Beyond the Learned Academy

Beyond the Learned Academy

The Practice of Mathematics, 1600–1850

Edited by
PHILIP BEELEY
Faculty of History, University of Oxford
AND
CHRISTOPHER D. HOLLINGS
Mathematical Institute, University of Oxford

OXFORD
UNIVERSITY PRESS

Great Clarendon Street, Oxford, OX2 6DP,
United Kingdom

Oxford University Press is a department of the University of Oxford.
It furthers the University's objective of excellence in research, scholarship,
and education by publishing worldwide. Oxford is a registered trade mark of
Oxford University Press in the UK and in certain other countries

Published in the United States of America by Oxford University Press
198 Madison Avenue, New York, NY 10016, United States of America

British Library Cataloguing in Publication Data

Data available

Library of Congress Control Number: 2023940942

ISBN 9780198863953

DOI: 10.1093/oso/9780198863953.001.0001

Printed and bound by
CPI Group (UK) Ltd, Croydon, CR0 4YY

Preface

The tremendous growth of the mathematical sciences in the early modern world was reflected contemporaneously in an increasingly sophisticated level of practical mathematics in fields such as merchants' accounts, instrument making, teaching, navigation, and gauging. Mathematics in many ways shaped the knowledge culture of the age, infiltrating workshops, dockyards, and warehouses, before extending through the factories of the Industrial Revolution to the trading companies and banks of the nineteenth century. While theoretical developments in the history of mathematics have been made the topic of numerous scholarly investigations, in many cases based around the work of key figures such as Descartes, Huygens, Leibniz, or Newton, practical mathematics, especially from the seventeenth century onwards, has been largely neglected. The present volume, comprising fifteen essays by leading authorities in the history of mathematics, seeks to fill this gap by exemplifying the richness, diversity, and breadth of mathematical practice from the seventeenth century through to the middle of the nineteenth century. Its appearance coincides with increased historical interest in the social and cultural milieus in which pre-modern and modern science were carried out, and will, it is hoped, facilitate and promote further investigations on the practice of mathematics itself.

Initial plans for this volume were drawn up by the editors during the lifetime of a three-year interdisciplinary project entitled 'Mathematical Culture in Restoration England: the life and letters of John Collins', hosted jointly by the Faculty of History and the Mathematical Institute of the University of Oxford, and generously funded by the Arts and Humanities Research Council (AHRC). A two-day conference held in York in April 2017 provided a number of participants—namely, Jim Bennett, Rebekah Higgitt, Sloan Despeaux, Benjamin Wardhaugh, Stefano Gulizia, Boris Jardine, and Mark McCartney—with the opportunity to explore themes which they have since developed into their contributions here. We should like to thank them sincerely for their perseverance and patience in what has turned out to be a long haul since those early deliberations on the topic.

Some time later, following the receipt of reviewers' reports on our original proposal, discussions with Daniel Taber and his then assistant at Oxford University Press, Katherine Ward, led to the suggestion that the scope of the present publication be considerably extended to take better account of European developments. We gladly signed up to this proposal and sent out further invitations to Ivo Schneider, Albrecht Heeffer, Margaret Schotte, Thomas Morel, João Domingues, Brigitte Stenhouse, and David Bellhouse. We should like to express our gratitude to all these colleagues for their willingness to join the team of contributors to this volume at a later stage in its conception and realization. Likewise, we should like to thank Dan Taber and a succession of his assistants, Katherine Ward, Charles Bath, John Smallman, and most recently Giulia Lipparini, not forgetting our anonymous referees, for their fulsome support and encouragement in bringing this publication to a successful conclusion.

Funding for the Collins project was originally awarded to Philip Beeley and Jacqueline Stedall. Sadly, the recurrence of Jackie's earlier illness meant she was scarcely able to begin working on her 'retirement project' before her premature death in 2014. Her place on the project was formally filled by Pietro Corsi, while Christopher Hollings, Jackie's successor in Oxford as lecturer in the history of mathematics, took on the task of co-organizing the York conference. All three are bound by their friendship and indebtedness to Jackie, a superb scholar who played a major role in establishing history of mathematics on a sound basis at Oxford. Fittingly, it is to her that this volume is dedicated.

Acknowledgements

Research for Chapter 4 was supported by funding from the Leverhulme Trust as part of the project Metropolitan Science: Places, Objects and Cultures of Practice and Knowledge in London, 1600–1800.

The research for Chapter 5 was partially financed by Portuguese Funds through FCT (Fundação para a Ciência e a Tecnologia) within the Projects UIDB/00013/2020 and UIDP/00013/2020.

Support for the research for Chapter 10 came from the project 'TacitRoots', PI Giulia Giannini, funded by the European Research Council (ERC) under the European Union's Horizon 2020 Research and Innovation Programme (GA n. 818098).

We thank the following libraries and archives for permission to use images from their collections: Admiral Long's Foundation; Ambrosian Library, Milan; Biblioteca Nacional de Portugal; Biblioteca Pública de Évora (Portugal); Biblioteca Nazionale Centrale, Florence; Bodleian Library, Oxford; British Museum; Christ Church Library, Oxford; Cuneiform Digital Library Initiative; Early English Books Online; History of Science Museum, Oxford; John Carter Brown Library, Providence, RI; Lincolnshire Archives; Maritiem Museum Rotterdam; Nationaal Archief, Den Haag; National Maritime Museum (UK); Penn Libraries, University of Pennsylvania; Sächische Landesbibliothek, Dresden; Science Museum, London; University of Oklahoma Libraries; Wellcome Collection; Whipple Museum of the History of Science, Cambridge. We are also grateful for the assistance offered by the Stadtarchiv Ulm.

Contents

PART III MATHEMATICAL PRACTITIONERS AND THEIR SCIENTIFIC MILIEUS

PART IV THE PRACTICE AND TEACHING OF MATHEMATICS

List of Figures

Notes on Contributors

Philip Beeley is Research Fellow and Tutor in the Faculty of History and Fellow of Linacre College, Oxford. A former president of the British Society for the History of Mathematics, his research covers mathematical correspondence networks in early modern Europe with a special focus on the history of analysis. He is also the editor (with the late Christoph J. Scriba) of the multi-volume *Correspondence of John Wallis* (1616–1703), published by OUP.

David Bellhouse is Professor Emeritus of Statistics at the University of Western Ontario. He holds a BA and MA from the University of Manitoba and a PhD from the University of Waterloo. His historical research interests include the history of probability, statistics, and actuarial science up to and including the nineteenth century. He has published two books related to the history of actuarial science: *Abraham De Moivre: Setting the Stage for Classical Probability and Its Applications* and *Leases for Lives: the Emergence of Actuarial Science in Eighteenth-Century England*. Recently, he has completed a biography of the father of statistical graphics, William Playfair.

Jim Bennett is Keeper Emeritus at the Science Museum, London. He works in the history of instruments, of astronomy, and of practical mathematics.

Sloan Evans Despeaux is Professor of Mathematics at Western Carolina University. Her research interests centre on the history of mathematics (especially the mathematics, mathematicians, and scientific journals in nineteenth-century Britain, and more recently, early twentieth-century America). She is secretary of the International Commission of the History of Mathematics, chair of the AMS HMATH Editorial Committee, and deputy editor of the *American Mathematical Monthly*.

João Caramalho Domingues is a lecturer at the Department of Mathematics of the University of Minho (Braga, Portugal). Since 2010 he has been a member of the *secretariado* (executive council) of the National Seminar for the History of Mathematics (an autonomous section of the Portuguese Mathematical Society). His research mostly focuses on eighteenth-century analysis and on mathematics in Portugal in the eighteenth century.

Stefano Gulizia is an intellectual historian specializing in the early modern period, with a focus on scholarly networks, print culture, and other themes. Trained as a classicist and philologist, he taught extensively in the US (after his PhD from Indiana University) and held fellowships in California, Oxford, Chicago, Montreal, Berlin, Wolfenbüttel, Bucharest, and Warsaw. He has joined the University of Milan as a member of TACITROOTS, and he is the editor-in-chief of the Scientiae Studies series at Amsterdam University Press.

Albrecht Heeffer is an engineer and philosopher based at the University of Ghent who works on the history of optics and the history and philosophy of mathematical practice. He has a special interest in cultures of mathematics in different time periods and in practices of non-scholarly and non-Western mathematics. He has published about sixty articles on the history of optics, emergence of symbolism in mathematics, Renaissance mathematics, and recreational mathematics.

Rebekah Higgitt has been Principal Curator of Science at National Museums Scotland since 2020. After completing her PhD at Imperial College London in 2004 she undertook postdoctoral research at the University of Edinburgh, was Curator of History of Science at Royal Museums Greenwich and a Senior Lecturer at the University of Kent's School of History. She has been a co-investigator on AHRC projects on the history of the Board of Longitude (2010–2015) and the scientific instrument trade (Tools of Knowledge, 2020–2023). She was also lead researcher on the Leverhulme Trust-funded Metropolitan Science project (2017–2020), which explored cultures of knowledge and practice within institutional sites in London in the seventeenth and eighteenth centuries. She is author of *Recreating Newton* (2007), co-author with Richard Dunn of *Finding Longitude* (2014), and, with Jasmine Kilburn-Toppin and Noah Moxham, of *Metropolitan Science* (forthcoming).

Christopher Hollings is Departmental Lecturer in Mathematics and its History at the Oxford Mathematical Institute and Clifford Norton Senior Research Fellow in the History of Mathematics at The Queen's College, Oxford. His research interests cover a range of topics in nineteenth- and twentieth-century mathematics.

Boris Jardine is a researcher at the University of Cambridge, based at the Whipple Museum of the History of Science. His research covers many aspects of the history of scientific instrumentation, especially the craft of instrument making and the uses of early-modern mathematical instruments. He is currently co-investigator on the AHRC-funded project 'Tools of Knowledge: Modelling the Scientific Instrument Trade, 1550–1914', which uses digital techniques to analyse local communities of artisans as well as long-term trends in the history of instrument-making.

Mark McCartney is senior lecturer in mathematics at Ulster University and a past president of the British Society for the History of Mathematics. His research interests include mathematical modelling, discrete chaos, and the history of mathematics and physics in the nineteenth and twentieth centuries.

Thomas Morel is a French historian of mathematics, working primarily on early modern practitioners. He completed his PhD at the Université de Bordeaux (2013) and has worked in Berlin and Lille. He is now professor of the history of mathematics and its teaching at the University of Wuppertal, and recently published *Underground Mathematics: Craft Culture and Knowledge Production in Early Modern Europe* (Cambridge University Press, 2022).

Ivo Schneider is emeritus professor of the history of science at the Universität der Bundeswehr, Munich. His research interests include the history of probability theory and its applications in physics, the use of measuring instruments by applied mathematicians, and philosophical influences upon mathematics.

Margaret E. Schotte is associate professor of history at York University (Toronto), where she teaches early modern history of science, technology, and history of the book. She holds a

PhD in the history of science from Princeton University (2014). Her prizewinning first book, *Sailing School: Navigating Science and Skill, 1550–1800* (Johns Hopkins University Press, 2019), is a comparative study of maritime expertise and training, with particular attention to the connections between classrooms, textbooks, and tacit knowledge. She has published on nautical instruments, logbooks, and imperial record-keeping.

Brigitte Stenhouse is a Lecturer in History of Mathematics at The Open University, UK. Through studies of the work of Mary Somerville (1780–1872), her research addresses the circulation of mathematics from France to Britain in the early nineteenth century and gendered barriers to mathematical communities. Current research looks broadly at the role of marriage in mathematics.

Benjamin Wardhaugh is based in Oxford and is a former fellow of All Souls College. He is the author of *Gunpowder and Geometry*, a biography of Charles Hutton, and the editor of Hutton's letters and library catalogue.

1
Introduction

Philip Beeley and Christopher D. Hollings

The second half of the seventeenth century witnessed a remarkable growth in the mathematical sciences in England, culminating in the publication of Isaac Newton's (1642–1727) *Principia* in 1687. This progress was reflected not only in the newly established Royal Society and in the work of those scholars who occupied one of three academic chairs, but also in an increasingly sophisticated level of practical mathematics in accountancy, commerce, navigation, and instrument making. It was in many ways through its diverse applications that mathematics came most to impact the lives of ordinary people. Although men like John Wallis (1616–1703) at Oxford or Isaac Barrow (1630–1677) at Cambridge made significant contributions to academic learning and to discussions in the wider Republic of Letters, their ideas seldom impacted those outside such lofty circles.

Nor, it must be said, did the latest developments in the mathematical sciences impact significantly on the undergraduate curriculum at England's universities. Although Oxford could boast two mathematical chairs founded by Henry Savile (1549–1622) in 1619, one for geometry and one for astronomy, their statutes were heavily indebted to Savile's humanist background, focusing almost entirely on the works of classical authors such as Euclid or Archimedes, Ptolemy, or Apollonius.[1] The mathematical chair at Cambridge, founded by the bequest of Henry Lucas (*c.*1610–1663) and first occupied by Barrow, was rather more open to contemporary developments.[2] Some college tutors also provided instruction in mathematics, but provision was uneven across both English universities.[3] Famously, the mathematician and instrument maker William Oughtred (1574–1660), formerly fellow of King's College, Cambridge, gave personal instruction in his rectory at Albury in Surrey to numerous scholars, including Seth Ward (1617–1689), Charles Scarburgh (1615–1694), and Robert Wood (1621/2–1685).[4] But the universities steadfastly resisted change. It was not until the end of the century that concrete proposals were drawn

[1] See William Poole, 'Sir Henry Savile and the Early Professors', in: *Oxford's Savilian Professors of Geometry*, ed. Robin Wilson, Oxford: Oxford University Press 2022, pp. 2–27.

[2] See Frances Willmoth, *Sir Jonas Moore: Practical Mathematics and Restoration Science*, Woodbridge: Boydell Press 1993, p. 8; Katherine Hill, 'Mathematics as a Tool of Social Change: Educational Reform in Seventeenth-Century England', in: *The Seventeenth Century* 12 (1997), pp. 23–36 esp. pp. 30–2; Walter William Rouse Bell, *A History of the Study of Mathematics at Cambridge*, Cambridge: at the University Press 1889, pp. 47–8 at pp. 100–3.

[3] See Philip Beeley, '"A designe Ichoate": Edward Bernard's Planned Edition of Euclid and its Scholarly Afterlife in Late Seventeenth-Century Oxford', in: *Reading Mathematics in Early Modern Europe: Studies in the Production, Collection, and Use of Mathematical Books*, ed. Philip Beeley, Benjamin Wardhaugh, and Yelda Nasifoglu, London: Routledge 2021, pp.192–229 esp. pp. 207–8.

[4] Jacqueline Stedall, 'Ariadne's Thread. The Life and Times of Oughtred's Clavis', in: *Annals of Science* 57 (2000), pp. 27–60.

up at Oxford, by the then professor of astronomy David Gregory (1659–1708), to introduce substantial reforms to the teaching of mathematics at that institution such as classes on fortification, hydrostatics, the use of globes, navigation, and notably through the medium of the English language.[5]

In London, the college created at the end of the sixteenth century through the legacy of the London merchant Thomas Gresham (1519–1579) provided an important forum for those interested in the mathematical sciences, with two of the original seven professorships likewise devoted to geometry and astronomy.[6] All of the Gresham professors were obliged to deliver their lectures informally in English as well as in the learned language of Latin, and a considerable number of incumbents of the mathematical chairs, including Edmund Gunter (1581–1626), Henry Briggs (1561–1630), and Samuel Foster (d.1652), had strong leanings towards the practical applications of their assigned fields.[7] In the absence of records of attendance, we can only speculate as to the make-up of audiences. While the college did not conceive itself as a teaching institution it did serve as a meeting place for men interested in the mathematical sciences and mechanical philosophy. Nowhere is its significance as such better reflected than in the fact that the idea for establishing 'a colledge for the promoting of Physico-Mathematicall Experimentall Learning', the future Royal Society, was devised following a lecture there by Christopher Wren (1632–1723) in November 1660.[8]

At a national level there was little in the way of formal instruction in mathematics available. Few grammar schools taught commercial subjects and what instruction in mathematics the two universities offered was largely restricted to Euclid's *Elements* and other classical texts. This educational deficit made itself increasingly apparent during the course of the century. England was steadily falling behind its Continental European competitors in areas where mathematical knowledge was required, such as navigation, shipbuilding, accountancy, or gunnery.[9] More perceptive contemporaries recognized that mathematics was of crucial importance to trade and the economic well-being of the land. Against this background, the Oxford-educated teacher of

[5] 'Dr Gregory's Scheme', in: *Private Correspondence and Miscellaneous Papers of Samuel Pepys 1679–1703*, ed. Joseph R. Tanner, 2 vols, London: G. Bell 1926, vol. II, pp. 91–4.

[6] Mordechai Feingold, 'Gresham College and the London Practitioners: The Nature of the English Mathematical Community', in: *Sir Thomas Gresham and Gresham College: Studies in the Intellectual History of London in the Sixteenth and Seventeenth Centuries*, ed. Francis Ames-Lewis, Aldershot: Ashgate 1999, pp. 174–88; Deborah E. Harkness, *The Jewel House: Elizabethan London and the Scientific Revolution*, New Haven, CT: Yale University Press 2007, pp. 119–20.

[7] See Charles Webster, *The Great Instauration: Science, Medicine, and Reform, 1626–1660*, London: Duckworths 1975, p. 353; Ian R. Adamson, 'The Administration of Gresham College and its Fluctuating Fortunes in the Seventeenth Century', in: *History of Education* 9 (1980), pp. 13–25, p. 19; Hill, 'Mathematics as a Tool', p. 27. On Gunter's design of instruments for use by seamen see Gerard L'E Turner, *Elizabethan Instrument Makers. The Origins of the London Trade in Precision Instrument Making*, Oxford: Oxford University Press 2000, pp. 67, 70–71, 255. See also Stephen Johnston, 'Mathematical Practitioners and Instruments in Renaissance England', in: *Annals of Science* 48 (1991), pp. 319–44.

[8] Marie Boas Hall, *Henry Oldenburg: Shaping the Royal Society*, Oxford: Oxford University Press 2002, pp. 54–9; Thomas Birch, *The History of the Royal Society of London for Improving of Natural Knowledge*, 4 vols, London: A. Millar 1756–7, vol. I, pp. 3–4; *The Record of the Royal Society of London*, 3rd edn, London: for the Royal Society 1912, pp. 7–9.

[9] Charles Wilson, *England's Apprenticeship, 1603–1763*, London: Longmans, Green & Co 1965, pp. 36–65.

mathematics John Newton (1621–1678) painted a correspondingly bleak picture in his *Cosmographia*, published posthumously in 1679, that

> four of the seven Liberal Arts, are almost wholly neglected, as well in both Universities, as in all Inferiour Schools; and setting aside the City of London, there are but few Places in this Nation, where a man can put his Son, to be well instructed in Arithmetick, Geometry, Musick and Astronomy; and even that Famous City was without a Publick School for Mathematical Learning, till His present Majesty was pleased to lay the Foundation; nay so averse are men in the general to the Arts (which are the support of all Trade) that without a high hand, it will be almost impossible, to make this People wise for their own good.[10]

The 'Publick School' to which Newton refers was the Royal Mathematical School, founded at Christ's Hospital in 1673 at the instigation of the surveyor-general Jonas Moore (1617–1679) and with the active support of Samuel Pepys (1633–1703) in his capacity as naval administrator.[11] The school's averred aim was to train each year forty poor boys in the arts of arithmetic and navigation in order that they might serve at sea. The course of study to be pursued under the school's mathematical master included such topics as the principles of geometry with the practice thereof, the division and proportional section of lines by the use of the diagonal scale and the rule of three in lines, decimal arithmetic, a general rule for finding latitude by the Sun or fixed stars, questions of plane sailing with the use of the plane sea chart, the use of logarithms and table of artificial sines and tangents, the use of Gunter's Scale, and the projecting of the sphere in circles or globe on a plane. At the conclusion of their instruction the boys were examined with a view to their proficiency, usually by a specially appointed examiner from Trinity House. Later reforms of the curriculum were overseen by some of the most esteemed mathematicians of the land, including Isaac Newton and Wallis.

Unfortunately, in practice things did not quite work out at the school as initially envisaged. New recruits were often lacking in basic knowledge, there were problems with discipline, and some boys succeeded in obtaining privileges in return for cash payment. Nor did all the mathematical masters employed by the school prove to be fit or entirely suitable for the task. Robert Wood arrived in January 1680/1 with excellent credentials, having studied at Eton and Merton College, Oxford as well as having been privately tutored by William Oughtred at Albury. However, not only was his teaching judged to be inadequate and discipline in class to be poor, but also Wood himself was frequently found to be absent, installing a deputy in his place. In the light of overwhelming evidence not in his favour, Wood resigned after just over a year in post in the spring of 1682.[12] His successor, Edward Paget (1652–1703) did not fare much better. Formerly a Fellow of Trinity College, Cambridge, he enjoyed a good reputation as mathematician and was appointed on Newton's recommendation. However, he turned out to be overly inclined to drink, and after two extended periods of leave,

[10] John Newton, *Cosmographia, or a view of the Terrestrial and Coelestial Globes, in a Brief Explanation of the Principles of plain and solid Geometry, applied to surveying and gauging a cask*, London: for Thomas Passinger 1679, sig. A4^r.

[11] Willmoth, *Sir Jonas Moore*, pp. 195–207.

[12] Ibid., p. 197.

ostensibly on health grounds, he was considered to be neglectful of his duties and resigned in 1695.[13]

The Royal Mathematical School responded to the need for better training of navigators and mariners, but its creation reflected in many ways what was already happening on the ground. It came at a time when mathematical learning was already taking hold in the metropolis. Promoted by self-declared philomaths, instrument makers, allied printers and booksellers, and innumerable private teachers, mathematics successfully permeated workshops, warehouses, dockyards, coffee houses, and taverns, and was disseminated by means of inexpensive printed books, leaflets, and letters. New means of making a living sprang up for those with little else by way of material or financial means. There was a growing need for accountants, diallers, gaugers, surveyors, and the like.[14]

Economically, England saw its principal competitor at this time in the United Provinces of the Dutch Republic. Here, the shortcomings in instruction and training in mathematics were leading to serious economic consequences, especially in the fishing industry. Dutch ships were extracting tons of herring from the waters washing English shores and could boast the largest merchant fleet in Europe. As has recently been argued, the sustained demonstration of Dutch commercial prowess had a more powerful effect on the English imagination than any other economic development of the seventeenth century.[15] The indefatigable promoter of mathematical learning John Collins (1626–1683) had himself witnessed the superiority of Dutch shipping vessels and the training of their crews during his involvement in the early years of the fifth Ottoman–Venetian war (1645–69). Even before then, the London merchant and law reformer Henry Robinson (1605–1673) had lamented that the Dutch could outstrip English shipbuilding efforts, despite their lack of timber.[16]

Such achievements were part of a broader picture. The natural philosopher William Petty (1623–1687), who served as an administrator in Ireland, recognized that Dutch prosperity rested essentially on its shipbuilding: by this means had not only the transportation of goods been promoted, but also ultimately the nation's commercial relations extended overseas, most visibly through the Dutch East India Company (Verenigde Oostindische Compagnie; VOC).[17] The political writer Roger Coke (1628–1704/7) went further and in a geometrically fashioned demonstration found eighteen reasons for the growth and increase of the Dutch trade above the English, of which two are particularly pertinent. Firstly, that the Dutch 'generally

[13] See Robert Iliffe, 'Mathematical Characters: Flamsteed and Christ's Hospital Royal Mathematical School', in: *Flamsteed's Stars: New perspectives on the life and work of the first Astronomer Royal (1646–1719)*, ed. Frances Willmoth, Woodbridge: The Boydell Press 1997, pp. 115–44 at p. 129.

[14] Philip Beeley, 'Practical Mathematicians and Mathematical Practice in Later Seventeenth-Century London', in: *British Journal for the History of Science* 52 (2019), pp. 225–48; Harkness, *The Jewel House*, pp. 97–103.

[15] Joyce Oldham Appleby, *Economic Thought and Ideology in Seventeenth-Century England*, Princeton, NJ: Princeton University Press 1978, p. 73. See also Jonathan Israel, *The Dutch Republic: Its Rise, Greatness, and Fall, 1477–1806*, Oxford: Clarendon Press 1995, pp. 617–18.

[16] Henry Robinson, *Englands safety in Trades Encreased*, London: E. P. for Nicholas Bourne 1641, p. 1: 'And yet the Hollanders who have no Timber at their owne growth, doe farre surpasse us in number of ships'.

[17] See William Petty, *Political Arithmetick or a discourse concerning, the extent and value of lands, people. Buildings, husbandry, manufactrure, commerce, fishery, [...]*, London: for Robert Clavel 1690, pp. 19–23.

breed their youth of both Sexes in the Studies of Geometry and Numbers, especially more than the English do', and secondly, and more broadly, that 'the Study of Geometry and Numbers, is the best Education for understanding Trade'.[18] Citing examples from classical antiquity, Coke proceeded to explain that these two studies constituted the foundation for all fortification, architecture, surveying, and measuring all bodies and surfaces, astronomy, navigation, geography, and so on.[19]

The contrast to the Netherlands in England was stark. Against the background of there being little or no formal provision of mathematics in schools, the path to mathematical knowledge, whether in the city or in the country, was seldom straightforward or cheap. However, learning some trades could provide a good start. In his autobiography, John Wallis tells us that it was his brother Henry (*c.*1620–1666), an apprentice draper, who first introduced him to

> the Practical part of Common Arithmetick in Numeration, Addition, Substraction, Multiplication, Division, The Rule of Three (Direct and Inverse,) the Rule of Fellowship (with, and without, Time,) the Rule of False-Position, Rules of Practise and Reduction of Coins and some other little things.[20]

Furthermore, Henry showed him these operations and rules 'by steps, in the same method that he had learned them: and I had wrought over all the Examples which he before had done in his book'.[21] This was the typical method that was found in most of the practical books at the time. Credibly, Wallis claims that following his studies and before his appointment to the Savilian chair in 1649, he was only able to pursue mathematics in his spare time, and that is to say, when he was not engaged in theological tasks such as those at the Westminster Assembly for which he was gainfully employed. Citing this common trope, he concludes his account thus:

> This was my first entry into Mathematicks, and all the teaching I had. But did afterward prosecute it, as a pleasing diversion at spare hours, as books of Arithmetick, or others Mathematical fell occasionally in my way, without any to direct me, what books to read, or what to seek of in what methode to proceed. For Mathematicks were not, at that time, (with us) looked upon as Accademical Learning; but as the business of Traders, Merchants, Sea-men, Carpenters, land-measurers, or the like; or perhaps some Alamanak-makers in London.[22]

[18] Roger Coke, *A Discourse of Trade. In two Parts. The first treats of the Reason of the Decay of the Strength, Wealth, and Trade of England. The latter, of the Growth and Increase of the Dutch Trade above the English*, London: for H. Brome and R. Horne 1670, p. 50; Appleby, *Economic Thought*, p. 77. On contemporary promotion of practical mathematics in the Netherlands see Fokko Jan Dijksterhuis, 'Duytsche Mathematique and the Building of a New Society: Pursuits of Mathematics in the Seventeenth-Century Dutch Republic', in: *Mathematical Practitioners and the Transformation of Natural Knowledge in Early Modern Europe*, ed. Lesley B. Cormack, Steven A. Walton, and John A. Schuster, Cham: Springer 2017, pp. 167–81.

[19] Coke, *Discourse of Trade*, pp. 69–75.

[20] Christoph J. Scriba, 'The Autobiography of John Wallis, F.R.S.', in: *Notes and Records of the Royal Society* 25 (1970), pp.17–46 at p. 26.

[21] Ibid.

[22] Ibid., pp. 26–7.

John Collins started off with an apprenticeship to the Oxford bookseller, Thomas Allam (17th century). After Allam's business failed, no doubt brought on by the effects of political turmoil on that university city, Collins managed to find work as clerk in the household of the then Prince of Wales in London, and was fortunate to encounter in his immediate superior one of the best constructors of sundials in the country, the legendary John Marr (*fl.* 1614–1647).[23] This was his first encounter with the practical side of mathematics. When he later served on board an English merchantman hired by the Venetians for their fight against the Turks in the eastern Mediterranean, Collins was able to study the rudiments of navigation in his spare time. Such practically acquired knowledge he would later put to use in his own publications, most notably, *Navigation by the Mariners Plain Scale new plain'd* and *Geometricall Dyalling*.[24] But like other practitioners, too, he was clearly concerned from the outset to acquire a broad range of skills. Probably by means of personal study he acquired a sufficiently good understanding of bookkeeping to enable him, already in 1653, to publish a manual on the topic, the *Introduction to Merchants-Accompts*. Reprinted a number of times later on, this book laid the foundation for his employment as an accountant in several government offices during the course of his lifetime.[25] By the time of the Great Fire, he had also forged ties to academic colleagues such as Wallis and Barrow, both of whom he regularly supplied with copies of the latest mathematical publications from abroad. As a man unerringly committed to promoting mathematics, Collins soon made himself indispensable to practical and academic milieus alike, and was in this respect respected and valued as an exceptional figure. Thus, he acted as midwife to numerous important books including Wallis's *Mechanica*, Barrow's *Lectiones opticae XVIII*, and the considerably augmented English edition of Johann Heinrich Rahn's (1622–1676) *Teutsche Algebra*, not to mention his repeated efforts to persuade Newton to publish some of his analytical writings.[26]

While some practitioners acquired their skills while being apprenticed to a master, in ways similar to Collins, others found the resources to engage the services of one of the army of teachers or 'professors of the mathematicks' who ran their own schools in places such as London, Liverpool, or Newcastle. One particularly notable case is that of John Kersey (1616–1677), a former servant to the eminent Denton

[23] John Collins, *An Introduction to Merchants-Accompts*, London: William Godbid for Robert Horne 1674, sig. B1r. See also Eva Germaine Rimington Taylor, *The Mathematical Practitioners of Tudor and Stuart England*, Cambridge: at the University Press 1968, pp. 203–4.

[24] John Collins, *Navigation by the Mariners Plain Scale new plain'd: or, a treatise of geometrical and arithmetical navigation*, London: Thomas Johnson for Francis Cossinet 1659; John Collins, *Geometricall Dyalling: or, dyalling performed by a line of chords only, or by the plain scale*, London: Thomas Johnson for Francis Cossinet 1659. See also Philip Beeley, 'To the publike advancement: John Collins and the Promotion of Mathematical Knowledge in Restoration England', in: *BSHM Bulletin* 32 (2017), pp. 61–74; William Letwin, *The Origins of Scientific Economics: English Economic Thought, 1660–1776*, London: Methuen 1963, pp. 100–13.

[25] See Natasha Glaisyer, *The Culture of Commerce in England, 1660–1720*, London and Woodbridge: The Royal Historical Society and The Boydell Press 2011, p. 115

[26] On the difficulties encountered in persuading stationers to undertake the printing of mathematical books see Adrian Johns, *The Nature of the Book: Print and Knowledge in the Making*, Chicago: University of Chicago Press 1998, pp. 448–54.

family of Buckinghamshire, who possibly with that family's help took instruction from the London teacher of mathematics John Speidell (1577–1649). Kersey subsequently went on to set up a mathematical school of his own in Covent Garden, where he was able to offer board to gentlemen wanting instruction in a wide range of topics, including arithmetic, merchants' accounts, algebra, geometry, navigation, dialling, the construction and use of mathematical instruments, and chirographie or 'The Art of accurate and exact Hand-writing, in the English and best Italique formes, by genuine Principles, and plain Demonstrations'.[27] By all accounts this school was successful, perhaps because it coincided with a trend on the part of young men from good families to seek careers in merchant adventuring, where sound knowledge of the art of reckoning was a prerequisite. Adam Martindale (1623–1686), a non-conformist minister originally from St Helens in Lancashire, offered instruction to young gentlemen and others at his house near Dunham in Cheshire on a wide range of mathematical skills, which included balancing accounts as well as other types of vulgar and artificial arithmetic, geometry, the principles of astronomy and navigation, and the art of dialling.[28]

The mathematician and landsurveyor William Leybourn (1626–1716) started out as a bookseller and printer in partnership with Robert Leybourn (17th century) in Cripplegate, London, specializing in scientific and mathematical books. By 1653 he had published *The Compleat Surveyor* under his own name, having previously brought out a pamphlet on the same topic under a pseudonym.[29] Dedicating the book to the Oxford-educated lawyer and mathematician Edmund Wingate (1596–1656), whom he addresses as his patron, Leybourn makes a point of indicating that all the instruments he employs could be bought from the instrument-maker Anthony Thompson (d.1665) in Smithfield.[30] A remarkable engraved diagram by Thompson of an instrument for converting money, weights, and measures to their decimal equivalents is found in the first edition of his *Arithmetick, Vulgar, Decimal, and Instrumental*, published in 1657.[31] Leybourn enjoyed a good public reputation, too. Shortly after London had been ravaged by the Great Fire, he was one of six practising mathematicians called upon by the lord mayor, aldermen, and council to survey the remains and

[27] Edmund Wingate, *Arithmetique made easie, or, A perfect Methode for the true knowledge and practice of Natural Arithmetique., according to the ancient vulgar way, without dependence upon any other Author for the grounds thereof.* Second Edition by John Kersey. London: J. Flesher for Philemon Stephens 1650, advertisement following p. 461.

[28] Adam Martindale, *The Country-Survey-Book: or Land-Meters Vade-Mecum. Wherein the Principles and practical Rules for Surveying of Land, are so plainly (though briefly) delivered, that any one of ordinary parts (understanding how to add, subtract, multiply and divide,) may by the help of this small Treatise alone, and a few cheap Instruments easy to be procured, Measure a parcel of Land, and with judgment and expedition Plot it, and give up the Content thereof,* London: for R. Clavel and T. Sawbridge 1692, advertisement following p. 195.

[29] William Leybourn, *The Compleat Surveyor: containing the whole Art of Surveying of Land, by the Plain Table, Theodolite, Circumferentor, and Peractor*, London: R. & W. Leybourn for E. Brewster and G. Sawbridge 1653.

[30] On Thompson see Taylor, *Mathematical Practitioners*, pp. 220–1; on Wingate see Mordechai Feingold, *The Mathematicians' Apprenticeship: Science, Universities and Society in England, 1560–1640*, Cambridge: Cambridge University Press 1984, pp. 171–3.

[31] William Leybourn, *Arithmetick, Vulgar, Decimal, and Instrumental. In three parts*, London: R. and W. Leybourn for G. Sawbridge 1657. The diagram is between pages 222 and 223.

prepare for the clearing and building of the city.[32] Six years later he was hired by the cartographer and impresario John Ogilby (1600–1676) to head a further survey. In his books, Leybourn often advertised his services as a surveyor and teacher of mathematics. In the second edition of his *Arithmetick*, in which he adds a fourth part on algebra to those three listed in the first edition, he advertises his services as teacher of mathematics, as land surveyor, and as dialler:

> If any Gentleman, or other Person, desire to be instructed in any of the Sciences Mathematical as Arithmetick, Geometry, Astronomy, the use of the Globes, Trigonometry, Navigation, Surveying of Land, Dialling, or the like; either at their own houses, his habitation, or such other convenient place as the Party shall direct, the Author hereof will be ready to attend them at times appointed.[33]

What probably allowed him to stand out against his competitors was his ability and willingness to carry out practical tasks himself. Thus, Leybourn also offered to survey land for building and to make measurements for master builders, carpenters, joiners, and the like. He also could provide the service of designing or building sundials or any dials whatsoever, fixed or movable, about the client's house or garden. And as a way of establishing trust, he lists the eminent London stationer and bookseller Nathaniel Brooke (*fl.* 1646–1676), sometime stationer to the Ordnance Office, and the highly regarded instrument maker Walter Hayes (*fl.* 1651–1692) from whom he might be heard.[34] The eco-system of merchants, mathematicians, tradesmen, booksellers, and instrument makers in early modern London flourished through the mutual dependency it engendered and the professional and moral support it offered.

If all else failed, prospective practitioners could always fall back on self-instruction. For this time-honoured path to betterment there were innumerable cheap and plain publications available, in which the author took his readers step by step through the basics of surveying, merchants' accounts, or practical geometry as if he were a friend or personal tutor.

Martindale, who wrote his *Country-Survey-Book* after he had ceased offering personal instruction, declares his motivation for going into print to have been for scientific and ideological reasons.[35] On the one hand, he had seen that men engaged in measuring land often lacked awareness of the fundamentals of geometry or had simply imbibed false principles in the course of their work, while on the other private tuition was usually prohibitively expensive and was in any case difficult if not impossible to obtain in the country. He conceived his book as a suitable remedy to both problems. Not only did it present no barriers to those with little or no foreknowledge,

[32] See Willmoth, *Sir Jonas Moore*, pp. 136–7. The result of the survey was *An Exact Surveigh of the Strees Lanes and Churches Contained within the Ruines of the City of London*, first completed in December 1666. The other mathematicians involved were John Leake, John Jennings, William Marr, Thomas Streete, and Richard Shortgrave.

[33] William Leybourn, *Arithmetick: vulgar, decimal, instrumental, algebraical. In four parts.* Third edition, London: S. Streater for George Sawbridge 1668, facing p. 1.

[34] Ibid.

[35] Ibid. Martindale begins his advertisement thus: 'The Author hereof useth in Winter and Spring Seasons to board young Gentlemen and others in his Habitation'. See also Glaisyer, *Culture of Commerce*, p. 111.

it was also small and affordable even for those who had little to spare. Importantly, too, he restricts himself to what the reader needs to know. As he writes in the preface:

> Mathematical-Schools, where better things might be learned, are very rare, and an able Artist to instruct one in private is hard, and chargeable to be procured. Excellent Books indeed there are in our English Tongue, Written by our famous Rathborn, Wing, Leybourn, and Holywell, to which may be added Industrious Mr. Atwells Treatise, and some parts of Capt. Sturmy's: But those I rather esteem fit to be read by an able Artist (towards his perfecting) than by a new beginner, for in the best of those Books he will find the most useful and plain Rules so intermixed with others that are less necessary, and more intricate, (though very excellent for their proper ends) and so many Curiosities touching Trigonometry, Transmutation of Figures, &c. which his business never calls for, for want of judgment to pick out that which fits his present purpose, and to study higher Speculations afterwards, he is apt to be confounded and discouraged; whereby it accidentally comes to pass that plenty, makes him poor. [...] I have therefore made my Book so little, that the Price can neither much empty the Pocket, nor the Bulk overfill it. And yet so plain, that I doubt not to be understood by very ordinary Capacities.[36]

A similar line is taken by John Mayne (*fl.* 1673–1675), about whom little is known apart from the fact that he styles himself 'Philo-Accomptant' and appears to have run a school in the south London parish of Southwark, where he offered instruction in all parts of arithmetic, the doctrine of plain and spherical triangles, as well as the use of globes, quadrant, sector, and other mathematical instruments. And this was not to mention shorthand, merchants' accounts, and 'fair writing'. Addressing his readers as 'thy friend', Mayne makes a point of being almost apologetic for publishing yet another book claiming to be 'a plain and easie Introduction to Arithmetick, Vulgar and Decimal', but he, too, sought to promote his *Socius Mercatorius* as an easily understandable, cheap alternative to paid tuition:

> My Design in this Work is, to render the Rules of those excellent Arts, which the Title-page pretends to, so plain and obvious, as that they may be easily apprehended without the Assistance of a living Master. And if there were nothing new in the whole, but the perspicuity of the Principles, and easiness of the Method (which out of civility to my self I must deny) yet those alone are sufficient to vindicate me in this Publication; and I hope thou wilt not be angry, that I am a Well-wisher to thy Understanding.[37]

Nor would it have been considered odd within the context of the milieus to which his book was directed that a teacher should seemingly offer a more affordable route to the knowledge that otherwise provided his main source of income. Apart from the ideal of promoting mathematical learning, to which men like Mayne, Collins, and

[36] Martindale, *Country-Survey-Book*, preface.

[37] John Mayne, *Socius Mercatoris: or the Merchant's Companion*, London: W. G. for N. Crouch 1674, sig. A3^r.

Martindale subscribed, books were an important means of advertising a teacher's skill while also drawing attention to other elements of the metropolis's mathematical eco-system: instrument makers, printers, engravers, booksellers, and so on. Collins pointed out in an advertisement for his services as an accountant that he could be heard of at Mr Robert Horne, the stationer.[38] Others like Michael Dary (1613–1679) advertised where the instrument they described could be obtained.[39]

Many prospective merchants are known to have received instruction at a writing school or from a private teacher of mathematics. John Verney (1640–1717) at first hoped to be taught by John Kersey who had long-established connections to that Buckinghamshire family, but eventually acquired his knowledge of accounting and arithmetic at the mathematical school run by a certain Mr Rich, before later being apprenticed to a Turkey merchant.[40] A well-known writing school was run by John Ayres (d.1704/9), a prolific author of manuals on handwriting and arithmetic, whose premises were in St Paul's churchyard in close proximity to numerous booksellers.[41] Charles Snell (1667–1733), author of various book-keeping texts and writing manuals, succeeded John Seddon (1643/4–1700) as master of Sir John Johnson's Free Writing School in Foster Lane, Cheapside, where he offered instruction in 'Writing in all the Hands of England, Arithmetick, and Merchants Accompts', recommending his services particularly 'for the convenience of Merchants and Tradesmen, who are desirous their apprentices should improve themselves in the above-said Arts'.[42]

Henry Robinson collaborated closely with the educational reformer Samuel Hartlib (*c*.1600–1662) and the Scottish preacher and ecumentist John Durie (1596–1680) in developing plans to reform the English state's schooling system. In the wake of the Revolution of 1649, Hartlib is known to have looked to him with a view to establishing his proposed college for the invention and the advancement of the mechanical arts.[43]

Not all academic or at least academically trained mathematicians valued their more practically orientated counterparts to such an extent. The aforementioned Robert Wood, sometime mathematical master at the Royal Mathematical School, is a case in point. Overly convinced of his own abilities, Wood had some twenty-five years preceding that endeavour followed Henry Cromwell (1628–1674) to Ireland and was involved in the plans of Oliver Cromwell's (1599–1658) son to establish a university

[38] John Collins, *Introduction to Merchants-Accompts*, To the Reader, Sig. B1^{v}.

[39] Michael Dary, *Dary's Diarie. Or, the description and use of a Quadrant*, London: by T. F. for George Hurlock 1650, sig. A3v: 'Note that this Instrument or any other Mathematicall Instrument, either for Sea or Land, is exactly made in Brasse or Wood, by Henry Sutton at Tower-hill neare the Posterne-Spring'.

[40] Susan E. Whyman, *Sociability and Power in Late Stuart England: The Cultural Worlds of the Verneys, 1660–1720*, Oxford: Oxford University Press 1999, pp. 41–3.

[41] See John Ayres, *The trades-mans Copy book, or, Apprentices Companion*, London: by the author 1688, p. 3, where the work is described as being 'chiefly intended and designed for Apprentices, Youths, and others employed in Business of Trade, and Merchandize'. See also his *The Penmans daily Practise*, London: by the author 1690, and Glaisyer, *Culture of Commerce*, p. 113, and Ambrose Heal, *The English Writing-Masters and Their Copy-Books, 1570–1800*, Cambridge: at the University Press 1931.

[42] Charles Snell, *The Tradesman's Director; or, a short and easy Method of Keeping his Books of Accompts*, London: Richard Baldwin 1697, advertisement facing p. 1.

[43] Webster, *Great Instauration*, pp. 210–13, 347–8, 363–7; See also Paul Slack, *The Invention of Improvement. Information and Material Progress in Seventeenth-Century England*, Oxford: Oxford University Press 2015, pp. 93, 100, 105.

college in Dublin.[44] In 1657, he was appointed professor of mathematics at Durham College by the Lord Protector himself, but there is no evidence that he actually took up that position. What is certain is that he was a member of the circle around Samuel Hartlib, to whom he boasted on one occasion that 'the two most eminent mathematicians of this age, viz. Mr Oughtred my ever honoured Master, & Dr Wallis were pleased to owne me in Epistles before some of their workes'.[45] Hartlib evidently valued Wood's judgement, and on one occasion in advance of the mathematician's visit to London sent him a paper on sundials by an unnamed gentleman. Wood's response is telling. He agrees with the gentleman that mathematical studies are more worthwhile than the philosophical disputes so valued in the universities, but suggests that practical mathematics are devoid of intellectual value:

> In Answer to your desires of seeing my thoughts concerning the Gentlemans Paper of Dials, before I come to London: 1st I easily accord with him That Mathematicks are better worth our studies then Academical Haeccieties: but for Practical Mathematicks alone, without their speculative Root in their Owners head to sustaine them, I look upon them but as cropt Flowers in a Pot, compared with those growing in the Garden.[46]

Despite manifest differences in their economic and social standing, academic mathematicians and their practically engaged counterparts alike were generally committed to a common aim of promoting the mathematical sciences. There was, for example, a common awareness that England lagged behind France, Italy, and the Low Countries, whether it be in the training of navigators or in innovative work in algebra. Collins in particular was led by this awareness to expend a considerable amount of his time, when not engaged in duties as accountant in chancery or later as manager of the Farthing Office, in seeking to obtain copies of the latest scientific publications from abroad.[47] Nor would these books by the likes of Pietro Mengoli (1626–1686), Christoph Grienberger (1561–1636), Pierre Fermat (1607–1665) or Jacques de Billy (1602–1679) be loaned exclusively to academic mathematicians, although books that were not written in English could sometimes present a barrier for practitioners. After he had successfully persuaded his friend, the mathematics teacher John Kersey, to embark upon writing a substantial introductory work on algebra in English, Collins regularly channelled books he had successfully procured from abroad in Kersey's direction. When the first part of the *Elements of Algebra* came out, in 1673, the author took the opportunity of acknowledging his considerable debt to Collins, describing

[44] Webster, *Great Instauration*, pp. 81–2, 240, 243.

[45] Robert Wood to Samuel Hartlib, 31 August/[10 September] 1659, Sheffield University Library, Hartlib Papers 33/1/63A–64B. See also Toby Barnard, 'The Hartlib Circle and the Cult and Culture of Improvement in Ireland', in: *Samuel Hartlib and Universal Reformation: Studies in Intellectual Communication*, ed. Mark Greengrass, Michael Leslie, and Timothy Raylor, Cambridge: Cambridge University Press 1994, pp. 281–97 esp. pp. 284–5.

[46] Robert Wood to Samuel Hartlib, 28 August/[7 September] 1657, Sheffield University Library, Hartlib Papers 33/1/27A–28B.

[47] See for example the mathematical books Collins sought to procure from Leibniz in 1673, listed in Philip Beeley, 'Leibniz and the Royal Society Revisited', in: *Leibniz's Legacy and Impact*, ed. Julia Weckend and Lloyd Strickland, New York: Routledge 2020, pp. 23–52 esp. p. 37.

him as 'an industrious promoter of the Mathematicks in general', pointing out, at the same time, that he had been 'a principal Instrument' of bringing the work to light.[48]

Collins provided assistance in similar ways to others, too. Thus, he supported the mathematical endeavours of Thomas Strode (d.1697), a land owner from Somerset, whose studies at Oxford had first been disrupted and then curtailed by the Civil Wars. With the help of the Cambridge-educated clergyman and natural philosopher John Beale (1608–1683), Collins carefully guided Strode's work on conic sections and combinatorics in the 1670s, lending him a considerable number of books by authors such as Gerard Kinckhuysen (1625–1666), Ismaël Boulliau (1605–1694), Bernard Frénicle de Bessy (*c.*1604–1674), and Barrow along the way. The fruits of these labours were an algebraic tract on conics and a treatise on combinatorics. But as ever Collins's support was not disinterested. Being keen to promote the study of algebra in England, he made no secret of the fact that in his view Strode should devote his energies to getting his work on conics published.[49] By the same score, he pointedly mentioned that there were already good contemporary publications on the other topic by the likes of Blaise Pascal (1623–1662) and G. W. Leibniz (1646–1716). Ironically, it was the treatise on combinatorics which in difficult times eventually found a publisher, while the analytic conics has survived only as an unpublished manuscript.[50] A later work by Strode on the topic of dialling, which appeared after Collins's death, reflected the enduring market for practical mathematics. Somewhat ashamedly, the author declares that he had chiefly composed the tract 'for some near Relations', but at the same time in evident reference to Collins and Beale he acknowledges 'That I received Seeds and Plants from some eminent Persons of the Royal Society, since deceased'.[51]

Given his background, it is perhaps not surprising that Strode was not completely comfortable with the idea of asserting his authorship of a work of such practical nature. A more representative stance is adopted by Kersey in the preface to the first part of his *Elements of Algebra*. Here the author first sets out the proud heritage of mathematics as the embodiment of demonstrative truth stretching back to ancient times, before then turning to its excellency and utility in a contemporary context. He thereby also bridges the aforementioned divide, but in a way that leaves no doubt that for him, as for Collins, that divide was largely artificial in the first place:[52]

> Nor are Arithmetick and Geometry excellent in themselves only, but highly esteem'd also for their manifold Utility, as well in the Employments of Men about Accompts,

[48] John Kersey, *The Elements of that Mathematical Art commonly called Algebra, expounded in four books*, 2 parts, London: William Godbid for Thomas Passinger and Benjamin Hurlock 1673–4, vol. I, sig. b3^r.

[49] John Collins to Thomas Strode, 24 October/[3 November] 1676, Cambridge University Library, MS Add. 9597/13/6, ff. 175^r–175a^v; Stephen Jordan Rigaud, *Correspondence of Scientific Men of the Seventeenth Century*, 2 vols, Oxford: at the University Press 1841, vol. II, pp. 453–5.

[50] Thomas Strode, *A Short Treatise of Combinations, Elections, Permutations & Composition of Quantities*, London: W. Godbid for Enoch Wyer 1678; Thomas Strode, 'Apollonius Analiticus', Oxford, Bodleian Library, MS Savile 43 (part one) and MS Savile 44 (part two).

[51] Thomas Strode, *A new and easie method to the art of dyalling*, London: for J. Taylor and T. Newborough 1688, 'To the Reader'.

[52] Michael Hunter appropriately describes Collins as 'a broker between eminent scholars and more practical men needing theoretical assistance'. See Michael Hunter, *Science and Society in Restoration England*, Cambridge: Cambridge University Press 1981, p. 96.

> Trade, Building, Measuring of Land, and divers other common Affairs, as in facilitating and enlivening divers other Noble Arts, for how can Harmonical Composition in Musick, or exact Measure and Proportion in Painting be perform'd, without the assistance of Arithmetick and Geometry.[53]

Kersey's *Elements of Algebra* proved to be highly successful, going through three or possibly four editions, while parts of the work continued to be published well into the eighteenth century, by which time it was included on the University's curriculum at Cambridge.[54] Neither the *Introduction to Algebra* by John Pell (1611–1685) and Thomas Brancker (1633–1676) nor John Wallis's *Treatise of Algebra* could make a similar claim.[55] But Kersey's work is significant in another way, too. By the time it came to writing the third book, devoted to number theory, Collins and Kersey appear to have decided jointly that the work should come to symbolize the new-found strength of English mathematics. Having obtained for Kersey copies of the very latest French publications by de Billy and Fermat, Collins went so far as to suggest to Beale that Kersey would top anything those two authors had produced.[56] That notion was of course completely unrealistic, but it sprang from that desire he shared with many if not most of his academic friends that English mathematics should not be found wanting when compared to other nations.

There was a strong social component to the work of mathematical practitioners on a local level. They formed a cohesive knowledge community which intersected closely with instrument makers, merchants, printers, and booksellers. Although evidence is sketchy, it is known that they organized themselves into mathematical societies or clubs that contrasted strongly with more illustrious and socially elevated counterparts to which they generally were not admitted.[57] It would not be until the following century that in the course of trends to popularize science and extend access to contemporary developments, societies, and journals would emerge that served as forums where mathematicians and the broader public could meet.[58] The trained seafarer Samuel Sturmy (1633–1669), best known as author of *The Mariners Magazine*, conveyed carefully collected data on tides and magnetic variation to the Royal Society with the aim of becoming a member, but was excluded from joining the fellowship. Although Collins was admitted, this was primarily so that he could employ his

[53] Kersey, *Elements of Algebra*, vol. I, sig. b2^{r}.

[54] See Ball, *History of the Study of Mathematics at Cambridge*, p. 95; Helena M. Pycior, *Symbols, Impossible Numbers, and Geometric Entanglements: British Algebra through the Commentaries of Newton's Universal Arithmetick*, Cambridge: Cambridge University Press 1997, p. 102.

[55] Johann Heinrich Rahn, *An Introduction to Algebra. Translated out of the High-Dutch into English by Thomas Brancker, M.A. Much altered and augmented by D[r] P[ell]*, London: William Godbid for Moses Pitt 1668.; John Wallis, A *Treatise of Algebra, both Historical and Practical*, London: John Playford for Richard Davis 1685.

[56] John Collins to John Beale, 20/[30] August 1672, Cambridge University Library, MS Add. 9597/13/5, ff. 83^{r}–85a^{v}, at f. 83^{v}; Rigaud, *Correspondence of Scientific Men*, I, pp. 195–204 at pp. 197–8: 'As to Diophantus, the late ones of Fermat now here to be had, Billy since it published a Booke called Diophantus Redivivus, but you will have that which is better than both in Kersey'.

[57] Beeley, 'Practical Mathematicians', pp. 243–4.

[58] See Larry Stewart, *The Rise of Public Science: Rhetoric, Technology, and Natural Philosophy in Newtonian Britain, 1660–1750*, Cambridge: Cambridge University Press 1992; Peter Clark, *British Clubs and Societies, 1580–1800: The Origins of an Associational World*, Oxford: Oxford University Press 2000, p. 74.

extensive knowledge of mathematical and natural philosophical publications in order to build up the institution's library. He was exempted from paying fees, because of the various administrative chores he was obliged to carry out, and always felt he was viewed as being socially inferior.[59] The only Fellow he is recorded as having proposed was his friend the Scottish mathematician James Gregory (1638–1675).[60] Nevertheless, he was involved in a number of notable publication projects under the auspices of the Society, including Wallis's *Treatise of Algebra*.[61]

Practical mathematics is thus at once part of a story that transcends the history of the universities of Oxford and Cambridge as well as that of institutions such as the Royal Society or Gresham College. While being a key element of the scientific movement of the seventeenth and eighteenth centuries, the various kinds of practitioning also enter in important ways the wider social and economic history of the country. Yet we should not see barriers where none existed. Despite limitations engendered on an institutional level, practical mathematicians often enjoyed close intellectual ties to their academic counterparts, just as there were clear efforts on the part of the Royal Society to embrace theoretical knowledge in a practical form. One thinks here of William Petty's (1623–1687) attempt to introduce the public to Galilean physics by showing its value in shipping, artillery, and building.[62] In similar vein, Collins is able to point out on one occasion that Adam Martindale's erstwhile country almanacs were esteemed by several members of the Royal Society as being 'very useful, especially for Country Affairs'.[63] As the mathematical intelligencer par excellence, Collins was one of the most important scientific correspondents of Wallis and Newton, not to mention his extensive epistolary commerce with James Gregory, Edward Bernard (1638–1697), and Isaac Barrow. There were others, too. Newton exchanged numerous letters with his friend the gauger and tobacco cutter Michael Dary, just as he later befriended the Welsh mathematician of humble background, William Jones (1675–1749). The mathematics teacher and surveyor Jonas Moore (1617–1679) was ultimately able to shake off his modest origins in Lancashire to receive a knighthood, become a friend of Christopher Wren and Robert Hooke (1635–1703), and be admitted Fellow of the Royal Society. When it came to practical mathematicians there were, it would seem, no hard and fast rules.

The essays collected in this volume are divided thematically. The first section looks at examples of how developments in mathematics impacted on navigation, seafaring, and warfare. It begins with a glance back at navigational techniques in the

[59] See Michael Hunter, *Establishing the New Science. The experience of the early Royal Society*, Woodbridge: The Boydell Press 1989, p. 11, p. 270; Hunter, *Science and Society*, pp. 72–3.

[60] Michael Hunter, *The Royal Society and its Fellows, 1660–1700*, 2nd edn, Oxford: The Alden Press for The British Society for the History of Science 1994, p. 65

[61] See Birch, *The History of the Royal Society*, vol. IV, pp. 4, 155, 166–7, 284.

[62] William Petty, *Discourse made before the Royal Society the 26. November 1674. Concerning the Use of Duplicate Proportion in sundry important particulars*, London: for John Martyn 1674. In his epistle to William Brouncker, sig. a3r, Petty writes: 'For that near half the whole Discourse relates to Shipping, Artillery, Fortresses, Seabanks, &c. which all concern his Majesties Service'; See also Hunter, *Science and Society*, p. 96.

[63] John Collins, 'To the Reader' in Martindale, *Country-Survey-Book*. For an example of Martindale's almanacs see *The Country Almanack for the Year, 1675. Suted to the several Capacities, Humours, and Occasions of Gentlemen, Scholars, Travellers, and Husband-men, &c.* London: F. L. for the Company of Stationers 1675.

Renaissance. Up to now a considerable amount of scholarly work has been done on the practice of mathematics in Renaissance Europe and the early decades of the seventeenth century. There are numerous reasons why this should have been so. The earlier period is at once marked out dramatically by voyages of exploration, in many ways made possible through the introduction of new or improved navigational instruments or the employment of ever more sophisticated nautical charts. At the same time, computational techniques were transformed, even revolutionized, through Napier's invention of logarithms.

In 'Mechanicall practises drawne from the Artes Mathematick', Jim Bennett considers the relationship between mathematical learning and navigational practice in the work of the Elizabethan explorer John Davis (*c.*1550–1605). Living at a time when there were a number of prominent mathematicians with experience at sea, including John Dee (1527–1608/9), Thomas Harriot (*c.*1560–1621), and Edward Wright (1561–1615), Davis is remarkable in that he sought to reconcile theoretical and practical aspects of navigation and thereby ultimately succeeded in establishing a sounder foundation for various navigational techniques than had previously existed. As Bennett points out, Davis achieved considerable fame already during his lifetime. Not only was a strait he passed during the course of an expedition to find the Northwest Passage named after him, but also the quadrant for measuring the Sun's altitude, more commonly known as the backstaff and by means of which the ship's latitude could be established. Equally if not more important historically, however, is the evidence of his navigational practice as revealed by his journals and his book *Seamans Secrets*, printed in 1595. Here for the first time we find bearings, distances, and other measurments set out in tabular form, with one column free for entering data on longitude. This measurement was of course not determinable at sea at that time, and is something Davis only referred to in the context of 'paradoxall' or great circle sailing. Importantly, he recognized that the practical means of following one's course when sailing in that fashion was to use a chart rather than a globe. Mercator's chart had been published in 1569 and was therefore available to Davis, but it fell to his assistant Wright to explain the mathematics on which its projection was based.

Margaret Schotte's chapter looks at another side of navigation, namely the formalizing of examinations seafarers had to undergo in the early modern era before they could take charge of a ship. Edward Wright also plays a role here, because he was hired by the English East India Company in 1614 to teach and examine its mariners. But although similar regulatory roles were given to mathematicians elsewhere in Europe, no precise pattern to the implementation and content of examinations is discernable. A good hundred years before Wright, in 1508, Spain introduced the requirement that its navigators be subjected to a formal exam. The explorer Amerigo Vespucci (1451–1512) was tasked with examining the use of maps, instruments, and knowledge of new techniques of celestial navigation to determine position—a clear contrast to the earlier predominance of coastline sailing and requiring awareness of the theoretical principles on which they were based.

Over time every state with an extensive maritime presence required that its pilots, masters, or mates demonstrated their abilities through examination, although in some cases, as with the Dutch East India Company, individual organizations established their own regulatory regimes. Schotte identifies five categories of practical or theoretical knowledge that to differing degrees might be examined: the use of instruments, the art of pilotage, the principles of cosmography, familiarity with geography, and the ability to calculate one's position. Of all the routines for assessing a navigator's ability, she suggests, this last category was most challenging: often involving a written test in mathematics developed in tandem with bespoke manuals designed for seafaring practice.

The focus of Rebekah Higgitt's chapter is again on nautical examinations, but this time with an eye to the examiners themselves, namely those employed by the august institution Trinity House, a charitable maritime guild chartered under Henry VIII. Originally situated near the docks in Deptford, Trinity House relocated to the City of London after the Restoration. This was a significant move, because the institution was now at the confluence of maritime and mercantile interests in the metropolis, with the Custom House and the Navy Office on the one side and the headquarters of overseas trading companies close by on the other side.

As Higgitt points out, examining the abilities of navigators became increasingly important after 1621, because Trinity House now had to certify ships' masters intending to work on the king's ships. In the second half of the century it went on to acquire additional roles in examining and certifying the boys who passed through the Royal Mathematical School. With examinations becoming increasingly text-based, it soon became necessary to appoint mathematicians to carry them out instead of the practically trained brethren of old. Most of the examiners are shown to have been private teachers of mathematics who advertised their services in books and journals and often formed part of a professional family dynasty. Sometimes boundaries were transcended. Thomas Weston (d.1728) served an apprenticeship under John Flamsteed (1646–1719) at the Royal Obervatory before establishing an academy for gentlemen and members of the nobility while being asked to assess new books and instruments submitted to Trinity House. He also wrote a treatise of arithmetic that was published posthumously.

The question of striking a balance between pure mathematics and practical applications is an important part of João Domingues's account of the development of military training for engineers in Portugal following the restoration of that country's independence from Spain in 1640. A pressing demand at that time was the construction of modern fortifications along the border to its Iberian neighbour for which nationally trained personnel was required. Out of the need to fulfil this demand emerged the Class of Fortification and Military Architecture under chief engineer Luís Serrão Pimentel (1613–1679), whose approach to training was straightforward: prospective engineers needed only sufficient mathematics to be able to design and build defences. Crucial changes to this approach came about through Manuel de Azevedo Fortes (1660–1749). Trained in Madrid and Paris, Azevedo Fortes sought to elevate the status of engineers and to this end brought about the transformation of the Class of Fortification to the Military Academy around 1720. He compiled a book entitled *O Engenheiro Portuguez* [The Portuguese Engineer], partly based on translations of French works by the likes of Jacques Ozanam (1640–1718) and Claude Dechales (1621–1678) covering topics such as trigonometry, surveying, and practical geometry. For him, Portuguese engineers should be trained broadly in the mathematics relevant to their military careers.

In the second section the focus is on the local societies that catered for the interests of amateur and professional mathematicians from the seventeenth century onwards and on what can be best described as the contemporary burgeoning of mathematical culture in Britain and Ireland. It begins with a contribution by Sloan Despeaux and Brigitte Stenhouse on so-called 'Questions for Answer' journals in north-west England and their philomath editors. The concept of these journals was not in itself novel; the *Ladies' Diary* of the early eighteenth century had already featured mathematical questions for answer as a polite endeavour suitable for its readership. Later on, non-university men, some of whom practised mathematics in their working lives as surveyors, teachers, or instrument-makers while others simply studied mathematics as an avocation unconnected with their employment, formed dedicated societies in industrial northern towns and cities like Manchester, Oldham, or Liverpool. Although these mathematical societies published no formal proceedings, their activities are recorded in journals such as *The Student*, the *Liverpool Apollonius*, and the *Preston Chronicle*.

Some contributors of solutions were former mathematical masters, while others like Thomas Wilkinson, a farmer from Blackburn, were self-taught. As Despeaux and Stenhouse point out, there was often a spirit of competition apparent on the part of those solving more demanding questions, some of which were drawn from the Senate House examinations at Cambridge. Less well-off contributors sometimes sold their solutions for cash so that their purported authors could take the glory. Journals like the *Philosophical Magazine* are known to have been read by the gentlemen of nineteenth-century science or fellows of London's learned societies, but generally they hesitated to grant working-class mathematicians the recognition they deserved. An exception is the self-taught mathematician and gas company employee Septimus Tebay, who so impressed gentlemanly readers of the *Preston Chronicle* with his solutions that they arranged for him to study at Cambridge where he subsequently excelled.

One of the most remarkable figures in late eighteenth and early nineteenth-century mathematics in Britain is Charles Hutton (1737–1823), a man of humble origins from the north-east who rose to become professor of mathematics at the Royal Military Academy, Woolwich, and Fellow of the Royal Society. Equally remarkable was his library which contained some three thousand mathematical books and a huge collection of scientific journals and pamphlets ranging from the Royal Society's *Philosophical Transactions* to the *Gentleman's Magazine* and the *Ladies' Diary*. In these two popular journals he had in early adulthood published numerous solutions to mathematical questions. A product of his maturity, the *Dictionary of Mathematics* (1795) is considered a masterpiece of the popularization of science.

Part of the reason for Hutton's extensive collection according to Benjamin Wardhaugh in his chapter on Georgian mathematicians was a profound cultural change that occurred around the middle of the eighteenth century when it became usual for pupils—and not just their teachers—to possess copies of their textbooks. Another important factor was that military academies like that at Woolwich and dissenting academies like that at Warrington took mathematics seriously. Thanks to his increasing national prominence, Hutton was able to establish a course-based programme of instruction at Woolwich and to export it through his textbooks and disciples to other institutions. His *Course of Mathematics*, which appeared in 1799, achieved considerable resonance and was popular with British overseas trading companies. Besides containing works of British mathematicians such as John Keill (1671–1721) and William Whiston (1667–1752), Hutton's library boasted a sizeable quantity of books intended for practitioners and the numerate trades as well as foreign publications by the likes of Leonhard Euler (1707–1783) and Pierre-Simon Laplace (1749–1827). Without doubt worthy of preservation, it was eventually put up for auction and dispersed along with Hutton's papers: a doleful tale of destruction, as Wardhaugh pertinently comments.

We stay in the late eighteenth and early nineteenth centuries for Christopher Hollings's chapter, but here the focus switches to the so-called literary and philosophical societies which sprang up around the British Isles and Ireland at that time, especially in industrialized towns and cities. These 'Lit & Phils' often had links to mechanics' institutes or museums, and such connections most likely would be reflected in the programme of lectures typically organized for members. Generally speaking their averred aim was to communicate scientific knowledge to an interested lay audience, but as Hollings discovered in the course of his investigation mathematical knowledge of an innovative nature tended to come up short. Most talks would have a local flavour, discussing perhaps the flora and fauna of the region, or nearby archaeological finds. It is not unusual to find lectures on practical mathematics on the programme of a Lit & Phil or perhaps remarks on the mathematics of ancient Egypt in the context of an archaeological presentation, but anything else such as a lecture by the Cheshire clergyman Thomas Kirkman (1806–1895) on group theory to the Manchester Literary and Philosophical Society in 1861 was truly exceptional.

But that society was exceptional, too, for its *Memoirs* also contain contributions from further contemporary mathematicians of note, namely Arthur Cayley (1821–1895) and James Cockle (1819–1895). As Hollings points out, although these papers were published alongside others for a more general audience on topics in medicine, technology, and natural history, they would in fact be intended for a wider scientific and mathematical community beyond Manchester. But if the Lit & Phils fell short when it came to promoting mathematics, the same cannot be said about another institution: the British Association for the Advancement of Science, founded in 1831. Prominent members, including David Brewster (1781–1868) and William Whewell (1794–1866), deplored the failure of government, universities, and the Royal Society to stimulate the advance of science in Britain. Already in its foundational year, the Association, aware that the Royal Society in particular was no friend of mathematics, commissioned a report on progress in mathematics with a particular view to Continental developments. Although not in time for the next meeting of the Association, that report, by George Peacock (1791–1858) of Trinity College, Cambridge, was presented in 1832.

Only in a few areas did new professional opportunities for mathematicians open up in the eighteenth century. One of these was in actuarial science, and in his chapter David Bellhouse takes us through the main developments up to the founding of the Institute of Actuaries in 1848. After first setting the scene with reference to early work on mortality tables by John Graunt (1620–1674) and William Petty, he notes that there was no proper actuarial pricing in eighteenth-century England even though attracting annuitants was essential to government already at this time for financing wars. Although Edmond Halley (1656–1742) suggested age-based pricing of government life annuities, he was ignored. At the same time, his computational methodology for calculating these annuities contained a weakness that Abraham De Moivre (1667–1754) was the first to resolve. Interestingly, he worked as a consultant from around 1739, using Slaughter's Coffeehouse in St Martin's Lane, London as his 'office'. De Moivre's former pupil James Dodson (*c.*1705–1757), who enjoyed close ties to London's mathematical community, carried on his work after his death and wrote an annuity book, the third volume of his *Mathematical Repository* (1748–55). However, all was not well with the market. When companies started offering annuity schemes in response to growing demand on the part of an expanding middle class, mathematicians such as Richard Price (1723–1791) and William Dale (18th century) were able to identify serious defects in their pricing.

In respect of life insurance things were even worse, as Bellhouse shows. In the early eighteenth century there were no premium schedules: the only advice given to undertakers was to become acquainted with the person to assess the individual risk. In Scotland, however, the mathematician Colin Maclaurin (1698–1746) devised a novel approach based on aggregate risk. The main promoters of life insurance schemes in the eighteenth century were merchants and businessmen, and mathematicians generally had no impact on the market, although Dodson did attempt to do so. When, in the early nineteenth century, several new insurance companies were formed they hired men with mathematical skills to carry out their pricing. But no professional body existed at that time to set up a system of examination and thus ensure standards. This only changed with the creation of the Institute of Actuaries.

The third section looks at mathematical practitioners and their scientific milieus. In 'Assembling the Scribal Self', Stefano Gulizia explores the role played by the Paduan virtuoso and collector Vincenzo Pinelli (1535–1601) and his renowned library in the scientific life of the Republic of Venice in the late Renaissance. Aptly describing the library as something like a public institution with open connections to the Venetian government, Gulizia points out that it was visited by patrons and collaborators alike, making use of its rich mathematical resources, including a substantial collection of instruments, often with booklet or broadsheet instructions on how they were to be employed, while many of the books understandably concerned naval or maritime themes. There is ample evidence that the library was frequented not just by scholars, but also by practitioners of mathematics, not a few of whom left traces of their 'scribal selves' in the form of notes, comments, or diagrams. Indeed, the picture that emerges is one of an amenable locus of artisanal and scholarly interaction and collaboration: the library was not so much a place for solitary reading as a site of writing and production by like-minded investigators. Visitors valued the range of its contents, stretching from astronomy and cosmology to fortification, optics, and cryptography. They would copy out papers or passages from books, enter notes of their own, and discuss topics of common interest, with Pinelli evidently overseeing everything and occasionally keeping records of what had transpired. Although as a wealthy patrician he did not visit warehouses or dockyards, he clearly aligned both speculative and practical mathematics with the interests of the state and correspondingly allowed his library to be used as a kind of clearing house of early modern mathematical science.

A considerable number of mathematicians can be classed as members of Pinelli's circle and often were among his closest friends. Famously, Galileo Galilei (1564–1642) drew on his collection, but others were perhaps more visible. Giuseppe Moleto (1531–1588), for instance, who had trained under the Sicilian mathematician Francesco Maurolico (1494–1575) and shared Pinelli's desire of promoting interest in ancient Greek mathematics. Moleto collaborated with Matteo Macigni (*c.*1510–1582), who possessed an extensive collection of Greek mathematical manuscripts, in producing a tract on reforming the Roman calendar that was given to Pinelli. Another member of the circle was Francesco Barozzi (1537–1604), who is credited with spurring a European revival of Proclus. Barozzi had close collaborative ties to Venetian shipyards and in particular to senator Giulio Savorgnan (1510–1595), commissioner to the arsenal, who oversaw a workshop for constructing mathematical instruments. Savorgnan, too, frequented Pinelli's house and is known to have received from its owner a set of wheels for some intended mechanical construction. Such scholarly circles were after all based on the principle of give and take.

Philip Beeley's chapter brings us back to the seventeenth century and seeks to throw light on the nature of the relationship between practical mathematicians and their academic colleagues in the universities, focusing particularly on the role of John Collins since this is best documented. In the 1650s he was to be found in London, leading what was in many ways a typical existence for a practical mathematician. Collaborating with the instrument maker Henry Sutton (d.1655), he wrote a number of books commissioned by him explaining how his instruments were to be used. Soon after the Restoration, however, in 1662, Collins established contact with the Cambridge-educated mathematician John Pell and asked him to provide a rigorously derived proof for an (undescribed) procedure he had previously only carried out practically. This was perhaps a reflection of the contacts Pell had already cultivated by this time to other mathematicians in the metropolis's practical milieus such as John Leake and Henry Bond (*c.*1600–1678). Not long afterwards, Collins would help see the *Introduction to Algebra* by Pell and Brancker through the press. When, at the height of the Great Plague in 1665, Collins decamped to Oxford, there was a somewhat similar meeting on his part with John Wallis. The London mathematician asked Wallis to provide a method for determining the volume of an elongated spheroid, this being a stereometrical problem of potential importance to gaugers, since the body concerned was equivalent to the kind of vessel that might be used to store wine. Not only did Wallis readily agree to deliver a solution, but he also showed Collins some of the treasures of the Bodleian Library, a privilege usually reserved for eminent visitors. Over the years, Collins and Wallis conducted a scientific correspondence that was increasingly collegial, but often with clearly defined roles: Collins would send Wallis mathematical news, especially of new books from abroad, while Collins occasionally posed some practical question to which the Savilian professor would patiently respond. There were however topics of common interest, especially algebra, on which they would meet on fairly equal terms.

As Beeley makes clear, books constituted an important part of Collins's correspondence with all his academic friends in Britain, because he was uniquely able to procure those publications from abroad which they desperately desired—and without which academic discussion in the country would have been greatly impoverished.

On the other hand, Collins expended a great deal of time and energy in his epistolary exchanges with Barrow and Newton, both of whom were notoriously reluctant to publish, in trying to get their work into print. On the whole there was a productive link between the two spheres of activity with a common aim of promoting the growth of mathematical knowledge. But differences were not ignored, either. Thus, Collins noted on one occasion in his exchanges with James Gregory (1638–1675) that the quadrature of the hyperbola, with which the Scottish mathematician was theoretically concerned, was also of importance to gaugers like his friend Michael Dary (1613–1679). Barrow for his part would occasionally point out to Collins his own limitations when it came to practical problems. In one instance, he implored Collins to adopt a more abstract style of writing, as this corresponded more to his own way of thinking.

Practically proven procedures also confront mathematical methods devised by scholars in Thomas Morel's contribution 'All of this was born on paper', in which he investigates the historical context in which geometry came to be employed successfully in the planning and development of long-distance tunnels for draining water out of metallic mines in Germany's Harz mountain region. This was a transformative development. Even at the turn from the fifteenth to the sixteenth century geometry was primarily used in the context of land surveying to establish the limits of mining concessions, while drainage tunnels were made on an ad hoc basis according to need, and not planned on paper. As Morel points out, until the end of the seventeenth century the applications of mathematics were fairly clearly defined: it was usual for mining maps to be drawn in order to plan extraction, while underground surveyors were in charge of linking mine shafts to existing drainage tunnels. By the beginning of the eighteenth century, there was, he finds, increasing trust in the power of geometry to plan extraction work, surveys and maps were regularly used, but in a piecemeal fashion. There was still the view that it was best to work over short distances in order to minimize risk of error. However, improvements in precision eventually led to the idea of fully mathematized planning of drainage tunnels in straight lines over long distances, with the first instance being the Deep George Tunnel in the kingdom of Hanover.

Mining engineers in the Clausthal-Zellerfeld region had come to the conclusion that only a new tunnel of this nature would be able to bring long-term relief to drainage problems. Importantly, precise maps of the mines were already available, so that geometrical expertise could be applied almost immediately. Being able to draw on a vast quantity of data collected over the previous century, the two surveyors tasked with planning, Samuel Gottlieb Rausch (d.1778) and Johann Christian Länge (d.1803), could restrict new on-site work to a minimum and thus keep costs low. They were however called upon by the mining council, with which they were in regular contact, to carry out a full quantitative survey of the different options before construction began. The final plans were then taken to London for approval by George III.

When the tunnel was finally inaugurated in 1799, after 22 years of building, it proved for the first time that it was possible to conceive major engineering projects from scratch using geometry. But it was not the result of abstract academic theories, certainly not of the kind that Abraham Gotthelf Kästner (1719–1800) presented in lectures at the University of Göttingen. Rather, the plans were drawn up by qualified practitioners who took into account the technical and local specificities of the Harz Mountains.

The final section of *Beyond the Learned Academy* is devoted to the practice and teaching of mathematics. Ivo Schneider takes us first to the southern German city of Ulm and considers Johannes Faulhaber's (1580–1635) path from schoolmaster to fortification engineer. Immediately, a stark contrast to mathematics teachers in early modern England becomes apparent, for a five- or six-year apprenticeship was prescribed for any man wanting to become a *Rechenmeister* and set up his own writing or reckoning school in that city. The number of teaching positions was limited to six at that time, and Faulhaber was fortunate that one of these became vacant in 1600. Once established, he had various means at his disposal to get a step ahead of competitors. Already in early years, he had a portrait produced with a clear intention to convey his professional status and abilities to potential clients. Hence, beneath the portrait the impressive list of his publications in German, noting they were also available in Latin, was cited. As a correspondingly attired teacher of reading, writing, and reckoning he cut a modest but respectable figure. However, the number and range of publications clearly suggested that his skills well exceeded those of his peers in neighbouring cities like Augsburg or Nuremberg. Other things were standard. As a teacher he was required to have a house and be married, because pupils lived in the teacher's household and were provided for. For these provisions the parents of a pupil paid a fee to the teacher, while the city paid him a modest salary for his services.

Rechenmeister sought competitive advantages over one another in other ways, too. It was not unusual for them to set their peers mathematical challenges, although few went as far as Faulhaber who in one instance boldly set a challenge to all the philosophers and mathematicians of Europe. The aim of such exercises, Schneider suggests, was to establish a kind of ranking, but in Faulhaber's case they also enabled him to build up an impressive network of correspondents. Problems and their solution were also an important part of the day-to-day business of teachers. In his first publication, the so-called *Lustgarten* of 1604, Faulhaber set out 160 mostly cubic problems, providing in each case at least one solution. The methods for arriving at solutions were not disclosed. Whoever wanted to grasp the way he had proceeded had to go to Faulhaber and receive instruction against payment. In this sense some arithmetic textbooks were intended to accompany personal instruction, while others were intended for self-instruction. Likewise, some clients were happy just to receive a solution to a problem they raised, while others desired knowledge of methods. Unscrupulous *Rechenmeister* could of course abuse this service by pretending that different instances of one method were in fact truly different.

One of the ways Faulhaber sought to stand out brought him into difficulty. He claimed that he was a prophet directly enlightened by God and through deliberations on biblical numbers, partly arising from his involvement with the Rosicrucian movement, suggested that he was able to decipher divine secrets. Unsurprisingly, these aspects of his professional life fell foul both with civic and ecclesiastical authorities and for a time he landed in prison. This setback did not last long however, so good was his reputation. And he was able to enhance it further by extending his expertise to fortification, contributing directly to the strengthening of Ulm's defences. He also spent time in Basel and in the Low Countries in this capacity, but eventually returned to Ulm where he was appointed engineer and earned a salary ten times greater than that he had had previously.

Questions of a different sort feature in Albrecht Heeffer's chapter on Sybrandt Hanszoon Cardinael (1578–1647). He reminds us at the start that notable mathematical scholars of the early modern period such as Johannes Kepler (1571–1630) and Claude Mydorge (1585–1647) looked down on men who used algebra, considering it as purists to be a tool for merchants and unsuitable for constructions in geometry. However, in the innumerable mathematical schools flourishing at that time on the Continent where arithmetic and geometry were taught to the sons of merchants, gaugers, and surveyors algebra was not to be found either, at least not on the curricula of these schools. And this despite the fact that many of the teachers employed by these schools were well apprised of the subject and even wrote treatises on it: Adam Ries (1492–1559) and Henricus Grammateus (1495–1525/6) in Germany, and Simon Stevin (1548–1620), Jan Stampioen (1610–1653), and Valentin Mennher (1521–1570) in the Low Countries all come to mind. Stevin, for example, founded an engineering school in Leiden and devoted a substantial part of his *L'Arithmetique* (1585) to the subject, but algebra was not to be found on his school's curriculum. Furthermore, when these authors wrote on topics such as engineering or surveying the title pages of their books stated explicitly that problems were solved without the use of algebra. Clearly, Heeffer suggests, they thought that this was the right way to do things.

Almost all the questions in Cardinael's book *Hondert Geometrische questien* [Hundred geometrical questions], published in Amsterdam in 1612, are set in a practical context with many of them concerned with tasks falling squarely in the domain of surveying. The author, who besides teaching mathematics practised as a surveyor and wine-gauger, was highly regarded in Dutch mathematical circles. His book was translated into German by Sebastian Kurz (1576–1659), a *Rechenmeister* in Nuremburg and friend of Faulhaber, while Thomas Rudd (1583?–1656) prepared the English translation. What makes it particularly remarkable is that, as Heeffer has established, the surveying techniques employed by Cardinael are clearly related to certain propositions in Book II of Euclid's *Elements*, but are even more closely aligned with surveyor formulations found in Old Babylonian mathematics, and are part of a tradition that can be identified in Fibonacci's *Practica geometriae*. The surveyor's geometry found in Cardinael's work is devoid of symbolism, Heeffer concludes, because it derives from an ancient tradition of geometrical algebra. The algebra which practical mathematicians distanced themselves from was symbolic algebra.

In Boris Jardine's 'The Life Mathematick' the focus switches back to early modern England and one of the pre-eminent teachers of mathematics in London, John Speidell (1577–1649), author of an important early work on logarithms as well as a number of elementary tracts including *A Geometricall Extraction*, published in 1617. Although relatively little has been known about him up to now, his son Euclid, who followed in his pedagogical footsteps, drew up a memoir entitled 'The Life of John Speidell', only recently discovered in Lincoln, that not only throws important new light on him but also on mathematical culture in the metropolis in general.

Previously known facts about John Speidell can be easily summarized. Apart from writing on logarithms, geometry, and arithmetic, he taught mathematics privately near Drury Lane and also for a time at Sir Francis Kynaston's (1587–1642) short-lived academy Musaeum Minervae, where it seems he taught geometry and its practical applications as part of a course deemed suitable for gentlemen. As we now learn from Euclid's memoir, his father Sebastian (d.1597) was a merchant from the Palatinate who settled in London around 1561. John Speidell began teaching around 1607 and invented a calculating scale that was made by Elias Allen (*c.*1588–1653) in the Strand and John Thompson (*fl.* 1609–1648) in Hosier Lane, Smithfield. Like Euclid later, John Speidell made instruments central to his teaching, probably in such a way that they constituted the beginning and end of instruction. The instruments both men used were probably aimed at aiding calculation and thus addressing the well-known deficits in numeracy at the time. Possibly John was responsible for other inventions, but despite his keen interest in instruments he never published a single description of one. However this fact, as Jardine indicates, does not allow us to draw any sound conclusion, as some teachers were reticent in publishing such information, when explaining the use could be reserved for paying clients. And there are in general problems about establishing the rightful ownership of a design when it comes to early modern mathematical instruments. Old instruments were often falsely described as being new inventions, or existing forms could be augmented and adapted so as to appear as an entirely new device. Possibly more important than authorship is the fact that John Goodwyn (d.1605?) and Edmund Gunter (1581–1626) both taught the use of the sector and likewise adapted the design so as to shift emphasis from geometry to arithmetic.

Even in his last book, *A Briefe Treatise for the Measuring of Glass, Board, Timber or Stone*, published in the 1640s, John Speidell advertised instruction in arithmetic. He had earlier taught his son the same subject 'as far as Trade might require' according to Euclid's memoir. Euclid was soon spotted and admitted to Westminster School by its legendary master, Richard Busby (1606–1695). After completing his schooling he continued to take instruction from his father, now in advanced arithmetic and geometry. After a spell working as a clerk he set up as a teacher of mathematics in Threadneedle Street, where Henry Sutton had his workshop. He apparently had Sutton make a calculating scale for the purpose. Before long he was close to John Collins and Michael Dary who shared with him the precariousness of living the life of a practical mathematician in later seventeenth-century London.

From early modern London we switch to nineteenth-century Belfast for the final chapter, looking at the father of Lord Kelvin, the somewhat neglected figure James Thomson senior (1786–1849), and the role he played in the development of mathematics at the Belfast Academical Institution. The author, Mark McCartney, points out that Thomson was initially self-taught, having studied *The Scholar's Guide to Arithmetic* by John Bonnycastle (1751–1821) and a number of books on dialling already as a young man. Indeed, he is reported as having constructed sundials at the family home in County Down, Ireland, already at the youthful age of eleven.

Progress, betterment, and the value of education became key parts of Thomson's life. His precociousness was recognized early on and when eventually he attended a privately run local school he soon rose through the ranks of his fellow pupils to become a teaching assistant. After delays to acquire the necessary knowledge of Greek and Latin, he attended Glasgow University from 1810, where among other awards he won the senior mathematics prize. Fortunately, his graduation coincided with the founding of a new academic institution in Belfast, pertinently named the Belfast Academical Institution, which combined a school with a college aimed at providing students with higher education equivalent to that of the Scottish universities. Thomson applied successfully for the post of teacher of mathematics and arithmetic at the school, but within a short time was offered the mathematical professorship at the college as well. He married and began raising a family around the same time, subsequently taking it upon himself to educate his children, while also spending his time at home writing books such as *A Treatise on Arithmetic in Theory and Practice* (1825). Having previously used the well-acquainted Bonnycastle as a textbook, Thomson soon replaced this with his own books instead. Although there were by now other good arithmetic books on the market, McCartney notes, Thomson's stood out on account of the extra factual content relating to daily business of many of the questions posed. He also authored books for the college such as his *Introduction to Differential and Integral Calculus* (1831).

Thomson was active in other respects, too. For example, he became an active member of the Belfast Literary Society, comparable to the 'Lit and Phils' in England, and read papers on topics in mathematics, astronomy, and geography. It is perhaps a further reflection on the widespread popular interest in science at the time that the local *Belfast Magazine and Literary Journal* not only reported on the mathematical examinations at the college but also published a selection of the questions that were set. Following the death of his wife, Thomson moved back to academia proper and took up the chair in mathematics at Glasgow, a post he held from 1831 through to the end of his life. He is remembered, however, particularly for his impact on the Belfast Institution, situated in what was then a rapidly growing town, where he served as an enthusiastic and respected mathematical educator.

These studies, covering a period from the late Renaissance through to the nineteenth century, reflect the richness and diversity of the mathematical practice taking place in Britain and in parts of Continental Europe at that time. We see examples of the erstwhile central role of noble patronage in supporting the mathematical sciences slowly being replaced by civic and state involvement, while practitioners in London and elsewhere developed and extended their ties to the mercantile environment in which they largely operated. With their financial means often limited, their existence correspondingly precarious, surveyors, diallers, teachers, and others drew crucial support from the milieus in which they lived and worked. Practitioners, stationers, and instrument makers collaborated closely, their activities were strongly interdependent, quite apart from the fact that they were often immediate neighbours. Over and above these local interdependencies, practitioners formed clubs and associations where they could exchange news and ideas pertaining to their day-to-day activities. The sheer number of inexpensive practical mathematical books that were printed since the late Renaissance is truly remarkable, yet only with the onset of conscious moves to popularize science in the eighteenth century were broader sections of the populace embraced by this burgeoning culture of mathematics. Teaching, too, underwent considerable changes from the privately run writing and arithmetic schools that were so prominent in seventeenth century England to what appears to have been more widespread instruction in grammar schools later on. Yet even in nineteenth-century Ireland it often came down to individual families to organize schools for the education of local children. A lot remains to be done before a more finely grained picture emerges, and it is to be hoped that these studies will serve to promote further work in an area of the history of mathematics that has tended to be neglected, especially when it comes to the seventeenth century and thereafter.

PART I
NAVIGATION, SEAFARING, WARFARE

2

'Mecanicall Practises Drawne from the Artes Mathematick'

The Mathematical Identity of the Elizabethan Navigator John Davis

Jim Bennett

'A most learned seaman and a good mathematician'

The relationship between learning and practice in early modern times is a well-worn debate among historians of mathematics. That generous and useful category 'mathematical practitioner', though for the most part constructed in later historical work,[1] can be applied to many authors of mathematical handbooks from the period, who were convinced that mathematics could have a transformative practical role. For them, mathematics could offer practical work beyond everyday techniques for calculation and drawing, could be adopted as an identifying, enabling, and reforming characteristic in a range of disciplines. To what extent the unmathematical practitioners were converted to this opinion is open to question.

This is not exactly the question addressed here, but something closely related. To what extent are the perspectives, categories, and issues regarding learning and practice that exercise historians reflected in the thinking of the practitioners themselves, mathematical or otherwise? We will look at a group of writers on navigation in late sixteenth-century England, while focusing particularly on one of their number, John Davis—the 'most learned seaman' and 'good mathematician' of the quotation that forms the title to this section.[2] Did Davis see the relationship between learning and practice as an issue that might concern him and, if so, what kind of analysis did he bring to it?

Early modern mathematicians, and scholars in general, have been open to the charge of being unfamiliar with the rigours of practical work and unrealistic about the relevance of their ideas and proposals. In 1581 the seaman author Robert Norman ridiculed his scholarly detractors, 'in their studies amongest their bookes'.[3] The

[1] E. G. R. Taylor, *The Mathematical Practitioners of Tudor & Stuart England*, Cambridge: Cambridge University Press 1954; E. G. R. Taylor, *The Mathematical Practitioners of Hanoverian England, 1714–1840*, Cambridge: Cambridge University Press 1966; Stephen Johnston, 'The identity of the mathematical practitioner in 16th-century England', www.mhs.ox.ac.uk/staff/saj/texts/mathematicus.htm#note3.

[2] See note 12 below.

[3] Robert Norman, *The Newe Attractive, Containyng a Short Discourse of the Magnes or Lodestone*, London: Iohn Kyngston for Richard Ballard 1581, sig. B.i^{v}.

charge can still be levelled today by historians.[4] Yet more English mathematicians of significance had experience at sea than might be expected. Thomas Harriot sailed to Virginia in 1585 to participate in the colonial venture of Walter Ralegh, and returned to England with Francis Drake the following year.[5] In 1589 Edward Wright joined a privateering expedition to the Azores, as a Captain in a fleet commanded by George, Earl of Cumberland, in the course of which he spent time with John Davis.[6] On an earlier voyage he had assisted in the evacuation of Ralegh's Virginia colony, and had probably encountered Harriot on the voyage home. These were prominent mathematicians who went 'beyond the academy' and experienced life at sea. Less well known was William Borough, who sailed with Richard Chancellor and was a pilot to the Muscovy Company before writing on navigation.[7] Robert Hues was a mathematician associated mainly with Oxford and a pupil in navigation of Harriot.[8] He had a range of maritime experience, including a circumnavigation with Thomas Cavendish, service under Wright in the Azores, and a voyage home from the Magellan Strait with Davis. Emery Molyneux sailed with Drake and possibly Davis, before becoming a reputable mathematical instrument maker.[9] Davis himself must be included, as a lifelong seaman who wrote a book, *The Seamans Secrets*, on the mathematical art of navigation.[10]

Davis is a promising candidate for a case study, because he has always figured strongly on both sides of any boundary we might construct between shipboard practice and mathematical learning. His contemporary *fame* derived from his achievements at sea, while a very significant part of his contemporary *reputation*, if we can make the distinction, rested on his technical knowledge of navigation within a broad command of seamanship. Further, this underlying technical knowledge was seen as a species of mathematics.

Two prominent eponymic memorials mark this duality: the Davis Strait and the Davis quadrant. The Davis Strait, an expanse of sea between Greenland and Baffin Island, was named after Davis on account of his explorations there in search of a Northwest Passage to the north Pacific and so onward to the East Indies. The Davis quadrant, today more commonly called the backstaff, was the early name for an instrument for measuring the altitude of the sun, the name and the instrument

[4] For a recent example, see Thomas Morel, '*De Re Geometrica*: Writing, Drawing, and Preaching Mathematics in Early Modern Mines', in: *Isis* 111/1 (2020), pp. 22–45.

[5] J. Roche, 'Harriot, Thomas (*c.*1560–1621), mathematician and natural philosopher', in: *Oxford Dictionary of National Biography*, Oxford: Oxford University Press 2004, https://doi.org/10.1093/ref:odnb/12379.

[6] A. Apt, 'Wright, Edward (bap. 1561, d. 1615), mathematician and cartographer', in: *Oxford Dictionary of National Biography*, Oxford: Oxford University Press 2004, https://doi.org/10.1093/ref:odnb/30029; Albert Hastings Markham, ed., *The Voyages and Works of John Davis the Navigator*, London: The Hakluyt Society 1880, pp. xxxvii, 65–92.

[7] Baldwin, 'Borough, William (bap. 1536, d. 1598), explorer and naval administrator', in: *Oxford Dictionary of National Biography*, Oxford: Oxford University Press 2008, https://doi.org/10.1093/ref:odnb/2915.

[8] Susan M. Maxwell, 'Hues, Robert (1553–1632), mathematician and geographer', in: *Oxford Dictionary of National* Biography, Oxford: Oxford University Press 2008, https://doi.org/10.1093/ref:odnb/14045.

[9] Susan M. Maxwell, 'Molyneux, Emery (d. 1598), maker of globes and ordnance', in: *Oxford Dictionary of National Biography*, Oxford: Oxford University Press 2008, https://doi.org/10.1093/ref:odnb/50911.

[10] John Davis, *The Seamans Secrets*, London: Thomas Dawson 1595; Helen M. Wallis, 'The First English Globe: a Recent Discovery', in: *The Geographical Journal* 117/3 (1951), pp. 275–90, see p. 279.

itself being most commonly used by English sailors in the seventeenth and eighteenth centuries. The Strait may be taken as a reminder of Davis's accomplished seamanship, the quadrant of his mathematical practice in navigation.

This is not an original observation. The editor of Davis's Hakluyt Society volume of 1880, A. H. Markham, opens his introduction with the claim that

> Among the distinguished English seamen of the sixteenth century, John Davis of Sandridge stands out conspicuously as one who, more than any other, united the qualities of a daring adventurer with those of a skilful pilot and a scientific navigator.[11]

Robert Dudley echoes the combination of seamanship and mathematics in his *Arcano del Mare*, 1646: 'Capitano Giovanni Davis Inglese era dottissimo marinero e buon matematico'.[12] As early as 1585, Davis was known as 'a man very well grounded in the principles of the Arte of Navigation'.[13]

As a sailor, Davis is best remembered for his three early voyages in search of a Northwest Passage, undertaken in 1585, 1586, and 1587, although he was often at sea later in his life, with voyages (including privateering) in the south Atlantic and the far East, eventually losing his life in a fight with Japanese pirates off the coast of Borneo.[14] The most challenging seas he encountered were around Cape Horn, but the navigational challenges were greatest in the high latitudes of the Atlantic in his early exploratory expeditions, accounts of which are preserved in Richard Hakluyt's *Principal Navigations*.[15]

The education that enabled Davis to become a successful author and to acquire sufficient mathematics to be considered expert in the art of navigation has been regarded as something of a puzzle. He must have had significant experience of the sea to be presented as the ideal commander of the exploratory expedition of 1585, where considerable personal investments were at stake. By that time his activities were linked to notable maritime families close to his boyhood home in Sandridge near Dartmouth, the Gilberts and the Raleghs, and he was a particular friend of Adrian Gilbert, brother of the famous Sir Humphrey. The chief supporter of these ventures was the London merchant Sir William Sanderson, who is best known in the history of mathematics

[11] Markham, *Voyages and Works*, p. i.

[12] Robert Dudley, *Dell' Arcano del Mare*, Florence: 1646, 1647; quoted from Markham, *Voyages and Works*, p. lv n. 1.

[13] Markham, *Voyages and Works*, p. 1.

[14] On Davis, see Markham, *Voyages and Works*; Clements R. Markham, *A Life of John Davis, the Navigator, 1550–1605: Discoverer of Davis Straits*, London: George Philip & Son 1889; A. McConnell, 'Davis, John (d. 1621), sailor', in: *Oxford Dictionary of National Biography*, Oxford: Oxford University Press 2008, https://doi.org/10.1093/ref:odnb/7284; David W. Waters, *The Art of Navigation in England in Elizabethan and Early Stuart Times*, Greenwich: National Maritime Museum 1978, pp. 201–12; Mary C. Fuller, 'Arctics of Empire: the North in Principal Navigations (1598–1600)', in: *The Quest for the Northwest Passage: Knowledge, Nation and Empire, 1576–1806*, ed. Frédéric Regard, London: Routledge 2016, pp. 15–29. A background account is Susan Rose, 'Mathematics and the Art of Navigation: The Advance of Scientific Seamanship in Elizabethan England', in: *Transactions of the Royal Historical Society* 14 (2004), pp. 175–84.

[15] Richard Hakluyt, *The Principall Nauigations, Voiages and Discoueries of the English Nation*, London: George Bishop and Ralph Newberie 1589; and second edition in three volumes, London: George Bishop, Ralph Newberie, and Robert Barker 1599–1600. Documents related to Davis are helpfully brought together in Markham, *Voyages and Works*; it will be convenient to refer to this edition, not least because the original edition of *The Seamans Secrets* is not paginated.

as the sponsor of the first English globes, a very large pair—terrestrial and celestial—made by the instrument maker Emery Molyneux, who was introduced to Sanderson by Davis. John Dee also took an interest in the projects of Davis and Gilbert, meeting them on several occasions between 1579 and 1583. These discussions, and others involving Secretary of State Sir Francis Walsingham, led to the granting of letters patent by the Queen authorizing the expedition, funded by a number of merchants, the principal being Sanderson.

Two small vessels, the *Sunshine* and the *Moonshine*, departed from Dartmouth in June 1585 on the first of the three voyages. Hakluyt published original journals of all three eventful voyages in the first edition of his *Principal Navigations* of 1589—quick work, since the third voyage took place in 1587. The first is described by the merchant John Janes, who went on the voyage to take care of the investors' interests in the role of a 'supercargo' of some kind, though without there being cargo for him to superintend. He confirms that Sanderson, who was his uncle, was the leader of the group of merchants involved, the chief projector, and that he 'commended vnto the rest of the companie one M. Iohn Dauis, a man very well grounded in the principles of the Arte of Nauigation, for Captaine and chiefe Pilot of this exployt'.[16]

The record of Hakluyt's Principal Navigations

The first recorded application of Davis's mathematical skill on the voyage was his making a chart based on a survey of the Scilly Isles, where the ships were delayed for twelve days by contrary winds: 'the Captaine did platte out and describe the situation of all the Ilands, rockes and harboroughs to the exact use of Navigation, with lynes and scale thereunto convenient'.[17] A chart drawn to scale, on the basis of geometrical survey, was a novelty in contemporary hydrography. Hakluyt's record of the third voyage also came from Janes, so it is not surprising that neither of these narratives has much navigational content. One of two accounts of the second voyage, however, is by Davis, and in Hakluyt's second edition of 1600 he prints Davis's influential tabular log or journal for the third voyage.

In Davis's account of the second voyage there is an episode that illustrates the traditional technique that he will call 'horizontal navigation' in *The Seamans Secrets*. He is in Baffin Strait in August 1586 and records:

> [...] the eighteenth of August we discovered land Northwest from us in the morning, being a very fayre promontory, in latitude 65 degrees, having no land on the South. Here we had great hope of a through passage.
>
> [...] The nineteenth of this moneth at noone, by observation, we were in 64 degrees 20. minuts. From the eighteenth day, at noone, unto the nineteenth at noone, by precise ordinary care, we had sailed 15 leagues South and by West, yet by art and

[16] Markham, *Voyages and Works*, p. 1.
[17] Ibid., p. 2; 'describe' here means 'draw'.

> more exact observation, we found our course to be Southwest, so that we plainely perceived a great current striking to the West.[18]

This means that on the 19th at noon he took an altitude of the sun and from that deduced his latitude. He also worked out his position from the progress made since the last determination of position, at noon on the 18th. This was done by 'dead reckoning', that is, assembling the record of the courses steered by the compass and the corresponding distances estimated or measured, and inferring a new position 'by account' or, as he puts it, 'by precise ordinary care'. This led to the conclusion that the course overall has been south by west, that is, one point of the compass to the west of south, and that they had made 15 leagues. His problem was that sailing 15 leagues south by west would not have brought him to the new-measured latitude of 64 degrees 20 minutes. In this divergence, the latitude is a definite measurement and Davis is confident of its accuracy. To reconcile the bearing and distance record with the measured latitude, he could adjust either direction or distance (or both to some extent) but he concludes that his course has in fact been south-west, that is, all of three further points to the west than he had inferred from dead reckoning.

Dead reckoning cannot reliably take account of a current: the seaman might make allowance for a suspected current but his measurements of speed and therefore of distance are only made with respect to the sea, not the seabed. The compass bearing indicates direction only with respect to the ship's prow, that is the 'heading', not the direction the ship might actually be taking under the influence of any side wind or current. On the other hand, the latitude fix does give a definite coordinate on the surface of the earth. Davis accounts for the discrepancy between these two positions—'by precise ordinary care' and 'by art and more exact observation'—by inferring 'a great current', that has taken him much farther to the west. This in turn would support his 'great hope of a through passage', perhaps a step in the discovery of a Northwest Passage.

In *The Seamans Secrets*, in addition to his eponymous quadrant (Figure 2.1), Davis published for the first time one of the fundamental tools of 'horizontal navigation', also known as 'plane sailing', namely the tabular journal or log: 'A Table shewing the order how the Seaman may keepe his accompts'.[19] Davis offers a short extract from his homeward passage from the Strait of Magellan in March–April 1593, 'to shew you after what sorte I haue beene accustomed to keepe my accomptes in my practises of sailing'. The table has columns for the date, the measured latitude, the course in the previous 24 hours, the distance in leagues, the wind direction, and a column for other records, mostly occupied by measurements of magnetic variation.

In 1600 Hakluyt published a much more extensive table from Davis, 'A Traverse-Booke', as an additional record of the third Northwest voyage of 1587.[20] Since Hakluyt

[18] Ibid., p. 27. I am grateful to Professor Mary Fuller for drawing my attention to this passage; I discuss it also in Jim Bennett, 'Instruments and Practical Mathematics in the Commonwealth of Richard Hakluyt', in *Hakluyt and Oxford*, ed. Anthony Payne, London: The Hakluyt Society 2017, pp. 35–52 esp. pp. 50–1.

[19] Davis, *Secrets* (no pagination); Markham, *Voyages and Works*, pp. 281–2. On Davis's journal, see Waters, *Art of Navigation*, pp. 203, 277, 282–3; Margaret Schotte, 'Expert Records: Nautical Logbooks from Columbus to Cook', in: *Information & Culture* 48 (2013), pp. 281–322 esp. pp. 289–90.

[20] Ibid., pp. 49–58.

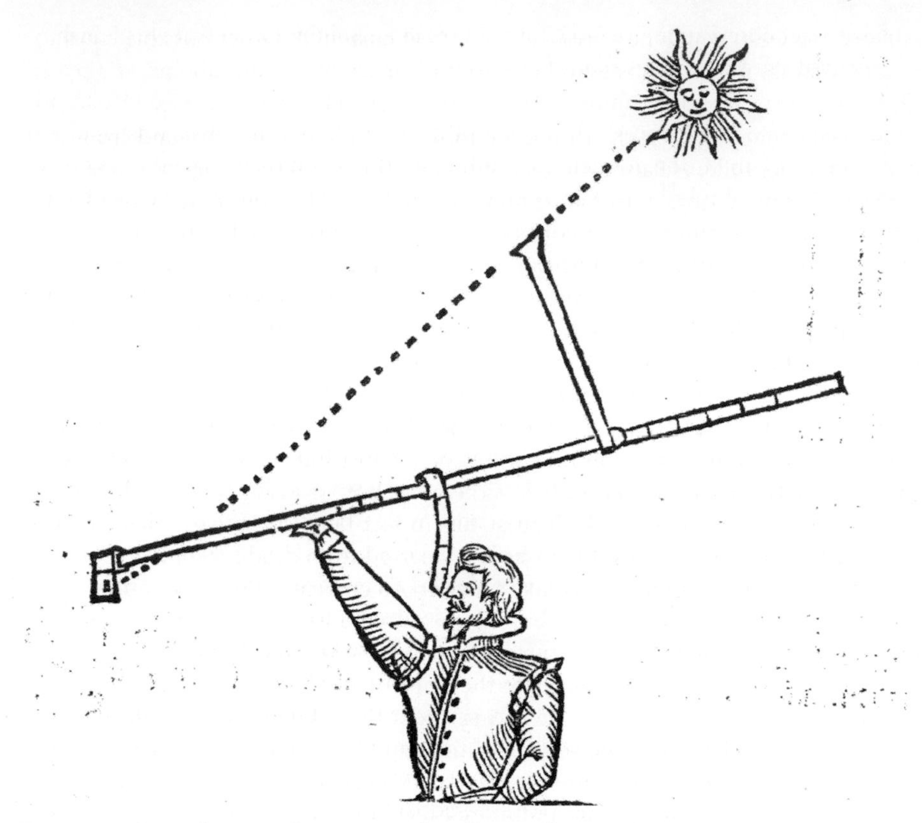

Figure 2.1 One of two configurations of quadrants illustrated in *The Seamans Secrets* for measuring the altitude of the sun without looking at it directly. Some modifications of this design would result in the standard backstaff or 'Davis quadrant'.

is more concerned with historical record than with teaching navigational practice, he prints the record of the entire voyage. The columns in this earlier example, though not in quite the same order, contain much the same information as the later one, except for an additional column for 'Howers', indicating not the time but the interval since the previous determination. In the 1587 journal, there is a line for each bearing and the period of time it was maintained; in 1593 these are reckoned up into the overall daily values from noon to noon.[21] In the 1587 journal the final column, headed 'The Discourse', is much more informative, especially in the later stages of the voyage, but the 1593 example was probably shortened and rationalized for instruction by removing the incidental detail.

'The Discourse' contains a note, 'The true course, distance, and latitude', whenever Davis has reconciled these parameters, which occurs only when he has measured a latitude. On one occasion this caused him to adjust his course by an astonishing six

[21] This is explained in Waters, *Art of Navigation*, p. 203.

points to the west, which he attributed to 'a great current'.[22] Other possible clues to the 'Passage' are noted, such as a tide measured vertically at four fathoms, at a place where there were 'an huge number of isles' and where 'a great whale passed by us, and swam West in among the isles'.[23] It was tempting to think that the whale was heading somewhere with intention. A week later, sailing south, they 'crossed over the entrance or mouth of a great inlet or passage, being 20 leagues broad, and situate between 62 and 63 degrees'.[24] From the estimate of size and especially the latitude, this was the eastern entrance to the (later named) Hudson Strait.

The publication of Davis's navigational journal in its tabular form, the first example to appear in print, is understood to have been an influential model and so a very significant contribution to basic navigational practice. The tabular log became fundamental to the daily work of seamen for several centuries, continuing through many developments in ever more sophisticated techniques. One column that does not appear in either of Davis's examples, but that would become standard, was longitude. Its absence was not, as may be thought, simply because longitude was difficult to find; this was just as difficult for long after it was being included in the tabular journal. If it could not be directly measured, it could be inferred from bearings and distances, that is, calculated by geometrical and later arithmetical dead reckoning. Longitude, as a value for seamen to calculate, seems not to have fallen within Davis's view of 'horizontal navigation', whose parameters were bearing and distance, reconciled whenever possible by the measurement of latitude. The result would be pricked on a chart, rather than entered as a longitude value in a tabular log. Perhaps this was a residue of the ancient bearing and distance technique before latitude finding was developed. At any rate, this absence would not survive for much longer, for Davis does mention longitude in the context of the 'paradoxall navigation' he would introduce in his major written work.

Davis on navigation in The Seamans Secrets

Two books by Davis were published in 1595, *The Seamans Secrets* and *The Worlds Hydrographical Discription*. The latter was an argument for the existence of a Northwest Passage. Dedicated to the Privy Council, it posits an ambitious geopolitical significance for the navigational enterprise of Davis and his associates. A Northwest Passage could transform the balance of the shipping and trading fortunes of European maritime nations, with England strategically positioned to flourish in a new order: 'her Majesties dominions should bee the storehouse of Europe, the nurse of the world, and the glory of nations, in yielding all forrayne naturall benefites by an easie rate'.[25]

The Seamans Secrets was part of this vision. The navigational lessons it taught were forged in the northwest, its techniques fashioned to meet the special navigational

[22] Ibid., p. 54.
[23] Ibid.
[24] Ibid., p. 55.
[25] Markham, *Voyages and Works*, p. 194.

challenges of high latitudes. Just as a new passage to the East would alter the geography of global trade, a new navigational practice would be needed to master it. Davis expressed it well in *Hydrographical Discription*:

> in that part of the world Navigation cannot be performed as ordenarily it is used, for no ordenarie sea chart can describe those regions either in the partes Geographicall or Hydrographicall, where the Meridians doe so spedily gather themselves togeather, the parallels beeing a verye small proportion to a great circle, where quicke and uncertayne variation of the Compasse may greatly hinder or utterly overthrow the attempt.[26]

Other elements in this prospect would be the 'paradoxal compass' (a chart on a polar projection mentioned by Dee and by Davis[27]), the globes of Emery Molyneux (financed by Sanderson, navigation by the globe being the new approach taught by Davis as suitable for high latitudes), the textbook by Hues on the use of the globes,[28] in particular those by Molyneux, and the mathematical account of the Mercator projection published by Wright in 1599. A new navigational practice would enable a new economy of maritime trade and power.

Davis was inclined to relate *The Seamans Secrets* to his experience in his northwest voyages; it was in these challenging waters that he honed the techniques he now shares with his readers. In the dedication to Charles Howard, Lord Admiral of England, he explained that:

> In those Northwest voyages where Nauigation must be executed in most exquisite sorte, in those attempts I was enforced to search all possible meanes required in sayling, by which occasion I haue gathered together this breefe treatise.[29]

'Sailing' here does not refer to wind-propelled travel at sea in its later, general sense, but to techniques of navigation, incorporating finding a position and setting a course, according to different clusters of geometrical constructions and observational and instrumental practices. Working in high latitudes obliged Davis, he says, to seek out 'all possible meanes', which he organizes, as we shall see, into three kinds of sailing.

The dedication also makes it evident that Davis was fully alert to the different areas of competence evident in his subject and their demarcation. In the terminology he adopts, they range from the 'speculative' to the 'mechanical'. With this concern, in response to his direct experience, his text becomes an appropriate study for the present volume. His boundary between what lies within and what beyond the

[26] Ibid., p. 197.

[27] Waters, *Art of Navigation*, pp. 209–12; Davis, *Secrets*, sig. A^{v}. The word 'compass' here refers to a complete circuit, containing all the degrees of longitude; the chart would be 'paradoxal' because it would show rhumb-lines or loxodromes spiralling to the pole—experienced as straight courses followed on a steady bearing, they appear as spiral on this chart. For Wright's doubts about this terminology, see Edward Wright, *Certaine Errors in Nauigation*, London: Valentine Sims [and W. White] 1599, sig. F2.

[28] Robert Hues, *Tractatus de globis et eorum vsu*, 2nd edn, London: Thomas Dawson 1594; Robert Hues, *Tractatus de globis et eorum usu: a treatise descriptive of the globes constructed by Emery Molyneux and published in 1592*, ed. Clements R. Markham, London: The Hakluyt Society 1889.

[29] Davis, *Secrets*, dedication, sigs I2v–I3; Markham, *Voyages and Works*, pp. 233–4.

academy is figuratively, but almost literally, the shoreline: 'neither haue I laide downe the cunning conclusions apt for Schollers to practise vpon the shore, but onely those things that are needfullye required in a sufficient Seaman'.[30] We shall choose several topics from *The Seamans Secrets* where the author had to address this boundary.

The moon and the tides

An occasion for Davis as author, where a mathematical topic of established scholarly interest touched on a routine procedure for practical calculation at sea, was the relationship between the motion of the moon and the times of tides. Although Davis had emphasized his concern with practice at sea and his own experience in that regard, and although he would eventually describe the actual procedures recommended to sailors, this is not where he began his account. Instead he opened with a simple lunar theory, of no value at all for knowing the times and characteristics of tides, but a 'speculative' introduction to all that would follow.

Davis's introduction to the moon's motion relies on the most basic elements of the Ptolemaic lunar theory. He distinguishes the moon's 'violent' from its 'natural' motion, the former being shared with the whole heavens, as 'the violent swiftnes of the diurnall motion of primum mobile', the latter being the moon's own contrary (eastward) motion in the zodiac.[31] This varies throughout her orbit and Davis gives figures in degrees per day for the minimum at apogee, maximum at perigee and a mean motion at two points between. He ascribes this variation (known as the moon's first anomaly) to 'the eccentricity of her Orbe wherein she moueth'.[32]

Is this of interest to a seaman? Even if it is, it will not help him calculate the tides. Davis rapidly changes gear and justifies a shift to common practice at sea. The lunar anomaly may take place in the zodiac, but the seaman calculates by his familiar compass card, even if this is an instrument for the horizontal plane:

> the Seamen for their better ease in the knowledge of tides, haue applied this the Moones motion, to the points, degrees, and minutes of the Compasse, whereby they haue framed it to be an Horizontal motion which sith by long practise is found to bee a rule of such certaintie, as that the errour thereof bringeth no danger to the expert Seaman, therefore it is not amisse to follow their practised precepts therein.[33]

(We shall see an example of this use of the compass card in registering the position of the moon when we come to consider Davis's instructions for finding the times of tides.)

Davis may have shifted his account to shipboard but he retains his commitment to a lively explanation. The moon's 'naturall motion' in the zodiac has a period of 27 days, 8 hours, but successive conjunctions with the sun, the interval of importance

[30] Ibid., sigs I3^{v} ff.
[31] Markham, *Voyages and Works*, p. 244.
[32] Ibid., p. 245.
[33] Ibid.

for the tides, are separated by 29 days, 12 hours, and 44 minutes.[34] This difference arises because, when the moon returns to the place of the former conjunction, 'she doth not finde the Sunne in that parte of the Ecliptick where she lefte him'. The sun's natural motion has taken him one degree eastwards for every day since their previous conjunction. No sailor needs to know this to calculate times of tides, but Davis clearly wants to promote some underlying understanding. At the same time, he is aware, of course, that the procedures used by seamen do not recognize the first lunar anomaly or the variation in the sun's motion (we would say apparent motion) through the year: 'the Seaman accompteth the Moones motion, to be vniforme in all places of the Zodiac'. A figure of 30 days 'reconed escry change and change', i.e. between successive new moons, is generally adopted, a practice Davis accepts, since the consequent errors are small:

> to alter practised rules where there is no vrgent cause, were a matter friuolous, which considered, I thinke it not amisse that we proceede therein by the same methode that commonly is exercised.[35]

In moving to accommodate seaboard practice, Davis even introduces the reader to some nautical terminology: 'in the Seaman's phrase, all the time of her application [from new to full moon] she is before the Sunne, and in the time of her separation [full to new] she is abaft the Sunne'.[36] 'Before' and 'abaft' here refer to the bodies' relative positions in the appearance of their movement across the sky, where the 'violent' component predominates.

Davis wants to take his readers all the way to calculating the times of tides from scratch and this he does, but at different stages he interrupts his instruction narrative, and perhaps undermines it, by offering the more usual assistance by way of pre-calculated tables and instruments to facilitate calculation routines. A more regular handbook would contain only these 'interruptions' (my term). As steps to finding the age of the moon, Davis tells his readers how to calculate the 'golden number', which he also calls the 'prime', for any year, and the 'epact' for the year, knowing the prime, and he illustrates these by worked examples. The prime, from 1 to 19, indicates the position of the year in the Metonic cycle of 19 years, after which the lunar phases occur on the same dates in the solar calendar as in year 1. The epact is the age of the moon on the first day of the given year. Having related both calculations, Davis provides a disc-shaped table (Figure 2.2) giving the prime (from January) and the epact (from March) for every year from 1593 to 1630, so covering two Metonic cycles. The circular shape emphasized the cyclical character of the relationship and setting two cycles of years in two circular rows demonstrated the repetition of the pattern.

> By this Table the prime and Epact may for euer be found, for when the yeeres be expired, you may begin againe, and continue it for euer at your pleasure.[37]

[34] Ibid.
[35] Ibid., p. 246.
[36] Ibid.
[37] Ibid., p. 249.

This 'interruption' removes any need for calculation.

Davis then explains how to find the age of the moon for any date, knowing the epact for the year. It would be useful at sea, as on land, to know when there might be moonlight, but Davis wants to press on to the real benefit to the seaman, 'the account of Tydes': 'to know the tyme of full Sea in any place at all such seasons as occasion shall require'.[38] For this he will need not only the age of the moon but the 'establishment' of the port in question, which is the time interval between the meridian passage of the moon and the next high tide on the days of new and full moon. Davis expresses this as the bearing of the moon by the compass at high tide, registered by one of the traditional 32 points of the compass, each being equivalent to 45 minutes of time. Davis then enumerates the procedure for finding the time of high tide, first arithmetically and then by a simple counting procedure using the compass card as an instrument of calculation. He goes a stage further, one typical of contemporary books of mathematical practice, by describing an instrument for performing these calculations, a

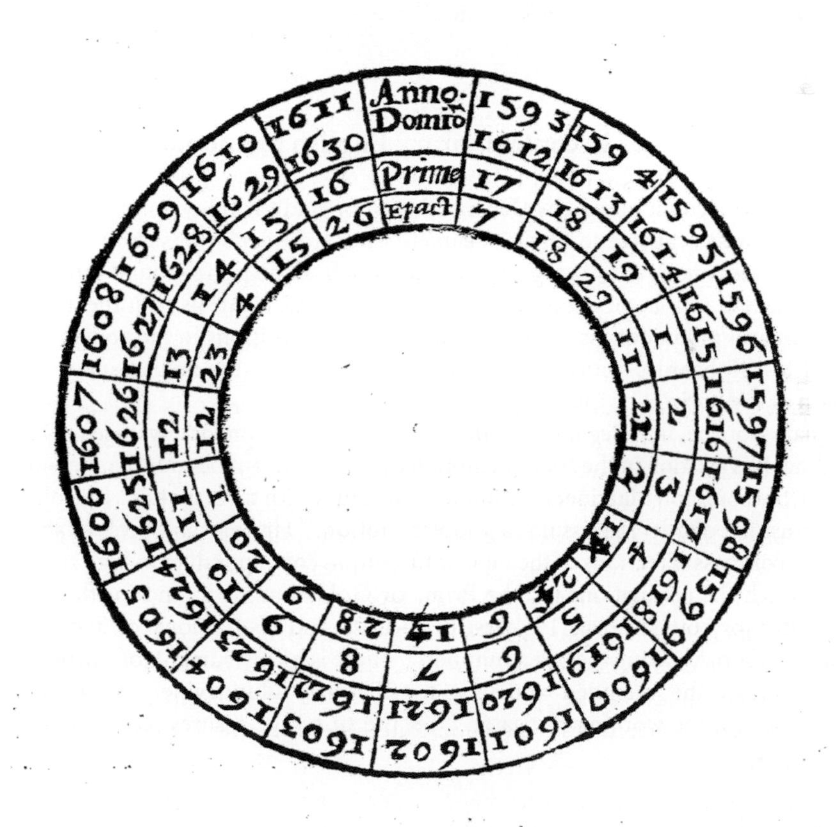

Figure 2.2 A circular table in *The Seamans Secrets* giving the golden number (the 'prime') and the epact for every year from 1593 to 1630.

[38] Ibid.

'Horizontall tyde Table', mentioning twice that this is intended for 'yong practisers in Nauigation'.[39] Whether he thinks that young men are drawn to new gadgets or older ones do not need them, he does not say.

It is worth noting that Davis himself employs what he describes as the seaman's register of establishments by the compass bearing of the moon at high tide. In the log reproduced by Hakluyt, at the position where they see the great whale Davis notes: 'In this place a S.W. by W. moone maketh a full sea'. He notes also that the magnetic variation is all of 30 degrees.[40]

It is not completely clear whether it was intended to include the tide calculator as a paper instrument on a page of the book. 'Heere followeth a very necessary Instrument', says Davis, but no copy in any edition seems to have one.[41] However there is a very complete description of a volvelle, whose base is a compass rose divided into both 32 points and 24 hours, subdivided to 15 minutes, above which are two moveable discs, one with a sun index and a lunar scale of 30 days, the other with an index for the moon. A set of worked examples explains how to find the time by the moon, the age of the moon, and the times of tides.[42]

The tides in contemporary English handbooks

How does the account offered by Davis of the moon and the tides compare with the books otherwise available to English seamen? Richard Eden's translation (1561) of Martin Cortes, *The Arte of Navigation*, covers much of the same ground but with a less precise and satisfactory account of the times of tides. He deals with the motions of the sun and moon but his volvelle for finding their positions in the zodiac is less attuned to shipboard practice in England, being based on the zodiac and not the division of the horizon by the compass rose.[43]

William Bourne, in *A Regiment for the Sea* (1574), has a similar coverage to Davis, including the variation in the zodiacal motion of the moon, but he adds that he 'would not wish the common Marriners to trouble themselues with these matters, but followe their accustomed order', i.e. assume a uniform motion.[44] His tide table gives establishments as compass directions of the new or full moon corresponding to high tide. He goes through the calculations for 'the Prime or Golden number' and for the epact, with a table for both covering 19 years, 1574–1592, and instructions for finding the age of the moon, knowing the epact number.[45] There is no instrument for calculation.

By contrast, the intended audience for Thomas Blundeville, *His Exercises* (1594), was not the working navigator.[46] Yet the title page assures readers that the

[39] Ibid., pp. 251–2.
[40] Ibid., p.55.
[41] Ibid., p. 251.
[42] Ibid., pp. 251–5.
[43] Martín Cortés, *The Arte of Nauigation*, London: Richarde Lugge 1572, ff. xxix–xxxii^{v}, xlviii–l.
[44] William Bourne, *A Regiment for the Sea, and Other Writings on Navigation*, ed. E. G. R. Taylor, Cambridge: Cambridge University Press for the Hakluyt Society 1963, pp. 179–80.
[45] Ibid., pp. 181–3, 174–81.
[46] Thomas Blundeville, *M. Blundevile his Exercises*, London: Iohn Windet 1594.

'furtherance' of the art of navigation is the central intention of the whole work and the last of six treatises assembled by Blundeville and written by himself and others is devoted to navigation as an 'arte'. The idea is that educating young gentlemen to understand the art, even if they will never practise it, will enlarge its appreciation and esteem. This makes the relationship between knowledge, art, and practice fundamental to Blundeville's project.

The *Exercises* as a whole was an eclectic assemblage, covering aspects of mathematics, astronomy, and cosmography, so the motion of the sun and moon had been treated before navigation was broached. Blundeville was then able to move quickly to the golden number, the epact and finding the age of the moon, and the days of new and full moon. He then covers many topics—the calendar, planetary ephemerides, instruments for navigation, the compass and variation, charts, dead-reckoning, and latitude finding by the pole star and by the sun at noon. Only then is there a chapter 'Of the Moone and of all her diuers motions', which treats the first anomaly in a manner very similar to Davis. Blundeville is aware that 'Mariners doe account the moouing of the Moone by the pointes of their Compasses' and that they treat this incorrectly as uniform.[47] Unlike Davis he does not add that this does not really matter in practice. Instead he takes the reader through the cycle of swift and slow motion according to the Golden Number and the moon's cycling place in the zodiac. He deals also in a descriptive way with her variation in ecliptic latitude, something not mentioned by Davis. Blundeville finally devotes two chapters to finding the times of tides, introducing the seaman's register of 'establishment' by the compass bearing of the new or full moon at high tide, and he describes an instrument, in principle like that of Davis, which he attributes to Michel Coignet.[48] Blundeville has left to the end the topic, namely tides, which Davis judged should come first. He echoes his title-page by an apology:

> I end this Treatise, praying all the learned Seamen not to be offended or greéued with mee for that I doe make young Gentlemen our owne Countrimen partakers of their most worthy knowledge, whereof the ignorant are not able to iudge, nor to yeélde them that prayse which they deserue.[49]

Davis's intention to offer 'onely those things that are needfully required in a sufficient Seaman' and Blundeville's target readership of 'young Gentlemen' helps explain why the former treated compass and tides first and the latter last.

Solar declination and latitude finding

A second instance for Davis where mathematical learning and shipboard practice require an accommodation concerns finding latitude from the meridian altitude of the sun. The observer's latitude is the angle from the equator to the zenith, found by measuring the maximum solar altitude with a cross-staff (or now with an instrument

[47] Ibid., p. 347.
[48] Ibid., pp. 349–50.
[49] Ibid., final page (not numbered).

for the back-observation), finding the complement of this angle and making an adjustment for solar declination, the sun's distance from the equator in its annual cycle.[50] The routine for the calculation was slightly different depending on the relative orientations of the pole, the zenith (equivalent to the observer's position), the equator and the sun, but Davis deals with these, as any account of the method was obliged to do. That would usually be considered sufficient, but Davis adds a section, 'Are these all the rules that appertain to the finding of the Poles height?', and additional material is introduced through his own experience at sea.[51] 'Those that trauell farre towards the north' might have two meridian transits of the sun in 24 hours and be making a measurement with the sun to their north and require yet another routine. This prompts a further thought. The 'Regiment' or declination table is calculated for a particular meridian. The possible use of two meridians on the same date is a reminder that the sun's declination changes continuously, 'therefore it is necessary to know the Suns declination at al times, and vpon euery point of the Compasse'. The occurrence of 'great fogges and mistes' obscuring the sun 'in my northwest uoiages' obliged Davis to resort to off-meridian observations.[52] Further, even in more temperate locations, longitude difference can be a consideration:

> as I haue by my experience found at my being in the Straights of Magilane, where I haue found the suns declination to differ fro' my regiment calculated for London, by so much as the Sunne declineth in 5 howers, for so much is the difference betweene the Meridian of London, and the Meridian of Cape froward, being in the midst of the said straights.[53]

Davis goes through a proportional calculation to find the sun's declination when crossing any meridian, based on the figures for the two adjacent days, a model that will serve for any case 'if you reade with the eye of reason, and labour to vnderstand with iudgement that which you read'.[54] He then outlines a 'most excellent' method for finding the declination for any time from ephemerides of the sun's longitude in the ecliptic, by a trigonometric calculation using sine tables, 'but because seamen are not acquainted with such calculations, I therefore omit to speake further thereof, sith this plaine way before taught is sufficient for their purpose'.[55]

Davis concludes with an instrument for these calculations, explained by worked examples, for the benefit of 'the yoong practiser',[56] and a new set of solar declination tables, calculated from the ephemerides of Johannes Stadius.[57]

We can consider Bourne and Blundeville in a brief comparison with contemporary English writers. The former does deal with sighting the sun on a northern meridian

[50] Markham, *Voyages and Works*, pp. 258–63. Davis mentions the possibility of having a cross-staff scale graduated directly in zenith distance, although the angle taken is from the horizon to the sun.

[51] Ibid., pp. 254–5.

[52] Davis does not take his readers through the calculations that would be required to find the latitude from these observations.

[53] Ibid., p. 265.

[54] Ibid., p. 267.

[55] Ibid.

[56] Ibid.

[57] The tables, extending over 12 pages in the original (one for each month), are not included in the Hakluyt Society edition, Markham, *Voyages and Works*.

in extreme latitudes and adjusts the figures in the *Regiment* accordingly but he does not generalize this for any longitude.[58] The latter explains how to find the solar declination for any position of the sun in the zodiac, 'Per tabulas Sinuum',[59] but does so in the astronomy section of his *Exercises* and does not return to this in his very basic account of finding latitude by the altitude of the sun in his chapters on navigation. There he reminds the reader of the earlier sections ('in my Spheare, in my Treatise of the Globes, and also in my treatise of the Astrolabes'), refers seamen to the relevant chapters of Bourne's *Regiment*, and reproduces a passage from Norman's *New Attractive*.[60] Treated as *navigation*, Davis offers the fullest account.

So Davis again seems, by comparison, particularly exercised by the tensions created by divergent ambitions: he wants to give navigators a grounded understanding of their mathematical art, while knowing from experience the limitations of mechanical practice.

Three kinds of navigation (or sailing)

It might have been expected that a writer of a practical, experience-based, how-to manual would be content with teaching a single methodology, what would today most likely be called 'best practice'. Davis however deals with three kinds of navigation or, as he also calls them, 'kinds of sayling'. He does not hide from the seaman that each kind has its advantages and limitations. Further, in this demarcation he is fully abreast of the art of navigation and well ahead of its practice. The three are, in his terminology, 'Horizontall paradoxall and great circle Nauigation'.[61]

Horizontal navigation is plane sailing, using a magnetic compass and a plane chart, keeping an account of the distance sailed on what bearing in every watch, and checking latitude by the measurement Davis has already explained. Only the latitude is reliable: 'the Pilot hath onely his height in certaintie'.[62] The bearing is unreliable: 'the corse is somewhat doubtful'; the distance estimate even worse: 'the distance is but barely supposed'. Davis explains how to make the calculations required but there are so many 'impediments', that is disturbing influences of wind, current, and tide, that art is inadequate. The needful skills 'are better learned by practice then taught by penne'.[63]

In horizontal navigation, therefore, art cannot provide a complete solution. Here Davis invokes the distinction between practice and art, with the latter demoted to the role of assistant: 'it is not possible that any man can be a good and sufficient pylot or skilful Seaman but by painful and diligent practise with the assistance of arte'.[64] As mentioned already, Davis reinforces his practical mission by showing sailors how to keep the continuous record of their sailing in a log and reproduces an extract from

58 Bourne, *Regiment*, pp. 223–6.
59 Blundeville, *Exercises*, ff. 52^v–53.
60 Blundeville, *Exercises*, ff. 345^v–346.
61 Markham, *Voyages and Works*, p. 314, see also pp. 239–40, 282.
62 Ibid., p. 275.
63 Ibid., p. 277.
64 Ibid. The punctuation in the first edition makes the practice / art distinction more emphatic: 'but by painefull & diligent practise, with the assistance of arte, whereby the famous Pylote may be esteemed worthy of his profession', Davis, *Secrets*, sig. G3.

his own journal. From the information recorded, the navigator can calculate overall direction and distance in 24 hours from the previous noon and plot a new position on his chart.

Horizontal navigation treats the sea as a flat surface, taking no account of the earth being a sphere, something of which the navigator is perfectly aware. Its use, with the plane chart, should be restricted to 'the coasting of any shore of country, or for shorte voyage'.[65] Anything more ambitious should accommodate the shape of the earth, so before dealing with the other two kinds of navigation, Davis introduces the globe. He has already mentioned 'paradoxall' and 'great circle' sailing but, given the limitations of the plane chart, or 'sea chart' as he calls it, he must explain the globe, the circles it displays, and the stand and accessories that make it an instrument of geometrical calculation.

The use of the globe

The introduction of the globe is such an important step for Davis that he uses it to transition to the second of the two 'books' of the *Seamans Secrets*. The globe will allow him, he promises, to deal with all three kinds of navigation. He begins, however, not with the 'globe' but the 'sphere', by which he sometimes means the geometrical figure but more often the armillary sphere. In referring to the 'globe', although Davis does not say so explicitly, the text is most easily understood with reference to what Dee called 'the Globe Cosmographicall', that is, a terrestrial globe with added circles relating to the heavens, notably the ecliptic but also considering the equator to have both celestial and terrestrial significance.[66] As Davis puts it: 'And further know that the Equator is the beginning of al terrestrial Latitude, and the declination of all celestial bodies'.[67] The cosmographical globe. with the addition of the spiral rhumb lines, was made, for example, by Gerard Mercator. It is significant that Davis introduced his second kind of sailing as 'paradoxall or Cosmographicall Nauigation'.[68]

When describing the planetary motions performed within the zodiacal band, we find again that unexpected terminology encountered with Davis's treatment of the moon. The motions of the planets in their individual periods from west to east are 'natural', but their diurnal motion from east to west is 'violent', imposed by the motion of the *primum mobile*. This is such an unusual conceit that it is worth quoting Davis in full:

> And note that this naturall motion of the Planets in the Zodiac, is from the West toward the East, the diurnall motion is violent, caused by the first mouer, or primum mobile, who in euery 24. houres doth performe his circular motion from the East to the West, carrying with him al other inferiour bodies whatsoeuer.[69]

[65] Markham, *Voyages and Works*, p. 272.

[66] John Dee, 'Mathematicall Praeface', in Euclid, *The Elements of Geometrie*, translated Henry Billingsley, London: Iohn Daye 1570, sig. b.iii.

[67] Markham, *Voyages and Works*, p. 292.

[68] Ibid., p. 239.

[69] Markham, *Voyages and Works*, p. 293.

'Natural' and 'violent' are, of course, the Aristotelian distinctions of motion on earth. The idea of a violent motion in the heavens is however profoundly contrary to Aristotle's thought and it is difficult to say where Davis may have found such a notion, unless it was one of his own.[70] Otherwise this part of the *Seaman's Secrets* seems comfortably in tune with the popular literature of cosmography.

When Davis comes to describe the navigational use of the instrument, he has in mind, of course, a *terrestrial* globe. The woodcut used by the printer to illustrate a solid globe, not now an armillary sphere composed of rings, is something of a puzzle (Figure 2.3). It has an accessory known as a 'circle of position', which Davis has to admit 'serueth to no great purpose for Nauigation'.[71] In fact its function was

Figure 2.3 The globe figure in *The Seamans Secrets.*

[70] On the impossibility of violent motion in the heavens, see Edward Grant, 'Celestial Motions in the Late Middle Ages', in: *Early Science and Medicine*, 2 (1997), pp. 129–48.

[71] Markham, *Voyages and Works*, p. 302.

astrological, being used for casting a horoscope, or 'setting a figure', and allocating the planets to the 'houses of heaven' for the time and place of the event in question. (The system of the houses was based on the horizon, so terrestrial location was important.) This was more relevant to the celestial globe than the terrestrial.

Once the reader has mastered the basic manipulation of the globe and its accessories, the three types of sailing can be demonstrated. Finding the distance between two places on the globe is straightforward. An opening of 'a paire circular compasses' (dividers with inward-curving legs) can be set against the degree scale at the equator (thus converting the chord of a great circle into degrees of arc) and allowing 20 leagues per degree (a league being three nautical miles) will give the distance. Covering that distance by sailing is a more complex matter.[72]

A globe rotates on north and south poles fixed to a vertical 'meridian ring', which is carried in a horizontal 'horizon ring' and where it can be turned so as to set these poles for latitude, this being the angle between the pole and the horizon. The course between two places is found by setting ('rectifying') the globe for the latitude of the starting point and then bringing this point to the meridian ring, where it will be at the zenith, 90 degrees from the horizon. A moveable and rotatable brass arc, the 'altitude quadrant', is clamped to the meridian ring at the zenith and its direction, when aligned with the destination and followed down to the horizon ring, will indicate the bearing of the destination from the current position. If these two places are reasonably close—Davis mentions a limit of 45 degrees—the navigator can treat the given bearing and distance as a 'Horizontall Corse' and set sail.[73]

By following the given bearing, however, he will veer farther and farther from his desired course. The meridians on the globe are not parallel, but converge towards the poles, and as the ship's track makes the same angle with them successively, it will increasingly deviate in a spiral path to one of the poles. As Davis says: 'if I saile vpon the Horizontall Corse, I shall never arriue vnto the same place'.[74] Horizontal sailing has its limitations.

The answer is frequent course correction. Every 20 or 30 leagues, as Davis recommends, reset the globe for the new latitude, bring the new position to the zenith, set the altitude quadrant through the destination and sail on the bearing indicated at the horizon ring. The sequence of course components will be close to a great circle—not only 'on target' but by the shortest course. As is usual, things will not be that simple—winds may not be favourable and the ship may have to tack and make way as well as possible, but at least the objective will be a sound one.[75]

'Paradoxall Navigation' involves sailing along a rhumb-line, maintaining a constant bearing. The rhumb-line course is longer than the great circle, but the technique is more straightforward, provided the navigator can choose the bearing to follow. As we have seen, this will not be the initial bearing pointing to the destination. Neither will the course be the straight one.

> Paradoxall Nauigation, demonstrateth the true motion of the Ship vpon any Corse assigned, in his true nature by longitude, latitude, and distance [...] by which motion

[72] Markham, *Voyages and Works*, p. 310.
[73] Ibid., pp. 311–13.
[74] Ibid., p. 312.
[75] Ibid., pp. 312–15.

> lines are described neyther circular nor straight, but concurred or winding lines, and are therefore called paradoxall, because it is beyond opinion that such lines should be described by plaine horizontall motion.[76]

Mercator's terrestrial or cosmographical globe had printed rhumb-lines distributed on its surface, the equivalent on a globe of the lines radiating from the compass roses on plane charts.[77] His world map or chart was constructed by a projection with parallel meridians and latitude lines increasingly spaced towards north and south, such that rhumb-lines were truly projected as straight lines. This meant that the navigator could set a straight rule between departure and destination and immediately find the 'paradoxall' compass bearing to follow. He would not, of course, be taking the shortest route.

Davis appreciates that the practical route to rhumb-line or paradoxal sailing (later called 'Mercator sailing') would be through a chart, being a much more manageable instrument than a globe, and he intends to rise to the challenge:

> I purpose (if God permit) to publish a paradoxal Chart with all conuenient speede, and so will discouer by the same at large, all the practises of paradoxall and great Circle nauigation, for vpon the paradoxall Chart it will best serue the Seamans purpose, being an instrumēt portable, of easie stowage and small practise, perfourming the practises of Nauigation as largely and as beneficially as the Globe in all respects.[78]

Although Mercator published his world chart in 1569, the mathematics on which the projection was based had not yet been explained, and this would be necessary for any general application of such charts to the practice of sailing. It would fall to Davis's associate Edward Wright to meet this need in his book of 1599, *Certaine Errors in Nauigation*. As mentioned above, he and Davis had served in 1589 as fellow captains in the fleet commanded by George, Earl of Cumberland, on his third voyage to the Azores. It was Wright, having taken the name 'Captain Edward Carelesse' for the occasion, who would write the account published in *Certaine Errors* and reprinted by Hakluyt, who described him as 'the excellent Mathematician and Enginier'.[79]

The first map on Mercator's projection published in England was the world chart in Wright's *Certaine Errors*, with the results of Davis's voyages to the northwest, as they had appeared on Molyneux's terrestrial globe of 1592.[80] On both we find 'Fretum Davis', 'Sandersens toure' (on the globe: 'Sandersons Tour') and 'A furious Ouerfale' (on the globe: 'A furious ouer fall')—a feature Davis had noted in his journal of the third voyage as, 'a mighty overfal, and roring, and with divers circular motions like whirlepooles'[81] A triangular relationship between Molyneux, Wright, and Davis connects the globe and the chart. Sanderson and Hakluyt are supporting links in this network. The impending globe was announced in the first edition of *Principal*

[76] Ibid., p. 315.

[77] In *Hydrographical Discription* Davis refers to 'Curious lyned globes to the right use of Navigation', ibid., p. 197.

[78] Markham, *Voyages and Works*, p. 315.

[79] Ibid., p. 60.

[80] Helen Wallis, 'England's Search for the Northern Passages in the Sixteenth and Early Seventeenth Centuries', in: *Arctic* 37 (1984), pp. 453–72 esp. pp. 464–7. See also Wallis, *First English Globe*.

[81] Markham, *Voyages and Works*, p. 56. This was at the entrance to what would become Hudson Strait.

Navigations, and the chart appears in some copies of the second. Molyneux may have sailed with Davis, who acknowledged in print that the globe (and so the chart) drew on his northwest voyages.[82] Davis had cartographic skill and experience of his own[83] and was well placed, through colleagues and associates, to hear about hopes and possibilities for solutions to inconsistencies in methods of sailing through the introduction of a new basis in projective geometry. Davis's remarks on this reflect the currency of such ambitions among his circle. A cartouche, presumably written by Wright, on the chart is an admirably succinct instruction for its use in the realization of paradoxal navigation.[84]

As well as his vision of the paradoxal chart, Davis imagines a very general shift to 'another knowledge of Nauigation, which so farre excelleth all that is before spoken'.[85] Here the calculations of spherical trigonometry would replace the graphical constructions of geometry: 'this sweete skill of sayling may well be called Nauigation arithmeticall, because it wholly consisteth of Calculations, comprehended within the lymit of numbers'. A lengthy paean to this future state of navigational practice, based on 'this heauenly hermonie of numbers', ends with an optimistic promise of a future account of its 'orderlye practise'—'to the best of my poore capacitie'.[86] This also was not realized by Davis but its basis was laid by Wright and its practical application made possible through the instrumentation of Edmund Gunter.

Speculative science, navigational art, and mechanical practice

The Seamans Secrets is the work of an author seeking to improve the practice of navigation through developments in mathematical art, while at the same time wanting to ground the work of practical seamen on a more secure grasp of the principles on which such developments were based. This was a challenging ambition. At several points we see Davis reaching the limits of its practical value, or beyond them, and retreating to something closer to customary practice. He had to accommodate also the current limits of his own theoretical knowledge, learnt at least in part through the discourse of his mathematical, maritime associates. Circumstances had brought him into relationships with the leading English navigational theorists of his day, Dee, Harriot, and Wright, but in this and in the wider group of mathematical practitioners, he was most prominently and permanently committed to a career at sea. His situation

[82] Wallis, *First English Globe*, pp. 178–80; Markham, *Voyages and Works*, p. 211.

[83] His early chart of the Scilly Isles is mentioned above; a more impressive achievement was his chart (not extant) of the Straights of Magellan, 'an exquisite plat', according to John Jane, Markham, *Voyages and Works*, pp.117–18. Davis mentions this work in *Hydrographical Discription*, saying of the Strait that 'I have described every creke therein', where 'described' almost certainly means drawn, ibid., p. 204. He also collaborated in the production of a chart of the English Channel, ibid., p. 280.

[84] Bennett, *Instruments and Practical Mathematics*, pp. 47–8. Sarah Tyacke suggests that the 'paradoxal chart' is nothing more than a polar chart with spiral rhumb lines, i.e. is the same as the 'paradoxal compass', Sarah Tyacke, 'All at Sea: Some Cartographical Problems in the North 1500–1700', *IMCoS Journal* 111 (2007), pp. 38–41.

[85] Markham, *Voyages and Works*, pp. 316–17.

[86] Ibid.

led to thoughtful reflection on roles and relationships relevant to our examination of 'the academy' and the world 'beyond'. For my concluding subtitle I have constructed Davis's taxonomy of the discipline of mathematical navigation as 'speculative science, navigational art and mechanical practice'; his own terminology would have been something like 'Theoricall speculations', 'Artes Mathematicke', and 'mecanicall practises'.[87]

At the very beginning of the dedication of *The Seamans Secrets*, addressed to Charles Howard, Baron Howard of Effingham, Elizabeth's Lord Admiral, Davis commends noble patronage of 'practises either speculative or mecanicall'.[88] As a good student of Dee, Davis had learnt the general notion of there being a range of fields of practice based on mathematics, but his taxonomy was developed from Dee's, where mathematical arts were derived from the mathematical sciences of arithmetic and geometry. Davis's own experience of mathematical practice obliged him to acknowledge an extension to three denominations. Speculative mathematics, though important, was not part of Davis's work; he placed his calling between mathematical art and mechanical practice. His seaborne activity comprised, so far as possible, 'mecanicall practises drawn from the Artes of Mathematicke'.[89] The accommodation he seeks is not between mathematical sciences and mathematical arts, but between mathematical arts (in his case navigation) and mechanical practice.

Davis exemplifies his principal categories with a few notable examples of men who occupy them admirably. England, he says, can boast of 'men of rare knowledge, singular explication, and exquisite execution of Artes Mathematicke', his examples being Thomas Digges, John Dee, and Thomas Harriot. Likewise, 'for the mecanicall practises drawn from the Artes of Mathematicke, our Countrie doth yeelde men of principal excellencie'. Here his examples are the globe-maker Emery Molyneux, the shipwright Matthew Baker, and the portrait miniaturist Nicholas Hilliard.[90]

In writing a navigation manual to capture his experience of sailing in northerly latitudes and to help realize a new commercial and geopolitical ambition for England, Davis found himself obliged repeatedly to accommodate the realities of shipboard practice. In this respect his work addressed the culture of the sea. At the same time, he needed to face in a different direction, to address his mathematical colleagues in a way that recognized and applied their categories of knowledge and skill. In spite of his regard for Dee, the 'Sciences, and Artes Mathematicall' of his famous 'Groundplat'[91] were not sufficient to capture Davis's shipboard experience. Instead of stopping at navigational art, Davis needed the explication extended to recognize the distinctive role of mechanical practice.

[87] Ibid., pp. 234–35.

[88] Ibid., p. 231.

[89] Ibid., p. 235. The spelling used in the title to this chapter is taken from the first edition, 1595.

[90] Ibid., pp. 234–5. In the preface to Robert Recorde, *The Pathvvay to Knowledg*, London: Reynold Wolfe 1551, we find: 'Carpenters, Caruers, Ioiners and Masons, | Painters and Limners with suche occupations, | Broderers, Goldesmithes, if they be cunning, | Must yelde to Geometrye thankes for their learning'.

[91] Dee, *Praeface*, 'the Groundplat'.

3
Navigation Examinations in the Early Modern Period

Margaret E. Schotte

'These are not matters for fools'

Navigators' examinations were challenging events in the careers of many early modern European sailors.[1] In the late 1760s, one determined Dutch East India Company (VOC) hand presented himself to no fewer than three different committees, hoping to pass the captain's examination.[2] Frans van Ewijk, who hailed from Alblasserdam just south of Rotterdam, was already a *schipper* (captain), sailing for the Delft chamber of the VOC in 1765. However, he ran into difficulties after a ruling by the VOC directors made examinations mandatory for ships' captains.[3] After a two-year voyage for Delft, Van Ewijk secured a new contract with the Rotterdam chamber. In October 1768, he presented himself in Rotterdam for the *schipper* examination—and was summarily failed. Van Ewijk was deemed unqualified to serve as captain. That winter, the company circulated a warning about his ineptitude. In a formal report, he blamed his household servants for distracting him with 'much confusion and trouble' at home, but this failed to persuade the chamber to grant him a second examination. Undaunted, he sought out a willing committee in the Amsterdam chamber. In the spring, they questioned him exactingly about theory and practice, almost certainly following the well-established series of topics that appeared in published model examinations: calculating the time of high tide, determining position from altitude observations, and recalling geographical details about sailing routes. This time, his answers proved sufficient for a favourable report, and by July 1769 Van Ewijk was finally employed again, as a captain for Zeeland.[4]

How did Captain van Ewijk prepare for these rites of passage? Did he avail himself of one of the crash courses offered by the many navigational teachers in Dutch

[1] The heading for this section is taken from Estienne Cleirac, *Us et costumes de la mer, divisées en 3 parties*, Bordeaux: G. Millanges 1647, p. 493.

[2] The *Vereenigde Geoctroyeerde Oostindische Compagnie*, hereafter VOC.

[3] On 1751 captain's examination (1766 in Zeeland), see C. A Davids, *Zeewezen en Wetenschap: De Wetenschap en de Ontwikkeling van de Navigatietechniek in Nederland Tussen 1585 en 1815*, Amsterdam: Bataafsche Leeuw 1986, pp. 294–6; C. A. Davids, 'Het navigatieonderwijs aan personeel van de VOC', in *De VOC in de kaart gekeken: Cartografie en navigatie van de Verenigde Oostindische Compagnie 1602–1799*, ed. Patrick van Mil and Mieke Scharloo, The Hague: SDU 1988, pp. 65–74 at p. 73.

[4] VOC 1.04.02 222 Zakenindex op de resoluties van de Heren XVII, 1737–1784, p. 748 (13.10.1768, 10.4.1769); 308 Zakenindex op de resoluties van de kamer Amsterdam, 1744–1788, pp. 514–15 (28.1 & 29.12.1768, 23.3.1769); due to Van Ewijk's 'onbekwaamheid', the second examination was 'naauwkeurig en Strict'; in March they found that he 'in de Theorie als practycq genoegen heeft gegeven'. See also Jaap R. Bruijn, *Schippers van de VOC in de achttiende eeuw aan de wal en op zee*, Amsterdam: Bataafsche Leeuw 2008, pp. 136–7.

port cities? What was the examination process like? How much mathematical knowledge did his examiners expect men like him to have? Like Van Ewijk, thousands of mariners across Europe subjected themselves with varying degrees of enthusiasm to licensing examinations for navigators. From the sixteenth century onwards, candidates spent considerable money and time acquiring sufficient mathematical and technical skills to satisfy the examiners. Sailors of all stripes were motivated to submit to these tests, since passing at least one examination typically led to higher wages and increased opportunities.[5] In some places they were a prerequisite for any legitimate work; in others, they were the first step toward being admitted to a school. As navigation became increasingly theoretical, examinations became a fixture in maritime communities. And yet, while similar interrogations were repeated in port towns and naval establishments across Europe, these important professional benchmarks have been generally overlooked by historians.

The evidence that survives from this widespread testing is scarce and uneven. Archival traces include terse instructions for French state examiners, and a limited number of passing certificates preserved in England, alongside bureaucratic protests from men who had failed their tests. As for the examinations themselves, a handful of model examinations appeared in Dutch textbooks, each with two or three dozen remarkably standardized questions about mathematics and geography. This very standardization has caused scholars to minimize the significance of the examinations, dismissing the questions as 'predictable'. Maritime historians also dispute the degree to which examinations were ever truly mandatory.[6] However, by analysing the examination records in conjunction with more abundant surviving educational materials such as textbooks and student manuscripts, it becomes apparent that these examinations had a much wider impact than has been recognized. Even if not all mariners presented themselves to take the examinations, anyone who took a navigation course learned the material that was tested on them.

Examiners in various regions and eras focused on different aspects of navigation and had different ideas about what qualified as mastery. After a brief overview of the institutions that introduced examinations, and why they considered these formal assessments to be useful, this chapter will discuss five types of examinations. These each focused on a separate body of knowledge essential to navigating, and consequently tested candidates in a distinctive way, asking them to physically demonstrate the use of instruments, recite information—in the form of memorized dialogues—about piloting, cosmography, and geography, or solve mathematical formulas to find their position. Once codified in print, the last of these, the mathematical VOC written

[5] The higher positions had a substantial pay differential: in seventeenth-century Netherlands, the three ranks of navigators earned from 26 to 40 or 50 guldens per month, or more than all but the captain, minister, surgeon, and the senior carpenter, and at least three times the average sailor's wages. L. M. Akveld and W. J. van Hoboken, eds., *Maritieme geschiedenis der Nederlanden: Zeventiende eeuw, van 1585 tot ca 1680*, Bussum: De Boer Maritiem 1977, vol. II, p. 141, Table IV.

[6] Bruijn, *Schippers van de VOC*, p. 136. C. A. Davids notes that naval examinations did not become mandatory in the Amsterdam Admiralty until 18 September 1749 (for *commandeurs* and *luitenants*), and across all the chambers the Netherlands until the 1780s; even after it was universally required, an 1851 survey of merchant captains revealed that 63% had *not* taken the examination. Davids, 'Het zeevaartkundig onderwijs voor de koopvaardij in Nederland tussen 1795 en 1875. De rol van het Rijk, de lagere overheid en het particuliere initiatief', in *Tijdschrift voor Zeegeschiedenis* 4 (1985), pp. 164–90 at p. 175.

examination, held sway for more than a century. A dialogue from the turn of the eighteenth century offers insights into how candidates may have fielded questions in the examination room, while evidence of the mathematical examinations appears in textbooks, student manuscripts, and fill-in-the-blank tests. This chapter will analyse the relationship between published manuals and examinations, and how these shaped the educational process. Far from being limited due to their 'predictable' nature, these examinations were very effective at instilling knowledge of arithmetic and geometry, trigonometry and tables, aiding maritime communities to embrace applied mathematics.

Chronology and logistics

Spain was the first nation to require its navigators to undergo a formal examination: on 6 August 1508, King Ferdinand appointed Amerigo Vespucci the first *piloto mayor* (principal pilot), assigning to him the responsibilities of examining the *pilotos* (navigators) who sailed to the West Indies, as well as approving their maps, instruments, and *derroteros* (rutters, sailing directions). The first official legislation describing this examination process dates from 1527.[7] The *Casa de la Contratación* (House of Trade) established its official school in 1552, placing university-educated instructors in charge of training and examining *pilotos* (navigators) who were not necessarily literate. The new techniques of celestial navigation—where mariners rely not on the visible coastline and their memories but on heavenly bodies like the sun and stars to determine position—were more abstract, and consequently not as easy to learn on board ship as an apprentice. To teach the theoretical underpinnings of this science, the cosmographers turned to the pedagogical approach with which they were most familiar from their own university educations: lectures, textbooks, and examinations. The teachers first adapted older textbooks, such as Sacrobosco's *De Sphaera* and Apian's *Cosmographia*, which taught the fundamentals of cosmography (describing the structure of the heavens and Earth). They then began writing their own nautical manuals with the same cosmographical focus. It was these Iberian manuals that carried that particular, theoretical version of the science of navigation north to other corners of Europe.

Every state with an extensive maritime presence invested in educating and certifying navigators.[8] Portugal followed Spain's lead and appointed a theoretically oriented *cosmografo mor* (chief cosmographer) in 1547. In 1558, after a hospitable visit to the *Casa de la Contratación* in Seville, the Dover sea captain Stephen Borough lobbied to be named England's first 'Cheyffe Pylott', proposing to undertake 'the examynacion and appoyntyng of all suche maryners . . . [that wished to become] a pilott or

[7] On the *piloto mayor*, see Alison Sandman, 'Cosmographers vs. Pilots: Navigation, Cosmography, and the State in Early Modern Spain', PhD dissertation, University of Wisconsin, Madison 2001, p. 97; and on the rules for examinations, p. 115, citing the *cedula* of 2 July 1527 (Seville, Archivo General de Indias [AGI] Patronato 251 R. 22).

[8] See Margaret Schotte, 'Sailors, States, and the Creation of Nautical Knowledge', in: *A World at Sea: Maritime Practices and Global History*, ed. Lauren Benton and Nathan Perl-Rosenthal, Philadelphia: University of Pennsylvania Press 2020, pp. 89–107.

master'.[9] While Queen Elizabeth I never signed Borough's contract, nor established the proposed school that Borough had modelled upon the *Casa de la Contratación*, by 1621 the medieval pilots' guild Trinity House had begun examining ships' masters for the navy.[10] The merchant marine briefly shared the impulse to educate and regulate sailors: in 1614 the English East India Company hired Edward Wright to teach and examine its mariners, but this arrangement lasted only two years.[11]

In France, a 1584 ordinance proscribed men from carrying out the duties of pilots or masters if they had not publicly proved their abilities in an examination. The legislation emphasized the 'noble sciences' that comprise the art of navigation, namely cosmography and mathematics. Recognizing how challenging it was to master all the listed fields, the Bordeaux lawyer Estienne Cleirac noted that 'these are not matters for fools, which is why the *Ordonnance* obliges those [men] to undergo an examination before becoming masters, pilots or mates'.[12] Examiners were appointed in 1615 and 1627, first in the port town of Dieppe, and then nationally.[13] In the United Provinces, in 1619 the Amsterdam Chamber of the VOC appointed prolific textbook author Cornelis J. Lastman the first examiner to examine *opper-* and *onderstuurlieden* (chief and assistant navigators). He soon obtained assistance in this considerable task from the cartographic publisher Willem J. Blaeu, and other chambers and the Admiralties gradually introduced their own examiners.[14]

With the exception of Spain, these early appointments left few traces of classroom practices or examination formats.[15] However, by the last decades of the seventeenth century officials had instituted the examinations that would eventually shape the training of most working navigators in northern Europe, from novice trainees to seasoned practitioners, as well as supervising officers in both the navy and the merchant marine.[16] More schools were established: a second Spanish institution, the Real Colegio Seminario de San Telmo, opened in Seville in 1681. In England, 1670 saw the founding of the Royal Mathematical School, the mission of which was to train young boys for service as masters in the navy. At the same time, in 1677, the Royal Navy implemented examinations for lieutenants. This surprising decision to subject elite officers to the same certification process as able seamen can be seen as an effort to level the playing field and admit to the navy men who were intelligent but not elite.[17]

[9] David W. Waters, *The Art of Navigation in England in Elizabethan and Early Stuart Times*, New Haven, CT: Yale University Press 1958, p. 515.

[10] Waters, *Art of Navigation*, p. 9 (charter from Henry VIII in 1514), pp. 106–7; records of the examinations themselves date from the 1660s; see TNA SP 46/136, 137 Admiralty Papers: Masters' certificates, 1660–73.

[11] Waters, *Art of Navigation*, pp. 243, 294.

[12] Cleirac, *Us et costumes de la mer*, pp. 491–3. The topics in which candidates must demonstrate competency include: 'Astronomie, Geometrie, Trigonometrie, Les Meteores [weather], L'Aritmetique, La Mechanique, Physique, Painture, Bon & ferme jugement'.

[13] Albert Anthiaume, *Evolution et enseignement*, Paris: E. Dumont 1920, vol. I, pp. 13–14.

[14] Bruijn, *Schippers van de VOC*, p. 135. For the most complete chronology of Dutch examiners, see Davids, *Zeewezen en Wetenschap*, Appendix II, pp. 398–405.

[15] The records of Spanish examinations survive only from 1568 on; Sandman, 'Cosmographers vs. Pilots', p. 104.

[16] Boundaries blurred between merchant and naval training, as many men sailed for both types of vessels.

[17] Robert Iliffe, 'Mathematical Characters: Flamsteed and Christ's Hospital Royal Mathematical School', in: *Flamsteed's Stars: New Perspectives on the Life and Work of the First Astronomer Royal, 1646–1719*, ed.

The French caught up quickly. Louis XIV's naval minister Jean-Baptiste Colbert oversaw the establishment of nearly a dozen schools for officers (including gentlemen *gardes de la marine*). To assess their 'talents and capacity', the state compelled local authorities and captains who happened to be in port to serve as examiners.[18] In the words of a mid-eighteenth-century examiner, strict testing was 'the only way to generate a spirit of emulation, and to judge their capacity with certainty'. He also reported on his classroom management: 'I give them not a moment of relaxation during the time we are together, and I work to interrogate them continuously about Pilotage [i.e. navigation] and Geometry, in order that they teach each other, and are forced to respond as best they can to the questions I pose'.[19]

The Netherlands, by contrast, did not set up state-funded schools. Instead, the long-distance trading companies, the East and West Indies Companies (VOC and WIC), relied on a decentralized system of independent schools to introduce their extensive labour force to the complex mathematics, record-keeping, and cartography required by their relentless commercial voyages.[20] The examiners were expected to publish textbooks, an unofficial prerequisite for their posts; the most successful of these were supplied on each merchant vessel sailing to Asia.[21] The companies and admiralties relied on these educators to disseminate the new mathematical techniques and standards of precision—not just to young men, but also to seasoned captains and other mariners.

All of these educational institutions had as part of their curriculum some type of examination.[22] It seems counterintuitive that a professional sailor could not—or must not—carry out his job until he had passed an examination. Although sailing and navigating had long been learned through formal or informal apprenticeships, the maritime world did not follow the conventions of artisanal guilds where an apprentice was required to produce some type of refined object, a masterpiece, at the completion of his training. Instead, as nautical training became increasingly connected to formal classroom learning from the sixteenth century onwards, students were expected to demonstrate their grasp of the material in a culminating examination, similar to the *disputatio* of the universities, a debate that tested both their knowledge of the subject and their dialectical skills.[23]

Less-elite educational institutions also expected students to give oral displays of their knowledge, although these focused more on memorization than disputation. In the first decade of the seventeenth century in Leiden, the 'Duytsche Mathematique' engineering school put new surveyors through an applied examination, about

Frances Willmoth, Woodbridge: Boydell Press in association with the National Maritime Museum 1997, pp. 115–44 at p. 122.

[18] Anthiaume, *Evolution et enseignement*, II, p. 86. Michel Vergé-Franceschi, *Marine et éducation sous l'ancien régime*, Paris: Editions du Centre national de la recherche scientifique 1991, p. 94.

[19] AN-Paris MAR/G92 'Mémoires et projets. Observatoire de Marseille . . .', 1701–1783, 1738, ff. 61^{v}–62.

[20] The *Geoctroyeerde West-Indische Compagnie*, hereafter WIC.

[21] Davids, 'Het navigatieonderwijs', p. 43. C. H. Gietermaker's *'t Vergulde Licht* (ed. princ. 1660) was later replaced by Pybo Steenstra's *Grond-beginzels der Stuurmans-kunst*, Amsterdam: G. Hulst Van Keulen 1779.

[22] The term 'examination' has myriad overlapping meanings. This chapter focuses on the 'process of testing a person's knowledge, skill, or competence' (*OED*), and will not touch on the Jesuit 'examen (of conscience)'. The Portuguese term '*exame*' means 'analysis'.

[23] D. D. Breimer et al., *Hora Est! On Dissertations*, Leiden: Universiteitsbibliotheek Leiden 2005, p. 26.

which few details are known.[24] Grammar and church schools alike relied on catechistic dialogues to help their pupils absorb—and hopefully retain—basic definitions and concepts.[25] Students were then expected to recite these texts in class. As a result, dialogues often functioned simultaneously as mnemonic tool and test material. Maritime educators embraced the form, incorporating numerous examples into nautical manuals well into the eighteenth century.[26] Some penned imaginative conversations between men with Latinate sobriquets ('Philomathes') or named ships for classical figures (Caesar and Pompey, Alexander and Darius).[27] Most dialogues, however, began as Euclid did: by defining standard terms. Alternating pairs of discursive queries and responses would, by the late seventeenth century, appear in mathematical texts as page after page of practice questions, usually published with answers.[28]

Examinations were initially designed to certify those men who were already working navigators, with the goal of regulating their income, and limiting or preventing foreigners from becoming overly familiar with a nation's trade secrets. (Prior to the mid-seventeenth century, locales with insufficient numbers of skilled men allowed naturalized foreigners to be pilots.)[29] The rite of passage was then extended to help train youth, as well as officers who needed to learn sufficient details about the theory and practice of sailing prior to taking command of men who had years of experience. Examinations might be held within or parallel to educational institutions, and could be used to increase the number of practitioners, or to limit them.[30] Early modern navigating professionals (and their instruments) were ostensibly subjected to examination at least once and often repeatedly during their careers, as part of the educational process, prior to employment, or upon their return from a voyage.[31]

In most instances, the certification process aimed to select desirable individuals, seeking attestations of character as much as intelligence. Future officers were expected

[24] E. Muller and K. Zandvliet, eds, *Admissies als landmeter in Nederland voor 1811*, Alphen aan den Rijn: Canaletto 1987, pp. 150, 154.

[25] Peter Burke, 'The Renaissance Dialogue', in: *Renaissance Studies* 3/1 (1 Mar 1989), pp. 1–12 at p. 3.

[26] In addition to those discussed below, see Garcia de Palacio, *Instrvción Návtica*, Mexico City 1587; many Dutch pamphlets were written as conversations, e.g. Heyndrick Reyersz., *Jaep en Veer, of stuurmans praetjen* [. . .], Amsterdam: Dirk Pietersz. 1622. French authors favoured the dialogue (or question-and-answer) format for their textbooks. Some were aimed not at uneducated practitioners but administrators, e.g. Nathaniel Boteler, *Six dialogues about sea-services between an high-admiral and a captain at sea*, London: for Moses Pitt 1685.

[27] Thomas Hood revised William Bourne's *Regiment for the Sea* (ed. princ. [1574?]) in 1592, final 21 ff. comprise 'The Marriners guide set forth in the form of a Dialogue' on sea charts.

[28] Marjolein Kool, *Die conste vanden getale: Een studie over Nederlandstalige rekenboeken uit de vijftiende en zestiende eeuw, met een glossarium van rekenkundige termen*, Hilversum: Verloren 1999, p. 229.

[29] Initially, experienced masters were excused, out of respect for their seniority rather than due to any conviction about their facility with the new techniques. S. Borough, cited in Waters, *Art of Navigation*, p. 513: 'not the olde and approvid good masters and marinors, but suche as dayly of youthe springeth vpp'; Bruijn, *Schippers van de VOC*, p. 136. On guarding against foreign interlopers, see Cleirac, *Us et costumes*, p. 493. On naturalized pilots, Sandman, 'Cosmographers vs. Pilots', p. 114 n. 55.

[30] In England, after the 1677 initiation of the lieutenant's examination, Secretary to the Admiralty Samuel Pepys was pleased to report that 'we have not half the throng of those of the bastard breed pressing for employment [. . .] they being conscious of their inability to pass this examination'. Dickinson, *Educating the Royal Navy*, Hove, Sussex: Psychology Press 2007, p. 11.

[31] Davids, *Zeewezen en Wetenschap*, p. 295. In England, the ease with which RMS boys passed was directly related to the school board's level of satisfaction with their instructor; see Iliffe, 'Mathematical Characters', pp. 122–4.

to be upstanding and diligent, while subordinate navigators needed to be 'men of probity and good morals'.[32] In addition to these subjective traits, candidates were required to meet specific objective qualifications. The respective governing bodies stipulated minimum (and in some cases maximum) ages, as well as a set number of years of service aboard ship.[33] For merchant and naval positions alike, time at sea was considered a valid proxy for experience, operating under the belief that one could not help but learn a specific set of practical skills, just by being on board.[34]

Having fulfilled the biographical requirements, candidates were then questioned by a panel of experts—but the type of questions and the format varied with time, location, and professional institution. From the first decades of the sixteenth century onwards, naval administrators wished to verify that aspiring navigators knew the ins and outs of pilotage and seamanship—how to handle a boat—and the same for the new instruments that were transforming navigational practice. Soon there were additional bodies of knowledge to master. Examinations began to incorporate questions connected to three distinct genres of nautical text: cosmographical, geographical, and mathematical. Each required the candidate to (re)produce knowledge in a different way, with varying relevance to the sailor's daily practice. Some examinations required hands-on demonstrations, but more commonly candidates would simply have to repeat memorized details and definitions. The latest type of examination to become popular required advanced mathematical calculations. Certain examinations combined all of these elements. While each of these tests required a degree of numeracy, the final one introduced and cemented important mathematical concepts—from logarithms to spherical trigonometry—well beyond the bounds of the ship's deck. It was this type of mathematical navigational examination that would turn out to have the longest legacy.

Demonstrating instruments

The pilot must have 'his astrolabe for the sun and quadrant for the north star and of both things know the use both in measuring the altitude and in adding or subtracting the declination of the sun'.[35]

The earliest type of examination, first conducted in Spain at the beginning of the sixteenth century, focused on the mariner's physical competence with charts and instruments. As these were new tools, it was important that *pilotos* be familiar

[32] *Ordonnance de Louis XIV, donnée à fontainebleau au mois d'Aoust 1681, touchant la Marine*, Paris: Denys Thierry 1714, p. 157.

[33] The exact ages varied over the years, and by country; for instance, English lieutenants were required to be at least 25 in 1677, but only 20 in 1703. (See TNA ADM 107/2/1-2 Lieut. Passing Certifs. 19 April 1703: 'prevent[ing] the Inconveniencies that happens by Persons comeing too Young to that Station' but volunteers could be aged 13–16.) The age for French candidates was more consistent: 25 in 1681; 24 in 1786; 30 was the maximum age for becoming an *enseigne*, and 50 was too old to take the captain's examination. See Danielle Fauque, 'Les Écoles d'hydrographie en Bretagne au XVIIIe siècle', in: *Mémoires de la Société d'histoire et d'archéologie de Bretagne* 78 (2000), pp. 369–400 at p. 388.

[34] By the end of the seventeenth century, most candidates were required to present a journal documenting their service, as the practice of keeping such a logbook was both concrete proof of their time at sea, and a form of training. On logbooks, see M. Schotte, 'Expert Records: Nautical Logbooks from Columbus to Cook', in: *Information & Culture: A Journal of History* 48/3 (2013), pp. 281–322.

[35] AGI Patronato 251 R.22 (2 July 1527), cited in Sandman, 'Cosmographers vs. Pilots', p. 116.

with how to operate them. The questions posed by examiners seem not to have been preserved in textual form, but we can find hints of the procedures in Spanish legislation as well as records of lieutenants' examinations in England's Royal Navy from the close of the seventeenth century. These two case studies, separated by one and a half centuries, confirm that manual dexterity was a considerable component of navigational expertise.

The earliest description of the Spanish examinations, a 1527 royal order, notes that each candidate was required to bring his astrolabe, quadrant, regiment, and sea chart to the licensing examination to be inspected, and also to demonstrate how to use them. He must be able to use a sea chart to plot a course and identify the major ports and dangers he would face in the journey. To ascertain that he was proficient at the new celestial methods of navigation—specifically using instruments and calculations to find latitude—the *piloto* needed 'his astrolabe for the sun and quadrant for the north star and of both things know the use both in measuring the altitude and in adding or subtracting the declination of the sun and the amount that the [north] star rises and falls [i.e., its offset from the celestial pole], together with knowledge of the hours at any time of the day or night'.[36] This examination foregrounded the use of key instruments. Candidates required some degree of numeracy for assessing angles along the graduated scale, and for reading the 'regiments' of the Sun and stars (astronomical tables)—but the instruments were designed to minimize the need for mathematical computation. The *pilotos* were then expected to answer additional questions, demonstrating their knowledge of the route to a particular island in the Caribbean, for example, as well as queries about cosmography and how to handle various emergency situations.

Across the Channel in England, there were separate examinations for the masters and lieutenants of the Royal Navy, and only one of these groups had to demonstrate physical mastery of their essential equipment. The masters, who typically went to sea as young teens and worked their way up over a decade or longer, were examined by 'senior brethren' from Trinity House. In light of their years at sea there was little need to verify the manual skills they would have routinely practiced. Their examiners were concerned with the masters' knowledge of geography—like the Spanish *pilotos*, they sought certification for a specific route—and their biographical bona fides. They were required to furnish testimonies from captains for whom they had sailed, confirming their reputation.[37] The Navy Board examiners, on the other hand, were less concerned about the personal backgrounds of the aspiring lieutenants. Instead, these young officers, who faced examinations from 1677 onwards, needed to prove that they were actually competent on board a ship.

Just as in Spain, the option to take an examination at the Navy Board was framed as a privilege, for which the gentlemen had to meet certain criteria to be 'entituled to ye benefitt of Examination'.[38] Unlike the masters, the aspiring lieutenants usually had only the minimum three years of experience at sea.[39] Therefore, their time in naval

[36] Sandman, 'Cosmographers vs. Pilots', pp. 115–17.

[37] TNA ADM 106/2908 e.g. 'a sober & carefull man fit to be warranted'.

[38] TNA ADM 107/1/1 Lieutenants' Passing Certificates 1690s–1702 (1701).

[39] N. A. M. Rodger, 'Commissioned Officers' Careers in the Royal Navy, 1690–1815', in: *Journal for Maritime Research* 3/1 (2001), pp. 85–129.

service took on great importance; lacking it could disqualify them. The lieutenants' passing certificates, which survive in considerable numbers from the 1690s into the nineteenth century, document the specific tasks the young gentlemen were expected to master. For example, the certificate of John Preake attests to the fact that he could 'splice rigging, take up a Reef in a Sail, take Observations, bring a ship to sayle and tack her in sayling'. More details are noted after Joseph Worlidge passed his examination on 17 March 1700: he could 'splice, knott, reife a saile, work a shipp in sailing, find the Prime, Epact, Moon's age, Shift his tydes, observe by sunn or star, keep a reckoning of the ship's way'.[40]

The first half of the lieutenant's examination, with its focus on sails, rope, and knots, may seem curiously plebeian for future officers. However, by asking these gentlemen to demonstrate some of a sailor's basic tasks, the examiners could easily see who had actually learned practical skills during their months-long voyages. Contemporary manuals did not include instructions; these details needed to be learned from other mariners. Although commissioned officers would not likely find themselves splicing rigging or hauling on sails, it was important that they understand how such tasks were done. With this aim in mind, the curriculum of the Royal Mathematical School had a similar emphasis on hands-on instrument use.[41]

The second half of the lieutenant's examination, as detailed on Worlidge's certificate, consisted of the calendrical information required to compute the times of the tides, as well as taking observations and charting the vessel's course. These standard duties involved basic arithmetic: addition and division to determine the current position of the moon and the year relative to the leap year; subtraction to account for the tides shifting back by forty-five minutes each day. There is no evidence that these were written examinations, or that the men were permitted to bring almanacs or tables with them. Instead, it seems likely that these candidates were expected to shift their tides on their thumbs—the long-standing computational aid—and sent home if they could not do this. A certain Capt. Dering, who was already dishonestly posing as a lieutenant, failed to cast the time of high tide at London Bridge correctly, and was immediately sent back to his ship and told to reapply at a later date.[42] Here we see the value placed on fundamental physical skills: whereas men tended to be given a second chance if they were missing a journal or letter from their captain, those unable to carry out basic computational tasks were failed unceremoniously.

If physical dexterity with instruments was important, a greater number of examinations focused on memorizing details for a range of subjects. Since these other aspects of navigational knowledge could be assessed through more traditional questions and answers, they left a more substantial documentary trail.

[40] TNA ADM 107/1/21 Mr John Preake, 23 December 1692. Preake satisfied all the requirements for seamanship, astronomical observation, and journal keeping, and his undoing was ultimately biographical rather than technical: his application was denied because his fifteen years of experience had been exclusively on merchant ships. ADM 107/1/192 midshipman Joseph Worlidge, 17 March 1700.

[41] See Margaret Schotte, *Sailing School*, Baltimore: Johns Hopkins University Press 2019, ch. 3.

[42] Samuel Pepys, *The Tangier Papers of Samuel Pepys*, ed. E. Chappell, London: Navy Records Society 1935, p. 131; Dickinson, *Educating the Royal Navy*, p. 11, n. 8. See also TNA ADM 107/1/8 Mr. W[illia]m Gill, 25 March 1692: 'something short in shifting his Tydes'.

Memorizing pilotage

Q. When your ship is out of dock, what is first to be done?
A. When the proper Officers have examined the limbers, I would level the hold, by laying the kentledge [ballast] from the fore-part of the fore hatchway to the after-part of the after hatchway.[43]

The second general category of examination—the dialogue or question-and-answer (Q & A) format—was a time-honoured way of assessing how well a candidate had memorized a specific body of material. Such examinations are the most diverse in topic: depending on the century and locale, the questions focused on pilotage and seamanship, cosmography, or most frequently, geography. Although the content varied based on time period and goal, these examinations were essentially recitations, a form of assessment familiar from oral examinations both medieval and modern.

In the early Spanish attempts to certify navigators, Spanish *pilotos* were asked about the 'art of pilotage'. (In Spanish and French, the term *piloto/pilote* refers to men responsible for 'large', open-water navigation, unlike the English association with 'small', coastal and harbour navigation.) Pedro Dias, a Spanish navigator, who in 1585 described his own examination at the *Casa de la Contratación*, recalled that he was expected to answer 'the most difficult questions' about the *arte de pilotaje*.[44] These questions, which were intended to confirm a *piloto*'s ability to manoeuvre a boat, do not appear in any published sixteenth-century document. As the examiners were themselves maritime men, they presumably already knew the points to interrogate.

It is possible, however, to draw inferences about the questions on this type of examination from a version published far later. John Adams' *The Young Sea-Officer's Assistant* (London, 1773) was a 79-page pocket guide to 'the substance of that examination, which every candidate for a commission in the East-India-Service, of the Navy, must necessarily pass'. Adams presented the information 'in the way of Question and Answer', because he viewed it 'as the most probable method of engaging the attention of young beginners, and imprinting the Replies and Observations in their memories [and they will benefit from] being pre-informed of what is to constitute their future Examination'.[45] The first thirty pages of the model examination focused on pilotage and seamanship: how to moor, which ropes and sails to use in specific locales, and general geographic directions. Although Adams saw educational value in dialogues, his decision to include a 24-line table on the final pages disrupts any idea that this matched the actual conversation in the examination room.

Adams's examination focuses on logistics and local knowledge. The second part consists of questions about working a ship 'in all difficult cases that may probably happen', the same sort of emergency preparedness that had concerned the *pilotos* at the *Casa de la Contratación*. Whereas the seventeenth-century English lieutenants' examination omitted pilotage and problem solving in preference to manual skills and a solid grounding in navigational tasks, by the latter half of the eighteenth

[43] John Adams, *The Young Sea-Officer's Assistant*, London: Lockyer Davis 1773, pp. 3–4.
[44] Richard Hakluyt, *Principal Navigations*, London 1600, III, pp. 862–4, 866–8, printed account by Pedro Dias, taken by Richard [Grenville] in 1585/86; reprinted in Waters, *Art of Navigation*, pp. 555–7.
[45] Adams, *Young Sea-Officer's Assistant*, p. vi.

century the emphasis had completely reversed. Now the 'Candidate for Commission' was expected to be conversant with the basics of seamanship, from loading cargo to mooring in particular harbours. Meanwhile the list of tested skills of the passing certificates had been abbreviated to 'can reef a saile, etc.'—a succinctness that calls into question the thoroughness of the examination procedures.[46] Examiners became increasingly anxious to verify the candidates' origins and age.[47] As the social status of aspiring Royal Navy officers grew more diverse, the promotion process shifted from verifying competencies to verifying identity, something much more difficult to prove in an examination.

Memorizing cosmography

> *If a man should aske you how many degrees are in the Compasse of the whole world: you may answer, there are 360. degrees, each degree being 15. Dutch miles, and 17. Spanish miles and a halfe.*[48]

Pilotage fell into the category of presumed knowledge, and was thus rarely treated in print. By contrast, cosmography, as explicated by the first Spanish navigation teachers, was an almost entirely textual subject. The textbooks they produced were not instruction manuals for daily tasks. Rather, the concepts taught within them were abstract, aiming to explain the sailor's position in the universe. Readers were expected to memorize a series of definitions: what is a pole, what is the zenith? Authors such as Martín Cortés and Pedro de Medina leavened their lengthy theoretical chapters with helpful diagrams.[49] With an eye to making the science of navigation more accessible to a general audience, Cortés penned two dialogues and Medina produced an adaptation of his work in dialogue form.[50] Although those particular texts circulated only in manuscript form, it did not take long for nautical writers beyond the Iberian Peninsula to harness the mnemonic power of dialogues.

Perhaps the earliest printed example of this type of cosmographical Q & A appeared at the back of the second volume of Jan Huygen van Linschoten's *Itinerario* (1595–96; see Figure 3.1).[51] While this catechistic dialogue has elements in common with some later examinations, there is no evidence that these questions were ever

[46] TNA ADM 107/4 Lieutenants' Passing Certificates, 1745–57.

[47] TNA ADM 107/40 Lieutenants' Passing Certificates, 1809, includes records of baptisms and affidavits for documents lost at sea.

[48] Jan Huygen van Linschoten, *His Discours of Voyages into ye Easte & West Indies Deuided into foure bookes*. London: J. Windet for J. Wolfe 1598, pp. 445–6, first question.

[49] Martín Cortés, *Breve compendio de la sphera y de la arte de navegar*. Seville: Anton Alvarez 1551; Pedro de Medina, Arte de navegar, Valladolid: F. Fernández de Córdoba 1545.

[50] See Martín Cortés, *Arte of Navigation. Conteynyng a Compendious Description of the Sphere, with the Makyng of Certen Instrumentes and Rules for Navigations* [...], London: Richard Jugge 1561, ed. David W. Waters. Delmar, NY: Scholars' Facsimiles 1992. Pedro de Medina, *Svma de cosmographía* (1561), ed. Juan Fernández Jiménez, [Valencia]: Albatros 1980.

[51] Jan Huygen van Linschoten, *Reys-gheschrift van de navigatien der Portugaloysers in Orienten*, Amsterdam: Cornelis Claesz 1595, pp. 132–4 (final pages of the second volume).

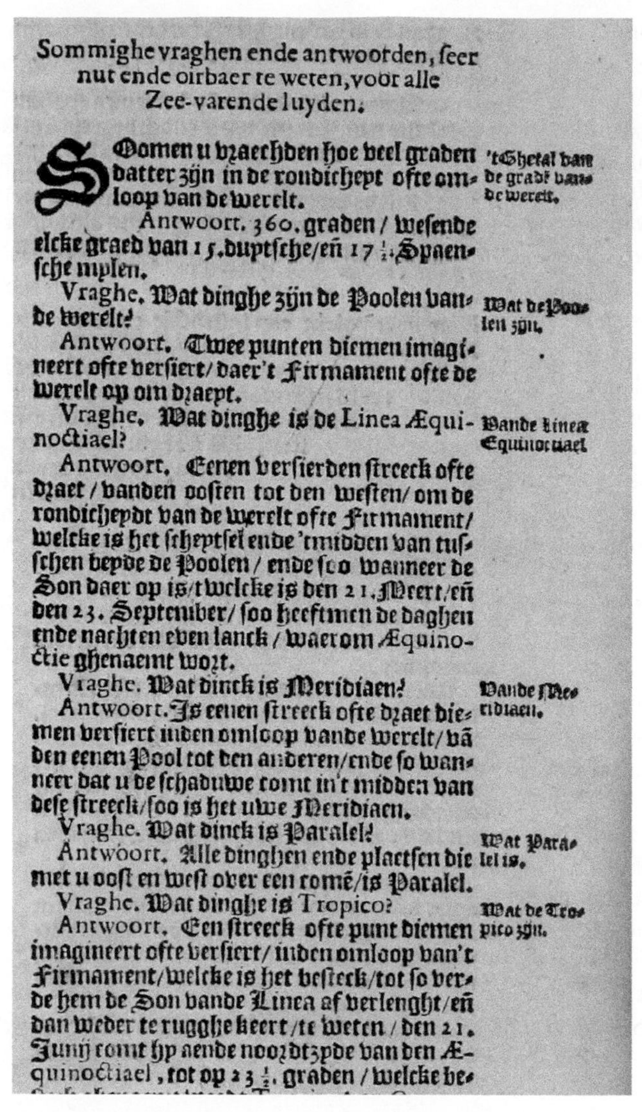

Sommighe vraghen ende antwoorden, seer
nut ende oirbaer te weten, voor alle
Zee-varende luyden.

Somen u vraechden hoe veel graden datter zijn in de rondicheyt ofte omloop van de werelt.

't Ghetal van de gradẽ vande werelt.

Antwoort. 360. graden / wesende elcke graed van 15. duytsche/ eñ 17 ½. Spaensche mylen.

Vraghe. Wat dinghe zijn de Poolen vande werelt?

Wat de Poolen zijn.

Antwoort. Twee punten diemen imagineert ofte versiert/ daer't Firmament ofte de werelt op om draept.

Vraghe. Wat dinghe is de Linea Æquinoctiael?

Vande Linea Equinoctiael

Antwoort. Eenen versierden streeck ofte draet / vanden oosten tot den westen/ om de rondicheydt van de werelt ofte Firmament/ welcke is het scheptsel ende 'tmidden van tusschen beyde de Poolen / ende soo wanneer de Son daer op is/ twelcke is den 21. Meert/ eñ den 23. September/ soo heeftmen de daghen ende nachten even lanck / waerom Æquinoctie ghenaemt wort.

Vraghe. Wat dinck is Meridiaen?

Vande Meridiaen.

Antwoort. Is eenen streeck ofte draet diemen versiert inden omloop vande werelt/ vã den eenen Pool tot den anderen/ ende so wanneer dat u de schaduwe comt in't midden van dese streeck/ soo is het uwe Meridiaen.

Vraghe. Wat dinck is Paralel?

Wat Paralel is.

Antwoort. Alle dinghen ende plaetsen die met u oost en west over een comẽ/ is Paralel.

Vraghe. Wat dinghe is Tropico?

Wat de Tropico zijn.

Antwoort. Een streeck ofte punt diemen imagineert ofte versiert/ inden omloop van't Firmament/ welcke is het besteck/ tot so verde hem de Son vande Linea af verlenght/ eñ dan weder te rugghe keert/ te weten / den 21. Junij comt hy aende noordtzyde van den Æquinoctiael, tot op 23 ½. graden / welcke be-

Figure 3.1 'Some questions and answers, very useful for all seafaring people': A dialogue summarizing the essentials of cosmography. Jan Huygen van Linschoten, *Reys-gheschrift*, Amsterdam: Cornelis Claesz, 1595.

posed to examination candidates.[52] Linschoten's lavish, three-volume set of travel accounts and navigational information was published by the prominent Amsterdam

[52] Jan Huyghen van Linschoten, *Itinerario, voyage ofte schipvaert naer Oost ofte Portugaels Indien. Deel 4 en 5*, ed. J. C. M. Warnsinck, The Hague: Martinus Nijhoff 1939, p. LXIX. Although Warnsinck suggests these are better suited to examination questions than a study guide, this hypothesis is unsupported.

publisher Cornelis Claesz. Claesz published a wide variety of texts, with something of a specialization in maritime material.[53] Linschoten spent several years in Goa, collecting numerous reports and observations, and then prepared his manuscript during a lengthy stopover in the Azores.[54] Because volume two of the collection, the *Reys-gheschrift*, contains considerable information about Portuguese voyages, it was rushed to print in 1595 to be ready for the first Dutch voyage to the far east.[55] It is not certain whether the three-page dialogue at the end of the volume was collected by Linschoten or Claesz. It may have come from a yet-to-be-traced Iberian cosmographical text, since the author provides the equivalent distances in Dutch as well as Spanish miles and mentions Havana.

This 1595 set of 'Certain questions & answeres very profitable & necessarie to be knowne by all Saylers' includes a dozen cosmographical definitions, a discussion of how to take altitude at different times, and explanations of the mariner's astrolabe and 32-point directional compass. With the definitions ranging from obvious to obtuse, the text focuses on summarizing rather than explaining. These jargon-laden passages would only be comprehensible to individuals familiar with the subject matter. Such lessons were not treated elsewhere in Linschoten's work but did appear in Medina and Apian. The dialogue is saturated with numbers: round numbers that facilitate simple mental calculations—10°, 90°, and 47° (the ecliptic, doubled)—are joined by special cases such as the equinoxes and high noon. The answers indicate that their author expected his readers to share his clear grasp of the physical model of the heavens, as well as a high level of numeracy.

Despite Linschoten's contention that 'all Saylers' could profit from this knowledge, his pricy folio volumes would not have been accessible for practitioners hoping to learn new techniques. However, this particular element of his text did reappear in a more affordable format. Jan van den Broucke, who operated a school in Rotterdam in the early years of the seventeenth century, published the same questions in his eclectic textbook.[56] Linschoten's dialogue was well suited to the classroom, fitting the traditional model of rote learning intended to cement concepts into the minds of students.

Numerous other educators produced similar cosmographical dialogues.[57] Daniel Newhouse, for instance, adapted the same set of sixteenth-century definitions. In his *The Whole Art of Navigation in Five Books*, published in 1685, he explains that his

[53] Günter Schilder, 'Cornelis Claesz (*c.*1551–1609): Stimulator and Driving Force of Dutch Cartography', in: *Monumenta Cartographica Neerlandica*, vol. 7. Alphen aan den Rijn: Canaletto/Repro-Holland 2003.

[54] Arun Saldanha, 'The Itineraries of Geography: Jan Huygen van Linschoten's 'Itinerario' and Dutch Expeditions to the Indian Ocean, 1594–1602', in: *Annals of the Association of American Geographers* 101/1 (2011), pp. 149–77.

[55] Saldanha, 'Itineraries', p. 157.

[56] Jan van den Broucke, *Instructie der Zee-Vaert*, Rotterdam: Abraham Migoen, 1610, f. 21r–v, 'Sommighe Vraghen ende Antwoorden, seer nut ende oirbaer te weten, voor alle Zee-varende persoonen'. On Van den Broucke's mnemonic strategies, see Schotte, *Sailing School*, pp. 55–7.

[57] See e.g. Cornelis Jansz. Lastman, *Beschrijvinge van de Kunst der Stuer-luyden* (ed. princ. 1642), Amsterdam: Symon Cornelisz. Lastman, 1657, where chapters 25–70 were divided into 118 questions, beginning with 'What is the art of navigation?' G. M. Oostwoud, *Vermeerderde Schoole der Stuurluyden*, Hoorn: Stoffel Jansz. Kortingh 1712, is also in dialogue form.

book 'is chiefly designed for Beginners, (although many Pilots may want it) . . . that is the reason that I chuse to make it by Dialogues'.[58] His discussion between 'a young Scholar' and 'his Tutor' opens with the ubiquitous question, 'What is Navigation?'[59] Soon after, curious about the list of tasks and instruments Newhouse lays out as the requirements for practising large and small navigation, the student inquires, 'Is this *all* that a Pilot should know?' The tutor responds, 'All good Pilots are obliged at least to know well what I have named, and to answer to it when examined about it'.[60] Not only does he confirm the existence of examinations, but his subsequent sentence reiterates the importance of instruments. While enthusiasts might 'perfect themselves' with advanced geometry, Newhouse suspected that working mariners would be more likely to skip mathematical tasks in favour of using the 'sinical quadrant or scale'.[61]

Samson Le Cordier included the same set of topics in his popular French manuals. Le Cordier also acknowledged the need to prepare for examinations, although he sceptically suggested that the examiners already knew who would pass before the tests even began.[62] Unlike Newhouse, Le Cordier viewed mathematical knowledge as essential: to undertake large navigation, 'it is à propos to have some knowledge of the Sphere, which is the foundation of Astronomical questions, Arithmetical knowledge, and some elements of Geometry'.[63] The publisher of later editions of Le Cordier's popular *Instruction des Pilotes* made some savvy adjustments to the original manual. From the 1748 edition onwards, the text was divided into alternating 'Questions et Réponses'.[64] The editor explained that the dialogue format was 'more familiar and more suited for beginners as an extended discourse ordinarily demands more attention'.[65] It is worth noting that the material about 'small and large navigation' was first presented under the heading 'Examen des Pilotes, et Maîtres et Pilotes'. However, in the 1754 edition, the same section was renamed as a 'General Instruction about Pilotage', with no other changes to the text.[66] Le Cordier's publisher may have wished to attract readers who did not intend to put themselves through a formal examination. These Q & A's facilitated the memorization of large quantities of information, but were less effective at helping navigators understand when or how to deploy that information at sea.

[58] Daniel Newhouse, *The Whole Art of Navigation in Five Books*, London: Printed for the author 1685, 'To the Reader'.

[59] Newhouse, *Whole Art*, Book 1, p. 1, 'Navigation is a Science or Art which contains certain Rules, absolutely necessary for every man to know that undertakes the Conduct of Ships from one Country or Harbour to another'.

[60] Newhouse, *Whole Art*, Book 1, p. 2.

[61] Newhouse, *Whole Art*, Book 1, p. 2.

[62] Samson Le Cordier, *Instruction des pilotes* (ed. princ. 1683), ed. M. Fouray, Havre de Grâce: P. J. D. G. Faure 1786, p. 187.

[63] Le Cordier, *Instruction des pilotes* 1786, p. 10.

[64] Le Cordier, *Instruction des pilotes* 1748, pp. 111–52, 'Examen des Pilotes, et Maîtres et Pilotes'.

[65] Le Cordier, *Instruction des pilotes* 1748, p. 7, Avertissement.

[66] Le Cordier, *Instruction des pilotes* 1754, p. 109.

Memorizing geography

Captain: What course should you (having sailed out from Texel) hold in order to pass the Strait of Dover freely and comfortably?
Navigator: Just a bit more southerly than a south-west heading.[67]

More immediately useful for daily practice were the examinations that encouraged navigators to memorize geographical details. The ability to recall the local harbours of small navigation (within sight of land), or the shifting currents and prevailing winds along specific routes required by each merchant company or navy, was of paramount importance for navigators well into the seventeenth century. Geographic questions and answers consequently appeared on many examinations. The hands-on Spanish examination discussed above included a geographical component, focusing on the course for which a navigator wished to be certified. English masters, and navigators (*stuurlieden*) for the Dutch merchant companies similarly needed to commit crucial details to memory.

The primary evidence of these geography-focused examinations appears in Dutch manuals. In his 1660 textbook, *'t Vergulde Licht der Zeevaert*, the prominent examiner and author Claas Hendriksz. Gietermaker introduced a pair of pioneering elements: the first published versions of the written and oral navigators' examinations. Both of these examinations proved extremely influential. The questions on the oral examination (*mond examen*) were entirely focused on geography. Once *stuurlieden* candidates had passed the written, mathematical component (analysed below), they were expected to prove their mastery of this geographic material.

Gietermaker presented a series of nineteen questions as an instructional dialogue, 'Consisting of various questions and answers between a *Schipper* and a *Stuurman*, relating to some courses, ocean bottoms, depths and shallows: The *Schipper* asks and the *Stuurman* answers'.[68] This dialogue appeared verbatim in the twenty-one editions of *'t Vergulde Licht* that appeared in the ensuing 125 years. (Gietermaker also included the exchange in his 1665 tract on trigonometry, and it was translated into French in the 1667 edition of his *Flambeau Reluisant*, despite its limited relevance to French navigators.[69]) Numerous Dutch textbook authors would follow Gietermaker's lead, publishing very similar versions of the *mond examen*, most distilled into six or seven questions. In his 1695 text, Claas J. Vooght indicated that such dialogues were more than classroom exercises: 'Before we break off, it pleases us to hereafter include the

[67] Claas Hendriksz. Gietermaker, *'t Vergulde Licht der Zeevaart*, ed. Frans vander Huips, Amsterdam: Hendrik Donkker 1677, p. 148, first question.

[68] Claas Hendriksz. Gietermaker, *'t Vergulde Licht der Zeevaert*, Amsterdam: J. van Keulen 1710, pp. 148–50.

[69] Gietermaker, *Driehoex-rekening bestaende in de verklaringe en ontbindinge der platte driehoecken . . . Met een discours tusschen een schipper en stuurman, aengaende de zeevaert*, [Amsterdam]: gedruckt voor den Autheur 1665; Gietermaker, *Le Flambeau reluisant, ou proprement Thresor de la navigation*, tr. J. Viret, Amsterdam: Henri Donker 1667. In recognition of the different system in France, the French translator/publisher omitted the *written* examination from the French edition.

Verbal questions (otherwise called the Oral Examination) which is always used by the VOC in question-and-answer fashion'.[70]

This standardized oral examination drilled candidates on their knowledge of the most common routes for VOC vessels. As the most pressing geographic question for the Company was how best to cross the Indian Ocean during the monsoon season, one might expect the examination to touch on this. Indeed, the company equipped every vessel with an instruction booklet on sailing to Java, and Gietermaker and others began printing the same information in their manuals.[71] But while the oral examination was presented as questions about sailing 'from Texel [north Holland] to the East Indies', the answers were in fact limited to adjacent European waters: Scotland, the English Channel, and south to Spain. While examiners placed importance on this local information, the details were assumed to be so familiar that they were not discussed in navigation manuals.

The standard geographical questions bear a strong resemblance to the information presented in rutters or *routiers*, the textual sailing directions that had their heyday before the rise of published nautical atlases. Compass bearings were more prominent than distances, and details about shallows and other hazards were described verbally.[72] Although navigators were becoming more comfortable with interpreting the visual information on nautical charts and atlases by the time Gietermaker published his *mond examen*, geographical knowledge continued to be described verbally. For the most important routes, or at least those closest to home, navigators needed to have the details in their memories rather than rolled up in a chart in the ship's chest.

The geographic examination remained remarkably stable until the mid-eighteenth century. Even at that point, new locations were added to the test but the dialogue format remained unchanged. Entrepreneurial instructor Pieter Holm published an updated *Mond Exame*, first in his 1748 textbook *Stuurmans Zeemeeter*, and then as a separate publication in 1759.[73] This test, like Gietermaker's, expected candidates to memorize depth soundings and compass headings, as well as lighthouse locations along the Dutch, English, and French coasts. The majority of the questions pertained to the North Atlantic (the Isles of Scilly, the Flemish Banks), but there were two questions about sailing to Suriname and the Caribbean.[74] On the penultimate question, No. 21, when the examiner asks about the distances in a six-leg traverse course, the published answer indicates that the candidate must have a chart or diagram of the same journey to Suriname in front of him in order to complete

[70] Claas Jansz. Voogt, *De Zeemans Wegh-Wyser*, Amsterdam: J. van Keulen 1695, p. 234.

[71] Heren XVII, *Instructie Van de Eygenschap der Winden, In het vaerwater Tusschen Nederlandt en Java*, Amsterdam: Paulus Mathysz. for the Oost-Indische Compagnie 1671. Gietermaker, *'t Vergulde Licht* 1677, pp. 1–8, separate pagination after p. 152.

[72] Pierre Garcie, *Le grand routier et pilotage et enseignement pour ancrer tant ès ports, havres qu'autres lieux de la mer*, Rouen: chez Jehan Burges 1521. Atlases followed this model, placing textual directions on the verso of the maps, e.g. Lucas Janszoon Waghenaer, *Spieghel der Zeevaerdt*, Leiden: Christophe Plantin 1584.

[73] Pieter Holm, *Mond-Exame* [sic] *Voor de Stierlieden van Het Schip Recht door Zee*, [Amsterdam]: the Author 1759.

[74] Holm, *Mond-Exame*, pp. 6–7, No. 18 (misnumbered as 19) and No. 19.

the question. From this internal evidence, Holm seems to have been teaching both VOC and WIC mariners. (In the eighteenth century, English treatises included similar details about the Channel—noting the depth in fathoms, the distinctive rocks on the ocean floor near Scilly—although not in dialogue form.[75]) These concise examination answers reveal just how much was left up to the navigator's accrued experience: given just five bearings, these men were expected to successfully cross the Atlantic to Suriname.

Calculating position

A captain, sailing in northern latitudes, shot the Sun in the ESE at 7 hours and 17 minutes after noon, observing a 12 degrees 0 min. southern declination. Question: at what latitude did this occur.
Answer, at ____ degrees ____ minutes ____ Latitude.[76]

Of all the different routines for assessing a navigator's abilities, the most challenging was the mathematical written test, which got right to the heart of the daily calculations required by celestial navigation. This style of test—deemed by the VOC and WIC to capture essential knowledge—was extremely influential, going on to shape coursework and subsequent textbooks in the Netherlands and beyond. Teachers initially experimented with different approaches to teaching celestial navigation.[77] However, once the first model examination was published in 1660, the curriculum in Dutch schools coalesced around those topics, with the end goal of helping as many mariners as possible to become applied mathematicians. The written examination developed in parallel with the mathematically focused nautical manuals of the seventeenth century. These, by Lastman, Gietermaker, Klaas de Vries, Jan van Dam, and others, aimed to help mariners grapple with logarithms and trigonometry, innovative mathematical tools that could simplify the process of finding one's position from the constantly moving celestial bodies. In order for navigators to take advantage of these powerful solutions, they needed to become comfortable with dense numerical tables and extensive computations. The best way to do this was through repetition. Consequently, students signed up for classes where teachers walked them through countless practice questions. Once they familiarized themselves with the different types of problems—time calculations, position, direction—and knew their way around the related tables, they were ready to face the written test.

In 1660, in addition to the pivotal *mond examen*, Claes Gietermaker published a set of thirty-six questions under the heading 'Toetse de Navigatie' (Navigation Test). He revisited a set of 'Questions for my disciples to practice', which he had published the year before, and placed new questions on the same topics at the halfway point

[75] John Hamilton Moore, *The Practical Navigator*, London: W. and J. Richardson 1791, pp. 295–6.
[76] Gietermaker, *'t Vergulde Licht* 1677, p. 152, question 35.
[77] See e.g. the creative and personalized practice questions in William Cuningham's *Tractaet des Tijdts*, Franeker: Dirck Jansz. Prins en Rombertum Doyma 1605, pp. 50–2.

of *'t Vergulde Licht*. To challenge his test-takers he increased the number of questions and removed the printed answers. His fidelity to these particular topics—from tides to Mercator charts—suggests that early seventeenth-century examinations were already being conducted in a similar sequence. Gietermaker effectively codified them for the ensuing century.[78] Every few decades the dates in the questions would be updated so that the students would be able to find the relevant astronomical and tidal information in current almanacs. The *toets* itself became a key part of those courses based on Gietermaker's textbook, appearing in virtually every extant manuscript based upon *'t Vergulde Licht* for at least one hundred years.[79]

Gietermaker's decision to expand familiar material and present it in examination format may have stemmed from a desire to increase his market share, or an awareness of student demand, but it must also have reflected an instructor's opinion of the best way to prepare for the type of test he administered for both the WIC and the VOC. In the next fifty years, at least half a dozen prominent textbook authors, many of whom also held examiner appointments, followed Gietermaker's example.[80] It became *de rigueur* for an examiner to produce his own version of the test as a calling card to demonstrate his competence to the companies or the Admiralty, as well as to offer candidates a better chance of passing his version of the test. (Dutch examiners were paid only when candidates passed.[81]) Most chose to include answers so readers could correct their practice exercises, converting the mock examination into a study guide.[82]

In almost every case, written examinations in the Gietermaker tradition opened with a multi-part question about the time of high tide in several key harbours. These examinations typically devoted at least one and as many as five questions to each of the following topics: courses and currents, plane charts, the core altitude observations (using the sun and stars, and how to work with two differing sun sights on the same day), two methods of determining magnetic variation, and a concluding section on Mercator charts. The topics echo those daily tasks tested in Spain and England, with the addition of questions about currents and more technical queries about magnetic variation, which the Dutch emphasized through the mid-eighteenth

[78] The examination remained the same up through the 1774 edition of *'t Vergulde Licht*, edited by Frans vander Huips. The WIC examinations do not survive in print, but since the two companies often shared examiners, the questions would have been very similar (except for the attention to the seasonal monsoons in the Indian Ocean).

[79] For an example from several manuscripts clearly based on Gietermaker, see J. W. Sleutel's 16-question practice exam, MMR H631: J. W. Sleutel, *Konstige oefeningen begrepen in drie boecken*, Hoorn 1675–7 at pp. 157–86.

[80] Gietermaker, Dirk Makreel, and Abraham de Graaf all published textbooks containing examinations within a year of their appointments as examiner. Claas de Vries and Jan Albertsz. van Dam also published examinations in conjunction with their tenure.

[81] There is evidence of conflicts of interest in almost every maritime community. For instance, after Abraham de Graaf died in 1714, no one was allowed to be an examiner and simultaneously run a private school. Bruijn, *Schippers van de VOC*, p. 137. Iliffe, 'Mathematical Characters', p. 121.

[82] The 31-question examination in Dirk Makreel's *Lichtende leydt-starre der groote zee-vaert*, Amsterdam: H. Doncker 1671, appeared without solutions, but Makreel himself went through copies and inked in many answers. A. de Graaf's 21-question exam, in the *De kleene Schatkamer*, Amsterdam: voor den Autheur by d'Erfg. van Paulus Matthysz 1688, was the first to provide printed answers. Answers also appear in C. J. Vooght, *De Zeemans Wegh-Wyser*, Amsterdam: J. van Keulen 1695, p. 227, ch. 31, 'Om de Examina, of d'Ondervragingh in de Stuurmans Konst wel te beantwoorden'.

century for its potential use for determining longitude.[83] By including extra questions in certain areas, authors indicated which topics their students should study most thoroughly. Which were more difficult and therefore required extra practice? Which were absolutely essential aboard ship, and which were less common problems? Gietermaker privileged the infrequent operations: latitude measurements from stars as opposed to the sun, magnetic variation from one observation rather than two, and charting courses using Mercator charts rather than plain charts. Others, like the Amsterdam Admiralty examiner Dirk Makreel, preferred the most familiar ones: on his test solar observations outweighed stellar ones, for example.[84] Textbook authors continued to debate whether sailors learned best by working their way through exhaustive treatments or whether it was preferable to offer reluctant readers a more streamlined course of study; the robust Dutch market offered options to learners of every stripe.

Concepts in conversation

Q. If you were to use your cross-staff to measure the Sun's altitude, could you do it properly in three different ways?
A. Yes, I would be able to do that.
...
Q. Well, tell us one [method]?[85]

How did mariners approach this material—in their classes as well as in the examination room? They were responsible for solving mathematical equations but also for understanding general concepts. We can glean insights about the latter from one unusual dialogue, essentially a discursive version of the written examination. Around 1701 an Amsterdam publisher, Jacobus Robyn, printed an anonymous sixteen-page pamphlet that consisted of an almanac and astronomical tables, with the second half of the volume devoted to a unique model examination (see Figure 3.2).[86] This dialogue is more imaginative than the rote catechistic dialogues in standard study guides. The candidate begins by laying out his credentials, explaining his aspirations and past experience: fifteen years at sea and voyages to most corners of the globe.[87] The examiner humorously pushes the *Jongman* for sufficient detail in his answers. When pressed about the same key navigational skills that appeared on other written examinations, the young mariner first gives a glib answer—perhaps the kind of thing an unprepared candidate would offer—before providing a thorough multi-part answer that satisfied the examiner. Unlike Gietermaker's test-takers or Linschoten's readers, this candidate was never asked to calculate a specific latitude or even to add simple round figures based on hypothetical locations on the globe, but his jocular

[83] A. R. T. Jonkers, *Earth's Magnetism in the Age of Sail*, Baltimore: Johns Hopkins University Press 2003.
[84] See Gietermaker, *'t Vergulde Licht* 1677, pp. 150–2; and Makreel's five solar and three stellar questions, *Lichtende leydt-starre*, pp. 192–4.
[85] *Nieuwe uytgereckende Taafelen, Wegens de Son, Maan en Sterren, Als meede een Almanach*, Amsterdam: Jacobus Robyn [1701], pp. 8–16, 'D'Examinateur, der Stuerlieden'.
[86] *Nieuwe uytgereckende Taafelen*, pp. 8–16.
[87] *Nieuwe uytgereckende Taafelen*, p. 8.

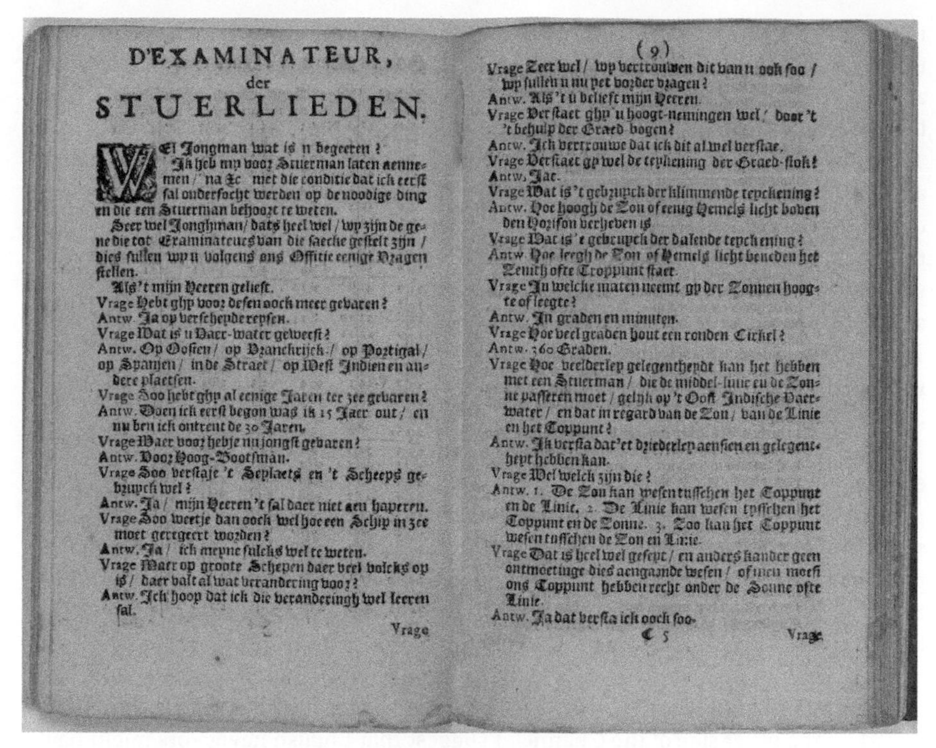

D'EXAMINATEUR,
der
STUERLIEDEN.

WEl Jongman wat is u begeeren?
Ik heb my voor Stuerman laten aennemen / na &c met die conditie dat ick eerst sal ondersocht werden op de noodige ding en die een Stuerman behoort te weten.
Seer wel Jonghman / dats heel wel / wy zijn de gene die tot Examinateurs van die saecke gestelt zijn / dies sullen wy u volgens ons Offitie eenige Vragen stellen.
Als 't mijn Heeren geliest.
Vrage Hebt ghy voor desen oock meer gevaren?
Antw. Ja op verscheyde reysen.
Vrage Wat is u Vaer-water geweest?
Antw. Op Oosten / op Vranckrijck / op Portigal / op Spanjen / in de Straet / op West Indien en andere plaetsen.
Vrage Soo hebt ghy al eenige Jaren ter zee gevaren?
Antw. Doen ick eerst begon was ik 15 Jaer out / en nu ben ick ontrent de 30 Jaren.
Vrage Waer voor hebje nu jongst gevaren?
Antw. Voor Hoog-Bootsman.
Vrage Soo verstaje 't Seplaets en 't Scheeps gebruyck wel?
Antw. Ja / mijn Heeren 't sal daer niet aen haperen.
Vrage Soo weetje dan oock wel hoe een Schip in zee moet geregeert worden?
Antw. Ja / ick meyne sulcks wel te weten.
Vrage Maer op groote Schepen daer veel volcks op is / daer valt al wat verandering voor?
Antw. Ick hoop dat ick die veranderingh wel leeren sal.
Vrage

(9)
Vrage Seer wel / wy vertrouwen dit van u ook soo / wy sullen u nu yet vorder vragen?
Antw. Als 't u believt mijn Heeren.
Vrage Verstaet ghy u hoogt-nemingen wel / door 't 't behulp der Graed-bogen?
Antw. Ick vertrouwe dat ick dit al wel verstae.
Vrage Verstaet gy wel de teykening der Graed-stok?
Antw. Jae.
Vrage Wat is 't gebruyck der klimmende teyckening?
Antw. Hoe hoogh de Son of eenig Hemels licht boven den Horison verheven is.
Vrage Wat is 't gebruyck der dalende teyckening?
Antw. Hoe leegh de Son of Hemels licht beneden het Zenith ofte Toppunt staet.
Vrage In welcke maten neemt gy der Sonnen hoogte of leegte?
Antw. In graden en minuten.
Vrage Hoe veel graden hout een ronden Cirkel?
Antw. 360 Graden.
Vrage Hoe veelderley gelegentheydt kan het hebben met een Stuerman / die de middel-linie en de Sonne passeren moet / gelijk op 't Oost Indische Vaerwater / en dat in regard van de Son / van de Linie en het Toppunt?
Antw. Ik versta dat 'et driederley aensien en gelegentheyt hebben kan.
Vrage Wel welck zijn die?
Antw. 1. De Son kan wesen tusschen het Toppunt en de Linie. 2. De Linie kan wesen tusschen het Toppunt en de Sonne. 3. Soo kan het Toppunt wesen tusschen de Son en Linie.
Vrage Dat is heel wel geseyt / en anders kander geen ontmoetinge dies aengaande wesen / of men moest ons Toppunt hebben recht onder de Sonne ofte Linie.
Antw. Ja dat versta ick oock soo.
C 5 Vrage

Figure 3.2 'D'Examinateur, der Stuerlieden': Talking through the technicalities of the written exam. *Nieuwe uytgereckende Taafelen*, Amsterdam: Jacobus Robyn, [1701], pp. 8–16.

answers displayed a clear understanding of cosmography and astronomy. When he was quizzed on his knowledge of instruments, the examiner was more interested in how the *Jongman* would manipulate the observed number than the technicalities of wielding the tool.[88] The manual side of things was once again assumed to be familiar from frequent use. The conversation concludes abruptly, without any closing remarks about whether the candidate passed, a concision due either to the pamphlet's sixteen-page format or the author's lack of behind-the-scenes knowledge of the official examination process.

In this atypical examination, comprehension is even more important than calculation. To earn praise for a question 'well answered', the *Jongman* needed to lay out all the possible cases for each type of problem, such as the three distinct positions of the sun with respect to the equator and the zenith.[89] These responses, set phrases that reveal how navigators were taught to analyse the situation, imparted a mental

[88] *Nieuwe uytgereckende Taafelen*, p. 9.

[89] If the *Stuurman* must measure the sun in East Indian waters, how many possible positions are there for the observer with regard to the sun and the equator? The youth correctly answers three: the sun could be between the zenith and the horizon; the horizon could fall between the zenith and sun; and the zenith could fall between the sun and horizon. He then must explain whether to add or subtract depending on the type of graduations on his cross-staff. *Nieuwe uytgereckende Taafelen*, p. 9.

image of the underlying geometry so that the correct formula could be applied. Even those candidates who simply parroted rote answers and relied upon printed tables aboard ship could not help but absorb a three-dimensional notion of the globe and the observer's place relative to the heavens, the tenets of cosmography.

Many of the basic definitions reappear from Linschoten's dialogue a century before, although Robyn's respondent is notably more comfortable with astronomical concepts. Like Linschoten's interlocutor, the questioner is interested in general cases rather than numerical answers, and in fact incorporates fewer numbers than Linschoten, avoiding even simple sums. (The sole quantities mentioned were right angles, the number of minutes by which the tide shifts, and the young man's age.) The author made the significant assumption that his readers' mathematical skills were sufficient to translate their comprehension of general principles into accurate calculations at sea.

What was the purpose of this mock examination? In a sea of conventional mathematical dialogues, this text stands out. The author may simply have intended to present the information from the written examination in a more engaging form, as a stylised conversation. If candidates wanted to prepare for the VOC *toets*, they could purchase the latest edition of Gietermaker. However, if they wanted a glimpse into the examination room, this dialogue imagined how a knowledgeable navigator would have responded to his interrogation.

The verbal rather than numerical expression of key concepts offers a window into other examination rooms as well. Returning to the inscrutable halls of the Navy Board on the other side of the Channel, I suggest that English navigators might have had similar responses to oral questions about altitude, instrument use, and other familiar subjects. Robyn's dialogue aligns with the topics not only of Gietermaker's written examination, but also those on the second half of the English lieutenants' certificates. However, where Dutch *stuurlieden* were given a specific set of dates and observations and required to compute position down to degrees and minutes using diagrams, logarithmic tables, and trigonometry, English lieutenants fielded different questions. Rather than expecting their candidates to carry out trigonometric computations mentally, British Admiralty examiners would have inquired about the best method of obtaining the relevant observations, or asked questions about common cases using round numbers, just as Linschoten's and Robyn's examiners did.

Evidence of mathematical practice

> *Difference of Latitude with the Longitude known*
> *Let there be a place, at N latitude 12 deg. 27 min., and longitude of 356 deg. 2 min.; and a second place at N latitude of 12 deg. 27 min. and longitude of 339 deg., 39 min. What is the course direction and distance?*[90]

[90] NL-HaNA 1.01.47.11 inv. nr. 11A Luytenant's Exam, Middelburg, 1767, p. 4, question 18, with blanks filled in by the examiner.

If the Robyn examination focused on concepts, the standard VOC examinations focused on obtaining precise numbers. Of all five types of navigation examinations, only this last leaves substantive archival evidence from the candidates. As they worked to master the theoretical side of navigation, mariners sought out lessons. When it came to large navigation, beyond sight of land, it was no longer sufficient for them to memorize procedural points about pilotage manoeuvres, instruments, or geographical details. Instead, they needed to learn how to use trigonometry to calculate distances from a known compass bearing, or changes in latitude from observations taken a few hours apart. They needed guidance in the use of logarithm tables and Mercator charts. Thus, in countless classrooms from the late seventeenth century until at least the early nineteenth century, diverse members of the Dutch maritime community produced manuscripts filled with nautical mathematics. These *schatkamers* ('treasure chests', named after Lastman's textbook) show how students approached these questions. Many of the surviving manuscripts were taken to sea by their creators, while others, judging by the fastidiousness of their lettering and decoration, were more likely created by gentlemen or merchants who did not pursue a nautical career.[91]

The navigator's primary task was to determine the ship's position. In order to derive a geographical coordinate, he needed to track the vessel's route (distance and direction), compare it with the previous known location, and (weather permitting) take a number of observations that would allow him to compute his current coordinates. The VOC written examination broke this complex task down into manageable components: how did one determine one's latitude from a vertical observation; triangulate between two observations to pinpoint a location; quantify the effect of the current on a ship; or predict the intersection point of two vessels each moving on a known course? Solving any of these problems could take as many as a dozen steps, often relying upon geometrical diagrams to visualize the respective positions of celestial features and the observer. Navigators needed to get comfortable consulting dense logarithmic and trigonometric tables, almanacs and astronomical tables, as well as rhumb line (*streek*) tables.[92] It was also crucial to memorize mathematical formulas, and know which one to apply at different times, for instance, to determine latitude when the vessel was south of the Equator. Navigators needed to work through such questions repeatedly so that they would be able to answer in a timely manner without making potentially dangerous errors.

Most student *schatkamers* contain at least one practice examination. In assigning problems to their students, teachers drew upon various popular textbooks if they had not published one themselves. Gietermaker's original *toets* appears in at least five manuscripts created between 1705 and 1774. In his 150-page workbook, one *stuurman*, Assuerus van Asson, worked through two and a half practice examinations,

[91] C. J. Lastman, *Schat-kamer des grooten seevaerts-kunst*, Amsterdam 1621. For a discussion and census of these manuscript *schatkamers*, see Schotte, *Sailing School*, ch. 4 and Supplemental bibliography. See also E. Crone, *Cornelis Douwes, 1712–1773: Zijn Leven en Zijn Werk; Met Inleidende Hoofdstukken over Navigatie en Zeevaart-Onderwijs in de 17de en 18de eeuw*, Haarlem: Tjeenk Willink & zoon 1941, ch. 2.

[92] Martin Campbell-Kelly, Mary Croarken, Raymond Flood, and Eleanor Robson, eds, *The History of Mathematical Tables: From Sumer to Spreadsheets*, Oxford: Oxford University Press 2003.

those of Gietermaker, Abraham de Graaf, and Jan Albertsz. van Dam.[93] In the latter half of the eighteenth century, other titles rose in popularity. The twenty-five-question test that Claas de Vries published in *Kunst der Stuurlieden* (1702) appears in four different, extensively decorated manuscripts.[94] Hendrik de Vos, whose charming sketches suggest he had first-hand experience on a whaling vessel, painted his spherical trigonometrical diagrams in vibrant colour. (Textbook authors and instructors could not agree if spherical trigonometry was a valuable tool, or if sailors were incapable of understanding its complexities. The VOC examinations could be solved without it.) These *schatkamers* offer intriguing evidence of regional differences and teacher preferences. Vos and another student, Dirk de West, answered the same questions from De Vries in contrasting ways. For question No. 2, about the direction of a current, they each placed the current in a different location on their diagrams. Vos first solved for the angle of the ship's course, and then computed the change in distance.[95] West, using Dutch rather than Latin terms (*hoekmaat*, abbreviation 'hmt', for *sinus*) came up with a slightly different course angle, but the same final distance.[96] Both used trigonometric tables with a radius of 100,000. In question No. 4, Vos solved for each leg of the course sequentially, whereas West used his diagram to solve the cumulative journey in a single equation. West, with his pristine, elaborately calligraphed manuscript, then skipped the five questions about altitude, 'due to their simplicity'.[97]

In other classrooms, certain lessons seem to have consisted only in copying out the entire examination—transcribing the questions and answers without pausing to solve them.[98] One sailor on the Danish merchantman *Scielland*, whose ship was seized by the English in 1798, filled several cartonnage-covered workbooks with nautical mathematics. In one modest notebook, he began transcribing a text entitled 'Grond Beginsels' (Basic Principles) in beautiful, flowing script, wasting not a centimetre of the page. Instead of the fundamentals of navigation that the title might lead us to expect, these 'basics' were in fact a geographic Q & A tailored to the route to Asia: What lighthouses are on the English coast? What is the best course from the Equator to the Cape [of Good Hope]? After a dozen pages his handwriting deteriorated, he ripped some pages out, and after copying a single line from a *Mond Examen* at the top of the volume's final page, this anonymous mariner stopped his exam prep mid-lesson.[99]

The *schatkamer* manuscripts not only demonstrate that students became familiar with the examinations in their classes, but also that they worked through hundreds of similar questions from the other sections of the textbooks. Certain mariners spent their winters ashore, carefully working through one question a week, while

[93] OTYA 16,275 Assuerus van Asson, 'Schatkamer Ofte Konst Der Stuurlieden', *c.*1705.

[94] HSM S.3003 J. Kok, 'Schatkamer ofte konst der stuurlieden', Kampen 1763 (carefully calligraphed and coloured, worn binding but no evidence of being taken to sea). MMR H891.2 [H. M. Hoffman?], Navigation workbook based on De Vries, 1758 (disbound, toned, but not waterstained). See also Vos and West below.

[95] MMR H632 Hendrik de Vos, 'Schatkamer', 1748–53.

[96] HSM S.0187 Dirk de West, 'Schatkamer', *c.*1760.

[97] West, 'Schatkamer', pp. 169, 170, '*ligtheyt der zaken*'.

[98] HSM S.0712 M. J. Grootschoen, Schatkamer, 1728, [ff. 100r–v].

[99] TNA HCA 30/763i, 'Grond Beginsels', *c.*1795. This text is not based on Steenstra's manual of the same title.

others took up old manuals years after they were published and proceeded to work through the questions and interleave their answers.[100] The *Scielland* navigator also had amongst his papers a manuscript workbook filled with questions from the works of examiner Pybo Steenstra. He drew careful circular diagrams as he learned the intricacies of spherical trigonometry.[101] By taking time to work through these examinations and textbooks, these mariners gained practice moving between the books of tables and their own calculations. If they erred, they or their teachers made careful ink corrections.[102] Through such repetition, these men became comfortable with advanced mathematical tools from logarithms to spherics. Just as there had been an expectation in England that lieutenants learn how to tie knots while at sea, here in these manuscripts we can see how these students got up to speed on their geometry and ratios; these practice examinations were simply a different mode of hands-on learning.

Evidence of the examination room

Both of these examinations, both oral and written, should be given to each person according to his abilities, and following the Model for all of the Exams, specified by the Executive Committee, printed at their orders, and following these instructions.[103]

With this understanding of the skills required to pass early modern navigation examinations, let us now return to Captain van Ewijk. What happened when he failed his first examination in Rotterdam in October 1768? We cannot know for sure, but based on his remarkable improvement the following spring, especially under more assiduous questioning, it appears that he put effort into studying for his test. Van Ewijk may well have signed up for lessons that winter, or perhaps even a cram course from one of the many independent instructors. Pieter Holm, for one, ran a popular school in Amsterdam. Although the standard course was six weeks of lessons at a cost of 36 florins, Holm also offered specialized 'exam instruction' for 6 florins.[104] As mentioned above, Holm excerpted the oral examination from his larger textbook and sold that separately to facilitate test preparation.

To some extent it is possible to make inferences about Van Ewijk's experience when it came time to reattempt the written *toets*. By the second half of the eighteenth century, VOC candidates took their test in a room with others. The examiner had strict instructions not to let anyone cheat: he must take care that those in the room were not allowed to share anything orally or in writing regarding numerical calculations

[100] e.g. MMR ARCH 8D21 G. M. Oostwoud, *Vermeerderde Schoole der Stuurluyden*, Hoorn: Stoffel Jansz. Kortingh 1712, with exercises interleaved by V. J. Warnar in 1806–7.

[101] TNA HCA 30/763i, 'Kunst der Stuurlieden' schatkamer, 1790–98. Steenstra, *Grond-beginzels der Stuurmans-kunst*, 1779, where the order of topics changed to emphasize trigonometry. See also Steenstra, *Uitgewerkt Examen*, Middelburg: P. Gillissen en Zoon 1781, which is an explanatory textbook rather than an examination.

[102] See e.g. Van Asson, 'Schatkamer', ff. 90^v, 92, etc.

[103] NL-HaNA VOC inv. nr. 5026 Heren XVII, *Instructie voor de Examinateurs der Stuurlieden by de edele Oost-Indische Compagnie*, 1793, p. 3, ch. 2, Art. 2.

[104] Ernst Crone, 'Pieter Holm en zijn Zeevaartschool', in: *De Zee* 52 (1930), pp. 489–97 at p. 493.

('Cyfferen').[105] The examiner's guidelines also included instructions about giving a second chance to candidates at risk of failing: providing their errors were simply from sloppy addition or incorrectly looking up 'Sines or Logarithms', they could be given new questions—but if the authorities judged they suffered from 'ignorance or faulty understanding of the rules', they would be failed.[106]

Van Ewijk would probably have received a copy of the written examination questions and a blank workbook. As the eighteenth century progressed, the company arranged to have fill-in-the-blank examination forms printed, a sign of their efficient information management system as well as a considerable number of examination candidates. The earliest of these printed examination booklets is that of examiner P. Warius, who had his twenty-three-question *Examen der Stierlieden* (Navigator's Examination) printed in Hoorn around 1730.[107] Another fill-in-the-blank exam, this one for 'Captains and Captain-Lieutenants of the VOC', survives from 1794—but that appears to be a study guide rather than an actual test. In one extant copy, the blanks for questions and answers were filled out in the same hand, with three of the thirty-one questions left blank, and textbooks are cited in footnotes for anyone needing additional information.[108] To streamline the test-setting process, in many questions the publisher printed pairs of words: North or South, under or over. With a stroke of the pen, the examiner or the candidate could eliminate one of the two options and then fill out the numerical portion of the question.

There are no traces of Van Ewijk's examinations from 1768–9, but a completed written test survives from just two years previously. A young German gentleman, Baron van Kinckel, sat down in Middelburg (Zeeland) to complete his lieutenant examination on 15 May 1767 (see Figures 3.3 and 3.4).[109] While there may have been distinctions between the administration of the Admiralty and VOC examinations, there were few substantive differences in the mathematical content. Van Kinckel's examination papers show how seriously test candidates took their assessments. The twenty-year-old baron did not rush, carefully copying out each question from the printed examination. Van Kinckel was equipped with a drawing compass, aware in advance that the examiners would ask, at question No. 20, that he 'Demonstrate all this with diagrams'.[110] He also had tables to hand, as the second question instructed that he use tide tables from Dirck Rembrandtsz. van Nierop's '*Nieropper Graad-boekjes* ... or another [volume]'.[111] For question No. 19, he was directed to use rhumb tables, and for question No. 20 and others, he relied on logarithmic and

[105] *Instructie voor de Examinateurs*, p. 4, ch. 2, Art. 5.

[106] *Instructie voor de Examinateurs*, p. 4, ch. 2, Art. 7.

[107] P. Warius, *Examen der Stierlieden*, Hoorn: Adriaan Brouwer *c.*1731.

[108] HSM S.4793(342) VOC Heren XVII. *Examen voor de Capitain en Capitain-lieutenants by de Oost-indische Compagnie*, [Amsterdam? 1794].

[109] NL-HaNA 1.01.47.11 inv. nr. 11A 'Examen of voorgegeeve Questien over de Navigatie', [1767]; inv. nr. 13 H. A. van Kinckel, Zeevaartkundige vraagstukken, examen voor luitenant, May 1767. Note: these images were incorrectly described in Schotte, *Sailing School*, pp. 144–5, figs. 4.19 and 4.20; the manuscript copy is not a draft but rather Kinckel's completed examination.

[110] 'Examen of voorgegeeve Questien over de Navigatie', p. 4, 'Dit alles figuurlijk te demonstreeren'.

[111] 'Examen of voorgegeeve Questien over de Navigatie', p. 1; *Nieuw Nieroper Graed-boeck*, Amsterdam: H. Doncker 1683. The 1794 VOC *Examen voor de Capitain* indicates that the candidate is allowed to consult the Tafel van Douwes [No. 8] and Steenstra's *Zeemans Almanach* [No. 27].

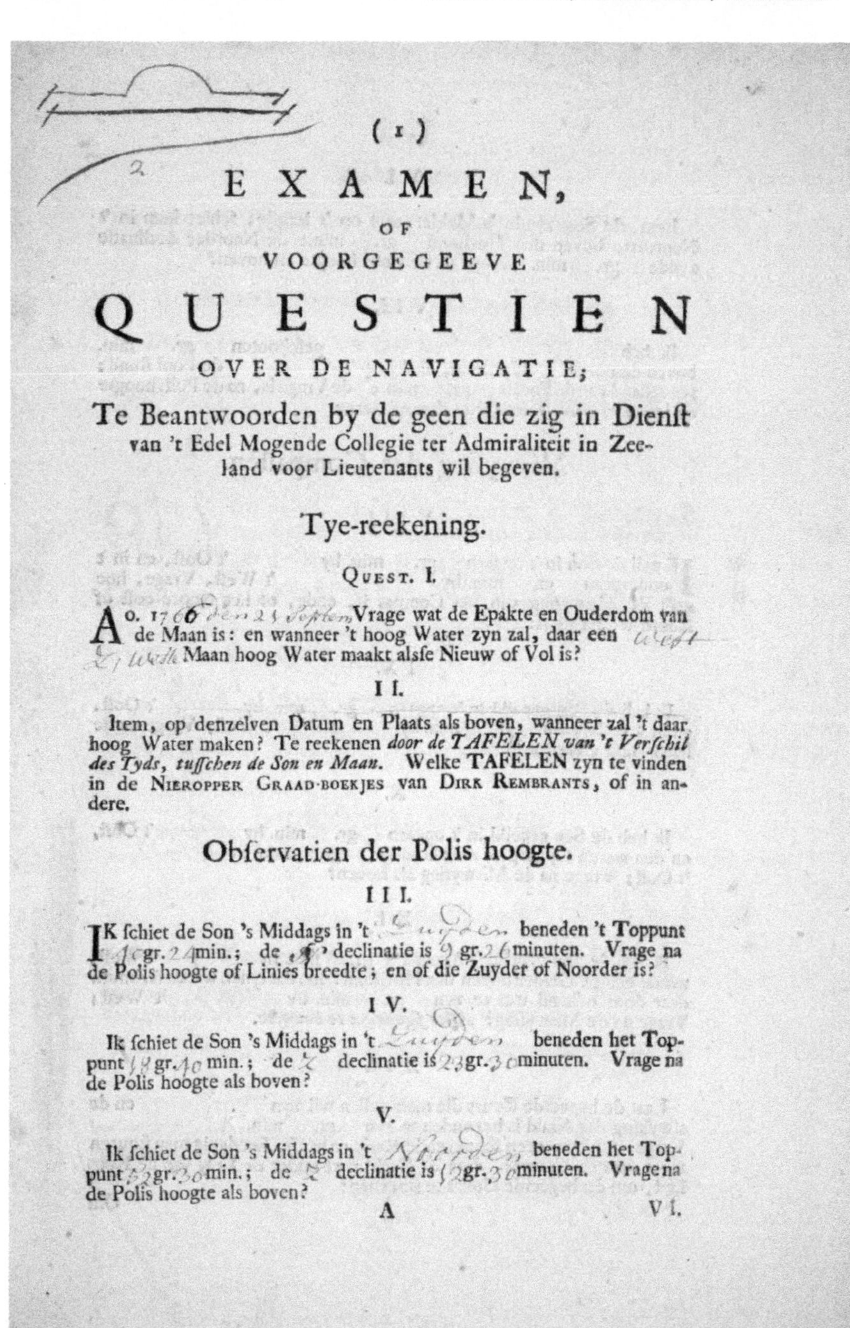

(1)

EXAMEN,

OF

VOORGEGEEVE

QUESTIEN

OVER DE NAVIGATIE;

Te Beantwoorden by de geen die zig in Dienſt van 't Edel Mogende Collegie ter Admiraliteit in Zeeland voor Lieutenants wil begeven.

Tye-reekening.

QUEST. I.

A°. 1766 den 25 Septem, Vrage wat de Epakte en Ouderdom van de Maan is: en wanneer 't hoog Water zyn zal, daar een West Maan hoog Water maakt alsſe Nieuw of Vol is?

II.

Item, op denzelven Datum en Plaats als boven, wanneer zal 't daar hoog Water maken? Te reekenen *door de TAFELEN van 't Verſchil des Tyds, tuſſchen de Son en Maan.* Welke TAFELEN zyn te vinden in de NIEROPPER GRAAD-BOEKJES van DIRK REMBRANTS, of in andere.

Obſervatien der Polis hoogte.

III.

IK ſchiet de Son 's Middags in 't Zuyden beneden 't Toppunt 41 gr. 24 min.; de ♈ declinatie is 9 gr. 26 minuten. Vrage na de Polis hoogte of Linies breedte; en of die Zuyder of Noorder is?

IV.

Ik ſchiet de Son 's Middags in 't Zuyden beneden het Toppunt 18 gr. 40 min.; de ♋ declinatie is 23 gr. 30 minuten. Vrage na de Polis hoogte als boven?

V.

Ik ſchiet de Son 's Middags in 't Noorden, beneden het Toppunt 83 gr. 30 min.; de ♑ declinatie is 12 gr. 30 minuten. Vrage na de Polis hoogte als boven?

A VI.

Figure 3.3 In Baron H. A. van Kinckel's *Luytenant's Exam* test booklet (Middelburg, 15 May 1767) the VOC examiner has inked in values for the preprinted examination questions.

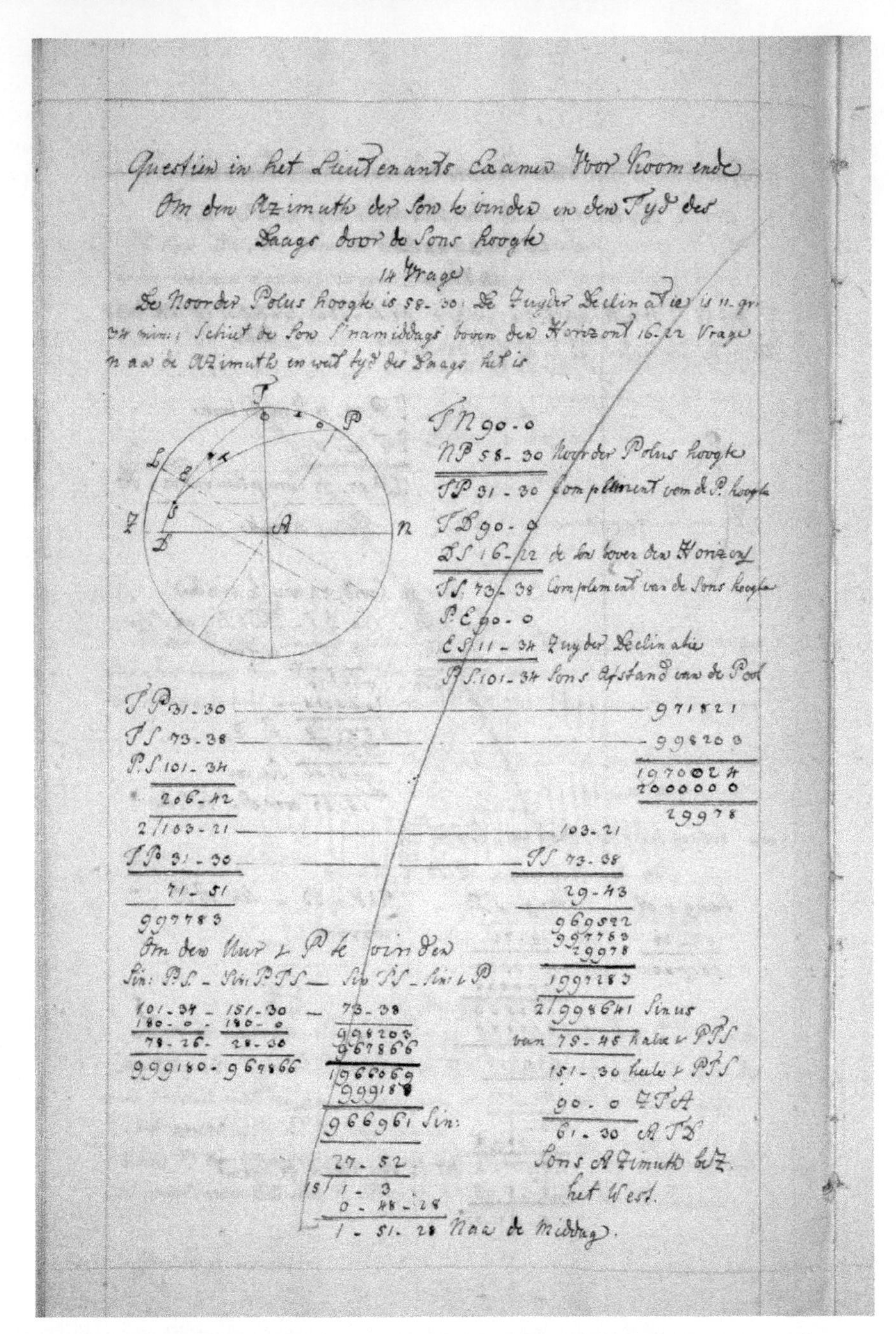

Figure 3.4 Kinckel has worked out meticulous answers, which the examiner who graded the test seems to have struck through.

trigonometry tables (radius 1,000,000).[112] Aside from a few minor corrections to his arithmetic, Van Kinckel completed the examination confidently, tackling problems about everything from compass variation to spherical trigonometry with no slips of the pen, doodles, or obvious errors. The young baron then went on to a respectable maritime career, securing commissions as second lieutenant and then captain before becoming a diplomat and possibly a British agent.[113]

As for Captain van Ewijk, once he passed the written examination, he would also have faced the oral examination. The geographic content, as we have seen, would likely have focused on local waters, with some questions about VOC routes further afield. In the later decades of the eighteenth century, this would change. By 1794 the 'Mond-Examen' had ballooned from half a dozen to twenty-three questions, with only the second-last of these about geography. Instead, hearkening back to the physical demonstrations of the sixteenth century, the main emphasis was on instruments. The very first question stated that the candidate 'shall present his own instruments, and position them on the chart; on the plain as well as on the Mercator chart'. From there the candidate was expected to explain the advantages and shortcomings of more than half a dozen instruments, old and new.[114] Van Ewijk, however, would not have had to put his technical expertise on display. Instead, any of the *mond examen* popular in the 1760s, from Holm, Van Dam, or Huips' edition of Gietermaker, would have been the classic geographical dialogue. For Van Ewijk to impress the Amsterdam committee with his competence, he needed to be able to find his way through tables and almanacs, be comfortable with the correct formulas, and know the appropriate terminology. Van Ewijk's career recovered; he would go on to sail for Zeeland, serving as captain of the *Burch* in 1769. That vessel got damaged in a storm and sought refuge in Canton. Van Ewijk makes one final voyage as captain of the *Europa*, returning from China in 1776.[115]

Conclusion

By analysing these five types of early modern examinations, we can see how the navigator's knowledge was conceived at various moments in time. At the beginning of the sixteenth century, most nautical instruments were unfamiliar, even to men who had spent years plying the coasts; therefore, their examiners required a physical demonstration of these new tools. Similarly, the theoretical cosmographical conception of the shape of the universe, from equator to pole, azimuth to zenith, was little known outside the universities. Given its potential utility for *pilotos* setting out to the far

[112] 'Examen of voorgegeeve Questien over de Navigatie', p. 4, question No. 19 'Door de Streek-tafel' (rather than solving this traverse course with a diagram).

[113] F. van der Horst, 'Daar donderd het canon van Vinckel aan de Theems!: Hendrik August, baron van Kinckel (1747–1821)', in: *Marinekapiteins uit de achttiende eeuw: Een Zeeuws elftal*, ed. J. R. Bruijn, A. C. Meijer, and A. P. van Vliet, Den Haag/Middelburg: Koninklijk Zeeuwsch Genootschap der Wetenschappen 2000, pp. 167–79.

[114] *Examen voor de Capitain*, pp. 12–13, e.g. question No. 12.

[115] J. R. Bruijn, F. S. Gaastra, I. Schöffer, with E.S. van Eyck van Heslinga, *Dutch-Asiatic Shipping in the 17th and 18th centuries*, digital edition, http://resources.huygens.knaw.nl/retroboeken/das, III, nos. 4052.3 & 7956.2, cited in Bruijn, *Schippers van de VOC*, p. 137.

side of the globe, cosmographical definitions featured in examinations well into the seventeenth century. On the other hand, the pilot's daily manoeuvres, the fundamentals of seamanship, were so familiar as to need no assessment—or classroom instruction.

The two types of knowledge that remained vital for navigators up through the close of the eighteenth century and beyond, even with the expansion of printed reference works such as atlases and almanacs, were geography and the calculation of positions. For these, maritime bureaucracies developed formulaic examinations to ensure the basic competency of their officers and navigators. We can reconstruct the Dutch testing regimen with considerable clarity, thanks to those written examinations that survive in the archives. This analysis can also offer suggestions about how to reconstruct practices elsewhere. In late eighteenth-century France, for instance, naval ensigns took part in challenging oral competitions where they were expected to solve trigonometry and other math problems in front of a public audience. The Royal Navy, for its part, expanded its reliance on both mathematical and geographical *written* examinations in the nineteenth century.[116]

The VOC's decision to allow their examinations to be published so extensively seems perplexing. Surely, they would have been more challenging if Gietermaker's canonical questions were not so widely available. And yet the very act of publishing the questions in popular textbooks, rather than keeping them secret, was a way of ensuring widespread competency. The men who worked through the codified textbooks would learn best practices and raise standards across the company. Instead of examinations functioning to exclude, they became a means of ensuring that a larger proportion of the crew knew the definitions and formulas.

Nor should these examinations be viewed solely as a summation of the lessons in the various textbooks, or even as the culmination of the navigators' training programme. For many learners they functioned in the opposite way: they were a starting point that shaped their education. If the mariners could learn enough to answer a particular set of 'predictable' questions, they could successfully take up their commission. And if they should run into difficulty, they could consult one of the many readily available textbooks, with their extensive sets of practice questions. It becomes clear that examinations and textbooks were closely interconnected, and that together they had a definitive impact not only on the way navigation was taught but also the way it was transmitted to the community at large. Inevitably, teachers started to adjust their courses to reflect the content and format of examinations. If there had originally been little curricular standardization in these independent schools, examinations quickly became the yardstick against which everyone was measured, even if not all students presented themselves to the examiners. Student workbooks filled steadily with examination questions, so even if these men went on to non-maritime careers, they would now be familiar with geometry and trigonometry, logarithms and tables.

From this range of examination types, we can also see how the focus of a sailor's memory changed over the centuries. There was no steady progress towards a more

[116] Anthiaume, *Evolution et enseignement*, II, p. 293. See e.g. NMM DRY/10, 'A Master's Examination for the Channel', 1780, which is a geographic dialogue; NMM HSR/U/17 Mathematics examination paper, 1858.

mathematical modernity. Illiterate *pilotos* already had a strong sense of numeracy in the sixteenth century: reading angles off their instruments, computing the time of high tide, and measuring depths in fathoms. But over time, as textbooks, atlases and almanacs became ubiquitous, it became possible to devote their prodigious memories towards things other than geography and pilotage. They could look at maps and consult printed instructions, leaving themselves free to carry out the computations their jobs required.

As part of their extensive preparation for these navigation examinations, early modern mariners and others turned to dialogues and practice questions to tackle increasingly sophisticated mathematics. When they stumbled over technical details, underprepared candidates like Captain van Ewijk and the presumptuous Captain Dering were sent back to their books to study further. Yet for every individual who was under-qualified, there were those who, like Dirk de West and Baron van Kinckel, addressed the material with ease. Whether the examinations demanded demonstration, memorization, calculation, or a mixture of these, the bar was set high enough to ensure that competent men—and not fools—took the helm.

4

Mathematical Examiners at Trinity House

Teaching and Examining Mathematics for Navigation in London During the Long Eighteenth Century

Rebekah Higgitt

Introduction

When Henry VIII put his seal to its charter, Trinity House was a charitable maritime guild in Deptford Strond with oversight of pilotage on the Thames.[1] As the traffic of the river and the wealth of maritime trade increased, so too did its jurisdiction and responsibilities. After the Restoration they transferred their business to Water Lane in the City of London, close to key locations such as Custom House, the Navy Office, the Ordnance, and the headquarters of overseas trading companies, including East India House.[2] Examination of the skills of navigators became an increasingly regular part of the Corporation's work when, in 1621, they added certification of ships' masters seeking to work on the king's ships to their regulation of Thames pilots. Because of these responsibilities, and because the Brethren of the guild were understood to be experienced and skilled ships' masters and captains, Trinity House went on to acquire the further roles of examining and certifying boys who passed through the Royal Mathematical School (RMS) at Christ's Hospital and men who hoped to be appointed as naval schoolmasters. However, the content of these examinations focused on the application of mathematics to navigation and so the Elder Brethren, who formed the government of Trinity House, found they required the assistance of mathematical examiners.

The RMS was founded by Charles II in 1673 on the instigation of Jonas Moore, Surveyor General of the Ordnance, and with the interested involvement of, among others, Samuel Pepys at the Navy Office, Isaac Newton in Cambridge, Robert Hooke, the Royal Society's Curator of Experiments, and John Flamsteed, appointed as the first Astronomer Royal in 1675.[3] It was to consist of up to forty poor scholars

[1] The most recent institutional history is Andrew Adams and Richard Woodman, *Light Upon the Waters: The History of Trinity House 1514–2014*, London: The Corporation of Trinity House 2013. See also Joseph Cotton, *Memoir on the Origin and Incorporation of the Trinity House of Deptford Strond*, London: J. Darling 1818; C. R. B. Barrett, *The Trinity House of Deptford Strond*, London: Lawrence & Bullen 1893; G. G. Harris, *The Trinity House of Deptford, 1514–1660*, London: Athlone Press 1969.

[2] On the Water Lane location, see Rebekah Higgitt, Jasmine Kilburn-Toppin, and Noah Moxham, 'Spaces and Geographies of Metropolitan Science', in: *Science Museum Group Journal*, 15 (2021) https://doi.org/10.15180/211506.

[3] On the RMS, see Clifford Jones, *The Sea & the Sky: The History of the Royal Mathematical School of Christ's Hospital*, Horsham: Christ's Hospital 2015; Frances Willmoth, *Sir Jonas Moore: Practical Mathematics and Restoration Science*, Woodbridge: Boyell Press 1993, pp. 195–207; Rob Iliffe, 'Mathematical

of Christ's Hospital, who were selected to be 'further educated in a Mathematical Schoole' at the King's expense. They were to be

> taught and instructed in the Art of Navigačon and the whole Science of Arithmatique until their age and competent proficiency in these parts of the Mathematiques shall have fitted and qualified them in the judgment of the Master of the Trinity House for the tyme being to bee initiated into the practices of Navigation and to bee bound out as Apprentices for seaven yeares to some Captaines or Comanders of Shipps.[4]

However, the text-based theoretical content proved, apparently unexpectedly, to be beyond the Elder Brothers' own training and regular practice. A mathematical gap was revealed that was dealt with by appointing and paying for a mathematician to carry out the examinations on their behalf. The debates about the balance of the curriculum between theory and practice, and the difficulties of finding masters for the school that could achieve results that satisfied all sides, have been discussed by historians.[5] However, this history has not yet been explored from the perspective of Trinity House, and nor has the role of the mathematical examiner, created in 1682.[6]

The succession of mathematical examiners (see table in Appendix 4.1), the majority of whom were private mathematical teachers, allowed Trinity House to fill the mathematical gap and carry out their obligation to the monarch who sponsored the scholars' education. From 1702, they also allowed the Corporation to fulfil an obligation to the Admiralty, by examining and certifying naval schoolmasters. These were men who went on to be appointed by the Admiralty to instruct Volunteers (those wishing to become warranted officers via a nomination by the Admiralty rather than through an apprenticeship to a commander) on board ships, 'not only in the theory but the practical part of Navigation' and 'the Art of Seamanship'.[7] The Trinity House mathematical examiners include some who are reasonably well known and others who are much harder to trace. There is no evidence that provides clear explanations of why they were selected or why they wanted to take on the role, although some can

Characters: Flamsteed and Christ's Hospital Mathematical School', in: *Flamsteed's Stars: New Perspectives on the Life and Work of the First Astronomer Royal, 1646–1719*, ed. Frances Willmoth, Woodbridge: Boydell Press 1997, pp. 115–44; N. Plumley, 'The Royal Mathematical School within Christ's Hospital: The Early Years—Its Aims and Achievements', in: *Vistas in Astronomy* 20 (1976), pp. 51–9; Nerida F. Ellerton and M. A. Clements, *Samuel Pepys, Isaac Newton, James Hodgson, and the Beginnings of Secondary School Mathematics: A History of the Royal Mathematical School Within Christ's Hospital, London 1673–1868*, Cham: Springer 2017.

[4] E. H. Pearce, *Annals of Christ's Hospital*, London: Methuen & Co. 1901, p. 101.

[5] Iliffe, 'Mathematical Characters'; Margaret E. Schotte, *Sailing School: Navigating Science and Skill, 1550–1800*, Baltimore: Johns Hopkins University Press 2019, pp. 95–113; Jason Grier, 'Navigation, Commercial Exchange and the Problem of Long-Distance Control in England and the English East India Company, 1673–1755', unpublished PhD thesis, York University 2018, pp. 29–85.

[6] With the exception of a brief article by a twentieth-century warden of Trinity House, which misses the first three holders of the position, chiefly focuses on one, and contains several errors: William Chaplain, 'William Mountaine, F.R.S., Mathematician', in: *The American Neptune* 60 (1960), pp. 185–90. This chapter answers Ellerton and Clements's question, 'Did Trinity House have persons with sufficient mathematical, navigational, and educational experience to be well-placed to evaluate the RMS students' learning and readiness for sea service?' (*Samuel Pepys*, p. 115). They did not believe this was always the case and suggest that there was a lack of teaching experience at Trinity House.

[7] H. W. Dickinson, *Educating the Royal Navy: Eighteenth- and Nineteenth-Century Education for Officers*, London: Routledge 2007, pp. 10, 13.

be conjectured by considering their other activities. Their published works, including advertisements for the teaching, publications, and wares that they or others offered, are revealing of their activities, careers, and connections.[8] Elsewhere, records of their lives and manuscript sources provide clues. As with many positions in the gift of Trinity House or other institutions at this period, family succession played a central role. Master-pupil and business relationships, often overlapping with family ones, were also evident, although these can be harder to confirm. None of the men discussed here received university training and, although in the second half of the eighteenth century the position was held by individuals who were, or became, Fellows of the Royal Society, this was not the case either before or after. While some seem to have sought out the role primarily for its potentially useful connections, others required the financial remuneration it offered, and even the security of a guild connection to support retirement and dependants.

Exploring these individuals as a group does not provide a particularly clear story about the changing status of mathematical navigation or education. At the beginning of the period we can see more advanced mathematics becoming significant in navigation—along with a shift away from the linkage of mathematics with astrology to more purely practical and/or natural philosophical applications—and, at the end, a broad incorporation of mathematical training and examinations into state-run institutions that bypassed Trinity House. There is some evidence of a change in the reputation of the Corporation toward the end of the eighteenth century, as the examiners began to advertise their connection in print. As a group, these men reveal the range and variety of lives and careers that were either wholly or partially supported by mathematical work, but their stories also show the close networks of teaching, publication, geography, family, and patronage that sustained them.

Turning mathematical

Initially it was a selection of Trinity House's Elder Brethren who examined the 'Mathematicall Boyes'. They were not active in advising on the RMS curriculum, although Samuel Pepys, who lacked mathematical knowledge and practical experience at sea but was both an Elder Brother and a governor of Christ's Hospital, took a close interest. It was soon obvious that it was essential to know what had been taught to determine what could or should be examined. In September 1676, the Court of Trinity House agreed that they would not examine the boys until the Mathematical Master at the RMS, John Leake, had certified their competencies.[9] However, the

[8] Such sources were, of course, key to E. G. R. Taylor's defining works, *The Mathematical Practitioners of Tudor and Stuart England*, Cambridge: Cambridge University Press 1954 and *Mathematical Practitioners of Hanoverian England, 1714–1840*, Cambridge: Cambridge University Press 1966. They have still proved revealing in more recent work, see e.g. Philip Beeley on the 'cohesive knowledge community' of mathematical authors and teachers, instrument-makers, printers, and booksellers in 'Practical Mathematicians and Mathematical Practice in later Seventeenth-Century London', in: *British Journal for the History of Science* 52/2 (2019), pp. 225–48.

[9] 27 September 1676, Court Minutes, MS 30004/005, p. 15, Archives of the Corporation of Trinity House, London Metropolitan Archives. Pepys more than once asked the Court 'to propose what parts

following year it was reported that the Elder Brothers 'have mett with difficulty in examining' the boys, and Pepys, as the then Master of Trinity House, reported a whole list of shortcomings to Christ's Hospital.[10] It was unclear whether the problem was with the curriculum, the teaching, or the examinations, and nor was it evident that ships' commanders wanted to take on mathematically trained apprentices. Pepys, at least, seems to have become increasingly critical of the mathematical skills of mariners generally and of the Elder Brethren in particular. Although he joined them in pushing for RMS teachers with experience of practical navigation, he also suggested that the Elder Brethren 'through want of use and overlooking their Books . . . are not themselves best fitted for Examining ye Children'.[11]

Leake's successor, Peter Perkins, seems to have succeeded in pleasing both the Elder Brethren and the governors of Christ's Hospital during his short tenure (see table of RMS Masters in Appendix 4.2). However, the pupils of his replacement, Robert Wood, were repeatedly found wanting at their examinations. This led to investigations at both institutions, with Trinity House asking Richard Norris, who Pepys referred to in his notes as 'Old Norris, our Mathematician', to comment.[12] Wood seems to have been found generally unsatisfactory, although he blamed 'a Tide' against him at Trinity House.[13] He may have had a point, given that Norris subsequently applied to replace him at the RMS. However, Norris's view, that the teaching was 'short of their [the pupils'] abilities, as well as of his majesty's design in this his Royal Foundation', prevailed.[14] Wood's replacement was Edward Paget, a Fellow of Trinity College, Cambridge, with Newton's backing but no practical experience of navigation. In Pepys's opinion, if he had not been 'superannuated' and no Latinist, Norris would have been better qualified for the position than Paget. He judged him to be 'equally if not more Largely certifyed-for' in his 'knowledge and Profession of teaching the Mathematicks noe less then experience in the Art of Navigation'.[15] If he is the same Richard Norris as the one from East Ham who drew up a will in 1695, his experience is unquestionable, as 'Mariner and superannuated officer of one of the ships of his Majesties Royall Navy'.[16]

Norris was probably the author of a 1685 pamphlet objecting to the Mercator map projection described in Edward Wright's *Certaine Errors in Navigation* (1599). On the title page he describes himself as 'Mariner' and, at the end of the text, he set out a list of items he intended to include in a future publication. To this he said he would 'annex an Examination, which all such Navigators ought to pass, that shall be judged capable

of Arithmetique and Mathematticks ought to be taught': 15 July 1676, 12 August 1676, Court Minutes, MS3004/005, pp. 7, 10.

[10] 15 October 1677, Court Minutes, MS30004/005, p. 65; Iliffe, 'Mathematical Characters', p. 122.

[11] Pepys Manuscripts on Christ's Hospital, quoted in Grier, 'Navigation', p. 63.

[12] 2 February 1681/2, Court Minutes, MS 30004/006, p. 34; Samuel Pepys, *Samuel Pepys's Naval Minutes*, ed. J. R. Tanner, London: The Navy Records Society 1926, p. 124.

[13] Grier, 'Navigation', p. 69.

[14] Norris's report is in the Pepys Library, quoted in Mordechai Levy-Eichel, '"Suitable to the Meanest Capacity": Mathematics, Navigation and Self-Education in the Early Modern British Atlantic', in: *The Mariner's Mirror* 103/4 (2017), pp. 450–65at p. 452.

[15] Pepys, *Naval Minutes*, p. 149; Memorandum of Pepys quoted in Iliffe, 'Mathematical Characters', p. 129.

[16] Will of Richard Norris, Prerogative Court of Canterbury, PROB 11/427/8 (The National Archives).

of taking Charge as Master of any of his Majesties or Merchants Ships, &c'.[17] Obviously Norris's knowledge and experience of the curriculum and examinations at Christ's Hospital and Trinity House could easily have been used to attract pupils or readers who were studying to gain such certification. He may also be the 'Industrious and Experienced Seaman, Richard Norris' who apparently wrote 'an Observation of the Tide, and how to turn out of the *Streights-Mouth* the Wind being Westerly', annexed to an account of five years' captivity in Algeria, *The Adventures of (Mr T.S.) An English Merchant*. This is a fictional narrative, based on geographical facts, and the annex may have been inserted to add plausibility to the tall, often erotic tales in the main text. If this were the case, it would suggest that his name carried some degree of recognition.[18]

After Norris had examined the mathematical scholars as part of his investigation for Trinity House, it was suggested that they should pay him for his pains. Further, it was agreed that the Corporation might 'admitt him into that Service' on a regular basis. As there were expected to be ten scholars ready to be examined each year, they would pay him £10. The Elder Brethren admitted that while they remained 'the most proper Judges of the Qualifications of those Children for Sea Affaires, yet they may not be so well acquainted with the School Ma:^{rs} Rules leading thereunto, w:^{ch} ought to be Supplyed by one whose dayly practice it is'.[19] This was deemed satisfactory, though Pepys remained scathing in private, noting that 'our present Tr[inity] H[ouse] are so conscious of their own disability to examine the Christ's Hospital boys [...] that they professedly allow 10^{l}. per annum out of the poor's money' to pay a mathematical teacher 'to do it for them in their presence'.[20] Norris was, however, removed within months by unanimous vote because of his continuing criticisms of Christ's Hospital and Paget. Trinity House, mindful of the king's interest in the RMS project, were concerned that 'his continuance to Examine the Said Children might be of ill Consequence both to the Hospitall & this House'.[21] The expertise of the mathematical examiner had to conform to the interests of the Corporation.

The Colson clan

When Norris was removed, a replacement was proposed and, after some discussion of whether he would be likely to accept the role, carried by vote. This was Lancelot Colson, or Coelson, who had already had a long career as a physician, astrologer, and chymist. A portrait frontispiece, which places him as 35 in 1662, describes him as 'Student in Astrologie & Physick' (Figure 4.1).[22] In his *Philosophia Maturata* (1668), on

[17] Richard Norris, *The Manner of Finding of the true Sum of the Infinite Secants of an Arch, by an Infinite Series*, London: Thomas James for the author 1685, p. 12.

[18] A. Roberts, *The Adventures of (Mr T.S.) An English Merchant*, London: Moses Pitt 1670, p. 247. This text is discussed in Gerald M. MacLean, *The Rise of Oriental Travel: English Visitors to the Ottoman Empire, 1580–1720*, Basingstoke: Palgrave Macmillan 2004, pp. 179–82, although Norris is not identified.

[19] 19 December 1682, Court Minutes, MS 30004/006, p. 71.

[20] Pepys, *Naval Minutes*, p. 401.

[21] 21 March 1682/3, Court Minutes, MS 30004/006, p. 86.

[22] Lancelot Coelson, *The Poor Man's Physician and Chyrugion*, London: printed by A.M. for Simon Miller at the Starre in St Pauls Churchyard 1656 (second edition 1663).

The Effigies of Lancelot Coelſon Student
in Aſtrologie & Phyſick. Ætatis 35. 1662.
John Dunſtall fecit.

Figure 4.1 Portrait of Lancelot Colson, used as a frontispiece to his 1663 *The Poor Man's Physician.*

'The Practick and Operative part' of philosophy for 'gaining the Philosophers Stone', he styled himself as 'Dr. in Phys. and Chym'. By the date his name appears in the minutes of Trinity House, he was regularly publishing an almanac called *Speculum Perspicuum Uranicum*, containing astronomical, astrological, medical, social, and historical information. Dedicatory letters and advertisements give his addresses: in 1670 'at the sign of the Royal-Oake in St. *Katherine's Lane* near the *Tower of London*', in 1673 the Still-Yard (or Distillers Yard) on Tower Hill under the same sign, in 1681 on a parallel street, the Postern (Postern Row), and 1686 on the street between, George Yard. His proximity to the Tower, the location of the Ordnance Office, is perhaps explained by a former military career and continued service to the Tower

Hamlets bands.[23] He was very well placed for Trinity House, just along Tower Street and down Water Lane.

An advertisement in his 1677 almanac informs the reader that from his house they could acquire 'that Great Antidote of the Ancient Philosophers, *Van Helmont*, *Paracelsus*, and *Crollius*, called by them (*Elixir Proprietaris*) the Greatest Cordial and only Medicine in the World for long and sound Life' for four shillings an ounce. He named various other vendors selling on his behalf and noted that 'my Honoured and Worthy Friend Mr. *William Lilly*' had mentioned his elixir in his own almanac.[24] Colson also offered an antiscorbutic pill, said to cure scurvy and venereal disease, a method of curing deafness in ten days, 'if curable', and worm powder. A letter to the reader in 1676 noted his having broken his leg over the last year, 'upon which it was very much reported abroad, both in City and Country that I was dead of the said hurt'.[25] He reassured readers that he was now cured and ready to receive customers. Even if exaggerated—it was a trick repeated after other bouts of ill health—this suggests a degree of renown. The running titles to his almanacs were 'Coelson' and the year, suggesting his close identification with these texts and how they were commonly referred to. Receiving patients in the mornings and visiting the sick in the afternoon, he probably had a wide acquaintance.

It does not appear that Colson was known for teaching mathematics but the Elder Brethren clearly believed he was capable of examining mathematical scholars. He was skilled in astronomical observation and calculation and seems to have been well known to astrologers such as Lilly and Richard Saunders.[26] It is possible that he was known to Trinity House through Sir William Warren, to whom the 1674 edition of his almanac was dedicated. A ship owner and timber merchant in Wapping, Warren had been a mentor to Pepys and went on to establish a near monopoly of supply to the Navy Board.[27] It may, equally, have been family connections that ensured he was in the orbit of mariners. Nathaniel Colson, who styled himself 'Student of the Mathematicks', produced multiple editions of *The Mariners New Kalendar* from the 1670s. This contained 'The Principles of Arithmetick and Geometry', including square and

[23] Born in Colchester, Colson was wounded fighting in Cromwell's Scottish campaign in 1650: Bernard Capp, 'Coelson [Colson], Lancelot (1627–1687?), astrologer and medical practitioner', in: *Oxford Dictionary of National Biography*, Oxford: Oxford University Press 2004, https://doi.org/10.1093/ref:odnb/5995.

[24] Lancelot Coelson, *Speculum Perspicuum Uranicum, or, An almanack for the year 1677*, [London: Company of Stationers 1677, p. 46]. In a later edition he indicated that it could be purchased from certain pubs and a stationer's shop: Louise Hill Curth, 'Medical Advertising in the Popular Press' in: *From Physick to Pharmacology: Five Hundred Years of British Drug Retailing*, ed. Louise Hill Curth, Aldershot: Ashgate 2006, pp. 29–48 at p. 44.

[25] Lancelot Coelson, *Speculum Perspicuum Uranicum, or, An almanack for the year of Christ 1676*, London: Company of Stationers 1676, [47].

[26] Capp, 'Coelson'; John Gascoigne suggests that Lancelot was the Mr Colson referred to in correspondence of Newton and Flamsteed in 1681 in relation to comet observations but it seems more likely that this was John Colson, discussed below: Gascoigne, 'Sensible Newtonians: Nicholas Saunderson and John Colson', in: *From Newton to Hawking: A History of Cambridge University's Lucasian Professors of Mathematics*, ed. Kevin C. Knox and Richard Noakes, Cambridge: Cambridge University Press 2003, pp. 171–204 at pp. 191, 203 n. 27. He also assumes that the Lucasian Professor John Colson was related to the London Colsons discussed here.

[27] J. D. Davies, 'Warren, Sir William (bap. 1627, d. 1695), naval contractor', in: *Oxford Dictionary of National Biography*, Oxford: Oxford University Press 2004, https://doi.org/10.1093/ref:odnb/58160.

cube roots, a range of astronomical tables, rules to calculate tides, methods of calculating latitude and longitude, descriptions of the use of cross-staff, backstaff, and nocturnal, problems in plain sailing and astronomy as well as a coastal rutter, channel soundings, and sailing directions for the principal harbours. His opening epistle 'To the Ingenious Mariner', signed 'thy Friend', claimed the contents would engage the novice and, beginning with arithmetic, were set out as a course of learning: 'I assure thee', he wrote, 'were I present to instruct thee, I could by no means render things more intelligible than I have here done'.[28]

Some editions of the *Kalendar* carried advertisements for another of the Colson clan, John Colson. In 1677 and 1688 these placed him at Marsh Yard, 'a little below the *Hermitage-Stairs* in *Wapping*', where he taught

> Mathematical Sciences, (viz.) *Arithmetick, Geometry, Algebra, Trigonometry, Navigation, Astronomy, Dialling, Surveying, Gauging, Fortification* and *Gunnery*, the Use of the Globes and other Mathematical Instruments, *Projection of the Sphere*, and other parts of the Mathematics.[29]

By 1693 John Colson had moved, closer to the City and just northeast of the Tower, to Prescot Street near Goodman Fields, from where his advertisements could state 'Youth Boarded'. By the 1697 edition, the *Kalendar* was printed for Richard Mount 'at the antient Shop at the *Postern* on *Tower Hill*; where you may have all sorts of Mathematical and Sea-Books', while the sixth edition of 1701 boasted an addition 'of *Mercator*'s Sailing, by JOHN COLSON, *Teacher of the Mathematicks*'. Throughout the eighteenth century Mount & Page continued to publish versions of this work—as the *Kalendar* and then the *Calendar*—under Nathaniel Colson's name and with his opening address to the Ingenious Mariner.

John Colson built up a considerable reputation as a mathematical teacher. From at least 1679 he had examined RMS pupils for Christ's Hospital, declaring whether they were ready to proceed to Trinity House and, if necessary, commenting on the teaching they had received. In this he worked, in the problematic year of 1681, alongside Flamsteed, whose correspondence also referred to his astronomical observations.[30] John Colson went on to become mathematical examiner of Trinity House but, frustratingly, it is not clear from the surviving records when this happened. Indeed, there is no entry in the Court Minutes to confirm whether or not Lancelot Colson actually took up the position in March 1682/3 though, equally, no indication that he did not. The clerk's marginal note by the minute mentioning Lancelot, 'Mr Colson

[28] Nathaniel Colson, *The Mariners New Kalendar*, London: printed by J. Darby for William Fisher, at the Postern-Gate near Tower-Hill; Robert Boulter, at the Turks-Head; and Ralph Smith at the Bible in Corn-Hill, near the Royal Exchange 1677, p. 3.

[29] Nathaniel Colson, *The Mariners New Kalendar* (1677), p. 163, and *The Mariners New Kalendar*, J. Darby for William Fisher at the Postern-Gate near Tower-Hill, Thomas Passenger, at the Three Bibles on London-Bridge, and Eliz. Smith, at the Bible in Corn-hill, near the Royal Exchange 1688, [p. 4].

[30] Iliffe, 'Mathematical Characters', pp. 124–5. Here and elsewhere there is confusion over whether examinations were being carried out for Christ's Hospital or Trinity House. For discussion of Colson's eclipse observations see *The Correspondence of John Flamsteed, the First Astronomer Royal*, ed. Eric G. Forbes, Lesley Murdin, and Frances Willmoth, 3 vols, Bristol: Institute of Physics Publishing 1995–2002, vol. I, p. 403; for discussion of comet observations see footnote 26.

admitted' suggests that he did. However, throughout the 1680s and 1690s, mentions of Colson lack a first name and there is no indication of change. No minute of John's appointment has been found, although there is a record of his having been admitted as a Younger Brother on 23 May 1692, when he appears to be referred to as 'our Mathematician'.[31] He remained in post until his death in 1716.

There was a strong endorsement of John Colson's teaching in the opening letter of the second edition of John Newton's *The English Academy*. Among mathematical teachers he noted first Paget and the RMS and then, although he said he did not know him personally, 'the well Accomplished Mr. *John Colson*, now living in *Goodmans-Fields*', for 'Report hath rendered him to the World a worthy Master and Teacher of that Science'. He noted that there were 'few School-Masters that can Teach these things' and even 'not many Tutors in either of our Universities that do'.[32] Colson's reputation was sufficient that he taught Jacob Bruce, later founder of an observatory and the Mathematical and Navigation School in Moscow, while he was part of Peter the Great's entourage on his 1698 Embassy to Europe. Colson was paid 48 guineas to train, lodge, and board Bruce for six months and, recorded as 'Ivan Kolsun', he dined with the Tsar on 7 February 1698, the day after Peter had visited Flamsteed in Greenwich.[33] Bruce is more famous as a practitioner of astrology, alchemy, and natural magic, which makes the presumed family link between John and Lancelot Colson of interest, although the fact that Bruce was a Jacobite would have sat badly with Lancelot's pro-Parliament and anti-Catholic views.[34]

While it is not known that Lancelot and John Colson were related, it seems probable. Likewise, it is not known if John had alchemical and astrological interests. His own publications were limited to the addition to Nathaniel Colson's *Kalendar* and revised editions of another popular work, *The Mariners Magazine* by Captain Samuel Sturmy, in 1679, 1684, and 1700. Here he was a 'Teacher of the *Mathematicks* in London', who undertook this task 'for the advantage of young Students', to whom he said he was 'an obliged servant for their Instruction and Information'.[35] John Colson's world was largely practical, pedagogical, and maritime. However, what might be a rather different image of John, possibly in conflation with Lancelot, appears in Ned Ward's satirical serial *The London-Spy* (1698–1700). After visiting the Tower of London, the Spy, a country gentleman guided by a native Londoner, finds himself in a 'remote part of the Town, which my Friend told me was as much *Incognito* to

[31] The first word is indistinct and overwritten, perhaps changed from or to 'a Mathematician': 23 May 1692, Court Minutes, MS30004/007, p. 277. There are no Court Minutes for the end of 1692 to 1705 and no Board Minutes surviving before 1685, when Lancelot Colson was appointed, or for 1689–98, when John Colson may have taken up the post at the same time as he was made a Younger Brother (as happened with Noor Colson, discussed below). Margaret Schotte discusses an examination carried out by John Colson on 29 March 1683, the report on which is preserved in the Pepys Library. She says this was carried out at Trinity House but on this date, shortly after Lancelot Colson's appointment, John Colson was more likely acting for Christ's Hospital and responding to Norris's criticisms: Schotte, *Sailing School*, p. 95.

[32] John Newton, *The English Academy, or, A Brief Introduction to the Seven Liberal Arts*, London: A. Milbourn for Tho. Passenger 1693, [n.p.].

[33] Robert Collis, *The Petrine Instauration: Religion, Esotericism and Science at the Court of Peter the Great*, 1689–1725, Leiden: Brill 2012, pp. 61–4.

[34] Capp, 'Coelson'.

[35] Samuel Sturmy, *The Mariners Magazine*, London: Anne Goodbid for William Fisher, at the Postern-Gate near Tower-Hill and five others 1679, [n.p.]

many thousands in *London*, as it was to me'. Here they wandered the streets and came across a doorway decorated with zodiac signs: 'we presently concluded no less than an Eminent *Conjurer*, or some strange foretelling *Star-Peeper*, could be Lord of the House, whose Door was so gloriously set of with such a Number of Constellations'. A malignant-looking figure then exited: 'one Eye look'd upwards and the other downwards, as if he was Stargazing with one Eye, and minding his way with the other'. On asking his companion the name of the place, the Spy reports 'he told me 'twas *Prescot-street*'.[36]

Back at Trinity House, we might assume Lancelot was still in position in 1685, when 'M.r Colson' complained that Paget was supplying imperfect certificates for the boys deemed ready for examination. Pepys again urged caution, noting 'what precaution was used & what ingenious persons were consulted in the contriveing that method', but Colson wished to 'draw up a forme of his owne' for consideration.[37] This certificate evidently indicated the curriculum for, on 25 August 1685, Colson submitted his proposed 'alterations and additions [...] to those heads the Hospitall Boyes were formerly examined upon'. These were considered by the Deputy Master, Captain John Hill, who found them 'more methodicall then those the Said Boyes had hitherto been examined upon', and Paget was persuaded to agree, though 'excepting against some p[ar]ticular heads or rather the wording thereof'. Colson suggested that they be proposed to Christ's Hospital, with 'Some explanacons therein, & p[ar]ticularly to add thereto the methods of keeping a Sea Jornall'.[38] It was not long after this that the Court 'Orderd That a paire of [Joseph] Moxons Globes glaz'd of about 10 or 12 inches Should be bought, being wanted upon the examination of the Children of the Mathematicall Foundacon in Christs Hospitall & other occasions'.[39] These shifts to more applied than purely book learning seem to be a response to concerns that former RMS scholars were not 'better qualified than others' for careers at sea. Pepys had discovered that some found themselves unemployed after their apprenticeships were concluded, despite 'the charge his Matie is at in their Breeding'.[40]

The relationship with Christ's Hospital was, presumably, eased by the fact that for many years John Colson held his role as an examiner for Christ's Hospital in parallel with the one at Trinity House. It also, as Pepys wrote in 1694, seemed 'to make the examination in either place the greater mockery'.[41] He resolved to put an end to what he saw as an irregularity but nothing changed: John was in both places in 1695 and 1708 when Trinity House again failed a number of boys in their examinations.[42] This created the oddity of John Colson, examiner for Christ's Hospital, declaring that the

[36] Edward Ward, *The London-Spy Compleat, in Eighteen-Parts*, London: Printed and sold by J. How 1703, pp. 317–19. This figure perhaps inspired Swift's speculative Laputians, with 'one of their Eyes turned inward, and the other directly up to the Zenith', who wear garments 'adorned with the Figures of Suns, Moons, and Stars': [Jonathan Swift], *Travels into Several Remote Nations of the World, By. Lemuel Gulliver*, London: printed for Benj. Motte 1726, vol. 2, part 3, p. 16.

[37] 18 August 1685, Court Minutes, MS 30004/007, p. 16.

[38] 25 August and 15 October 1685, Court Minutes, MS 30004/007, pp. 19, 35.

[39] 18 February 1685/6, Court Minutes, MS 30004/007, p. 65.

[40] 8 April 1685, Court Minutes, MS 30004/006, p. 190. On 'applied' and 'hands-on theory' at RMS, see Schotte, *Sailing School*, pp. 106–13.

[41] Pepys, *Naval Minutes*, pp. 401–2.

[42] Ellerton and Clements, *Samuel Pepys*, pp. 76, 116,

boys were qualified and ready to be examined, and John Colson, examiner for Trinity House, apparently failing them. However, two or three Elder Brethren were present at examinations in this period—described, variously, as supervising Colson in his work or as being assisted by him in theirs—and these appear to be cases in which they, rather than their mathematical examiner, raised objections. This mix of personnel and responsibilities led to the 1708 clash, which seems to have been the last. The complaint was made by Captain John Merry, a warden of Trinity House and a governor of Christ's Hospital, who had commanded an East Indiaman and became an investor and eventually Deputy Governor of the Hudson's Bay Company.[43] He said that he had sat through the examinations of five boys at Trinity House over several days and found them 'more ignorant in their business than any others that have of late years come before the Brothers of Trinity House'. This was despite Colson, as examiner for Christ's Hospital, having found them 'competently knowing' a couple of months earlier.[44] The then RMS teacher, Samuel Newton, protested about the process, and the inability of those present to recognize the 'divers ways' in which particular operations could be undertaken or taught, but he ultimately resigned.[45] He was replaced by James Hodgson, a former Royal Observatory assistant, and thereafter the Elder Brethren seem to have been content. It is perhaps the case that they relied increasingly on their mathematical examiners and made a tactical decision to withdraw from the examination room.

More typically, things went smoothly, as well they might, but a different problem arose in 1707, when Colson was reprimanded and suspended after a complaint that he had refused to examine a prospective naval schoolmaster unless he paid five guineas if qualified, or a greater sum 'if not so thoroughly qualified for a Teacher as he ought to be'. Colson denied the charge but did not satisfy the Brethren or 'behave himself so respectfully as he should have done'. The 'Court being of Opinion That the taking of Money by such undue Practices ought never to be Countenanced', they voted to dismiss him. When informed, Colson offered his apologies and asked them to 'take into Consideration his past trouble in Examining Teachers, since he apprehended his Sallary of £20 p[er] ann:[u] was given him only for Assisting in the Examination of the Boys from Christs Hospitall'.[46] The Elder Brethren were unsure how to proceed. They debated Colson, and the problem of being without a mathematical examiner, at the following Court before deciding to consult Admiral George Churchill, their previous Master and a Lord of the Admiralty. He declined to be involved, despite the work being done for the Admiralty, and the Elder Brethren weighed up the request of Thomas Weston, who had recently completed his indentured term as assistant to Flamsteed at the Royal Observatory, to replace Colson. This led to debate and

[43] Merry's biographical epitaph in Hampstead is recorded in William Toldervy, *Selected Epitaphs*, 2 vols, London: printed for W. Owen 1755, vol. II, pp. 149–50; see also his Hudson's Bay Biographical Sheet, Archives of Manitoba, https://www.gov.mb.ca/chc/archives/_docs/hbca/biographical/m/merry_john.pdf

[44] Excerpts from the Minutes of the Committee of Almoners, 8 December 1708 and 16 September 1708, in Ellerton and Clements, *Samuel Pepys*, p. 76.

[45] Excerpts from the Minutes of the Committee of Almoners, 9 June 1708 and 8 December 1708, in Ellerton and Clements, *Samuel Pepys*, pp. 75–7.

[46] 27 August 1707, Court Minutes, MS 30004/008, p. 62. I have not been able to establish when the salary for the Mathematical Examiner increased from the £10 offered in early 1683.

a ballot, which was 15 to 1 in favour of Colson, who was reappointed. Interestingly, it prompted a resolution that a ballot box should be used in all future elections.[47]

The following February Colson wrote to ask 'that the Corp:° would be pleas'd to take it into their Consideration his past Services, as Assistant in the Examination of about a 100 Persons for Math:[le] Teachers on board her Maj:[ts] Fleet, which he thought deserved no less than 100 pounds'. However, the Brethren considered this 'an unreasonable Summe' and so Mr Noyes, the clerk, was 'order'd to return him his Letter again and that no regard should be had thereto'. Colson, evidently waiting outside, returned the letter with 'a Submissive Postscript' asking that they might give him whatever they considered appropriate. After more debate it was resolved to add £10 per annum to the £20 'already given him for his Constant attending & assistance in Examining the Children if the Royall Foundation in Christs Hospital'. This was 'in full Consideration of his assisting in the Examination of both Teachers and hospitall Boys', on which it is clear the Elder Brethren thoroughly relied.[48]

John Colson's death in early 1716 meant a successor was required. As was often the case with positions appointed by Trinity House—buoy or light keepers, agents, overseers of the Ballast Wharf, and so on—personal recommendation, family connection, and skilful lobbying were key. On this occasion the Master, Vice-Admiral James, Earl of Berkeley, had written to the Deputy Master, Captain John Hazelwood, to indicate that he was 'pleas'd to recomend M.[r] Noor Colson the Son to Succeed his said Father in the aforemention'd Station'. This was unanimously agreed, 'in respect to his Lordship', and he was appointed 'the Assistant Mathematical Examiner to this Corporation, with the same Salary of 20*l* p annum as was allow'd to his Father'. At the same meeting he was sworn in as a Younger Brother.[49] Little is known of the unfortunate Noor Colson. In 1719 he retired from his position because of his 'distracted Condition', though the advantage of his having formalized his connection with Trinity House was evident in his being offered a pension of 10 shillings a month.[50] He must have continued his father's school after this date, for in 1738, presumably after his death, he was referred to as 'Late of Prescod Street, Goodman's Fields' and a 'Teacher of Mathematics'. He had, however, become an insolvent debtor and his creditors were petitioning for the assignment of his goods.[51] If he was the Noor Colson, Mariner, who married Elizabeth Olson in 1728, he was then already in Fleet Prison.[52]

Thus ended the Colsons' more than three-decade connection with Trinity House, during which time the Corporation's role in certifying competence in mathematical navigation was solidified. Though incidental to Lancelot's main business, it perhaps brought him additional connections and customers. For John it was central to his role and identity as a mathematical teacher and, even as the work increased, the opportunities for training pupils to pass exams may have been lucrative. Even if his

[47] 19 November 1707 and 19 December 1707, Court Minutes, MS 30004/008, pp. 65, 68–9.
[48] 5 February 1707/8, Court Minutes, MS 3004/008, pp. 70–1.
[49] 20 April 1716, Court Minutes MS 3004/008, p. 184.
[50] 22 April 1719, Court Minutes MS 30004/008, p. 201.
[51] Petition of the Creditors of Noor Colson, Surrey Quarter Sessions, QS2/6/1738/Mid/48 (Surrey History Centre), details from The National Archives: https://discovery.nationalarchives.gov.uk/details/r/cdd6be0d-5e10-4397-9d32-35874ddb7243.
[52] Fleet Register, Clandestine Marriages, The National Archives, RG 7/85 (via Find My Past).

extra £10 payment stopped him asking for examination fees or bribes, he surely would have gained from attracting pupils that he could promise to certify. Noor did, perhaps, manage the Prescot Street school effectively for a number of years but ill health or misfortune ensured that it was neither a flourishing business nor supported by publication and advertisement. Trinity House had been happy to trust consanguinity and their Master's patronage but found they were soon seeking a replacement.

The Greenwich Academy

Thomas Weston evidently had useful connections providing news from London and considered the Trinity House position to be advantageous, for he was again immediately on hand to petition to replace a Colson.[53] He was this time in a significantly different position: although in 1707 he had completed exemplary training at Greenwich, by 1719 he was running a successful academy, known for its mathematical and navigational training.[54] This had become sufficiently established and recognized for selected pupils of the Greenwich Hospital School, the sons of naval pensioners, to be sent to him for further instruction. He provided something akin to the RMS within Christ's Hospital, likewise foregrounding mathematics as part of a charitable maritime education. From this position, Weston seems to have taken over Colson's dual role of examining RMS scholars for both Trinity House and Christ's Hospital.[55] He was thus the adjudicator of the effectiveness of the teaching of his former colleague, Hodgson, with whom he had overlapped for some years at the Royal Observatory. It is perhaps unsurprising that reports were glowing and examinations proceeded satisfactorily.

Flamsteed had particularly appreciated Weston as a draughtsman but his apprenticeship at the Observatory also trained him in mathematics and observation.[56] By 1715 he was considered 'well-qualified' to teach the Greenwich Hospital School boys.[57] An indication of his early curriculum is apparent from his 1716 request to Greenwich Hospital for a pair of globes, a *Treatise of the Globes* (probably that by John Senex), Hodgson's *Theory of Navigation* and Euclid.[58] Weston's academy primarily catered for private pupils, however, and an advertisement of February 1727/28 claimed its suitability for 'Young Noblemen and Gentlemen'. A wide range of subjects were covered by up to 13 masters: 'Writing, Arithmetick, Merchants Accompts, or the Italian Method of Book-Keeping, Foreign Exchanges, the Mathematicks (in English,

[53] 22 April 1719, Court Minutes MS 30004/008, p. 201.

[54] On Weston's teaching, see Kim Sloan, 'Thomas Weston and the Academy at Greenwich', in: *Transactions of the Greenwich and Lewisham Antiquarian Society* 9/6 (1984), pp. 313–33 and 'The Teaching of Non-Professional Artists in Eighteenth-Century England', unpublished PhD thesis, University of London 1986, pp. 72–7, 82–90.

[55] Ellerton and Clements (*Samuel Pepys*, p. 136) mention Weston as an 'external expert' reporting positively on Hodgson to the Christ's Hospital Committee of Almoners in 1726.

[56] Francis Baily, *An Account of the Revd. John Flamsteed, the First Astronomer Royal*, London: printed by order of the Lords Commissioners of the Admiralty 1835, pp. 64, 65.

[57] Kirby, 'Early Greenwich Schools', p. 232.

[58] Sloan, 'Thomas Weston', p. 318.

Latin or French) Short-hand, Drawing, Fencing, Musick and Dancing', all taught in 'the most rational Way', as well as English, Latin, Greek, Hebrew, French, Italian, High-Dutch, and Spanish. Weston also advertised 'frequent Courses of Philosophical Experiments' with 'Explanatory Lectures concerning them', as well as lectures in 'Geometry, Geography and Astronomy three Days a Week'. The school boasted 'an excellent apparatus of Instruments (Geometrical, geographical, Astronomical and Philosophical)'.[59] His portrait, from 1723, shows him with mathematical instruments and an astronomical diagram and handwritten lecture (Figure 4.2).

This portrait appeared as a frontispiece to the published version of the short works for teaching drawing, writing, and arithmetic that Weston used at his Academy.[60] His more significant *Treatise of Arithmetic* was offered to '*Young Students*' and all '*Those* who are desirous of being *rationally* skill'd in the *Science* of ARITHMETIC', although dedicated to 'The Young Gentlemen of the Academy in Greenwich, For whose USE it was Originally COMPOS'D'. In his preface, Weston outlined a theory of effective education with practical outcomes. He emphasized the 'Universal Usefulness of my Subject' and claimed that he aimed at plain rules so that 'All Mankind' might 'in some measure' understand arithmetic by following reasoned steps. He and Hodgson, like their master Flamsteed, criticized rote learning, though Weston wrote that neither 'Working by heart' nor being taught 'Practice' alone was sufficient.[61]

Most of Weston's work for Trinity House was identical to that of Colson. It is interesting to note, though, that on at least one occasion he was consulted by the By-Board of Trinity House to assess an astronomical instrument. This had been submitted by Captain Jacob Rowe to the consideration of the Elder Brethren, who, both at the larger Courts and the smaller By Boards, often assessed and passed judgement on books, instruments, and ideas brought before them by projectors. It appears that this one was beyond their expertise and so they turned to Weston, who was familiar with practical and theoretical astronomy and its navigational applications.[62] This was at a period before the Commissioners of Longitude met regularly to consider such submissions and so the Admiralty had approached Newton, who was one of the Commissioners, directly. While he approved the theory, he deferred practical assessment to Trinity House.[63] Weston thus seems to have stood at midpoint between the academician and the practitioners.

[59] Quoted in Sloan, 'Thomas Weston', pp. 321–2.

[60] Thomas Weston, *A Copy-Book Written for the Use of the Young-Gentlemen at the Academy in Greenwich*, [London] 1726; *Drawing-Book compos'd for the use of the young gentlemen at the Academy in Greenwich*, [London 1726]; *Veteris arithmeticae elementa: sive De symbolicis et practicis partibus arithmeticae, ab antiquis hebraesis, graecis et romanis usurpatae ... tractatus: in usum studiosae juventutis in Academiâ Grenovici* [...] [London 1726]; *A Treatise of Arithmetic, in Whole Numbers and Fractions*, London: published by John Weston, printed for J. Hooke 1729.

[61] Weston, *Treatise of Arithmetic*, preface.

[62] 27 November 1725, Board Minutes, MS 30010/4. On Trinity House's assessment of instruments, schemes and publications, see Rebekah Higgitt, Jasmine Kilburn-Toppin, and Noah Moxham, *Metropolitan Science: London Sites and Cultures of Knowledge and Practice, 1600–1800*, Bloomsbury, forthcoming, Chapter 5. This instrument was probably the fluid quadrant described in the first part of Jacob Rowe, *Navigation Improved: In Two Books*, London: printed for John Hooke 1725.

[63] Draft letter from Isaac Newton to the Lords Commissioners of the Admiralty, 26 August 1725, Cambridge University Library, MS.Add.3972, https://cudl.lib.cam.ac.uk/view/MS-ADD-03972/59.

Figure 4.2 Thomas Weston. Mezzotint by J. Faber, junior, 1723, after M. Dahl.

Weston's *Treatise of Arithmetic* was published posthumously in 1728 by his brother John, who had taken over the Academy.[64] He also petitioned to succeed to the role of Trinity House's mathematical examiner. This time the position and its vacancy were

[64] Sloan, 'Teaching', p. 89.

sufficiently well known to attract three petitioners, the others being identified as Mr Hazelden and Mr Ham.[65] The former was undoubtedly Thomas Haselden, who, in a posthumously published address to his readers, said he had been 'Educated in the Theory of Navigation, almost from my Childhood' before gaining practical experience from sixteen years in the merchant navy. He must have passed a Trinity House examination himself, for he became a naval schoolmaster before setting up as a private teacher. In 1722 he described himself as 'Late *Teacher* of the *Mathematicks*, to his *Majesty's Volunteers*, in the *Royal Navy*' and signed his opening letter to Edmond Halley, by then Astronomer Royal, 'From my SCHOOL, near Wapping Old Stairs'.[66] This text defended the now widely used Mercator projection over a globular projection that was being promoted by Henry Wilson, who said it had been approved by Halley. It was one sortie within a dispute between Haselden and Wilson about the best means of drawing charts that were accurate, easy to produce, and useful for navigators.[67]

The dispute was described as 'sharp' in 1723 and it was evidently well known within mathematical and philosophical circles in London. It potentially pitted Haselden against Halley and the several other undertakers of the project, including the publisher and chart-maker John Senex and engraver John Harris.[68] It is possible that its aftermath played a part in Trinity House's 1728 decision, although Haselden's career ultimately flourished. Shortly after he lost out to John Weston, a longitude scheme that he had sent to John Machin was read to the Royal Society.[69] In 1730 he published a translation of the *Leçons de mathematiques* by Joseph Privat de Molières, and was then advertising himself as near Union Stairs in Wapping, where 'young Gentlemen are boarded and taught the Mathematics, and Merchants Accompts'. He still noted his former Royal Navy role, suggesting that the accreditation involved, as well as service to the navy, were useful testaments to his ability and character.[70] Haselden's most significant publication was *The Seaman's Daily Assistant*, which appeared in many editions to the end of the century.[71] By January 1740, when he was elected a Fellow of the Royal Society—proposed as 'well versed in Mathematicks and

[65] 7 August 1728, Court Minutes, MS 30004/009, p. 66.

[66] Thomas Haselden, *The Seaman's Daily Assistant*, London: printed for W. and J. Mount, T. Page, and Son 1761, prefatory letter; *The Description and Use of that most Excellent Invention Commonly call'd Mercator's Chart*, London: printed for the author 1722, title page and p. xii. He may also have passed examination as an RMS scholar: see Dickinson, *Educating the Navy*, p. 35.

[67] Wilson's projection was used in charts published the *Atlas Maritimus and Commercialis* (1728), and had been trailed in advertisements from about 1718. On the dispute, see Katy Barrett, 'The Wanton Line: Hogarth and the Public Life of Longitude', unpublished PhD thesis, University of Cambridge 2014, pp. 34–8.

[68] The comment was made by William Blundel, who sought attention for his own projection, quoted in Barrett, 'The Wanton Line', p. 37. Haselden clearly knew these men personally: see his *Description and Use*, p. i.

[69] 'A method for finding the Longitude in a Letter from Mr Thomas Haselden to Mr Machin', 27 August 1728 (read 6 February 1728/9), Register Book Original, RBO/14/33 (Royal Society).

[70] Joseph Privat de Molières, *Mathematic lessons, for the use of students in the mathematics and natural philosophy*, trans. Thomas Haselden, London: printed for John Clarke 1730. He also revised and extended John Darling's *The Carpenter's Rule Made Easy* for an edition of 1727 and published a revised edition of Isaac Barrow's Euclid (1732). He was then near the Hermitage in Wapping.

[71] The earliest edition I have seen is dated 1761, long after Haselden's death, though the book was advertised in 1749 in Colson's *The Mariner's New Kalendar* and presumably was in print during his lifetime, given his authorship of a prefatory letter.

Astronomy' by Machin and eight others—he was Master of the Royal Naval Academy at Portsmouth.[72]

John Ham was a teacher of mathematics based in Hatton Garden, who, like Haselden, had qualified as a naval schoolmaster.[73] He, too, was involved in issuing and adding to a range of revised publications, although largely after his application to Trinity House. In 1729 he published a new edition of Henry Coggeshall's *The Art of Practical Measuring*, to which he added a method of using Scamozzi's Lines for measurement in carpentry. He described himself there as 'Master of the Mathematical School at the Chapel in *Hatton-Garden, Holborn*' three days a week and a private teacher to 'Gentlemen at their own Houses' the other three. He taught 'Mathematicks in Theory and Practice, *viz.* Arithmetiçk Vulgar and Decimal, with all the usual of Rules, Merchants Accompts, or the true *Italian* Method of Book-keeping'. He also offered algebra, geometry, and trigonometry, with applications to mensuration, gauging, surveying, navigation, geography, astronomy, dialling, gunnery, and fortification, as well as use of the globes.[74] Ham's preface advertised his knowledge of mathematical literature (mentioning texts by Richard Towneley, Jonas Moore, Edmund Wingate, and Thomas Everard) and the wares of the instrument maker Thomas Heath, based on the Strand. An illustration of the lineages of mathematical publication is Ham's revision, with added appendix, of Samuel Cunn's revised edition of John Keil's Euclid. He was by then based at Great Kirby Street in Hatton-Garden.[75]

Both Haselden and Ham, then, seem to have engaged more significantly with the wider world of mathematical publishing than did John Weston, although he did see his brother's *Treatise of Arithmetic* (1729, and a second edition in 1736) and translation of Galileo's *Dialogues on Two New Sciences* (1730 and 1734) to press. While Haselden seems to have had the more eminent career, he was perhaps then better known for his dispute with Wilson than for his teaching and authorship. Ham was yet to establish his publishing career. Weston therefore had the advantage of the connection to the substantial Greenwich Academy as well as a fraternal bond to the man who had served Trinity House effectively for nearly a decade. The Elder Brethren voted between the three: 'their Names were Order'd to <be> affixed on the Balloting box which being done and Each Brother had put in his ball there were Nineteen found to be for M^{r}. Weston, five for M^{r}. Hazelden and none for M^{r}. Ham'.[76] Weston's business at his Academy, and the greater distance of his home from Water Lane, could have been a disadvantage. Given the emphasis on examinations taking place at Trinity House—a way of ensuring Corporate ownership of the increasingly lone work of the appointed examiner—he was perhaps fortunate that almost immediately after his appointment it was agreed that he might examine prospective schoolmasters 'at

[72] Thomas Haselden Election Certificate, Royal Society, EC/1739/12.

[73] His examination by Thomas Weston is recorded on 1 November 1721, Board Minutes, MS 30010/4, [n.p.].

[74] Henry Coggeshall, *The Art of Practical Measuring, Easily Perform'd, by a Two-Foot Rule*, London: printed for Richard King 1729.

[75] Euclid, *Euclid's Elements of Geometry, from the Latin translation of Commandine*, London: printed for Thomas Woodward 1733. He also produced a fourth edition translation of Christiaan Huygens's *Of the Laws of Chance*, London: printed for B. Motte and C. Bathurst 1738.

[76] 7 August 1728, Court Minutes, MS 30004/009, p. 66.

his own house at Greenwich' in 1728.[77] The work of examining RMS scholars, more predictable and more prestigious, remained tied to the building.

When John Weston died in mid-1744, the Elder Brethren were again faced with a decision in appointing a successor, as 'the Several Petitions of M^{r}. James Rossam, M^{r}: W^{m}: Mountaine & M^{r}: John Collier, praying respectively to Succeed him, were read'. The decision for Rossam was unanimous, and he was appointed at the usual salary of £20 per annum.[78] Again they followed the line of succession of a family business, for Rossam had taken over the Greenwich Academy and was married to the Westons' sister, Sarah. Little else is known of him, for he fell ill and died within months of his brother-in-law. Another family member took over the Academy—Reverend Francis Swinden, who had previously taught there and was related to Rossam—but he does not seem to have sought appointment by Trinity House.[79]

One of the unsuccessful candidates at the previous election left a positive impression, and will be discussed in the next section. The other, Collier, was another former 'Teacher of the *Mathematicks* to the Gentlemen *Voluntiers* in the ROYAL NAVY' who published *Compendium Artis Nauticæ* in 1729 and, in a pamphlet advertising this, referred to himself as '*Navigator, Surveyor*, and *Teacher of* MATHEMATICKS'.[80] The *Compendium* was dedicated to Lord Archibald Hamilton, a naval officer, MP and Commissioner of the Admiralty under whom Collier had served. He stated that he had 'for many Years, carefully Instructed several Gentleman *Voluntiers* and others in the *Royal-Navy*', which had led him to construct his various tables and methods. He also included a letter from William Whiston, as former professor of mathematics at Cambridge, that commended him and the ease of his methods. A concluding advertisement stated that he was surveyor at Swaffham in Norfolk and ready to undertake land surveys and produce plans 'in a curious New-print, with a Scale, Compass, and all other useful Embelishments'. The *Compendium* found its place on the regular lists in maritime and mathematical works advertised by Mount & Page on Tower Hill in the 1740s but was not enough to trump the existing link to Rossam.

As mathematical examiners for Trinity House between 1719 and 1744, the masters of the Greenwich Academy assessed the products of Hodgson's teaching. It is therefore interesting to note that his apparently impressive tenure as master of the RMS, and smooth relationship with examiners, ran into problems in around 1745. Ellerton and Clements, who credit Hodgson with making the school run efficiently and for establishing an advanced curriculum for secondary mathematical education, suggest that as he aged his pedagogical powers weakened. This accounted for the fact that when William Brakenridge, a Scottish mathematician and cleric, examined

[77] 21 August 1728, Board Minutes, MS 30010/4. On the Corporate attitudes of Trinity House to examinations, knowledge, and practice, see Higgitt, Kilburn-Toppin, and Moxham, *Metropolitan Science*.

[78] 7 July 1744, Court Minutes, MS 30004/010, p. 180.

[79] John Weston left a £400 investment and the remainder of his estate to his sister, Sarah, and brother-in-law James Rossam, as well as £50, his mathematical instruments, and his books to Rossam, who was also an executor, PROB/11/734/106 (TNA). Swinden was brother-in-law to Rossam's son, so perhaps married to Rossam's daughter: Sloan, 'Thomas Weston', pp. 325, 333 n. 44; Sloan, *Teaching*, p. 109.

[80] John Collier, *Compendium Artis Nauticæ. Being the Daily Practice of the Whole Art of Navigation*, London: sold by J. Harbin, B. Motte, F. Simons, W. Meadows, S. Goodwin, S. Fitzer, C. Digby, and E. Baldwin 1729; John Collier, *A Letter to the Practisers, Promoters, and Learners of Navigation*, London: printed by W. Pearson for the author 1730. On Collier see Dickinson, *Educating the Navy*, p. 20.

the RMS scholars for Christ's Hospital on 4 April 1745 he found that although they 'performed the mechanical parts well' they were 'defective in the science and do not answer my expectations'. Further questions were raised and it was found necessary to appoint an assistant master, John Robertson.[81] This period, then, saw an interruption to the Greenwich connection, stemming from the linked training of Hodgson and Thomas Weston under Flamsteed, and passing through their teaching at Greenwich and Christ's Hospital. It also seems finally to have done away with the practice, under Colsons and Westons, of having the same individual examine RMS boys on behalf of two different institutions.

Gainsford Street, Shad Thames

On 5 January 1744/5, there was only one petitioner, William Mountaine, who had previously lost out to Rossam. He was a 'Teacher of the Mathematicks at Shad Thames', who 'was Elected accordingly, at the usual Salary of Twenty pounds a year'.[82] Mountaine was born in the early years of the eighteenth century in Clint, near Ripley in Yorkshire, but little is known of his education there. He was in St Olave, Southwark, by the time he married in 1732 and he was evidently well set-up as a teacher by 1744 in the neighbouring riverside area of Shad Thames, in the parish of St John Horselydown. An advertisement in the first of many editions of his *Seaman's Vade-Mecum* stated

> The CLASSICS, WRITING, ACCOMPTS, BOOK-KEEPING after the *Italian* form, NAVIGATION, ASTRONOMY, the USE of the GLOBES and other Branches of the MATHEMATICS are regularly Taught, and young Gentleman Boarded, by *William Mountain*, In *Gainsford-Street*, near *Shad-Thames, Southwark*.[83]

Mountaine was prolific in his publications, nearly all of which were printed for Mount & Page. These included *The Practical Sea-Gunner's Companion* (1744, 1747) and *The Seaman's Vade-Mecum* (1744, 1756, 1761, 1778, and posthumous). He produced revisions of James Atkinson's *Epitome of the Art of Navigation* (1744), Andrew Wakley's *The Mariner's Compass Rectified* (1753), Henry Wilson's *Navigation New Modelled* (1777) and many editions of Nathaniel Colson's *The Mariner's New Calendar* from at least 1753. The year 1744, when he first petitioned Trinity House and produced several publications, was evidently a turning point in his career.

Mountaine had other ambitions, however. He collaborated with a Wapping-based teacher of mathematics, James Dodson, to produce a new chart of geomagnetic

[81] Quoted in Ellerton and Clements, *Samuel Pepys*, p. 140. Brakenridge seems to have been a regular examiner for the school thereafter: Jones, *The Sea & the Sky*, pp. 166–7, 169.

[82] 5 January 1744/5, Court Minutes, MS 30004/010, p. 193.

[83] William Mountaine, *The Seaman's Vade-Mecum*, London: printed for W. Mount and T. Page 1744, [n.p.]. On Mountaine see Maurice Edward Ogborn, *Equitable Assurances: The Story of Life Assurance in the Experience of The Equitable Life Assurance Society 1762–1962*, London: George Allen & Unwin 1962, pp. 77–80 and Andrea Ives, *Admiral Long's Foundation & Burnt Yates School: 250 Years of History*, Harrogate: printed for the author 2014, pp. 128–9.

variation in 1745. This sought to update the observations that Halley had published as a potential longitude solution at the beginning of the century, drawing on data from logbooks and journals of the Royal Navy and the East India and Royal African Companies.[84] Mountaine and Dodson's promotional pamphlet noted both authors as teachers, while another, advertising tables updated to 1756, styled both as Fellows of the Royal Society. The two men used the Society's *Philosophical Transactions* to disseminate and promote their work in 1753 and Mountaine continued the campaign after Dodson's death. Mountain also published a 'Defence of the Mercators Charts' there in 1763, which referred back to the 1720s dispute between Haselden and Wilson, championing the former after another critique of the Mercator projection.[85] Mountaine, described as 'A Gentleman well skill'd in Mathematicks, Natural Philosophy and most branches of Polite Literature', had been elected in 1751. Still based in Gainsford Street, he was supported by Hodgson and Robertson at the RMS, Gowin Knight (a physician, known for his research into magnetism and invention of magnetic instruments), John Colson (the Lucasian Professor), and Joseph Ames (a bibliographer and antiquary).[86] He was able to support Dodson's election in 1755, also proposed by the Society's president, the Earl of Macclesfield.[87] Dodson had recently, perhaps with Mountaine's assistance, succeeded Robertson as master of the RMS. Conversely, it was through Dodson, and their mutual connections with Gowin Knight, who was its first chair, that Mountaine became involved with The Society for Equitable Assurances. Dodson had left his papers and his shares in the Society to Mountaine and it was to him that they seem to have turned for mathematical support.[88]

Mountaine also remained close to Robertson, who moved from the RMS to become headmaster of the Royal Naval Academy in 1755 and, subsequently, the Royal Society's clerk and librarian. In the 1772 edition of his *Elements of Navigation*, Robertson promoted Mountaine and Dodson's work, suggesting that the government should pay for the publication of regularly updated charts. After Robertson's death in 1776, Mountaine published a description of his improved Gunter's scale, a copy of which he presented to Trinity House.[89] Robertson's son, who succeeded him at the Royal

[84] William Mountaine and James Dodson, *An account of the methods used to describe lines, on Dr. Halley's chart of the terraqueous globe, shewing the variation of the magnetic needle*, London: William Mount and Thomas Page 1746, p. 4. Later versions reported that they had also gained access to Hudson's Bay Company logs, possibly facilitated by Mountaine's Trinity House connections.

[85] Barrett, 'The Wanton Line', pp. 38–9.

[86] He was elected 14 March 1750/1: Election Certificate, Royal Society, EC/1750/16.

[87] On Mountaine and Dodson's variation charts see Barrett, 'The Wanton Line', pp. 41–2 and, for an example of the use and testing of their tables and chart at sea, see Jim Bennett, 'Mathematicians on board: introducing lunar distances to life at sea', in: *British Journal for the History of Science* 52/1 (2019), pp. 63–83; pp. 68–71; Election Certificate, Royal Society, EC/1754/17.

[88] Ogborn, *Equitable Assurances*, pp. 37, 80–1; on Dodson see also G. J. Gray and Anita McConnell, 'Dodson, James (*c.*1705–1757), mathematician and actuary', in: *Oxford Dictionary of National Biography*, Oxford: Oxford University Press 2004, https://doi.org/10.1093/ref:odnb/7756.

[89] W. F. Sedgwick, rev. Anita McConnell, 'Robertson, John (1707–1776), mathematician', in: *Oxford Dictionary of National Biography*, Oxford: Oxford University Press 2004, https://doi.org/10.1093/ref:odnb/23802; Barrett, 'The Wanton Line', p. 43; William Mountaine, *A Description of the Lines Drawn on Gunter's Scale, as Improved by Mr. John Robertson*, London: Nairne & Blunt 1778. Blunt also attended Trinity House to present the scale itself: 4 April 1778, Court Minutes, MS 30004/013, p. 11.

Society, received a £50 legacy in Mountaine's will.[90] Mountaine had accumulated enough money and property to be generous to a large number of friends, relations, and charitable causes. He was a gentleman, who had taken on responsibilities such as becoming a Justice of the Peace and a governor of St Thomas's Hospital. His will bequeathed his significant book collection, some 560 volumes largely on mathematics, navigation, and science, to the Burnt Yates School, Clint, for which he had helped to generate support during his lifetime.[91] He requested that the books be catalogued and given labels, 'having a neat Scroll around and this description: "The Gift of William Mountaine, F.R.S to Clint School"'. He also left money to be invested to pay for clothing and books for poor pupils, and for a librarian to take care of the books and 'other materials'. The latter included a pair of 17-inch globes, two telescopes, his 1744 and 1756 'variation Charts neatly coloured and in rolling fframes canvassed', together with his pamphlets, a manuscript chart of the south seas on vellum, a 9-foot and a 3-foot refracting telescope, other optical and mathematical instruments, an octagonal concave mirror, and portraits of him and his wife Mary (Figure 4.3), and of the king and queen.[92]

Figure 4.3 Portraits of William and Mary Mountaine by Joseph Highmore, bequeathed by him to the Burnt Yates School, Clint, Yorkshire.

[90] Sedgwick, 'Robertson, John'; Will of William Mountaine, PROB/11/1053/120.

[91] Ogborn, *Equitable Assurances*, p. 81; Mountaine, contacted by the Rector of Ripley, William Gawthorpe, in 1755, as someone who had been 'promoted in the world by means of their education in this parish', persuaded another London-based native of the area, Admiral Robert Long, to support the school: Ives, *Admiral Long's Foundation*, pp. 5–13, quote at p. 4.

[92] PROB/11/1053/120. Sadly only the portraits of the Mountaines remain at the school. The catalogue of Mountaine's books is in Ives, *Admiral Long's Foundation*, pp. 141–8.

We have a rare description of Mountaine's examination of a naval schoolmaster, and a glimpse of his home, in the journal of Nathan Prince. Prince had been a scholar and tutor of mathematics at Harvard for twenty years but, after losing his position for alleged intemperance, and failing to establish a mathematical school in Boston, was on his way to missionary work in the Caribbean. He came to London in 1747 to be ordained and to be retrospectively certified as a naval schoolmaster, a post he had filled on his voyage over as the only way to afford his passage.[93] Advised by the Trinity House doorkeeper to visit Gainsford Street to discuss the examination content, Prince was, on presentation of a letter from the lieutenant of his ship, received kindly by Mary Mountaine and invited to dine. She entertained him in one room, and they went down to dinner before William arrived. After eating, the two men went up to Mountaine's room to discuss the examination; there was sufficient trust between the two for Mountaine to lend the relevant texts.[94] The examination took place on two mornings, necessarily at Trinity House and on Court days, when Elder Brethren and the Deputy Master would be in the building. On each occasion, Prince met Mountaine at Gainsford Street before the two proceeded together over the river to Water Lane. Just the two men were present at the examination, which was a discursive process in which Mountaine sometimes prompted or explained alternative approaches. It was unusual that a university man should seek this certification and Prince had already been told that they would make allowance for deficiencies in one area if the candidate proved sufficient in others.[95] They skipped the theory of the globe, since he had 'taught it to scores in New England'.

The result was a foregone conclusion, with Mountaine completing his testimonial before the examination was finished.[96] He, however, declined the offer of a fee but told Prince that he should pay the Secretary 15s for the certificate and tip the doorkeeper.[97] Chaplain claimed that 'there is good evidence that a fee attached to every examination', from which the examiners benefited, though did not outline what this evidence was.[98] In fact, as we have seen, John Colson had been admonished and temporarily dismissed when it was found that he had been taking such fees, although he later received his supplementary payment in recognition that it was a particularly busy period.[99] Mountaine petitioned successfully in 1745 for the same additional £10 per annum as Colson had received in 1707–1714, citing the 'great Increase of his Business by the War (Especially since the R.^{t} Hono:^{ble} the Lord's Comr.^{s} of the Admiralty have Invited School-Masters into the Navy)'.[100] Prince's journal suggests Mountaine kept his side of the bargain and neither charged individual fees nor accepted them

[93] William L. Sachse, 'The Journal of Nathan Prince, 1747', in: *The American Neptune* 16 (1956), pp. 81–97; Laura Morris, 'Biographical Note on Prince', Harvard Library (2011), https://hollisarchives.lib.harvard.edu/repositories/4/resources/4037.

[94] Sachse, 'Journal', p. 87. Mountaine lent Prince Hodgson's *A System of the Mathematics* (1723, 2 vols) and Archibald Patoun's *A Compleat Treatise of Practical Navigation Demonstrated* (1734).

[95] Sachse, 'Journal', pp. 94, 87.

[96] Sachse, 'Journal', pp. 87, 89–90, 92–3.

[97] Sachse, 'Journal', p. 93. Prince's successful examination was recorded on 27 June 1747 in the Board Minutes, MS 30010/011, p. 80.

[98] Chaplain, 'William Mountaine', p. 186.

[99] 27 August 1707 and 5 February 1707/8, Court Minutes, LMA MS 30004/008, pp. 62, 70.

[100] 16 March 1744/5, Court Minutes, LMA MS 30004/010, p. 304.

when offered. His salary reverted to £20 in 1747 but again increased in 1756, because 'the Business of Mathematical Examinations was increas'd about Equal to what it was at any time during the Late War', and 1778, because of 'the additional trouble in the present Times'.[101] It is interesting to note that the occasion of war meant an increase in numbers rather than an overlooking of naval schoolmasters.[102]

Mountain's self-descriptors in his published works indicate that, first and foremost, he saw himself as a mathematical teacher. His will gave a major legacy to one school and smaller ones to others. It also noted that he was a member of the Society of Schoolmasters. While he had clearly been keen to gain the appointment at Trinity House, and kept it over more than 30 years, it does not seem to have been a significant part of his public persona for most of that time. It was not until 1778 that he added 'Mathematical Examiner to the Honourable Corporation of Trinity House of Deptford-Strond' to his authorship credentials, where it remained in posthumous editions. Trinity House had, it seems, gained sufficient prestige among mathematical readers to be worth mentioning. Certainly it had become increasingly wealthy, particularly from its monopoly in the supply of ballast to Thames shipping, and its Elder Brothers included wealthy merchants and individuals with improving instincts or scientific interests. The Deputy Master at this period, Captain Sir William James, was a baronet, member of the Bombay Marine, a Director of the East India Company, MP, and Governor of Greenwich Hospital, who, along with other Brothers, began Trinity House's first formal experiments with lighthouse illumination.[103]

Within a month of Mountaine's death, petitions from potential replacements were invited. It appears that, for all the Corporation's increasing status, only one was received, for on 5 June it was reported that the Court had appointed Gideon Fournier 'to be Mathematical Examiner to this Corporation'.[104] Fournier is not generally known to historians, or known as a mathematician. He was, however, a neighbour to Mountaine, demonstrating again the significance of personal connections. He was one of five friends chosen by Mountaine as the executors of his will and received £50, plus another £10 each for him, his wife, and daughter for mourning. Mountaine also bequeathed him 'my case and apparatus of the small artificial Magnets by Dr Knight and my tripod Quadrant'. While it was a John Davis, a more significant beneficiary, who received his large magnets, and his papers on the construction of his variation charts, Fournier was evidently known to Mountaine as a man of science. In 1783, indeed, he became a Fellow of the Royal Society. Identified

[101] 7 January 1748 and 4 September 1756, Court Minutes, MS 30004/001, pp. 42, 270; 11 April 1778, Court Minutes, MS 30004/013, p. 13.

[102] Naval schoolmasters were not well remunerated and had been seen as rare and/or inefficient. However, they have been shown to be numerous and in many cases capable: Dickinson, *Educating the Royal Navy*, pp. 15–20. The fact that Haselden, Ham, and Collier chose to advertise their experience of this role suggests that it was seen positively within the practical mathematics community. In 1702–5, well before Colson's busier period, Trinity House issued 62 certificates for competent schoolmasters: F. B. Sullivan, 'The Naval Schoolmaster During the Eighteenth Century and the Early Nineteenth Century', in: *The Mariner's Mirror* 62 (1976), pp. 311–26 at p. 317.

[103] T. H. Bowyer, 'James, Sir William, first baronet (1722–1783), naval officer and director of the East India Company', in: *Oxford Dictionary of National Biography*, Oxford: Oxford University Press 2004, https://doi.org/10.1093/ref:odnb/14626; Higgitt, Kilburn-Toppin, and Moxham, *Metropolitan Science*, Chapter 5.

[104] 31 May 1779, 5 June 1779, Court Minutes, MS 30004/013, pp. 54, 57.

as of Middle Temple he was said to be 'well versed in many branches of polite and useful literature, particularly mathematics'. He was nominated by, among others, James Horsfall, a mathematician and the librarian of Middle Temple, and William Wales, then master of the RMS.[105]

Other traces of Fournier show that he, too, lived in Gainsford Street. He was there in 1766, identified as a 'Gentleman', when a man was accused of stealing his railings, and in 1764, when *The General London Guide* listed him as a magistrate of the Police Office at Union Hall in Southwark.[106] The Union Hall Police Office was one of six established in 1792 by the Middlesex Justices of the Peace Act. Fournier, who had taken an oath as a Justice of the Peace for Surrey in 1778, was named for the new position by Lord Onslow, Lord Lieutenant of Surrey, and also 'earnestly' recommended as a 'proper Person' for the role by the MP for Southwark, Paul Le Mesurier, based on 'the very intimate Connection I have had these eight Years past with the Borough of Southwark'. Le Mesurier also recommended another appointment, Benjamin Robertson, and said he had known both for over 20 years and believed them to be 'strictly unexceptionable for Character Ability and Attachment to His Majesty's Persona and Government'. Fournier's ability was testified to by his being a barrister and Robertson was said to be 'extremely reputable, as a Gentleman of general Knowledge and a considerable Freeholder in the County'.[107] Robertson, who maintained a large botanical garden in Stockwell, requested his release from his existing position at the Navy Office in order 'to devote himself entirely to the new Southwark police'.[108] On his death in 1813, Fournier was reported to be 'chief magistrate of the county of Surrey'.[109]

In 1791, two years after a significant reorganization of their own examinations of masters and pilots with the creation of an Examining Committee, the Brethren reminded themselves that the certification they had returned to Christ's Hospital was 'Sign'd by the Deputy Master and Elder Brethren, notwithstanding their Examination by our Mathematical Examiner, is not attended by any of the Brethren, which in the Opinion of the Court appears to be very inconsistent'. They resolved

> That (tho' we are perfectly satisfied with the Abilities and Assiduity of M.r Fournier our present Mathematical Examiner) All future Examinations of these Children, as well as Mathematical teachers in the Navy, be held in the Presence of the Committee

[105] Gideon Fournier, Election Certificate, EC/1783/01.

[106] Catalogue entry for Surrey Quarter Sessions https://discovery.nationalarchives.gov.uk/details/r/a3930d57-0d95-48d6-a0da-ba701783da7d.

[107] Catalogue entry for Oaths of Justices of the Peace, 1778, https://discovery.nationalarchives.gov.uk/details/r/C3646052; Fournier's was the first name on a list appended to a letter from Lord Onslow to Evan Nepean, 7 July 1792, HO 42/21/22, ff. 50–2; letter from Paul Le Mesurier to Henry Dundas, 17 July 1792, HO 42/21/59, ff. 129–130 (digitized copies available via The National Archives). Born in Guernsey, Le Mesurier was a merchant, shipowner, and Director of the East India Company: W. R. Meyer, 'Le Mesurier, Paul (1755–1805), merchant and politician', in: *Oxford Dictionary of National Biography*, Oxford: Oxford University Press 2008, https://doi.org/10.1093/ref:odnb/16428.

[108] *Edinburgh Magazine*, 15 (1800), p. 509; letter from Robert Barclay to Evan Nepean, 20 July 1792, enclosing a letter from Fournier, HO 42/21/69, ff. 151–4 (digitized).

[109] Fournier's death was recorded as having taken place 'lately' at the end of the list of deaths for 1812 in the *Annual Register... For the Year 1812*, London: printed for W. Otridge and Son and others 1813, p. 183, his burial, aged 70, was recorded on 30 January 1813 at St John Horsleydown.

for Examining Masters of Men of War, or some of them, or of some Elder Brethren whom they may engage in Case the Duty of the Committee shall not permit; and that such Examinations be sign'd by the Elder Brethren attending, as well as the Mathematical Examiner.[110]

Thereafter, one Brother, occasionally two, joined Fournier in the examinations. Usually it was Captain Thomas Brown, who continued to perform this service with Fournier's well-qualified successor, although the process seemed to have been dropped again by the 1810s. Fournier was, nevertheless, recognized as deserving of the usual increment for the 'Extra Trouble' that came in time of war, with a 'Gratuity of Twenty Guineas' being added to a salary that was now £30 per annum. Ongoing warfare led in 1794 to an increased salary of £50 replacing the gratuity.[111]

In the 1720s and 1740s the position of mathematical examiner to Trinity House was sufficiently desirable—whether for the additional income or for the connections it could offer—for there to have been a degree of competition. In William Mountaine the Elder Brethren found someone who served for many years and who enjoyed a reputation as a gentleman, natural philosopher and Fellow of the Royal Society, along with that of a teacher of mathematics. The role he played for Trinity House was, ultimately, deemed significant enough to be noted on the title page of his publications. However, this was not a sign that it had become professionalized. While Gideon Fournier may have been able to examine naval schoolmasters and RMS scholars, he did not undertake this as part of a wider career of mathematical teaching and publishing. He presumably took an interest in Mountaine's magnetic researches, but although he became a Fellow of the Royal Society, he does not seem to have been active beyond supporting the elections of three Fellows, two of whom were barristers and the other an Elder Brother.[112]

Dr Maskelyne's recommendation

Fournier's immediate successors at Trinity House suggest that the role of mathematical examiner now more clearly shifted to one that honoured the ability of the holder, and that the Elder Brethren might take external advice to fill. When ill health and infirmity led Fournier to request permission to resign, the application of Andrew Mackay (Figure 4.4) to succeed him was provisionally agreed, 'he being well known to several of the Elder Brethren, and otherwise respectably recommended'.[113] Soon, 'having perform'd the Duty for several Months past to much satisfaction, and he producing Testimonies from D.r Maskelyne Astronomer Royal, and several other eminent Mathematicians and Astronomers, to his Knowledge of the Mathematical

[110] 3 November 1791, Court Minutes, MS 30004/015, p. 73. On the Examining Committee see Cotton, *Memoir*, pp. 174–6.

[111] 2 January 1794, Court Minutes, MS 30004/014, p. 193.

[112] Election Certificates for John Morgan, Royal Society, EC/1785/21; Sir Arthur Leary Piggott, Royal Society, EC/1786/24; Joseph Huddart, Royal Society, EC/1791/23.

[113] 4 April 1805, Court Minutes, MS 30004/015, pp. 362–3.

Figure 4.4 Andrew Mackay, stipple engraving by James Heath after a drawing done in Aberdeen by A. Robertson.

Sciences, Navigation &c'., Mackay was officially appointed with a salary of £50.[114] While largely self-taught, Mackay had been an unpaid assistant at the observatory of Marischal College at the University of Aberdeen, where he had supportive patrons. He impressed Nevil Maskelyne with his navigational treatise in 1787, subsequently published as *The Theory and Practice of Finding the Longitude at Sea or on Land* (Aberdeen, 1801).[115] He was a Fellow of the Royal Society of Edinburgh and held an honorary LLD, though his correspondence with Maskelyne shows his desire for election to the Royal Society of London and to gain paid employment. They discussed his becoming an assistant at Greenwich and a Board of Longitude-sponsored observer to accompany Matthew Flinders's Australian circumnavigation but both seemed insufficiently rewarding.[116] In 1804 Mackay moved to London and set up as a teacher.

An advertisement of 1808 showed him teaching 'Mathematics, Natural Philosophy, Navigation, Lunar Observations, Geography, Ship-building, Naval Tactics, Architecture, Fortification' and more from George Street off Trinity Square on Tower Hill (Figure 4.5). Trinity Square was a new location, named for the grand new Trinity House built there at the end of the eighteenth century in classical style. However, Mackay's was in fact the same street as that in which Lancelot Colson had lived, then known as George Yard, more than a century earlier. In this advertisement Mackay presented himself as a prolific mathematical and scientific author and as examiner not just to Trinity House but also to the East India Company and Christ's Hospital.[117] Thus the Trinity House examiner was, once again, examining the RMS scholars twice, and also offering certification, probably in the lunar-distance method of finding longitude, for candidates undertaking East India Company examinations. In 1808 Mackay signed a handwritten volume of glowing testimonials for his book, which included the dedication to Trinity House that was to appear in the new edition.[118] Sadly, just over a year later, as his career appeared to be ready to flourish, he died and was buried at All Hallows Barking on Tower Street.[119] His widow, Margaret, was left with 'Eight very young Children, in the deepest Distress' and without means to pay for the funeral. The Elder Brethren agreed to offer her a pension of £30 per annum during her widowhood, 'to be applied toward the Support of herself and Infant Family'.[120]

[114] 4 July 1805, Court Minutes, MS 30004/015, p. 379.

[115] Alexi Baker, '"Humble Servants," "Loving Friends" and Nevil Maskelyne's Invention of the Board of Longitude', in: *Maskelyne: Astronomer Royal*, ed. Rebekah Higgitt, London: Hale Books 2014, pp. 203–28; pp. 223–4; David Gavine, 'Mackay, Andrew (1758–1809), teacher of navigation', in: *Oxford Dictionary of National Biography*, Oxford: Oxford University Press 2004, https://doi.org/10.1093/ref:odnb/17552.

[116] See Nevil Maskelyne's correspondence with Andrew Mackay, MKY/8, National Maritime Museum, available at Cambridge Digital Library: https://cudl.lib.cam.ac.uk/view/MS-MKY-00008/1

[117] Advertisement, Letters relating to Andrew Mackay's teaching activities and publications, National Maritime Museum MKY/9/4 (Cambridge Digital Library: https://cudl.lib.cam.ac.uk/view/MS-MKY-00009/11). These papers include an 1805 certificate by Mackay for a 'Mathematical Teacher in His Majesty's Navy' (MKY/9/5) and an examination plan for naval schoolmasters (MKY/9/8).

[118] Andrew Mackay, 'Some Public Testimonials in Favour of the First Edition of the Theory and Practice of Finding Longitude', MKY/9/9 (https://cudl.lib.cam.ac.uk/view/MS-MKY-00009/27).

[119] He was buried on 8 August 1809, apparently without a memorial, Joseph Maskell, *Collections in Illustration of the Parochial History and Antiquities of the Ancient Parish of Allhallows Barking, in the City of London*, London: Bryan Corcoran & Co. 1864, p. 108.

[120] 1 September 1809, Court Minutes, MS 30004/016, pp. 125–6.

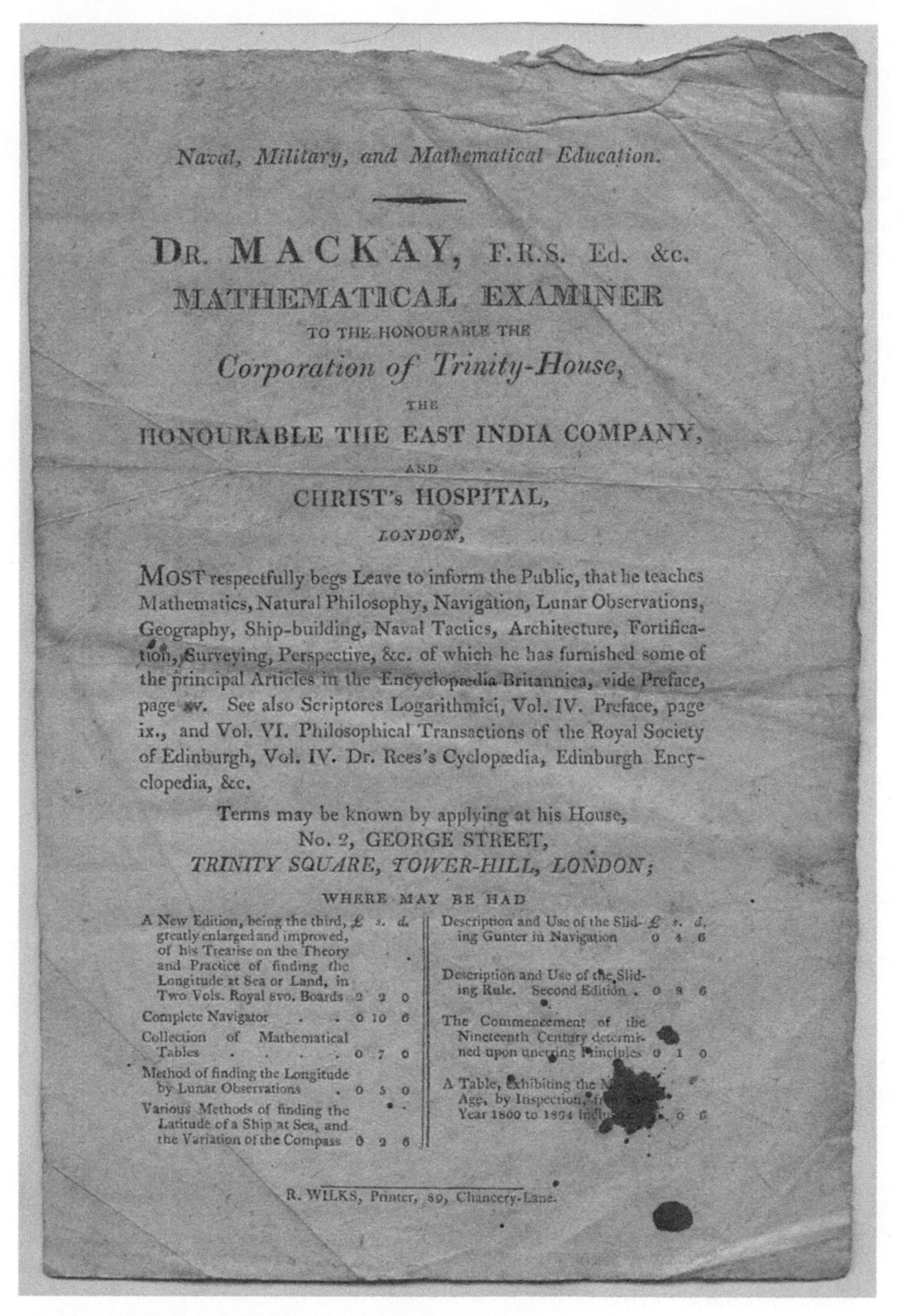

Figure 4.5 Advertisement for Andrew Mackay.

The patronage of Nevil Maskelyne also appears to have been a factor in the appointment of Mackay's successors.[121] In 1809 there were five petitioners, John Crosley, George Sanderson, Benjamin Workman, J. F. Dessiou, and John Epps. The first three, at least, had connections to the Astronomer Royal. For Crosley this had been for many years, as he had been appointed an observatory assistant from 1789 to 1792, as a Board of Longitude observer on expeditions to the northwest Pacific and Australia, and as a *Nautical Almanac* computer. He was also president of the Spitalfields Mathematical Society.[122] Workman, who had been a naval schoolmaster after being dismissed from his position as an instructor in mathematics at the University of Pennsylvania for radicalism, had undertaken computing work for the *Nautical Almanac* and other Board of Longitude tables in 1808 and 1809, but found himself in prison for debt by 1810. He petitioned the Board in a dire position, despite receiving a small amount of money from Maskelyne toward his food and having pawned books belonging to the Board. He tried again in 1818, when he also enclosed an advertisement for his teaching any 'Young Gentlemen designed for entering any of the Royal Military or Naval Colleges' or ladies looking to become governesses.[123]

While the well-known Crosley was appointed subsequently, it was George Sanderson who received the position in 1809, possibly at the specific request of the Astronomer Royal. Sanderson appears in Maskelyne's records in 1801–2 as receiving 'into his hand' books that his son Thomas required for computing work.[124] George himself was a tailor but considered by some to be 'one of the first mathematicians in England'.[125] An 1839 speech by Olinthus Gregory indicated the role of patronage in the appointment, as well as holding up to members of the Woolwich Institution an example of the benefits to individuals and society of 'the acquisition of knowledge by humble individuals'. Although unnamed, Sanderson was described as

> a tailor, who was an excellent geometrician, and who discovered curves which had escaped the sagacity of Newton; with whom Dr. Hutton, Bishop Horsley, Dr. Maskelyne, and Baron Maseres delighted to converse on mathematical subjects, yet who laboured *contentedly* at his trade till nearly 60 years of age, when, by the recommendation of his scientific friends, he was first appointed master of Neale's Navigation School, and afterwards Nautical Examiner at the Trinity House.[126]

Sanderson died in 1814 and Crosley was appointed in his stead, apparently without competition. A petition was submitted on behalf of Sanderson's 'two deserted

[121] On Maskelyne as patron to practical mathematical roles, see Rebekah Higgitt, 'Equipping Expeditionary Astronomers: Nevil Maskelyne and the Development of "Precision Exploration"', in: *Geography, Technology and Instruments of Exploration*, ed. Fraser MacDonald and C. W. J. Withers, Basingstoke: Routledge 2015, pp. 15–36, and Mary Croarken, 'Nevil Maskelyne and his Human Computers', in: *Maskelyne: Astronomer Royal*, pp. 143–4.

[122] Currently the most useful short biography, with links to other sources, is on Wikipedia: https://en.wikipedia.org/wiki/John_Crosley.

[123] Croarken, 'Nevil Maskelyne', pp. 154–5; Nevil Maskelyne, 'Diary of Nautical Almanac Work', Royal Greenwich Observatory Archives, RGO 4/324, p. 34^r (Cambridge Digital Library: https://cudl.lib.cam.ac.uk/view/MS-RGO-00004-00324/69); Papers of the Board of Longitude, 'Petitions and memorials', RGO 14/12, pp. 486^r–494^r (https://cudl.lib.cam.ac.uk/view/MS-RGO-00014-00012/381).

[124] RG0 4/324, p. 26^r (https://cudl.lib.cam.ac.uk/view/MS-RGO-00004-00324/53).

[125] *The Quarterly Visitor* 2 (1814), Acknowledgements.

[126] *The Mechanics Magazine* 32 (1839), p. 106.

Children', and the Court ordered that £15 a year for each be allowed to Thomas Sanderson 'toward their Board and Education, till they can be put out to Service'.[127]

Conclusion

Most of the lives highlighted in this chapter were built around mathematical teaching, supplemented and advertised by forays into print and via institutional connections. They were well acquainted with the methods and curricula required, helping Trinity House to fill its 'mathematical gap' and fulfil its responsibilities to the monarch and the Admiralty. They reveal the overlapping networks that linked friends and family businesses with publishing houses, the merchant and royal navies, a maritime guild and newly arising institutions such as the RMS, Royal Observatory, Board of Longitude, and Royal Naval Academy. These teachers and their publications met an evident demand, and their role as examiners was one of potential national importance, but it was a precarious world. Many of those who applied to become examiner badly needed a regular source of income to supplement more piecemeal work and, while £20 was not a huge sum, it equalled that paid to assistants at the Royal Observatory. Working for Trinity House also had the potential benefit—for those such as the distracted Noor Colson, the widowed Margaret Mackay or orphaned Sanderson children—of future pensions or alms. Even those with successful careers must have appreciated security against unforeseen misfortune. The position also put the holders in contact with a range of people who might help their businesses and interests. Teachers might gain private pupils hoping to pass examinations, or be recommended by Elder Brethren within their spheres of influence, including the Navy Office, Admiralty, and East India Company. Such connections might lead to book sales, patronage, and access to knowledge and resources that would support the development of new resources for and approaches to navigation.

Despite all this, Trinity House did not feature in the advertisements and publications of the mathematical examiners before the final decades of the eighteenth century, by which time it was a wealthier and more influential corporation. Mid-century, experience of having taught volunteers on Royal Navy vessels was highlighted by some, if not the Trinity House certification that allowed them to do it. Many worked hard to piece such careers together, and successful schools and publishing enterprises could be passed to family members. Throughout the eighteenth century, Trinity House seems to have privileged family and personal connections in making their appointments, perhaps seeing sons as bred to the business, just as mariners and lighthouse keepers usually were. Being in the right place at the right time probably helped Thomas Weston and Gideon Fournier into the job but, for a few years in the early nineteenth century, the patronage of Nevil Maskelyne may have been the crucial factor, as he sought to reward what he saw as deserving merit.

It was a close friend of Maskelyne's, Patrick Kelly, who succeeded Crosley in 1817. He was a teacher of mathematics and navigation, who ran the successful Finsbury Square Academy for many years, and it is possible that his application was actively solicited. The Board Minutes record that there had been three individuals who put

[127] 6 January and 3 November 1814, Court Minutes, MS 30004/016, pp. 336, 354.

themselves forward on 2 October but they deferred discussion until the following week, by which time an application had also been received from Kelly, who was unanimously recommended to the Court.[128] Indicating that the appointment did honour to both sides, Kelly recorded his affiliation with Trinity House in his publications until his resignation in 1838 and it was often mentioned posthumously.[129] That resignation was probably due to illness or old age (he was then in his eighties) but the fact that it preceded his death, and was accompanied by an application from his son, suggests that he hoped he could control succession. However, when considering Kelly junior's request, it was stated that 'it appears to this Court that the duties of Mathematical Examiner, have very materially decreased' and that 'it is not therefore necessary to take any immediate steps' as a result of the senior Kelly's resignation.[130]

Patrick Kelly was thus the last of the sequence of Mathematical Examiners appointed by Trinity House since 1682. It is not clear that the duties had decreased, but it seems that they were increasingly being shared by the Examining Committee, and we find that Elder Brothers were, once again, joining their Mathematical Examiner to carry out examinations of the RMS scholars. This is testament to a growing confidence in the mathematical abilities of a well-schooled subset of the Brethren, and of the importance of the Examining Committee. They seem, however, to have been over-confident because, in 1845, one Boulter Bell was appointed 'to assist the Examining Committee'.[131] The salary, and the work, had increased in recognition of the greater mathematical content of training for other roles. Thus Bell was to receive £50 for examining masters and mates in the merchant navy, £50 for masters and pilots of Royal Navy vessels, and £20 for RMS boys. In the following year, he examined 90 candidates.[132] However, Trinity House's role in examining all but pilots ended soon after, when the Board of Trade instituted compulsory examinations in 1850, and education and examination within the Royal Navy underwent significant change. It spelled a second end for the role of Mathematical Examiner at Trinity House, and a particular relationship between the Corporation and the community of mathematical teachers in London and its neighbouring riverside communities.

[128] 2 October 1817, Board Minutes, MS 30010/023, pp. 249–50 records applications from Lawrence Gwynne, a former Naval Schoolmaster and master of the RMS, William Garrard, an assistant at the ROG and teacher at RMS, and, as in 1809, the hydrographer J. F. Dessiou. Gwynne had undertaken a Trinity House examination of a naval schoolmaster while Crosley was ill: 4 September 1817, Board Minutes, MS 30010/023, p. 241. Kelly's recommendation was made 9 October 1817, Board Minutes MS 30010/023, p. 252 and his appointment confirmed 6 November 1817, Court Minutes, MS 30004/017, p. 63.

[129] e.g. Kelly 'held for many years the office of Mathematical Examiner to the Trinity House' (*Gentleman's Magazine* 172 (1742), p. 434). His resignation was acknowledged by Trinity House on 6 February 1838, Court Minutes, MS 30004/020, p. 303.

[130] 6 February 1838, Court Minutes, MS 30004/020, p. 303. Rev. A. P. Kelly was born and schooled at his father's academy, a Senior Optime at Cambridge and for many years vicar at St John the Baptist, Hoxton. His obituary appears in *The Gentleman's Magazine* (March 1865) pp. 373–5.

[131] Records show he was born in Berwick-upon-Tweed in 1801, died in 1866, and was listed on the register of Merchant Seamen for 1835–1836. On the 1851 and 1861 census he was based in Fulham, first listed as a Captain in the Merchant Navy and then as an Examiner in Navigation for the Port of London.

[132] 7 October 1846, Court Minutes, cited on the Trinity House History blog, https://trinityhousehistory.wordpress.com/2014/10/08/on-this-day-in-trinity-house-history-7-october/.

Appendix 4.1

Mathematical Examiners to Trinity House

Name	Dates	Appointed	Other roles
Richard Norris	d. 1695?	19 Dec 1682	mariner, teacher
Lancelot Colson/ Coelson	1627–1687?	21 Mar 1682/3	medical practitioner, almanac writer
John Colson	d.1716	By 12 May 1692	teacher
Noor Colson	d.1738?	20 Apr 1716	teacher, mariner?
Thomas Weston	d.1728	22 Apr 1719	teacher
John Weston	d.1744	7 August 1728	teacher
James Rossam	d.1744	7 July 1744	teacher
William Mountaine	*c.*1705–1779	5 January 1744/5	teacher, Justice of the Peace, FRS
Gideon Fournier	*c.*1742–1813	5 June 1779	barrister, magistrate, FRS
Andrew Mackay	1760–1809	4 April 1805 (prov.) 4 July 1805 (conf.)	astronomical assistant, teacher
George Sanderson	d.1814	1 September 1809	tailor, teacher
John Crosley	1762–1817	6 January 1814	astronomical assistant, observer, computer
Patrick Kelly	1756–1842	6 November 1817	teacher

Source: Court and Board Minutes of Trinity House (MS 30004, 30010), plus other works cited previously.

Appendix 4.2

Mathematical Masters of Christ's Hospital, 1673–1800

Name	Dates	Years of appointment
John Leake (or Leeke)	unknown	1673–1677
Peter Perkins	? –1680	1678–1680
Robert Wood	1621?–1685	1681–1682
Edward Paget	1652?–1703	1682–1695
Samuel Newton	unknown	1695–1708
James Hodgson	1678?–1755	1709–1755
John Robertson	1707–1776	(Assistant Master from 1747) 1755–1755
James Dodson	*c.*1705–1757	1755–1757
Daniel Harris	? –1775	1757–1775
William Wales	1734–1798	1775–1798
William Dawes	1762–1836	1799–1800
Lawrence Gwynne	1772–1854	1800–1813

Source: J. I. Wilson, *A Brief History of Christ's Hospital*, London: John van Voorst 1842, pp. 55–6, plus other works cited previously and genealogical records.

5
What Mathematics for Portuguese Military Engineers?
From the Class of Fortification to the Military Academy of Lisbon

João Caramalho Domingues

What is meant by 'mathematics'?

This chapter explores the question of what mathematics was taught at the Class on Fortification and its successor institution, the Military Academy, in Lisbon in the seventeenth and eighteenth centuries. Here, and in most of the chapter, the word 'mathematics' is used in an anachronistic way: what is meant by the term is what we would nowadays recognize as mathematics, which in the context of these military schools constituted propaedeutic material. However, in the period under consideration the meaning of 'mathematics' was much wider, and contemporaries of these institutions would have classified all the subjects taught there as 'mathematical'. The main character playing a role in this chapter, Manuel de Azevedo Fortes, delivered, in 1739, an 'academical oration' in the presence of the king, João V, praising mathematics and his majesty's benevolence towards mathematics.[1] The parts of mathematics referred to in this oration included (roughly in order of appearance) geography, astronomy, gnomonics, hydrography, optics, dioptrics, perspective, civil architecture, arithmetic, and geometry. More systematically, in a book published in 1744, Azevedo Fortes divided mathematics, as was then usual, into 'pure mathematics' and 'mixed mathematics'; pure mathematics in turn was divided into arithmetic and geometry, while mixed mathematics, which was 'applied to the knowledge of natural things, called physics by the philosophers',[2] had several parts, of which the main ones were cosmography, geography, hydrography or nautics, mechanics, statics, optics, catoptrics, dioptrics, pyrotechnics or artillery, and military and civil architecture.[3] This distinction may allow us to say that in this chapter we will examine what pure mathematics, in the seventeenth- and eighteenth-century sense, was taught at

[1] Manoel de Azevedo Fortes, *Oraçaõ Academica, que pronunciou Manoel de Azevedo Fortes, na presença de Suas Magestadas, hindo a Academia ao Paço em 22 de Outubro de 1739*, no place: no publisher 1739. Also reprinted in Luís Manuel A. V. Bernardo, *O Projecto Cultural de Manuel de Azevedo Fortes*, Lisbon: Imprensa Nacional-Casa da Moeda 2005, pp. 239–45.

[2] 'A mathematica mixta, he aquella, que se aplica ao conhecimento das cousas naturaes, que os Filosofos chamaõ Fisica': Manoel de Azevedo Fortes, *Logica Racional, Geometrica, e Analitica*, Lisbon: Jozé Antonio Plates 1744, p. 2 (third pagination).

[3] Ibid.

the Class on Fortification or Military Academy. But we would be hard pressed to decide whether practical geometry was part of pure or mixed mathematics. Be that as it may, the reader should be warned of this deliberately anachronistic use of the term 'mathematics'.

Luís Serrão Pimentel and the Class on Fortification in the seventeenth century

After sixty years of Iberian union, Portugal restored its independence in 1640 and fought the Restoration War against Spain until 1668. This made it necessary to build modern fortifications along the border between the two countries. At first, foreign engineers were hired, particularly from France and the Low Countries.[4] But it being desirable, especially in the context of war, to have Portuguese engineers, a Class on Fortification and Military Architecture was set up in Lisbon, with Luís Serrão Pimentel (1613–1679) as its first teacher.[5] In 1673, Pimentel was appointed Chief Engineer of the Realm (*Engenheiro mor do Reino*), in recognition both of his many years of teaching 'mathematics, fortification, and castrametation' and of his service in the war.[6] In the army, Pimentel reached the rank of lieutenant general of artillery. In addition, Luís Serrão Pimentel also held the office of Chief Cosmographer of the Realm (*Cosmografo mor do Reino*),[7] which entailed teaching mathematics and navigation, examining prospective pilots, and other technical duties.[8]

Now, what mathematics associated with fortification did Luís Serrão Pimentel teach? There were no precise regulations, but there are some clues on what was taught. The most important piece of evidence is a book written by Pimentel, *Methodo Lusitanico de desenhar as Fortificaçoens* (Lusitanian method of designing fortifications), printed in the year after his death (Figure 5.1).[9] As the title indicates, the bulk of this

[4] Antónia Fialho Conde, 'Alentejo (Portugal) and the Scientific Expertise in Fortification in the Modern Period: the Circulation of Masters and Ideas', in: *The Circulation of Science and Technology: Proceedings of the 4th International Conference of the ESHS*, ed. Antoni Roca-Rosell, Barcelona: Societat Catalana d'Història de la Ciència i de la Tècnica 2012, pp. 246–52.

[5] It is usually stated that a school, called by historians *Aula de Fortificação e Arquitectura Militar*, was created in 1647, with Luís Serrão Pimentel being appointed immediately. However, there was probably never any formal creation of a school, but simply an order for Pimentel to teach fortification, and apparently this happened only in 1654: a royal document appointing Pimentel to 'teach [in Lisbon] the art of fortification' is dated 9 October 1654. See Christovam Ayres de Magalhães Sepulveda, *Historia Organica e Politica do Exercito Português: Provas*, 17 vols, Lisbon: Imprensa Nacional, Coimbra: Imprensa da Universidade 1902–1932, vol. VIII, p. 548.

[6] Sepulveda, *Historia do Exercito Português: Provas*, vol. VIII, pp. 531–4.

[7] Officially from 1671 onwards, but since 1641 he had often performed the duties of the office, replacing the official holder António de Mariz Carneiro on many occasions. See Nuno Alexandre Martins Ferreira, 'Luís Serrão Pimentel (1613–1679): Cosmógrafo Mor e Engenheiro Mor de Portugal', master's dissertation, Lisbon: Faculdade de Letras da Universidade de Lisboa 2009, p. 8.

[8] Ferreira, 'Luís Serrão Pimentel', pp. 49–65.

[9] Luís Serrão Pimentel, *Methodo Lusitanico de desenhar as Fortificaçoens das Praças Regulares, & Irregulares, Fortes de Campanha, e outras obras pertencentes á Architectura Militar*, Lisbon: Antonio Craesbeeck de Mello 1680.

Figure 5.1 Frontispiece of Luís Serrão Pimentel's *Methodo Lusitanico.*

work is a treatise on fortification.[10] In its preface, Pimentel claims that the book is organized in such an easy and practical manner that it is not necessary for soldiers to know geometry or arithmetic beyond multiplication and division in order to learn

[10] For more details on Pimentel's *Methodo Lusitanico*, see Antónia Fialho Conde and M. Rosa Massa-Esteve, 'Teaching Engineers in the Seventeenth Century: European Influences in Portugal', in: *Engineering Studies*, 10/2–3 (2018), pp. 115–32.

how to design all kinds of fortifications. However, some arithmetic, geometry, and trigonometry is included, although apparently only as optional material. The first, larger part of the book gives 'practical rules', in such a manner that it is not necessary to 'attend to the foundations of the rules'; in the second part, about half the size of the first, Pimentel gives 'the reason, or demonstration' for each rule he thought necessary to prove, this being for the benefit of 'theoreticians'.[11] For example, in the first part he teaches how to draw regular polygons up to 20 sides by using a pattern given in a figure;[12] in the second part he gives the construction of this pattern. This construction is by approximation and trial and error, but the second part also includes some proper mathematical proofs, citing Euclid and the practical geometries of Christoph Clavius and André Tacquet. There are also four appendices, two of which are mathematical. One is on 'practical plane trigonometry',[13] preceded by an introduction on decimal fractions.[14] Because of their extensive use in trigonometry, logarithms are also addressed, while several theorems from Euclid about triangles are enunciated, as well as some trigonometrical theorems (such as the law of sines), but without proofs. This constitutes a practical trigonometry and the emphasis is on how to solve triangles, with plenty of worked-out examples. The final sections (about one fifth of the appendix) are about calculating the areas of triangles and other polygons, and the volumes of some solids. The other appendix is a short summary of geometrical problems and theorems:[15] thirteen problems (geometrical constructions, but sometimes also with numerical analogues), followed by thirty-three theorems, without proofs, almost all of them from Books I, III, and VI of Euclid's *Elements*. Occasionally, Pimentel alludes to other mathematical notes he wrote (but which were not printed) and lectured upon: on spherical trigonometry[16] (probably related to his duties as Chief Cosmographer, rather than as Chief Engineer and teacher of fortification), on practical geometry,[17] and possibly on speculative, that is, theoretical geometry.[18]

In short, in answer to the question of what mathematics Luís Serrão Pimentel taught in his capacity as teacher of fortification, we can say with certainty decimal arithmetic, plane trigonometry, practical geometry, and possibly also theoretical geometry. Except for the last subject, all the mathematics taught had direct applications to fortification. And, apparently, all this was optional, for Pimentel tried to make the basics of fortification accessible to soldiers with only elementary notions of arithmetic operations.

Luís Serrão Pimentel's immediate successor as teacher of fortification was his son, Francisco Pimentel. Most likely, Francisco Pimentel did not introduce any noteworthy changes as far as content was concerned. Manuscript lecture notes up to the

[11] Pimentel, *Methodo Lusitanico*, 'Proemio', unpaginated.
[12] Pimentel, *Methodo Lusitanico*, p. 8.
[13] 'Trigonometria Practica Rectilinea', Pimentel, *Methodo Lusitanico*, pp. 547–644.
[14] 'Practica da Arithmetica Decimal, ou Dizima', Pimentel, *Methodo Lusitanico*, pp. 548–57.
[15] 'Compendio de alguns problemas da Geometria practica, & Theoremas da especulativa', Pimentel, *Methodo Lusitanico*, pp. 645–66.
[16] Pimentel, *Methodo Lusitanico*, p. 547.
[17] Pimentel, *Methodo Lusitanico*, pp. 650, 654.
[18] Pimentel, *Methodo Lusitanico*, p. 666.

early eighteenth century show a long-lasting influence of the elder Pimentel.[19] Significant changes would occur only slightly later, with the appointment of a new Chief Engineer, Manuel de Azevedo Fortes.

The Jesuits and fortification

The Society of Jesus had a dominant presence in Portuguese education from the late sixteenth century up to 1759. The Jesuit institution that most concerns us here is the College of Santo Antão in Lisbon, where public lectures in mathematics (in the wider sense outlined above), the so-called *Aula da Esfera* (Class on the sphere), were established around 1590.[20] According to the bibliographer Diogo Barbosa Machado, Luís Serrão Pimentel studied mathematics at the College of Santo Antão as well as with the then Chief Cosmographer, Valentim de Sá.[21] At first, the emphasis of lectures at the *Aula da Esfera* was on matters related to navigation. But, after 1640, in the context of the Restoration War, military subjects gained ground, including fortification.

During the seventeenth century, most teachers of mathematics in Portuguese Jesuit colleges, including Santo Antão, were foreigners. Their list overlaps with that of foreign engineers mentioned above. One such engineer was Jan Ciermans, a former pupil of Gregory of Saint-Vincent, who went to Portugal in 1641, taught at the *Aula da Esfera*, and built several fortifications, before dying in battle in 1648.[22] However, in 1692, the General of the Jesuits ordered a fundamental reform of mathematical studies in the Portuguese province, after complaints were made that they had reached a particularly low level. As a result of this reform a number of Portuguese Jesuits were trained in mathematics and in the eighteenth century (before their expulsion in 1759) almost all teachers at the *Aula da Esfera* were Portuguese.

Of course, the students at the *Aula da Esfera* were not only supposed to learn fortification. This was only one subject in a much broader mathematical curriculum. At least from the 1692 reform onwards they would certainly receive instruction in Euclid's *Elements*. Arithmetic and trigonometry were probably also always obligatory, as well as astronomy. In the 1730s, Manuel de Campos, a teacher at the *Aula da Esfera*, published a version of Euclid's *Elements* and a textbook on trigonometry which were specifically to be used there. At the end of the trigonometry text, he mentioned that applications would appear in two further treatises, one on the sphere (that is, on astronomy and geography) and the other on optics, but these were never published. And many other subjects, unrelated to fortification, could also be covered, although probably not on a fixed basis. For instance, Inácio Vieira left several manuscript treatises on a range of subjects such as

[19] Margarida Tavares da Conceição, 'A teoria nos textos portugueses sobre engenharia militar: o *Engenheiro Portuguez* e os tratados de fortificação', in: *Manoel de Azevedo Fortes (1660–1749): Cartografia, Cultura e Urbanismo*, ed. Mário Gonçalves Fernandes, Porto: Gedes 2006, pp. 35–55 at p. 41.

[20] Henrique Leitão, 'Jesuit Mathematical Practice in Portugal, 1540–1759', in *The New Science and Jesuit Science: Seventeenth Century Perspectives*, ed. Mordechai Feingold, Dordrecht: Springer 2003, pp. 229–47.

[21] Diogo Barbosa Machado, *Bibliotheca Lusitana*, vol. III, Lisbon: Ignacio Rodrigues 1752, p. 133.

[22] Leitão, 'Jesuit Mathematical Practice in Portugal', p. 237.

optics, perspective, astronomy, hydrography, and even chiromancy, that in all likelihood resulted from his teaching at the *Aula da Esfera* in the early eighteenth century.

Manuel de Azevedo Fortes

Manuel de Azevedo Fortes (1660–1749) is a very interesting figure in Portuguese intellectual history, who made important contributions to engineering, cartography, and philosophy (Figure 5.2). He was an early example of what Portuguese historiography calls *estrangeirados*: Europeanized intellectuals, aligned with the Enlightenment, who saw Portugal as being culturally backward and peripheral, and sought to modernize the country.[23]

Figure 5.2 Portrait of Manuel de Azevedo Fortes, from the frontispiece of the first volume of *O Engenheiro Portuguez*.

[23] Ana Carneiro, Ana Simões and Maria Paula Diogo, 'Enlightenment Science in Portugal: The *Estrangeirados* and their Communication Networks', in: *Social Studies of Science* 30/4 (2000), pp. 591–619.

Azevedo Fortes was born in Lisbon as an illegitimate child, and there are no clues about his mother other than that she was a 'reputable Portuguese lady'.[24] His father was most probably Jean Frémont d'Ablancourt (1625–1693), a French protestant who was chargé d'affaires in Lisbon between 1659 and 1664.[25] What is certain is that Azevedo Fortes had as benefactor a Frenchman who supported him and paid for his education which took place outside of Portugal from when he was around ten years of age. At that time, he went to Spain and studied at the Imperial College of Madrid, a Jesuit institution, before proceeding to study philosophy at the University of Alcalá de Henares. Afterwards, he went to Paris to study philosophy, theology, and mathematics at the Collège du Plessis. It appears that his studies in France had a deeper impact on him than those in Spain, for he became a Cartesian and remained so for the rest of his life. He spent some time in Italy, where he had a temporary position for six years teaching philosophy at the University of Siena.

When his position at Siena ended, his presumed father and benefactor having died, Azevedo Fortes, who was by now already in his thirties returned to Portugal. In 1696 he was appointed substitute teacher of fortification, filling in when necessary for Francisco Pimentel, and, in 1698, he was made captain engineer. He then embarked on a military career, taking part in the campaigns of the War of the Spanish Succession. He served as governor of the border town of Castelo de Vide for several years, and in 1735 reached a rank roughly equivalent to that of a modern brigadier.

Even as governor of Castelo de Vide, Azevedo Fortes kept some connection to the Class on Fortification in Lisbon; it is not clear when he became the official holder of the post of teacher there, but it must have been before 1712, for at this point he had his own substitute to do the actual teaching.[26] In 1719, Azevedo Fortes was appointed Chief Engineer of the Realm. As we will see, in the following years he did teach the Class on Fortification, now known as the Military Academy. He published part of his lectures in 1728–29 as *O Engenheiro Portuguez* (The Portuguese engineer);[27] the first volume of this work will be examined below.

Alongside his technico-military career, Azevedo Fortes moved in intellectual circles. He took part, in 1718–19, in learned meetings at the house of the count of Ericeira (known as the Portuguese Academy), where he lectured on modern logic in contrast with that of the ancients; and he was a founder member of the first official Portuguese learned society, the Royal Academy of Portuguese History, established in 1720. In this Academy he was charged with studying Portuguese geography, and specifically with map drawing. Moreover, Azevedo Fortes wrote specially for this purpose a 'Treatise on the easiest and most precise manner of making maps'.[28] This

[24] 'mulher Portugueza bem reputada': Jozé Gomes da Cruz, *Elogio Funebre de Manoel de Azevedo Fortes*, Lisbon: Jozé da Silva da Natividade 1754, p. 2.

[25] João Carlos Garcia, 'Manoel de Azevedo Fortes e os mapas da Academia Real da História Portuguesa, 1720–1736', in: *Manoel de Azevedo Fortes (1660–1749): Cartografia, Cultura e Urbanismo*, ed. Mário Gonçalves Fernandes, Porto: Gedes 2006, pp. 141–73; pp. 142–3.

[26] Dulcyene Maria Ribeiro, 'A formação dos engenheiros militares: Azevedo Fortes, Matemática e ensino da Engenharia Militar no século XVIII em Portugal e no Brasil', doctoral thesis, São Paulo: Faculdade de Educação da Universidade de São Paulo 2009, pp. 79–82.

[27] Manoel de Azevedo Fortes, *O Engenheiro Portuguez*, 2 vols, Western Lisbon: Manoel Fernandes da Costa 1728–9.

[28] Manoel de Azevedo Fortes, *Tratado do Modo o mais facil, e o mais exacto de Fazer as Cartas Geograficas*, Western Lisbon: Pascoal da Sylva 1722.

was intended to serve as a guide for teams of engineers that would cover the country producing accurate maps using modern techniques. But, much to his chagrin, the Royal Academy never showed any interest in this project, either because they did not wish to finance it or because many members of the Academy were old-fashioned intellectuals who were content with introducing unsystematic corrections to older maps.[29]

In his later years, Azevedo Fortes gave lessons in philosophy to the brother of the king, D. António, to whom he dedicated his last book, the *Logica Racional, Geometrica, e Analitica* (Rational, geometrical, and analytical logic).[30] The first part of this book, devoted to rational logic, is the first treatise of logic written in Portuguese rather than in Latin. It is mostly a Cartesian work, heavily influenced by the Port-Royal logic, although open, in particular aspects, to compromise with the work of other authors such as John Locke.[31] The second and third parts, which are mathematical, will be briefly discussed later.

The status of military engineers and the passage from Class on Fortification to Military Academy

Manuel de Azevedo Fortes was involved in a controversy over the status of military engineers with António do Couto de Castelo Branco, the author of a book on military topics (*Memorias Militares*).[32] In the first volume of the *Memorias*, published in 1719, Castelo Branco stated that engineering positions were less prestigious than other posts.[33] From his description of the duties of an engineer it appears that he himself had a very limited view of engineers, seeing them as sappers or a kind of manual labourers. Furthermore, Castelo Branco stuck to a very traditional conception that viewed military hierarchy as reflecting social origin and valour in arms, but not technical expertise.

Azevedo Fortes's views were diametrically opposed to these. In 1720, one year after being appointed Chief Engineer of the Realm, he published an address to the king with his proposals regarding engineers and military academies.[34] He was concerned about the low prestige of engineers and complained of Castelo Branco's ideas, one of the consequences of which was that too few engineers were trained and that the applicants were often not well suited.[35] He proposed reviving a royal decree from 1701 which purportedly created three classes on engineering in Portuguese provinces

[29] Garcia, 'Manoel de Azevedo Fortes e os mapas'.

[30] Azevedo Fortes, *Logica*.

[31] Bernardo, *Projecto Cultural*.

[32] The following account of this controversy and of Azevedo Fortes' views on the role of engineers in the military is indebted to Renata de Araújo, 'Manoel de Azevedo Fortes e o estatuto dos Engenheiros Portugueses', in: *Manoel de Azevedo Fortes (1660–1749): Cartografia, Cultura e Urbanismo*, ed. Mário Gonçalves Fernandes, Porto: Gedes 2006, pp. 15–34.

[33] 'Estes postos tem menos reputação que os mais', cited in Araújo, 'Manoel de Azevedo Fortes e o estatuto', p. 19.

[34] *Representação feita a S. Majestade que Deus guarde [...] Sobre a forma e direcção, que devem ter os Engenheiros [...]*, reprinted in Bernardo, *Projecto Cultural*, p. 231–8.

[35] There was also another, more mundane, aspect to this low prestige: engineers often received a lower pay than other officers of similar ranks. Azevedo Fortes, after becoming Chief Engineer, managed to improve this situation. Ribeiro, '*A formação dos engenheiros militares*', pp. 57–8.

bordering Spain, provided for lectures on fortification and artillery for soldiers who were not destined to become engineers, and also established a preference in promotion to be given to those who had attended these lectures and proved their worth to the teacher. However, these initiatives had been largely ineffectual. In particular, only one of the three new classes was actually established, and even that had very little success.

Azevedo Fortes insisted that those proposed classes be created outside Lisbon, and especially that men who had studied in military academies be preferred when it came to promotion. This was not only a matter of elevating the prestige of engineers by making this career more attractive, but also of improving the competence of all officers: all cadets and officers should in his view be encouraged to attend those lectures, which should cover not only fortification but 'all that pertains to war', including military evolutions (movements of troops), geometry, attack and defence of fortified places, castrametation, and artillery. According to Azevedo Fortes's proposal, ultimately almost all officers of infantry and artillery should be engineers by training, even if only a few actually became engineers by profession. This approach was associated with an idea of engineer training that went beyond the mere teaching of fortification, but included almost all military doctrines. It also meant that promotions should be associated with merit rather than social background, although Azevedo Fortes took the cautious approach of encouraging nobles to acquire the desirable technical expertise.

A detail that probably reflected Azevedo Fortes's project was the adoption of the name Military Academy. It should be pointed out that there was never an official change of name, as there had not been a fixed official name for the Class on Fortification in the first place, but the most commonly used appellation changed around 1720. Dulcyene Ribeiro, who has consulted many manuscript documents of this period such as those detailing the appointment of engineers, noticed that before 1720 the expressions used revolved around *Aula de Fortificação* (Class on Fortification). But after that year the name *Academia Militar* (Military Academy) becomes dominant.[36] It seems that the first occurrence of 'Military Academy' was precisely in the address to João V discussed above. This is the expression used also, for instance, in the 1732 decree mentioned below. 'Academy' was probably chosen because it sounded both more distinguished and more institutional than 'Class'. On the other hand, 'Military Academy' suggests a broader scope of instruction than the expression 'Class on Fortification'.

The controversy with Castelo Branco continued for many years. In the second volume of *Memorias Militares*, published in 1731, Castelo Branco mentioned Azevedo Fortes's complaint about his disregard for engineers, but he reaffirmed his earlier opinion. He saw his position as being reinforced by the fact that Azevedo Fortes himself had not had success in reviving the 1701 decree on the creation of engineering classes. In 1732, finally, a new royal decree was published renewing the command to establish the classes planned in 1701, now promoted to 'military academies' and providing for promotion of those officers and soldiers who made 'special progress' in these military academies. The delay from 1720 to 1732 shows

[36] Ribeiro, 'A formação dos engenheiros militares', pp. 99–100.

that there was resistance to Azevedo Fortes's project (as we will see below, the text of the decree was ready in 1729, but it was signed only in 1732). In 1733, Azevedo Fortes's final reply to Castelo Branco appeared. It took the form of a 271-page book under the title *Evidencia Apologetica e Critica sobre o primeyro, e segundo Tomo das* Memorias Militares (Apologetic and critical demonstration on the first and second volumes of *Memorias Militares*), and although ostensibly written by the students of the Lisbon Military Academy, it was possibly penned by Azevedo Fortes himself. In the *Evidencia Apologetica* many passages from *Memorias Militares* were heavily criticized, even ridiculed, such as numerous mistakes in translation from the French. After the 1732 decree, it would appear that Azevedo Fortes was feeling triumphant.

Mathematics for the Military Academy in Azevedo Fortes's books: practical geometry and trigonometry

In this and the following section we will examine the mathematics present in Manuel Azevedo Fortes's teaching at the Military Academy. As we will see, he did not compose any mathematics texts himself, but instead produced translations with only minor adaptations from French books. However, what is interesting is the choice of authors and material, as well as the order in which the material was presented, for all these things reflect the kind of mathematics Azevedo Fortes saw as being fit for instruction at the Military Academy.

The most obvious and most accessible source for the mathematics in Azevedo Fortes's teaching programme at the Military Academy is his book *O Engenheiro Portuguez*, which explicitly derives from the notes of lectures he delivered there. However, as we shall see below, it only covers part of the mathematics that Azevedo Fortes actually taught.

The first of the two volumes of *O Engenheiro Portuguez*, published in 1728, is presented as a treatise on practical geometry, as 'the science that teaches how to divide and measure continuous quantity according to its extension'.[37] It contains three books, namely on longimetry (the measure of lengths), planimetry (the measure of areas), and stereometry (the measure of volumes), these being the standard main parts of practical geometry.[38] There is also an appendix on plane trigonometry. For most of the time in those three books Azevedo Fortes follows closely a seventeenth-century French handbook, *La Geometrie Pratique* by Jean Boulenger, with notes by the prolific mathematical author, Jacques Ozanam.[39]

The book on longimetry actually begins with some definitions (accessible and inaccessible lines, point, line, and so on) that are absent from Boulenger's book, followed

[37] 'a sciencia, que ensina a dividir, e medir a quantidade continua segundo a sua extençaõ': Azevedo Fortes, *O Engenheiro Portuguez*, vol. I, p. 2.

[38] Dominique Raynaud, 'Introduction', in: *Géométrie pratique: Géomètres, ingénieurs et architectes, XVIe—XVIIIe siècle*, ed. Dominique Raynaud, Besançon: Presses universitaires de Franche-Comté 2015, pp. 9–20 at p. 15.

[39] Jean Boulenger, *La Geometrie Pratique*, new edition with notes and a 'Traité de l'Arithmetique par Geometrie' by Jacques Ozanam, Paris: Michel David 1691.

by a short discussion of Portuguese units of measure. Azevedo Fortes would later return to his preference for the Portuguese span (*palmo*) over the foot (*pé*), which Portuguese engineers had adopted under French influence. The reason for this was partly patriotic and partly mathematical: 1 fathom (*braça*) was precisely 10 spans, and he felt this bit of decimal measure ought to be put to good use.[40] This is followed by an introduction to decimal arithmetic, taken from Boulenger. Boulenger strongly believed, and Azevedo Fortes clearly agreed, in operating with decimal fractions and converting only in the end to common units of measure, which usually were not decimal. This conversion could be performed using the rule of three, but Boulenger and Azevedo Fortes recommended a simple graphical converter. Naturally, this gave only an approximate result, but that was not of much concern for practical purposes; it was enough that the errors did not bring a noticeable loss to anyone.[41] Next came a chapter on geometrical constructions, such as to draw a straight line, to draw a perpendicular, or to find the centre of a circle, etc. Azevedo Fortes observed that some of these constructions had been more fully explained in his treatise on maps. In this chapter we also first see how much Boulenger, and even more so Azevedo Fortes, have in mind operations relating to the physical terrain. To bisect a line, Boulenger proposes the classical Euclidean construction, explaining that on the physical terrain instead of using a compass one uses a string with pegs. Ozanam, in a note, observes that the Euclidean construction may be too difficult depending on the terrain, and so proposes extending the string along the line and taking its half by joining its extremities. Azevedo Fortes cites only Ozanam's note. This book then finishes with the measurement of lines properly speaking.

The book on planimetry again starts with some definitions (plane surface, circle, ellipse, etc.) that are absent from Boulenger's book. But Azevedo Fortes then follows Boulenger and Ozanam in the calculation of areas of triangles, squares, rectangles, and so on, up to a segment of a parabola, or the surface of a sphere. Proofs are presented just as quick arguments, often relying on and simply referencing Euclid's *Elements* or works of Archimedes. Azevedo Fortes then adds three chapters on subjects not addressed by Boulenger. The first is on the division of lands and the second contains a critical report on how fields were customarily measured in Portugal. Azevedo Fortes points out that the division of lands belongs more to measurers, that is, land surveyors, than to engineers, but he argues that any practical geometer should have knowledge of it.[42] We should note that the royal decree of 1732, mentioned in the previous section, called for measurers of civil works to be instructed in that part of practical geometry in military academies. A summary of this decree appears in the second volume of the *Engenheiro Portuguez*, including the provision about civil measurers.[43] That decree was ready and awaited only the royal signature. The final chapter deals with the transformation of areas, for example the construction of a square equal to a given circle, but does not contain demonstrations.

[40] Azevedo Fortes, *O Engenheiro Portuguez*, vol. II, p. 54.
[41] Azevedo Fortes, *O Engenheiro Portuguez*, vol. I, pp. 9, 50.
[42] Azevedo Fortes, *O Engenheiro Portuguez*, vol. I, p. 144.
[43] Azevedo Fortes, *O Engenheiro Portuguez*, vol. II, pp. 433–4.

The book on stereometry again starts with definitions absent from Boulenger's book, of the pyramid, prism, polyhedron, and so on, but proceeds to deal with the calculation of volumes, up to and including the volumes of ramparts or walls. At that point Boulenger's text finishes, but Azevedo Fortes continues with other topics such as the customary measurements of civil works, the construction of regular solids (including some we would not classify as 'regular', such as a pyramid with square base), and a long chapter on instruments used by engineers, both on paper and on physical terrain (Figure 5.3). The final chapter is devoted to the drawing of military maps.

For the appendix on plane trigonometry, Azevedo Fortes follows part of a treatise on trigonometry by Jacques Ozanam, originally published together with a revised version of Adriaan Vlacq's trigonometrical and logarithmic tables.[44] It starts with basic definitions (arc of circle, degree, complement and supplement of an arc, chord, right sine and versed sine, and so on), followed by nine theorems that establish relations between trigonometrical quantities (the first is the classical statement that the chord of an arc is double the sine of half the arc, while the second is the Pythagorean trigonometric identity). The next section explains how to construct trigonometric tables, using those relations together with results from Euclid's *Elements* (Books IV and XIII) on the sides of regular polygons. Next comes a section on logarithms, which as usual at the time are defined as numbers in an arithmetic progression corresponding to others in a geometric progression—base 10 is adopted. After some propositions, a calculation of the logarithm of 9, with 7 decimal places and requiring 26 iterations, is given as an example for the construction of tables. Subsequently, several simple problems are solved using tables, both trigonometrical and logarithmic. The final section is about solving triangles and comprises five theorems, including, as would be expected, the law of sines. Additionally, there are four problems, only one of which is in Ozanam's text, and logarithms are employed to simplify the calculations.

The second volume of *O Engenheiro Portuguez*, published in 1729, is a treatise on fortification, and includes an appendix on weapons of war. We will not dwell on this volume, but only note a short passage that is relevant for the mathematical curriculum. Azevedo Fortes included a chapter 'on the accomplishments that engineers must possess',[45] where he states what in his opinion it is indispensable for an engineer to know:

> Arithmetic, Euclid's Elements, Practical Geometry, Trigonometry, fortification, attack and defence of fortified places; the use of Mathematical instruments belonging to his profession; the method to make topographical plans and maps, with their profiles, elevations, and fronts, and how to draw them; and he should not ignore artillery.[46]

[44] Jacques Ozanam, *La Trigonometrie rectiligne et spherique*, Paris: Claude Jombert 1720, pp. 1–59.

[45] 'Das partes, que devem ter os Engenheiros': Azevedo Fortes, *O Engenheiro Portuguez*, vol. II, pp. 423–38.

[46] 'a Arithmetica, os Elementos de Euclides, a Geometria Pratica, a Trigonometria, a fortificaçaõ, ataque, e defença das Praças; o uso dos instrumentos da Mathematica pertencentes à sua proficaõ; o methodo de tirar as plantas, e cartas topographicas com seus perfis, elevações, e fachadas, e o modo de as desenhar; e naõ deve ignorar a Artilheria': Azevedo Fortes, *O Engenheiro Portuguez*, vol. II, pp. 427–8.

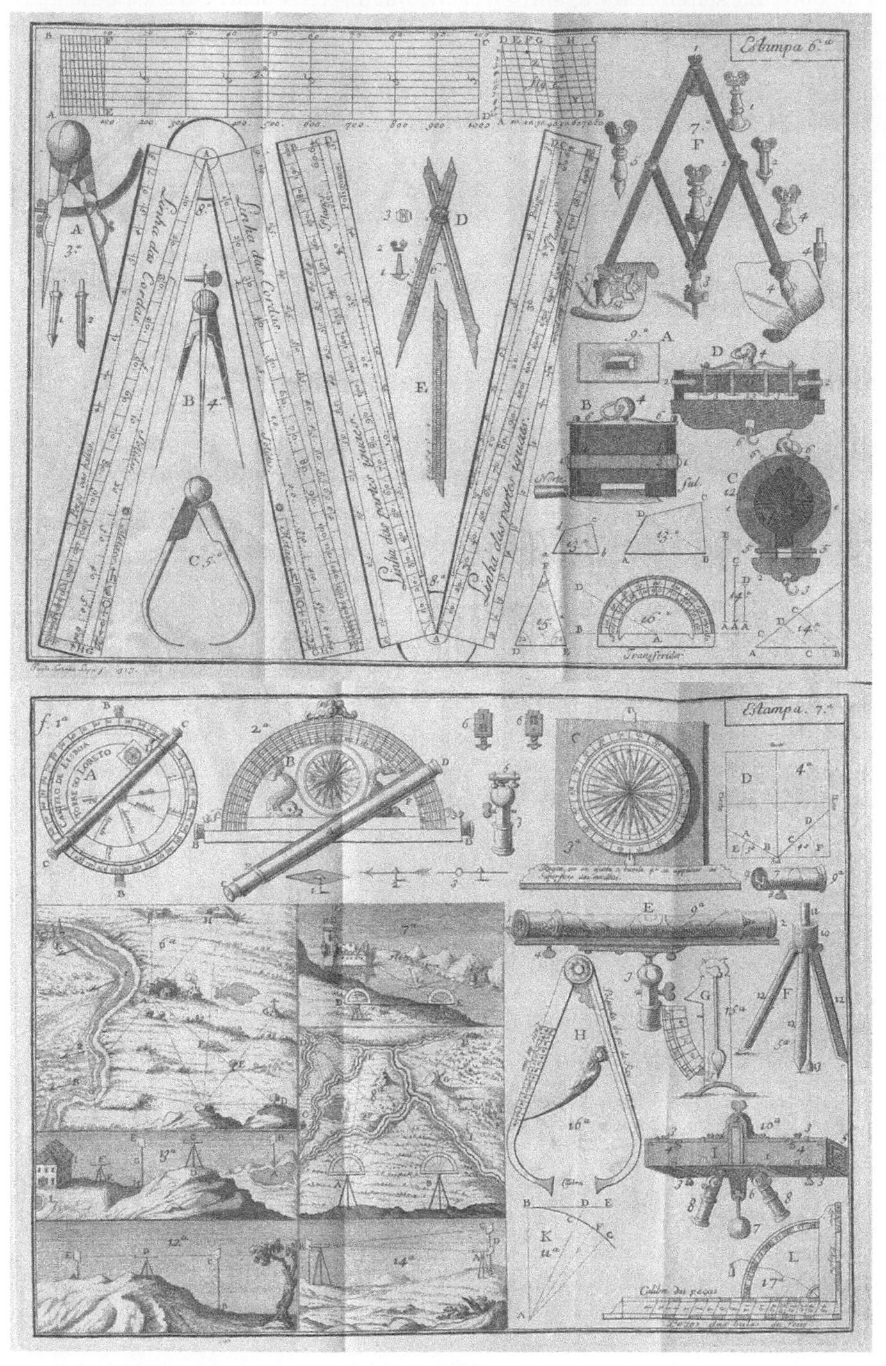

Figure 5.3 Plates on the use of instruments by engineers from the first volume of *O Engenheiro Portuguez*.

It is possible that basic arithmetic, in integers, was a pre-requisite to enter the Military Academy, and as we saw decimal arithmetic was covered in *O Engenheiro Portuguez*. However, this quotation indicates that Euclid's *Elements* was taught at the Military Academy despite being absent from that work. As we shall see, even more pure mathematics, further advanced arithmetic, and algebra were taught at the Military Academy, even though they were not considered indispensable.

It must be mentioned that Azevedo Fortes published another book with substantial mathematical content: the *Logica Racional, Geometrica, e Analitica* (already mentioned above), which came out in 1744. While the first part of this work is a text on philosophy, namely logic, the second and third parts on geometrical and analytical logic respectively are incomplete translations of two textbooks by the French Oratorian priest Bernard Lamy, one devoted to geometry and the other to arithmetic and algebra. In the foreword to the *Logica*, Azevedo Fortes says he is writing for everyone, wishing the work to be useful to the nation as a whole. While it is true he addresses specifically engineering officers, for whom he argues knowledge of logic is very necessary, these officers would probably already have completed the course of the Military Academy. But when he justifies himself writing in Portuguese rather than in Latin, he mentions having thought not only of military officers but also of Portuguese ladies.[47] Moreover, the second part of the *Logica*, on geometry, is entirely different from the introduction to geometry that was taught at the Military Academy in the 1720s: the former is a translation of Lamy's Cartesian *Elemens de geometrie* (1685), while the latter was a version of Euclid's *Elements* (see below). The third part of the *Logica* is an abridged translation of another textbook by Lamy that, in a complete translation, was taught at the Military Academy (again, see below); however, this (partial) connection between the third part of the *Logica* and the teaching at the Military Academy is never made explicit, and can be inferred only by comparison with manuscript lecture notes.

Mathematics for the Military Academy in Azevedo Fortes's manuscript lecture notes: Euclid's *Elements* and 'magnitude in general'

The analysis of a few contemporary manuscripts will allow us to complete the picture of the mathematics taught by Manuel Azevedo Fortes at the Military Academy.

We will first look at a manuscript kept at the public library of Évora, containing three texts, and which has obvious points of contact with *O Engenheiro Portuguez*.[48] Its title page very clearly gives the author[49] as being Manuel Azevedo Fortes and the date as 1724 (see Figure 5.4). The first text in this manuscript, and by far the longest, is a version of Euclid's *Elements*, more precisely of Books I to VI, XI, and XII, a very common selection of Euclidean content. It is essentially a translation of Jacques Ozanam's

[47] Azevedo Fortes, *Logica*, 'Antiloquio', unpaginated.

[48] Manoel de Azevedo Fortes, *Geometria Espiculativa; Trigonometria Espherica; Modo de riscar e dar aguadas nas plantas melitares*, Biblioteca Pública de Évora, cód. Manizola 258.

[49] Author of the content; the physical author of the manuscript was probably a student.

Figure 5.4 Title page of Manuel de Azevedo Fortes' manuscript *Geometria Espiculativa; Trigonometria Espherica; Modo de riscar e dar aguadas nas plantas melitares.*

version, published in the first volume of his *Cours de Mathematique*,[50] which in turn is based on the version by the French Jesuit Claude François Milliet Dechales.[51] These are liberally edited versions, substituting some of Euclid's proofs and omitting those propositions that were deemed unnecessary, while keeping the original numbering. Thus, in Book I, Proposition 8 follows on directly from Proposition 6, because Proposition 7 has been omitted, the reason for this omission being that an alternative proof of Proposition 8 makes it unnecessary.

In the seventeenth and eighteenth centuries the theory of parallel lines and the theory of proportions were often rewritten, and in various ways, as their Euclidean variants were regarded as being not clear enough. The case of Dechales, Ozanam, and

[50] Jacques Ozanam, *Cours de Mathematique*, vol. I, Paris: Jean Jombert 1693.

[51] Claude François Millet Dechalles, *Les Elemens d'Euclide*, new edition, Paris: Estienne Michallet 1690. Ozanam also published a revised and augmented edition of Dechalles's version as a free-standing volume, but that is distinct from the version included in his *Cours*.

Azevedo Fortes provides no exception to this rule. In their versions of the *Elements*, the theory of parallel lines is based on an alternative axiom, namely that all perpendiculars (that is, all perpendicular segments) between two parallel straight lines are equal. Both Dechales and Ozanam, and likewise Azevedo Fortes, state Euclid's original axiom, now known as Euclid's parallel postulate, but they express the opinion that it is not evident enough and instead use this alternative axiom based on equal perpendiculars to prove the lemma that a perpendicular to one of two parallels is also perpendicular to the other one.[52] They then use the axiom together with this lemma to prove Proposition I, 29, which as is well known is the first proposition in Euclid's *Elements* that requires the parallel postulate. Later they use Proposition I, 34 (on the equality of opposing sides and angles of a parallelogram) to prove the statement of Euclid's parallel postulate.

More radical changes occur in Book V on the theory of proportions.[53] Euclid's definition of sameness of ratios for general magnitudes, based on equimultiples, was often regarded as too complicated; the corresponding definition for numbers in Book VII (numbers are proportional when the first is the same multiple, or the same part, of the second that the third is of the fourth) seemed much simpler. A common trend was to try to adapt the latter definition to general magnitudes. Apart from its simplicity, this definition was also coherent with the tendency for algebraization of mathematics—a tendency that blurred the classical Greek distinction between magnitudes and numbers. Both Ozanam and Azevedo Fortes followed an approach that went back to the even more liberal version of the *Elements* by the aforementioned Flemish Jesuit Tacquet.[54] Azevedo Fortes defined equal or similar ratios as 'those where antecedents [equally] contain, or are equally contained in their consequents'.[55] This is a partial translation of Ozanam's version of Tacquet's definition, but Ozanam helpfully also explains that it is the same as saying that 'the antecedent of one ratio contains as many times whatever aliquot part of its consequent, as the antecedent of the other ratio contains a similar aliquot part of its consequent'.[56] However, Azevedo Fortes also uses this characterization in terms of aliquot parts in proofs, similarly to Ozanam and Tacquet. Because Propositions V, 1 to V, 6 are auxiliary results specific to the method of equimultiples, Tacquet, Ozanam, and Azevedo Fortes all omit them. Ozanam and Azevedo Fortes also omit Propositions V, 19, V, 20, and V, 21, which are deemed to be 'useless'.[57]

A particular trait of Ozanam's version of Euclid's *Elements*, also present in Azevedo Fortes's translation, is the systematic indication of the uses for each proposition. Sometimes the only use is in proving other propositions, but sometimes there are

[52] Dechalles claims to use the definition of parallel lines, rather than the axiom, but that is because his wording of the definition actually assumes the axiom.

[53] What follows is just a very short summary. A full comparison between Azevedo Fortes, Ozanam, and other authors on the theory of proportions was made by Maria Elisabete Barbosa Ferreira, 'Teoria(s) de Proporções em Portugal na Primeira Metade do Século XVIII', master's dissertation, Braga: Escola de Ciências da Universidade do Minho 2013.

[54] Andreas Tacquet, *Elementa geometriæ planæ ac solidæ*, 2nd edn, Antwerp: Jacob van Meurs 1665.

[55] 'Razois iguais, ou semelhantes saõ aquellas em q.[e] os antecedentes conthem, ou saõ igualm.[te] contheudos nos seus consequentes', Azevedo Fortes, *Geometria Espiculativa*, f. 114.

[56] 'l'Antecedent d'une Raison contient autant de fois quelque partie aliquote que ce soit de son Consequent, que l'Antecedent de l'autre Raison contient une semblable partie aliquote de son Consequent', Ozanam, *Cours*, I, p. 268 (second pagination).

[57] Tacquet includes Proposition 19, but omits other propositions.

practical uses as well. A simple example is provided by Proposition I, 13 which asserts that if a straight line falls upon another, these lines make either two right angles or angles equal to two right angles. It is used to prove the next proposition and several others, but it can also be used to measure an inaccessible angle such as the interior angle *ABC* of a wall, by measuring the exterior angle *CBD* and subtracting it from 180° (see Figure 5.5).[58]

There is one aspect of Azevedo Fortes's version of the *Elements* that does not derive from Ozanam's. Azevedo Fortes includes an appendix to Book III on the angle of contact (the angle between a circle and a tangent straight line), which is not present in Ozanam's edition. As is well known, there had been controversies about the status of such 'horn angles', which are less than any rectilinear angle and thus, if accepted as magnitudes, would seem to contradict the Eudoxus–Archimedes principle. In this appendix, Azevedo Fortes presents the arguments put forward by Galileo and his disciple Vincenzo Viviani maintaining that the angle of contact is not truly an angle. Galileo had in fact given three arguments, two of them based on the consideration of the circle as a regular polygon with infinite sides and the other on a sort of intermediate value property, to the effect that the angle of contact is null, and not divisible. Viviani for his part had questioned whether Euclid had truly considered non-rectilinear angles, suggesting that the few references to them in the *Elements* might be interpolations.[59]

The second text in the Évora manuscript is presented as a treatise on trigonometry and practical geometry, but in fact it includes only trigonometry. It is probable that it was to be continued but that there was insufficient space for the remaining text. The first book is on plane trigonometry and is almost identical to the appendix that would be published four years later in the first volume of *O Engenheiro Portuguez*. However, unlike what Azevedo Fortes would do there, in this manuscript he also includes a book on spherical trigonometry. Almost half of this is taken up with a

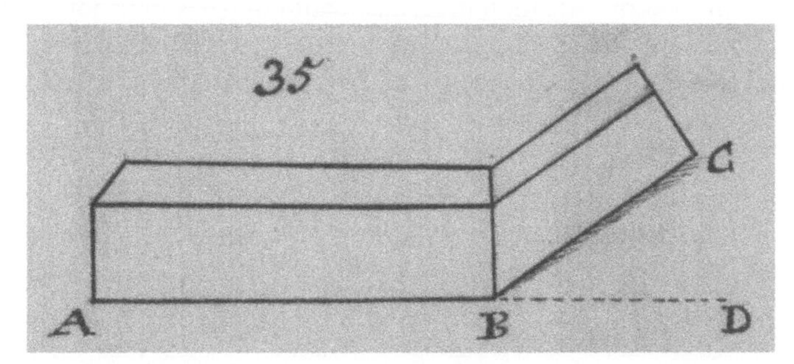

Figure 5.5 Diagram for the 'use' of Proposition I, 13 from Azevedo Fortes, *Geometria Espiculativa*.

[58] Azevedo Fortes, *Geometria Espiculativa*, ff. 20–21.

[59] This appendix has been reported upon by Catarina Alexandra Pereira Mota, 'A história do conceito de reta tangente em Portugal: um estudo desde o século XVIII até à matemática moderna', doctoral thesis, Braga: Escola de Ciências da Universidade do Minho 2018, pp. 95–101.

summary on the sphere in the geographical–astronomical sense. He sets out that there are three sorts of sphere: the celestial sphere, the terrestrial sphere, and the armillary sphere. The third represents both the first and the second, and is recommended as a tool for readers to use. This part is followed, after he has outlined some principles of spherical geometry, by a translation of the section on spherical trigonometry in the treatise by Ozanam from which Azevedo Fortes had already drawn his material on plane trigonometry.[60] Ozanam finishes with some problems that apply spherical trigonometry to astronomy, but Azevedo Fortes does not include these. The result is a very theoretical exposition of spherical trigonometry, with seventeen theorems but no problems. The final text in this manuscript is a short treatise on drawing and painting military architectural plans.

A noteworthy detail in the Évora manuscript is the fact that Euclid's *Elements* is presented as the 'second treatise' and the treatise on trigonometry (and purportedly on practical geometry) as the 'third treatise'.[61] This begs the obvious question: what was the first treatise which was to come before Euclid's *Elements*? The answer appears in the first sentence of Azevedo Fortes's short prologue to the *Elements*:

> Although in the treatise of magnitude in general we have given the elements for all sciences, and have treated much of what Euclid teaches in his books of Elements, for the good understanding of geometry, both speculative and practical, let us seek to explain in a clear, easy, and brief manner the first six books of Euclid's Elements, 11th and the 12th [...].[62]

So, what was this 'treatise of magnitude in general', which opened the mathematical course of the Military Academy? Its text can be found in three manuscripts kept in the Portuguese National Library in Lisbon, bearing the title *Elementos das Mathematicas, ou Tractado da grandeza em geral* (*Elements of mathematics, or treatise of magnitude in general*).[63] It is a translation of a textbook by Bernard Lamy entitled *Elemens des Mathematiques ou Traité de la Grandeur en general*.[64]

[60] Jacques Ozanam, *La Trigonometrie*, pp. 60–91.

[61] The much smaller final treatise on military architectural plans is not numbered.

[62] 'Ainda, q[e] no tratado da grandeza em geral temos dado os Elementos p.[a] todas as sciencias, e tratamos m.[ta] parte do q[e] Euclides ensina nos seus Livros dos Elementos, com tudo p.[a] a boa intelligencia da Geometria assim especulativa, como pratica, procuremos explicar de hum modo claro, facil, e breve, os seis primr.[os] L.[os] dos Elementos de Euclides, e o 11, e 12 [...]': Azevedo Fortes, *Geometria Espiculativa*, f. 1.

[63] [Manoel de Azevedo Fortes], *Elementos das Mathematicas, ou Tractado da grandeza em geral*, Biblioteca Nacional de Portugal, cód. 1861; cód. 5194//1; cód. 6205//17 (incomplete). That this is the text taught by Azevedo Fortes is not so obvious as in the case of the Évora manuscript, but it is a safe conclusion: cod. 5194 contains two texts, the first being a copy of this treatise and the second a copy of the same text of Euclid's *Elements* found in the Évora manuscript; in this codex, a title page for Euclid's *Elements* contains the indication 'dictated by [name crossed out, replaced with Joze Sanches da Silva]' and the date 1739, clearly altered from 1722; it is practically certain that the original name was Manoel de Azevedo Fortes, who taught/dictated this in 1722, that José Sanches da Silva was among the listeners and produced this manuscript, and that he reused it in 1739, when he had to teach the same subject.

[64] Bernard Lamy, *Traité de la Grandeur en general*, Paris: André Palard 1680; later editions with the title *Elemens des Mathematiques ou Traité de la Grandeur en* general, 2nd edn, Paris: André Palard 1689; 3rd edn, Florentin Delaulne 1704; 4th edn, Paris: Nicolas Pepie 1715.

Philosophically, Lamy was a Cartesian with connections both to Pierre Malebranche and to the Port-Royal group. Besides this tract on magnitude in general he wrote the aforementioned *Elemens de geometrie* (1685), a textbook on geometry which instead of following Euclid followed a Cartesian 'natural order' from the simple to the complex.[65] Other publications included a treatise on mechanics, another on perspective, and several works on philosophy, biblical studies, and morality.[66]

Lamy's treatise of magnitude in general comes from a particular seventeenth-century Cartesian tradition that promoted *mathesis universalis*, understood as the science of magnitude in general. Under this concept was understood all that could increase or decrease regardless of being continuous or discrete magnitudes or quantities. *Mathesis universalis* was considered to be more fundamental than geometry or arithmetic, and was therefore deemed the proper introduction to mathematics. It was also usually identified with symbolic algebra or, to give it a more historically correct name, specious algebra, invented by François Viète in the late sixteenth century. Since algebra taught how to operate with quantities represented by letters, it was possible to regard these as abstract, or general, magnitudes.[67]

The preface to Azevedo Fortes's version begins thus:

> Having to begin a short course on the parts of mathematics most needed for a man of war, it seemed to me necessary to start with the elements of this noble science, and [...] although commonly it is Euclid who is followed, especially in the Spains [*sc.* in the Iberian Peninsula], I believe father Lamy [...] excels in order, method, and clarity all those who wrote until his time. His treatise of magnitude in general is capable of disposing, developing, and perfecting the understanding [...] and it may be called the Logic of Mathematics.[68]

He proceeds to translate or paraphrase some parts of Lamy's own preface, but omitting the Oratorian priest's apology of mathematics and religious arguments. Euclid's *Elements* was disapproved of as an introduction to mathematics or sciences for two reasons. The first is that Euclid only treated a particular kind of magnitude, namely that of geometrical bodies. The properties of those bodies were 'certainly more composite and more difficult to know than those of magnitude in general'.[69] Furthermore,

[65] On this 'natural order', see Évelyne Barbin, 'On French heritage of Cartesian geometry in Elements from Arnauld, Lamy and Lacroix', in: *'Dig where you stand' 5. Proceedings of the fifth International Conference on the History of Mathematics Education*, ed. Kristín Bjarnadóttir et al., Utrecht: Freudenthal Institute 2015, pp. 11–28.

[66] François Girbal, *Bernard Lamy (1640–1715): étude biographique et bibliographique*, Paris: Presses universitaires de France 1964.

[67] '"Mathesis universalis" as the discipline of symbolic algebra among other mathematicians': Chikara Sasaki, *Descartes's Mathematical Thought*, Dordrecht: Springer 2003, pp. 394–8.

[68] 'Havendo de dar principio a hũ curso compendiozo das p.^{tes} da Math. mais necessr.^{as} a hũ homem de guerra, me paresseo precizo começar pl.^{os} elem.^{os} desta nobre sciencia, e [...] ahinda q^{e} mais ordinr.^{a} m.^{e} se segue a Euclides, principal m.^{e} nas Hespanhas; acho porém q^{e} o P.^{e} Lamy [...] excede em ordem, meth., e clareza a todos os q^{e} escreveraõ athe o seu tempo. O seu tract. da grand.^{a} em g.^{l} he capaz de dispor, formar, e aperfeiçoar o entendim.^{o}, [...] e se póde chamar [...] a Logica das Mathematicas': [Azevedo Fortes], *Elementos*, cód. 1861, f. 1–1^{v}.

[69] 'cujas propried.^{es} saõ certa m.^{e} mais compostas e mais deficeis de conhecer do q^{e} as da grand.^{a} em g.^{al}': [Azevedo Fortes], *Elementos*, cód. 1861, f. 1^{v}.

it was 'very important for those who begin the study of mathematics to accustom themselves to make use of pure understanding, without intervention from the senses or the imagination which are the cause of many errors'.[70]

The second reason is that Euclid, like the other ancient geometers, had not kept a 'natural order', but instead had mixed different matters together. Of course, the ancients had stuck to the deductive order, as Lamy explained in a passage omitted by Azevedo Fortes.[71] But Lamy had sought to lead his readers from what they knew to what they did not know, keeping different matters apart, and striving to take the shortest paths. In other words, he pursued a Cartesian and analytic order instead of a synthetic one.[72]

As far as content was concerned, the title page of Lamy's *Elemens* indicates that it 'comprises arithmetic, algebra, and analysis' and this was a fair description, analysis here being understood as a reference to the analytic method. Both Lamy's original work (from the third edition onwards) and Azevedo Fortes's version of it are divided into seven books plus four shorter appendices, the latter in Lamy forming actually an eighth book.

The first book begins with some opening remarks on the nature of magnitude and the distinction between continuous and discrete magnitudes, the latter having separate parts. Since we can conceive 'at least in thought the parts of continuous quantity as distinct and separate',[73] discrete magnitude is held by Azevedo Fortes to be more general so that what is taught about discrete quantity or numbers also applies to continuous quantity. These remarks are followed by a short chapter on writing numbers and by a list of nine general principles or axioms such as 'the whole is greater than a part', or 'two magnitudes equal to a third are equal to one another'. There is a section on the four basic arithmetic operations, and another on the same operations on magnitudes noted by letters of the alphabet. For instance, the product of multiplying $B + D$ by $X - Z$ is $XB + XD - ZB - ZD$, whereby such 'arithmetic with letters' is called 'algebra'.

The second 'book' addresses powers and extraction of roots, but includes also a final section on permutations, with repetition (called 'combinations') and without repetition (called 'changes of order').

The third and fourth books are on ratios. In the third book Azevedo Fortes introduces ratios and geometrical proportions, and differences and arithmetical proportions. Clearly, Lamy wished to give a clear idea of ratios, while at the same time avoiding complex definitions, necessarily a desperate task. After much discussion and explanation, ratio is defined by Azevedo Fortes as 'the manner in which a magnitude contains, or is contained in another'.[74] All examples are in integers, although the surd ratio is defined as a ratio that is not that of a number to a number. There are hints of a

[70] 'He m.[to] importante aos q[e] principiaõ o estudo das Math. o acustumaremse a fazer uzo do entendim.[to] puro sem intervençaõ dos sentidos, ou da imaginaçaõ, q[e] saõ cauza de m.[tos] erros': [Azevedo Fortes], *Elementos*, cód. 1861, f. 2.

[71] Lamy, *Elemens*, 4th edn, xiii.

[72] See footnote 65.

[73] 'podemos perceber ao menos pl.[o] pençam.[to] como destintas, e separadas as p.[tes] da quantid.[e] continua': [Azevedo Fortes], *Elementos*, cód. 1861, f. 4[v].

[74] 'o modo com q[e] huã grand.[a] conthem, ou he contheuda de outra': [Azevedo Fortes], *Elementos*, cód. 1861, f. 79[v].

characterization using similar aliquot parts, but this was apparently too complicated. A definition should be 'short and clear', he notes.[75] The chapter on geometrical proportions includes the rule of three, the rule of company, and the rule of false position. The fourth book is shorter and deals with compound ratios.

The fifth book is on fractions, that is, on arithmetic with fractions.[76] Because fractions are ratios for Lamy and Azevedo Fortes, arithmetic operations are performed indiscriminately on fractions and ratios. It appears that up to the second section of this book Azevedo Fortes followed the second edition of Lamy published in 1689, while afterwards he followed the third edition of 1704.

The sixth book is on incommensurable magnitudes. Existence is proven by giving the examples of the square root of 18 and the cube root of 24; for the former, having proven that, given a continued geometrical proportion $H\,.\,L\,.\,M$, if the ratio H: M is not expressed by square numbers, then L is incommensurable with both H and M, it is enough to consider the geometrical proportion $1\,.\,x\,.\,18$, and notice that the ratio 1 to 18 cannot be expressed by square numbers, so that x (the square root of 18) is incommensurable with both 1 and 18. The last section deals with arithmetic operations on incommensurable magnitudes, which really means on roots and on 'binomials' (expressions like $A + \sqrt{B}$, A and $\sqrt{B}$ being incommensurable) and 'multinomials' ($A + \sqrt{B} + \sqrt{C}$).

The seventh 'book' is 'on the method to solve questions, or problems', that is to say, on the analytic method. This is identified with the process of finding equations that express the conditions of the problems, and then solving them. After giving rules to find and simplify such equations, there is a list of 30 problems solved using first-degree equations. The final chapter addresses the solution of second-, third-, and fourth-degree equations; the quadratic formula is first enunciated and proven to be correct, and then also derived by completing the square; cubic equations, in depressed form (the general method to remove the second term of equations had already been given), are solved using a substitution proposed by Pierre Varignon. An abridged solution for quartic equations is also given, but it is wrong, as it assumes that they can always be reduced to biquadratic equations (supposedly by applying the method to remove the second term also to the fourth term).

As for the appendices, the first one in Azevedo Fortes's version of the text[77] addresses permutations and combinations, while the second is on harmonic proportions.[78] The third appendix deals with logarithms presented as numbers from an arithmetic progression corresponding to others in a geometric progression, as well as with the construction and use of logarithm tables.[79] The last appendix is a short and

[75] 'Huã defin. deve ser curta, e clara': [Azevedo Fortes], *Elementos*, cód. 1861, f. 91.

[76] In the Portuguese version, on 'broken' (i.e. fractional) numbers.

[77] In the original version the order of the appendices is precisely the reverse of this.

[78] In the third edition Lamy moved the section on permutations to the new eighth 'book', which is a collection of four appendices with the title 'Supplement to the Elements of Mathematics'. Because in the second 'book' Azevedo Fortes was following the second edition, he included that section there; but in the appendices he was following the third edition; in this case, he included the passages on permutations that were new in the third edition.

[79] This matter is later repeated, but with much more detail, in the treatise on trigonometry present in the Évora manuscript and as an appendix in *O Engenheiro Portuguez*. Azevedo Fortes acknowledges this

elementary introduction to the 'arithmetic of infinities'. It starts with several simple results about the arithmetic progression of natural numbers, followed by a number of results about the progression of odd numbers. We are then asked to conceive an infinity of terms interposed between each pair of consecutive numbers in the progression of natural numbers, keeping it as an arithmetic progression. The appendix finishes with a geometrical result: a pyramid having no perceptible difference from an infinity of squares whose sides are in arithmetic progression piled one upon another, its volume (as the sum of those squares)[80] is computed to be one third of the circumscribed parallelepiped (Figure 5.6).

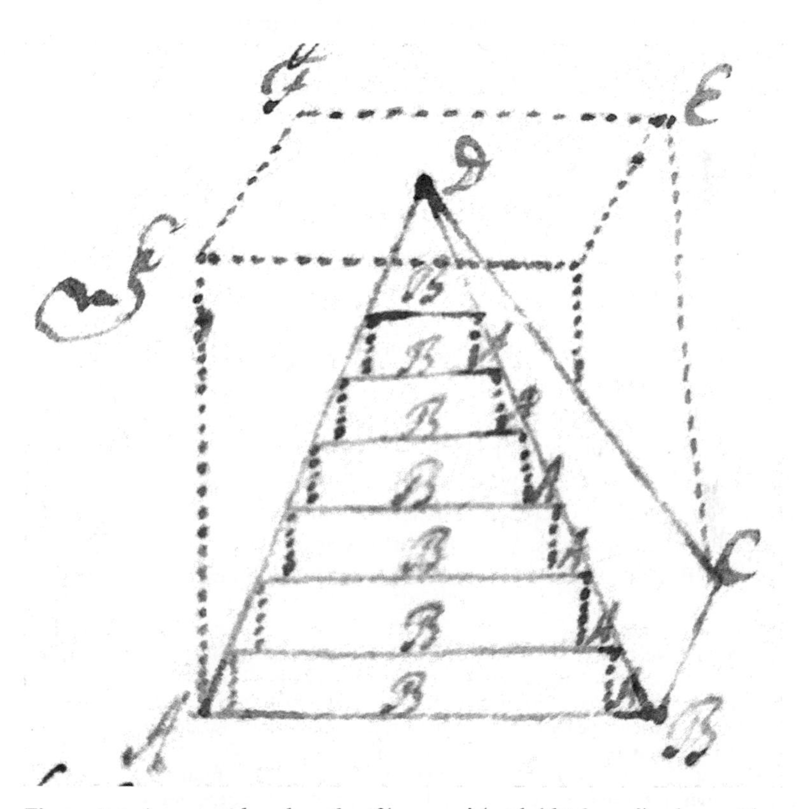

Figure 5.6 A pyramid and a pile of 'squares' (with 'thickness') whose sides are in arithmetic progression.

repetition: 'Although in a treatise that I have dictated on magnitude in general I have spoken of these tables [...]' ('Ainda que em hum tratado, que ditey da grandeza em geral faley destas taboadas [...]'): *O Engenheiro Portuguez*, I, p. 475; a similar sentence appears in the Évora manuscript.

[80] This language of 'squares' suggests the use of indivisibles, but before being made infinite in number they are supposed to have 'thickness', so that we are really dealing with infinitesimally thin parallelepipeds.

Concluding remarks

By taking the manuscripts we have examined together with *O Engenheiro Portuguez* in the light of a passage from the 1732 decree on military academies, we have been able to reconstruct the sequence of subjects taught at the Military Academy by Manuel de Azevedo Fortes, and probably also by his successors,[81] as follows:

1. 'magnitude in general' (arithmetic and algebra);
2. speculative geometry (Euclid's *Elements*);
3. trigonometry (plane, always; spherical, occasionally?);
4. practical geometry;
5. mathematical instruments belonging to engineering;
6. topographical plans and maps;
7. fortification, attack and defence of fortified places;
8. artillery;
9. castrametation.[82]

Practical geometry may have been taught before trigonometry. The order from item 5 to item 8 is based on that which is found in *O Engenheiro Portuguez*, but cannot be stated with certainty. Furthermore, there is no information on the proper position of item 9. It is possible that some other subjects, particularly of a military nature, such as movements of troops, even if they were not deemed indispensable for engineers, were also taught.[83]

One may ask why Azevedo Fortes published *O Engenheiro Portuguez*, a book that covers some but not all of the subjects in this list? Two partial answers may be conjectured. First, it is possible that with a view to the new military academies that were to be created by the 1732 decree, he wished to standardize some practical aspects, such as measurement practices. Azevedo Fortes was less concerned with what particular version of Euclid's *Elements* would be taught in each academy. Second, we can

[81] Besides recalling what was said in footnote 63. about José Sanches da Silva repeating in 1739 the lectures he had heard in 1722 about 'magnitude in general', we may add that there is another manuscript, dated 1732–34, with a somewhat different version of the treatise of magnitude in general, probably an adaptation produced by Filipe Rodrigues de Oliveira, then substitute at the Military Academy; this manuscript has been studied by Ribeiro, 'A formação dos engenheiros militares', pp. 146–69.

[82] The decree of 1732 on military academies prescribed that disciples from these academies intending to become engineers would be examined on encampments and entrenchments, as well as attacks, measurements, plans and maps, and fortifications: Araújo, 'Manoel de Azevedo Fortes e o estatuto', p. 33.

[83] Apart from the short introduction to the 'sphere' in the book on spherical trigonometry of the Évora manuscript, there is no indication that astronomy was ever taught at the Military Academy. Accordingly, the cartographical techniques taught by Azevedo Fortes focused on large- and medium-scale maps and did not include measurements of latitudes and longitudes. Those techniques sufficed for the European territory of Portugal, but not for larger areas. Thus, it is no wonder that when in the 1720s it was felt necessary to draw maps of the interior of Brazil, Azevedo Fortes and the military engineers he trained were not involved. Instead, the cartographical expedition that was sent in 1729 was headed by two Jesuit priests, the Italian Domenico Capassi and the Portuguese Diogo Soares: André Ferrand de Almeida, '"Arrumar as terras, os rios e os montes": os jesuitas matemáticos e os mapas do Brasil meridional, 1720–1748', in: *Manoel de Azevedo Fortes (1660–1749): Cartografia, Cultura e Urbanismo*, ed. Mário Gonçalves Fernandes, Porto: Gedes 2006, pp. 99–122.

surmise that in publishing a book about engineering, he sought to influence higher powers towards having the king, to whom the book is dedicated, sign the decree creating new military academies and thereby improve the prospects of their students. For this political purpose those technical subjects would have sufficed.

Considering the list proposed above, we may safely draw some conclusions regarding the mathematical subjects to be taught. If we compare these to the mathematical content of Luís Serrão Pimentel's lectures, and even more so to the limited mathematics that was mandatory to learn fortification according to Pimentel's method, it is obvious that Azevedo Fortes's mathematical curriculum was far more comprehensive.

It is also striking that nearly all of the mathematics that we can identify as having been taught by Pimentel in the Class on Fortification had direct applications to fortification, while Azevedo Fortes lectured on a considerable number of topics that did not. Indisputable examples of this wider scope include instruction on the nature of the angle of contact and the arithmetic of infinities. Furthermore, the whole structure and characteristics of the 'treatise of magnitude in general' is clearly more directed to 'disposing, developing, and perfecting the understanding' than to practical applications.

This does not mean that Pimentel was less of a mathematician than Azevedo Fortes. Pimentel was certainly capable of writing for theoreticians and we know he lectured on other subjects that were not really suitable for engineers such as spherical trigonometry. Moreover, he appears to have had sound knowledge of classical geometry, and was interested in specious algebra, considered at the time to be a forward-looking field.

Nor does this mean that Azevedo Fortes for his part neglected applications. Rather, it seems that he tried to strike a balance between a very 'pure' introduction to mathematics and what he considered to be serious applications that were to come later on. This may be seen not only in the practical geometry of *O Engenheiro Portuguez*, where concerns with operations on the physical terrain are very much present, but already in the instruction he provided on Euclid's *Elements*, where the uses of propositions prepare the ground for applications.

Once again, it should be stressed that while Azevedo Fortes was not the intellectual author of these works, it was he who chose them, and did so deliberately. It is significant, for instance, that he chose the version of Euclid's *Elements* found in Ozanam's *Cours de Mathematique*, whose title page claims that it contains 'all the parts most useful and most necessary *for a man of war*, and for those who wish to perfect themselves in mathematics', but that he did not take practical geometry or trigonometry from that work, even though he took trigonometry from another publication by Ozanam.

What comparisons between the different mathematical levels of Azevedo Fortes's and Pimentel's lectures suggest is that the two men pursued quite different projects. Pimentel was charged with teaching fortification, and he sought to train engineers, understood as technicians of fortification, as most straightforwardly as possible. Azevedo Fortes on the other hand had a grander plan: to enhance the prestige of engineers, and to educate military officers, giving them training not only in fortification, but also in other military subjects. He did not expect all the men he taught to become engineers. While he certainly sought to cover all relevant technical aspects, he also sought to develop the understanding of his students.

PART II
PROFESSIONS, SOCIETIES, AND CULTURES OF MATHEMATICS

6
Mathematical Men in Humble Life
Philomaths from North-West England as Editors of 'Questions for Answer' Journals

Sloan Evans Despeaux and Brigitte Stenhouse

Introduction

By the 1710s, working-class men from the North of England began to establish informal societies in order to pursue avocational interests in mathematics. In 1718, the Manchester Mathematical Society was founded by working-class men with strong interests in geometry. In 1794, the Oldham Mathematical Society, not far from Manchester, was similarly established.[1] Thomas Turner Wilkinson (1815–1875) a mathematician and historian of his native north-west England, considered these two societies the catalysts for what he called the Lancashire School of Geometers.[2]

This group of mathematically able working-class men gained a reputation among their local communities as well as within the wider mathematical community of Britain. Their visibility in Lancashire is evidenced in the work of Manchester-based author Elizabeth Gaskell, who gave these mathematicians an honourable mention in her 1848 novel *Mary Barton*. Gaskell described

> a class of men in Manchester [...] whose existence will probably be doubted by many, who yet may claim kindred with all the noble names that science recognises. I said 'in Manchester', but they are scattered all over the manufacturing districts of Lancashire. In the neighbourhood of Oldham there are weavers, common hand-loom weavers, who throw the shuttle with unceasing sound, though Newton's 'Principia' lies open on the loom, to be snatched at in work hours, but revelled over in meal times, or at night. Mathematical problems are received with interest, and studied with absorbing attention by many a broad-spoken, common-looking, factory-hand.[3]

A year later, Thomas Stephens Davies (1794–1851), whom we will meet properly below, commented that the concentration of geometry questions posed in the Senate House exams at the University of Cambridge was gradually increasing. That this

[1] J. W. S. Cassels, 'The Spitalfields Mathematical Society', in: *Bulletin of the London Mathematical Society* 11 (1979), pp. 241–58 at pp. 253–4.

[2] Thomas Turner Wilkinson, 'The Lancashire Geometers and Their Writings', in: *Memoirs of the Literary and Philosophical Society of Manchester* 11 (1854), pp. 123–57 at p. 127.

[3] Elizabeth Cleghorn Gaskell, *Mary Barton: A Tale of Manchester Life*, vol. 1, London: Chapman and Hall 1848, p. 55. Gaskell similarly notes the expertise of working-class botanists who could be met in Lancashire; these botanists are treated in Anne Secord, 'Science in the Pub: Artisan Botanists in Early Nineteenth-Century Lancashire', in: *History of Science* 32/3 (1994), pp. 269–315.

increase resulted from the influence of the Lancashire geometers is suggested by Davies' remark that mathematicians working on geometry in Cambridge 'will reap the field which the poor weavers of Lancashire have grown'.[4]

Why did these Lancashire men, well outside of the university sphere, take an interest in mathematics? Some of them could be categorized as mathematical practitioners who used mathematics in their day-to-day work, such as surveying, teaching, and instrument making. Growth in trade since Elizabethan times encouraged men to acquire practical mathematical knowledge. Besides mathematical practitioners, however, many members of these societies could be characterized as 'philomaths', that is, people who studied mathematics as an avocation unconnected to their employment.

On the occasion of the inaugural meeting of the British Association for the Advancement of Science held in York in 1831, Plymouth-based mathematician George Harvey marvelled that there

> existed a devoted band of men in the North of England, resolutely bound to the pure and ancient forms of geometry, who in the midst of the tumults of steam-engines, cultivated it with unyielding ardour, preserving the sacred fire under circumstances which would seem from their nature most calculated to extinguish it. In many modern Publications, and occasionally in the Senate-House Problems proposed to the Candidates for Honours at Cambridge, questions are to be met with derived from this humble but honourable source.[5]

Harvey felt it natural to imagine that the taste for what he called 'mechanical combinations' would have prevailed among the philomaths, given that they lived in a machinery-rich environment. However, as he had to concede, geometrical topics reserved for 'the very greatest minds' in ancient times 'were here familiar to men whose condition in life was, to say the least, most unpropitious for the successful prosecution of such elevated and profound pursuits'.[6]

For his part, Wilkinson speculated why these philomaths had a fascination in particular with geometry:

> The weaver at the loom,—the farmer in the field,—the mechanic in the shop,—or the miner in the drift, is too much occupied by *manual* labour to be able to *write out* long *analytical* investigations, but each can contemplate and deduce at pleasure the properties of a geometrical diagram, either *actually* constructed or mentally conceived. The farmer and the miner soon acquire the power of depicting vivid *mental* representations of the constructions necessary for their geometrical inquiries, and are thus enabled to carry on their processes of deduction even when buried in the mine or following the plough. The weaver and the mechanic can sketch their diagrams on

[4] Chetham's Library, Manchester. Manuscripts/1/215, A.3.93, T. S. Davies to T. T. Wilkinson, 24/02/1849.

[5] George Harvey, 'Extract of Letter', in 'Scientific Transactions of the Meeting: Tuesday Morning', in: *Report of the First and Second Meetings of the British Association for the Advancement of Science*, London: John Murray 1833, pp. 58–9.

[6] Ibid.

a slate, and thus pursue their favourite studies whilst their hands and feet are almost instinctively engaged in their monotonous operations [...].[7]

Other motivations for the educational pursuits of these workers are suggested by the intellectual historian Jonathan Rose:

Weavers and lead miners were well-paid and had short work hours: six hours a day for miners, four days a week for weavers. Weavers had to be literate for their work, and mining companies wanted an educated work force. Both trades had a history of friendly society activity and self-education.[8]

Although Harvey in his account of 'the very remarkable circumstance of the Geometrical Analysis of the ancients having been cultivated with eminent success in the northern counties of England' was unable to trace the 'true cause for this remarkable phænomenon', he was nonetheless able to cite proof of its existence 'from a variety of periodical works'.[9] In fact, although the mathematical societies to which these philomaths in the North of England belonged unfortunately published no formal proceedings, a remarkable record of their mathematical activities often appeared in 'question for answer' journals. This chapter will explore the central role played by these journals in the motivations and the networks of mathematical communication of this seemingly unlikely group of mathematicians. In particular, we will focus on the editors of three journals from one region from north-west England, a region including the cities of Manchester and Liverpool. The editors of these three journals, *The Student* (1797–1800), the *Liverpool Apollonius* (1823–4), and the *Preston Chronicle* (1844–5), provide a case study of how mathematics could be pursued well outside of the university sphere as a means of sociability, friendly competition, and social mobility.

'Questions for Answer' Journals as a means of philomathic communication

Since the early eighteenth century, mathematical 'questions for answer' appeared as regular features in English journals and almanacks. The most well-known and long-lived English almanack with such a feature was the *Ladies Diary*, which published almost 2000 mathematical questions and their answers between 1707 and 1840.[10]

The mathematical journalistic genre that was so successfully developed by the *Ladies Diary* was soon copied by other almanacks, which included mathematics alongside enigmas, poetry, and ephemerides, as well as journals exclusively devoted to mathematics. Over a century after the *Ladies Diary* first began publishing

[7] Wilkinson, 'The Lancashire Geometers', p. 130.

[8] Jonathan Rose, *The Intellectual Life of the British Working Classes*, New Haven, CT: Yale University Press 2001, p. 59.

[9] Harvey, 'Extract of Letter', p. 58.

[10] Joe Albree and Scott H. Brown, '"A Valuable Monument of Mathematical Genius": The Ladies' Diary (1704–1840)', in: *Historia Mathematica* 36 (2009), pp. 10–47 at p. 40.

mathematical questions for answer, University of Edinburgh Professor John Playfair (1748–1819) described the periodicals that comprised this genre as follows:

> In these, many curious problems, not of the highest order indeed, but still having a considerable degree of difficulty, and far beyond the mere elements of science, are often to be met with; and the great number of ingenious men who take a share in proposing and answering these questions, whom one has never heard of any where else, is not a little surprising. Nothing of the same kind, we believe, is to be found in any other country.[11]

This journalistic tradition, in Playfair's opinion, provided proof that 'a certain degree of mathematical science, and indeed no inconsiderable degree, is perhaps more widely diffused in England, than in any other country of the world'.[12] To give a sense of the pervasiveness of this genre, we can cite the 1929 'Notes on Some Minor English Mathematical Serials' in which Raymond Archibald surveyed 43 question for answer journals from the eighteenth and nineteenth centuries.[13] Davies, writing in 1850, gave a sense of both the extent and rareness of the older journals: 'Although at one period of our life we took great pains to make a collections of the *periodicals* which, during the last century, were devoted wholly or partially to mathematics, yet we could never even approximate towards completeness'.[14]

Historians of mathematics: Thomas Stephens Davies and Thomas Turner Wilkinson

Much of what is known today of the Lancashire Geometers and their engagement with mathematical periodicals comes from the tireless chronicling of Thomas Stephens Davies and Thomas Turner Wilkinson.

Davies started out as a private tutor in Bath, before being employed as one of the mathematical masters at the Royal Military Academy, Woolwich in 1834. He was a prolific contributor to periodicals such as the *Leeds Correspondent* (1814–23) and the *Lady's and Gentleman's Diary* (1841–71), and acted as editor for the first volume of *The Mathematician* (1843–50).[15] Wilkinson was himself a quintessential member of the Lancashire School of Geometers. He grew up near Blackburn (around 25 miles north-west of Manchester) and, after leaving school to provide labour on his father's

[11] John Playfair, '*Traité de Méchanique céleste*', in: *Edinburgh Review* 11 (1808), pp. 249–84 at p. 282.

[12] Ibid.

[13] Raymond Archibald, 'Notes on Some Minor English Mathematical Serials', in: *The Mathematical Gazette* 14 (1929), pp. 379–400. More details on 34 of these periodicals are given in Sloan Evans Despeaux, 'Mathematical Questions: A Convergence of Mathematical Practices in British Journals of the Eighteenth and Nineteenth Centuries', in: *Revue d'histoire des mathematiques* 20 (2014), pp. 5–71.

[14] Pen-and-Ink [T. S. Davies], 'On the Cultivation of Geometry in Lancashire', in: *Notes and Queries* 57 (1850), pp. 436–8 at p. 436.

[15] For more biographical information on Davies see June Barrow-Green, '"A Senior Wrangler among Senior Wranglers": Ellis's Mathematical Education', in: *A Prodigy of Universal Genius: Robert Leslie Ellis, 1817–1859*, ed. Lukas M. Verburgt, Cham: Springer 2022, pp. 21–49.

farm, taught himself mathematics through the study of books sourced for him by an uncle.[16] Wilkinson began submitting solutions to questions posed in the *Ladies Diary* in 1837, and subsequently became a contributor to the *Preston Chronicle* (to which we will return later), the *York Courant* (1828–46), and the *Educational Times* (1847–1923), among many others.

The prolific contributions Davies and Wilkinson made, were spread across numerous disparate periodicals, magazines, and journals; the contributions included solutions to mathematical puzzles, longer mathematical papers, and, pertinent to this chapter, historical essays. It was through Wilkinson's submission of mathematical exercises to the *Mechanics Magazine* that he first became acquainted with Davies.[17] The two men subsequently became friends and frequent correspondents; three volumes of letters written by Davies to Wilkinson between 1848 and 1850 are held today at Chetham's Library in Manchester.

In 1848, Wilkinson's first entries in his series of historical articles on eighteenth- and nineteenth-century periodicals appeared in the *Mechanics Magazine*. Within five years, he had written over forty articles in which he surveyed twenty periodicals that were either exclusively mathematical, or contained a significant amount of mathematical content. Many of these periodicals featured contributions by mathematicians who were identified with the Lancashire School of Geometers, or indeed, as we will see in this chapter, were edited and run by such mathematicians. On one occasion Davies complimented Wilkinson on the quality of his historical studies, saying that 'Mr. Wilkinson has shown himself to possess so many of the qualities *essential* to the historians of mathematical science, that we trust he will continue his valuable researches'.[18] Regarding the working-class geometers of Lancashire specifically, Wilkinson and Davies published numerous biographies of individuals, alongside more general essays about the geometry that was cultivated by these men.[19]

That Davies and Wilkinson took their work as historians of mathematics seriously is frequently evidenced in their correspondence. A major difficulty was divining the true identities of contributors whose work was hidden behind one or more pseudonyms. In early 1850, Wilkinson and Davies frequently discussed their investigations of Thomas Leybourn's (1770–1840) *Mathematical Repository*, which ran from 1795 through to 1835. Davies arranged to visit one of the main contributors to the journal, Mark Noble, who had served as mathematical master at the Royal Military College from 1806 to 1820, and in 1850 was living just a few miles away from Davies in London. Davies felt that it was 'very important that the anonymous should be cleared up' in his and Wilkinson's works on mathematical periodicals, and

[16] William Alexander Abram, 'Memorial of the Late T.T. Wilkinson, F.R.A.S, of Burnley', in: *Transactions of the Historic Society of Lancashire and Cheshire*, 3rd ser. 4 (1875), pp. 77–94.

[17] Wilkinson, quoted in Abram, 'Memorial', p. 88.

[18] [Davies], 'On the Cultivation', p. 436.

[19] Examples of their writings include: Thomas Stephens Davies, 'XXIX. Geometry and Geometers No. II', in: *The London, Edinburgh, and Dublin Philosophical Magazine and Journal of Science* ser. 3, 33/221 (1848), pp. 201–6; Thomas Turner Wilkinson, 'On the origin and progress of the study of geometry in Lancashire', in: *Notes & Queries* 34 (1850), pp. 57–60; Thomas Turner Wilkinson, 'An account of the Life and Writings of the Late Henry Buckley', in: *Transactions of the Historic Society of Lancashire and Cheshire*, new ser. 3 (1863), pp. 115–28; and other articles referenced throughout this chapter.

hoped that Noble would be able to help.[20] Other contributor's identities were divined using a copy of the *Repository* on which authors' names had been written down 'from Leybourn's dictation' next to their pseudonymous contributions; Leybourn identified himself as 'Samuel Thornaby'.[21] At other times, identities were proposed for pseudonymous works based on their style or content.[22]

As quoted above, Davies and Wilkinson invested a significant amount of their time in sourcing as many original issues and volumes of mathematical periodicals as they could. They shared their own copies between each other where necessary, consulted volumes held in the library of the Royal Society of London, and requested support from publishers of their acquaintance. Even when original copies could be found, they were not always complete. For example, Wilkinson wished to know why the final issue of a periodical he refers to only as 'Turner' was printed in Wrexham. But when Davies was eventually able to get hold of a copy of this issue the title page was missing. As a result, he could neither illuminate why this printing location was chosen, nor even confirm that it was.[23]

The seriousness with which Davies and Wilkinson took their historical investigations can perhaps partially be explained by their viewing this work as part and parcel of their mathematical endeavours. In his 1849 article on *The Mathematician*, Wilkinson makes specific reference to work done by Davies on generalizing a theorem regarding properties of trapezia. The published version of the article has an extensive footnote directing the reader to further work done by Davies on this topic in Leybourn's *Repository*, and exhorting the importance of someone furthering this work by preparing a classification of the properties of certain types of quadrilaterals.[24] This footnote was in fact proposed by Davies himself on reading a manuscript copy of the article; he wished 'to persuade somebody or other to take up the subject' as he would liked to have done so himself but did not have the time.[25] Thus, although their articles were ostensibly historical surveys of periodicals, many of which had long since ceased publication, Davies and Wilkinson nevertheless saw this work as having the potential to direct, encourage, and inspire mathematical research in their present day.[26]

Alongside advertising the mathematics practised by the Lancashire geometers, Davies certainly saw himself as an advocate for the mathematicians themselves. After the death of John Swale (1775–1837), a mathematical editor from north-west England who will be discussed below, Davies embarked on the substantial project of analysing and circulating a large collection of manuscripts that had been given to him by Swale's son. Davies' intention was to 'claim for Mr Swale his place as a geometer'

[20] Chetham's Library, Manchester, Manuscripts/1/215, A.3.94, Davies to Wilkinson 14/01/1850.

[21] Chetham's Library, Manchester, Manuscripts/1/215, A.3.94, Davies to Wilkinson 31/03/1850.

[22] Chetham's Library, Manchester, Manuscripts/1/215, A.3.92, Davies to Wilkinson 23/11/1848.

[23] Chetham's Library, Manchester, Manuscripts/1/215, A.3.93, Davies to Wilkinson 20/02/1849.

[24] Thomas Turner Wilkinson, 'Mathematical Periodicals (continued)', in: *The Mechanics' Magazine* 50 (1849), pp. 5–9.

[25] Chetham's Library, Manchester, Manuscripts/1/215, A.3.92, Davies to Wilkinson 28/12/1848.

[26] For more on historical work as part of mathematical practice, see Nicolas Michel and Ivahn Smadja, 'Mathematics in the Archives: Deconstructive Historiography and the Shaping of Modern Geometry (1837–1852)', in: *British Journal for the History of Science* 54/4 (2021), pp. 423–41.

and posthumously 'obtain for [him] his proper place in the scientific pantheon'.[27] Although Davies died before his project reached completion, an article on Swale that he had written for his long-running series 'On Geometry and Geometers' was posthumously inserted in the *Philosophical Magazine*.[28] Further notes on the Swale manuscripts were subsequently contributed by Wilkinson to the same publication.

Notably, the *Philosophical Magazine* was contributed to and read by the gentlemen of nineteenth-century science—the professors at Oxford and Cambridge, and the fellows of London's elite learned societies. It is thus clear that Davies was not so much concerned with building Swale's reputation in Lancashire, but rather of extending it to the metropolis and to the upper echelons of society.

In his letters to Wilkinson, Davies makes clear the hesitancy that gentlemen and their prestigious scientific societies had in giving recognition to the working-class geometers of Lancashire. At the end of 1850, he wrote to Wilkinson: 'Do you know any particulars of Henry Clarke of Salford? He afterwards became LLD: though he was one of those whom Banks rejected as F.R.S because of his social status—"being only a school master"'.[29] That same year, Wilkinson was successfully elected a Fellow of the Royal Astronomical Society (RAS) thanks to assistance from Davies and mathematician Augustus De Morgan (1806–1871), who was then the Secretary of the RAS. Davies appears to have taken it upon himself to organize the two signatures of current Fellows required to propose Wilkinson from 'personal knowledge'. He wrote to Wilkinson asking if the latter could name any Fellow with whom he was personally acquainted. Davies assured Wilkinson that the acquaintance need not be intimate, and indeed enclosed a letter from De Morgan that he suggested Wilkinson could show to any potential nominator to 'convince him of the estimate formed of you by the Secretary', which would assure them that their signature was needed only for a formal requirement rather than to personally assure Wilkinson's suitability for the RAS.[30] It is intriguing that even though Wilkinson had established a reputation or cultivated sufficient social connections to impress De Morgan with his mathematical abilities, yet the requirement for even just a second signature of a current fellow provided Wilkinson with difficulties in the process of his election to this socially elite society. Ultimately, Wilkinson was nominated by De Morgan and John William Whittaker (*c.*1790–1854).[31] Although not sharply divided by social class, the networks of correspondence and recognition between mathematicians in nineteenth-century Britain were evidently heavily shaped by it.

Building on the historical works of Wilkinson and Davies, we will now turn to three mathematical periodicals and their editors to illuminate the networks and practice of this so-called Lancashire School of Geometers.

[27] Chetham's Library, Manchester, Manuscripts/1/215, A.3.94, Davies to Wilkinson 01/10/1850.

[28] Thomas Stephens Davies, 'LXXVI. Geometry and Geometers No. VII', in: *The London, Edinburgh, and Dublin Philosophical Magazine and Journal of Science*, ser. 4, 1/7 (1851), pp. 536–44.

[29] Quotation marks in the original. Chetham's Library, Manchester, Manuscripts/1/215, A.3.94, Davies to Wilkinson 04/12/1850. F.R.S. refers to 'Fellow of the Royal Society' and the Banks referred to is Joseph Banks (1743–1820), who was then president of the Royal Society. Clarke was most likely Henry Clarke (1743–1818), a mathematician and schoolmaster who was based in Lancashire for most of his life.

[30] Chetham's Library, Manchester, Manuscripts/1/215, A.3.94, Davies to Wilkinson 18/04/1850.

[31] Thank you to Sian Prosser at the RAS for providing me with the election certificate of Wilkinson.

The Student (1797–1800)

In 1797, Liverpool philomath John Knowles began an annual journal entitled *The Student* '[t]o inspire active EMULATION, to supply rational AMUSEMENT, and to diffuse useful KNOWLEDGE'.[32] In the preface of the journal's first number, Knowles situates the goals of the journal against a background of commerce and competition:

> By the assistance of philosophy, of chemistry, and of the mathematics, the powers of nature are made subservient to the health and happiness of mankind. From the same sources is chiefly derived whatever is great or valuable in arts, manufactures, navigation, and commerce. From the admiration of excellence naturally arises a desire to excel: And this generous emulation, in well formed minds, is productive of the ablest efforts and of the most benevolent sentiments. Such a happy disposition is sometimes excited and always improved by a liberal and candid competition for preeminence; to which, all those who wish to receive or to communicate information are respectfully invited.[33]

This 'candid competition for preeminence' among the participants in question for answer journals was quite different from the genre's roots. In fact, mathematical 'questions for answer' began as a 'polite' endeavour suitable for ladies. For example, mathematical questions submitted to the *Ladies Diary* were initially posed in rhyming verse. However, beginning in the 1720s, the *Diary* changed

> from a forum where women's 'geometrical, algebraical, astronomical and philosophical' skills were displayed and praised to one nominally dedicated to women and to mathematics, but in which polite banter had been displaced by an increasingly skill-oriented, even confrontational, discourse.[34]

Instead of rhyming verse, questions were set in plain prose and accompanied with figures and equations. As a result, 'women dropped out of its mathematical dialogue for more than twenty years, resurfacing [by mid-century] only in very reduced numbers and in a much more competitive context'.[35]

Davies, in his account of the Lancashire School of Geometers, recognized the spirit of competition inherent in the activity of solving mathematical questions for answer:

> the spirit of emulation did something; from the belief that *insertion was an admitted test of superiority*, it was as much an object of ambition amongst these men to solve the 'prize question' as it was by philosophers of higher social standing to gain the 'prize' conferred by the *Académie des Sciences*.[36]

[32] Title Page, in: *The Student*, no. 1 (1797).
[33] [John Knowles], 'Preface', in: *The Student*, no. 1 (1797).
[34] Shelley Costa, 'The "Ladies' Diary": Gender, Mathematics, and Civil Society in Early-Eighteenth-Century England', in: *Osiris*, 2nd series, 17 (2002), pp. 49–73 at p. 70.
[35] Ibid., pp. 71–2.
[36] [Davies], 'On the Cultivation', p. 437.

The editor of *The Student* pledged to distribute all of the profits from the sale of the journal in prizes for the next year.[37] In most journals of this genre, prizes usually consisted of free copies of the publication. While *The Student* did give out free copies as prizes,[38] it also gave direct disbursements of money.

Besides competition, Knowles wanted his journal to encourage cooperation and the formation of something similar to working groups:

> If small societies, consisting of six or eight persons, were formed for the purpose of making the experiments related in the following pages, and debating on the queries, it would greatly increase the pleasure of each, and reduce his expense to a mere trifle.[39]

The Student consisted of six parts: grammar, polite and useful arts, natural and experimental philosophy (what would now be called physics), chemistry, geometry and mathematical correspondence, and poetry (consisting of enigmas, rebuses, charades, and French excerpts for translation). While short articles appeared in the first number,[40] by the second, the entire journal consisted of queries, questions, and answers provided by contributors. Prize questions appeared in each section. Knowles was competent in the content of all six sections: Wilkinson recounts that Knowles 'was then a schoolmaster and lecturer on Natural Philosophy; he was also well versed in the classics, and possessed a fair knowledge of chemistry and mathematics'.[41]

Of the six sections, 'geometry and mathematical correspondence' was the most extensive: this section took up from nearly a third to almost a half of all the pages in each of the four numbers of the journal. In a subsection of the first number called 'Mathematical Correspondence', solutions were given to questions first proposed in the last issue of an almanack, the *British Diary* for 1796. This almanack had only survived for nine annual issues and had begun its first issue with questions posed in the defunct *British Miscellany*.[42] This passing of the mathematical torch from journal to journal was a common practice and indicates the short life expectancies of journals in this genre.[43]

[37] [Knowles], 'Preface'.

[38] For example, Question 33 posed in the Polite and Useful Arts section of the fourth number of *The Student* offered a prize of six copies of the journal: 'What is the cheapest and most effectual method of bleaching flannel without giving it any disagreeable smell?' (*The Student*, no. 4 (1800), p. 15).

[39] [Knowles], 'Preface'.

[40] There was also a calendar of festivals and feasts for the upcoming year of 1798. This kind of section was common in almanacks, but it was dropped by *The Student* after the first number.

[41] Thomas Turner Wilkinson, 'Biographical Notices of some Liverpool Mathematicians', in: *Transactions of the Historic Society of Lancashire and Cheshire*, new ser. 2 (1862), pp. 29–40 at p. 33.

[42] Archibald, 'Notes', p. 389.

[43] For more on almanacks and passing the mathematical torch, see Sloan Evans Despeaux, 'Connected by Questions and Answers: The Milieu of Mathematical Editors of English Commercial Journals, 1775–1854', in: *Circulation des mathématiques dans et par les journaux: Histoire, territoires, publics*, ed. Hélène Gispert, Jeanne Peiffer, and Philippe Nabonnand, forthcoming. The run of journals each following on from the other could become quite long: for example according to Davies, the *Scientific Receptacle* commenced with the questions left unsolved in the *Leeds Correspondent*, which itself began with questions from the *Enquirer*, which in turn began with questions from the *Quarterly Visitor* (Chetham's Library, Manchester, Manuscripts/1/215, A.3.92, Davies to Wilkinson 11/03/1848) and Wilkinson, 'Mathematical Periodicals (continued)', p. 8.

While he provided some of the journal's material under his pseudonyms, *Non Sibi* and *N. Selwon*, Knowles was joined on the pages by active local mathematicians including John Swale, James Wolfenden (1754–1841), and William Hilton (1772–1826). Wolfenden and Hilton were both employed as handloom weavers, but Hilton studied under Wolfenden on Sundays. Wolfenden was self-taught, having been taken out of school after only one week, 'the bobbin-wheel and the loom being considered much more profitable employments than learning to read'.[44] With the help of his grandfather and of local mathematician Jeremiah Ainsworth (1743–1784), co-founder of the Oldham Mathematical Society, Wolfenden began to pose and answer mathematical questions in almanacks in 1781. Wolfenden continued to contribute solutions and questions (many of them prize questions) for almost sixty years to mathematical journals, including to *The Student*. Wolfenden gave up his hand loom at the age of 62 and began tutoring students in mathematics; nonetheless he spent the end of his life in poverty. In fact, in 1841, some members of the Manchester Literary and Philosophical Society collected funds to provide for Wolfenden and to later erect a memorial at his grave.[45] Wolfenden's case was not unique: in 1843 the 'Society for the Relief and Encouragement of Scientific Men in Humble Life' was established in Manchester to provide support for working-class men engaged in science, and specifically in mathematics.

Unlike Wolfenden, Hilton had been able to obtain some elementary education at a local school, but was trained to work as a weaver for his father. He succeeded in combining both activities. While he learned mathematics from Wolfenden on Sundays, Hilton 'carried his studies to his work at the loom, for it is said that often when the picking-stick should have been going, William was found sitting aside from his work, thinking and figuring out some abstruse problem'.[46] He began contributing mathematical problems and solutions to a variety of questions for answer journals, including *The Student*. When Knowles died after the second number of that journal, Hilton took over as editor. Issue number three lists his name on the title page as 'W. Hilton, Teacher of the Mathematics'.

Hilton continued the recurring section of the journal called 'Lineal Sections' where 'methods are exhibited of determining *Two right Lines* by having certain relations given'. These methods were deemed to be useful points of reference, because they could serve 'to render many solutions more elegant and concise'.[47] He also continued another recurring section of *The Student*, called 'Modern Geometry', which concerned a triangle with inscribed and circumscribed circles. One figure (our Figure 6.1), reproduced in subsequent numbers of the journal with the same labelling but ever more complicated, was the point of reference for a variety of geometrical propositions and demonstrations by different contributors. By the fourth and final number of *The Student*, the figure included 34 different labelled points. Davies called

[44] Thomas Turner Wilkinson, 'Memoir of James Wolfenden, of Hollinwood', in: *Mechanics Magazine* 50 (1849), pp. 387–93 at p. 387.

[45] Ibid., p. 392.

[46] Philander, 'William Hilton, The Lancashire Mathematician', in: *Manchester City News Notes and Queries*, no. 96 (1882), pp. 337–8.

[47] 'Lineal Sections', in: *The Student*, no. 1 (1797), p. 38.

this section on 'Modern Geometry' a 'little work of great merit' that had, because of available copies, been transcribed 'from MS. copies at *third* or *fourth hand*'.[48]

In his musings on the question of why weavers from the North of England pursued geometry, Davies referred to this type of diagram (Figure 6.1):

> Algebraic investigation required writing: but the weaver's hands being engaged he could not write. A diagram, on the contrary, might lie before him, and be carefully studied, whilst his hands and feet may be performing their functions with an accuracy almost instinctive. Nay more: an exceedingly complicated diagram which has grown up gradually as the result of investigations successively made, may be carried in the memory and become the subject of successfully peripatetic contemplation.[49]

Signs of financial trouble appeared by the time Hilton took over the editorial helm of *The Student*. The journal had yielded some profits, which Hilton distributed to those who had contributed material to each number. However, he lamented:

> If some Gentlemen, who took large quantities of those numbers, had remitted the money for them, a much larger distribution [of profits] would have been made, if it be not done before the publication of the next Number their Names will then be made public.[50]

While no final notice was made in the fourth and last number of the journal, possibly *The Student* succumbed to similar economic woes shared by many of the journals in this genre. Another possible reason for the journal's demise is Hilton's move from Saddleworth to Liverpool, according to Wilkinson 'about at the close of the last century. He was first an assistant in some school—then a clerk in a merchant's office—and finally a dealer in cotton, by which he realised several thousand pounds'.[51]

An account of Hilton by someone writing under the pseudonym Philander gives a slightly different timeline. After Hilton solved a difficult problem on the tides that

> had been going the round of the papers for solution [...] a Mr. Knowles arrived at the home of William in his carriage, and took him in triumph out of his loom-house [...] to Liverpool. It was arranged that William should attend for a time the Grammar School at Liverpool. He afterwards became a schoolmaster in Liverpool, calculating the Tide Tables. After this he appears to have held a situation in connection with the port of Liverpool, and knowing the progress and extent of the cotton trade in his native neighbourhood [...] he set up business as a cotton merchant [...] William Hilton, out of a poor weaver lad, became one of the wealthiest merchants in Liverpool.[52]

[48] [Davies], 'On the Cultivation', p. 438.

[49] Ibid., pp. 436–7.

[50] [W. Hilton], [Editor's note], in: *The Student*, no. 3 (1799), p. 72.

[51] Wilkinson, 'Biographical Notices', p. 33.

[52] Philander, 'William Hilton', pp. 337–8. There are grounds for questioning the complete accuracy of this account, because it also recounts that '[i]t is said that the friendship between Mr. Knowles and Hilton continued for many years, but that on account of some difference of opinion on a mathematical subject the friendship was broken off'. However, Knowles died after the second number of *The Student*.

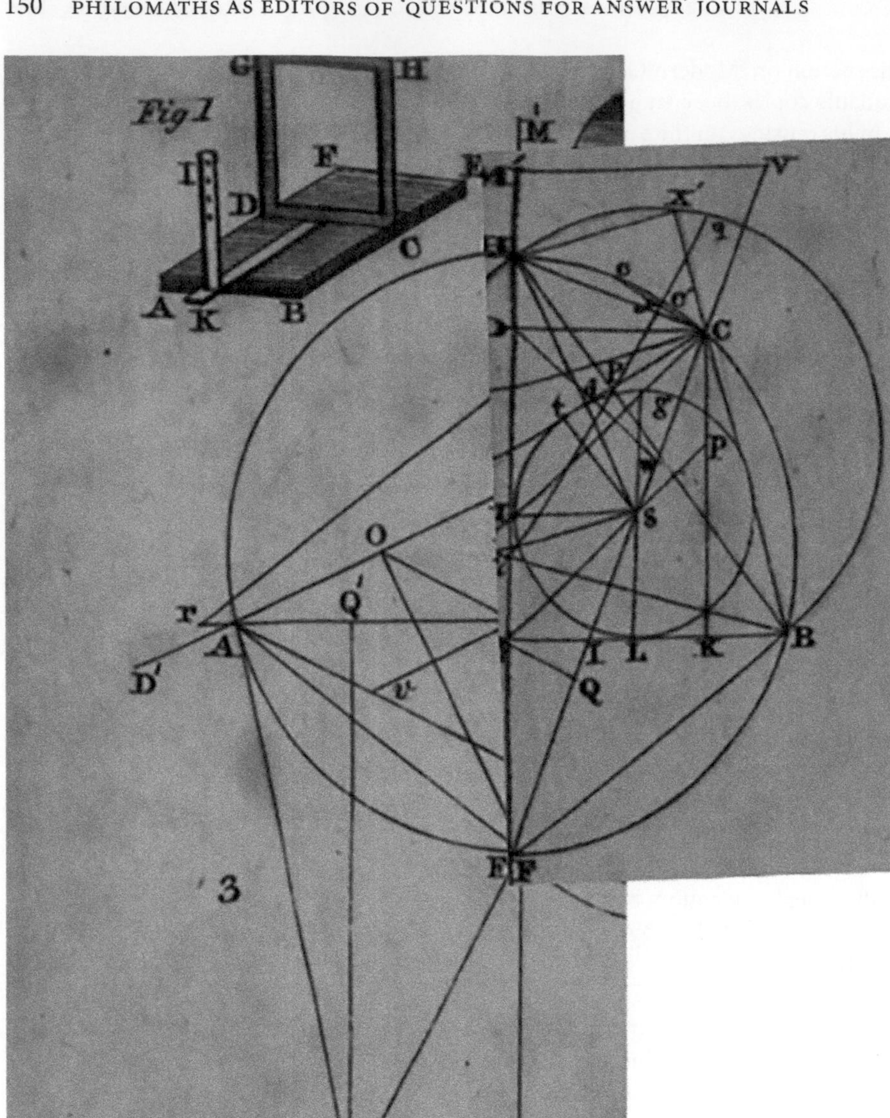

Figure 6.1 The labelled figure from the 'Modern Geometry' section of *The Student*. This figure was a pull-out and available scans of *The Student* only captured portions of the diagram. The version above is an amalgam of the diagrams from two different numbers. The figure, reproduced in subsequent numbers of the journal with the same labelling but ever more complicated, was the point of reference for a variety of geometrical propositions and demonstrations by different contributors.

Hilton would have been one of many people in Lancashire in the eighteenth and nineteenth centuries who made their fortune through the trade of enslaved African people, or in adjacent markets. By the end of the eighteenth century, Liverpool controlled around 60% of the British slave trade, and thousands of people living in the surrounding areas were employed in shipbuilding, sugar-refining, and textile production.[53] Even after the ostensible abolition of the slave trade in Britain in 1807, the labour of enslaved peoples in the Americas was vital to the local economy, with Liverpool receiving 90% of the total raw cotton imported to Britain in 1830.[54] Once this raw cotton had been manufactured into cloth in the mills in and around Manchester, it was then exported back to southern American states, bringing in further revenue for the Lancashire merchants.

Hilton's fortunes thus took a very different turn to those of *The Student* and his teacher, Wolfenden. Whether his move to Liverpool was a direct consequence of his activity in question for answer journals or not, his work with mathematical questions clearly prepared him for teaching, the first step in his new career path.

Liverpool Apollonius (1823–24)

One of the regular contributors to *The Student*, John Henry Swale (1775–1837), began his mathematical journey in Yorkshire, where he became a valued and recognized contributor to many question for answer journals. Swale befriended a number of mathematical journal editors, including Hilton and Leybourn. He moved to Liverpool in 1810, where he established an academy and continued his interactions with mathematical editors.[55]

In 1823, Swale joined the ranks of mathematical editors by launching the *Liverpool Apollonius*. As part of his motivation, he indicated his bias towards geometry and against the analytic methods of the Continent being adopted in Britain:

> the Editor presents, to junior Geometricians in general, to those of Liverpool in particular, and to all promoters of the pursuits of Intellect; the first number of the APOLLONIUS: a periodical work, intended to furnish a page of record for the productions of Genius: to supply the Curious with useful and ennobling subjects of inquiry: to induce habits of THINKING: to encourage the prosecution of mathematical and physical Science: to familiarize THE STUDY OF GEOMETRY [...] NEWTON lamented the neglect of Geometry in his day: and, at the present, the fascinating and profound Analysis of La Grange, seems to occupy (exclusively) the attention of English Mathematicians.[56]

[53] Marika Sherwood, *After Abolition: Britain and the Slave Trade Since 1807*, London: I. B. Taurus & Co. 2007, p. 28.

[54] Ibid., p. 31. See also Jessica Moody, *The Persistence of Memory: Remembering Slavery in Liverpool, 'Slaving Capital of the World'*, Liverpool: Liverpool University Press 2020.

[55] Thomas Turner Wilkinson, 'Memoir of the Late J. H. Swale', in: *Mechanics Magazine* 56 (1852), pp. 194–6, 206–9, 224–6 at pp.196, 206–7.

[56] J. H. Swale, 'Advertisement', in: *Liverpool Apollonius*, no. 1 (1823). Swale dedicated his first number to Thomas Leybourn, by then firmly established as a Mathematical Master at the Royal Military Academy.

The two numbers of the *Apollonius* covered a variety of mathematical areas and contained reprints of memoirs by, amongst others, the contemporary textbook writers Charles Bossut (1730–1814) of Paris and Samuel Vince (1749–1821) of Cambridge; the eighteenth-century French authority on the history of mathematics, Jean Etienne Montucla (1725–1799); and the eighteenth-century Scottish mathematicians Colin Maclaurin (1698–1746) and Matthew Stewart (1717–1785). However, Swale's predominantly geometrical focus is reflected by the fact that original geometry articles occupied over one-third of the work. In this regard, Wilkinson noted that the

> whole of the geometrical papers are by Mr. Swale himself, and are everywhere characterised by his usual elegance, originality, and fertility of invention. All of them possess a peculiar value to the student of pure geometry, and the historical interest attaching to many of the problems discussed will always render the Apollonius one of our most esteemed mathematical periodicals.[57]

Swale provided historical details as well as his own opinions about various mathematical schools in his geometrical contributions:

> To inscribe, in a given Circle, a Polygon of six Sides, so that the sides shall tend to, or pass through the given Points P,Q,R,S,T,V... Geometers have considered this Problem, especially the general Case, as very difficult. It appears that CASTILON first published a laborious, synthetic Solution of it, in the 'Berlin Memoirs' for 1778: and LAGRANGE, in the same volume, has given another Solution, entirely analytical. Also, LEXELL, at the request of EULER, gave, in the 'Petersburgh Memoirs', a Construction of Lagrange's Formula; but failed in attempting to construct the inscribed Quadrilateral. An Inference from this, would place the GEOMETRY of the FRENCH ANALYSTS on a puerile and retrograde Scale![58]

Besides its original articles, the *Apollonius*, like other mathematical question for answer journals, also included junior and senior mathematical problem sections. Swale himself vigorously contributed to the problem section under his pseudonym, Apollonius.[59]

The first number of the *Apollonius* contained nineteen junior and senior questions, and more than twice that many were added to the second number; however, the solutions to this second set would remain in manuscript form after the early failure of the journal.[60] One factor cited in the ruin of this enterprise was Swale's decision to print

He dedicated his second number to the American mathematician and mathematical journal editor, Robert Adrain, '[a]s a public expression of esteem for his worth and talents'. Adrain in turn contributed 'A View of the Diophantine Analysis' (*Liverpool Apollonius*, no. 2 (1824), pp. 86–91) to the second number. The promised continuation of this article did not appear because of the journal's early death. For more on Adrain, see David E. Zitarelli, *A History of Mathematics in the United States and Canada*, vol. 1: 1492–1900, Providence, RI: MAA Press 2019, 122–9.

[57] Thomas Turner Wilkinson, 'Mathematical Periodicals', in: *Mechanics Magazine* 58 (1853), pp. 306–7, 327–328 at p. 307.

[58] J. H. Swale, [Geometrical Construction #4], in: *Liverpool Apollonius*, no. 2 (1824), pp. 51–2.

[59] Wilkinson, 'Mathematical Periodicals', p. 327.

[60] Wilkinson noted that '[s]everal of the more simple problems in the two courses have since been discussed in the pages of the *Gentleman's Diary*, the *Mathematical Repository*, and the *Educational Times*,

a 55-page series of 'Letters on the Newtonian System' by a Mr. Bar. Prescot.[61] Swale explained this initiative to his readers in this way:

> [t]hough the Editor does not profess a coincidence in all Mr. P's opinions and principles, yet, as an Advocate for unshackled, and even bold, Inquiry, (the inalienable Birth-right of Man) he feels no hesitation in extending the pages of the Apollonius, as a channel of communication; and recommending this interesting subject to the consideration of the curious.[62]

Unfortunately, Swale admitted that the publication of Prescot's work pre-empted other mathematical articles. This extensive publication also necessitated an increase in the price of the journal. Promising to return the price to normal in future numbers, Swale continued, explaining that '[t]he delay and defects of the present Number may derive some apology from continued Indisposition; which suspends all intellectual pursuit'.[63] In his letters, Prescot deemed Newton's 'system, from beginning to end [...] [as] altogether supposititious and imaginary'.[64] Prescot's system of the world, on the other hand,

> shall in no wise be at variance with the senses, nor, consequently, with God's revealed history of the Creation: such as shall enable us to form all our computations for practice, in Astronomy and Navigation, directly from the real appearances and the true distances of the Sun and Moon, moving in circular orbits; and in which will, consequently, be rejected, the expedients of eccentrics and epicycles, the deformity of elliptical orbits, and the monstrous doctrines of the Newtonian perturbations and derangements.[65]

Calling Newton's physics 'monstrous' was never destined to find acclaim among the British public, who generally venerated Newton as its greatest scientist.[66] In the tenuous environment of commercial periodicals, one editorial misjudgement by Swale may have outweighed all of his previous labours with the *Apollonius*, and no further numbers were printed.

At the end of 1823, Swale's health took a downward turn and he was forced to leave his position at the academy he had founded in Liverpool. Although the academy was flourishing at the time and could possibly have been sold for a tidy profit, Swale was reluctant to do so owing to his 'scruples respecting the propriety of such transfers'.[67]

but the most interesting, and at the same time the most difficult portion, of this selection [...] has never yet found its way out of the extensive MS. Collections of the gifted Editor' (Wilkinson, 'Mathematical Periodicals', p. 328).

[61] Wilkinson, 'Mathematical Periodicals', p. 307.

[62] J. H. Swale, 'To Correspondents &c.', in: *Liverpool Apollonius*, no. 2 (1824).

[63] Ibid.

[64] Bar. Prescot, 'Letters on the Newtonian System', *Liverpool Apollonius*, no. 2 (1824), pp. 132–86, p. 152.

[65] Ibid., pp. 176–7.

[66] Prescot, however, was not the only one to criticize Newton at this time. For example, Augustus De Morgan's father-in-law, William Frend, was also critical of Newton's ideas: Kevin C. Knox, 'The Revolting Propositions of Newtonian Mechanicks': Natural Philosophy and the Trial of William Frend', in: *Enlightenment and Dissent* 17 (1998), pp. 126–53.

[67] Wilkinson, 'Memoir of the Late J. H. Swale', p. 225.

A few years later, he lent £1,200—an enormous sum of money at the time—to a close relative. Swale never saw this money again, and in a note found amongst his papers after his death, he lamented that after toiling for 33 years he was unable to provide his children with financial security: 'the poverty to which I and my children are reduced by plundering villains has prostrated all my enjoyments and hopes!'[68]

Nevertheless, Swale was able to continue benefiting from the sociability cultivated by periodicals amongst mathematicians in the North of England. When his good health returned, he travelled often about the country, renewing his personal acquaintances with notable working-class geometers who he had first met on the printed page.

Preston Chronicle (1844–45)

Septimus Tebay (1820–1897) represents another example of a self-taught mathematician from north-west England who used the pages of question-and-answer journals as a means for not only self-improvement but also social mobility. According to an 1856 account in his local newspaper, *The Preston Chronicle*, we read:

> Mr. Tebay was originally a labourer in a mechanics' shop in Preston, and had received scarcely more than the ordinary education of that class. After he had been thus employed for some years, he was upon one occasion attracted by a work on one of the lower departments of mathematics at an old book stall, and purchasing it, it formed for some time the amusement and occupation of his evenings. He speedily made himself master of its contents, and he then pursued the study into the higher branches. His fondness for the pursuit became with him a passion and he soon became known by his contributions to the Ladies' Diary, and other mathematical publications, as a master of the exact sciences. Application to his studies brought upon him a serious illness, which incapacitated him for some time from attending to either work or books.[69]

After recovering from this illness Tebay worked for a few more years at Preston Gas Company while he continued to teach himself mathematics.[70] He also continued to contribute to a variety of mathematical publications, including the mathematics column (1828–46) of a regional newspaper, the *York Courant*. In 1845, Tebay established a similar column in the *Preston Chronicle*. During its two-year run, this column published twenty-seven mathematical questions, a third of which were authored by Tebay himself.

For example:

> Question 25: By Mr. Septimus Tebay, Preston
>
> Let two circles be described to touch a given circle internally and externally at the same point, and also to touch a right line passing through its centre; then the

[68] Ibid., p. 226.

[69] 'University Honours', in: *Preston Chronicle*, issue 2266, Saturday 2 February 1856: https://link.gale.com/apps/doc/Y3207445121/BNCN?u=lancs&sid=zotero&xid=44a262c8. The venerable annual almanack the *Ladies Diary* was by then renamed *The Lady's and Gentleman's Diary.*

[70] Ibid.

> difference of the reciprocals of the radii of these two circles will be of an invariable magnitude, whatever be the position of the line. Required the demonstration.[71]

While he wrote many of the mathematical questions himself, Tebay's column also got support from local mathematicians. For example, Question 28 was posed by Wilkinson:

> Question 28: By Mr. T. T. Wilkinson, Burnley
>
> If the product of any two quantities be equal to four times the square of a third, the sum of the first two quantities will not be less than four times the third. Required the proof.[72]

Looking back on this period, Wilkinson recounted that Tebay occasionally joined the meetings of a group of mathematicians from Blackburn, around ten miles from Preston.[73] According to 'a very distinguished mathematician' in the *British Weekly*,

> Tebay used to walk over, a distance of eight or nine miles, to discuss with us [sc. Blackburn mathematicians] the questions proposed in the Diary. His knowledge of mathematics was wider than ours, and I rather think we learnt more from him than he did from us. We often walked with him half way back to Preston.[74]

The cultivation and circulation of new mathematical knowledge found in the pages of periodicals was thus complemented by in-person interactions between the contributors. Much as they would compete against each other for the privilege of having a solution published or the award offered for a prize problem, these same men actively sought each other's company for collaboration and social interaction.

In the early 1850s, Tebay was given the opportunity to study at the University of Cambridge, the university in nineteenth-century Britain that was by far the most well respected for the mathematical education it provided. Thanks to the moral and financial support of a group of gentlemen, Tebay was able to leave paid employment and devote three years to the study of higher mathematics in preparation for sitting the prestigious Mathematical Tripos examination. Only a few weeks before the final exams in 1856, Tebay was however taken ill and had to leave Cambridge to convalesce; he returned in slightly better health and was ranked 27th wrangler. All students who sat the Tripos were ranked by their examination results, with the title of Senior Wrangler given to the student with the highest marks. Results of these examinations were frequently reported by newspapers across Britain, and indeed an announcement of Tebay's success appeared in the *Preston Chronicle*.[75] The short announcement conveys palpable pride that a townsman of Preston had been so distinguished at Cambridge, noting that Tebay was frequently placed alongside the Senior Wrangler in the

[71] 'Mathematics', in: *The Preston Chronicle*, issue 1703, Saturday 19 April 1845.

[72] 'Mathematics', in: *The Preston Chronicle*, issue 1713, Saturday 28 June 1845. This question was solved by Tebay.

[73] T. T. Wilkinson, quoted in Abram, 'Memorial', p. 88.

[74] 'Rambling Remarks', in: *British Weekly: A Journal of Social and Christian Progress*, 1 July 1897, p. 184.

[75] 'University Honours'.

smaller college examinations, and that he achieved the title of wrangler even while suffering ill-health. After obtaining his Cambridge honours, Tebay went on in 1857 to become a Headmaster at Rivington Grammar school in Lancashire, where he stayed until 1875.[76]

While at Cambridge, Tebay engaged in a heated exchange with Wilkinson on the pages of the *Mechanics Magazine.* At the heart of their argument were dual accusations of plagiarism. In his article surveying the 'question for answer' journal *The Northumbrian Mirror* (1837–41), Wilkinson noted the editors' complaints that the authors of solutions might not be the actual solvers of the problem, and went on to provide a concrete example: 'The prize in one of our most valued annual publications has recently been awarded to a Lancashire gentleman, whose solutions are well known to have been furnished to him by a friend at Cambridge'.[77] Tebay, assuming that he was the 'friend at Cambridge', wrote a scathing letter to the editor in the next volume of the *Mechanics Magazine*, stating that he only clarified to the 'Lancashire gentleman' the wording of the prize problem he had posed in the *Lady's and Gentleman's Diary* for 1853.[78] He then accused Wilkinson of sharing solutions with a mutual friend, 'that is, [they] made mutual exchanges, in order that both might make a greater score than either could separately'.[79] The *Lady's and Gentleman's Diary* awarded prizes of free copies of the *Diary* for both prize problems (such as the one Tebay authored) as well as for 'general mathematical answers'.[80] Tebay continued:

> The system of exchange alluded to above, I believe to have been practised to a considerable extent among the present degenerate race of 'Lancashire mathematicians', of the glory of whose ancestors so much has been said. Such, however, is the general wreck, that, with one or two solitary exceptions, there does not remain a single spark of that sterling genius which characterises the labours of Butterworth, Smith, Swale, and Wolfenden.[81]

In fact, in the 1873 account of scientists in 'humble life', Manchester journalist James Cash described John Butterworth (1774–1845) of Haggate, near Oldham, as being annoyed but 'obliged by his poverty to answer mathematical questions brought to him by persons who wished to insert them in newspapers as their own, and who paid him sixpence or a shilling for his services'.[82] In his retort, Wilkinson wrote that 'I am in possession of undeniable evidence that as much as £2 have been paid for a single solution'. He pointed out that 'it would have been something more than simply ridiculous for me to have pointed out the system of deception' had he also been a part

[76] Wm. Fergusson Irvine, *A Short History of the Township of Rivington in the County of Lancaster with some Account of the Church and Grammar School*, Edinburgh: Ballantyne Press 1904, p. 122.

[77] Thomas Turner Wilkinson, 'Mathematical Periodicals', in: *Mechanics Magazine* 59 (1853), pp. 528–9 at p. 529.

[78] Tebay had posed problem 1868: 'Show how three billiard balls must be struck simultaneously, so that each ball shall *just* touch the other two' (*Lady's and Gentleman's Diary for 1853* (1852), p. 78).

[79] Septimus Tebay, 'To the Editor of the Mechanics' Magazine', in: *Mechanics Magazine* 60 (1854), pp.158–9 at p. 159.

[80] See for example *Lady's and Gentleman's Diary for 1854* (1853), p. 64.

[81] Tebay, 'To the Editor', p. 159.

[82] James Cash, *Where there's a will there's a way. Or, Science in the cottage: An account of the Labours of Naturalists in Humble Life*, London: Robert Hardwicke 1873.

of it. While offended by Tebay's accusation, Wilkinson pledged to 'continue to expose the system as occasion occurs'.[83] Tebay, in the final volley of this back-and-forth affair wrote that 'I hope Mr. Wilkinson does not mean to charge me with having received pecuniary aid in the way he alludes to'. He then cited a letter to himself from Wilkinson in which the latter asks for hints on prize problems. Instead of an F.R.A.S. (Fellow of the Royal Astronomical Society), Tebay said that Wilkinson's title 'would *sound* much more consonant, in the present instance, if written F.A.R.S.'[84]

Conclusion

In his evaluation of the 'Modern Geometry' section of *The Student*, Davies lamented that

> The inquiries have in several cases been successfully followed up in those different periodicals which are principally devoted to mathematics, but in so unconnected a form (which is unavoidable in those works) that they are comparatively little known, and their relations are as little perceived.[85]

He then reflected that

> [s]pecial inquiries are prosecuted with great vigour and acumen; but we look in vain for system, classification, or general principles [...] in truth, it must be confessed to be a vice [...] almost universal amongst English geometers; and even in the geometry of the Greeks themselves, the great object appears to have been 'problem-solving' rather than deduction and arrangement of scientific truths. The modern French geometers have, however, broken this spell.[86]

Wilkinson lamented that '[i]solation and promiscuous arrangement are indeed, among the characteristics of the Geometry of the Lancashire School'.[87] In fact, the publication format available to these mathematicians—questions for answer—circumscribed the reach, coherence, and direction of their research programme in geometry.

Appearing in print in these question for answer journals brought satisfaction and local renown to those who approached mathematics purely as an avocation. Moreover, it brought acceptance and job prospects, especially in military schools, to those who hoped their mathematical attainments would help them climb the social ladder.

[83] Thomas Turner Wilkinson, 'Mathematical Periodicals', in: *Mechanics Magazine* 60 (1854), pp. 182–3.
[84] 'F.A.R.S.' being a pun on the word 'farce'. Septimus Tebay, 'To the Editor of the Mechanics' Magazine', in: Mechanics Magazine 60 (1854), p. 200.
[85] [Davies], 'On the Cultivation', p. 438.
[86] Ibid.
[87] Wilkinson, 'The Lancashire Geometers', p. 140.

7

Collection, Use, Dispersal

The Library of Charles Hutton and the Fate of Georgian Mathematics

Benjamin Wardhaugh

Introduction

June 1816 saw the sale at auction of the remarkable mathematical library of Charles Hutton (1737–1823). Hutton's had been one of the leading voices in English writing about mathematics for two decades, and the dispersal of his collection of mathematical books attracted regretful comment from various quarters. It is fortunate for the historian that we possess a detailed inventory, since his library provides an unusual snapshot of the mathematical culture of which he was part, and which was—like the library itself—in the process of being swept away during the early decades of the nineteenth century. This chapter examines Hutton's activities as a collector of mathematical texts from various sources, his work as a disseminator of what he learned from those texts, and the circumstances that led to the dispersal of his library and the replacement of his mathematical culture.

Hutton as collector

Hutton's origins lay in Newcastle upon Tyne and its environs; his family was a humble one and it is likely that he worked in the coal pits for a period during his childhood.[1] Nevertheless, he early acquired a reputation as a buyer and collector of books; obituarists apparently relying on his personal reminiscences spoke of his enthusiastic purchase of books of ballads even in childhood,[2] while another anecdote relates his destruction of this material under the influence of devotional tracts and his

[1] See Benjamin Wardhaugh, *Gunpowder and Geometry: The Life of Charles Hutton, Pit Boy, Mathematician and Scientific Rebel*, London: William Collins 2019, and the evidence cited there, especially John Bruce, *A Memoir of Charles Hutton*, Newcastle 1823. Personal communication from Alyson Pigott at Tyne and Wear Archives and Museums indicates that the paybill seen by Bruce can no longer be located.

[2] Anon, 'Charles Hutton', in: *Public Characters*, 10 vols, London, 1799–1809, vol. 2, pp. 97–123 at pp. 100–1, 103. See Anon, *A Catalogue of the Entire, Extensive and Very Rare Mathematical Library of Charles Hutton, L.L.D.*, [London 1816]; subsequent references to items owned by Hutton are to this (printed) catalogue unless otherwise specified. A more complete version of Hutton's library catalogue is now available online: see http://www.benjaminwardhaugh.co.uk/HuttonsBooks/. Relevant items in the catalogue include Allan Ramsay's *Poems* (1761) and a collection of Scots poems of before 1600 (1761); also the *Gentle Shepherd* (Newcastle, 1760), *Caledonian Miscellany* (Newcastle, 1762) and *Tim Bobbin's Toy-Shop opened* (Manchester, 1763).

conversion to Methodism.[3] His library as catalogued in 1816 contained clutches of items that clearly related to certain periods of his life: two English grammars from the 1760s indicating a period of anxiety about his English skills around the time he first became a published author, for instance;[4] a number of books from the 1780s onwards relating to property law and finance, from the period when Hutton was acquiring land and erecting buildings of his own.[5] From the period of his widely publicised dispute with the Royal Society in 1784 he kept copies of the pamphlets published for and against him and his friends.[6] He also kept copies of other works critical of him, including those of Thomas Saint and Robert Woodhouse from the years around 1810. Hutton's more general immersion in the print culture of his period is witnessed by long series of periodicals: over 200 volumes in total. At one time or another he subscribed to the *Gentleman's Magazine*, the *Monthly Magazine*, the *Monthly Review*, the *Philosophical Magazine*, and the short-lived *Projector*, as well as to the *Philosophical Transactions of the Royal Society* and the *Transactions* of the American and the Edinburgh Philosophical Societies. He acquired copies of general reference works such as the *Encyclopaedia Britannica*, the *Encyclopedia Londoniensis* and the *Spectator*. He also owned about sixty pamphlets and papers produced by the Newcastle Literary and Philosophical Society, reflecting his particular interest in that institution and his continued connection with Newcastle after his move to Woolwich in the 1770s.[7]

But it is naturally as a collection of mathematical books that his library is of most interest. He acquired these books during a period of intense mathematical education from his teenage years onwards, and continued to acquire them by purchase, subscription and gift until (at least) the end of 1815 when his library was catalogued. At that time, the mathematical and scientific items numbered something like three thousand. The coverage of instructional works is conspicuous. Hutton, having learned mathematics himself, spent a decade delivering school-level instruction followed by forty years as mathematical professor at the Royal Military Academy (RMA); a friend spoke of his 'never-failing love of the act of communicating knowledge by oral instruction'.[8] He appears to have continued to acquire instructional books on mathematics throughout his career. Hutton's career and his book-buying years very largely fell after the watershed of the mid-century, when it became usual for each pupil—rather than each teacher—to own a copy of a cheap mathematical textbook. Francis Walkingame had been the pioneer in this regard, pressing for the practical advantages; Hutton followed him, writing that each scholar 'ought' to 'have a printed book', and indeed imagining them 'having the book always in their pockets, to get off any rules or tables

[3] Eneas Mackenzie, *A Descriptive and Historical Account of Newcastle-upon-Tyne*, Newcastle 1827, p. 560.

[4] Lowth's *Introduction to English Grammar*, 1763 and 1769 editions.

[5] Mortimer's *Every man his own Broker* (1785); Paul/Wilson on *Laws relating to Landlords and Tenants* (1791), Paul's *Parish Officer's Guide* (1793) and Sugden on the *Sale of Estates* (1809).

[6] Peter Pindar's works in three volumes (Dublin 1792). In total Hutton had 19 copies of the *Narrative of the Dissensions* (London, 1784) and *Appeal to the Royal Society* (London, 1784), suggesting, indeed, that he had a hand in their production.

[7] Anon, *Catalogue of a Miscellaneous Collection of Books: being the valuable and scientific library of the late Dr. Olinthus Gregory ... which will be sold by auction by Messrs. Southgate and Son ... on Thursday, March the 17th, 1842 and following day*, [London 1842], p. 9.

[8] Anon, [Obituary of Charles Hutton], in: *The Mathematical Repository* (N.S.) 5 (1830), pp. 187–96, at p. 194.

when they are out of school'.[9] Pupils at elementary level would almost all have prepared a 'cyphering book' consisting of extracts and worked examples;[10] dozens if not hundreds of these would have been prepared under Hutton's eye, but this is one type of mathematical text that does not appear in his library catalogue.

At a somewhat higher level, Hutton had the textbooks of Moore and Ward, and Emerson's *Cyclomathesis*. These were apparently intended for use in the grammar schools to provide students with the grounding for the mathematics they would study at the English or Scottish universities. Nineteenth-century official figures would show that only about 70% of those in public schools and only about a third of those in private schools were taught arithmetic.[11] But many specialist private schools specialized either in mathematics or in preparation for military training: the latter was conceived as involving a significant amount of mathematical work. Cyphering books which surely originated in these contexts are numerous, and show pupils continuing a study of algorithmic calculation through fractions, decimals, and the more advanced techniques of proportional reasoning, as well as, in some cases, the study of mensuration and the techniques of surveying and navigation.[12]

The military academies, another important and much-emulated educational site, on the whole took mathematics seriously: more and more so during the eighteenth century. The example of the Royal Mathematical School at Christ's Hospital seems to have been important nationally in setting both the style of the mathematical teaching at military schools and the particular content of what was taught, even perhaps down to the order in which topics were treated; a recent historian of the institution writes of its stranglehold over British mathematical curricula.[13] Other military schools appointed distinguished mathematicians as professors, and they in turn did much for the visibility of mathematics nationally and of British mathematics internationally. Hutton in particular was able to gain quite a lot of power over the curriculum at the RMA, where he was Professor of Mathematics from 1773 to 1806, and once he had established a mathematically based course at that institution his national prominence enabled him to export it to other institutions through his textbooks and his disciples.[14]

The English universities retained a commitment in principle to teaching at least part of Euclid's *Elements* as part of the arts curriculum. Daniel Waterland's *Advice to a Young Student*, written around 1706 and first published in 1740, listed the

[9] Charles Hutton, *The School-master's Guide: or, a complete system of practical arithmetic, adapted to the use of schools*, Newcastle 1764, pp. ii, iii.

[10] Nerida Ellerton and M. A. Clements, *Rewriting the History of School Mathematics in North America 1607–1861: The Central Role of Cyphering Books*, Dordrecht: Springer 2012; John Denniss, *Figuring it Out: Children's Arithmetical Manuscripts, 1680–1880*, Oxford: Huxley Scientific Press 2012.

[11] Denniss, *Figuring it Out*, p. 4; see also A. G. Howson, *A History of Mathematics Education in England*, Cambridge: Cambridge University Press 1982.

[12] Ashley Smith, *The Birth of Modern Education: The Contribution of the Dissenting Academies, 1660–1800*, London 1954, p. 250.

[13] Nerida Ellerton and M. A. Clements, *Samuel Pepys, Isaac Newton, James Hodgson, and the Beginnings of Secondary School Mathematics: A History of the Royal Mathematical School Within Christ's Hospital, London 1673–1868*, Cham: Springer 2017.

[14] W. D. Jones, *Records of the Royal Military Academy*, 2nd edn, Woolwich: Royal Artillery Institution 1895, pp. 18, 20 (entrance exam), 23, 28. Cambridge: University Library, RGO 4/187/18 (Charles Hutton to Nevil Maskelyne, Woolwich, 19 December 1793) gives some hints about how Hutton handled the students at Woolwich.

books necessary for university study in England. The mathematical books listed include Euclid's *Elements*, the *Arithmetic, Trigonometry* and *Astronomy* of Webster Wells, Philippe de La Hire's *Conic sections*, and the *Trigonometry* and *Optics* of Isaac Newton. On astronomy and mathematical natural philosophy Waterland mentioned works by William Whiston, John Keill, and David Gregory. Hutton owned them all. Although delivery of this curriculum was reportedly patchy by mid-century, college lecturers in mathematics continued to be appointed, certain colleges introduced examinations in mathematics, and editions of the *Elements* intended for university use continued to be produced and therefore presumably to be purchased. By the late century the mathematical tripos (properly named the Senate House Examination) was taking shape at Cambridge,[15] with its culture of intensive study with a tutor and repetitive private rehearsal of mathematical proofs and techniques.[16]

Finally, some of the Dissenting Academies took the view that mathematics was of increasing importance in the modern world and went out of their way to teach a good deal of it. On the whole they taught algebra and calculus and their applications, rather than Euclid. Pupils at the dissenting academies were (relatively) conspicuous in the mathematical periodicals, and graduates of those academies filled some of the key roles in British mathematical life, particularly towards the end of the Georgian period. But other dissenting academies took an opposite line, reasoning that mathematics in particular or abstraction in general were dangerous distractions from the serious business of intellectual life, and refusing to have anything to do with them. The result was that the influential class of nonconformist intellectuals and ministers included a wide range of levels of mathematical attainment, from the very highest proficiency to practically nil.[17]

As well as textbooks, Hutton had an ample store of the mathematical periodicals. His library contained about 170 bound volumes of almanacs, going back to the early seventeenth century. Hutton himself sometimes disparaged this material; in a letter he wrote mockingly of their prognostications as 'belonging to a Science far above my pitch'.[18] But the almanacs had a close cousin in the philomath journals, which were much more highly regarded and important to British mathematical culture.

The original and arguably the best of these journals was the *Ladies' Diary*, founded in 1704 and finally terminated in 1871.[19] Formally, it was an annual almanac, the year's

[15] John Gascoigne, *Cambridge in the Age of the Enlightenment: Science, Religion and Politics from the Restoration to the French Revolution*, Cambridge: Cambridge University Press 1989, pp. 8, 183.

[16] Andrew Warwick, *Masters of Theory: Cambridge and the Rise of Mathematical Physics*, Chicago: University of Chicago Press 2003, p. 37.

[17] Smith, *Birth of Modern Education*; Irene Parker, *Dissenting Academies in England, their Rise and Progress and their Place among the Educational Systems of the Country*, Cambridge 1914; David A. Reid, 'Science and Pedagogy in the Dissenting Academies of Enlightenment Britain', PhD thesis, University of Wisconsin–Madison 1999.

[18] Charles Hutton to Robert Harrison, 19 March 1781. Printed in Sidney Melmore, 'Some Letters from Charles Hutton to Robert Harrison', in: *The Mathematical Gazette* 30 (1946), pp. 71–81, at pp. 78–9; also in Benjamin Wardhaugh (ed.), *The Correspondence of Charles Hutton (1737–1823): Mathematical Networks in Georgian Britain*, Oxford: Oxford University Press 2017, letter 15.

[19] Teri Perl, '*The Ladies' Diary or Woman's Almanack*, 1704–1841', in: *Historia Mathematica* 6 (1979), pp. 36–53; Shelley Costa, 'The *Ladies' Diary*: Society, Gender and Mathematics in England, 1704–1754', PhD thesis, Cornell University 2000; Shelley Costa, 'The *Ladies' Diary*: Gender, Mathematics, and Civil Society in Early-Eighteenth-Century England', in: *Osiris* 17 (2002), pp. 49–73; Joe Albree and

calendar filling the first 24 pages; set up initially as an almanac aimed at women it quickly became—by the choice of its contributing readers—a well-known forum for setting and answering mathematical problems. These ranged from relatively trivial puzzles and calculations to questions whose answers required formal proof in Euclidean or algebraic style. The *Diary* found many imitators; during the eighteenth century there were at least thirty more or less long-lived periodical publications containing similar material.[20] The *Gentleman's Diary* adopted a very similar format to the *Ladies'* but was subtly differentiated in its content; at least during the middle part of the century it contained a higher proportion of Euclidean material, reflecting the fact that Euclid was part of the education of a gentleman but not of a lady: the *Ladies' Diary* by contrast contained a higher proportion of algebra.[21] The two converged in their interests by the end of the century, and indeed in 1841 merged into the *Ladies' and Gentleman's Diary*. The total number of individuals participating actively by sending in solutions to the philomath journals was probably at any moment several hundred or a few thousand, while those participating passively as readers may have numbered in the tens of thousands. Some provincial newspapers (the *Newcastle Courant* was one) also printed mathematical problems and solutions.

Correspondence from the period gives a clear impression of the enthusiasm with which readers engaged with the problems in the periodicals, circulating material cribbed from proof sheets, exchanging gossip about who had solved (or claimed to be able to solve) what, and whose solutions had been included or excluded.[22] Mary Croarken notes that 'for about one-third of the [Royal Observatory] assistants the *Ladies' Diary* was an important part of their mathematical education'.[23] Hutton himself would state in the 1770s that the *Ladies' Diary* had 'made more Mathematicians within the last 60 or 70 Years, than half the Books that have been written upon the Subject';[24] Thomas Leybourne wrote in 1801 that 'there are scarcely any mathematicians of eminence in this country, who will not readily acknowledge, that in one part or other of their scientific course, they have been assisted by such works';[25] in 1823 the periodicals were noted as being 'exceedingly instrumental in exciting and augmenting a love of literature and science among the middle classes of society in England'.[26] Hutton's library bears witness to all of this, containing a complete set of the *Ladies' Diary* whose acquisition he stated cost him 'much trouble', as well as sixty issues of the *Gentleman's Diary* among other periodicals.

Scott H. Brown, '"A Valuable Monument of Mathematical Genius": *The Ladies' Diary* (1704–1840)', in: *Historia Mathematica* 36 (2009), pp. 10–47.

[20] Niccolò Guicciardini, *The Development of Newtonian Calculus in Britain 1700–1800*, Cambridge: Cambridge University Press 1989, p. 115.

[21] Shelley Costa, presentation at workshop on Georgian mathematics, All Souls College, Oxford, December 2015.

[22] Cambridge: University Library, RGO 35/115: David Kinnebrook to his father, 29 October 1795.

[23] Mary Croarken, 'Astronomical Labourers: Maskelyne's assistants at the Royal Observatory, Greenwich, 1765–1811', in: *Notes and Records of the Royal Society of London* 57 (2003), pp. 285–98, at p. 296.

[24] *Public Advertiser* (London, England), Thursday 4 July 1771, advertisement for Hutton's *Miscellany*; cf. *Public Characters*, vol. 2, p. 99.

[25] Thomas Leybourne, 'Preface', in: *The Mathematical Repository* 2 (1801), pp. v–vi at p. v.

[26] Olinthus Gregory, 'Brief Memoir of the Life and Writings of Charles Hutton', in: *Imperial Magazine* 5 (March 1823 [obituary dated 1 February]), pp. 201–27 at p. 205.

Another large category in Hutton's library was the professional mathematical books: books by and for mathematical practitioners and members of the numerate trades. Gaugers, surveyors, builders, and accountants were among those whose need for mathematical rules along with numerical tables and ready reckoners was catered for by an extensive printed literature, including works aimed at both adult and child learners.[27] Hutton had textbooks on all of these subjects. Navigation was a practice that became increasingly numerate during the century; the advent of the *Nautical Almanac* in 1767 created an expectation that at least some navigators would learn how to take and process lunar observations so as to find their longitude at sea.[28] Some of the skills were much the same as those involved in reading any almanac or extracting information from any table of numbers, but others were much more specialized, including the use of the quadrant and the detailed computational recipe that had to be followed. Training was provided by the Board of Longitude at certain ports in an attempt to create a sufficiently numerate body of navigators to make use of the new almanac. Hutton possessed fifty issues of the *Nautical Almanac*, as well as the three volumes of the 'tables requisite' for its use.

Hutton also seems to have acquired printed copies of practically everything that could reasonably be described as British mathematical research from the period of his own career. He had copies of the nearly 100 volumes of the *Philosophical Transactions*, for instance, which by his own estimate contained a good deal of mathematics and mathematical science.[29] He had the works of Nevil Maskelyne, John Bonnycastle, William Emerson (17 volumes), James Ferguson, Benjamin Franklin, Olinthus Gregory, John Leslie, Benjamin Martin, John Muller, Joseph Priestley, and Thomas Simpson. Mathematical novelties were produced by practitioners, by mathematics teachers, and particularly by the elite mathematical teachers at the RMA and the Royal Military College (RMC): the former acquired a strong reputation for applied and the latter for pure research by the end of the century. And there were the mathematical professors at Gresham College, at Oxford and Cambridge, and at the Scottish universities and Dublin, all of whose employment envisaged at least the possibility of doing new mathematical work. Many of these books would have been presented to Hutton by their authors; some were acquired by subscription (Hutton subscribed to at least 29 books, including works of Antonio Mario Lorgna, Margaret Bryan, Joseph Priestley, and John Dalton).[30]

Hutton—as did many of his contemporaries—certainly also used correspondence and the circulation of manuscripts as sources of mathematical text. From the mid-1770s onwards his correspondence was of notable size throughout Great Britain and Ireland. I have estimated elsewhere that his total correspondence is unlikely to have

[27] E. G. R. Taylor, *The Mathematical Practitioners of Hanoverian England, 1714–1840*, Cambridge: Cambridge University Press 1966.

[28] Mary Croarken, 'Providing Longitude for All: The Eighteenth Century Computers of the Nautical Almanac', in: *Journal of Maritime Research* 4 (2002), pp. 106–26; Mary Croarken, 'Tabulating the Heavens: Computing the Nautical Almanac in 18th-Century England', in: *IEEE Annals of the History of Computing* 25/3 (2003), pp. 48–61.

[29] See below of Hutton's *Abridgement* of the *Transactions*.

[30] Ruth Wallis and Peter John Wallis, *Index of British Mathematicians. Part III, 1701–1800*, Newcastle upon Tyne: University of Newcastle upon Tyne 1993, p. 72.; cf. Peter Wallis and Ruth Wallis, *Mathematical Tradition in the North of England*, Durham: NEBMA 1991, p. 13.

numbered fewer than 10,000 items,[31] and at the time his library was catalogued in 1815–16 it is likely that many of those items were still present in his house, although they were not included in the catalogues. Hutton occasionally acknowledged his informants in print, on matters ranging from methods of gauging casks and solutions of recreational problems to the provision of detailed data in experiments on gunpowder. Conversely, at least once in his publications he requested improvements to his books to be sent to him or his publishers, thereby acknowledging the possibility of readers' agency in getting the details right.[32] Early in his London career he sent drafts of his papers to Nevil Maskelyne for comment, and more than once he acknowledged the 'generous advice and assistance' of the Astronomer Royal as well as his contribution of particular details or passages.[33]

Hutton collected antiquarian mathematics, too: he had a small collection of editions of Newton's *Principia* and a larger one—twenty-odd copies—of Euclid's *Elements*; he had printed mathematical books from the sixteenth and seventeenth centuries including classics such as works by François Viète and Niccolò Tartaglia as well as more obscure items.[34] He also acquired a number of manuscripts including a medieval Bible on vellum, a manuscript copy of *Elements* I–VI, and works on mechanics, gunnery, and astronomy. Smaller items acquired through personal correspondence included papers of Henry Cavendish on cometary orbits and of Francis Maseres on problems including the motion of a vibrating line.[35] Major Edward Williams of the Royal Artillery forwarded Hutton accounts of his scientific experiments in Quebec;[36] various series of experiments on gunnery also reached Hutton from Gibraltar and elsewhere.[37] Through his hands passed the books and papers of Edward Rollinson in 1773; Hutton purchased part of the papers of John Robertson which in turn included papers of William Jones; he also seems to have seen the papers of John Whitehurst.[38]

[31] Wardhaugh, *Correspondence*, p. xxviii.

[32] See Benjamin Wardhaugh, 'Rehearsing in the Margins: Mathematical Print and Mathematical Learning', in: *The Palgrave Companion to Mathematics and Literature*, ed. Robert Tubbs, Alice Jenkins, and Nina Engelhardt, London: Palgrave Macmillan 2021, pp. 553–67.

[33] Cambridge: University Library, RGO 4/187/11: Charles Hutton to Nevil Maskelyne, 27 June 1785; printed in Wardhaugh, *Correspondence*, letter 31. Charles Hutton, *Mathematical Tables*, London 1785, dedication.

[34] Cambridge: Trinity College Library, R.1.59: Miscellaneous mathematical manuscripts by Charles Hutton and others, contains at 64^{r}–66^{r} some of Hutton's summaries and notes on his older mathematical books.

[35] Cambridge: Trinity College Library, R.1.59, ff. 168, 170: Henry Cavendish to Charles Hutton, 7 July 1779; printed in Wardhaugh, *Correspondence*, letter 10; also ff. 155^{r}–165^{r} ('On finding the Orbits of Comets in a Parabola from 3 Observations. (By the Hon[oura]ble Henry Cavendish) (Copied June 1779.) C.H.'), 90^{r}–105^{v} ('Mr Maseres's Problem of a Vibrating Line', followed by further notes of Hutton and Cavendish on the solution of the problem).

[36] Charles Hutton, 'Experiments on the Expansive Force of Freezing Water, made by Major Edward Williams of the Royal Artillery, at Quebec in Canada ... Communicated in a letter from Charles Hutton ... to Professor John Robison, General Secretary of the Royal Society of Edinburgh', in: *Transactions of the Royal Society of Edinburgh* 2 (1790), pp. 23–8. Reprinted in *The Literary Magazine and British Review* 6 (January 1791), pp. 20–2.

[37] London: Royal Artillery Museum ('Firepower': closed at the time of writing; materials seen in February 2016), MD/913/2 ('Experiments in bursting ?Cochorn shells in a case in Queens Lines Gibraltar. 13th January 1781'., apparently communicated by a Captain Seward).

[38] Anon, *A Catalogue of the Curious Mathematical, &c Books of the Late Mr. Edw. Rollinson* [London, 1775]; Cambridge: Trinity College Library, MS R.1.59 52^{v}–63^{v} (items copied from Robertson's papers by

Hutton's correspondence extended overseas; we have his testimony that he was in touch with his 'worthy friend' Christiaan Damen in Leiden concerning Arabic mathematics.[39] And in view of what would later happen to British mathematics it is of interest to notice how far his collection of mathematical texts included non-British materials. Foreign mathematical works of the eighteenth century were numerous in his library, from Jean le Rond d'Alembert to Pierre Varignon: there were more than forty authors and more than eighty volumes, plus foreign serials totalling more than fifty volumes. He had multiple volumes of such advanced writers as Alexis Clairaut, Leonhard Euler, Jérôme Lalande, and Pierre-Simon Laplace. Hutton could read and write French,[40] and had dictionaries and/or grammars also in Italian, Spanish, Dutch, Greek, and German; he owned works in Latin and French, and occasionally Italian and German, plus one Dutch Bible and one Greek New Testament. He admired Continental mathematics; remarking on a recent French version of William Gardiner's tables he wrote that 'it is but justice to remark the extraordinary spirit and elegance with which the learned men and the artisans of the French nation undertake and execute works of merit'.[41] His admiration of French culture also played out in the placement of his youngest daughter Charlotte—and probably her sister Eleanor—at a convent school in France for a time in the 1780s.[42] The exact reasons for the decision are nowhere recorded, and the girls were of course brought safely home once political trouble began (his library contained a 1781 plan of Paris, very probably in connection with this sojourn, though it is not known that Hutton himself ever left England).

Hutton was not exceptional: recent research is increasingly tending to show that mathematical material both from the centres of research at Paris and St Petersburg, and from further afield, was coming into Britain, although it is certainly true that its impact was limited.[43] The Royal Society, for instance, collected (mainly by gift) a large quantity of foreign mathematical publications; the British reviewing journals did their best to keep pace with Continental scientific works including mathematical works, although they were hampered at times by the practical conditions of war. Similarly, the commercial scientific journals of William Nicholson, Alexander Tilloch, Thomas Thomson, and others aimed to 'provide readers with access to the scientific discoveries of other countries',[44] and summaries or extracts from many French works

Hutton; see also London: British Library, Loan 96 RLF 1/315/2: Patrick Kelly and Charles Hutton to the Committee of the Literary Fund, April 1814 on behalf of Robertson's daughter Ann Coppard); Hutton, *Tables*, p. 117; Wallis and Wallis, *Index*, p. 6.

[39] Charles Hutton, *A Mathematical and Philosophical Dictionary*, London: Johnson 1795–6, vol. 1, p. 67.

[40] See letters 24 and 120 in Wardhaugh, *Correspondence* for specimens (one certain, the other less so) of Hutton's written French.

[41] Hutton, *Tables*, p. 41.

[42] Anon, [Obituary of Charlotte Hutton], in: *The Gentleman's Magazine* (October 1794), pp. 960–1; Portsmouth History Centre, Vignoles Papers, Letter 127 (Eleanor Wills (née Hutton) to Charles Blacker Vignoles, December 1816).

[43] Jonathan R. Topham, 'Science, Print, and Crossing Borders: Importing French Science Books into Britain, 1789–1815', in: *Geographies of Nineteenth-Century Science*, ed. David N. Livingstone and Charles Withers, Chicago: University of Chicago Press 2011, pp. 311–44 at p. 312; Alex D. D. Craik, 'Mathematical Analysis and Physical Astronomy in Great Britain and Ireland, 1790–1831: Some New Light on the French Connection', in: *Revue d'histoire des mathématiques* 22 (2016), pp. 223–94, *passim*.

[44] Topham, 'Science, Print, and Crossing Borders', p. 320.

on subjects like heat, light, and electricity appeared in British journals.[45] Hutton's colleagues such as Peter Barlow, Olinthus Gregory, and John Bonnycastle were thus able to acquire an extensive knowledge of French work on practical subjects such as the strength of materials, magnetism, and ballistics.[46]

Foreign correspondence, however, is extremely rare in Hutton's surviving letters. We have just one letter from Edme-Sébastien Jeaurat (who visited Hutton, probably in 1781) and one from Laplace, as well his own statement that he was in touch with Damen.[47] Out of more than fifty known correspondents that is not many, and gives an impression that Hutton was keener to acquire print rather than manuscript material from overseas.

Hutton's interests as a mathematical historian ranged beyond Europe, and his library had copies of Edward Strachey's and John Playfair's works on Indian algebra and astronomy. He reported that he had seen specimens of algebraic works from India, 'both in the native language and in Persian translations' with partial translations into English, at the hands of Samuel Davis, one of the directors of the East India Company;[48] he also saw translations sent to England by Strachey. He went so far as to copy out by hand some of Reuben Burrow's articles in the *Journal of Asiatic Researches* on astronomical subjects.[49]

Hutton as interpreter and writer

Up to now I have attempted to give a sense of the range of mathematical texts being produced in, or imported into, later eighteenth-century Britain and consumed there. I have also sought to give a sense of the scale and range with which Charles Hutton collected specimens of those texts in his remarkable library. Hutton did not acquire mathematical texts in print and manuscript merely to improve his own mind, however, nor indeed to fuel his own *viva voce* teaching. He also worked over many years as an author of his own mathematical texts, in which he re-presented much of what he had assimilated from others. One of his most notable functions, indeed, was as a gatekeeper through whom historical and foreign mathematics passed before reception by a British public. When his library was sold, he disposed of 66 volumes of which he was the author or editor, as well as more than 150 offprints and pamphlets; more than 500 further volumes of his own works were disposed of after his death.

Hutton's first book was *The School-Master's Guide*, 'an Attempt to introduce a regular and rational Method of Teaching this most necessary Science [sc. arithmetic] into the generality of Schools, and to ease the Masters of part of the Labour which

[45] Craik, 'Mathematical Analysis', p. 233; cf. Guicciardini, *Development*, p. 119.

[46] Guicciardini, *Development*, p. 108.

[47] Letters 16 and 122 in Wardhaugh, *Correspondence*: respectively Washington D.C., Smithsonian Institution, MS 752 A (Edme-Sébastien Jeaurat to the Misses Hutton, 20 October 1781) and *Philosophical Magazine* 1/56 (November 1820), pp. 321–2 (Pierre-Simon de Laplace to Charles Hutton, 11 September 1820).

[48] Charles Hutton, *Tracts on Mathematical and Philosophical Subjects*, 3 vols, London: Rivington et al. 1812, vol. 2, Tract 33 (History of algebra), p. 153; cf. p. 163.

[49] Cambridge: Trinity College Library, R.1.59, 124^{r}–153^{v}: 'Extracts from the Asiatic Researches Vol. 1 & 2'.

necessarily attends their Business without such a Help'.[50] Hutton's love for ordering, reordering, and improving what he found in the works of others was evident throughout, particularly in his 'Definitions delivered in a Manner proper for Children to copy from' and the preface 'containing some Hints towards a proper Method of teaching this useful Art'.[51] He used the textbook form to present an authoritative and deceptively cosmopolitan *persona*, referring repeatedly to long-distance travel, to international trade, and to the large-scale purchase and sale of various commodities and investments. Individual problems derived from a long line of predecessors, who were not acknowledged.[52] The book was emphatically not meant for self-study: Hutton's gnomic explanations would usually have required the elucidation of a teacher in order for an uninitiated student to make any sense of them.

Hutton went on to notable success with his 1770 account of mensuration, which more resembled a reference book or encyclopaedia than either a practical manual or a schoolbook. The range of sources was wide, long lists of ancient and modern names were cited, and there was a notable historical dimension.[53] A cut-down version for schools appeared in 1786.[54]

The major textbook of Hutton's maturity, though, was the *Course of Mathematics*, which appeared in 1799 and was intended first for the use of the cadets at the RMA. He noted in an introduction that he and his colleagues had been accustomed to teach from several books by several authors 'selecting a part from one and a part from another': this was a hassle and an expense, and was far from ideal because of the non-uniformity of the material.[55] The *Course* thus exemplified Hutton's activity as a collector and digester of a range of mathematical sources into uniform wholes for the use of his contemporaries. As originally published, the *Course* covered arithmetic, algebra, geometry, applications, and fluxions,[56] though subsequent additions and re-editions did much to upset its arrangement.

With respect to philomath culture, Hutton had been a contributor from his early adulthood, writing problems or solutions for the *Ladies' Diary*, the *Gentleman's Diary*, and *Martin's Magazine of the Sciences* during the 1760s. Between 1771 and 1775 he edited a new reprint of the problems that had appeared in the *Ladies' Diary* since its inception.[57] The work was largely one of transcription rather than editing as such, although merely to have assembled a complete set of old issues of the *Diary* was a notable achievement. Furthermore, Hutton included in the edition new mathematical correspondence, with original essays and problems, amounting to around 15% of the total; Hutton himself was said to have written at least some of this material

[50] Charles Hutton, *The School-Master's Guide*, Newcastle: for the author 1764.

[51] *London Chronicle*, issue 1140 (10–12 April 1764): advertisement for Hutton, *Guide*.

[52] On the longevity of this tradition and its contents see Frank J. Swetz, *Mathematical Expeditions: Exploring Word Problems across the Ages*, Baltimore: Johns Hopkins University Press 2010.

[53] Charles Hutton, *A Treatise on Mensuration, both in theory and practice*, Newcastle and London; T. Saint 1770, p. xviii.

[54] Charles Hutton, *The Compendious Measurer: being a brief, yet comprehensive, treatise on mensuration and practical geometry*, London: G. G. and J. Robinson and R. Baldwin 1786.

[55] Charles Hutton, *A Course of Mathematics*, 2 vols, London: for G. G. and J. Robinson 1798, vol. 1, p. iii.

[56] Howson, *Mathematics Education*, p. 68.

[57] Charles Hutton (ed.), *The Diarian Miscellany*, 5 vols, London: Robinson and Baldwin 1775.

under pseudonyms.[58] After this apprenticeship, Hutton's dominance of the philomath world became complete in 1773 when, following the death of Edward Rollinson, he was asked to take over as regular editor of the *Ladies' Diary*; he subsequently took on the editorship of all but two of the annual almanacs produced by the Stationers' Company of London, employing assistants 'to make most of the calculations & certainly all the Prognostications'.[59] Thus the essential question-and-answer material for the *Diary* passed through his hands and underwent his selection and arrangement. He saw the number of contributors more than double, and went so far as to produce an annual 'supplement' in order to accommodate more of the mathematical material. The *Gentleman's Diary*, meanwhile, came under the editorship of Charles Wildbore in 1780 at Hutton's recommendation.[60]

Hutton's public-facing work also apparently included reviewing mathematical books for the London periodicals. Three obituarists reported the fact, and one review survives in manuscript (under the pseudonym 'N. Bosworth'), although signed examples in print have not turned up.[61] Editing the works of deceased colleagues was a related activity, and it was another area in which Hutton's rich collections of printed and manuscript mathematics came into their own, as a resource for selection, arrangement, and dissemination. He edited posthumous works of William Emerson, Thomas Simpson, John Robertson, and Benjamin Robins, in most cases having acquired or borrowed portions of their papers and in the case of Emerson at least with his agreement during his lifetime: Hutton already had 'the care' of a new edition of Emerson's *Algebra* in 1779, three years before his death.[62]

The pinnacle of Hutton's editorial work was a second project ostensibly similar to his re-edition of the *Ladies' Diary*, but whose context made it something very different: the 'Abridgement' of the *Philosophical Transactions* that appeared between 1803 and 1809.[63] Three editors were named on the title pages of the series, but it was widely reported that Hutton undertook 'the general editorship and correction of the press, of the whole'[64] and that biographical articles and translations as well as abridgements of articles concerning mathematics and the mathematical sciences were all his

[58] *Public Characters*, vol. 2, p. 105.

[59] Charles Hutton to Robert Harrison, 19 March 1781, printed in Melmore, 'Some Letters', pp. 78–9 and in Wardhaugh, *Correspondence*, letter 15; cf. Reuben Burrow's statement that Hutton 'does not know how to make an Almanack': T. T. Wilkinson, 'Mathematics and Mathematicians, the journals of the late Reuben Burrow', in: *London, Edinburgh, and Dublin Philosophical Magazine*, 4th series, 5 (1853), pp. 185–93, 514–22; 6 (1853), pp. 196–204 at p. 187. Obituaries of Henry Andrews in 1820 stated that Andrews computed the *Nautical Ephemeris* and *Moore's Almanack* for the Stationers' Company for more than 40 years, never receiving more than £25: *Monthly magazine, or, British register* (December 1820), pp. 480–8 (obituary of Henry Andrews); cf. Stationers' Company of London, *The Records of the Stationers' Company 1554–1920* (microfilm series: Cambridge, 1985), Series I, Box B, folder 6, item ii (draft agreement between the Stationers' Company and Henry Andrews dated 13 September 1788).

[60] *Public Characters*, vol. 2, p. 113; Charles Hutton to Robert Harrison, 31 May 1780, printed in Melmore, 'Some Letters', pp. 77–8; also in Wardhaugh, *Correspondence*, letter 14.

[61] London: Royal Artillery Museum ('Firepower'), MS 913/6.

[62] Charles Hutton to Robert Harrison, 4 August 1779, printed in Melmore, 'Some Letters', pp. 74–6 and in Wardhaugh, *Correspondence*, letter 12.

[63] Charles Hutton, George Shaw, and Richard Pearson (eds), *The Philosophical Transactions... abridged*, 18 vols, London 1803–9.

[64] Anon, Obituary (1830), p. 193.

work. The quantity of text and the huge sum he was said to have been paid (£6000),[65] on the other hand, clearly indicates that assistants were again involved. Once more, Hutton's library came into play: his own collection of old issues of the *Transactions*, his biographical notes and his knowledge about mathematicians and others and his large store of their letters including contributions—whether published or not—to the *Ladies' Diary*.

The *Abridgement* was a particular opportunity to put Hutton's own slant on the work of the Royal Society, an institution with which he was still at odds after his catastrophic, widely publicised falling-out with its president Joseph Banks in 1784.[66] Critics commended the even-handedness of the project,[67] but it is difficult to be quite convinced. The volumes were internally arranged by subject, and mathematics and mechanical philosophy were the first two categories (contrast the 1787–91 abridgement of the *Transactions* printed in Paris, in 14 subject-specific volumes, which did not devote a volume to mathematics).[68]

Hutton's public-facing work also included perhaps the most enduring of his publications and the most important use to which his magnificent library was put during its existence: the 1795 *Dictionary of Mathematics*.[69] An immense project of reputation-building on both his own behalf and that of mathematics in general, and of British mathematics in particular, it filled two very large volumes and ranged across the whole gamut of the mathematical culture Hutton knew: from the contents of textbooks and the incremental innovations of the philomaths, through the biographies and foibles of the philomath authors themselves, recent practical and technical work, and the latest in both British and Continental mathematical research. There were probably few mathematical volumes in Hutton's library not used at least indirectly in its pages, whose compilation took him ten years of labour.[70]

On the Continental side, the *Dictionary* contained much bibliographical and biographical information to orient the British reader, with outlines of the methods and results of Jean le Rond d'Alembert, Leonhard Euler, and Joseph-Louis Lagrange. Hutton cited the output of the major journals in Paris, Berlin, St Petersburg, and Leipzig, as well as works of Marin Mersenne, Jérôme de Lalande, Christian Wolfius, Nicolas-Louis de la Caille, Gua, Girolamo Saladini, Giambatista Beccaria, and Charles Messier. It was possible to carp at individual omissions (there were no biographies of Ruggiero Boscovich, Jean Étienne Montucla, or Giuseppe Torelli),[71] and some would point out that Hutton's account was short of technical details, but on the whole it was thought that Continental work was amply covered. One reviewer even complained of

[65] Anon, [Obituary of Charles Hutton], in: *The Edinburgh Annual Register* 16 (December 1823), pp. 328–31, p. 330.

[66] Benjamin Wardhaugh, 'Charles Hutton and the "Dissensions" of 1783–84: Scientific Networking and its Failures', *Notes and Records of the Royal Society* 71 (2017) 41–59.

[67] Anon, [Review of Hutton et al. (eds), *Abridgement*, vol. 1], in: *The British Critic* (1803), p. 540.

[68] Jacques Gibelin (ed.), *Abrégé des Transactions Philosophiques de la Société Royale de Londres*, Paris: Buisson 1787–91.

[69] Hutton, *Dictionary*.

[70] See Cambridge: University Library, White b.8 (a single page advertisement for Hutton, *Dictionary*, with letter of Charles Hutton to David Stephenson, 7 February 1795 attached); also printed in Peter John Wallis, *Newcastle Mathematical Libraries: William Armstrong, Charles Hutton and Others, Northern Notes* 4 (1972: supplement), University of Durham, pp. 10–11 and in Wardhaugh, *Correspondence*, letter 70.

[71] Anon, [Review of Hutton, *Dictionary*], in: *Critical Review* (November 1796), pp. 302–5 at p. 305.

too much of Pierre Louis Maupertuis and Alexis Claude Clairaut;[72] another claimed that the *Dictionary* was little more than a translation of the mathematical parts of Denis Diderot's *Encyclopédie*.[73]

Hutton's achievement in the *Dictionary* was the more remarkable in that the wider world of scientific popularization tended very largely to take a non-mathematical approach to its subjects: scientific lecturers on the whole studiously avoided the use of mathematics in their presentations of the Newtonian world, and some lecturers were notoriously ignorant of mathematics.[74] The *Dictionary* was far from being a work of 'polite science' as it was normally understood.[75]

Hutton also, and finally, produced mathematical writings whose purpose was originality rather than re-presentation of what was already known. On the boundary of this category are his two volumes of mathematical tables, which he presented as a decisive improvement over any that had come before and were adopted and paid for with some enthusiasm by the Board of Longitude.[76] The second of these volumes, the table of logarithms, was the occasion for a lengthy history of logarithms and their calculation and use, into which Hutton put much of what he had learned from his years of collecting and reading on the subject. This was his first major historical project; a second would be the history of algebra which was published both separately and as part of the *Dictionary*, and on which he claimed to have spent two years of work, bringing together summaries, translations, and selections from the many relevant works in his library, that spanned medieval, Renaissance, and modern treatments of the topic.[77]

In a similar calculatory vein was Hutton's work on Maskelyne's project to determine the gravitational attraction—and hence the density and mass—of the earth using astronomical observations. Hutton spent nearly a year performing the calculations, transforming a ground survey into an estimate of the local gravitational field strength at two points in order to interpret the observed deflections of a plumb line from vertical.[78] The project was a clear illustration of Hutton's strengths as a mathematical technician; it was Maskelyne who received a Copley medal for the work,

[72] Anon, [Review of Hutton, *Dictionary*], in: *The Monthly Review* (1798), p. 185.

[73] Anon, [Review of Hutton, *Dictionary*], in: *English Review* 28 (July 1796), pp. 14–19 at pp. 18, 19.

[74] Hutton, *Tracts* (1812), vol. 3, Tract 38 (Miscellaneous problems), p. 379: a story about the lecturer James Ferguson.

[75] Alice N. Walters, 'Conversation Pieces: Science and Politeness in Eighteenth-Century England', in: *History of Science* 35 (1997), pp. 121–54 at pp. 122, 127–8.

[76] Hutton, *Tables*, p. 40; Mary Croarken, 'Tabulating', p. 55 (list of items to be sent to *Nautical Almanac* computers and comparers, September 1799).

[77] Hutton, *Tracts* (1812), vol. 2, Tract 33; also in Hutton, *Dictionary*, vol. 1, and cf. the history of the binomial theorem in Charles Hutton, *Tracts, Mathematical and Philosophical*, London 1786, Tract 6, pp. 67–75.

[78] Nevil Maskelyne, 'A Proposal for Measuring the Attraction of Some Hill in this Kingdom by Astronomical Observations', in: *Philosophical Transactions* 65 (1775), pp. 495–9; Nevil Maskelyne, 'An Account of Observations Made on the Mountain Schehallien for Finding its Attraction', in: *Philosophical Transactions* 65 (1775), pp. 500–42; Charles Hutton, 'An Account of the Calculations Made from the Survey and Measures Taken at Schehallien, in Order to Ascertain the Mean Density of the Earth', in: *Philosophical Transactions* 68 (1778), pp. 689–788; Charles Hutton, 'On the Calculations for Ascertaining the Mean density of the Earth', in: *Philosophical Magazine* 38 (1811), pp. 112–16; Charles Hutton, 'On the Mean Density of the Earth', in: *Philosophical Transactions* 111 (1821), pp. 276–92, reprinted in *Philosophical Magazine* 58/279 (1821), pp. 3–13.

though Hutton liked later to present his own role as an active, intellectual one. When a commemorative medal was struck in the final years of his life it showed an emblem of 'weighing the world'.[79]

Hutton's own more original research papers favoured topics in the manipulation and summation of series, including extensions of work by Euler, John Landen, and Stephen Hales.[80] Over many years he carried out experimental work on the force of fired gunpowder, and he received the Copley medal himself for one of his papers on the subject, shortly before his break with the Royal Society.[81] His work on the analysis of bridge designs, although it found a mixed reception, was the occasion in its later versions to synthesize what he had learned from William Jones and John Robertson, in both cases using their autograph manuscripts which Hutton had acquired over the years.[82]

Hutton's textbooks and his public-facing work were notably successful, and he was able in 1812 to ascribe his fairly substantial fortune to the 'liberal encouragement of the Public', meaning at least in part the sale of his books.[83] The *Guide*, *Mensuration*, logarithm tables and *Course* all continued to be reprinted for many years after his death; the *Guide* and the tables until nearly the end of the nineteenth century. The *Course* remained in use at the RMA and several sister institutions until the 1840s, its importance to training and examination several times reaffirmed.[84] In an 1841 edition it was stated that nearly 30,000 copies of the work had so far circulated.[85]

As with the presence of foreign works in his library, there is particular interest in the degree to which Hutton's own works travelled internationally. Approbation in America was provided by certain teachers who used the *Course*, notably Robert Adrain at West Point, whose list of the greatest mathematicians included Blaise Pascal, Gottfried Wilhelm Leibniz, the Bernoullis, Emerson, Robert Simpson, Hutton, and Samuel Vince.[86] There were American printings of extracts from the *Guide* up to 1824, and by chance a manuscript has survived entitled 'Book Keeping by single entry Extracted from the works of Charles Hutton', prepared by one William Mahan in 1827.[87] The *Course* meanwhile was used at West Point up to about 1823;[88] its

[79] Anon, *Tribute of Respect to Charles Hutton, LL.D. F.R.S. &c. &c.*, [London 1822].

[80] Hutton, *Tracts* (1786), Tract 1, p. 2; Tract 2, p. 34; Tract 6, pp. 73–4.

[81] Charles Hutton, 'The Force of Fired Gun-Powder, and the Initial Velocities of Cannon Balls, Determined by Experiments', in: Philosophical Transactions 68 (1778), pp. 50–85; further discussions in Hutton, *Tracts* (1786) and Hutton, *Tracts* (1812).

[82] Hutton, *Tracts* (1812), vol. 1, Tract 1, p. 89; cf. Tract 2 (a paper by George Dance on London Bridge) and Tracts 3 and 4 (papers from the Robertson papers).

[83] Hutton, *Tracts* (1812), vol. 1, p. x.

[84] Jones, *Records*, pp. 59, 60, 65, 69, 96, 100.

[85] Hutton, *Course* (1841 edition, ed. T. S. Davies), preface.

[86] Julian L. Coolidge, 'Robert Adrain and the Beginnings of American Mathematics', in: *The American Mathematical Monthly* 33 (1926), pp. 61–76, p. 75; cf. Frank J. Swetz, 'The Mystery of Robert Adrain', in: *Mathematics Magazine* 81 (2008), pp. 332–44 at p. 340.

[87] James Mulhern, 'Manuscript Schoolbooks', in: *The Journal of Educational Research* 32 (1939), pp. 428–48 at p. 443, describing an item in the collection of the Historical Society of Pennsylvania.

[88] V. Frederick Rickey and Amy Shell-Gellasch, 'Mathematics Education at West Point: The First Hundred Years', in: *Convergence* (2010): https://www.maa.org/press/periodicals/convergence/mathematics-education-at-west-point-the-first-hundred-years, esp. 'Hutton and the Notebooks'.

final American edition was printed in 1831. An American work extracted from the *Dictionary* also appeared in 1817.[89]

In British India the *Course* enjoyed a particular popularity at the hands of the East India Company and the Royal Bengal Artillery; a Gujarati translation was published in Bombay in 1828, and this was followed by Arabic, Sanskrit, and Marathi versions over the next twelve years; extracts in Urdu appeared in 1848.[90] But perhaps more surprising is that Hutton's work on ballistics was able to find acceptance in France. A few Continental journals had noticed his reprint of the *Ladies' Diary*, his tables, *Dictionary*, and *Course*,[91] and it is possible to find the occasional French or German citation of his research. But it was the work on gunpowder that found most reception here. Hutton's 1786 paper on the subject was translated into French in 1791–2, though not published; a report circulated that Lagrange himself was closely interested in the work.[92] Hutton's long, final discussion of gunpowder, published in English in 1812, was published in a French translation, as well as being reviewed, extracted, and cited.[93] A British reviewer could boast in 1822 that the French 'eagerly possess themselves of every essay, investigation, and experiment of Dr. Hutton on the subject, as soon as it is made public'.[94]

Hutton's ability to export his ideas through correspondence seems to have been much less important. Evidence of such direct contact is limited to single letters from Jeaurat and from Laplace, the suggestion that he was in touch with Damen, and a passage in Montucla's *Histoire* which may suggest contact with Hutton or his circle.[95]

The export of Hutton's books, it should be said, was not exceptional for his period. The *Philosophical Transactions* were sent to foreign members of the Royal Society in a number of countries, carrying with it a fair amount of mathematics. Ward's *Young Mathematician's Guide*, successful enough in Britain to run to fifteen editions, was translated into French in 1756 under the title *Le Guide des jeunes mathématiciens*. This translation was in fact part of a collection of mathematical works published in French by Charles Antoine Jombert; the set also included works of Newton and Colin Maclaurin. Ward's *Guide* also went to America, and is said to have been studied at

[89] Nathan S. Read, *An Astronomical Dictionary: compiled from Hutton's Mathematical and Philosophical dictionary*, New Haven 1817.

[90] J. F. Blumhardt, *Catalogue of the Library of the India Office*, vol. 2, part 5: *Marathi and Gujarati Books*, London 1908, pp. 89, 91, 245–6; Charles Ambrose Storey, *Persian Literature: a bio-bibliographical survey* 2/1 (1927), p. 19.

[91] *Journal Encyclopedique* (September 1776), pp. 355–7 (*Diary*); *Journal Encyclopédique* (October 1786), p. 179 (*Tables*); *Journal encyclopédique ou universal* 62 (October 1786), p. 339 (*Tables*); *Allgemeine Literatur-Zeitung* (November 1796), pp. 489–500 (*Dictionary*).

[92] Anon, [Obituary of Charles Hutton], in: *Monthly Magazine* 55 (March 1823), pp. 137–42 at p. 139; Louis-Bernard Guyton de Morveau, 'Notes', appended to J. B. J. Delambre, 'Notice sur la vie et les ouvrages de M. Malus, et de M. le Comte Lagrange', in: *Mémoires de la classe des sciences mathématiques et physiques de l'institut mperial de France. Année 1812. Première partie* (Paris, 1814), pp. lxxviii–lxxx; essay review including the *Tracts* in *The British Review* (1822), p. 300.

[93] Cf. J. Madelaine, [Review of Hutton, *Nouvelles expériences d'artillerie*], in: *Journal des sciences militaires* 5 (1826), pp. 350–79 and Brett D. Steele, 'Military "Progress" and Newtonian Science in the Age of Enlightenment', in: *The Heirs of Archimedes: Science and the Art of War through the Age of Enlightenment*, ed. Brett D. Steele and Tamera Dorland, Cambridge, MA: MIT Press, 2005, pp. 361–90 at p. 373.

[94] Essay review including the *Tracts*, in *The British Review* (1822), p. 299.

[95] Montucla had an English informant when he was working on books 3 and 4 of his *History*, with information about Newton and who had apparently seen the Pemberton papers. This could have been Hutton, though there are other candidates and no direct proof (personal communication by Niccolò Guicciardini).

Harvard in its early days. Another international traveller was Simson's 1756 edition of Euclid's *Elements*, which was widely used across the English-speaking world and eventually received translations into Arabic and various Indian languages.

It is notable that the *Ladies' Diary* also received some attention abroad, with French reviews of Leybourne's collection of its problems in 1817 remarking favourably on the mathematics it contained and the mathematical culture it represented. A fair case can be made, indeed, that during the whole of the eighteenth century British mathematics enjoyed at least modest success as a product for export.

Equally, certain British-trained individuals were able to make careers as teachers outside Britain, including Robert Adrain at West Point and William Marrat who went to New York 'carrying with him letters of recommendation from Dr. Hutton and Dr. [Olinthus] Gregory; and he is now a Professor of Mathematics in one of the colleges of that State'.[96] Walter Minto's trajectory was similar, taking him eventually from Scotland to America where he taught mathematics at Princeton. Britain was thus a serious player on the international mathematical stage, both exporting and importing quantities of texts and people. Expressions of self-confidence can fairly easily be found, and it was not entirely an idle boast when in 1784 Samuel Horsley claimed the country had in the person of Charles Hutton 'one of the greatest mathematicians in Europe'.

Dispersal

I have represented Hutton and his library as something like a sieve—or perhaps a better comparison would be a mangle—through which almost the widest possible range of mathematical text from eighteenth-century Britain and Europe passed, and which produced as its output a series of publications that themselves covered a wide range of genres, were addressed to a range of audiences, and successfully travelled beyond the British Isles. Hutton retired from his position at the RMA in 1807, and his books moved along with his other possessions from his house on Woolwich Common to his new residence on Bedford Row.[97] The new location was just a few streets away from the British Museum, and reasonably close to the heart of London's intellectual and scientific life. Hutton appears to have been a sociable man,[98] but his correspondence does not at this or at any other period specifically mention either lending out books from his library or granting access to it to his friends and colleagues. (The closest is a mention of books exchanged between him and Reuben Burrow at the Royal

[96] [Olinthus Gregory], 'A Review of Some Leading Points in the Official Character and Proceedings of the Late President of the Royal Society', in: *Philosophical Magazine*, series 1, no. 56 (1820), pp. 161–74, 241–57 at p. 251.

[97] Jones, *Records*, p. 57 (26 June 1807); the address is consistently given as 34 Bedford Row in Hutton's surviving correspondence, although 36 appears in the manuscript catalogue of his library (see below).

[98] Portsmouth History Centre, Correspondence and papers of C. B. Vignoles, Letter 65 (C. B. Vignoles to Mary Griffiths): 'literary friends' frequented Hutton's house during Vignoles's time there. Cf. Catherine Hutton, *Reminiscences of a Gentlewoman of the Last Century*, Birmingham 1891, p. 178: at his home in Bedford Row Hutton 'was constantly visited by an extensive circle of friends'.

Observatory in 1773.)[99] Already in 1802, a printed appreciation of Hutton had styled his library as one of the best mathematical libraries ever assembled in Britain.[100]

The exact sequence of events leading up to the library's dispersal is unclear. It appears that Hutton determined to dispose of his library during late 1815. Statements were made in print that he intended to retire into the country;[101] a later account had it that Hutton 'by reason of his advanced age, formed a determination to relinquish the habits of a student and the active pursuits of an author'.[102]

Our only detailed account of what happened next is from the pen of Olinthus Gregory, Hutton's protégé and supporter, who was writing in a polemical context shortly after the death of Joseph Banks, with whom Hutton had fallen out long before. He stated that Hutton formed the wish to sell his library to the British Museum, and that the trustees of that institution responded positively. He proposed to abide by a valuation decided upon by one representative from each party; the Museum appointed an officer to inventory the books, who reported favourably. Hutton then wrote to Banks, who was one of the trustees, in Lincolnshire, 'fearing he might take offence if not apprized of what was going on [...] and to express his hopes that the proposal would be approved'. There was, it was said, no reply, but within two weeks Banks was in London 'and busily employed among the other governors of the British Museum in dissuading them from the purchase'. Negotiations ceased.[103]

The outlines of this story find confirmation in Hutton's correspondence, where he wrote that his planned sale to the Museum had been 'cruelly prevented' by his 'old implacable enemy' Banks,[104] and even of Banks's 'triumph in disappointing me' in the matter.[105] But Hutton cannot be called an impartial observer of Banks's conduct any more than can Gregory, and I know of no other evidence that Banks continued to feel such animosity towards Hutton, three decades after their dispute at the Royal Society. It is at least possible that others at the Museum were involved in the decision not to take the mathematical books, or that the combination of mis-step and misunderstanding looked different to other eyes.

Once the sale to the Museum had fallen through, Hutton determined instead to sell the books at auction. I am at a loss to account for this decision: Hutton possessed a fairly sizeable personal fortune and cannot plausibly have needed the money. In private correspondence he stated that the library 'would prevent me from chusing [*sic*] another residence' and that 'I shall have little or no further use for it'; but in the event he neither moved from Bedford Row nor ceased to publish mathematical work, including new editions of his books and a new paper in the *Philosophical Transactions*. Hutton also stated with uncharacteristic harshness that the library 'could be

[99] London: UCL Library Services, Special Collections, MS Graves 23/3/5: Reuben Burrow to Charles Hutton, 24 September 1773, printed in Wardhaugh, *Correspondence*, letter 4.

[100] *Public Characters*, vol. 2, p. 112.

[101] Anon, *A Catalogue* (1816): title page; cf. 'Literary Intelligence', *The Literary Panorama* 4 (June 1816), p. 432.

[102] [Gregory], 'A review', p. 248.

[103] [Gregory], 'A review', p. 248 (no direct evidence of the independent inventory is known).

[104] Charles Hutton to John Bruce, first half of 1816; extract printed in Bruce, *Memoir*, pp. 32–3; also in Wardhaugh, *Correspondence*, letter 115.

[105] Charles Hutton to John Bruce, 22 March 1822; printed in Bruce, *Memoir*, pp. 39–42; also in Wardhaugh, *Correspondence*, letter 129.

of no use to any of my family, after my death':[106] but that still does not explain why its dispersal had to take place during his lifetime.

A manuscript catalogue, consisting of an alphabetical list of the books, was made by Hutton's daughter Isabella, listing 2,193 items in 3,315 volumes.[107] This was a draft, with poor bibliographic data in a number of cases, and underwent some revisions in Isabella's hand as well as some additions in Hutton's. Tick marks were added—presumably by Hutton—indicating items to be kept rather than sold. Many of the ticked items were non-mathematical ones, though the list was not in any event a complete list of the books Hutton had in his house: sheet music, for instance, is conspicuous by its absence, despite references in letters to pianos (plural) and music-making in the household; novels likewise are not listed in any quantity, though references in the correspondence indicate that both Hutton and his daughter read them.[108]

There is no evidence to show when this first catalogue was made. A fair copy was taken by Hutton himself in November 1815.[109] Now styled a 'Catalogue of Doctor Hutton's Mathematical Library', it quite consistently omitted the items ticked in the draft, although there were some second thoughts and some further revision of the list. This time there was no count made of the items (by my count there were 1,749).

Finally, and independently of the existing catalogues, Hutton's books were catalogued in print by the firm of Leigh and Sotheby. Copies of the printed catalogue were distributed nationally; the sale took place over six days from 11 June 1816, at the auctioneers' establishment on the Strand.[110] This final catalogue listed 1,841 lots of books: something of a reduction compared with the first manuscript catalogue, although since the books were now organized by day of sale and by physical format it is difficult to correlate or compare them in detail. Still, the number compared favourably with the 757 lots of Maskelyne's library sale in 1811 or the 1,421 of John Playfair's in 1819.[111] It is typical of the obscurity of the whole tale that there were sold with Hutton's books nine lots of mathematical instruments, described as formerly the property of Benjamin Franklin, including a theodolite, a microscope, a sextant, geometrical models, and two telescopes.[112] We have no other evidence of a direct connection between the two men (they had acquaintances in common in the Royal Society in the 1770s), and it seems impossible to say whether Hutton himself had actually owned Franklin's instruments during the more than forty years since the latter had left England.

After the sale, regret was expressed from more than one side, with the Literary and Philosophical Society in Newcastle stating that had it only known of the unfolding

[106] Charles Hutton to John Bruce, first half of 1816; extract printed in Bruce, *Memoir*, pp. 32–3; also in Wardhaugh, *Correspondence*, letter 115.

[107] Yale University: Lewis Walpole Library, LWL Mss Vol. 54, part 1.

[108] Portsmouth History Centre, Vignoles Papers, Letter 295 (C. B. Vignoles to Mary Vignoles, 4 August 1823), with a reference to 'one of the pianos'; Catherine Hutton, *Reminiscences*, pp. 182–3 (letter to Isabella Hutton, 6 November 1822).

[109] Yale University: Lewis Walpole Library, LWL Mss Vol. 54, part 2.

[110] Anon., *A Catalogue* (1816), title page.

[111] Christa Jungnickel and Russell McCormmach, *Cavendish: The Experimental Life*, 2nd edn, s.l., 1999, p. 323.

[112] Anon, *Catalogue* (1816), p. 80.

situation, it would have stepped in and purchased the library complete.[113] In fact, it was able to secure only a few volumes. It seems certain that other friends and colleagues attended the sale, too.

In fact, the sale of printed books was the most visible but not the only and perhaps not even numerically the largest dispersal of material from Hutton's library, if manuscript materials are taken into account. Just fifteen manuscripts were sold in 1816, including items which may have been Hutton's own work such as 'Experiments on Gunnery' as well as others such as copies of works by Euclid and by Flamsteed. A huge quantity of manuscript material, as well as many printed books, was evidently dispersed by other routes.[114]

The evidence for this is patchy. Some items were certainly destroyed in accidents during Hutton's lifetime. These included his manuscript lectures on natural philosophy and some of his experimental equipment, the latter succumbing to a fire at the RMA in 1802.[115] Notes for an enlarged treatise on bridges was also apparently lost,[116] and there is no trace of the calculations and map relating to his work on Schiehallion: frustratingly so, since the map appears to have been the first to have included lines of equal height, qualifying Hutton as the inventor of contour lines in that sense.[117] It is quite possible that some manuscript material was left behind at the RMA on Hutton's retirement, where it was subsequently affected by another fire. By contract he was required to return to the Stationers' Company relevant papers when he ceased to compile almanacs in 1818, but the Company's archive does not in fact seem to contain more than a few isolated documents relating to Hutton's long period of work.

Hutton had a reputation for methodical habits, and some of the surviving letters and manuscripts do possess numbers or labels added in his later life: but it is clear in fact that he treated some manuscripts carelessly. In two cases, printed books were sold which had letters (from Maskelyne and from the bishop of Clonfert) pasted into them.[118] Two letters, according to later endorsements, found their way into the archives of third parties.[119] Hutton also had the habit of cutting up manuscript sheets—particularly of calculations, but in at least one case a letter was affected—in order to reuse their reverses.[120]

A group of manuscripts including material on gunnery was given to his protégé Gregory in November 1818,[121] and the collection of pamphlets from Newcastle was

[113] Charles Hutton to John Bruce, 22 March 1822; printed in Bruce, *Memoir*, pp. 39–42; also in Wardhaugh, *Correspondence*, letter 129. Charles Hutton to John Bruce, 8 May 1817; printed in Bruce, *Memoir*, pp. 33–4; also in Wardhaugh, *Correspondence*, letter 116.

[114] Anon, *Catalogue* (1816), p. 79.

[115] Wardhaugh, *Correspondence*, p. xxviii.

[116] Anon, Obituary (1830), p. 190.

[117] K. Rann and R. S. Johnson, 'Chasing the line: Hutton's contribution to the invention of contours', in: *Journal of Maps* 15 (2019), pp. 48–56.

[118] Anon, *Catalogue* (1816), pp. 30, 70.

[119] Cambridge: University Library, RGO 35/92 (Nevil Maskelyne to Charles Hutton, 20 June 1796, 1797, or 1798); an endorsement states that this letter found its way into Waring's papers. London: Wellcome Collection, MS 5270 no. 70 (Patrick Kelly to Charles Hutton, 24 December 1821, fragment); this letter was endorsed later by Catherine Hutton and had probably come into her possession.

[120] Cambridge: Trinity College Library, R.1.59, fols. 52–63 are mostly written on cut-up sheets of calculations; fol. 56 is a fragmentary draft letter in French, possibly in Hutton's hand.

[121] London: the Royal Artillery Museum ('Firepower'), MS 913/5, item 551(1); MS 913/3; MS 913/1, item 1.

presented to Gregory after his death.[122] Gregory also acquired other manuscripts, including Hutton's calculations relating to the division of the quadrant, his selections from Archimedes and Pappus, and the treatment of the vibrating line by Maseres, as well as a pendulum. Gregory's books, in turn, were dispersed at auction after his death in 1842: the gunnery material remained at the RMA, but the other manuscripts of Hutton's were sold. They passed through the hands of Joseph Clinton Robertson (1788–1852), founder and editor of the *Mechanic's Magazine*, and of J. S. Davies, a mathematical master at the RMA. He left them to his son Charles Butler Davies, a fellow of Trinity College, Cambridge, and the volume of Hutton's manuscripts was given to his college in 1873 (where it remains).[123] This trajectory may be typical of the complex paths of individual items; it is exceptional in being fully documented. Charles Babbage also acquired some of Hutton's books, perhaps indirectly, and they are now in the Crawford Collection at the Royal Observatory in Edinburgh.[124] Similarly, Augustus De Morgan acquired several printed and manuscript items of Hutton's from Gregory's sale and at other public sales, including a manuscript translation from Tartaglia, and these remain in the Senate House Library in London.[125]

In April 1835, the London scientific bookseller John Weale issued a catalogue indicating that he had made substantial acquisitions from the libraries of Maskelyne, Horsley, and Hutton (as well as William Phillips and Richard Heber).[126] It is impossible to make any surmise about those books in his catalogue for which provenance is not specifically given, but of the more than 1,000 volumes listed, sixteen are specifically identified as having been Hutton's.[127] Where these books had been during the twelve years since Hutton's death, or the twenty years since the sale of his library, we do not know.

What was not sold or dispersed during Hutton's lifetime passed to his descendants. His will left virtually everything to his eldest daughter Isabella, and family letters show that what she received included a large number of books. Five hundred copies of Hutton's 1812 *Tracts* are mentioned: some were given away or sold individually;

[122] Anon, *Catalogue* (1842), p. 9.

[123] Cambridge: Trinity College Library, MS R.1.59: 4th initial unfoliated leaf, r, and bookplate.

[124] Eric G. Forbes, 'Collections II: The Crawford Collection of Books and Manuscripts on the History of Astronomy, Mathematics, etc., at the Royal Observatory, Edinburgh', in: *British Journal for the History of Science* 6 (1973), pp. 459–61 at p. 459.

[125] London: Senate House Library, MS 235 (translation from Tartaglia by Charles Hutton), librarian's annotation on fol. 1r; MS 913B/3/1 (xiv) (letter of John Playfair to Charles Hutton, 12 December 1782), endorsement on fol. 2^{v}; MS 913B/3/1 (xv) (letter of John Leslie to Charles Hutton, 14 October 1795), endorsement on fol. 2^{v}. See also Karen Attar, 'Augustus De Morgan (1806–1871), his Reading, and his Library', in: Mary Hammond (ed.), *The Edinburgh History of Reading: Modern Readers*, Edinburgh: Edinburgh University Press 2020, pp. 62–82.

[126] John Weale, *A catalogue of Books, on the Sciences: Astronomy, Mathematics, Natural Philosophy, &c; With some added that are curious and miscellaneous; chiefly from the libraries of Rev. Nevile Maskelyne, D.D., Astron. Royal and F.R.S.; Bishop Horsley, F.R.S., &c.; Dr. Charles Hutton, LL.D. F.R.S., &c.; William Phillips, F.R.L. and G.SS.; and Richard Heber, esq. On Sale, By John Weale, (Scientific and Architectural Bookseller, 59, High Holborn, London.)*, [London] 1835.

[127] There were works by Apollonius, d'Alembert, d'Auvergne, Gautier, Guarini, Heron, Descartes, Borelli, Huygens, Kiel, Maupertuis, Fine, Ptolemy, Porphyry, Sacrobosco, Euclid, Saul, Michell, Templehof, Ferguson, Simson, Voltaire, and Young: some were bound together.

most were sold in bulk in 1830.[128] There was also an unknown number of copies of his treatise on bridges,[129] and at least a few copies of his other works. Some family papers remained with the archive of his grandson Charles Blacker Vignoles, now at the Portsmouth Library and Archive; some personal items including a journal—seen by Gregory in 1823—appear to have vanished.[130] Vignoles mentioned an 'accident',[131] but deliberate pruning is also possible. The fate of what must have been a large collection of mathematical letters is particularly obscure: only fifty-seven letters addressed to Hutton have been located to date.

Today, Hutton's books and papers are scattered over more than two dozen archives, with none holding more than a few volumes: the largest group appears to be the seven books held by the Senate House Library. Some are certainly in private hands;[132] many may have been destroyed.

The dispersal of Hutton's library and papers—a doleful tale of destruction, gift (sometimes disorganized), and sale—is an apt if an imperfect metaphor for the destiny of the mathematical culture it in many ways reflected.

British mathematics in Hutton's lifetime was to all appearances succeeding in its own terms, and there seems nothing feigned about the pride and enthusiasm expressed by Hutton and his colleagues; in a letter to Hutton, Maskelyne wrote of the achievements of 'Lyons, Emerson, Landen, Waring; the last of whom is [...] the author of some of the greatest discoveries in Algebra; algebraic curved lines, infinite series, increments and fluxions'.[133] This was a dynamic, active, and creative mathematical world, whose work seemed important and worthwhile to its authors and readers. The *Nautical Almanac*, for instance, was no trivial innovation in what mathematics could do or how it could do it; it was no trivial innovation in the fields of the organization of calculation or the large-scale printing of accurate mathematical tables. New proofs of the binomial theorem and new results about the convergence of series; new methods for the calculation of trigonometric ratios; and, of course, practical outcomes such as improvements in artillery or the determination of the density of the earth: all of these were the proud achievements of a culture that felt no need to apologise for itself.

Compared with Continental methods, partisans of British mathematics could point to the important role it gave to physical analogy and to geometrical and physical intuition.[134] In their view 'in point of intellectual conviction and certainty, the

[128] London: British Library, MS Add. 58,203 (Diary of C. B. Vignoles, 1824), 15r (week of 15 March); MS Facsimile *920 (1–3), 514 (27 December 1830); Portsmouth History Centre, Vignoles Papers, Letter 514.

[129] Portsmouth History Centre, Vignoles Papers, Letter 526; London: British Library, MS Facsimile *920 (1–3), 526 (21 July 1821).

[130] Gregory, 'Memoir', pp. 220, 221; Mackenzie, *Newcastle-upon-Tyne*, p. 560.

[131] Portsmouth History Centre, Vignoles Papers, Letter 751 (C. B. Vignoles to the Marquis of Northampton, 25 March 1841); London: University College archives, MS Galton 2/4/1/2/9 (C. B. Vignoles to Francis Galton, 17 November 1865).

[132] Peter John Wallis, *Newcastle Mathematical Libraries*, pp. 12–13; a copy of the 1815 (second edition) *Dictionary* sold during 2019 contains a long letter to J. B. Wise (Raymond V. Giordano, pers. comm.).

[133] Cambridge: University Library, RGO 35/92 (Nevil Maskelyne to Charles Hutton, 20 June 1796, 1797, or 1798), 2.

[134] Niccolò Guicciardini, 'Dot-Age: Newton's Mathematical Legacy in the Eighteenth Century', in: *Early Science and Medicine* 9/3 (2004), pp. 218–56 at p. 255; cf. Philip C. Enros, 'The Analytical Society

fluxional calculus is decidedly superior to the differential and integral calculus'.[135] The production of fluxional textbooks continued throughout the century and included the celebrated account by Maclaurin as well as works by Emerson and Simpson;[136] Hutton's own works regularly assumed familiarity with fluxional methods; his *Conics* contained a table of common fluxions and fluents for the use of students.[137]

Nevertheless, it could not escape the attention of a man as widely read as Hutton, nor of a culture as widely read as that of the Georgian mathematicians, that in Paris, Berlin, St Petersburg, and other Continental locations, things were done very differently. There, the status of original researchers in mathematical fields was high, and they enjoyed a protected leisure that had no equivalent for their British counterparts.[138] This small, privileged group had, over the decades, transformed the Leibnizian calculus into Eulerian analysis, stripping it of its geometrical roots.[139] The new methods were powerful but difficult, and it is hard to form an accurate estimate of how many British mathematicians at any given date would have been able to read the latest Continental research with full understanding. It is reported that by the 1770s Edward Waring at Cambridge 'found it virtually impossible' to follow Euler's 'use of partial differential equations and the calculus of variations to tackle problems in mechanics'.[140] Something like a language barrier came to exist, sustained by the belief of each side in the superiority of its methods as well as the understandable reluctance of teachers and expositors to introduce unfamiliar new notation to their students and readers.[141] Peter Barlow, for instance, in various works including encyclopaedia articles showed clear acquaintance with Continental works, but retained the language of fluxions.[142] Much the same was true of Hutton; a necessary feature of the activity of gatekeeper figures like him was that they translated into fluxional terms what could be so translated and left out what could not. The accounts of Continental figures in his *Dictionary*, for instance, often conspicuously lack technical details. I do not believe that that lack reflected incapacity on the part of Hutton to read and understand at least a good proportion of their work; it did reflect the deficiencies of the mathematical

(1812–1813): Precursor of the Renewal of Cambridge Mathematics', in: *Historia Mathematica* 10 (1983), pp. 24–47 at p. 38.

[135] Guicciardini, *Development*, p. 113, citing Gregory in the preface to the 1836–7 edition of Hutton's *Course*.

[136] Colin Maclaurin, *A Treatise of Fluxions: In Two Books*, Edinburgh 1742; Thomas Simpson, *The Doctrine and Application of Fluxions: Containing (besides what is common on the subject) a number of new improvements in the theory: and The Solution of a Variety of New, and very Interesting, Problems in different Branches of the Mathematicks*. Part I, London: Printed for J. Nourse 1750; William Emerson, *The Doctrine of Fluxions: not only explaining the elements thereof, but also its application and use in the several parts of mathematics and natural philosophy*, London: Printed for J. Richardson 1757.

[137] Charles Hutton, *Elements of Conic Sections: with select exercises in various branches of mathematics and philosophy. For the use of the Royal Military Academy at Woolwich*, London: Printed for J. Davis 1787, p. 171.

[138] Guicciardini, 'Dot-Age', p. 252; Enros, 'The Analytical Society', p. 41, quoting a letter of Herschel to Bromhead of 19 November 1813; Warwick, *Masters of Theory*, p. 35; also Jeremy Gray, 'Overstating their Case? Reflections on British Mathematics in the Nineteenth Century', in: *Bulletin of the British Society for the History of Mathematics* 21 (2006), pp. 178–85.

[139] Guicciardini, 'Dot-Age', p. 241; see Warwick, *Masters of Theory*, p. 34.

[140] Warwick, *Masters of Theory*, p. 34; cf. Guicciardini, 'Dot-Age', p. 246.

[141] Warwick, *Masters of Theory*, p. 75: Guicciardini, *Development*, p. 108; David Philip Miller, 'The Revival of the Physical Sciences in Britain, 1815–1840', in: *Osiris* 2 (1986), pp. 107–34 at p. 109.

[142] Guicciardini, *Development*, p. 113.

language in which he felt constrained to write, and probably the deficiencies in understanding he imputed to those for whom he was writing. For periods between the 1770s and the 1810s, too, France in particular was a hostile country, and the French associations of analysis did something to discourage British mathematicians from its use: in 1815 the patriotism of Robert Woodhouse was questioned because of his use of the French mathematical notation.[143] For all these reasons, it remained possible into the 1820s to become senior wrangler with little or no facility in Continental mathematical methods.

It has been clearly documented, though it probably bears repeating, that none of this means that British mathematical culture was in its 'dot-age' or that there was anything self-evidently wrong with it. Its practitioners show no sense of feeling that they were working in a field that had been exhausted or that their location—physical or intellectual—was a backwater. Very substantial numbers of Continental books were coming into Britain, and if they needed to be digested and translated before they could be presented beyond a narrow circle of specialists, no one during the years from mid-century up to the 1790s seems to have judged that that was a problematic state of affairs. British mathematicians did not feel—and had no very obvious reason to feel—they were cut off from the mainspring of mathematics.

Around 1800, however, things changed. A few whispers could be heard in the late 1790s: Woodhouse, reviewing the *Dictionary*, noted that Hutton had 'scarcely announced' the foreign method of analysis and hinted that a proper English work on the principles of fluxions was needed, as well as implying delicately that the English might be wrong not to take the Continental methods seriously.[144] But the watershed was the publication of Laplace's *Traité de mécanique céleste*, beginning in 1799. Early British reviews in 1799 and 1803 praised the work as an instance of French prowess but regretted its purely analytic character, its lack of diagrams. Suggesting a need for geometrical interpretation, they spoke to the existence of a language barrier which made the work less than perfectly comprehensible even to the best British mathematicians.[145] Woodhouse, reviewing it for the *Monthly Review*, judged it necessary to transliterate the word 'differentiation' and explain that it meant 'putting an equation or expression in fluxions'.[146] And in January 1808 Playfair, in what has become a much-quoted passage, wrote in the *Edinburgh Review* that

> a man may be perfectly acquainted with every thing on mathematical learning that has been written in this country, and may yet find himself stopped at the first page of the works of Euler or D'Alembert. He will be stopped, not from the difference of the

[143] Anon., 'Mr. Woodhouse on the rectification of the hyperbola', in: *Gentleman's Magazine* 85 (1815), pp. 18–22 at pp. 18–19, quoted in Harvey Becher, 'Radicals, Whigs, and Conservatives: The Middle and Lower Classes in the Analytical Revolution at Cambridge in the Age of Aristocracy', in: *British Journal for the History of Science* 28 (1995), pp. 405–26 at p. 410; cf. Warwick, *Masters of Theory*, p. 67.

[144] Anon., [Review of Hutton, *Dictionary*], in: *The Monthly Review* (1798), pp. 184–201, pp. 364–83 at pp. 193, 194–5; see also Howson, *History of Mathematics Education*, p. 232.

[145] Jonathan R. Topham, 'Science, Print, and Crossing Borders: importing French science books into Britain, 1789–1815', in: *Geographies of Nineteenth-Century Science*, ed. David N. Livingstone and Charles Withers, Chicago: University of Chicago Press 2011, pp. 311–44 at p. 326.

[146] Topham, 'Science, Print, and Crossing Borders', p. 326; for the attribution see B. C. Nangle, *The Monthly Review, Second Series, 1790–1815: indexes of contributors and articles*, Oxford: Clarendon Press 1955, vol. 2, p. 74; also Craik, 'Mathematical Analysis', p. 243.

> fluxionary notation, (a difficulty easily overcome), nor from the obscurity of these authors, who are both very clear writers, especially the first of them, but from want of knowing the principles and the methods which they take for granted as known to every mathematical reader. If we come to works of still greater difficulty, such as the *Méchanique Céleste*, we will venture to say, that the number of those in this island, who can read that work with any tolerable facility, is small indeed. If we reckon two or three in London and the military schools in its vicinity, the same number at each of the two English Universities, and perhaps four in Scotland, we shall not hardly exceed a dozen; and yet we are fully persuaded that our reckoning is beyond the truth.[147]

Playfair had been active as a promoter and improver of Hutton's own work on the density of the earth and air resistance, and was no enemy to British mathematics in general. His critique—and the blame he gave to the English universities and the Royal Society—weighed the heavier as a result. There followed a wave of criticism, in which the virtues of British mathematics were re-read as vices.[148] Thus it was now 'unnatural' to found the calculus on motion or velocity, which applied properly only to dynamics.[149] The fluxional notation, it was said, had been preserved merely out of sentimental attachment to the work of Newton.[150] The diversity of versions of 'Newtonianism' was itself no longer a sign of richness but of fragmentation.[151] The high technical standard of the Cambridge examinations denoted sterility, and the gradual improvements and discoveries of the philomaths were 'trifling':

> A mathematical production, above the level of school-practitioners, finds little encouragement in this country; to enable a book to sell, it must be trifling; it must reduce all rules to mere mechanical operations; it must in fact be suited to the taste of solvers of problems, and not to investigators: – we have more of the former class, and fewer of the latter, than any empire in Europe.[152]

John Toplis, Herschel, and even Gregory also took to the press with accounts of the 'decline' of mathematics in Britain.[153] Babbage in 1830 was still lamenting at book length 'the decline of science in England'.[154]

[147] John Playfair, '*Traité de Méchanique Céleste* [review]', in: *The Edinburgh Review* 11 (1808), pp. 249–84 at p. 281. See also Amy Ackerberg-Hastings, 'John Playfair on British Decline in Mathematics', in: *Bulletin of the British Journal for the History of Mathematics* 23 (2008), pp. 81–95.

[148] See Topham, 'Science, Print, and Crossing Borders', p. 328 on Herschel's review of translations of the *Mecanique Céleste*.

[149] Florian Cajori, 'Discussion of Fluxions: from Berkeley to Woodhouse', in: *The American Mathematical Monthly* 24 (1917), pp. 145–54 at p. 154.

[150] Craik, 'Mathematical Analysis', p. 264; Robert Woodhouse, [Review of Lagrange, *Theorie des fonctions*], in: *The Monthly Review* 28 (1799), pp. 481–99 at p. 487.

[151] See Guicciardini, 'Dot-Age', pp. 219, 223.

[152] Anon., 'Lagrange's "A Treatise upon Analytical Mechanics..."', in: *The Monthly Review or Literary Journal* 78 (October 1815), pp. 211–13; see Craik, 'Mathematical Analysis', p. 243.

[153] John Toplis, 'On the Decline of Mathematical Studies, and the Sciences dependent upon them', in: *Philosophical Magazine* 20 (1805), pp. 25–31; Olinthus Gregory, *A Treatise of Mechanics*, London 1806, vol. 1, preface; J. F. W. Herschel, *Memoir of Francis Baily, Esq.*, London 1845.

[154] Charles Babbage, *Reflections on the Decline of Science in England, and on Some of its Causes*, London 1830.

Several of those who wrote in this vein had—unlike Playfair—personal axes to grind. Some were younger mathematicians keen to show that they were more up to date than their older predecessors. Some, like Gregory, were enemies of Banks and wished to blame or at least embarrass him and the Royal Society;[155] others bore grudges against the universities.[156] Others again blamed the circle around Hutton for its now long-lived failure to cooperate with the Royal Society.[157] The rhetoric of the short-lived Analytical Society based at Cambridge was particularly exaggerated and self-serving.[158] Yet the judgement of these critics took hold, and the identification of a 'dot-age' in British mathematics has persisted to the present day.

The positive outcomes have been well documented. A wave of importation and translation of Continental books took place during the first two decades of the nineteenth century, introducing the differential notation to British mathematicians. A translation of Sadi Carnot's *Réflections* appeared in the *Philosophical Magazine* in 1800–01 using differential notation, even though its translator preferred fluxions;[159] John Colson and John Hellins translated Maria Gaetana Agnesi's textbook in 1802.[160] Gregory kept up with the works of Gaspard Monge, Jean Nicolas Pierre Hachette, and Claude-Louis Navier; his 1806 *Treatise of Mechanics* and 1807 translation of René Just Haüy's *Traité élémentaire de physique* were outcomes.[161] Toplis in 1814 published Book 1 of the *Méchanique céleste* as *A Treatise upon analytical mechanics*.[162] The Analytical Society completed a translation of Silvestre François Lacroix's calculus textbook in 1816, with a further volume in 1820.[163] Cambridge tutors meanwhile mastered the Continental notation and began to teach it.[164]

Periodicals and reference works became an important venue for the new mathematics. The fourth (1801–10) edition of the *Encyclopaedia Britannica* introduced much Continental mathematics, and William Wallace's article on 'Fluxions' for the Edinburgh *Encyclopedia* of 1815 was the first complete account of the calculus in English to use Continental notation.[165] The *Philosophical Transactions* carried three of James Ivory's papers on topics related to the *Mécanique céleste* between 1809 and 1812.[166] From 1806 the new series of the *Mathematical Repository*, edited by

[155] David Philip Miller, 'Between Hostile Camps: Sir Humphry Davy's presidency of the Royal Society of London, 1820–1827', in: *British Journal for the History of Science* 16 (1983), pp. 1–47.

[156] Gascoigne, Cambridge in the Age of the Enlightenment, p. 6; Miller, 'The Revival', p. 108.

[157] David Philip Miller, 'Sir Joseph Banks: An Historiographical Perspective', in: *History of Science* 19 (1981), pp. 284–92 at p. 289.

[158] Elizabeth Garber, 'On the Margins: Experimental Philosophy and Mathematics in Britain, 1790–1830'. in: *The Language of Physics: The Calculus and the Development of Theoretical Physics in Europe, 1750–1914*, Boston: Birkhäuser 1999, pp. 169–206 at p. 191.

[159] Guicciardini, *The Development*, p. 119.

[160] W. Johnson, 'Contributors to Improving the Teaching of Calculus in Early 19th-Century England', in: *Notes and Records of the Royal Society of London* 49 (1995), pp. 93–103 at p. 99.

[161] Albree and Brown, 'A Valuable Monument', p. 26; Ivor Grattan-Guinness, *Convolutions in French Mathematics, 1800–1840: From the Calculus and Mechanics to Mathematical Analysis and Mathematical Physics*, Basel: Birkhäuser 1990, pp. 437–8; cf. the letter of John Gough published in *The Mathematical Repository* 2 (1809), p. 7; Guicciardini, The Development, p. 113.

[162] Craik, 'Mathematical Analysis', p. 272.

[163] Warwick, *Masters of Theory*, p. 68.

[164] Warwick, *Masters of Theory*, p. 49; also Alex D. D. Craik, *Mr Hopkins' Men: Cambridge Reform and British Mathematics in the Nineteenth Century*, London: Springer-Verlag 2007.

[165] Guicciardini, *Development*, p. 120; Craik, 'Mathematical Analysis', pp. 242, 249.

[166] Maurice Crosland and Crosbie Smith, 'The Transmission of Physics from France to Britain: 1800–1840', in: *Historical Studies in the Physical Sciences* 9 (1978), pp. 1–61, p. 17.

Thomas Leybourne of the RMC, became a particularly important site for translations of shorter pieces of Continental mathematics as well as announcements of recent works available in Britain.[167] The teachers of the RMA and RMC were highly visible in its pages, and its contents included extracts from Lagrange, Legendre, and Euler; differential and integral notation were used.[168] Although the first volume of the *Repository*'s first series, in 1795, had been dedicated to Hutton (and gave more than a hint of being the mathematical rival to the *Philosophical Transactions* Hutton had once wished for),[169] the publication had now become one in which he and his style of mathematics were scarcely at home.

In the end, British mathematicians adopted Continental notation and methods wholesale, and formed a new set of research agendas influenced, though not dictated, by what they found in imported Continental works. The details varied in different locations.[170] In the Scottish universities, and at Sandhurst, Laplacian astronomy was taken up as a research interest, though teaching changed little. At Dublin, teaching reform centred on applied topics including Lagrange's work;[171] at Cambridge the focus was more on pure mathematics and algebra. Indeed, at Cambridge the teaching reform initially failed, but during the 1820s the foundation of the Classical Tripos and the solidification of a mathematical Tripos provided an occasion for a new start.[172] In respect of research at Cambridge, the Analytical Society was not the beginning of French mathematics or French notation in Britain (as they and others sometimes later claimed), but instead reinforced the acceptance of Lagrange's calculus and began a focus on the calculus of operators and functional equations, which lasted until the days of Arthur Cayley and George Boole.[173] Arguably this involved only a limited appreciation of Augustin-Louis Cauchy's rigorization of calculus, and thus paradoxically left British mathematicians once again somewhat isolated.[174]

Metropolitan mathematical practitioners were on the whole somewhat later adopters of these novelties; as late as 1816, Barlow could criticize Herschel and Babbage as too fond of 'a sort of parade' common in France which made 'a great display of intricate and almost unintelligible formulae, without the least consideration of their application to any purpose of real utility'.[175] But the RMA and the other military academies were by then following French engineering work; and pure topics such as Legendre's elliptic integrals and Lagrange's algebraic foundation of calculus

[167] *Mathematical Repository* 2 (1809), p. 64 and 3 (1814), p. 51: lists of foreign works. See also Brigitte Stenhouse, 'Mary Somerville's Early Contributions to the Circulation of Differential Calculus', in: *Historia Mathematica* 51 (2020), pp. 1–25, and for developments in the *Ladies' Diary* Albree and Brown, p. 27.

[168] Guicciardini, *Development*, pp. 108, 117.

[169] *Mathematical Repository* 1 (1795; 2nd edn 1799), pp. iii, iv; cf. Wardhaugh, *Gunpowder and Geometry*, pp. 120–1.

[170] Guicciardini, *Development*, p. 141.

[171] Maria Panteki, 'William Wallace and the Introduction of Continental Calculus to Britain: A Letter to George Peacock', in: *Historia Mathematica* 14 (1987), pp. 119–32, p. 124; Guicciardini, *Development*, pp. 132–3.

[172] Warwick, *Masters of Theory*, pp. 67, 76; cf. Johnson, 'Contributors', p. 96; Enros, 'The Analytical Society', pp. 24, 26; Guicciardini, *Development*, p. 135.

[173] Guicciardini, *Development*, p. 136.

[174] Guicciardini, *Development*, p. 138.

[175] David Philip Miller, 'The Royal Society of London 1800–1835: A Study in the Cultural Politics of Scientific Organization', PhD thesis, University of Pennsylvania 1981, p. 119, citing a review by Herschel in *The Monthly Review* 81 (1816), p. 393.

were being assimilated by a roll call of teachers including Wallace, Ivory, Hellins, John Brinkley, Woodhouse, Bonnycastle, and William Spence.[176]

In 1840 the use of Hutton's *Course* was discontinued at the RMA.[177] The last textbooks of fluxions to be reprinted were those of William Dealtry in 1816 and Vince in 1818.[178] There thus came to be a generation of British mathematicians who never learned fluxions, but acquired analysis directly from French works: to whom, therefore, older British works were inaccessible. Arguably, it would take the foundation of the Astronomical Society in 1820 and the London Mathematical Society in 1865 to bring the process of professionalizing British mathematics to anything near completion. But by then Georgian mathematics and its culture were, like Hutton's library, quite gone.

[176] Guicciardini, *Development*, p. 135.

[177] Jones, *Records*, p. 102; cf. Panteki, 'William Wallace', p. 124; Crosland and Smith, 'The Transmission of Physics', p. 18.

[178] Craik, 'Mathematical Analysis' p. 240.

8

Mathematics at the Literary and Philosophical Societies

Christopher D. Hollings

Introduction

In April 1861, a Cheshire clergyman, the Reverend Thomas Penyngton Kirkman (1806–1895), delivered a mathematical lecture before the Manchester Literary and Philosophical Society (MLPS). The title under which the lecture was subsequently published in the Society's *Memoirs* was 'On the Theory of Groups and many-valued Functions'.[1] In this, the first of several papers on the topic, Kirkman was employing the word 'group' in the sense in which it had been used by the precocious French mathematician Évariste Galois (1811–1832) some thirty years earlier:[2] as the name for a collection of permutations (or rearrangements) of some set of objects which has the property that if we take two such permutations and apply them to the set of objects in succession, then the overall permutation also belongs to the original collection. During the second half of the nineteenth century, this notion of a 'group' was subsequently developed into something rather more abstract, in which form it is a central object of study in modern mathematics. The details of this abstract theory and its development need not concern us here[3]—suffice it to say that even in the considerably less abstract form in which the idea was employed by Kirkman, this was a topic that would probably have been far-removed from the mathematical experience of many in his audience. Indeed, it seems likely that the novelty of the mathematics, not to mention the idiosyncratic nature of Kirkman's notation and terminology (which may have presented a challenge even to contemporary mathematicians), would have left his listeners rather bemused.[4]

[1] T. P. Kirkman, 'On the theory of groups and many-valued functions', in: *Memoirs of the Literary and Philosophical Society of Manchester*, 3rd ser., 1 (1862), pp. 274–398.

[2] For a comprehensive English edition of Galois's mathematical writings, see: Peter M. Neumann, *The Mathematical Writings of Évariste Galois*, Zürich: European Mathematical Society 2011.

[3] See instead: Hans Wussing, *Die Genesis des abstrakten Gruppenbegriffes*, Berlin: Deutscher Verlag der Wissenschaften 1969; English translation by Abe Schenitzer: *The Genesis of the Abstract Group Concept*, Cambridge, MA: MIT Press 1984.

[4] On Kirkman and his mathematics, see: W. W. K[irkman], 'Thomas Penyngton Kirkman [obituary]', in: *Memoirs and Proceedings of the Manchester Literary and Philosophical Society, Fourth Series* 9 (1895), pp. 238–43; Alexander Macfarlane, *Lectures on Ten British Mathematicians of the Nineteenth Century*, Mathematical Monographs 17, New York: John Wiley & Sons 1916, pp. 122–33; Stella Mills, 'Thomas Kirkman–The Mathematical Cleric of Croft', in: *Manchester Literary and Philosophical Society: Memoirs and Proceedings* 120 (1977–80), pp. 100–9; N. L. Biggs, 'T. P. Kirkman, Mathematician', in: *Bulletin of the London Mathematical Society* 13 (1981), pp. 97–120; Robin J. Wilson, 'Kirkman, Thomas Penyngton (1806–1895), mathematician and philosopher', in: *Oxford Dictionary of National Biography*,

If we examine the table of contents for the volume of the MLPS's *Memoirs* in which Kirkman's paper appeared (Figure 8.1),[5] we find a wide range of topics, including optics, steam power, glaciers, lighthouses, sun spots, the detection of diabetes, and microscopy, to name but a few. Meteorology is a dominant theme, represented by the lectures numbered X, XVII, XX, XXI, and XXIII; astronomy also emerges as a prominent subject, with three lectures (XI, XII, XXIV). We notice also that Kirkman's was not the only mathematical lecture to have been recorded in this volume: we find also James Cockle's 'Supplementary Researches in Higher Algebra' (delivered in November 1859) and Arthur Cayley's 'On the Δ-faced Polyacrons, in reference to the Problem of the Enumeration of Polyhedra' (October 1860). As in the case of Kirkman's, the topics of both Cockle's and Cayley's lectures would have seemed rather abstruse to a non-specialist audience. Indeed, these three mathematical lectures appear to have been of a rather different type from the rest of the Society's programme, whose other topics, as indicated above, tended to be rather more 'practical' in character: either experimental, or else of a more accessible nature (for example, astronomy). This is not to say that the other lectures were not technical—tables of figures appear in several of the lectures in their written-up form—but there is nothing else like the dense formalism and algebra of Kirkman. Moreover, if we browse adjacent volumes of the Manchester *Memoirs*, we find that this pattern repeats itself there: the majority of lectures delivered to the Society were on largely accessible topics, but the programme was interspersed with mathematical lectures (by Kirkman and others[6]) that would surely have been understandable only to a small section of the MLPS's membership.[7] The appearance and, indeed, reception of such lectures in the MLPS's programme warrants further investigation.

The Manchester Literary and Philosophical Society, founded in 1781 and still going strong,[8] is one of the oldest examples of this type of organization,[9] the 'Lit &

Oxford: Oxford University Press 2004, https://doi.org/10.1093/ref:odnb/51577; Caroline Ehrhardt, 'Tactics: In Search of a Long-Term Mathematical Project (1844–1896)', in: *Historia Mathematica* 42/4 (2015), pp. 436–67.

[5] This volume contains the texts of lectures delivered to the Society between October 1859 and April 1861.

[6] Indeed, Cockle's paper in the 1862 volume is a follow-up to one in the preceding volume of the *Memoirs*.

[7] The society did have a dedicated 'Physical and Mathematical Section', and, indeed, lectures XI, XII (astronomy), and XVII (meteorology) of the volume under consideration are marked out explicitly as having been read before this section; however, no such indication is given for any of the mathematical lectures.

[8] See https://www.manlitphil.ac.uk/.

[9] On the history of the Manchester Literary and Philosophical Society, see: R. Angus Smith, *A Centenary of Science in Manchester: in A Series of Notes*, London: Taylor and Francis 1883; James Bottomley, 'Notes on the Early History of the Literary and Philosophical Society', in: *Proceedings of the Manchester Literary and Philosophical Society* 25 (1885–86), pp. 3–7; Francis Nicholson, 'The Literary and Philosophical Society 1781–1851', in: *Memoirs and Proceedings of the Manchester Literary and Philosophical Society (Manchester Memoirs)* 68 (1923–24), pp. 97–148; C. L. Barnes, *The Manchester Literary & Philosophical Society*, Manchester: [s.n.] 1938; Donal Sheehan, 'The Manchester Literary and Philosophical Society', in: *Isis* 33/4 (1941), pp. 519–23; H. J. Fleure, 'The Manchester Literary and Philosophical Society', *Endeavour* 6 (1947), pp. 147–51; Arnold Thackray, 'Natural Knowledge in Cultural Context: The Manchester Model', *The American Historical Review* 79/3 (1974), pp. 672–709; Robert Kargon, *Science in Victorian Manchester: Enterprise and Expertise*, Manchester: Manchester University Press 1977; Chris E. Makepeace, *Science and Technology in Manchester: Two Hundred Years of the Lit. and Phil*, Manchester: Manchester

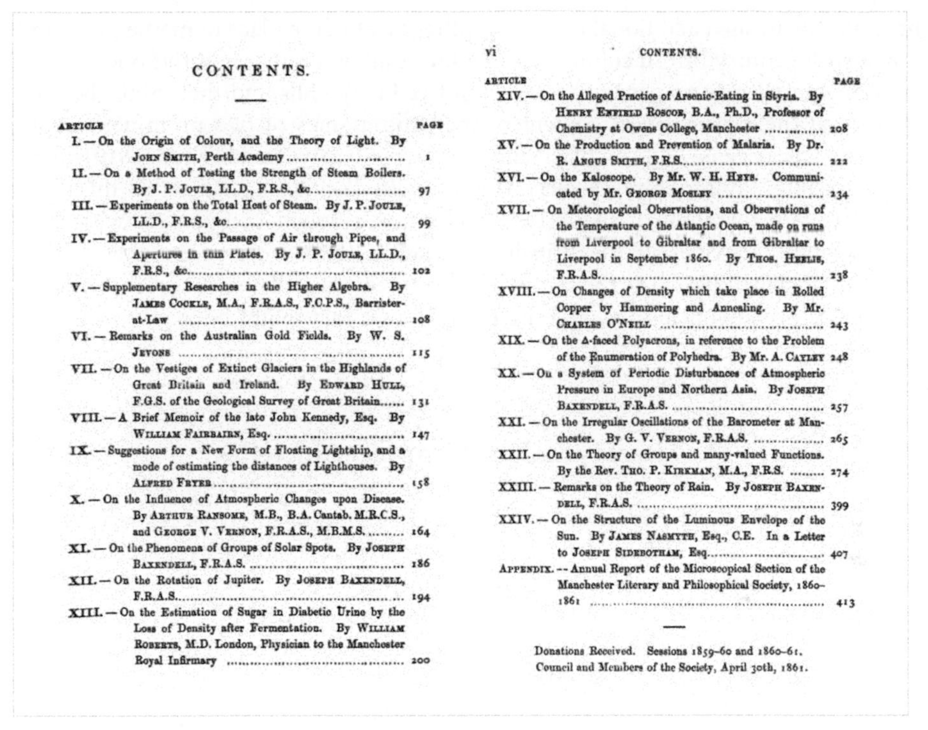

CONTENTS.

vi CONTENTS.

Figure 8.1 Contents pages of volume 1 (1862) of the third series of the *Memoirs of the Literary and Philosophical Society of Manchester.*

Phils', concerned with the communication of primarily scientific subjects to a general audience, and also with the study of topics of local interest: flora, fauna, local history, archaeology, and so on. As we will see, it is not unusual to find at least a little (practical) mathematics within the lecture programmes of these societies, but pure mathematics of the kind discussed by Kirkman is considerably rarer, though by no means incompatible with the usual aims of these societies. In this chapter, we begin the process of surveying the ways in which mathematical topics were handled by the Lit & Phils of the nineteenth century. The present investigation is very much a preliminary one, and certainly does not exploit all of the available resources[10]—it is included here as a pointer towards future lines of research. We adopt a somewhat naive and broad-brush approach to the activities of the Lit & Phils in order to draw general conclusions that might in the future be tested in the cases of specific societies. We hope

Literary & Philosophical Publications 1984; Jon Mee and Jennifer Wilkes, 'Transpennine Enlightenment: The Literary and Philosophical Societies and Knowledge Networks in the North, 1781–1830', in: *Journal of Eighteenth-Century Studies* 38/4 (2015), pp. 599–612; Jon Mee, 'The Transpennine Enlightenment, 1780–1840', in: *Annual Report of the Yorkshire Philosophical Society* (2016), pp. 65–71. Details of the society's archives may be found at https://discovery.nationalarchives.gov.uk/details/c/F103212.

[10] In particular, the present chapter is based entirely on published sources, a great wealth of which have been left to us by the various societies dealt with here. However, a fuller and more nuanced story can only emerge with the full exploitation of the often very comprehensive archives that have been preserved by these societies.

nevertheless to shed additional light on another facet of the place of mathematics in nineteenth-century British culture—even if the result is a rather negative one.

We begin with some general remarks on the Lit & Phils and their aims, before looking at the mathematical content of the programmes of two prominent such societies: the Leeds Philosophical and Literary Society (established 1819) and the Yorkshire Philosophical Society (founded in York in 1822). After this, we return to consideration of the MLPS, and draw a rough comparison with the mathematical activities of the British Association for the Advancement of Science (established 1831). At the end of the chapter, we pull the material together with some broad conclusions, and indicate avenues that might be explored in the future.

The Literary and Philosophical Societies

The societies that are the subject of the present chapter form an extremely varied collection of organizations, with a variety of different names. Some styled themselves 'Philosophical Societies', while others were 'Literary and Philosophical' or 'Philosophical and Literary'. In addition, there were 'Scientific Societies', 'Natural History Societies', 'Microscopial Societies', and all possible combinations of these labels and others. In the interests of simplicity, we will employ the blanket term 'Literary and Philosophical Society', or 'Lit & Phil' for short, throughout this chapter whenever referring to such societies in general.

Whatever its name, and whatever its particular disciplinary biases (to which subject we will return in a moment), each of these societies had a broadly similar set of aims: to promote the general education in, appreciation of, and participation in a range of disciplines that might normally have been seen as more-or-less academic.[11] The need for such organizations was felt in particular in those British towns and cities that did not have universities—which, at the beginning of the nineteenth century, was of course the vast majority. Among the earliest such societies to be formed, in the final decades of the eighteenth century, were the Manchester and Newcastle Literary and Philosophical Societies (in 1781 and 1793, respectively).[12] These were followed by many others, some very short-lived, over the course of the nineteenth century,

[11] For general remarks on such societies, see: Paul Elliott, 'The Origins of the "Creative Class": Provincial Urban Society, Scientific Culture and Socio-Political Marginality in Britain in the Eighteenth and Nineteenth Centuries', in: *Social History* 28/3 (2003), pp. 361–87; Paul Elliott, 'Towards a Geography of English Scientific Culture: Provincial Identity and Literary and Philosophical Culture in the English County Town, 1750–1850', in: *Urban History* 32/3 (2005), pp. 391–412. See also: Steven Shapin and Arnold Thackray, 'Prosopography as a Research Tool in History of Science: The British Scientific Community 1700–1900', in: *History of Science* 1 (1974), pp. 1–28; R. J. Morris, 'Clubs, Societies and Associations', Chapter 8 in: *The Cambridge Social History of Britain, 1750–1950*, ed. F. Thompson, Cambridge: Cambridge University Press 1990, 395–444; Peter Clark, *British Clubs and Societies 1580–1800: The Origins of an Associational World*, Oxford: Clarendon Press 2000; William C. Lubenow, *'Only Connect': Learned Societies in Nineteenth Century Britain*, Woodbridge: Boydell Press 2015.

[12] On the history of the Newcastle Literary and Philosophical Society, see: Robert Spence Watson, *The History of the Literary and Philosophical Society of Newcastle-upon-Tyne (1793–1896)*, London: Walter Scott Ltd 1897; Derek Orange, 'Rational Dissent and Provincial Science: William Turner and the Newcastle Literary and Philosophical Society', in: *Metropolis and Province: Science in British Culture, 1780–1850*,

although the rate of formation of new societies slowed considerably in the twentieth century.[13] A list of examples of Literary and Philosophical Societies may be found in Table 8.1.[14]

Societies of this type were formed throughout the British Isles and Ireland, although they seem to have been more common in England than elsewhere, with a particular concentration in the north of England, especially in recently industrialized locations, where newly wealthy middle classes with more free time, many of them religious nonconformists to whom university education was effectively closed, sought the betterment that a broader education might bring them.[15] Perhaps because modern science and technology were the source of this wealth, these were subjects that received particular attention in the activities of the societies. The presence of nineteenth-century amateur scientists, particularly among the clergy, gave a boost to those subjects also. The activity common to all of these societies was the organization of public lectures on a great range of subjects, which were often published in the society's *Memoirs*, *Transactions*, or similar publication; at the very least, we are able to gain an impression of lecture programmes (usually only titles, but occasionally abstracts or summaries by members of the audience) from the *Annual Reports* that many of these societies issued. As well as running lectures, several Lit & Phils hosted the more informal *conversazioni*,[16] and societies of this type were often linked to museums or to libraries.[17] As we shall see, both the Yorkshire Philosophical Society and the Leeds Philosophical and Literary Society, for example, were formed in connection with museums, and the Newcastle Literary and Philosophical Society

ed. Ian Inkster and Jack Morrell, London: Hutchinson 1983, 205–30; Mee and Wilkes, 'Transpennine enlightenment'; https://www.litandphil.org.uk/.

[13] See, for example, the comments in: Samuel J. M. M. Alberti, 'Natural History and the Philosophical Societies of Late Victorian Yorkshire', in: *Archives of Natural History* 30/2 (2003), pp. 342–58 at pp. 351–2. The formation of such societies did not stop entirely: the Macclesfield Literary and Philosophical Society, for example, was established in 2006 (https://www.litandphilmacc.org.uk/). Other existing societies, long moribund but never formally wound up, have been reinvigorated in recent years: for example, those of Hebden Bridge (https://hblitandsci.org.uk/) and of Belfast (http://www.belfastsociety.org/).

[14] Other lists of relevant societies may be found in: SDUK, *Report of the State of Literary, Scientific, and Mechanics' Institutions in England*, London: Society for the Diffusion of Useful Knowledge 1841; J. W. Hudson, *The History of Adult Education, in which is comprised A Full and Complete History of the Mechanics' and Literary Institutions, Athenæums, Philosophical, Mental and Christian Improvement Societies, Literary Unions, Schools of Design, etc., of Great Britain, Ireland, America, etc.*, London: Longman, Brown, Green & Longmans 1851, pp. 222–38; R. M. MacLeod, J. R. Friday, and C. Gregor, *The Corresponding Societies of the British Association for the Advancement of Science 1883–1929: A Survey of Historical Records, Archives and Publications*, London: Mansell 1975; Alberti, 'Natural History', p. 358; Martyn Austin Walker, '"A Solid and Practical Education within Reach of the Humblest Means": The Growth and Development of the Yorkshire Union of Mechanics' Institutes 1838–1891', PhD thesis, University of Huddersfield 2010, Appendices. A list that also includes similar such societies outside England may be found in: 'Societies, Literary and Scientific', *The Penny Cyclopædia of the Society for the Diffusion of Useful Knowledge*, vol. XIII, 1842, pp. 174–6.

[15] For early reflections on the benefits to be gained from the formation of such societies, see: Thomas Gisborne, 'On the benefits and duties resulting from the institution of societies for the advancement of literature and philosophy', in: *Memoirs of the Literary and Philosophical Society of Manchester* 5/1 (1798), pp. 70–88.

[16] Samuel J. M. M. Alberti, 'Conversaziones and the Experience of Science in Victorian England', in: *Journal of Victorian Culture* 8/2 (2003), pp. 208–30.

[17] See, for example: Anthony Burton, 'Lit and Phil Museums in the North-West', in: *Manchester Memoirs: being the Memoirs and Proceedings of the Manchester Literary and Philosophical Society* 155 (2016–17), pp. 101–8.

Table 8.1 Non-exhaustive list of Literary, Philosophical and Scientific Societies founded in the British Isles between the late eighteenth century and the early twentieth. Only societies whose name suggests the potential for mathematical interests are included in the list; thus, for example, the Chester Society of Natural History, Literature and Art is omitted. An asterisk in the third column indicates that the society still exists today.

Society	Founded	Wound up
Barnsley Literary and Philosophical Society	?	?
Bath Philosophical Society	1779	*c*.1815
Bath Royal Literary and Scientific Institution	1824	*
Belfast Natural History and Philosophical Society	1821	*
Bradford Scientific Association	1875	?
Bristol Literary and Philosophical Society	?	?
Canterbury Philosophical and Literary Institution	1825	?
Cork Literary and Scientific Society	1820	*
Derby Literary and Philosophical Society	1808	1816
Derby Philosophical Society	1783	?
Exeter Literary and Philosophical Society	?	?
Royal Philosophical Society of Glasgow	1802	*
Halifax Philosophical Society	?	?
Hebden Bridge Literary and Scientific Society	1904	*
Hampstead Scientific Society	1899	*
Hull Literary and Philosophical Society	1822	*
Leeds Philosophical and Literary Society	1819	*
Leicester Literary and Philosophical Society	1835	*
Limerick Literary and Philosophical Society	1840	?
Liverpool Literary and Philosophical Society	1812	*c*.1935
Manchester Literary and Philosophical Society	1781	*
Newcastle-upon-Tyne Literary and Philosophical Society	1793	*
Norwich Philosophical Society	1812	?
Plymouth Athenaeum	1812	*
Portsmouth and Portsea Literary and Philosophical Society	1823?	1831?
Rochdale Literary and Philosophical Society	1878	?
Scarborough Philosophical Society	1827	1853
Scarborough Philosophical and Archaeological Society	1853	*c*.1948
Sheffield Literary and Philosophical Society	1822	1932
Southport Literary and Philosophical Society	1880	?
Royal Institution of South Wales	1835	*
Wakefield Literary and Philosophical Society	1826	1838
Warrington Literary and Philosophical Society	1870	*
Whitby Literary and Philosophical Society	1823	*
Yorkshire Philosophical Society	1822	*

now survives only as the library that was part of its original foundation; similarly, the Whitby Literary and Philosophical Society now exists only as a museum.[18] Experimental demonstrations sometimes featured in the lectures organized by the Lit & Phils, and indeed some societies of this type modelled themselves closely on the Royal Institution, founded in London in 1799.[19]

In noting the educational aims of the Lit & Phils in general, we must observe that such societies often had links to nearby Mechanics' Institutes, bodies whose goal was the education of the working classes. The role of mathematics in the programmes of these institutions was rather more fundamental and systematic, and so in the interests of narrowing the scope of the present chapter, we omit the Mechanics' Institutes from further discussions.[20] On a related note, we also gloss over the subtle differences in membership that existed between distinct societies: some, usually those with close connections to Mechanics' Institutes, were open quite widely to members of the local community, whereas others were much more exclusive.[21] The accessibility of these societies to women was similarly varied, and would bear further investigation.

To return briefly to the question of disciplinary biases exhibited by the Lit & Phils, we note that it was rather common for nascent societies enthusiastically to profess an interest in all (scientific) branches of knowledge, as we shall see shortly. Such a broad scope, however, does not appear to have been sustainable in most cases, either because of the difficulties of securing suitable lecturers, or more simply because the societies were led by the interests of their prominent members, often in connection with the building up of museum collections, whose subjects necessarily required a greater focus. This is a general issue to which we will return in the concluding section, with particular reference to mathematics, but before that we will first explore the activities of the Lit & Phils through the discussion of specific examples.[22]

[18] https://whitbymuseum.org.uk/. On the history of the Whitby Literary and Philosophical Society, see: Horace B. Browne, *Chapters of Whitby History, 1823–1946: The Story of Whitby Literary and Philosophical Society and of Whitby Museum*, Hull: A. Brown & Sons 1946; Claire Loughney, 'Colonialism and the Development of the English Provincial Museum, 1823–1914', PhD thesis, University of Newcastle 2006, Chapter 5.

[19] For example, the Royal Institution of South Wales: http://www.risw.org/.

[20] See instead: Hudson, *History of Adult Education*; Walker, '"A Solid and Practical Education"'; Martyn Walker, *The Development of the Mechanics' Institute Movement in Britain and Beyond: Supporting Further Education for the Adult Working Classes*, London: Routledge 2017.

[21] Falling into the latter category, for example, we have the Yorkshire Philosophical Society, which in its early history was criticized by an anonymous correspondent to the *Yorkshire Observer* (no. 12, 18 January 1823, p. 92) for being more like a 'Yorkshire Aristocratical Society'.

[22] We take this opportunity to direct the interested reader towards the available histories of a few other Lit & Phils that are not cited elsewhere in the present chapter: R. P. Sturges, 'The Membership of the Derby Philosophical Society, 1783–1802', in: *Midland History* 4/3 (1978), pp. 212–29; Paul A. Elliott, *The Derby Philosophers: Science and Culture in British Urban Society, 1700–1850*, Manchester: Manchester University Press 2009; C. L. Boltz, *Seventy Five Years of Popular Science: Record of the Hampstead Scientific Society, 1899–1974*, London: Hampstead Scientific Society 1974; Stephen A. Shapin, 'The Pottery Philosophical Society, 1819–1835: An Examination of the Cultural Uses of Provincial Science', in: *Science Studies* 2 (1972), pp. 311–36.

The Leeds Philosophical and Literary Society

We first take the Leeds Philosophical and Literary Society (LPLS), founded in 1819.[23] This society will provide us with an indication of the very limited extent to which mathematics appeared in the programme of a typical Lit & Phil, and will also illustrate a point that we will explore in the concluding section: the biases in the interests of such societies towards those of their prominent members. The LPLS remains an active society to this day, with the aims, indicated on its website,

> [t]o promote the advancement of science, literature and the arts in the City of Leeds and elsewhere, and to hold, give or provide for meetings, lectures, classes, and entertainments of a scientific, literary or artistic nature.[24]

The book-length history of the LPLS, written by its then Honorary Secretary Lt.-Col. E. Kitson Clark in the early 1920s, shortly after the society's centenary,[25] points to a letter published in the *Leeds Mercury* in September 1818 as having provided the impetus for the creation of the society. The letter's anonymous author, 'Leodiensian', bemoaned the fact that, despite the number of philanthropic institutions already existing in Leeds, the city could

> boast of no Society for the promotion of intellectual and literary improvement, nor any which might afford opportunities to our youth for the increase of their knowledge [...].[26]

The author went on to call for the foundation of such a society, outlining the benefits, principally to 'the young men of Leeds', that its existence would bring, commenting that

> [t]he name of the Society I consider quite unimportant; Literary, Debating, Reasoning, Athenæum, or any other expressive of the thing it represents. Its purpose should be the discussion of subjects historical, literary and philosophical.[27]

During the following weeks, further correspondence in a similar vein appeared within the pages of the *Mercury*, broadly in favour of the formation of such a society,

[23] On the history of the LPLS, see: E. Kitson Clark, *The History of 100 Years of Life of the Leeds Philosophical and Literary Society*, Leeds 1924; E. D. Steele, 'The Leeds Patriciate and the Cultivation of Learning, 1819–1905: A Study of the Leeds Philosophical and Literary Society', in: *Proceedings of the Leeds Philosophical and Literary Society, Literary and Historical Section* 16/9 (1978), pp. 183–202; Loughney, *Colonialism*, Chapter 8; Mark Steadman, 'A History of the Scientific Collections of the Leeds Philosophical and Literary Society's Museum in the Nineteenth Century: Acquiring, Interpreting & Presenting the Natural World in the English Industrial City', PhD thesis, University of Leeds 2019, Chapter 1. The society's archives are held in Special Collections (SC/LPLS) at the University of Leeds Library; see https://explore.library.leeds.ac.uk/special-collections-explore/617562.

[24] https://www.leedsphilandlit.org.uk/.

[25] Clark, *History of 100 Years of Life*.

[26] Quoted in Clark, *History of 100 Years of Life*, p. 5.

[27] Ibid., p. 6.

but with concerns expressed as to whether the city could sustain one.[28] The letter-writers also debated the subjects that such a society might cover in the lectures that it arranged:

> [t]he attention of the members might be equally devoted to literature and natural philosophy, the two subjects alternately occupying the consideration of the meeting.[29]

However, one suspects from the outset a slight bias towards 'philosophical' subjects, which, it was suggested, might come to dominate the society's activities; literature was presented elsewhere merely as a possible break from science:

> [t]he introduction of literature would be a relief to the severe and dry study of science which again would be returned to with fresh zest after a literary discussion.[30]

The idea of forming such a society evidently gained momentum during the final months of 1818, and on 11 December 'a Meeting of Gentlemen' took place at the Court House in Leeds, at which it was decided to establish 'a Philosophical Institution for Leeds and its Vicinity' subject to the resolution that

> this Institution shall be denominated the Philosophical and Literary Society of Leeds, in which subjects shall be discussed, that include all the Branches of Natural Knowledge and Literature (strictly so called), but excluding all topics of Religion, Politics, and Ethics.[31]

Further general meetings of the new society were held during the following year. One of the early decisions made, in May 1819, was to purchase land for the construction of a new building for the society. Matters moved remarkably quickly: £6,100 was raised in subscriptions, and the corner stone of the building was laid in July 1819 by Benjamin Gott (1762–1840), a leading wool merchant in Leeds, and an enthusiastic driving force during the formation of the LPLS. He had also presided at the general meeting of the society in January 1819 at which the form that the LPLS ought to take was decided upon; based on decisions made there, the society's first officers and council were elected in November that year. Its first president was John Marshall (1765–1845), a local manufacturer who went on to become Member of Parliament for Yorkshire (1826–1830). The society's laws were finally enacted at a meeting in January 1820, although the formal business of the LPLS was not deemed to have begun until

[28] These concerns may have been well founded: an earlier Philosophical and Literary Society had been established in Leeds in 1783, but it does not appear to have survived the decade, having succumbed to internal disagreement as to the subjects that it ought to cover; see: Clark, *History of 100 Years of Life*, pp. 2, 11. The city would also eventually have a Mechanics Institute: A. D. Gardner and E. W. Jenkins, 'The English Mechanics' Institutes: The Case of Leeds 1824–42', in: *History of Education* 13/2 (1984), pp. 139–52.

[29] Quoted in Clark, *History of 100 Years of Life*, p. 8.

[30] Ibid.

[31] Ibid., p. 11. See also the introduction to the first volume of the society's *Transactions* (1837).

its first general meeting of 1822, held shortly after the society had taken possession of its new building, Philosophical Hall, on the corner of Park Row and Bond Street in Leeds. On the occasion of this meeting, on 6 April, an introductory address was given by one of the society's Honorary Secretaries, the physician Charles Turner Thackrah (1795–1833).[32] The text of this address gives an indication of the views of at least one member as to the early positioning of the society.[33]

The newspaper correspondence that sparked the foundation of the LPLS had made explicit mention of several subjects of a scientific nature: anatomy and natural history, for example, along with the much more broadly defined natural philosophy from which the society took part of its name. Thackrah's address expanded greatly upon this list, and it is here that we find the first explicit mention of mathematics as a possible subject of interest to the society. Indeed, the speech, later described as 'a dignified and inspiring discourse',[34] extolled the virtues of all forms of learning, the 'love' of which 'distinguishes man from every other class of the animal kingdom'.[35] The tone of the discourse, and perhaps also that of the LPLS as a society, was set on the first page, where Thackrah spoke of man's 'thirst for improvement' and of the pursuit of science as a hallmark of civilization, going on in later pages to cite the examples of ancient Egypt, Greece, and Rome, before arriving finally at nineteenth-century Britain. Thackrah praised Leeds for having established in the LPLS a body that might exploit 'the advantage to be obtained by the union of talent, and the conversational diffusion of knowledge',[36] and went on to outline the different spheres of learning that it might pursue. He defended natural philosophy against critics who condemned it for having promoted too sceptical an attitude, and argued instead for its 'numerous and important benefits' for mankind;[37] in his view

> classical studies, inordinately pursued, have often prevented the development of Science. They have been cultivated with undue eagerness, and at the expense of Knowledge much more important.[38]

Moreover, Thackrah was very clear about what knowledge he deemed to be important:

> Of the studies of Philosophy, however, none produce results so stable, satisfactory, and splendid, as those of *Natural Knowledge*, the *Mathematics* and *Mechanical Sciences*. Here are objects capable of employing all the energies of the mind, and of gratifying its ardent curiosity.[39]

[32] On Thackrah, see: J. Cleeland and S. Burt, 'Charles Turner Thackrah: A Pioneer in the Field of Occupational Health', in: *Occupational Medicine* 45/6 (1995), pp. 285–97.

[33] Charles Turner Thackrah, *An Introductory Discourse Delivered to the Leeds Philosophical and Literary Society, April 6, 1821*, Leeds: Philosophical and Literary Society 1821.

[34] Clark, *History of 100 Years of Life*, p. 25.

[35] Thackrah, *Introductory Discourse*, p. 1.

[36] Ibid., p. 2.

[37] Ibid., p. 26.

[38] Ibid., p. 14.

[39] Ibid., p. 21; emphasis in original.

Given that this was evidently a rousing address delivered to the nascent society, we might therefore expect to find Thackrah's three italicized subjects as the major themes of the early years of the LPLS. However, this was not so, as we can see from the data presented in Figure 8.2, concerning the topics of lectures delivered to the society between 1819 and 1831. It is not surprising that the heading 'Science, Chemistry, Geology, Entomology' (category C in Figure 8.2) is one of the main, if somewhat variable, themes of the lectures, though a perusal of the list of titles of these lectures[40] indicates that chemistry, geology, and entomology were the dominant disciplines under this heading, with the more general 'science' covering only the very occasional further lecture on topics such as optics.[41] Subjects that might be construed as 'mathematical' did occupy the society's attention on occasion, but not necessarily under the heading of 'science'—we will return to this point below. In reference to the distribution of topics in Figure 8.2, Clark observed that 'music and art which curiously do not appear to enter into the scheme of Mr. Thackrah are not forgotten',[42] and indeed, we see that 'Literature, Art, Pædogogy, Ethics' (category B) formed the other prominent theme for the society. Moreover, the spread of topics that the LPLS went on to cover during its early decades appears to have gone far beyond the expectations of some of its early members, who supposed that the society would focus on local studies, even though this specialism had played only a small role in the discussions surrounding its formation:

> Experience had proved that the Members had not been, as probably it was originally anticipated, principally occupied with the investigation of local geology, or the fauna and flora of the district. The meetings of the Society had on the other hand been attended by a miscellaneous audience rather than mainly of students of science, and the result had been that the work had developed rather in the popularization and dissemination of knowledge than in local investigation.[43]

An important aspect of the LPLS that we have not yet mentioned is its museum, which was a part of the original plan for the society, along with the establishment of a library, both of them to be accommodated in the society's new building.[44] Philosophical Hall with the museum therein was eventually transferred to the ownership of Leeds City Council in 1921, and the museum collections may now be found within the holdings of the current Leeds City Museum.[45]

The LPLS's museum was developed originally from donations by members, an important early acquisition being the private natural history collection of the surgeon John Atkinson (1787–1828), who served as the LPLS's first curator and librarian.

[40] As found on pp. viii–xvi of the first volume of the society's *Transactions*.

[41] For example, 'On Coloured Shadows', delivered by William Hey in December 1822.

[42] Clark, *History of 100 Years of Life*, p. 27.

[43] Clark, *History of 100 Years of Life*, p. 80, quoting from the society's *Annual Report* for 1869–1870.

[44] For a very detailed treatment of the LPLS's museum, see: Mark Steadman, *Scientific Collections*; see also Loughney, *Colonialism*, Chapter 8. Steadman (Chapter 1) argues that the place of the museum within early planning at the LPLS was in fact rather marginal.

[45] See https://museumsandgalleries.leeds.gov.uk/leeds-city-museum/ and Peter Brears, *Of Curiosities & Rare Things: The Story of Leeds City Museums*, Leeds: Friends of Leeds City Museums 1989. The LPLS's library was donated to the Brotherton Library of the University of Leeds in 1936.

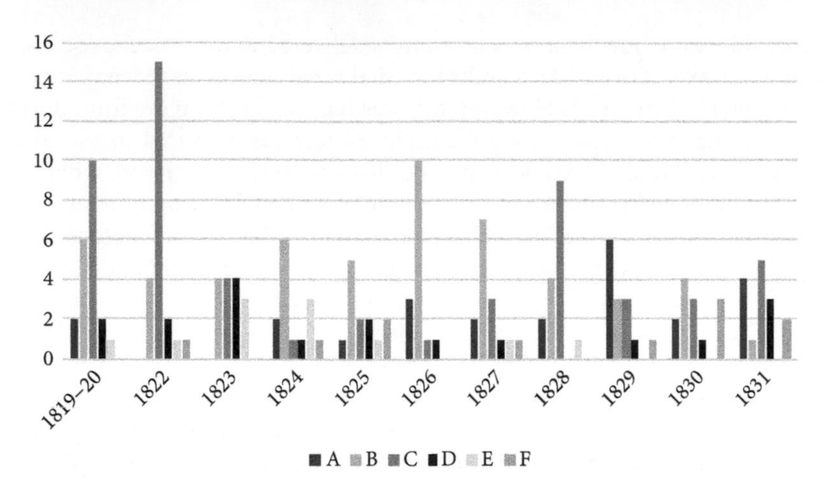

Figure 8.2 Distributions of topics of lectures delivered to the Leeds Philosophical and Literary Society 1819–31, according to data drawn from Clark, *History of 100 Years of Life*, p. 27, arranged there under the following categories: A = 'Archæology, History'; B = 'Literature, Art, Pædagogy, Ethics'; C = 'Science, Chemistry, Geology, Entomology'; D = 'Physiology, Biology', E = 'Geography, Travel, Ethnography'; F = 'Economics'.

Indeed, the museum's focus on natural history was strengthened further when the society proceeded to create the post of salaried curator, the first holder of which, the entomologist Henry Denny (1803–1871), who held the post for 45 years, was especially active in building up these collections. Many of the specimens acquired during this time can still be seen in Leeds:[46] for example, the skeleton of a prehistoric hippopotamus discovered at Armley,[47] and a stuffed Bengal tiger, purchased in 1862.[48] Another notable acquisition during the museum's early years was that of two Egyptian mummies,[49] one of which remains in the Leeds City Museum.[50] Indeed, as we shall discuss further below, the burgeoning field of Egyptology was a major interest for members of the early LPLS. Unsurprisingly, however, mathematics had little role to play in the society's museum.

We pick up now on the observation made above that a scattering of mathematical topics did in fact appear within the lecture programme of the LPLS during the first sixty years of its history. The limited available details of these lectures are summarized in Table 8.2. We note, however, that there is nothing here in the nature of the

[46] Adrian Norris, 'Leeds City Museum—Its Natural History Collections', in: *Journal of Biological Curation* 1 (1993), pp. 29–39.

[47] https://news.leeds.gov.uk/news/leeds-museums-and-galleries-object-of-the-week-the-armley-hippo.

[48] https://news.leeds.gov.uk/news/leeds-museums-and-galleries-object-of-the-week-the-leeds-tiger.

[49] William Osburn Jnr, *An Account of an Egyptian Mummy, presented to the Museum of the Leeds Philosophical and Literary Society, by the late John Bladys, Esq.*, Leeds: Philosophical and Literary Society 1828. See also: https://russellcroft.net/urban-history-leeds/the-leeds-mummy/.

[50] See https://news.leeds.gov.uk/news/leeds-museums-and-galleries-object-of-the-week-coffin-of-nesyamun. The other mummy was lost to an air-raid during the Second World War which also severely damaged Philosophical Hall, leading to its eventual demolition in 1965.

technical lectures that we observed at the MLPS. Many of these lectures delivered in Leeds might best be described as meta-mathematical, and several were probably placed under the heading of 'pedagogy', rather than 'science'. In the cases of those lectures on mechanical topics, one suspects that a qualitative rather than quantitative approach would have been taken—most particularly in the case of the experimentally illustrated lecture of the celebrated astronomer and science-popularizer Robert Stawell Ball (1840–1913) during the 1875–76 session. Indeed, Ball also lectured to the society (on astronomical topics) during the 1881–82, 1883–84, and 1891–92 sessions;[51] LPLS Honorary Member Sir Thomas Edward Thorpe (1845–1925) later noted that '[n]o lecturer was more popular than the late Sir Robert Ball, who managed

Table 8.2 Lectures on mathematical topics (some only slightly so) delivered before the Leeds Philosophical and Literary Society between 1823 and 1886; compiled from Clark, *History of 100 Years of Life*, pp. 154–92.

Year	Speaker	Title
1823–1824	John Dalton FRS, Manchester	A Course of Lectures on the Principle of Mechanics and Meteorology (six lectures)
1827–1828	John Darby	The Advantages to be derived from Mathematical and Philosophical Knowledge
1828–1829	William Osburn Jr.	The System of Numerical Notation, and on the Calendar of the ancient Egyptians
1829–1830	Rev. William Turner, Halifax	The Nature and Extent of Mathematical Evidence
1830–1831	Rev. James Acworth	The Application of Mathematical Reasoning to Metaphysical and Moral Science
1834–1835	Rev. Joseph Holmes	The Motions of Bodies generally acted upon by Forces tending to a point, and particularly with reference to the Solar and Lunar Theories
1837–1838	J. G. Marshall, Vice-President	The Nature of Statistic Enquiries, and their Relations to other Branches of Science, and to the Purposes of Social Improvement
1842–1843	R. D. Chantrell	The Geometric Principles of Gothic Architecture
1852–1853	Rev. H. R. Reynolds	Some of the Scientific and Practical Applications of the Doctrine of Probabilities
1853–1854	William Osburn Jr.	The Mode of Computing time in Ancient Egypt
1858–1859	James Yates, Statician [*sic*] (read by T. W. Stansfeld)	Illustrations of the Metrical System of Weights and Measures
1869–1870	William Farr MD FRS DCL	International Decimal Coinage
1875–1876	R. S. Ball LLD FRS, Professor of Astronomy, University of Dublin	The Laws of Motion, with Experimental Illustrations
1885–1886	C. V. Boys	Mechanical Aids to Calculation

[51] Clark, *History of 100 Years of Life*, pp. 190–1, 195.

to introduce humour into the driest of technical subjects'.[52] Regardless of whether Ball's or any of the other lecturers' talks had any substantial mathematical content, the lectures listed in Table 8.2 still only represent fourteen contributions over more than sixty years to a society that typically arranged 15–20 lectures per year. It is interesting to note in passing that the history of mathematics had a small presence, via the Egyptological interests of the Honorary Secretary and antiquarian William Osburn (1793–1875);[53] these lectures were presumably categorized as 'history', 'archaeology', or 'ethnography'.

Thus, over the first 100 years of its existence, the LPLS appears to have had little to do with mathematics, despite Thackrah's emphasis on the subject's 'stable, satisfactory, and splendid' results. We will explore the reasons for this in more detail in our concluding section. As the society entered its second century, however, changes were afoot. As we noted above, it was around this time that the LPLS transferred the ownership of its building and museum to Leeds City Council and the society became a registered charity, with the aims listed at the beginning of this section. It was also during the 1920s that the LPLS began publication of a new journal, *Proceedings of the Leeds Philosophical and Literary Society*, which appeared in two sections, 'Literary and Historical' and 'Scientific'. We will not attempt to treat this journal in any depth here, since it would take us well beyond the scope of the present book, but we observe that at least the first volume of the scientific section of the *Proceedings* contains a number of papers, all of them either written or communicated by members of the mathematics department of the University of Leeds, whose titles suggest overtly mathematical topics.[54] This sudden appearance of technical mathematics in the activities of the LPLS at such a late date warrants future investigation.

The Yorkshire Philosophical Society

Our second example of a Lit & Phil is provided by the Yorkshire Philosophical Society (YPS), founded in York in 1822.[55] We will see in this society a number of differences from the LPLS, particularly with regard to the subjects covered, but also several

[52] Quoted in Clark, *History of 100 Years of Life*, p. 141.

[53] On Osburn and his interests, which mostly concerned the connections between Egypt and the Bible, see: Morris L. Bierbrier, *Who Was Who in Egyptology*, 4th rev. edn, London: Egypt Exploration Society 2012, p. 410. Osburn's 1828–29 lecture on Egyptian numeration must have contained cutting-edge material, since the Egyptian scripts had not long been deciphered at this point.

[54] For example: 'The 10 nodes of the Hessian of the cubic equation' by William P. Milne (January 1926) or 'A 15-ic resolvent of the binary sextic' by J. Lamb (January 1927).

[55] On the history of the YPS, see: J. Kenrick, 'A Retrospect of the Early History of the Yorkshire Philosophical Society', in: *Annual Report of the Council of the Yorkshire Philosophical Society for MDCCCLXXIII*, York, 1874, pp. 34–44; A. D. Orange, *Philosophers and Provincials: The Yorkshire Philosophical Society from 1822 to 1844*, York: Yorkshire Philosophical Society 1973; Derek Orange, 'Science in Early Nineteenth-Century York: The Yorkshire Philosophical Society and the British Association', in: *York 1831–1981: 150 Years of Scientific Endeavour and Social Change*, ed. C. H. Feinstein, York: William Sessions, Ebor Press in association with the British Association for the Advancement of Science (York Committee) 1981, pp. 1–29; David Rubinstein, *The Nature of the World: The Yorkshire Philosophical Society, 1822–2000*, York: Quacks Books 2009; Sarah Sheils, *From Cave to Cosmos : A history of the Yorkshire Philosophical Society*, York: Yorkshire Philosophical Society 2022; see also the various articles in the society's *Annual Report* for 1971. The society's archives are held at the Borthwick Institute for Archives at the University of York (reference code YPS): see https://borthcat.york.ac.uk/index.php/yps as well as the resources available at https://www.ypsyork.org/resources/yps-archive/.

similarities in the initial formation of the society, most notably in the provision of a museum and a library. Moreover, the YPS is another Lit & Phil that still exists and still organizes a full programme of activities, with the stated aim

> to promote interest in and understanding of natural sciences, archaeology and antiquities together with any technological developments affecting them and any social issues which these may raise.[56]

We note the quite narrow focus adopted here, which will be contrasted with statements that the society made about itself during its earlier history.

The historian A. D. Orange, writing about the formation of the YPS, points to one event in particular to which the foundation of the society may be traced: the discovery of a cave full of prehistoric animal bones at Kirkdale in North Yorkshire in July 1821.[57] An early account of this discovery was provided by the Oxford geologist William Buckland (1784–1856) in a paper to the Royal Society, in which he noted the presence of the teeth and bones of 'Elephant, Rhinoceros, Hippopotamus, Bear, Tiger, and Hyæna, and sixteen other animals',[58] which he saw as being relics of the biblical Flood.[59] For many amateur collectors, the bones quickly became sought-after objects 'to furnish their geological cabinets'.[60] However, newspaper coverage of the discovery had fostered a general recognition of its importance, and the loss of vital specimens to private collections (whose owners for reasons of discretion remained undisclosed) was decried—for example, by Philip Francis Sidney, editor of the *Yorkshire Observer*, in November 1822:

> At Kirkdale, very recently, some fine Antediluvian Remains were discovered and seized upon by individuals. To whom do they now belong? [...] I am informed that some of the finest specimens are in the hands of________ and of________. Do they consider them private property? I do presume to hope that a sense of public spirit will interpose, and induce all parties concerned to reunite the specimens [...].[61]

Sidney was concerned also for the preservation of artefacts linked to York's long history, and proposed the foundation of a 'museum and school of arts' that would provide a repository for these, as well as the Kirkdale bones and other items. Although Sidney's call was not taken up directly, it did form part of a wider newspaper correspondence, similar to that which we saw in the case of the LPLS, concerning the possible establishment in York of a 'literary society [...] for *free* and *friendly* discussion upon the various subjects that science and literature present'.[62] Consciousness

[56] https://www.ypsyork.org/about-yps/objectives/.

[57] Orange, *Philosophers and Provincials*, p. 7.

[58] William Buckland, 'Account of an assemblage of fossil teeth and bones of elephant, rhinoceros, hippopotamus, bear, tiger, and hyæna, and sixteen other animals; discovered in a cave at Kirkdale, Yorkshire, in the year 1821: with a comparative view of five similar caverns in various parts of England, and others on the Continent', in: *Philosophical Transactions of the Royal Society of London* 112 (1822), pp. 171–236.

[59] A. D. Orange, 'Hyaenas in Yorkshire: William Buckland and the Cave in Kirkdale', in: *History Today* 22/11 (1972), pp. 777–85.

[60] Orange, *Philosophers and Provincials*, p. 7.

[61] *Yorkshire Observer*, no. 1 (2 November 1822), p. 4.

[62] Quoted from the *Yorkshire Courant*, 1821 by Orange, *Philosophers and Provincials*, p. 8; emphasis in Orange.

of the fact that societies of this type were already being formed in other towns and cities (the newspapers mention Hull and Sheffield) appears to have aroused civic pride.[63]

The YPS had its very modest beginnings in a meeting on 7 December 1822 of three men in Lendal, a street in the centre of York: William Salmond (1769–1838), a retired army officer and amateur geologist, James Atkinson (1759–1839), a retired surgeon, and Anthony Thorpe (1759–1830), a solicitor. Salmond had been involved in Buckland's investigation of the Kirkdale cave, and, along with the other two, had amassed a personal collection of fossilized bones from there. The discussion of the collation and housing of these collections appears to have been the purpose of the meeting, which was followed by another a week later, to which the churchman Rev. William Venables Vernon (1789–1871), son of the Archbishop of York, was invited (but which Thorpe apparently did not attend). A printed prospectus for the society that was to administer the collections had come out of the first meeting, and this had been circulated to several sympathetic individuals, who were all now elected to membership of what was to be called the 'Yorkshire Philosophical Society', with Atkinson as its secretary, and Salmond and Vernon as other members of its organizing committee. A second printed prospectus was also agreed:

> It is proposed to establish, at York, a Philosophical Society, and to form a Scientific Library, and a Museum. – *The Library* will consist principally of [...] Books of Reference, and in particular of the Transactions of the Philosophical Societies, and Journals of Science published in different parts of the World, to be consulted in the Reading-Room by Members living in York, and sent out to those who reside at a distance. The *Museum* will be open to every subject of scientific curiosity: but it is chiefly designed to be a Repository of Antiquities, in which the vicinity of this City abounds; and of Geological Specimens of which no district in England can furnish a greater variety than the County of York. – A nucleus of the geological part of the Museum, has already been formed by the contribution of a few Individuals, who have given a collection (the most perfect that is anywhere to be found) of the valuable Fossil Remains, lately discovered in the Cave of Kirkdale.[64]

Thus, geology was to be a clear focus for the society's museum from the outset—and, by extension, for the society itself. Other early statements make it clear, however, that this was not the only topic to be considered: '[t]he general object of the Society, is the promotion of Science in the district for which it has been instituted'.[65] In a copy of the YPS's *Objects and Laws* for 1825, we find geology referred to as '[o]ne of the Society's principal objects';[66] much discussion is also given over to natural history, another

[63] On the Hull Literary and Philosophical Society, see: http://www.hull-litandphil.org.uk; on the Sheffield Literary and Philosophical Society, see: William Smith Porter, *Sheffield Literary and Philosophical Society: A Centenary Retrospect 1822–1922*, Sheffield: J. W. Northend 1922.

[64] Quoted in Orange, *Philosophers and Provincials*, p. 10.

[65] *Objects and Laws of the Yorkshire Philosophical Society: with the Annual Report for MDCCCXXIII*, York, 1824, p. 5.

[66] *Objects and Laws of the Yorkshire Philosophical Society, instituted December, MDCCCXXII*, York, 1825, p. 6.

subject with a clear representation in the society's museum holdings. By 1829, we find the bold statement that 'there is NO PART of real knowledge, whether of art or nature, to which [the society] does not wish to apply itself', although this is paired with the qualification that 'LOCAL information, and whatever materials of information relate to Yorkshire, are its principal objects'.[67] Thus, although mathematics would not, in principle, be excluded from the activities of the society, it is clear that it was far from being a priority, and would easily be squeezed out by the greater focus on geological and natural historical topics. We note also that the YPS was firmly from the start a 'philosophical' society, rather than 'literary and philosophical'.

The YPS held its first official meeting in March 1823 in rooms purchased by a group of members for that purpose in Low Ousegate in York. Its first order of business, like the LPLS before it, was the question of finding suitable accommodations for its projected museum and library. The structure of the society was also settled that year, with Vernon being unanimously elected the YPS's first president. Even in the absence of a permanent home for a library and museum, the society appointed a librarian, as well as curators with responsibilities for 'geology, fossil and recent conchology', mineralogy, comparative anatomy, entomology and ornithology, and antiquities and coins. From its first volume for 1823, the YPS's *Annual Report* listed donations both to the library and to the museum, arranged according to the areas of responsibilities of the curators. By the end of the decade, the society had amassed collections consisting of many thousands of items.

By 1825, the YPS had identified the site for its new permanent home: a piece of land beside the River Ouse known as Manor Shore, adjacent to the ruins of St Mary's Abbey and to King's Manor, the former home of the Council of the North. Subscriptions were raised for the construction of a new building on the site which would house not only the society's museum but also accommodation for the keeper, a library, a lecture theatre, and a basement laboratory. However, legal difficulties over the formal acquisition of the land caused delays in the construction, which nevertheless began even before the receipt by the society of ten acres of land by royal grant in 1828.[68] The grant was made on the condition that part of the site be used to establish a botanic garden, a project which the society embarked upon during the 1830s, creating the forerunner of the current Museum Gardens.[69] The new building on the site, the Yorkshire Museum, was completed by 1830, and officially opened in February that year to much acclaim.[70] In a similar way to the museum of the LPLS, the Yorkshire Museum was eventually given over to York City Council in 1961, and is now managed by the York Museums Trust.[71]

Given the interests of its founders, it is unsurprising to find that scientific communications delivered to the society during its early years were often on geological topics,

[67] *Annual Report of the Council of the Yorkshire Philosophical Society for MDCCCXXIX*, York, 1830, pp. 4–5.

[68] The process appears to have been smoothed somewhat by a number of Vernon's cousins, who were well placed within certain government departments; see: Orange, *Philosophers and Provincials*, Chapter 4.

[69] Peter J. Hogarth and Ewan W. Anderson, *'The Most Fortunate Situation': The Story of York's Museum Gardens*, York: Yorkshire Philosophical Society 2018.

[70] See, for example, Orange, *Philosophers and Provincials*, p. 22.

[71] https://www.yorkshiremuseum.org.uk/.

but as the membership grew and as more external speakers were invited,[72] other subjects began to appear, and during the mid-1820s, the society organized courses of lectures on specific topics: for example, the *Annual Report* for 1824 records lectures on experimental chemistry and on zoology.[73] Once the society had taken possession of its new building in 1830, the lecture programme expanded still further—but not, it seems, to include mathematics: the LPLS may have received only a few mathematical communications, but the YPS had even fewer. Indeed, if we browse the society's *Annual Reports* to the end of the nineteenth century, we find only four overtly mathematical contributions to meetings of the society (probably short communications, rather than full lectures), all of them by the same honorary member, the Rev. William Taylor (1790–1870):[74]

- 'Drawing and Description of a Peculiar Method of Generating Curves' (November 1828);
- 'Notice of a Simple Method of Measuring Certain Inaccessible Distances' (March 1831);
- 'Notice of the Construction of a New Perpetual Sunday Calendar' (March 1831);
- 'A Demonstration of Euclid III.31' (January 1841).[75]

These four communications represent just a small selection of Taylor's wider contributions to the YPS, which also embraced, for example, improvements to clockwork mechanisms, and experiments with coal-gas and electrical apparatus.[76] Indeed, when describing the disciplinary biases of the YPS, Orange observes that

> [t]he physical sciences were largely ignored except by the Rev. William Taylor, an honorary member much given to invention, who exhibited and enlarged on a variety of devices mechanical and electrical.[77]

Taylor also appears frequently in the YPS's *Annual Reports* in connection with donations made by him to the society's museum and library. Like his communications, these were also quite eclectic, comprising, for example, fossils, minerals, coins, plant specimens, heraldic drawings, a 'new discharging Electrometer',[78] and a 'Piece of

[72] For example, William Osburn, whom we met in the section on the LPLS, delivered a lecture to the YPS in February 1828 with the title 'On the Hieroglyphical Inscriptions accompanying a Mummy in the collection of the Leeds Philosophical and Literary Society' (*Annual Report of the Council of the Yorkshire Philosophical Society for MDCCCXXVIII*, York, 1829, p. 33).

[73] *Annual Report of the Council of the Yorkshire Philosophical Society for MDCCCXXIV*, York, 1825, p. 4.

[74] *Annual Report of the Council of the Yorkshire Philosophical Society for MDCCCXXVIII*, York, 1829, p. 34; *Annual Report of the Council of the Yorkshire Philosophical Society for MDCCCXXXI*, York, 1832, p. 23; *Annual Report of the Council of the Yorkshire Philosophical Society for MDCCCXL*, York, 1841, p. 22.

[75] Euclid III.31: 'In a circle the angle in a semicircle is right, that in a greater segment less than a right angle, and that in a less segment greater than a right angle; and further the angle of the greater segment is greater than a right angle, and the angle of the less segment less than a right angle'. (T. L. Heath, *The Thirteen Books of Euclid's Elements*, 3 vols, Cambridge: Cambridge University Press 1908, vol. II, p. 61.)

[76] For example: 'Drawing and Description of a New Compensation Balance-Wheel for a Watch' (February 1828); 'Account of New Experiments on the Combustion of Coal-Gas' (February 1829); 'On Some New Modifications of Electrical Apparatus' (December 1830).

[77] Orange, *Philosophers and Provincials*, p. 29.

[78] *Annual Report of the Council of the Yorkshire Philosophical Society for MDCCCXXX*, York, 1831, p. 39.

protected Telegraphic wire used for Submarine purposes'.[79] Taylor's donations to the society's library included treatises on gaslight and optics, a twenty-year run of the *Philosophical Transactions of the Royal Society*, and, perhaps more interestingly for our present purposes, a copy (and subsequently the copper plates) of his own embossed version of Euclid's *Elements*, intended for the teaching of geometry to the blind.[80] The YPS's *Annual Report* for 1828 singled out this donation for special mention:

> Too little attention has hitherto been paid in this country to the means of communicating the advantages of education to the blind, especially in scientific attainments; and in the work now alluded to, a successful attempt has been made to supply part of that defect, by publishing the diagrams of Euclid in a *tangible* form.[81]

Indeed, the work appears to have enjoyed a brief vogue during the 1830s,[82] and was cited on Taylor's certificate of election to the Royal Society in 1836.[83] The education of the blind is a subject in which Taylor was particularly interested: he was the first superintendent of the Wilberforce Memorial School for the Blind (in York) and later founded the Worcester College for the Blind Sons of Gentlemen.[84] He also devised a mechanism, known subsequently as the Taylor Arithmetical Slate or Taylor Arithmetical Frame,[85] for teaching arithmetic to the blind, an example of which we also find among the lists of donations to the YPS.[86]

Thus, we see that what little mathematics found its way into the activities of the YPS was confined to topics related to education, or to others that might be described as mathematical curios (for example, Taylor's perpetual Sunday calendar). Once again, we see nothing in the same vein as the mathematics that appeared in the programme of the MLPS. In addition to those mentioned above, a handful of mathematical books were donated to the YPS's library over the years: for instance, its earliest library catalogue listed Andrew Motte's English translation of Newton's *Principia*;[87] Taylor

[79] *Annual Report of the Council of the Yorkshire Philosophical Society for MDCCCLI*, York, 1852, p. 22.

[80] William Taylor, *The Diagrams of Euclid's Elements of Geometry, in an Embossed or Tangible Form, for the use of Blind Persons who Wish to Enter upon the Study of that Noble Science*, York, 1828. See: *Annual Report of the Council of the Yorkshire Philosophical Society for MDCCCXXVIII*, York, 1829, p. 51; *Annual Report of the Council of the Yorkshire Philosophical Society for MDCCCXXX*, York, 1831, p. 40.

[81] *Annual Report of the Council of the Yorkshire Philosophical Society for MDCCCXXVIII*, York, 1829, p. 23; emphasis in original.

[82] It is praised, for example, in James Wilson, *Biography of the Blind: or the Lives of Such as Have Distinguished Themselves as Poets, Philosophers, Artists, &c.*, Birmingham: J. W. Showell 1838, p. 283.

[83] EC/1836/04. We note that William Venables Vernon (now appearing under the name William Vernon Harcourt, owing to a family inheritance—see: Orange, *Philosophers and Provincials*, p. 14) was the first proposer on the certificate.

[84] Amanda Bergen, 'A Philosophical Experiment: The Wilberforce Memorial School for the Blind c.1833–1870, in: *European Review of History: Revue européenne d'histoire* 14/2 (2007), pp. 147–164; D. Bell, *An Experiment in Education: The History of Worcester College for the Blind, 1866–1966*, London: Hutchinson 1967.

[85] See, for example, G. F. Meyer, 'Devices in Mathematics', in: *American Association of Instructors of the Blind: Twenty-Seventh Biennial Convention, held at Watertown, Massachusetts, June 23 to 27, 1924*, pp. 202–8.

[86] '"[A] simple Contrivance for teaching arithmetic to the blind"': *Annual Report of the Council of the Yorkshire Philosophical Society for MDCCCXXVI*, York, 1827, Appendix, p. xvi.

[87] *Objects and Laws of the Yorkshire Philosophical Society: with the Annual Report for MDCCCXXIII: Regulations and Catalogue of the Library: List of Officers etc. and General List of Members*, York, 1824, p. 39.

later gave a copy of Sacrobosco's *Libellus de Sphæra*, the polymath Charles Babbage (1791–1871) donated his *Table of Logarithms*, and the YPS also received a copy of Abraham Sharp's 1717 *Geometry Improv'd*. Most of the items in the society's possession, however, related to its core interests of geology and natural history, and one wonders how much use these mathematical books saw.

By way of concluding this section, we allow ourselves another brief trip beyond the true scope of the present book, and into the twentieth century. At various points in the YPS's *Annual Reports* for the first decades of that century, we find reaffirmations of the society's main interests. For example, in 1912, on the occasion of the opening of the refurbished lecture theatre in the Yorkshire Museum, the geologist T. G. Bonney (1833–1923) gave a lecture entitled 'Development of Education', in which he noted the value of learning mathematics and classics, but asserted that it is the natural sciences that '[train] us in accurate observation and sound induction'.[88] Ten years later, on the YPS's centenary, a congratulatory message from the Royal Society of Edinburgh pointed to the 'two fields of activity in which the work of the Society has achieved special distinction, viz.: Archæology and Geology'.[89] By the 1960s, in the wake of the loss of its museum,[90] and perhaps also in anticipation of the impending foundation of the University of York in 1963, the YPS gave a clear statement of its aims: namely, those quoted at the beginning of this section—to promote 'natural sciences, archaeology and antiquities'. Thus, although the society had made rhetorical gestures towards, and given some limited attention to, other subjects during its earlier history, it remained largely true to the intentions of its founders.

The Manchester Literary and Philosophical Society

We return at last to the question of the appearance of mathematics at the MLPS. The previous two sections have given us an impression of the minimal handling of mathematics at other Lit & Phils, and so we give here a contrasting preliminary survey of the representation of this subject at the MLPS.

As we have already noted, the MLPS was one of the first Lit & Phils to be founded, in 1781. A history of the society,[91] written on the occasion of its centenary by the chemist Robert Angus Smith (1817–1884), who had served both as the society's secretary (1852–57) and as its president (1864–66), emphasizes the influence of the physician and health reformer Thomas Percival (1740–1804) in the formation of the society, which grew out of the meetings of scientific men (predominantly medics) that took place at Percival's house in Manchester. As Percival put it himself in the Preface to the first volume of the society's *Memoirs* (printed in 1785):

> Many years since, a few Gentlemen, inhabitants of the town, who were inspired with a taste for Literature and Philosophy, formed themselves into a kind of weekly

[88] *Annual Report of the Council of the Yorkshire Philosophical Society for MCMXII*, York, 1913, pp. 3–4.
[89] *Annual Report of the Council of the Yorkshire Philosophical Society for 1922*, York, 1923, p. 40.
[90] Rubinstein, *Nature of the World*, p. 110.
[91] Smith, *Centenary of Science*.

> club, for the purpose of conversing on subjects of that nature. These meetings were continued, with some interruption, for several years; and many respectable persons being desirous of becoming Members, the numbers were increased so far, as to induce the founders of the Society to think of extending their original design. Presidents, and other officers were elected, a code of laws formed, and a regular Society constituted, and denominated, THE LITERARY AND PHILOSOPHICAL SOCIETY OF MANCHESTER.[92]

The principal founders of the society, alongside Percival, were two other Thomases: the Unitarian minister and educational reformer Thomas Barnes (1747–1810), and the surgeon Thomas Henry (1734–1816). In line with its origins, the spirit of the early society was that knowledge should be spread through conversation (though lectures of a more formal nature were also introduced from the start):

> Men, however great their learning, often become indolent, and unambitious to improve in knowledge, for want of associating with others of similar talents and acquirements: Having few opportunities of communicating their ideas, they are not very solicitous to collect or arrange those they have acquired, and are still less anxious about the further cultivation of their minds.—But science, like fire, is put in motion by collision.—Where a number of such men have frequent opportunities of meeting and conversing together, thought begets thought, and every hint is turned to advantage. A spirit of inquiry glows in every breast. Every new discovery relative to the natural, intellectual, or moral world, leads to a farther investigation; and each man is zealous to distinguish himself in the interesting pursuit.[93]

Thus, the MLPS began life as a rather different organization from the LPLS and the YPS: there was no museum to motivate the society's foundation.[94] The subjects taken up by the society would therefore not be skewed towards those connected with museum holdings. The MLPS also began as a much more exclusive society than the others would be: its membership was to be limited to 50 (although this rule was later relaxed), who would be selected by ballot; honorary members residing outside Manchester were permitted,

> provided no Gentleman be recommended, who has not distinguished himself by his literary or philosophical publications; or favoured the Society with some paper, which shall have received the approbation of the Committee of Papers.[95]

The first formal meeting of the MLPS took place in March 1781 in a back room of the Unitarian Cross Street Chapel, which the society used until 1799, when it moved

[92] *Memoirs of the Literary and Philosophical Society of Manchester*, vol. I, Warrington: for W. Eyres and T. Cadel 1785, p. vii.
[93] Ibid., pp. vi–vii.
[94] Provision for a library was, however, written into the MLPS's regulations: ibid., p. xv.
[95] Ibid., p. xi.

to new premises in George Street. It occupied these until their destruction in the Manchester Blitz of 1941, which also saw the loss of the society's library.[96]

From the start, an eclectic mix of permissible subjects was written into the MLPS's laws:

> That the subjects of conversation comprehend Natural Philosophy, Theoretical and Experimental Chemistry, Polite Literature, Civil Law, General Politics, Commerce, and the Arts. But that Religion, the Practical Branches of Physic, and British Politics, be deemed prohibited; and that the Chairman shall deliver his *Veto*, whenever they are introduced.[97]

The injunction against religion is of course one that we have already seen in the establishment of the LPLS, whilst that against British politics might be regarded as an effort to maintain harmony within the society, since a number of its members were involved in the Anti-Corn Law League.[98] The explicit mention of chemistry is noteworthy, since this is a subject that received a great deal of attention from the society. Indeed, John Dalton (1766–1844), the so-called 'father of (modern) chemistry', was a prominent member of the MLPS from the time of his arrival in Manchester in 1794, eventually serving as both its secretary (1800–1809) and its president (1816–1844), and conducting much of his research in a laboratory in the MLPS's George Street premises;[99] the society still awards a Dalton Medal for contributions to science.

Dalton was not the only nationally prominent professional scientific figure to be closely involved with the activities of the MLPS. We might also mention, for example, the physicist James Prescott Joule (1818–1889) later in the nineteenth century.[100] Indeed, the society had a number of high-profile contributors from its early years: the second volume of its *Memoirs* (covering 1783–85) features meteorological contributions from Benjamin Franklin (1706–1790), for instance. In contrast to the activities of the other Lit & Phils, the impression given by the MLPS in its early years is that of a much more academic organization—perhaps because of its rather exclusive beginnings. The MLPS received contributions not only from local luminaries, but also from scholars at the universities and elsewhere. Its *Memoirs* are reminiscent of the *Philosophical Transactions of the Royal Society*. All of this points towards a situation much more conducive to the inclusion of mathematics.

In the introduction to the present chapter, we mentioned Thomas Penyngton Kirkman, the prolific mathematician Arthur Cayley (1821–1895), and James Cockle (1819–1895) as being three mathematical contributors to the MLPS, but they were not the only ones. The papers delivered to the society in its early years were light

[96] Sheehan, 'Manchester Literary and Philosophical Society'.

[97] *Memoirs of the Literary and Philosophical Society of Manchester* I, pp. xii–xiii.

[98] Paul Pickering and Alex Tyrell, *The People's Bread: A History of the Anti-Corn Law League*, London: Leicester University Press 2000, p. 226.

[99] Indeed, it has been asserted that Dalton and the MLPS were 'virtually synonymous': Howard M. Wach, 'Culture and the Middle Classes: Popular Knowledge in Industrial Manchester', in: *Journal of British Studies* 27/4 (1988), pp. 375–404 at p. 378. Volume 13 of the third series of the society's *Memoirs* (1856) is given over entirely to an account of Dalton's life and work.

[100] Volume 6 of the fourth series of the society's *Memoirs and Proceedings* (1892) is given over entirely to an account of Joule's life and work.

on mathematics, with only a few isolated examples, the first being 'Some Properties of Geometrical Series explained in the Solution of a Problem, which hath been thought indeterminate' by the physician John Rotheram,[101] writing from Middlesex, in December 1787.[102] The 1790s, and the first decades of the nineteenth century, saw a scattering of papers on mechanics,[103] but pure mathematics did not reappear until the 1840s, with two communications from the mathematics teacher Robert Rawson (1814–1906),[104] who was elected to membership of the MLPS in January 1845; he went on to present a total of thirteen papers to the society between 1844 and 1884, mostly relating to mechanics, and to the solution of differential equations. The years following Rawson's election to membership of the MLPS coincidentally also saw the election of other figures who were to make mathematical contributions to the society: the Rev. Robert Harley (1828–1910) in 1849,[105] Kirkman (to honorary membership) in 1852, and Cayley (similarly to honorary membership) in 1859.[106] Rawson, Harley, and Kirkman, together with Cockle, who did not formally become an honorary member of the MLPS until 1870, were largely responsible for the explosion in mathematical papers that were presented to the society between the 1850s and the 1890s: up to the end of the nineteenth century, Harley contributed eleven papers (mostly on the solution of algebraic equations), Cockle fifteen (also on algebraic equations),[107] and Kirkman twenty-five (on combinatorics and permutation groups, plus some on a philosophical theme—see Table 8.3);[108] Cayley's tally was a more modest three.[109]

[101] Apparently, this was the John Rotheram (*c.*1750–1804) who was eventually to become Professor of Natural Philosophy at the University of St Andrews (Isobel Falconer, 'Rotheram, John (*c.*1750–1804), natural philosopher', in: *Oxford Dictionary of National Biography*, Oxford: Oxford University Press 2004, https://doi.org/10.1093/ref:odnb/24153), based on evidence of his address given elsewhere (*The Naval Chronicle*, London, 1805, p. 470).

[102] *Memoirs of the Literary and Philosophical Society of Manchester*, 3 (1790), pp. 330–6.

[103] See volumes 4 and 5 of the first series of the society's *Memoirs*, and volumes 2 and 5 of the second.

[104] R. Harley, 'Robert Rawson [obituary]', in: *Proceedings of the London Mathematical Society* 4 (1907), pp. xv–xvii; R. Harley, *Brief Biographical Sketch of Robert Rawson*, London: James Clarke 1910. See also: *Monthly Notices of the Royal Astronomical Society* 67 (1907), pp. 234–6.

[105] E. B. E[lliott], 'Obituary Notice: Robert Harley', in: *Proceedings of the London Mathematical Society, Second Series* 9 (1911), pp. xii–xv; P. A. M[acMahon], 'Robert Harley, 1828–1910', in: *Proceedings of the Royal Society* 91 (1915), pp. i–v; H. T. M. Bell and Maria Panteki, 'Harley, Robert (1828–1910), mathematician and Congregational minister', in: *Oxford Dictionary of National Biography*, Oxford: Oxford University Press 2008, https://doi.org/10.1093/ref:odnb/33715.

[106] Cayley was not the only internationally prominent mathematician to be elected to honorary membership of the MLPS; others included (up to the end of the nineteenth century): Paul Appell, 1894; Francesco Brioschi, 1892; Gaston Darboux, 1892; J. Williard Gibbs, 1892; J. W. L. Glaisher, 1894; Charles Hermite, 1892; Felix Klein, 1892; Leo Königsberger, 1894; Sophus Lie, 1894; Gösta Mittag-Leffler, 1895; Edward John Routh, 1889; George Salmon, 1889; George Gabriel Stokes, 1851; J. J. Sylvester, 1861; P. G. Tait, 1868.

[107] R. H[arley], 'Sir James Cockle [obituary]', in: *Memoirs and Proceedings of the Manchester Literary and Philosophical Society, Fourth Series* 9 (1895), pp. 215–228. On Cockle, see also: A. R. Forsyth, 'James Cockle [obituary]', in: *Proceedings of the London Mathematical Society* 26 (1895), pp. 551–554; R. Harley, 'James Cockle [obituary]', in: *Proceedings of the Royal Society of London* 59 (1896), pp. xxx–xxxix; J. M. Bennett, *Sir James Cockle: First Chief Justice of Queensland, 1863–1879*, Sydney: Federation Press 2003; T. A. B. Corley and A. J. Crilly, 'Cockle, Sir James (1819–1895), lawyer in Australia and mathematician', in: *Oxford Dictionary of National Biography*, Oxford: Oxford University Press 2004, https://doi.org/10.1093/ref:odnb/5788.

[108] K[irkman], 'Thomas Penyngton Kirkman', pp. 242–3.

[109] R. F. G[wyther], 'Arthur Cayley [obituary]', in: *Memoirs and Proceedings of the Manchester Literary and Philosophical Society, Fourth Series* 9 (1895), pp. 235–7. On Cayley more generally, see: Tony Crilly,

Table 8.3 The contributions of the Rev. Thomas Penynton Kirkman to the Manchester Literary and Philosophical Society; taken from K[irkman], 'Thomas Penyngton Kirkman', pp. 242–3. *Paper listed by title only.

1848	On Mnemonic Aids in the Study of Analysis
1851	On Linear Construction
1853	On the Representation and Enumeration of Polyedra
1854	On the k-partitions of N
1857	On the 7-partitions of X
1857	On the triedral partitions of the X-ace, and the triangular partitions of the X-gon
1858	General Solution of the Problem of the Polyedra
1858	On the absurdity of Ontology, and the vanity of Metaphysical Demonstrations; illustrated by reference to Professor Ferrier's Institutes of Metaphysic
1858	New Formula in Polyedra
1859	On the j-nodal k-partitions of the R-gon
1859	On the Partitions and Reticulations of the R-gon
1861	On the Theory of Groups and many-valued Functions
1861	Theorems on Groups
1862	On Non-Modular Groups
1863	On Maximum Groups
1863	The Complete Theory of Groups, being the Solution of the Mathematical Prize Question of the French Academy for 1860
1864	On the Relation of Force to Matter and Mind
1868	Note on 'An Essay on the Resolution of Algebraic Equations, by the late Judge Hargreave'
1868	On the Solution of Algebraic Equations
1868	Note on the Correction of an Algebraic Solution
1872	Once again the Beginning of Philosophy*
1891	On the number and formation of many-valued Functions of $x_1x_2x_3 - -x_n$, which of any degree can be constructed upon any Group of those elements, with exhibition of all the values of the Functions
1891	The 143 six-letter Functions given by the first transitive maximum group of six letters, with full exhibition of the values of the Functions
1893	On the k-partitions of R and of the R-gon
1893	On the k-partitions of the R-gon

As we have noted, these various figures were not the only mathematical contributors to the MLPS, but they were certainly the main ones, forming a mathematical clique attached to the society: it was not uncommon for there to be an interdependence between their papers.[110] Harley in particular played an important role as the only

Arthur Cayley: Mathematician Laureate of the Victorian Age, Baltimore: Johns Hopkins University Press 2006.

[110] For example, Harley's 'On the Rev. T. P. Kirkman's Method of Resolving Algebraic Equations' of October 1868 (*Proceedings of the Literary and Philosophical Society of Manchester* 8 (1868–9), pp. 4–20), inspired by a letter from Cockle to Harley. On some of these connections more generally, see: Tony Crilly, Steven H. Weintraub, and Paul R. Wolfson, 'Arthur Cayley, Robert Harley and the Quintic Equation: Newly Discovered Letters 1859–1863', in: *Historia Mathematica* 44/2 (2017), pp. 150–69.

member of this group to be a non-honorary member of the society: it fell to him to communicate the mathematical papers submitted by non-members (for example, those of Cockle until 1870).

One thing that becomes clear when we peruse the papers written for the MLPS by these various men is that we have strayed into a much more academic world: the presence of Arthur Cayley is perhaps the clearest indicator of this. These papers were not local and isolated mathematical curios, such as the few mathematical communications to the LPLS and the YPS, but technical contributions to a wider literature and mathematical community extending beyond Manchester. The papers by Harley and Cockle on the solution of algebraic equations, for example, form a part of the extensive, but little-studied, British reception of and response to Niels Henrik Abel's (1802–1829) 1824 proof of the unsolvability in radicals of the general quintic equation.[111] Moreover, none of these figures confined their work to the MLPS's *Memoirs*: they also published in the few other journals that were open to mathematics in nineteenth-century Britain: for example, the *Philosophical Magazine*, the *Quarterly Journal of Pure and Applied Mathematics*, and the *Philosophical Transactions of the Royal Society*.[112] Kirkman's outlook (along, of course, with Cayley's) was broader still: although he does not appear to have published in any foreign journals, he did famously submit a solution to a prize problem set by the Paris Academy in 1858.[113] Thus, the mathematical contributors to the MLPS were members of a broader national, or even international, community of mathematicians both within and outside academia. We have given here only a very brief indication of the mathematics hosted by the MLPS, as part of our discussion of the Lit & Phils, but the role played by the society and its journal in the mediation of mathematical research within this wider network would be an interesting topic of further investigation.[114]

What is rather curious and coincidental about Kirkman, Cockle, and Cayley is that they all died in the same year (1895), and have obituary notices within the same volume (namely, volume 9) of the MLPS's *Memoirs and Proceedings*, which (among other things) list their contributions to the society. A further and more detailed study of mathematics at the MLPS should look at how much mathematics remained in the programme of the society following the demise of Kirkman and Cockle in particular. A rough scan through the society's publications for the final decades of the nineteenth century indicates that a handful of new mathematical contributors were appearing

[111] See, for example: Henrik Kragh Sørensen, 'The Mathematics of Niels Henrik Abel: Continuation and New Approaches in Mathematics During the 1820s', PhD thesis, University of Aarhus 2002, §6.9; Crilly, Weintraub, and Wolfson, 'Arthur Cayley, Robert Harley and the Quintic Equation'.

[112] See, for example: Sloan Evans Despeaux, 'The Development of a Publication Community: Nineteenth-Century Mathematics in British Scientific Journals', PhD thesis, University of Virginia 2002; Sloan Evans Despeaux, 'A Voice for Mathematics: Victorian Mathematical Journals and Societies', in: *Mathematics in Victorian Britain*, ed. Raymond Flood, Adrian Rice, and Robin Wilson, Oxford: Oxford University Press 2011, pp. 155–74.

[113] As well as Kirkman's, the Academy received two further solutions from Émile Mathieu (1835–1890) and Camille Jordan (1838–1922), but decided not to award the prize to any of them. In his account of the affair (and of his solution) for the MLPS, Kirkman's aggrieved tone is clear: 'The Complete Theory of Groups, being the Solution of the Mathematical Prize Question of the French Academy for 1860', in: *Proceedings of the Literary and Philosophical Society of Manchester* 3 (1863–64/1864–65), pp. 133–52.

[114] The place of its journal within nineteenth-century British mathematical publishing has already seen some study: Despeaux, 'Development of a Publication Community', Chapter 2.

within this 'mathematics-friendly' environment established by Kirkman and the others: we mention, for example, Joseph John Murphy (1827–1894), who contributed papers on symbolic logic, and R. F. Gwyther (1852–1930), who wrote on mechanics and other topics. However, as we have already gone far beyond the scope of the present book, we must end our discussion of the MLPS here.

The British Association for the Advancement of Science

Before we attempt to make any general conclusions about the Lit & Phils and their engagement with mathematics, we first take a brief detour in order to draw a broad comparison between their activities and those of the British Association for the Advancement of Science (BAAS), founded in 1831 with thc goal of promoting the pursuit of science in Britain.[115] This short diversion will enable us to introduce a related question for future research. We must note from the outset that this comparison is not necessarily a fair one, for the BAAS was a society of a somewhat different type from the Lit & Phils, particularly by being national in its scope, and by having the goal of promoting government support of the sciences. In the 'Objects and rules' prepended to the published reports of its first two meetings, the BAAS's goals are given as follows:

> —To give a stronger impulse and a more systematic direction to scientific inquiry,
> —to promote the intercourse of those who cultivate Science in different parts of the British Empire, with one another, and with foreign philosophers,
> —to obtain a more general attention to the objects of Science, and a removal of any disadvantages of a public kind, which impede its progress.[116]

The BAAS grew not out of a local concern for personal betterment through education, but from a much broader worry about the state of the sciences in Britain. Perhaps in recognition of its fundamental position within a systematic development of the sciences, mathematics played a rather more prominent role in the activities of the BAAS than of the typical Lit & Phil.

The impetus for the foundation of the BAAS came from a number of directions, but the credit is conventionally assigned to a small number of scientific figures of the age who had all expressed their dissatisfaction at the failure of the government, the

[115] On the history of the BAAS, see: O. J. R. Howarth, *The British Association for the Advancement of Science: A Retrospect 1831–1921*, London: British Association for the Advancement of Science 1922; A. D. Orange, 'The British Association for the Advancement of Science: The Provincial Background', in: *Science Studies* 1 (1971), pp. 315–29; A. D. Orange, 'The Origins of the British Association for the Advancement of Science', in: *British Journal for the History of Science* 6/2 (1972), pp. 152–76; A. D. Orange, 'The Idols of the Theatre: The British Association and its Early Critics', in: *Annals of Science* 32/3 (1975), pp. 277–94; Jack Morrell and Arnold Thackray, *Gentlemen of Science: Early Years of the British Association for the Advancement of Science*, Oxford: Clarendon Press 1981; Jack Morrell and Arnold Thackray (eds), *Gentlemen of Science: Early Correspondence of the British Association for the Advancement of Science*, Camden Fourth Series, vol. 30, London: Royal Historical Society 1984.

[116] *Report of the First and Second Meetings of the British Association for the Advancement of Science; at York in 1831, and at Oxford in 1832: including its Proceedings, Recommendations, and Transactions*, London: John Murray 1833, p. ix.

universities, and the Royal Society to stimulate the progress of science in Britain. Charles Babbage famously wrote of 'the decline of science in England',[117] but it is the Edinburgh-based scientist David Brewster (1781–1868) whom A. D. Orange describes as having been 'the most constructive voice' in the debate.[118] Inspired by the recent success of the *Deutsche Naturforscher Versammlung* in the raising of public awareness of science in the German states, Brewster proposed the formation of a similar body in Britain. He aimed to launch the projected association by organizing a gathering of men of science in the summer of 1831. Owing in part to his prior contacts with the YPS (he had been elected an honorary member the year before), Brewster selected York as the location for this meeting, describing the city in a letter to John Philips, secretary of the YPS, as 'the most centrical city for the three kingdoms'.[119] The YPS, and its president, the Rev. William Vernon Harcourt,[120] therefore played a key role in the formation of the BAAS, which did indeed hold its inaugural meeting in the Yorkshire Museum in September 1831. In the interests of saving space, however, we will not go into a detailed discussion of the association's foundation here.[121]

From its inception, the BAAS was to be a peripatetic association that held annual meetings in different locations across the British Isles, with the proviso that London be specifically avoided.[122] By taking its gatherings in particular to the industrial cities of the north of England, some members (notably Babbage) hoped that the BAAS would serve as a meeting point of abstract and practical sciences. However, despite much rhetoric about the association bringing science to the country at large, the vast majority of its early prime movers were figures with well-established university positions. Indeed, it might be said that the BAAS was quickly hijacked by academics, who were perhaps more likely to have access to the kinds of national networks needed to run such an organization: after its inaugural session in York in 1831, its next four annual meetings were held in the university cities of Oxford (1832), Cambridge (1833), Edinburgh (1834), and Dublin (1835). As Morrell and Thackray comment, the association's 'provinciality [...] was rapidly eroded'.[123] Only when the BAAS's university-educated and socially prominent core had firmly established its control of the association did the meetings move to industrial and commercial centres elsewhere in the country, starting with Bristol in 1836.

The preponderance of academics among the organizers of the BAAS is probably the major reason for one clear difference between the BAAS and most Lit & Phils: the prominent position given to mathematics, a subject with which at least any Cambridge-educated figure could not help but be familiar. By 1836, after one or two

[117] Charles Babbage, *Reflections on the Decline of Science in England, and on Some of its Causes*, London: Fellowes 1830.

[118] Orange, *Philosophers and Provincials*, p. 32. See also: J. B. Morrell, 'Brewster and the Early British Association for the Advancement of Science', in: *'Martyr of Science': Sir David Brewster 1781–1868*, ed. A. D. Morrison-Low and J. R. R. Christie, Edinburgh: Royal Scottish Museum 1984, pp. 25–9.

[119] Brewster to Philips, 23 February 1831, quoted in Morrell and Thackray, *Gentlemen of Science: Early Correspondence*, p. 34.

[120] Previously referred to here as William Venables Vernon—see the comment in note 83.

[121] See instead the sources already cited on the history of the BAAS; in particular, on the role played by the YPS, see: Orange, *Philosophers and Provincials*, Chapter 4.

[122] This injunction was eventually broken a century later: the BAAS held its 1931 meeting in London.

[123] Morrell and Thackray, *Gentlemen of Science: Early Years*, p. 104.

small variations during its formative years, the activities of the BAAS were organized under the following seven sections:[124]

A. Mathematical and physical science;
B. Chemistry and mineralogy;
C. Geology and geography;
D. Zoology and botany;
E. Medical science;
F. Statistics;
G. Mechanical science.

The categories varied over the following years and decades, but mathematical science remained at the top of the list throughout, and was one of the most active sessions at BAAS meetings: Morrell and Thackray describe the Section A of the mid-1830s as 'bursting'.[125] The combination of the physical sciences with topics of a purer mathematical nature was justified by reference to the broader outlook that a general mathematical knowledge brought.[126] Academics who were heavily involved in the BAAS, such as the polymath William Whewell (1794–1866), saw to it that the mathematical subjects were placed at the top of a 'hierarchy of science', owing to their reliance on deductive rather than inductive reasoning; the inductive approach to science that had been promoted by Harcourt during the preparation for the inaugural meeting of the BAAS was relegated to a secondary position, and '[t]he Association's edicts on proper science came increasingly from Trinity College, Cambridge, not from the Yorkshire Museum'.[127] Indeed, Whewell, a Fellow of Trinity from 1817 and Master from 1841, seems to have been rather dismissive of the Lit & Phils, and adopted a somewhat imperious attitude in relation to their involvement in the activities of the BAAS.[128]

That mathematics was on the minds of the first organizers of the BAAS can be seen from the correspondence exchanged by Harcourt, Whewell, and Brewster in late 1831 concerning the search for an appropriate person to compile a report for the next meeting on recent progress in mathematics, particularly Continental mathematics, behind which British mathematics was deemed to lag.[129] Reports were also sought on other subjects (mainly those linked to Section A), and quickly became a fixture of the early BAAS meetings. The Rev. George Peacock (1791–1858), Fellow of Trinity College, Cambridge, was eventually commissioned to write 'a Report on recent progress of Mathematical Analysis, in reference particularly to the differential and integral calculus'.[130] Although the report was not completed in time for the 1832 BAAS meeting, it was presented the following year, taking up 168 pages of that

[124] Ibid., Table 11.

[125] Ibid., p. 260.

[126] See, for example, the remarks of James McCullagh (1809–1847), president of Section A at the Cork meeting of 1843: *Report of the Thirteenth Meeting of the British Association for the Advancement of Science; held at Cork in August 1843*, London: John Murray 1844, 'Notices and abstracts', p. 1.

[127] Morrell and Thackray, *Gentlemen of Science: Early Years*, p. 267.

[128] Ibid., p. 303.

[129] Morrell and Thackray, *Gentlemen of Science: Early Correspondence*, pp. 87, 103, 111, 115–6, 119.

[130] *First Report of the Proceedings, Recommendations, and Transactions of the British Association for the Advancement of Science*, York: Thomas Wilson and Sons 1832, p. iii.

year's printed transactions (the next longest report, on the physiology of the nervous system, accounted for 34 pages).[131] Peacock's report emphasized the need for further mathematization of the physical sciences, placing mathematics above experimentation. Other such Section A reports, many written by Cambridge-based academics, similarly laid a great emphasis on mathematics.[132] It is clear that, through the activities of many of its university-based members,

> [t]he mathematical and physical sciences enjoyed the commanding heights of the Association's intellectual economy. While the meetings of Section A were more arcane and less crowded than those devoted to geology or statistics, the interests at stake were closer to the heart of the Association's affairs.[133]

Alongside the specially commissioned survey reports, BAAS meetings also featured communications on the researches of members, and here too we find mathematics: famously, for example, William Rowan Hamilton's 1844 lecture on quaternions.[134] We don't attempt here to give a full survey of mathematics at the BAAS, nor a quantitative comparison with the representation of other subjects, but merely note the strong presence of mathematics. This strikes quite a contrast with the situation that we have seen at the Lit & Phils.

One of the reasons for the emphasis on mathematics at the BAAS may have been the fact that Britain's existing scientific association, the Royal Society, had gained a reputation at this time for being no friend to mathematics, for a range of convoluted reasons.[135] This is why we find mathematicians engaged so heavily in the activities of the (Royal) Astronomical Society during these decades.[136] In 1865, the London Mathematical Society was formed, and soon became a forum not only for mathematicians in the capital, but also those from the nearby university cities.[137] We might therefore expect the place of mathematics at the BAAS to have diminished—but this is not the case. Skipping to the BAAS's 1880 meeting in Swansea, for example, we still find a substantial number of technical mathematical communications appearing in Section A.[138] Moving further along, into the twentieth century, this remains the case, to varying degrees. Further discussion of these would of course take us too far away

[131] George Peacock, 'Report on the recent progress and present state of certain branches of analysis', in: *Report of the Third Meeting of the British Association for the Advancement of Science; held at Cambridge in 1833*, London: John Murray 1834, pp. 185–352.

[132] Morrell and Thackray, *Gentlemen of Science: Early Years*, p. 484.

[133] Ibid., p. 466.

[134] *Report of the Fourteenth Meeting of the British Association for the Advancement of Science; held at York in September 1844*, London: John Murray 1845, 'Notices and abstracts', p. 2.

[135] Stemming in part from the events described in: Benjamin Sutherland Wardhaugh, 'Charles Hutton and the "Dissensions" of 1783–84: Scientific Networking and its Failures', in: *Notes and Records of the Royal Society* 71/1 (2017), pp. 41–59.

[136] See, for example: Rebekah Higgitt, 'Why I don't FRS Mail Tail: Augustus De Morgan and the Royal Society', in: *Notes and Records of the Royal Society* 60/3 (2006), pp. 253–9.

[137] Adrian C. Rice, Robin J. Wilson, and J. Helen Gardner, 'From Student Club to National Society: The Founding of the London Mathematical Society in 1865', in: *Historia Mathematica* 22 (1995), pp. 402–21; Adrian C. Rice and Robin J. Wilson, 'From National to International Society: The London Mathematical Society, 1867–1900', in: *Historia Mathematica* 25 (1995), pp. 185–217; John Heard, *From Servant to Queen: A Journey through Victorian Mathematics*, Cambridge: Cambridge University Press 2019.

[138] *Report of the Fiftieth Meeting of the British Association for the Advancement of Science; held at Swansea in August and September 1880*, London: John Murray 1881, pp. 473–88.

from the timeframe of the present book, so we conclude this section by observing that a detailed and wide-ranging study of the place of mathematics within the activities of the BAAS would surely provide further insight into the cultures of mathematics in Britain during the nineteenth and twentieth centuries.

Concluding remarks

Throughout this chapter, the elephant in the room has been the general character of mathematics, and its suitability for live presentation to a general audience. This posed a problem even for the London Mathematical Society in its early years,[139] and for the Royal Society mathematics was, as Wardhaugh puts it, 'something of a cross [...] to bear'.[140] Meetings of the society consisted of secretaries reading out papers sent in by Fellows, as 'a sort of literary re-creation of experimental or observational work done far away'.[141] However,

> mathematics never lent itself to that kind of performance. If a paper was dense with algebra, if it needed its geometrical diagrams in order to be understood, or if—worse yet—it was actually about mathematics rather than its practical uses, there were real difficulties. Even the most skilful reader could scarcely make such things readily comprehensible, and not all Fellows were remotely interested in any case. Some Fellows persisted in sending mathematical papers to the Royal Society, but it was increasingly felt that by doing so they were causing a problem, breaking the rules of a world that was still to some degree about 'polite' discussion of accessible topics.[142]

Indeed, the idea of mathematics as 'antisocial behaviour' has a bearing on the attitude towards mathematics at the early-nineteenth-century Royal Society that we noted in passing in the preceding section. Turning elsewhere, we can gain an impression of what happens when a mathematical talk is delivered in a forum in which it is neither expected nor welcome by looking to the Highfield Astronomical and Meteorological Society, of Halifax, Yorkshire, a society not listed in Table 8.1 because its title does not suggest that it would have anything to do with mathematics. Indeed:

> At each meeting there would usually be an address on an astronomical (or frequently geological) topic by a member, but occasionally an invited outside speaker would do the honours. Thus, in 1860 November, a certain Mr Bowman gave a talk on mathematics, but overestimated the staying power of the members and left the minute-taker utterly bewildered.[143]

[139] Heard, *From Servant to Queen*, p. 77.

[140] Benjamin Wardhaugh, *Gunpowder and Geometry: the Life of Charles Hutton, Pit Boy, Mathematician and Scientific Rebel*, London: William Collins 2019, p. 98.

[141] Ibid., p. 98.

[142] Ibid., pp. 98–9.

[143] David Sellers, 'An Early Astronomical Society: Highfield Astronomical and Meteorological Society, Halifax', in: *The Antiquarian Astronomer: Journal of the Society for the History of Astronomy* 13 (June 2019), pp. 42–9 at p. 45.

It is easy to imagine that a technical mathematical lecture delivered to other Lit & Phils would be similarly received, the MLPS being a notable exception.

We have of course considered only three examples of Lit & Phils, so it would be instructive briefly to cast the net a little wider to see what was happening at other such societies. We have already mentioned the Newcastle Literary and Philosophical Society, for instance. If we peruse the historical volume written shortly after the society's centenary, we do find some lectures relating to mathematics, but of a meta-mathematical character similar to those that we observed at the YPS:[144]

- 'The importance of the study of mathematics, more especially with regard to their application in the Discoveries of Modern Astronomy' (two lectures, 1838);
- 'The Philosophy of Mathematics' (six lectures, 1863–1864).

All of these lectures were delivered by the Rev. James Snape (1815–1880) of Newcastle's Royal Grammar School. It seems likely that these lectures are reflective of the extent of the engagement with mathematics of the Newcastle Literary and Philosophical Society, but a detailed study of the society's annual reports and archives would be required to confirm this.

Looking into accounts of other societies, we find the same eclectic mixes of subjects as at the societies we have considered in greater detail. For example, physics, chemistry, geology, zoology, geography, engineering, physiology, botany, and education are identified as the topics that were of particular interest to the Bradford Scientific Association; its journal was organized into sections on anthropology, botany, geology, and zoology.[145] A similar spread is indicated for the Leicester Literary and Philosophical Society, although here the society was evidently open to philosophy, for we find a lecture in January 1871 by none other than Thomas Penyngton Kirkman, entitled 'The First Steps of the Cartesian Philosophy',[146] presumably related to his 1876 book *Philosophy without Assumptions*,[147] as well as to an 1872 communication to the MLPS (see Table 8.3). But there is no evidence of mathematics having been treated by the Leicester society. This stands in contrast to the Liverpool Literary and Philosophical Society,[148] whose *Proceedings* contain eight papers by Kirkman between 1875 and 1889, all on subjects broadly similar to those of the communications that he was making to the MLPS at this time. A more detailed study is needed to determine whether the Liverpool society, like its Manchester counterpart, hosted other mathematical contributors. Even if this is the case, however, it would

[144] Watson, *Literary and Philosophical Society of Newcastle-upon-Tyne*, pp. 238–9, 272.

[145] H. J. M. Maltby and W. P. Winter, *Fifty Years of Local Science, 1875–1925; A Record of Fifty Years of Work Done by Members of the Bradford Natural History and Microscopical Society and the Bradford Scientific Association*, Bradford, 1925.

[146] Frederick Barnes Lott, *The Centenary Book of the Leicester Literary and Philosophical Society*, Leicester: [s.n.] 1935, p. 69.

[147] Thomas Penynton Kirkman, *Philosophy without Assumptions*, London: Longmans, Green, and Co. 1876.

[148] Gordon W. Roderick and Michael D. Stephens, 'Approaches to Technical Education in 19th Century England, Part III: The Liverpool Literary and Philosophical Society', in: *The Vocational Aspect of Education* 23/54 (Spring 1971), pp. 49–54; Arline Wilson, 'The Cultural Identity of Liverpool, 1790–1850: The Early Learned Societies', in: *Transactions of the Historic Society of Lancashire and Cheshire* 147 (1997), pp. 55–80.

probably only bring the number of Lit & Phils engaging with mathematics to a total of two.[149]

So what determined the varied disciplinary profiles of these societies? We have seen in the example of the YPS that the intentions of the founders could, quite understandably, shape the direction that a Lit & Phil took. However, a society, like the LPLS, that made ambitious and sweeping statements about its intentions was unlikely to be able to follow through, simply because it could not expect to sustain expertise and interest in all branches of learning with only a small pool of members to draw upon. The disciplines covered by a Lit & Phil would therefore be skewed naturally towards the interests of its prominent members and, where appropriate, to the subjects that were relevant to its museum. Thus, we see in Leeds not only a major bias towards natural history, but also a continuing interest in Egyptology.[150] At the YPS, the initial focus on geology and natural history remained throughout. We might therefore argue that the YPS was from the start much more realistic in its disciplinary positioning.

To bring the discussion back to our main focus, we must now try to sum up our preliminary findings concerning the presence of mathematics at the Lit & Phils. It has become clear that the type of high-level mathematics that was being presented to the MLPS would have been inappropriate at most other societies. The first factor to which to point concerns the educational backgrounds of the participants in the societies: by the very nature of the Lit & Phils, the vast majority of their members had not attended university, and would therefore not have had the background knowledge to comprehend the type of mathematics communicated to the MLPS by Kirkman and others.[151] Of course, it would be far too simplistic to suggest that this is the only relevant factor: Robert Rawson, for example, whom we mentioned as a mathematical contributor to the MLPS, did not attend university, and was in fact largely self-taught in mathematics.[152] On the whole, therefore, any interest in mathematics would not have been in the details of the mathematics itself, but in its practical applications, for example to engineering, which could in fact be discussed without invoking the mathematics.

[149] There have been suggestions that George Boole delivered mathematical lectures to the Cork-based Cuvierian Society, but the evidence for this is apparently lacking: Desmond MacHale, *George Boole: His Life and Work*, Dublin: Boole Press 1985, p. 125.

[150] Indeed, the LPLS's connection to Egyptology outlived its apparent instigator, William Osburn: Clark (*History of 100 Years of Life*, p. 115) records a 1901 donation to the society by the Egypt Exploration Fund of artefacts from a recent excavation. Moreover, during the nineteenth century, this interest had spilled beyond the society into the city more generally, with the design (in which Osburn was involved) and construction of the Egyptian-style Temple Mill in Holbeck: Chris Elliott, *Egypt in England*, Swindon: English Heritage 2012, pp. 265–77.

[151] We note that from the 1860s onwards there were a number of pronounced local initiatives in England, mainly voluntary, aimed at founding university colleges, most of which began as science colleges; see, for example: Margaret Gowing, 'Science, Technology and Education in England in 1870', in: *Oxford Review of Education* 4 (1978), pp. 3–17.

[152] Indeed, Rawson appears to have belonged to the older tradition of the 'philomaths', which we have not mentioned here. He was certainly a contributor to the mathematical question-and-answer journals, such as *The Ladies' Diary*, about which much has been written. See, for example: Peter and Ruth Wallis, *Mathematical Tradition in the North of England*, Durham: NEBMA, 1991; Sloan Evans Despeaux, 'Mathematical Questions: A Convergence of Mathematical Practices in British Journals of the Eighteenth and Nineteenth Centuries', in: *Revue d'histoire des mathématiques* 20/1 (2014), pp. 5–71.

The question of whether a Lit & Phil had a museum is another relevant consideration: as we have observed, the subjects that a society would generally touch upon were firmly linked to the holdings of its museum. Although the idea of a mathematical museum is not an unusual one in the modern world,[153] it would not have had a place in the more traditional establishments of the Lit & Phils and their collections of tangible objects. In this respect, we see why natural history was such an easy subject for the Lit & Phils to cover, and why this was a popular amateur pursuit in nineteenth-century Britain more generally.[154] The emphasis that was often placed on *local* natural history is also an important point to note. In engaging in local studies—not just natural history, but also archaeology, geology, antiquarianism, and others—the Lit & Phils filled a niche not already occupied by university academics. But in this context there is no such thing as 'local mathematics'.[155] Nor (for those societies that modelled themselves on the Royal Institution) was mathematics a subject in which people might become engaged via experimental demonstrations.

Thus, it appears that, despite the pretensions that were sometimes made towards the pursuit of *all* knowledge, there was no automatic place for mathematics at the Lit & Phils. The only means by which it might find its way into their programmes in any substantial manner, as for instance with Osburn and Egyptology in Leeds, was via the personal interest of an individual: we may point, for example, towards William Taylor in York, or James Snape in Newcastle. As we have seen, however, Taylor remained an isolated example of a mathematician at the YPS, and apparently did not inspire the pursuit of mathematics in any of his fellow members. His own interests were certainly quite eclectic and far from being confined to mathematics anyway.

At the MLPS, the quite extensive coverage of mathematics was not introduced merely by an interested individual, but rather by a cluster of mathematical figures, together with some outside academic input (Cayley). Moreover, thanks to the much wider scholarly outlook of the MLPS (it appears to have spent a little less time on local studies, for instance), the society begins to emerge from the sources as one which, paradoxically, it is inappropriate to consider solely within the framework of the Lit & Phils, since it evidently had a rather different character. By a similar token, its journal must also be treated differently from those of the other Lit & Phils, certainly in the mathematical context.[156] Thus, the suggestion with which we began this chapter—that Kirkman's mathematical contributions to the MLPS were somehow misplaced[157]—now seems somewhat misguided: the MLPS's *Memoirs* did indeed

[153] See, for example: https://www.mathematikum.de/.

[154] Samuel J. M. M. Alberti, 'Amateurs and Professionals in One County: Biology and Natural History in Late Victorian Yorkshire', in: *Journal of the History of Biology* 34 (2001), pp. 115–47; Alberti, 'Natural History'.

[155] There could, however, be 'local *history* of mathematics', as shown, for example, by the (reasonably technical) article: T. T. Wilkinson, 'The Lancashire Geometers and Their Writings', in: *Memoirs of the Literary and Philosophical Society of Manchester* 11 (1854), pp. 123–57.

[156] Indeed, this has already been acknowledged by the inclusion of the MLPS's *Memoirs* as the only Lit & Phil journal to be considered in Despeaux, 'A Voice for Mathematics'.

[157] However, a paper read by Kirkman to the Historical Society of Lancashire and Cheshire in May 1857 does still seem oddly placed: 'On the perfect partitions of $r^2 - r + 1$', in: *Transactions of the Historical Society of Lancashire and Cheshire* 9 (1856–7), pp. 127–42.

cover a great variety of subjects, but mathematics no longer appears as an oddity among them, rather as just one further aspect of the society's activities. A much more detailed study of the place of mathematics within the MLPS, and the relationship of this to the mathematics being pursued elsewhere in Britain at that time, would certainly be worthwhile.

9
The Evolution of Actuarial Science to 1848

David R. Bellhouse

Introduction

The early development of actuarial science is closely tied to developments occurring in Britain and Ireland from the late seventeenth century onwards. There was interest in insurance schemes outside of Britain during the time period I consider (1693 to 1848) and within Britain prior to the late seventeenth century. It is, however, within Britain where most of the major mathematical developments occurred and where the market for life insurance and life annuities exploded. Seemingly paradoxically, these developments initially occurred outside the budding insurance industry. Prior to about 1770, the developments in Britain were motivated by life contingent contracts related to property. This chapter describes the developments in actuarial science up to 1848, the year when the Institute of Actuaries was founded.

On the European continent, there were many developments in the eighteenth and nineteenth centuries related to actuarial science. Some were concerned with models to describe the probability of surviving beyond certain ages. Others were devoted to the numerical calculation of mortality tables, or survivor probabilities, using local records. By comparison to what had happened in Britain, these developments were scattered and had little impact compared to what was happening in Britain.

I have described in more detail many of the developments up to about 1800 in my book *Leases for Lives: Life Contingent Contracts and the Emergence of Actuarial Science in Eighteenth Century England*.[1] Much of the source material that I have used in this article is reprinted, with insightful commentary, in a 10-volume series *History of Actuarial Science* by Steven Haberman and Trevor Sibbett.[2] Both are fully qualified actuaries, the former in academia and the latter in industry.

The seed that germinated

When one door closes, another one opens. Such was the case with Edmond Halley (1656–1742). Fourteen hundred kilometres to the east in Breslau, now Wrocław, a Protestant pastor named Caspar Neumann (1648–1715) collected data from parish records in the city to look at the question of climacterics. This was the idea that there

[1] David R. Bellhouse, *Leases for Lives: Life Contingent Contracts and the Emergence of Actuarial Science in Eighteenth-Century England*, Cambridge: Cambridge University Press 2017.

[2] Shelby Haberman and Trevor A. Sibbett, *History of Actuarial Science*, 10 vols, London: W. Pickering & Chatto 1995.

was a periodic pattern to human mortality.[3] Based on ideas from antiquity that persisted into the seventeenth century, there should be peaks in the number of deaths at ages 7, 14, 21, and so on, for example. Neumann had collected data that might answer this question. After extracting data from five years of the Breslau parish registers, he tabulated the number of deaths at each age in each year over the five years. When the data reached the Royal Society in London, in 1693, they were scrutinized by the fellows during one of their meetings. Thereupon it was decided that climacterics were a 'groundless conceit'.[4] Halley took the data home and played with them, and in the process discovered another application altogether, this one related to political arithmetic. From the 1660s to the early 1690s John Graunt (1620–1674) and William Petty (1623–1687) wrote on this topic, and Halley thought that he could add to what they had written. Using the data and a simple smoothing thereof, Halley constructed a population table for the City of Breslau showing the number in the city alive at each age. The total number in the table would show the population size of Breslau, something that Graunt and Petty could only roughly estimate for London. Realizing that this table could also be used as a life, or mortality, table (the terms are often used interchangeably), he showed, among a few other things, how the table could be applied to calculate the value of a life annuity at any given age for one, two or three lives, and to calculate the value of one-year term insurance, also at any given age of a single life.[5]

Though not stated explicitly by Halley, since he was looking to value annuities on three lives, one of his motivations for working on the mortality table was most likely the valuation of leases for lives. These were leases which fell out of favour by the end of the eighteenth century, on which the length of a lease was determined by the lifetimes of three persons named on the lease, typically to the death of the last survivor. When one of the lessees died, that person could be replaced by making a payment to the landlord thus restoring the lease to three lives. The actuarially proper way to value the addition to the lease would be the value of last survivor annuity on three lives less the value already paid in, the value of a last survivor annuity on two lives. There was a traditional way to value the lease. Equating one life to seven years, a lease on three lives was valued as the equivalent of a lease for a fixed term of 21 years. When one of the lessees died, adding another life would be equivalent to adding seven more years to the end of the lease. For this valuation, like mortgage tables today, there were tables by various mathematical writers that could be used to come up quickly with the desired value.

The traditional method of pricing leases for lives carried over to the method of financing the wars of the joint sovereigns William and Mary against the French. Beginning in 1693, money was raised by selling life annuities, called Exchequer annuities, with the price set at seven times the annual payment from the annuity. Using his life annuity valuations based on the Breslau table, Halley commented how advantageous it would be to buy Exchequer annuities especially for teenagers and those in

[3] David R. Bellhouse, 'A New Look at Halley's Life Table', in: *Journal of the Royal Statistical Society. Series A (Statistics in Society)* 174 (2011), pp. 823–32.

[4] Royal Society of London: *Journal Book*, 18 January 1693.

[5] Edmond Halley, 'An Estimate of the Degrees of the Mortality of Mankind, Drawn from Curious Tables of the Births and Funerals at the City of Breslaw; With an Attempt to Ascertain the Price of Annuities upon Lives', in: *Philosophical Transactions* 17 (1693), pp. 596–610.

their twenties. Halley's suggestion for age-based pricing of government life annuities was ignored throughout the eighteenth century. The issue there was not proper actuarial pricing, but attracting enough annuitants to finance the wars. It was stimulating demand that drove this market.

Again, though not stated explicitly, Halley's motivation for suggesting that his table could be used to set the price for a one-year term insurance policy probably came from some insurance practices of the day. There were no formal life insurance companies; the first of that type was established in 1696. Prior to that year, and perhaps well before, a person seeking life insurance would find another person or group of people willing to underwrite the insurance. There were no premium schedules. The only advice offered to the underwriters was that it was best to become acquainted with the person seeking insurance in order to assess the risk.[6] This approach to life insurance persisted into the eighteenth century even after other formal life insurance companies were established. Beginning in 1721, the London Assurance Corporation offered term insurance, usually for one year.[7] A committee of the company interviewed those applying for insurance to assess the risk. The premiums usually ranged from £5 to £5 10s per £100 of insurance and went up in higher risk situations when, for example, the insured was a soldier or was over 60 years of age. At £5 or more per hundred, the premium levels were conservatively high and so there was little risk of being unable to meet any claims.

Life Annuity Contracts in the Eighteenth Century

The problem with Halley's methodology for life annuities is computational. The price of a life annuity is the average value of all future payments. Supposing that the payments are made at the end of each year for life, the price is expressed in the simple formula

$$\sum_{t=1}^{\infty} P_t \times p_t,$$

where P_t is the present value of the payment to be made t years in the future and p_t is the probability of making that payment, i.e. the probability of surviving to t years in the future. The use of the symbol ∞ in the formula is a fiction made for convenience. There is an upper limit to the human life span so that at some point and beyond, the value of p_t is 0. The calculation of P_t depends on the rate of interest to be used and p_t is calculated from a mortality table. As might be easily guessed from the formula, evaluation of a life annuity in Halley's day required many time-consuming arithmetical calculations by hand. Make a simple change in the interest rate (the legal rate of interest changed from 6% per annum when Halley was writing to 5% in 1714) and many hours lay ahead in re-evaluating the same annuity. It was an

[6] G. Malynes, *Consuetudo; vel, Lex Mercatoria; or, the Ancient Law-merchant, in Three Parts*, London: Basset 1686, p. 115.

[7] B. Drew, *The London Assurance: A Chronicle*, London: The London Assurance 1928.

issue that would always be present until the invention of mechanical calculators, with a truly satisfactory solution coming only with digital computers in the mid-twentieth century.

Motivated by life-contingent contracts related to property, Abraham De Moivre (1667–1754) came up with a practical solution to Halley's computational problem in 1725.[8] For an annuity on a single life, De Moivre assumed that the number of survivors decreased linearly over the age span of a population. This reduced the valuation of a life annuity to one based on fixed-term annuities, whose values could be obtained from already existing tables. This simple assumption reduced the calculation time from several hours to several minutes. De Moivre justified his assumption by looking at Halley's population, or life, table. For a wide interval of ages beginning around age 30, the number of survivors did decrease in a linear fashion, at least approximately. For joint life annuities, De Moivre made two incompatible assumptions. Assuming that the number of survivors decreased exponentially with age and that the lifetimes of the people in question were independent, De Moivre expressed the value of a joint life annuity as a function of single life annuities. Then he used the linear survivor function to value a single life annuity. This result could be applied directly to the valuation of leases for lives.

With these simple assumptions, De Moivre solved a wide variety of annuity problems. These include for two or three lives:

- joint survivor annuities paying until the first death of one of the lives;
- last survivor annuities paying until the last death among the lives;
- reversionary annuities paying to one or more lives (either joint or last survivor) beginning on the death of the third life;
- succession annuities paying to one person and then to another on the first person's death.

These annuities can all be put in a context involving property. De Moivre's solutions to these problems were popular. By the time of the author's death in 1754, De Moivre's book *Annuities Upon Lives* had gone through four editions in England and one in Ireland. He also included material on annuities in the last two editions of his *Doctrine of Chances*.[9]

Let me demonstrate one of the annuity concepts and how it relates to property by using Jane Austen's *Pride and Prejudice* as a model. The novel opens with:

> It is a truth universally acknowledged, that a single man in possession of a good fortune, must be in want of a wife.

Mr Bennet, the father of the heroine Elizabeth, held an estate worth £2,000 a year. He held the estate for his lifetime so that the estate income to him was a life annuity. On

[8] Abraham De Moivre, *Annuities upon Lives, or, The Valuation of Annuities upon any Number of Lives, as also, of Reversion*, London: W. Pearson 1725.

[9] Abraham De Moivre, *The Doctrine of Chances: Or, a Method of Calculating the Probability of the Events in Play*, 2nd edn, London: Woodfall 1738; 3rd edn, London: Millar 1756.

his death the estate would pass to his closest male heir, in this case the pompous Mr Collins who was in want of a wife. After a failed proposal to Elizabeth, Collins married Elizabeth's best friend, Charlotte Lucas. In technical terms, Collins held the reversion on the estate. In case he wanted to borrow against his expectations, or even to sell his interest in the estate, the value to Collins would be the value of a life annuity of £2,000 a year payable after the death of Mr Bennet, provided that Mr Collins survived Mr Bennet. Given the ages of Bennet and Collins, De Moivre could easily have calculated Collins's expectations.

De Moivre was not the first to consider the assumption that the number of survivors in a population decreased linearly with age. Nicolaus Bernoulli (1687–1759) beat him to that distinction. Bernoulli's assumption shows up in his doctoral thesis of 1709.[10] Here Bernoulli was trying to complete some of his uncle Jakob's (1655–1705) unfinished work on the application of probability theory to economics and politics. Jakob Bernoulli's work on probability appeared posthumously in 1713. Nicolaus Bernoulli's work on annuities had little impact. The most likely reasons are that it was written in Latin and not widely available, and that the system of land ownership on the Continent did not provide the fruitful motivation for life annuity valuations that it did in Britain.

There was an even earlier attempt at modelling survivor functions that also had little impact. Near the end of his life, Johan de Witt (1625–1672) carried out valuations for annuities offered by the Dutch state to fund their military operations against Louis XIV of France.[11] For each of four age intervals, Witt assumed differing rates of mortality between the intervals but constant rates of mortality within an interval. Witt was killed in a riot about a year after he made his recommendations on annuity valuations. Like state annuities in Britain twenty years later, the actuarial valuation was ignored in favour of something simple to increase demand for the state annuities. Unfortunately, his work was relegated to the state archives and left relatively unknown for more than a century and a half.

Another British mathematician, Thomas Simpson (1710–1761) had a different solution to the problem of onerous life annuity calculations. He spent considerable time making these onerous calculations and producing tables for the values of single life annuities at 3, 4, and 5% annual interest. He also produced annuity tables for two and three lives, both joint and last survivor, where the lives have the same age at issue. Then he produced a set of approximations to handle the cases when the ages of the annuitants are different. The work appeared in Simpson's *Doctrine of Annuities and Reversions*, published in 1742.[12]

The mortality table at the base of Simpson's annuity calculations was one that Simpson adapted from a mortality table based on mortality experience in the City of

[10] Nikolaus Bernoulli, *Dissertatio Inauguralis Mathematico-juridica de Usu Artis Conjectandi in Jure*, Basil: Johannis Conradi 1709 An English translation by Thomas Drucker is in Haberman and Sibbett, *History of Actuarial Science*, vol. I, pp. 187–96.

[11] An English translation of de Witt's *Waerdye van Lijf-renten naer proportie van Los-rentenis* is in F. Hendriks, 'Contributions to the History of Insurance, and of the Theory of Life Contingencies, with a Restoration of the Grand Pensionary De Wit's Treatise on Life Annuities', in: *The Assurance Magazine* 2 (1852), pp. 222–58.

[12] Thomas Simpson, *The Doctrine of Annuities and Reversions: Deduced from General and Evident Principles*, London: J. Nourse 1742.

London. As a larger city with more crowding and disease, London had greater mortality than Breslau. Consequently, a life annuity calculated from the London table would have a smaller value than one based on the Breslau table. With greater mortality, fewer payments on average would be made to London annuitants than Breslau annuitants, or an equivalent community. Conversely, a London resident with an insurance policy would, on average, die sooner, and the beneficiary paid sooner, than a policyholder in Breslau so that life insurance in London should be costlier than in Breslau. It mattered which mortality table was used and to what purpose.

The valuation of leases for lives remained bound by tradition over the course of the eighteenth century. What became bread and butter problems for the mathematical consultants were non-standard life contingent contracts where there was no tradition to rely upon and no expertise among the contracting parties other than vague guesses about the values. What was requested of the consultant depended on the mathematical sophistication of the client. A client with very little mathematical ability might request the valuation of a simple contract. Where the mathematicians excelled, and were valued for their work, was in the valuation of more complex life contingent contracts.

De Moivre began working as a consultant for the valuation of life annuities by 1739, and probably much before.[13] His 'office' was Slaughter's Coffeehouse in St Martin's Lane, and it was there he met his clients. Slaughter's was known as a meeting place for artists and also for chess aficionados such as De Moivre, as well as a place to read the daily newspaper. Later in life, De Moivre's former pupil, James Dodson (*c.*1705–1757) assisted his teacher in making annuity valuations at Slaughter's. On De Moivre's death, Dodson continued with these consulting activities. Dodson went on to write his own annuity book, *The Mathematical Repository*.[14]

Here are some examples of life annuity questions posed to mathematicians from which they obtained some additional income. I have given the consulting mathematician, the year of the consultation, and a brief description of the problem.

- Thomas Simpson—1751:[15] Question put to him by his friend Thomas Blake. 'How many Year's purchase must be deducted from ye price of an Estate in Fee Simple, in consideration of ye joint Lives of a Woman 57 Years old, and of a Woman 37 Years old; both of them but moderately healthy?'
- De Moivre's methodology—1752:[16] A dragoon in the British army owned the reversion on an estate held by a man and his son (and heir). Wanting to purchase a commission in the army, the dragoon sold his reversion to someone else in order to raise the money. The man and his son died within a month of the sale and so the dragoon sued, claiming the reversion was undervalued. He, or his lawyer, used De Moivre's methodology to obtain the valuation. The High Court of Chancery ruled in favour of the defendant since there was no fraud

[13] Bellhouse, *Leases for Lives*, pp. 169–70.

[14] James Dodson, *The Mathematical Repository*, vol. 3, London: John Nourse 1755.

[15] Columbia University Library. David Eugene Smith Collection of Historical Papers Series I: Catalogued Correspondents Box 21 Thomas Simpson.

[16] F. Vesey, *Cases Argued and Determined in the High Court of Chancery: In the Time of Lord Chancellor Hardwicke, from the Year 1746–47, to 1755*, London: W. Strahan and M. Woodfall 1773.

involved in the sale, but awarded court costs to the dragoon since the defendant had obtained such a good deal.

- Richard Price—1777:[17] John Arthur Worsop inherited an estate on which there were conditions. Should he die without a male heir, the estate would go to a cousin. Should both die without male heirs, the estate would go to another cousin. Worsop wanted to get married and needed a marriage settlement. This would involve a piece of his property set aside from the estate that would support his bride should he predecease her. On her death, the property would revert to the whole estate. Because of the conditions on his inheritance, he could not set aside any property for this purpose. What he had to do in order to get married was to create a fund, priced by Price, which would pay his cousins the expected lost income should either come to inherit with the marriage settlement in effect.
- John Rowe—1782:[18] Exeter Cathedral was renting a house to a man aged 60. The cathedral personnel wanted to know from Rowe what should the valuation change to when two persons, one of whom was 20 years of age and the other 15, were added to the lease. The age difference between the original lessee and those added was considered too great to use the traditional method to value a lease for three lives.

A person knows that his or her ideas really have been accepted when lawyers take notice; and then the courts also take notice and rule on cases using that person's methodology. Between 1751 and 1774, De Moivre's and Simpson's methologies relating to life annuities appear in at least four court cases.[19] During the 1770s, lawyers sometimes called on mathematicians as proto-actuaries to value life contingent contracts that were disputed before the courts. After several such cases, the courts recognized that there was a problem with valuing life annuities. On the open market, there were several prices that could be attached to a life annuity. Which one should the courts choose? The market price could depend on whether the person was buying or selling the annuity; there was a price differential. Further, when buying an annuity, was the annuity on the purchaser's life or the life of someone else? When selling an annuity, did the seller have property as collateral in case the seller could not meet some future annuity payment? Courts like to have rules to follow, or to guide them; and there was not a standard market price. This led to a landmark ruling in 1787 in the case of *Heathcote v. Paignon* heard in the High Court of Chancery.[20] The court came down favouring a 'fair price' for an annuity—that is, one that was actuarially based. This was done so that every property dispute of this type would not come

[17] Institute and Faculty of Actuaries Archives RKN 43915. Described in Bellhouse, *Leases for Lives*, pp. 204–5.

[18] F. Hendriks, 'The Case Book of John Rowe, of London and Exeter, from 1775 to 1790. Edited from the original MS., with an Introductory Notice' in: *The Assurance Magazine and Journal of the Institute of Actuaries*, 7(1857), pp. 136–48.

[19] Ciara Kennefick, 'The Contribution of Contemporary Mathematics to Contractual Fairness in Equity, 1751–1867', in: *The Journal of Legal History* 39/3 (2018), pp. 307–39.

[20] W. Brown, *Reports of Cases Argued and Determined in the High Court of Chancery: During the Time of Lord Chancellor Thurlow, of the Several Lords Commissioners of the Great Seal, and of Lord Chancellor Loughborough, from 1778 to 1794*, London: W. Clarke and Sons 1819.

before Chancery. Actuarial calculations for life contingent contracts now became the legal norm for these kinds of contract disputes.

In England, actuaries or the earlier proto-actuaries, evaluated almost all insurance contracts using methods based on individual risk. For an individual under consideration, the actuary calculates the expected value of all future payments, either an insurance benefit or a series of annuity payments. The expected value of all premiums charged is set to cover these future payments. This results in having expected income equal to expected costs. The key word here is 'expected'. There could be variation around these expectations. By having a large number of policies the risk is reduced, so that the average observed difference between income and cost is approximately equal to the expected difference.

During the 1740s the Scots took a different approach when setting up the Scottish Ministers' Widows' Fund. The approach used was to model aggregate risk. Like individual risk assessments, in any pension fund for a group of individuals there are income and cost to consider. The aggregate risk approach is to value the total fund. In any year, the value of the fund is the observed value from the previous year plus interest earned on the fund, plus the value of premiums paid into the fund, minus the payments made from the fund. Three Scotsmen were involved in the establishment and design of the Widows' Fund: Colin Maclaurin (1698–1746), Robert Wallace (1697–1771), and Alexander Webster (1707–1784).[21] It was the mathematician Maclaurin who carried out the aggregate risk modelling.[22] Assuming a fixed number of ministers, where ministers who die during a year are immediately replaced, Maclaurin could easily model income. On the expense side, Maclaurin took a statistical approach. He used church registers to estimate the number of widows that would be paid each year and the fraction of those widows who would die in the year. Maclaurin set the annual premiums for the living ministers so that the income would cover expenses. The Scottish Ministers' Widows' Fund operated successfully until 1993. The reason that it was successful for about 250 years was that Maclaurin's initial assumptions of a relatively stable income flow and a stable mortality rate among the widows either held up or changes were well accommodated by later actuaries.

The crisis of 1771 over improperly valued life annuity contracts

As the middle class grew, this change was accompanied by a growing desire from many in that class to obtain some form of financial security in old age. It was a security that was not generally available unless it was attached to income from owning property. There were two kinds of products that interested the purchasing public. The first was a widow's pension. This would pay a certain amount annually to a woman on the death of her husband, in essence a reversionary annuity. The second was a straight pension. The potential pensioner paid into a fund for a number of years, and at a

[21] Christopher G. Lewin, *Pensions and Insurance before 1800: A Social History*, East Linton: Tuckwell 2002.

[22] *The Collected Letters of Colin MacLaurin*, ed. Stella Mills, Nantwich, Cheshire: Shiva 1982, pp. 105–10.

certain age quit paying in and began taking out annual lifetime payments. In annuity jargon, this is a type of deferred life annuity. The proper pricing for both these products was well known to many in the mathematical community, but ignored by all but one of the promoters of these products.[23]

The earliest off the blocks were two different societies: The Laudable Society for the Benefit of Widows in 1761 and The Law Society for the Benefit of Widows in 1766. Only one of them, the Laudable Society, made it beyond the drawing board. It must have been a success in the eyes of the public. By the end of the decade after the Laudable Society began its operations, this new society led the way in opening the taps to a torrent of annuity society formations: in 1769, The Amicable Society of Annuitants; in 1770, The Provident Society, The Society of London Annuitants, The Equitable Society of Annuitants, The Westminster Union Society, The London Union Society; in 1771, The Consolidated Society, The Public Annuitant Society, The Friendly Society of Annuitants; and, in 1772, The Rational Society. None of these later societies consulted with any mathematicians knowledgeable in annuities about their pricing systems.

Only the Law Society for the Benefit of Widows made an attempt to consult with mathematicians to put its pricing on a firm foundation. And that was at the same time the cause for its stillbirth. Initially, the society contacted Richard Price (1723–1791), a Nonconformist clergyman, mathematician, and philosopher. He thought that the premiums that the Law Society was planning to charge were insufficient to cover the promised benefits. Then the Law Society consulted with John Rowe (d. 1792), a little-known mathematician who was adept at annuity pricing. His premium calculations, based on Thomas Simpson's most recent mortality table published in 1752, were larger, by about a factor of three, than what the Law Society was proposing. As a result the proposal quickly died.

As he had done with the Law Society for the Benefit of Widows, Richard Price exposed the other annuity societies for their severe shortcomings in premium pricing. This time, it was done in public. His exposé appeared in 1771 in a lengthy publication, *Observations on Reversionary Payments*.[24] William Dale (18th century), who was a member of the Laudable Society, made his own criticisms of these societies in another publication, *Calculations Deduced from First Principles*, published in 1772.[25] Dale had worked as a household steward to the Duke of Beaufort. Price has become famous for his work on annuities and insurance, while Dale is now more or less forgotten. Dale's problem in this respect was that he carried out a very lengthy technical analysis of these annuity societies. For his part, Price was already well known at the time of his work. He also wrote in a style that was easily accessible to the general public. Both men showed that these annuity societies, if they continued in the manner in which they had started, would run out of money in a few years and would be unable to meet their annuity commitments to their members.

[23] The events up to and around the crisis are described in Bellhouse, *Leases for Lives*, pp. 124–53.

[24] Richard Price, *Observations on Reversionary Payments: On Schemes for Providing Annuities for Widows, and for Persons in Old Age; on the Method of Calculating the Values of Assurances on Lives; and on the National Debt*, London: Cadell 1771.

[25] William Dale, *Calculations Deduced from First Principles: In the Most Familiar Manner by Plain Arithmetic, for the Use of the Societies Instituted for the Benefit of Old Age*, London: J. Ridley 1772.

Following on from Price's lucid and devastating criticisms, the bubble burst and most of these annuity societies ceased operations. At least one of the societies, the Friendly Society of Annuitants, wrote to Price asking for his opinion on their premium structure. The reply was sufficiently negative that the Friendly Society was dissolved in the same year that *Observations on Reversionary Payments* was published.

The one major holdout was the firstborn, The Laudable Society for the Benefit of Widows. The directors of the Laudable Society were a responsible lot. Convinced of Price's dire predictions, the directors, in 1772, requested advice from Price and, to be free of perceived bias, two other mathematicians. Advice came in the form of a 60-page report that ultimately recommended an increase in the Society's fees—a doubling of the entry fee and a 60% increase in annual premiums. The consultants also recommended that those who were unwilling to pay the increase and thereby leave the Society should be paid 15 shillings on the pound for their prior investment. This did not go over well with many of the membership and so they hired another mathematician, a rogue consultant named Benjamin Webb (18th century). When first consulted, Webb agree with Price and the other two mathematicians, but later did an about-face. He consulted the Laudable Society's books and concluded that the Society was in good financial shape. In later meetings of the membership there was a vote favouring reform followed by a vote overturning that reform vote. The exasperated directors decided to petition Parliament to dissolve the Society. It was met with a counter-petition by some of the Society's members. The end result was that Parliament asked another mathematician, Edward Waring (*c.*1735–1798), who was Lucasian Professor of Mathematics at Cambridge, to weigh in on the dispute. Waring made his own recommendations for reform and these were adopted by Parliament. What Waring neglected to include in his calculations, however, was the effect of inadequate funding of the Society over the thirteen years since its inception in 1761. Waring's reform lasted only seven years, at which time another reform had to be made to bring the Society's books back to order.

One beneficiary of this crisis was the Society for Equitable Assurances on Lives and Survivorships. It was founded in 1762 and used sound mathematically based product pricing for both life insurance and life annuities. During the heyday of these insolvent annuity societies, the Equitable Society was not competitive in the annuities market. However, when the bubble burst, the Equitable Society thrived, eventually taking over the business of one of the struggling survivors that had hit the shoals of financial reality.

Life insurance in the eighteenth century

Mathematicians such as De Moivre and Simpson showed no interest in the problem of pricing life insurance. Their books are devoted solely to the pricing of life annuities. At the time, these mathematicians were more closely tied to the patronage system of the day; and the patronage system centred on men of property. Also, the chief economic engine at the time was agriculture, again tied to property. The promoters and proprietors of life insurance schemes were merchants and businessmen operating outside, or on the edges of, this patronage loop. It is no surprise that the two paths did not meet, so that these mathematicians and their ideas had very little penetration into the life insurance and life annuity markets.

From the beginning of the eighteenth century, there was an increasing interest among the British populace, particularly in London, in purchasing life insurance. One of the earliest and longest lasting companies, the Amicable Society for a Perpetual Assurance Office, was founded in 1706.[26] The Amicable took a simple approach to their product, an approach that minimized their risk. When it was founded, the Amicable Society offered a form of whole-life insurance, or insurance that is there for the life of the insured person, not just one year. With fluctuations in the number of deaths every year, the proprietors of the Amicable could be confronted with losses, possibly substantial ones. The solution was to fix the premium and vary the death benefit. There was an annual premium of £6 4s regardless of age. Based on a maximum of 2,000 policyholders, the death benefit in the business year was set at £10,000 divided by the number of deaths in the year. The risk to the company was reduced to attracting 2,000 policyholders to the company in any year. It was a successful strategy that continued late into the eighteenth century. With the changing insurance climate at the end of the century and a number of new companies sprouting up, the Amicable Society switched to age-based premiums in 1807.

Among the mathematicians, it was James Dodson who tried to enter the insurance market. Similar to the mathematical expression for an annuity valuation, there is a similar expression to evaluate an insurance policy. This is, in its simplest form,

$$\sum_{t=1}^{\infty} D_t \times d_t,$$

where D_t is the present value of the death benefit to be made t years in the future when death occurs, and d_t is the probability of making the death payment, which in this case is the probability that death occurs in the tth year in the future. Like life annuities, these calculations could be quite onerous especially for whole life insurance. For a one-year term insurance, the calculation is simply $D_1 \times d_1$.

Dodson had been dead for five years by the time the Society for Equitable Assurances on Lives and Survivorships opened its doors in 1762. It was the first insurance company to have age-based pricing of its products based on proper mortality considerations. Prior to his death, Dodson laid the foundations for the Equitable Society, both mathematical and corporate. On the corporate side, he brought together the initial investors behind the Equitable Society. On the mathematical side, he put together an extensive set of tables that provided pricing for various kinds of life annuities and life insurances. In doing this, he recognized the issue of different mortality experiences for different locations and situations. For his calculations, Dodson considered three different scenarios. In the first he looked at the Bills of Mortality for London and tried to capture the maximum mortality experience over a 22-year period. Then he loosely followed Thomas Simpson's mortality table that was related

[26] Amicable Society, *The Charter of the Corporation of the Amicable Society for a Perpetual Assurance-office; Together with the By-laws thereunto Belonging*, London: George Sawbridge 1710.

to mortality for the City of London. Finally, he applied, again loosely, De Moivre's linear survival curve assumption to obtain his third table. This gave him three mortality experiences—worst, middling, and best—that could be applied to whatever situation was deemed appropriate.

Edward Rowe Mores (1730–1778) took over the leadership of the Equitable Society and oversaw the birth of the Society after five years of gestation. During his tenure, the clerks, called actuaries for the first time in the insurance industry, probably determined the premiums to be charged for standard products offered by the Society. Mores appears to have been the one who worked on the pricing of non-standard requests for policies. Here is one example of a non-standard policy whose pricing can be attributed to him:

> An individual borrowed £500 from a friend in order to purchase a promotion. The loan would be repaid in monthly installments over five years. The individual also borrowed an additional amount from the friend to cover a single premium to buy an insurance policy that would continue the monthly payments on the loans if the individual died within the five years.[27]

Mores worked out what the premium should be in order to repay the total debt to the friend.

Within the Equitable Society there was dissatisfaction among the members over the issue of how the original promoters, Mores among them, should be compensated for their original investment in the Society. In a struggle that culminated in 1768, Mores resigned from his roles in the Society. This meant that someone else needed to handle the non-standard insurance contracts. The Equitable Society's choice was Richard Price. Between August 1768 and March 1771, Price corresponded with Equitable's actuary John Edwards (d.1774), who worked mainly as a clerk and had limited mathematical experience. Price answered a number of Edward's pricing questions regarding some specific life annuity and life insurance contracts. Moreover, he remained in a close relationship with the Equitable Society until his death in 1791. After Edwards fell ill in 1773, Price suggested that his nephew, William Morgan (1750–1833), could take on the mathematical side of the actuarial work. Edwards died early in 1774 and was replaced by John Pocock (d.1775) of unknown mathematical abilities. Morgan was made an assistant actuary at the Equitable in February 1774 and promoted to chief actuary in February 1775, following Pocock's death. From this point onward the actuary, in the modern sense, at the Equitable Society handled the pricing of all policies.

Before bowing out completely to his nephew, Price made two more contributions to actuarial science. One was a new mortality table, the Northampton Table published in 1783.[28] It became the basis for the Equitable's policy pricing, as well as for subsequent insurance companies. The other was a report to the Equitable Society on

[27] Edward Rowe Mores, *A list of policies and other instruments of the Society as well general as special*, London: s.n. 1764.

[28] Richard Price, *Observations on Reversionary Payments: On Schemes for Providing Annuities for Widows, and for Persons in Old Age; on the Method of Calculating the Values of Assurances on Lives; and on the National Debt*, London: Cadell 1783.

how to determine the Society's financial position. Both of these contributions later inspired substantial work in actuarial science.

In the 1771 edition of *Observations on Reversionary Payments*, Price presented a mortality table that he had constructed based on the parish registers of All Saints, Northampton from 1735 to 1770. Similar to Halley and to those who had followed him in table construction, Price tabulated the number of deaths, or burials, in different age intervals: 0 to 2, 2 to 5, 5 to 10, 10 to 20, 20 to 30, and so on. Finding that there were more burials than christenings, Price had to adjust his raw numbers to account for migration into the parish. Once this was done, he distributed his deaths within the age intervals to deaths at yearly ages. For example, he found 80 deaths in the age interval 20 to 30 and so assigned 8 deaths to each of the ages 20, 21, 22, and so on up to 29. When there was a change in the average number of deaths per year going from one age interval to another, Price smoothed out the transition. Ten years later, after working with the Equitable Society, Price decided to construct another table from the Northampton data specifically for the Equitable using methodology similar to 1771. The table was published in the 1783 edition of *Observations on Reversionary Payments*. Over the intervening ten years between 1771 and 1781, there were an additional 1,000 deaths approximately, and yet the table shows an approximate tenfold increase in each of the entries in the table. This was probably done to increase the precision of the calculations for annuity and insurance valuations. This table became the standard in the insurance industry for the next thirty years.

Determining the financial position of a life insurance company is not straightforward. Money is flowing into the company through premiums and immediate expenses are paid. But there are much more than immediate expenses. For a recently bought life insurance policy, the beneficiary may not receive the death benefit for several years or even a decade or two. From the premiums coming in, enough money must be set aside and invested to meet these future payments. Price had three suggestions, two of which were later put into effect and have become standard in insurance practice. The first practical suggestion was to compare the mortality experienced by a company with what is expected from the mortality table used to price the insurance. If the observed number of deaths over the year is less than what would be expected from the mortality table, then the company is in good financial shape, otherwise not. The second suggestion required much more work, especially when calculations are done by hand. This suggestion was to calculate the present value of all expected future benefit payments on the company's books and compare it to the present value of all expected premium incomes for policies in force. One would want the expected income from premiums to be greater than the expected death benefits to be paid.

William Morgan worked indefatigably for the Equitable Society. Following his uncle's suggestion, Morgan calculated the mortality experience of the Equitable Society on an annual basis. The great amount of work involved in carrying out Price's second suggestion is indicated by the fact that Morgan carried out these calculations about once every ten years during his tenure at the Equitable Society, which lasted until 1830. In addition to this work, after Price produced the Northampton table in 1781 for the Equitable Society, Morgan spent months calculating the values of various annuity and insurance contracts by age at issue.

Price had some—albeit short-lived—influence on the development of the insurance industry in France. In 1786, prior to the French Revolution, financiers Étienne Clavière (1735–1793) and Jean de Batz (1754–1822) started a fire insurance company. Two years later, they decided to offer life insurance and annuity products. Pricing of these products given in their prospectus is based on Price's Northampton table.[29] The company, as well as Clavière, died in the Revolution. It was not until after the Bourbon restoration that life insurance companies started up again in France.

Life insurance in the nineteenth century

An explosion in the number of life insurance companies began at the turn of the eighteenth to the nineteenth century. During the entire span of the rest of the eighteenth century up to 1792, there were only four life insurance companies of any significance: The Amicable Society (1706), The London Assurance Corporation (1721), The Royal Exchange Assurance (1721), and The Equitable Society (1762). The start of the exponential growth of life insurance companies is described in a short guide to the life insurance industry, written by Francis Baily (1774–1844) who was working as a stockbroker at this point in his career. The new era in life insurance began in 1792 with the establishment of the Westminster Society, followed by the Pelican in 1797. By 1807 eight additional companies had been formed. Baily was critical of many of these new companies.[30] The main model for these new companies was provided by the Equitable Society, which charged age-based premiums for a variety of life products. The new companies all followed suit with premiums very similar to the Equitable; many had identical premiums. That was the nub of the problem for Baily. The Equitable Society was a mutual company: in essence, the policyholders owned the company so that any surplus or profits were returned to them. Almost all the new companies, however, were formed as joint stock companies so that profits were returned to the stockholders, not the policyholders. To a policyholder, insurance from these new companies would be more costly overall than from the Equitable. By 1850, nearly 210 new companies had been established.[31] More than 60% of them failed. Of the failures, more than 15% of them essentially had done no business. Some of the failing companies were absorbed by other companies; most were not.

As a stockbroker, Baily was keenly interested in the full range of life contingent contracts, from leases for lives to life annuities and life insurance. He wrote on all these topics. Of his standing in this general community, one biographer wrote: 'During his long career on the stock exchange Baily became a powerful figure in defining financial practices—in particular those appropriate for annuities and assurance societies'.[32]

[29] Compagnie royale d'assurance, *Prospectus de l'établissement des assurances sur la vie*, Paris: Lottin l'aîné et Lottin 1788.

[30] Francis Baily, *An Account of the Several Life-assurance Companies Established in London: Containing a View of their Respective Merits and Advantages*, London: Richardson 1810.

[31] M. A. Black, *Chronological List & Statistical Chart of the Life Assurance Associations Established in the United Kingdom from 1706 to 1863 Showing where they Are, when & how they Disappeared*, London: Edward Stanford 1864.

[32] William J. Ashworth, 'Baily, Francis (1774–1844)', in: *Oxford Dictionary of National Biography*, Oxford: Oxford University Press 2007.

Baily put his recommendations for good financial practice in the insurance industry into print by writing the first comprehensive book on life insurance mathematics, or what is now usually called life contingencies. The book, *The Doctrine of Life-annuities and Assurances*, was first published in 1810 with a second edition appearing in 1813.[33]

What Price suggested to determine the financial position of an insurance company, and Morgan calculated, was the expected future liabilities of the company minus the expected future income. For any individual policy, this would be the expected future insurance benefit less the expected future premium payments. In actuarial parlance, this is given the name 'prospective reserve'. Mathematically, this is equivalent to what is called the 'retrospective reserve'. Looking back in time, the retrospective reserve is the accumulated value of the premiums paid in up until the current time, less the expected accumulated cost of insurance to the present time. The retrospective reserve can also be viewed as the surrender value of the policy. After paying in for several years, for whatever reason the policyholder may decide to terminate the policy. The retrospective reserve, or mathematically equivalently prospective reserve, is what is owed to the policyholder. Neither Price nor Morgan published any formulas for the value of the reserve, prospective or retrospective. The first algebraic expression for the reserve or surrender value appears in Baily's *Life-annuities and Assurances*.[34]

The British insurance industry saw the birth of another new company, the Sun Life Assurance Society, in June 1810. The new company hired a 34-year-old mathematician, Joshua Milne (1775–1851), as its actuary.[35] Milne became quickly engrossed in mortality statistics that he thought might help in the pricing of insurance products. In this context, he came across the work of Dr John Heysham (1753–1834) of Carlisle. In a letter to Heysham, written on 12 September 1812, Milne noted: 'it appears that the inhabitants of your city surpass in longevity those of any place (so far as I am informed) for which a similar table has yet been constructed'.[36]

That letter was the beginning of a two-year correspondence between Milne and Heysham about questions of mortality in Carlisle. They discussed among other things how to determine the population of Carlisle and the effect of various diseases on mortality, including the impact of Jenner's vaccine for smallpox. At the end of this lengthy correspondence Milne wrote to Heysham on 22 March 1814: 'I have reason to believe that the law of mortality in your two parishes has, for the last thirty years, been very nearly the same as the average of the kingdom'.[37]

Based on the data accumulated and his deliberations with Heysham, Milne published, in 1815, the Carlisle Table as part of his book *A Treatise on the Valuation of*

[33] Francis Baily, *The Doctrine of Life-annuities and Assurances: Analytically Investigated and Explained, Together with Several Useful Tables Connected with the Subject and a Variety of Practical Rules for the Illustration of the Same*, London: J. Richardson 1810.

[34] Baily, *Life-annuities and Assurances*, pp. 456–61.

[35] W. A. S. Hewins and Robert Brown, 'Milne, Joshua (1776–1851)', in: *Oxford Dictionary of National Biography*, Oxford: Oxford University Press 2008.

[36] Henry Lonsdale and Joshua Milne, *The Life of John Heysham, M.D. and His Correspondence with Mr. Joshua Milne Relative to the Carlisle Bills of Mortality*, London: Longmans, Green, and Co. 1870, p. 137.

[37] Lonsdale and Milne, *Life of John Heysham*, p. 156.

Annuities and Assurances on Lives and Survivorships.[38] The new table became the standard for use in insurance companies for over fifty years.

Extensive tables based on particular mortality data and a given rate of interest became the key to actuarial calculations. However, mortality tables in themselves were not enough. What was needed was a set of tables from which a few entries could be extracted and simple calculations made to determine the value of any annuity or insurance contract. An initial investment of a great deal of labour in the construction of the tables would result in substantial labour savings further down the road. William Dale had led the way in this idea in 1772 by calculating an extensive table that could be used to calculate variations on a life annuity for someone aged 50. In addition to a straight life annuity, there could be an upper bound on the age to which the annuity is paid or a lower bound after 50 on the age at which the annuity begins. The idea was expanded upon and simplified in the nineteenth century. Several hands, both in England and on the Continent, either rediscovered or massaged the idea until it came to full fruition with the actuary Griffith Davies (1788–1855). In 1821, the newly formed Guardian Assurance Company called on Davies for actuarial advice, and two years later proceeded to appoint him as actuary to the company. Motivated by William Morgan's work at the Equitable Society and by his new position, Davies published *Tables of Life Contingencies* in 1825. One of the tables, Table XIII, is a much simplified and easier-to-use version of what Dale had already done. In the table, calculated from ages 0 to 96 using the Northampton mortality table, there are five columns of numbers from which all life annuity and life insurance calculations can be made at 3% interest.[39] This approach to actuarial calculations, called 'commutation functions', continued until the penetration of digital computing into insurance companies during the second half of the twentieth century. Davies's work is also an example of early collaboration among actuaries in different insurance offices. Davies acknowledged help in his table calculations from actuaries in four other insurance companies.[40]

Up to and including the publication of the Carlisle Table, mortality tables in England were constructed from records such as parish registers or bills of mortality, for which the age at death is given. Table construction relies on a key assumption, that of a stationary population. These are populations for which the number of births equals the number of deaths and the death rate by age remains constant over time. In a relatively short period of time the assumption of the constant death rate is usually met. With migration or with a declining or increasing population, this assumption could become problematic. Table constructors as far back as Halley were aware that such assumptions were necessary. Halley, for example, checked to see that births and deaths in Breslau were approximately equal in number. Simpson adjusted a mortality table for the City of London in order to take into account immigration.

Wesley Stoker Barker Woolhouse (1809–1893) was faced with a new type of mortality data. After leaving a position at the National Almanac Office, located at the Royal

[38] Joshua Milne, *A Treatise on the Valuation of Annuities and Assurances on Lives and Survivorships*, London: Longman, Hurst, Rees, Orme, and Brown 1815.

[39] Griffith Davies, *Tables of Life Contingences*, London: Longman and Co. 1825.

[40] Davies, *Tables of Life Contingences*, p. viii.

Observatory in Greenwich, he had taken a new job, in 1839, as the first actuary at the National Loan Fund Society.[41] This was also the first position for Woolhouse in that particular line of work. The Loan Fund Society was established, in part, to fill a perceived gap in provision for working people. Policyholders had their premiums deposited into a fund that would eventually be used to purchase a deferred annuity, essentially a pension.[42] There was also the option, under certain restrictions, of borrowing up to two thirds of the deposited amount. Should the policyholder be unable to make premium payments, there was a provision for dispersing the funds that had accumulated in the account. Soon after he commenced working for the Society, Woolhouse was given the task of the analysis of mortality data taken among the British officers serving in the East India Company's army, a total of 12,000 soldiers, over approximately an eighty-year period. This was a 'population' with significant migration; soldiers came and went. To analyze the data, Woolhouse came up with the concept of what is now called 'exposure to risk'.[43] This is the time interval for which a soldier would be exposed to the risk of death only while serving in the army. To calculate this interval for any soldier, it would be necessary to obtain the date of entry into the army, as well as the age at entry, and the date of exit. Exit could be due either to death or to leaving the army or to termination of the mortality study. The exposure to risk is the time interval from the date of entry to the date of exit. The mortality table is then constructed based on the ages of the soldiers, the number of deaths among them, the number of survivors, and the exposure to risk.

De Moivre's model that the number of survivors decreases linearly by age stood for several years. With newer and better mortality tables, however, it became obvious that De Moivre's model no longer applied to many of these tables. Beginning with the 1783 edition of *Observations on Reversionary Payments*, edited after Price's death by his nephew, both Richard Price and William Morgan criticized the linear assumption, but made no attempt to replace it with a better model. Exactly 100 years after De Moivre first put forward his model, the actuary Benjamin Gompertz (1779–1865) suggested, in 1825, a more accurate mathematical expression for human mortality, this time applicable to ages 10 or 15 and above. To this day, Gompertz's model continues to be used as a reasonable model for human lifetimes.

Like De Moivre, Gompertz came by his model empirically.[44] He looked at some mortality tables including Price's Northampton table from 1783, Joshua Milnes's Carlisle table from 1815 and Antoine Deparcieux's table of French mortality from 1746.[45] In each table Gompertz looked at l_x, the number living at age x for $x = 10$ or 15 and then for $x = 20$, 25, and so on. He subsequently calculated the logarithm of the number living at each of these ages. Gompertz denoted these values by L_{10}, L_{15},

[41] G. H. R., 'The Late W. S. B. Woolhouse', in: *Journal of the Institute of Actuaries*, 31 (1894), pp. 362–5.

[42] *First report of the directors at the annual meeting of proprietors, held on the 13th day of May 1840, of the National Loan Fund Life Assurance and Deferred Annuity Society*. Accessed by Google Books.

[43] Wesley S. B. Woolhouse, *Investigation of Mortality in the Indian Army*, London, 1839.

[44] Benjamin Gompertz, 'On the Nature of the Function Expressive of the Law of Human Mortality, and on a New Mode of Determining the Value of Life Contingencies', in: *Philosophical Transactions of the Royal Society of London* 115 (1825), pp. 513–83.

[45] Milne, *Treatise*, p. 564 and Antoine Deparcieux, *Essai sur les probabilités de la durée de la vie humaine*, Paris: Guerin 1746.

L_{20}, and so on. Finally, he calculated $L_{10} - L_{15}$, $L_{15} - L_{20}$, and so on, and found these series of differences were in approximate geometric progression. This led him to postulate that the number living at a particular age x could be expressed in mathematical form as

$$l_x = dg^{q^x}$$

in Gompertz's notation, where d, g, and q are constants to be determined. For the Northampton table, Gompertz found $d = 8441$, $g = 0.7404$, and $q = 1.0261$. Figure 9.1 shows the Northampton table from ages 15 and on, along with Gompertz's curve and a linear approximation. The Gompertz curve fits well until about the age of 70 and provides a much better fit than the linear approximation.

Earlier in the eighteenth century there was another attempt beyond De Moivre to formulate a model for the survivor distribution or the number of survivors at each age in a population. Johannes Nikolaus Tetens (1736–1807) was a professor of mathematics and philosophy at Kiel University in Germany. In the 1780s he was asked to consult on the operation of a widows' fund in the principality of Calenberg, which was governed by the House of Hanover.[46] Tetens's task was to compare the observed number of deaths to the expected number. Motivated by this task, and the data he was given, Tetens put forward a survivor distribution that was conditional on attaining a certain age. Once that age was reached the number of survivors decreased exponentially with age. This has some similarities to Gompertz's distribution of survivors.

In the first decades of the nineteenth century, several new companies hired men with demonstrated mathematical skills to carry out the pricing of their insurance

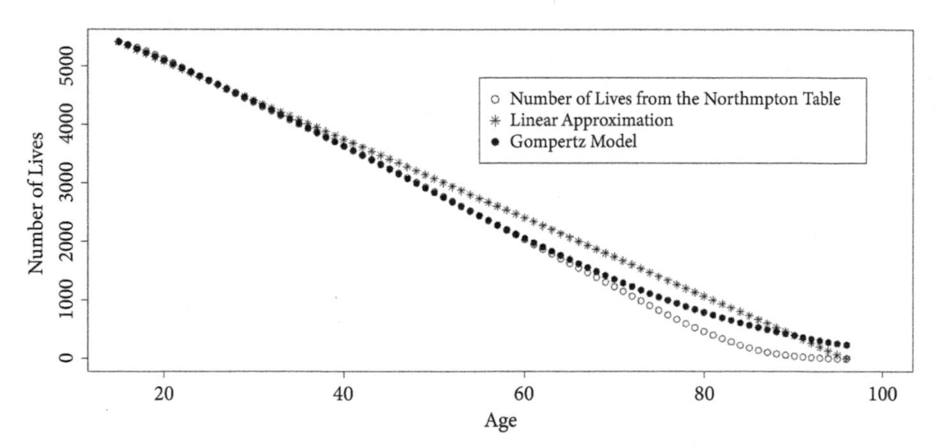

Figure 9.1 The Gompertz Model and the Linear Approximation for the Northampton Mortality Table.

[46] Niels Keiding, 'The Method of Expected Number of Deaths, 1786–1886–1986' in: *International Statistical Review*, 55 (1987), pp. 1–20.

products.[47] But there was no professional society for actuaries and no system of examinations to set standards. Those companies seeking to demonstrate the soundness of their products, and provide evidence of their own, chose men with 'proper' mathematical credentials. Of the early nineteenth-century actuaries mentioned thus far Baily, Gompertz, and Woolhouse were fellows of the Royal Astronomical Society, and Davies was given a certificate of actuarial competency by William Morgan. Gompertz was also a fellow of the Royal Society. All that is known of Joshua Milne is that he 'received an education grounded in mathematics and languages'.[48] During this time, many companies also began training actuaries in house. One example of such in house training is provided by the actuary Charles James Ansell (1794–1881) who began working for Atlas Fire and Life Assurance Company in 1808 as a junior at the age of fourteen.[49] Two years later he was taken on as a member of staff and in 1823 he was made actuary of the Atlas. By the 1850s the example of Ansell had become the norm.[50] The reason for this development is that actuaries began to fill two roles within a company: mathematician as actuary and actuary as company manager. As a manager, the actuary oversaw the day-to-day operation of the company. This was not a position that a 'parachuted-in' mathematician could necessarily fill well. It proved better instead to take on young clerks who showed mathematical promise: let them learn the ropes of an insurance company operation, nurture them in the necessary mathematics, and then move them into the role of actuary.

The collaborative approach between actuaries in different companies that Griffith Davies instituted in 1825 when calculating his *Tables of Life Contingencies* was expanded upon significantly in the late 1830s and early 1840s. Realizing that the population of those seeking insurance might have a different mortality experience from those in the general population or the populations in Carlisle and Northampton upon which earlier tables were based, a group of actuaries met in March 1838.[51] They decided to carry out a mortality study of insured lives that differentiated on the one hand between males and females, and on the other hand according to whether the insured policy was sold from an office located in a town or in the country or in Ireland. A group of seventeen insurance offices agreed to submit data from their companies covering policies in force between 1822 and 1837. It took until 1843 to complete the study, because some offices did not submit their data until 1841. When the data did come in, the actuaries working on the data had to contend with different classification schemes used by different companies (standardization of data remains a problem even today when trying to combine diverse databases). Essentially, the data arrived in the form that Woolhouse saw for the East India Company's army. And so Woolhouse's methodology was used to obtain mortality tables for males,

[47] Timothy Alborn, 'A Calculating Profession: Victorian Actuaries among the Statisticians', in: *Science in Context*, 7 (1994), pp. 433–68.

[48] W. A. S. Hewins and Robert Brown, 'Milne'.

[49] 'Obituary Notices of Fellows Deceased', in: *Proceedings of the Royal Society of London* 34 (1882), pp. vii–viii.

[50] Timothy Alborn, 'Quill-Driving: British Life-Insurance Clerks and Occupational Mobility, 1800–1914', in: *Business History Review* 82 (2008), pp. 31–58.

[51] Charles Ansell, *Tables Exhibiting the Law of Mortality: Deduced from the Combined Experience of Seventeen Life Assurance Offices*, London: J. King 1843.

females, town policies, country policies, and Irish policies. With the exception of Irish mortality, which was different from the rest of the United Kingdom, the tables were combined into a general table. Of the eight members on the committee that carried out the mortality study, five have been mentioned here: Ansell, Davies, Gompertz, Milne, and Woolhouse.

The findings were published, but probably only in limited circulation, by the end of April 1843 under the lengthy title *Tables exhibiting the law of mortality: Deduced from the combined experience of seventeen life assurance offices, embracing 83,905 policies, of which 40,616 are distinguished by denoting the sex of the lives assured, and by classing them into town, country and Irish assurances.* A total of sixty companies, including the seventeen that had participated in the study by submitting data, subscribed to the publication. A reporter for the newspaper *The Era* received a copy of the published findings and used this copy to comment on the problems faced by the consumer in purchasing insurance. There were what he called 'old-established and reputable offices' and 'cheap offices'. After looking at the tables, the reporter concluded with warnings similar to those of Price in 1771:

> What, therefore, between the increased mortality as shown by these tables, the low premiums taken to insure business, carelessness in the selection of lives, and the low rate of interest at which investments can be secured, it follows that, sooner or later, ruin, absolute and irretrievable, must be the lot of all cheap offices. Are we wrong then, and can we be accused of taking an invidious part, when we call for a Parliamentary inquiry to stay the plague which will involve thousands of families in hopeless ruin.[52]

Parliament did consider the issue and subsequently passed the Joint Stock Companies Act of 1844.[53] This Act required all joint stock companies, including insurance and annuity companies, to be registered and to provide a balance sheet. However, many insurance and annuity companies were able to evade the Act's provisions and problems in the insurance industry continued. In the four years 1845 to 1848, a total of 48 new insurance companies were formed, but they lasted on average less than ten years.

Jenkin Jones (1811–1866), an actuary from the National Mercantile Life Assurance Society, took the new life table and carried out some extensive calculations to make sets of tables that evaluated various kinds of annuities and insurances at different rates of interest. Established only in 1837, the year that marked the end of the collaborative mortality study, the National Mercantile Life was one of the companies that neither participated in the study nor subscribed to its resulting publication. Jones's tables were a prodigious undertaking on three counts. First, Jones did the calculations himself. But then he talked to a young colleague, Evan Owen Glynne (1824–1846?), based at another non-participating and non-subscribing company, also established in 1837. Glynne, who was recently fresh out of Merchant Taylors' School, repeated most

[52] *The Era*, 30 April 1843.

[53] Philip M. Booth, '"Freedom with Publicity"—the Actuarial Profession and United Kingdom Insurance Regulation from 1844 to 1945', in: *Annals of Actuarial Science* 2 (2007), pp. 115–45.

of the calculations independently.[54] Once he was finished, the results of Jones and Glynne were compared. Finally, Jones consulted Joseph Cleghorn who was Griffith Davies's assistant at Guardian Assurance. Cleghorn had computed the same tables for Guardian Assurance; they were compared to Jones's and Glynne's tables and found to be in agreement.

After corresponding with William Thomson (1813–1883), an actuary in Edinburgh, Griffith Davies called a meeting of a number of London actuaries from various life companies, twenty six in all. He also invited Thomson to attend, with a special reason in mind: he was to tell the assembled group about an association of actuaries in Edinburgh that had been meeting for about fifteen years.[55] The men gathered in London on 15 April 1848. At the end of their meeting the London actuaries resolved:

> That it appears desirable that those connected with the management of Life Assurance Institutions should have occasional opportunity of meeting together and consulting on subjects of mutual interest.
>
> That a Committee of not exceeding Ten be formed to consider the best mode of carrying out the same and to report thereon to a future meeting.[56]

Not much happened until two months later. With the proliferation of life insurance companies, many with questionable foundations, and with the need for proper actuarial skills to keep the industry on an even keel, an expanded group of sixty-six actuaries met on 10 June 1848. Deciding that they needed some kind of association of actuaries, they appointed a committee to prepare a plan for the association with rules and regulations. The committee met several times and by the end of the month came up with their recommendations.

The committee went well beyond the April resolution to form an official group that would meet together to discuss subjects of mutual interest.[57] With the proposed name of 'Institute of Actuaries', the committee laid out the purpose of the new body. It included the original resolution of 'meeting together and consulting on subjects of mutual interest'. What was altogether new was the method of how to become a member. There were to be two classes of membership: fellow and associate fellow. After proposing a method of 'grandfathering' in practising actuaries that would occur by 1 September 1848, the committee suggested a series of examinations that would lead to fellowship. Anyone wishing to be an associate of the Institute needed to be proposed by two fellows and then elected by ballot at one of the monthly meetings of the Institute. Once an associate obtained a position as an actuary in a life insurance company, that individual could proceed to the next level of fellowship through a set of examinations. There were to be four examinable topics: mathematical theory, vital statistics, computation and construction of life tables, and bookkeeping and office

[54] Charles J. Robinson, *A Register of the Scholars Admitted into Merchant Taylors' School*, vol. 2, Lewes: Farncombe and Co. 1883, p. 266.

[55] See George M. Low, 'The History and Present Position of the Faculty of Actuaries in Scotland', in: *Transactions of the Faculty of Actuaries* 1/1 (1901–3), pp. 3–16.

[56] Archives of the Institute and Faculty of Actuaries of the United Kingdom. History of the formation of the Institute of Actuaries, 1848, RKN 1279.

[57] Archives of the Institute and Faculty of Actuaries of the United Kingdom, RKN 1279.

routine. Those who were grandfathered in were senior practitioners recognized for their actuarial abilities and contribuions. They bypassed the proposed examination process. After the proposal was adopted at a meeting of actuaries on 8 July 1848, the Institute of Actuaries soon became the formal body for the training and continuing education of actuaries in England. Within weeks of its formation, one hundred and forty-eight individuals had applied for membership.[58]

[58] *Post Magazine*, 26 August 1848.

PART III
MATHEMATICAL PRACTITIONERS AND THEIR SCIENTIFIC MILIEUS

10
Assembling the Scribal Self
Gian Vincenzo Pinelli's Circle and Mathematical Practitioners in the Veneto, *c.*1580–1606

Stefano Gulizia

Introduction: files and numbers

Shortly after March 1584, as one gathers from internal evidence, the Jews of Padua addressed a petition to the government to avert confinement in the ghetto. This text in Italian is stern and no-nonsense; it culminates in a dire caveat against the high economic cost of setting up an enclosure.[1] The anonymous legal drafter makes smart remarks on the advantages of 'spying' versus 'locking', both of which sound eerily Foucauldian to our ears.[2] The Jews, he concludes, have always been loyal to the city. That the petition carried a solid argument can be deduced from the fact that it was not till 1602 that a formal ghetto was established in the Jewish district of Padua.[3] On the other hand, perhaps precisely because discussion among city officials was so intense in those years, this corporate document was culled from a variety of sources and considered a kernel of 'useful' knowledge by the local collector Gian Vincenzo Pinelli (1535–1601), who ordered the main scribe in his household to execute a fair transcription. Right next to the Jewish plea, Pinelli added a 'Note on the gems' in his own (coarse) hand, which first develops as a synoptic treatment, and later morphs in a fully alphabetical fashion. Pinelli's 'Note' is essentially a lexicographical exercise in mineralogy, which finds many counterparts in his papers, a vast *Nachlass* now housed at the Ambrosian Library in Milan.[4]

[1] Milan, Biblioteca Ambrosiana (subsequently: BAM), MS R 122 sup., ff. 111ʳ–113ᵛ.

[2] The key terms are *spiare* and *serrare* (BAM, MS R 122 sup., f. 113ʳ).

[3] See Benjamin Ravid, 'The Venetian Government and the Jews', in: *The Jews of Early Modern Venice*, ed. Robert C. Davis and Benjamin C. Ravid, Baltimore: Johns Hopkins University Press 2001, pp. 3–30, and Ariel Viterbo, 'Gli ebrei a Padova nel Settecento', in: *Ramhal: Pensiero ebraico e kabbalah tra Padova ed Eretz Israel*, ed. Gadi Luzzatto Voghera and Mauro Perani, Padua: Esedra 2010, pp. 13–23.

[4] As examples of Pinelli's high proficiency in the lexicographical practices, I cite three different manuscripts: (i) BAM, MS I 231 inf., ff. 25ʳ–52ᵛ, 85ʳ–103ᵛ, where Pinelli takes his cue from Dioscorides to produce a botanical *sylva* on loose folio pages cut in half that greatly expands on such a foundational text and includes discussions of method; (ii) BAM, MS D 290 inf., which reads out like an alphabetical commonplace of colour-related terms, dominated but not exhausted by classical references taken from Pliny and Gellius; and (iii) BAM, MS D 152 inf., especially ff. 8ʳ and 18ʳ, a letter of 1589 in which Pinelli, in order to facilitate the job of those who have to prepare an edition of the historian Carlo Sigonio, chiefly the Jesuit Alessandro Caprara who mediated between the printers and Pinelli's household, explicitly imagines that his private 'wish list' could function as a preliminary *errata corrige*. On (ii), see Valentina Pugliano, 'Ulisse Aldrovandi's Color Sensibility: Natural History, Language and the Lay Color Practices of Renaissance *Virtuosi*', in: *Early Science and Medicine* 20 (2015), pp. 358–96.

By accretion, the same codex acquired supplementary bits and parcels of knowledge, some of which were ostensibly plotted to fit an individual line of inquiry—by Pinelli himself or one of Cardinal Borromeo's library assistants, after the purchase and rebinding of the original collection.[5] Others, it would seem, were accidentally enmeshed in this vibrant information order based on miscellanies and their paper technologies.[6] One of these, a computational insert of noticeably different paper-length and watermarks compared to what stands materially on either side to it, was produced by one of Pinelli's closest friends in Padua, and possibly at his insistence too: the mathematician Giuseppe Moleto (1531–1588).[7] Like Pinelli, Moleto was born in Southern Italy; and he had also trained with the preeminent Sicilian mathematician

[5] Apart from Adolfo Rivolta, *Catalogo dei Codici Pinelliani dell'Ambrosiana*, Milan: Tipografia Arcivescovile 1933, who offers the first and still indispensable orientation on the reconfiguration of Pinelli's Latin papers in the hands of Borromeo's entourage, we are also well informed on the Greek interests pursued in this intellectual circle, which are discussed especially by Marcella Grendler, 'A Greek Collection in Padua: The Library of Gian Vincenzo Pinelli (1535–1601)', in: *Renaissance Quarterly* 33 (1980), pp. 386–416, Anna Meschini, *Michele Sofianòs*, Padua: Liviana 1981, and Anna Gialdini, 'Fonti codicoogiche e archivistiche per la ricostruzione della biblioteca di Michael Sophianos', in *Miscellanea Graecolatina II*, ed. Federico Gallo and Lisa Benedetti, Rome: Bulzoni 2014, pp. 287–323. On the making and undoing of the collection, see Anthony Hobson, 'A Sale by Candle in 1608', in: *The Library* 26 (1971), pp. 215–33, Massimo Rodella, 'Fortuna e sfortuna della biblioteca di Gian Vincenzo Pinelli: la vendita a Federico Borromeo', in: *Bibliotheca* 2 (2003), pp. 87–125, and Anna Maria Raugei, *Gian Vincenzo Pinelli e la sua biblioteca*, Geneva: Droz 2018, who unconvincingly downplays the importance of science and philosophy. Since Pinelli does not feature in Michael Hunter (ed.), *Archives of the Scientific Revolution: The Formation and Exchange of Ideas in Seventeenth-Century Europe*, Woodbridge: Boydell Press 1998, the best interpretation attempted so far remains Angela Nuovo, 'The Creation and Dispersal of the Library of Gian Vincenzo Pinelli', in: *Books on the Move: Tracking Copies through Collections and the Book Trade*, ed. Giles Mandelbrote et al., London: British Library 2007, pp. 39–67. For more on how editorial choices can affect the way in which a scientific legacy is read and understood, albeit with an Anglocentric approach, see Elizabeth Yale, 'The Book and the Archive in the History of Science', in: *Isis* 107 (2016), pp. 106–15, L. Daston (ed.), *Science in the Archives: Pasts, Presents, Futures*, Chicago: University of Chicago Press 2017, and V. Keller, A. M. Roos, and E. Yale (eds), *Archival Afterlives: Life, Death, and Knowledge-Making in Early Modern British Scientific and Medical Archives*, Leiden: Brill 2018.

[6] By 'paper technologies', I do not consider only those practices relevant—like lists, diagrams, tables, and observational notebooks—which have been underlined as organizational tools at the service of a specific natural philosophy, post-Baconian or otherwise, but also traditional habits such as marginalia or epistolary précis that come from a variety of fields, including the history of reading, diplomacy, politics, and the pharmacy countertop. In short, my view is that scribal enumeration in its innumerous guises performs high intellectual duties, but originates in everyday cut-and-paste methods, and was seen (by Pinelli) as a response to the cluttering of objects, not to an anxiety for 'information overload'. See Christoph Hoffmann, 'The Pocket-Schedule: Note-Taking as a Research Technique', in: *Reworking the Bench: Research Notebooks in the History of Science*, ed. F. L. Holmes, J. Renn, and H.-J. Rheinberger, Dordrecht: Springer 2003, pp. 183–202, Anke te Heesen, 'The Notebook: A Paper Technology', in: *Making Things Public: Atmospheres of Democracy*, ed. B. Latour and P. Weibel, Cambridge, MA: Harvard University Press 2005, pp. 582–9, Matthew Eddy, 'Tools for reordering: Commonplacing and the space of words in Linnaeus' Philosophia Botanica', in: *Intellectual History Review* 20 (2010), pp. 227–52, Isabelle Charmantier and Staffan Müller-Wille, 'Worlds of paper: an introduction', in: *Early Science and Medicine* 19 (2014), pp. 379–97, and Filippo De Vivo, 'Archives of Speech: Recording Diplomatic Negotiation in Late Medieval and Early Modern Italy', in: *European History Quarterly* 46 (2016), pp. 519–44.

[7] On Moleto, who named Pinelli as his executor, bequeathing in the process all his mathematical instruments, see Antonio Favaro, 'Amici e corrispondenti di Galileo Galilei: Giuseppe Moletti', in: *Atti del Real Istituto Veneto di Scienze Lettere ed Arti* 77 (1918), pp. 47–118, W. R. Laird, *The Unfinished Mechanics of Giuseppe Moletti: An Edition and English Translation of his Dialogue on Mechanics, 1576*, Toronto: University of Toronto Press 2000, Alessandra Fiocca, 'Giuseppe Moleto (1531–1588), matematico al servizio dei Gonzaga e della Repubblica di Venezia', in: *Contributi di scienziati mantovani allo sviluppo della matematica e della fisica*, ed. Fabio Mercanti e Luca Tallini, Cremona: Monotipia Cremonese 2001, pp. 111–29, Manuela Bragagnolo, 'Geografia e politica nel Cinquecento: La descrizione di città nelle carte di Gian

Francesco Maurolico (1494–1575), from whom he took the idea of reviving interest in Greek mathematics by preparing to publish new editions of Alexandrian textbooks. Moleto attended to this task mostly in his Paduan years, when he was a public professor in the discipline for eleven years,[8] and where he worked on a compilation of *Tabulae Gregorianae*, which was printed in Venice in 1580.

Perhaps in the context of that volume, or by individual solicitation by Pinelli himself, who might have tracked down one of his customary 'pet projects', Moleto sent the collector a gift copy of a small theoretical essay on the emendation of the Roman calendar.[9] Naturally, to bind somebody in favour or friendship is not by itself remarkable for an actor embedded in a scholarly network, not even if the nature of the intelligence itself was made heavy by technical analysis, as in the case of Moleto's booklet. The reform of the calendar year, moreover, remains one of the most nervous controversies in the mathematical republic of letters in this period.[10]

The archival handling of Moleto's computational booklets supplies a strong context for Pinelli's own reading of mathematics. This discipline is not found in isolation. *Mathesis* is configured as one of the main loci of mechanical as well as pedagogical apprenticeship,[11] and indeed is valued in early modern Venice for its ability to facilitate cross-cultural exchange. The same result could be obtained by looking at Pinelli's massive files on hydraulic energy and water flow, which are conspicuous for seeking a mathematical solution to what was habitually perceived as a technical problem of engineering.[12] In this and similar cases, it is best to remember that Pinelli's library was a public institution, with an open connection to Venice's government.[13] Near

Vincenzo Pinelli', in: *Laboratoire Italien* 8 (2008), pp. 163–93, and Stefano Gulizia, 'The Philosophy of Mathematics in Gian Vincenzo Pinelli's Papers', in: *Bruniana & Campanelliana* 25 (2019), pp. 459–74.

[8] Adriano Carugo, 'L'insegnamento della matematica all'Università di Padova prima e dopo Galileo', in: *Storia della cultura veneta*, ed. Girolamo Arnaldi and Manlio Pastore Stocchi, Vicenza: Neri Pozza 1984, vol. 4/II, pp. 151–99. As is well known, Moleto was replaced by Galileo in 1592 and at that juncture, from Pinelli's viewpoint, it would have been hard to tell them apart, whereas the other two tenured readers in this period, namely, Francesco Barozzi (1537–1604) and Pietro Catena (1501–1577), had done the most to elevate mathematics from adjunct teaching to the medical syllabus and both had a distinguished role in the Paduan controversies on the epistemic status of mathematical demonstrations, which I can only mention briefly; see Paolo Mancosu, 'Aristotelian Logic and Euclidean Mathematics: Seventeenth-Century Developments of the *Quaestio de certitudine mathematicarum*', in: *Studies in the History and Philosophy of Science* 23 (1992), pp. 241–65, Paolo Palmieri, 'Mental Models in Galileo's Early Mathematization of Nature', in: *Studies in History and Philosophy of Science* 34 (2003), pp. 229–64, and the bibliography cited in Gulizia, 'The Philosophy of Mathematics', pp. 464–71.

[9] BAM, MS R 122 sup., ff. 280^r–287^v; many aggregative structures that directly reflect Pinelli's decision making, and pre-date Borromeo, have an identification number (in this file *Y—9) which descended from the organizational technique of the library in Padua, based on numbered shelves, for which see Nuovo, 'The Creation and Dispersal', pp. 49–52.

[10] C. P. E. Nothaft, 'A Sixteenth-Century Debate on the Jewish Calendar: Jacob Christmann and Joseph Justus Scaliger', in: *Jewish Quarterly Review* 103 (2013), pp. 47–73.

[11] I resonate here with an argument expressed with admirable economy by Richard J. Oosterhoff, 'Tutor, Antiquarian, and Almost a Practitioner: Brian Twyne's Readings of Mathematics', in: *Reading Mathematics in Early Modern Europe: Studies in the Production, Collection, and Use of Mathematical Books*, ed. P. Beeley, Y. Nasifoglu, and B. Wardhaugh, London: Routledge 2021, pp. 151–66, namely that pedagogy generated scholarship too.

[12] See BAM, MS R 99 sup., ff. 1^r–70^v, with notes on hydrography attributed to the late 1550s, BAM, MS G 121 inf., a treatise on method between mathematics and logic, and P. Ventrice, 'Ettore Ausonio matematico dell'Accademia veneziana della Fama', in: *Ethos e cultura: Studi in onore di Ezio Riondato*, ed. Ezio Rionato, Padua: Antenore 1991, pp. 1135–54.

[13] I. MacLean, *Episodes in the Life of the Early Modern Learned Book*, Leiden: Brill 2020, p. 101.

the end of his life, Pinelli had become savvy enough as an empirical actor to extensively annotate an artillery treatise or to evaluate Tycho Brahe's proposal to move his astronomical research to the lagoon.[14] This chapter naturally presupposes the consolidated notion of a centrality of Pinelli's archive in the European republic of letters, but shifts the attention away from prosopography and book trading, that is, from the supplementary activities of scholar-agents, as well as from most of Pinelli's merits and accomplishment as a collector.[15] It concentrates, in its stead, on the 'atomic' level of scholarly practice, namely, assemblages, workshops, marginalia, and so on. This approach sheds a better light on the role of mathematical practice, and reveals how science, print culture, and antiquarianism have remained mutually exclusive in Pinelli's case. In the majority of occasions, historians wrote of his library as an impressive yet static arsenal of information that was mobilized by diverse actors, not as a site which exercised and negotiated its agency, identity, and power.

Pinelli's contemporaries, either patrons or collaborators, flocked to his house to write, not read, books. This explains why the library, such as we have it in Milan, deploys more than one *telos*. And while mathematics is something productive, not only reflective, of certain social and material realities, the discipline operated beyond academic boundaries and was a 'minor' pursuit re-collected, through scribal *desiderata*, alongside a larger concern for ethnography, religious dissent, or warfare.[16] As a result, this chapter also assumes that Pinelli's collection was more heterogeneous and politically dangerous than we might otherwise think, and especially that it functioned as a scientific clearing house before the actual cleansing occurred thanks to Cardinal Borromeo's entourage.[17] Some fragments of this earlier stage of the archive emerged—for example, due to the Arabic grammar by Ibn al-Hajib, which was produced for the *Typographia Medicea* in 1592 and later resurfaced at the National Library in Rome as an eccentric Oriental piece owned by Pinelli but not included in the Ambrosian purchase.[18] No one better than Venetian officials was aware of Pinelli's importance:

[14] For the first example, see BAM, S 85 sup., ff. 255^r–275^r, with Pinelli's notes on the military architect Carlo Teti (1529–1589); on Tycho, see Luisa Pigatto, 'Tycho Brahe and the Republic of Venice: A Failed Project', in: *Tycho Brahe and Prague: Crossroads of European Science*, ed. J. R. Christianson et al., Frankfurt: H. Deutsch 2002, pp. 187–202, and Giovanni Pizzorusso, 'Francesco Ingoli: Knowledge and Curial Service in 17th-Century Rome', in: *Copernicus Banned: The Entangled Matter of the anti-Copernican Decree of 1616*, ed. N. Fabbri and F. Favino, Florence: Olschki 2018, p. 160.

[15] As P. N. Miller, *Peiresc's Europe: Learning and Virtue in the Seventeenth Century*, New Haven, CT: Yale 2000, p. 20, writes, Gassendi spoke of a transfer of Pinelli's 'heroic qualities' onto his successors.

[16] For Pinelli's complementary gathering of sensitive materials, see J.-M. Philo, 'English and Scottish Scholars at the Library of Gian Vincenzo Pinelli (1565–1601)', in: *Renaissance and Reformation* 42 (2019), pp. 51–80, and 'Henry Savile's Tacitus in Italy', in: *Renaissance Studies* 32 (2017), pp. 687–707.

[17] A synopsis of the *Bibliotheca Pinelliana* after the death of its owner in 1601 has three stages. (i) Five inventories are drawn, until 1608; three of which are partial, all intersecting Paolo Gualdo's memories. (ii) After the heirs decide to transfer the library to the family castle in Naples, Turkish pirates sink the cargo off the coast of Ancona; 11 of 33 crates made in Padua are lost, including all the mathematical instruments, while parchment pages float for days until the fishermen use them to bolster their windows. (iii) In 1602, emissaries of Cardinal Federico Borromeo barely defeat the two competing bidders—the Jesuits of Naples and the Duke of Urbino—and acquire what remains of Pinelli's set, which is then shipped to Milan, via Genoa; within this new arrangement, Borromeo privileges the manuscripts, with lasting consequences on our scholarship.

[18] See Nuovo, 'The Creation and Dispersal', p. 53, R. Cassinet, 'L'aventure de l'édition des Éléments d'Euclide en arabe par la Société Typographique Médicis vers 1594', in: *Revue Française d'Histoire du*

after his death, they confiscated all the papers and carefully sifted them to decide what could have been brought back to the general public.[19]

Shifting the priority from extracting contents to interrogating the logic embedded in the collection allows one to see Pinelli's archive less as a mirror and more like a chessboard of cultural options.[20] Arguably, Pinelli saw a convergence of his note-taking techniques towards the humble, ordinary labour of double-entry bookkeeping, herbals, and pharmaceutical ledgers. Such a concourse of high and low reinforced the scholarly appeal around these techniques. The rest of this chapter performs three main functions. First it presents a series of examples and definitions that become crucial tools for rethinking existing approaches to record-keeping.[21] Secondly, it sketches the contextual frameworks within which Pinelli's mathematical and scientific collections must be set. Thirdly, it describes how we could interpret archival documents as taking their bearings from the agency of their owner, rather than from a storage. The challenge of these tasks lies in the nature of the evidence being primarily scribal. Beyond that, we are also unaccustomed to seeing Pinelli's papers or transactions as anything but a record of encounters, rivalries, and collaborations; their enduring value rests in the information they combine and select, not contain. As in the library of Sir Robert Cotton (1571–1631),[22] another late Renaissance collector, the truly creative dynamics in Pinelli's circle are pivoted on the sharing of quality transcriptions among co-investigators.

Aristotelian bricolage

It is ironic, or perhaps symptomatic, that despite the emphasis placed on a mathematical *instauratio* in the Veneto, or the philological restitution of key texts calibrated to end in the printing press, the discipline's revival is punctuated by a survival of handwritten ephemera or manuscript drafts. Our view of indexing and practical scribblings is dominated by a learned and obsessively bibliophilic preference. We see these tools and methods as a mitigating complement to the late sixteenth-century explosion of new printed editions, or as a document of social interactions whose archival usefulness is *both* limited in time *and* subordinated to a specific readerly task such as ecclesiastic censorship. Yet, although Pinelli, like a wealthy patrician, avoided personal mingling in the smelly stalls of fisheries and spice warehouses, mercantile

Livre 62 (1993), pp. 5–51, and C. Reimann, "Ferdinando de" Medici and the *Typographia Medicea*', in: *Print and Power in Early Modern Europe, 1500–1800*, ed. N. Lamal et al., Leiden: Brill 2021, pp. 220–38.

[19] Paul F. Grendler, *The Roman Inquisition and the Venetian Press, 1540–1605*, Princeton, NJ: Princeton University Press 1975, p. 288.

[20] Alexandra Walsham, 'The Social History of the Archive: Record-Keeping in Early Modern Europe', in: *Past and Present*, Supplement 11 (2016), pp. 9–48.

[21] Compared with the case studies of Hunter, *Archives of the Scientific Revolution*, particularly Boyle and Huygens, the general effect with Pinelli is not the pruning of magic or 'superstitious' interests that, some editors felt, were at odds with the serious gentlemanly ethos of the theoreticians, but a flattening of Pinelli's actual proficiency in a virtuoso's spectrum and the effacing of a style of reading and writing that was truly collective and already leading to an ideal, 'many-headed' natural history.

[22] I agree with Kevin Sharpe, *Sir Robert Cotton, 1586–1631: History and Politics in Early Modern England*, Oxford: Oxford University Press 1979, pp. 48–83, that the impetus behind collecting, in this age, is utilitarian, not a disinterested concern for scholarship.

and artisanal contexts in his writing always compete for space alongside forms of humanist training.[23]

Pinelli saw himself as someone entirely absorbed by the type of rough immediacy that derived from the epistemic field of Aristotelian *historia*,[24] with the important difference that assembling data was to him a worthy goal by itself, and already theory-laden,[25] not a temporary repository to allow the gradual elaboration of theory. He pursued collaborative ideals. Often, though, even those trusted correspondents who outranked him in Venetian society, like Alvise Mocenigo, saw the blunt side of his natural impatience.[26] For these and multiple other reasons, we should be wary of operating with Pinelli as in a system of centralized, closed relationships; and it would be reasonable to see him as a multisided, Zilselian man more than a fully Gesnerian type, with whom he still shared several traits.[27]

Pinelli was surrounded by, and transcribed from inside his household, an enormous number of specimen lists, lectures, or experimental reports. These were items where a distinction between materiality and textuality is not always clear-cut.[28] Mathematics, too, was part of these 'centrifugal' forces at play within Pinelli's collection. We know that hunting there was prized among fellow savants[29] or that he may have

[23] This is why Pinelli's papers are neither a classic example of an aristocratic diary, for which see Christiane Neerfeld, *'Historia per forma di diaria': La cronachistica veneziana contemporanea a cavallo tra il Quattro e Cinquecento*, Venice: Istituto Veneto di Scienze, Lettere ed Arti 2006, nor a simple *zibaldone* of the kind found on shop counters, an example of which is discussed by Suzanne B. Butters, *The Triumph of Vulcan: Sculptors' Tools, Porphyry, and the Prince in Ducal Florence*, Florence: Olschki 1996, vol. 2, pp. 454–9.

[24] Long understood, and perhaps minimized, as a mere building block or preliminary stage of the requirements of a 'philosophical' natural history, the Renaissance genre of *historia*, pursued by many actors in Pinelli's inner circle, already had stable epistemic and material characteristics; see Gianna Pomata and Nancy G. Siraisi (eds), *Historia: Empiricism and Erudition in Early Modern Europe*, Cambridge, MA: MIT Press 2005.

[25] Silvia Manzo, 'Probability, Certainty and Facts in Francis Bacon's Natural Histories', in: *Skepticism in the Modern Age*, ed. José R. Maia Neto, Gianni Paganini, and John Christian Laursen, Leiden: Brill 2009, pp. 123–38, persuasively argued that natural data, collected in Latin, are not free and detached from theory; see also Dana Jalobeanu, 'Core Experiments, Natural Histories and the Art of *Experientia Literata*: The Meaning of Baconian Experimentation', in: *Society and Politics* 5 (2011), pp. 88–103.

[26] Many of the 105 letters sent by Pinelli to Mocenigo, preserved in BAM, MS G 272 inf. and covering the period 1570–98, betray signs of impatience and hurry, also visible in the handwriting. This is the case, in particular, of an exchange regarding Moleto and astronomical ephemerides (f. 139^{r-v}), in reaction to a printed edition of 1582, but the examples can easily multiply. Remarkable in this manuscript, which I am studying elsewhere, is also the systematic juxtaposition of high and low, a planetary debate and the purchase of socks in Flemish wool (f. 61^{r}) or the mention of barrels of salted fish together with Thomas Savile's notes on Polybius (f. 41^{r}).

[27] The book inventory of 1604 registers two copies of Gessner's prohibited *Bibliotheca Universalis*, which however we do not possess and evidently did not survive the religious scruples of Pinelli's daughter-in-law; Nuovo, 'The Creation and Dispersal', p. 61.

[28] V. Pugliano, 'Specimen Lists: Artisanal Writing or Natural Historical Paperwork?', in: *Isis* 103 (2012), pp. 716–26.

[29] As is witnessed by Paolo Gualdo, *Vita Ioannis Vincentii Pinelli, Patricii Genuensis*, Augsburg: Markus Welser 1607, p. 27: 'Non adeo facile aditus praebebat ad hanc suam bibliothecam inspiciendam. Iis vero, qui illaudata periergia ducti, nullius pretii homines, ilam videre affectassent, omnino interclusam volebat: sufficiere dictitans commonstrari illis imagines illustrium virorum, quarum decurias aliquot in domesticis conclavibus asservabat'. For brokerage and display of collections, see Barbara Furlotti, 'Connecting people, connecting places: antiquarians as mediators in sixteenth-century Rome', in: *Urban History* 37 (2010), pp. 386–98.

inspected a now-rare copy of Diophantus'*Arithmetica*,[30] but the bulk of his daily mathematical transactions extended from cosmology and astronomy to the art of town fortification, as from optics and music to foreign military dispatches and even cryptography.[31] These topics were not dealt with in depth. Pinelli's method could be described as miscellaneous as a result of its outer filing, and beyond labelling his sensibility truly foregrounded the 'incremental fact gathering' that is associated with this category of material artefacts.[32] Our evidence, however, does not support the claim that this intellectual group encouraged a focused yet disjointed form of attention. It must be stressed that Pinelli and his friends did not see what they did as unruly in terms of philosophical method. Rather, they saw a Peripatetic science of nature that was so open and engaging that even experimental activity or the alchemical laboratory were perceived as beneficial to it.[33] In some ways, this Aristotelian bricolage was unique to the Veneto, with its humanist conception of geomorphology, local traditions, and antiquarian studies,[34] while in other, no less robust, aspects the communicative strategies were not self-contained and there was a considerable overlap with Continental European trends.

Moleto's calendrical booklet, from which we started here, was guided by a principle of *trust in tradition* that we see at work elsewhere in Pinelli and in his papers on the same subject. A perfect companion to this piece is another booklet, which contains notes on the calendric reform written by Matteo Macigni (*c.*1510–1582).[35] Macigni was an active mathematician and collaborated with Moleto on the aforementioned *Tabulae Gregorianae*; he enjoyed a structured readership inside the school

[30] Wolfenbüttel, Herzog August Bibliothek, MS Gud. Gr. 1.

[31] The note contained in BAM, MS R 110 sup., ff. 306^r–321^r, consists of the inventory of the private library of the obscure mathematician Agostino Amadi, which included 200 manuscripts, around 1500 printed books, and musical instruments; Amadi's collection is also cited in the civic travelogue of Francesco Sansovino, *Venetia città nobilissima et singolare*, Venice: Sansovino 1581, f. 258^r. In 1588, Amadi wrote a treatise on ciphers that emphasized its algebraic underpinnings: Piero Lucchi, 'Un trattato di crittografia del Cinquecento: le Zifre di Agostino Amadi fra cultura umanistica e cultura dell'abaco', in: *Matematica e cultura*, ed. Michele Emmer, Milan: Springer Italia 2004, pp. 39–50, Hannah Marcus and Paula Findlen, 'Deciphering Galileo: Communication and Secrecy before and after the Trial', in: *Renaissance Quarterly* 72 (2019), pp. 953–95. To an extent, Pinelli's interest in Amadi responds to a social enlargement of Renaissance professions, including the master of ciphers: I. Iordanou, 'The Professionalization of Cryptology in Sixteenth-Century Venice', in: *Enterprise & Society* 19 (2018), pp. 979–1013.

[32] On miscellanies, see R. Yeo, *Notebooks, English Virtuosi, and Early Modern Science*, Chicago: University of Chicago Press 2014, pp. 37–68, and A. Vine, *Miscellaneous Order: Manuscript Culture and the Early Modern Organization of Knowledge*, Oxford: Oxford University Press 2019.

[33] This is especially recognizable in Pinelli's extensive patronage of Ettore Ausonio (*c.*1520–1570), who combined optics, mathematics, and Lullian alchemy, and of whom Pinelli preserved, among many other files, an experimental diary from the 1540s, enclosed in a Roman missal and numbered as a result of heavy study (BAM, MS Q 120 sup., ff. 42^r–143^v). In a rare, autobiographical vignette Pinelli described himself humbly as a student of the Aristotelian commentary tradition (BAM, MS G 69 inf., f. 212^v), which in that context approximates being a student of the 'secrets of nature'; historians, in particular scholars of Della Porta, have erected too sharp a boundary between natural magic and Plinian 'catalogues', as if to suggest that one tradition was meant to supplant the other, when in fact they were coextensive. See Brian W. Ogilvie, *The Science of Describing: Natural History in Renaissance Europe*, Chicago: University of Chicago Press 2006.

[34] In the inaugural talk of a cycle of private lectures on Aristotle's meteorology, which took place in the summer of 1564, the philosopher Francesco Piccolomini suggested that citizens of Venice should study the 'chemical arts' of the weather because of their city's configuration, sitting within all natural elements (BAM, MS D 138/1 inf., f. 2^r).

[35] BAM, MS R 113 sup., ff. 28^r–34^v.

of philosophy at the University of Padua but he also participated in two important academies of his day, the Accademia degli Animosi, founded in Bologna in the late 1540s, and the Accademia degli Infiammati, operating in the same years but in Padua, whose cultural programme may be described as a blend of Petrarch and Aristotelianism.[36] Most importantly, Macigni was a private collector, with an extensive range of Greek manuscripts,[37] which adds yet another reason why he was one of Pinelli's privileged collaborators.[38] In his rejection of Scaliger and recent Protestant scholarship on the calendar Macigni insists on how a long chain of well-founded beliefs ought to be a necessary foundation.[39] And it is striking for the mathematical culture of Pinelli's circle that every research agenda or innovation, including those contradicting an Aristotelian *scientia*, is anchored by an ideal of epistemic modesty.

Inside Pinelli's archive

Many patricians of Pinelli's time or lineage shared his enthusiasm for the mathematical arts. Some of them became collectors, while others became patrons; others still made an extra step by translating unavailable textbooks from Greek. This is the case of Francesco Barozzi (1537–1604), one of four main teachers of mathematics in Padua before Galileo's tenure.[40] Barozzi singlehandedly spurred a European revival

[36] Heikki Mikkeli, 'The cultural programmes of Alessandro Piccolomini and Sperone Speroni at the Paduan Accademia degli Infiammati in the 1540s', in: *Philosophy in the Sixteenth and Seventeenth Centuries: Conversations with Aristotle*, ed. Constance Blackwell and Sachiko Kusukawa, London: Routledge 1999, pp. 76–85.

[37] Macigni's extraordinary rare catalogue is recorded in Giacomo Filippo Tomasini, *Bibliothecae Patavinae manuscriptae publicae et privatae*, Udine: Schiratti 1639, p. 109, a publication financed by the Roman *nunzio* in Venice; on this aspect of Pinelli's activity and the importance of mathematics, see Angela Nuovo, 'Gian Vincenzo Pinelli's Collection of Catalogues of Private Libraries in Sixteenth-Century Europe', in: *Gutenberg-Jahrbuch* 82 (2007), pp. 129–44, and the seminal article of Jean Irigoin, 'Les ambassadeurs à Venise et le commerce des manuscrits grecs dans les annees 1540–1550', in: *Venezia centro di mediazione tra Oriente e Occidente (secoli XV–XVI). Aspetti e problemi*, ed. H. G. Beck, M. Manoussacas, and A. Pertusi, Florence: Olschki 1977, pp. 399–415.

[38] Apart from Macigni's epistolary opinion on the calendar and the *Index librorum* taken from his personal collection, Pinelli kept four letters in his personal archive: three from the period 1572–76 in BAM, MS S 109 sup., and a fourth one, of 11 December 1576, in BAM, MS R 94 sup., ff. 169^{r}–170^{v}, which features in a fascinating mathematical collage. In that manuscript, to simplify matters and without elevating the empirical succession of 'building blocks' to the codex's overall rationality, we can clearly see Pinelli's desire (i) to assemble official state documents of the Republic of Venice along with theoretical excerpts from an unpublished treatise by Giuseppe Moleto himself, and (ii) to juxtapose the replies of Macigni to his queries in 1576 first with annotations from Euclid on how to use hydraulic machines (f. 171^{r-v}) and then with a disciplinary division of Euclidean geometry that was probably intended as a supplementary aid to academic teaching (f. 173^{r-v}). In addition, Pinelli's papers also preserve Macigni's unpublished astronomical treatise, entitled *Theorica orbium et motuum caelestium*, which occupies roughly sixty folio pages in BAM, MS B 274 suss. (incidentally, a call number suggestive of Borromeo's intervention). On this genre, see Adam Mosley, 'Objects of Knowledge: Mathematics and Models in Sixteenth-Century Cosmology and Astronomy', in: *Transmitting Knowledge: Words, Images, and Instruments in Early Modern Europe*, ed. Sachiko Kusukawa and Ian MacLean, Oxford: Oxford University Press 2006, pp. 193–216.

[39] The key passage is 'la lunga esperienza ci ha fatto vedere' (BAM, MS R 113 sup., f. 32^{v}). It is customary in our secondary literature to emphasize the ecumenical character of the meetings in Pinelli's household, which was friendly to the Jesuits, although eclectic as an information-gathering organ: see Aldo Stella, 'Galileo, il circolo culturale di Gian Vincenzo Pinelli e la *Patavina libertas*', in: *Galileo e la cultura padovana*, ed. Giovanni Santinello, Padua: Cedam 1992, pp. 307–25.

[40] See P.L. Rose, 'A Venetian Patron and Mathematician of the Sixteenth Century: Francesco Barozzi (1537–1604)', in: *Studi veneziani* 1 (1977), pp. 119–78, Marjorie Nice Boyer, 'Pappus Alexandrinus', in:

of Proclus.[41] And like the philosopher Francesco Patrizi, he should be seen as a typical product of that colonial and Adriatic society of Venice's *stato da mar* whose contribution to the rise of a 'mathematization of nature' as a shared scientific paradigm is still dimly understood.[42] In 1629, the Earl of Pembroke, William Herbert, bought Barozzi's Greek collection for the Bodleian Library, and those valuable manuscripts, mostly commissioned in Cyprus and displaying a deft interplay of text and diagram within the picturing space of the scribal *mise en page*,[43] disappeared once and for all from Padua's 'local' knowledge. Before this dizzying geographical shift however, and after two decades of quiet training in the Venetian islands, Barozzi had used many samples to accompany his bid for a tenure and found in the successful state-funded position solid grounds for optimism about his Euclidean project.

Barozzi had many supporters and counterparts in the lagoon. Some of them are still demanding attention from historians, like senator Giacomo Contarini (1536–1595), who was a Commissioner of the Arsenal and one of the main mathematical authorities in and beyond the shipyard and naval depot of Venice, while the fame of others, including Giulio from Thiene (d.1588), faded over time. Contarini was engaged in the promotion of the mechanical arts and supervised an actual workshop of instruments next to today's Palazzo Grassi, built in the eighteenth century, in Campo San Samuele (Figure 10.1), one of the few squares in the city with an unimpeded access

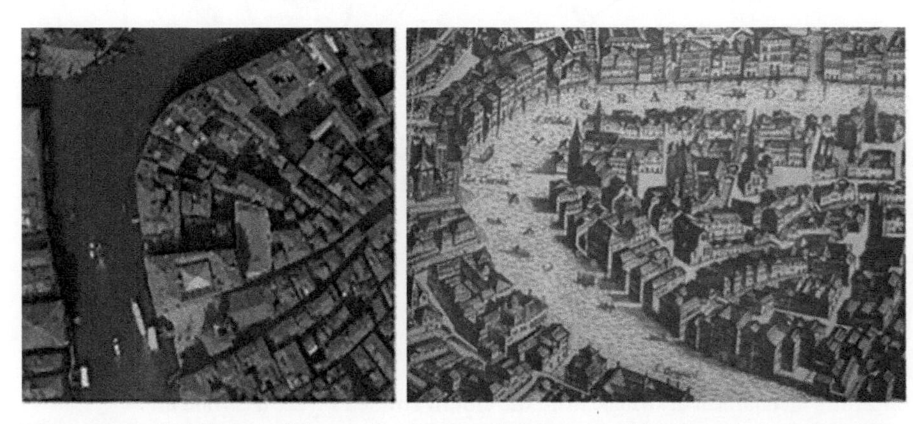

Figure 10.1 Aerial view of the mechanical workshop owned by Giacomo Contarini in Campo San Samuele, Venice.

Catalogus Translationum et Commentariorum, ed. P. O. Kristeller and F. E. Cranz, Washington DC: Catholic University of America Press 1969, vol. 2, pp. 205–13, and W. R. Laird, 'The Scope of Renaissance Mechanics', in: *Osiris* 2 (1986), pp. 43–68.

[41] Barozzi's 1560 Latin translation of Proclus' *Commentary to Euclid* was a masterpiece of scholarship; G. Claessens, 'Imagination as Self-Knowledge: Kepler on Proclus' *Commentary on the First Book of Euclid's Elements*', in: *Early Science and Medicine* 16 (2011), pp. 182–3.

[42] See David Holton (ed.), *Literature and Society in Renaissance Crete*, Cambridge: Cambridge University Press 1991, and Alberto Bardi, 'Scientific interactions in colonial, multilinguistic and interreligious contexts: Venetian Crete and the manuscript *Marcianus latinus* VIII.31 (2614)', in: *Centaurus* 63 (2021), pp. 339–52. On Patrizi, see Stefano Gulizia, 'Francesco Patrizi da Cherso and the anti-Aristotelian tradition: interpreting the *Discussiones Peripateticae* (1581)', in: *Intellectual History Review* 29 (2019), pp. 561–73.

[43] Barozzi's notes on Aristotle's *Physics*: Oxford, Bodleian Library, MS Barocci 79, f. 9^r.

to the Grand Canal. As for Thiene, he was a military man and also captain of mercenaries, a figure altogether similar to the engineer Giulio Savorgnan (1510–1595), who was in charge of military defences.[44] Barozzi was naturally associated with this expertise, as was Pinelli, and received an elliptical compass designed by Thiene,[45] which may not have differed much from the sketches drawn by Contarini himself for an assortment of gilded mathematical instruments that we possess thanks to a manuscript of the Bodleian (Figure 10.2). Both Robert Goulding and William Poole have commented that Henry Savile (1549–1622) began his mathematical career with a strong theoretical bent, but that a 'sympathy for the practical' took progressively hold of his attention in the later years.[46] Given Savile's own acquaintances in Padua and Venice, our evidence encourages us to view Pinelli as one inclined to a similar conflation of speculative and practical mathematics. For sure, optics, gnomonics, geography, and navigation are all well attested in Pinelli's papers.[47] But one notices,

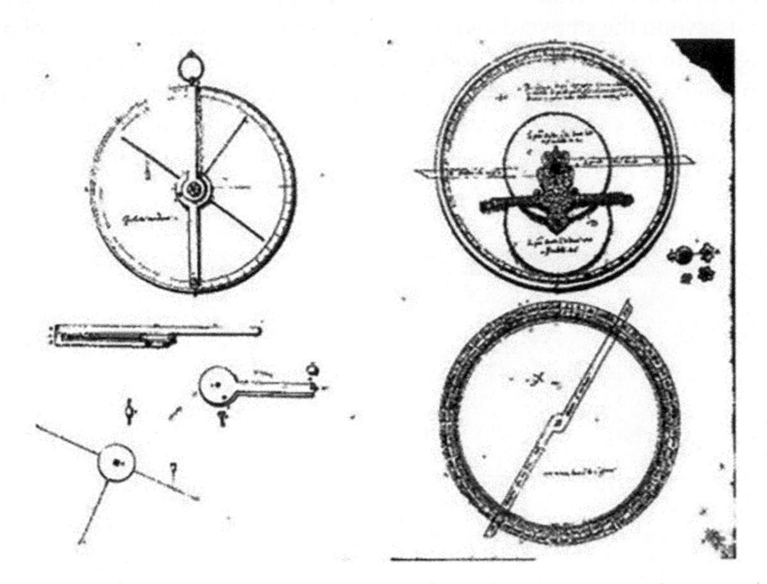

Figure 10.2 Visual elaboration, with mirroring of two different pages, of Giacomo Contarini's mechanical drawings, from the album entitled *Figure d'istromenti matematici* (*c*.1590).

[44] A. Manno, 'Giulio Savorgnan: machinatio e ars fortificatoria a Venezia', in: *Cultura scienze e tecniche nella Venezia del Cinquecento: Atti del convegno internazionale di studi Giovanni Battista Benedetti e il suo tempo*, ed. Antonio Manno, Venice: Istituto veneto di scienze, lettere e arti 1987, pp. 227–45; W. Panciera, 'Giulio Savorgnan e la costruzione della fortezza di Nicosia (1567–1570)', in: *La Serenissima a Cipro*, ed. E. Skoufari, Rome: Viella 2013, pp. 131–42.

[45] Nuovo, 'Pinelli's Collection of Catalogues', p. 141.

[46] See Robert Goulding, 'Numbers and Paths: Henry Savile's Manuscript Treatises on the Euclidean Theory of Proportion', and William Poole, 'The Origin and Development of the Savilian Library', in: *Reading Mathematics in Early Modern Europe*, pp. 33–61, 167–91.

[47] For navigation, one of the best examples known to me in Pinelli's archive is BAM, MS D 332 inf., ff. 5^r–6^r, which contains a brief 'Discorso intorno alla materia de' legnami', with annotations on the necessity of wood in Venice and riverine transport through barges in the Adige and Brenta; in terms of methods, Pinelli's notes are culled from Polybius and Hero of Alexandria, among others, and he shows a typical

as well, that while a fair number of his collaborators tend to look at technical and engineering matters as something of a notorious duty within the framework of what the Most Serene Republic asked of its city elite, Pinelli maintains overall a firm Aristotelian schedule. He builds and improves upon the medieval genre of the *quaestio*,[48] which he probably perceives primarily as a way to focus on specific passages and eschew cover-to-cover reading, but whose 'dialogic' orientation is turned inside out—refracted, as in a Dellaportian optical experiment, in the epistolary culture of Pinelli's circle.

This is how *and* why, therefore, Pinelli was often capable of generating a kind of 'galaxy' of multivocal responses built around discrete segments of textual knowledge, which were presented in the usual order of a given textbook but habitually avoided too strict a separation between print and scribal domains. The mathematical or natural historical information that was conveyed differed to some extent both from traditional, philological *adversaria*,[49] which under normal circumstances were directly attached to the text from which they were produced or else separated only by paper inserts,[50] and from *commonplace notes*, which were copied separately and sorted under thematic or topical headings made by an individual user to facilitate retrieval.[51] As a blueprint for collaborative projects based on the art of excerpting, sharing and storing, Pinelli's note-taking displays some uncanny similarities with the Greek style of 'problem-solving',[52] which in turn was epitomized by the *Problems*, a collection of dubious authenticity within the Aristotelian corpus, but genuinely appreciated and studied in depth in Pinelli's circle, almost as a premise of the epistemic movement.[53]

Unsurprisingly, Pinelli's friends took inspiration from him on multiple occasions. Some declared their debt by using the appropriate window in the book trade and some, like Galileo and Peiresc, eulogized the Paduan virtuoso only in

tendency to organize this 'file' synoptically, with something tabular, resembling yet not entirely identical to Ramist products. It is difficult to extract Pinelli's opinion on Ramus. A tantalizing note in a MS where Niccolò Tartaglia's view of the art of fortification is followed by Pinelli's notes on Plutarch and Bodin, mentions him, suggesting something worthy of praise, but the obscurity of the context is too serious to be conclusive: 'fortasse inter opera P. Rami [...] laudari possit' (BAM, MS I 186 inf., f. 184^{v}).

[48] Brian Lawn, *The Rise and Decline of the Scholastic 'Quaestio Disputata', With Special Emphasis on Its Use in the Teaching of Medicine and Science*, Leiden: Brill 1993.

[49] Liv Ingeborg Lied and Marilena Maniaci (eds), *Bible as Notepad: Tracing Annotations and Annotations Practices in Late Antique and Medieval Biblical Manuscripts*, Berlin: De Gruyter 2018, pp. 1–9.

[50] One should consider that extensive sections of Pinelli's own copy of the Aldine edition of Aristotle, in Greek, were systematically interfoliated to facilitate annotations and their sharing; e.g. BAM, INC 374/8 (with intense work on *De partibus animalium*).

[51] Ann Blair, 'Note-Taking as an Art of Transmission', in: *Critical Inquiry* 31 (2004), pp. 85–107.

[52] It is undeniable, based especially on Pinelli's marginalia to the Aristotelian corpus preserved in his Aldine edition of the 1490s, that he was particularly responsive to the medical side of the Greek problems; see Robert Mayhew (ed.), *The Aristotelian* Problemata Physica*: Philosophical and Scientific Investigations*, Leiden: Brill 2015.

[53] Since the topic is vast, I limit myself to two examples. (i) BAM, MS M 41 sup. is a beautiful Greek codex owned by Pinelli, which also contains the spurious *Problemata* of Alexander (ff. 130^{r}–140^{v}). (ii) BAM, MS I 231 inf., already mentioned in note 4, is a miscellany organized by Pinelli and supposedly quite close to the material state where he wanted things to be; at some juncture, Pinelli writes a terminological comparison between Euclid and Proclus (f. 120^{r}) and he proceeds with a different philological *collatio*, this time concerning Aristotle's collection of *Problems*, for which he makes a list of variants taken from a 'codex Basiliensis' (ff. 121^{r}–122^{v}). In short, (i) is part of Pinelli's elite culture of collecting, and (ii) shows how he also envisaged a constructive role for the 'problem' as a unit of scientific assembling.

hindsight,[54] depicting an ultimately ambiguous frame—perhaps drawing from nostalgic memories—as if to suggest that the key significance of the library had been to serve as a postal relay for itinerant *érudits* such as they saw themselves at the time. Or better still, such as the dynamic scale of intellectual life in the Italian peninsula was in the 1620s and 1630s, when things slowly moved out of Venice into the Transalpine world, the Levant, as well as towards Florence and Rome.[55]

By 1600, as we will shortly see, all bookish or scientific purveyors who had any intellectual commerce with Pinelli took in obvious consideration that solicitation by him implicitly carried within itself a polite request for compliance with this idiosyncratic filing system.[56] Antiquarian traditions and protocols per se cannot entirely do justice to the cognition of Pinelli's archive; in a fundamental way the collection was adapting to science and philosophy the political storage which patricians in the Veneto favoured.[57]

The historical actors who are registered in Pinelli's archive, including many mathematical practitioners, were part of a courtly culture marked by elaborate rituals, but inside the warm, intellectual gatherings of that unique Paduan household,[58] instead of co-opting social power, they were seemingly persuaded to leave their traces with prudent and minimalistic irony. What we possess is, mainly, their 'scribal self'. Pinelli himself was not a prolific correspondent, not at any rate in comparison with men like Henry Oldenburg (1619–1677) or Samuel Hartlib (*c*.1600–1662), who refigured the Comenian ideal by taking upon themselves a role of central intelligence with Continental natural philosophers.[59] Yet, it must be conceded, Pinelli was sufficiently well versed in this mode of communication to convince everyone to speak like confidential

[54] In the former case, the pharmacist Bartolomeo Maranta dedicated a book to Pinelli in 1559, reflecting his lifelong passion for botany and, incidentally, the only disciplinary setting in which it was socially acceptable to credit Pinelli in public: B. Maranta, *Methodi cognoscendorum simplicium libri tres*, Venice: Valgrisi 1559; for Galileo's well-known elegiac feeling towards his Paduan years, it is worth considering the vast uncertainty it generated in our historiography: see William A. Wallace, 'Randall Redivivus: Galileo and the Paduan Aristotelians', in: *Journal of the History of Ideas* 49 (1988), pp. 133–49; Nick Wilding, *Galileo's Idol: Gianfrancesco Sagredo and the Politics of Knowledge*, Chicago: University of Chicago Press 2014, pp. 28–9, notes that the strange fate of Pinelli's circle, as the centrepiece of Paduan intellectual brokerage, was to instigate its own eclipse.

[55] Peter N. Miller, *Peiresc's Mediterranean World*, Cambridge, MA: Harvard University Press 2015, pp. 320–1.

[56] Not all the 'epistemic genres' handled or mediated by Pinelli were codified in college. We can presume that when the philosopher Federico Pendasi offered his thoughts, in a letter, on the difference between Aristotle and Plato (BAM, MS D 226 inf., ff. 117^{r}–122^{r}) he was indeed trampling on solid academic ground: a form of doxography, companion to teaching, or a *quaestio* in its own rights. But when Clusius sent bulbs from Northern Europe (as in BAM, MS R 95 sup., f. 97^{r-v}), he simply adapted his speaking persona to a botanist engaging a colleague who also attended to a garden. On idiosyncratic filing protocols, see Elizabeth Yale, *Sociable Knowledge: Natural History and the Nation in Early Modern Britain*, Philadelphia: University of Pennsylvania Press 2016, p. 9.

[57] Dorit Raines, 'L'archivio familiare strumento di formazione politica del patriziato veneziano', in: *Accademie e biblioteche d'Italia* 64 (1996), pp. 5–36.

[58] 'Unique' in distinction, but certainly not isolated in the natural philosophical culture of the time: Paolo Sarpi, one of Pinelli's key associates, was also able to combine his duties as a Servite with a fully experimental programme: Ron Naylor, 'Paolo Sarpi and the first Copernican tidal theory', in: *British Journal for the History of Science* 47 (2014), pp. 661–75.

[59] Philip Beeley, 'A Philosophical Apprenticeship: Leibniz's Correspondence with the Secretary of the Royal Society, Henry Oldenburg', in: *Leibniz and His Correspondents*, ed. Paul Lodge, Cambridge: Cambridge University Press 2004, pp. 47–73.

insiders, rather than as scholars.[60] The smallest units of knowledge circulation in Pinelli's circle, I would finally claim, can indeed be as small as citations, but are neither commonplaces nor scholia. In many respects, the correspondence topics fit the *focal points* of the patron's investigation, not clear-cut academic disciplines. We may presume that for some folder to have survived until now, it evidently had Pinelli's approval. And although he was not always able to mediate the initial contact or to be alerted without aid to the first appearance of new printed editions,[61] it was almost certainly he who supervised the record-keeping or envisaged how certain research questions could be mutually reinforcing and ought to be bound together, however random that might seem to the onlooker.

To return for now to Barozzi's perspective, the final function of his bid in Padua was not entirely within the academic discourse; in other words, he knew that his patrons were eager collectors and kept private libraries. Their auxiliary design was to house mathematical instruments alongside books in a *continuum* of hands-on preoccupations.[62] Barozzi, it turns out, knew his audience well. In fact, he was not the first 'colonial' citizen who tried this strategy. In the course of the sixteenth century, Patrizi's case is remarkable in as much as he tried, unsuccessfully, to sell a stock of Greek manuscripts in Venice which he reckoned rare enough to elicit a sizeable financial return. After the transaction in the lagoon failed, they ended up being absorbed into the collection of Philip II in Spain.[63] Patrizi fashioned himself as a master of *prisca theologia* and a Neoplatonist. It makes perfect sense that he would complement this perception by speculating on a vast folder of enigmatic titles, given that his intended Venetian recipients had long held an interest in Arabic science and letters, as well as in Byzantine matters *qua* Byzantine, that is to say, as a living heritage of Hellenism. It also became clear on the strength of this comparison that Barozzi wanted to test his intellectual network while still living outside Venice, to collect ideal subscriptions to his mathematical ideas and thus to pave the way for such publication projects as Proclus on Euclid to be realized. In this light, the new Euclidean geometry promised by Barozzi did not unfold like an abstract template; it was received in a distinctively 'cluttered' intellectual space that demonstrates the important role played by aristocratic circles like Pinelli's. The centrepiece of this system is the household economy, in its ability to shift between humanist retreat and busy workshop.

[60] Considering the importance of intercepting or replicating communication in Pinelli's library, it is important to recall here a self-representation by one of the circle's members, Sagredo, who according to Wilding, *Galileo's Idol*, p. 84, 'portrays himself as a passive, but privileged amanuensis of international news'.

[61] Once again, Pinelli's correspondence with Mocenigo, cited in note 26, reveals how the celebrated Paduan collector was often at the receiving end of bibliographical updates.

[62] For libraries as 'experimental space', where a patron could assemble data provided by practitioners, see D. Bertoloni Meli, *Thinking with Objects: The Transformation of Mechanics in the Seventeenth Century*, Baltimore: Johns Hopkins University Press 2006, pp. 14–15, and M. Henninger-Voss, 'Comets and Cannonballs: Reading Technology in a Sixteenth-Century Library', in: *The Mindful Hand: Inquiry and Invention from the Late Renaissance to Early Industrialization*, ed. L. Roberts, S. Schaffer, and P. Dear, Amsterdam: Royal Netherlands Academy of Arts and Sciences 2007, pp. 11–33.

[63] Maria Muccillo, 'La biblioteca greca di Francesco Patrizi', in: *Bibliothecae Selectae: Da Cusano a Leopardi*, ed. Eugenio Canone, Florence: Olschki 1993, pp. 73–118, and Margherita Palumbo, 'Books on the Run: The Case of Francesco Patrizi', in: *Fruits of Migration: Heterodox Italian Migrants and Central European Culture, 1550–1620*, ed. Cornel Zwierlein and Vincenzo Lavenia, Leiden: Brill 2018, pp. 45–71.

Galilean scholarship intimated that a chief innovation during his Paduan period consists in the use of his city address as a boarding house, on which, in turn, is predicated the mathematician's refusal to work as a lonely theoretician and his desire to mingle in a 'trading zone' of artisan epistemology.[64] Actually, if one chooses to study the everyday routine of Pinelli's home and cares to ignore the anticlerical bias of Antonio Favaro, for whom Pinelli was nothing more than a convenient enabler,[65] it would be plain to see that Galileo here was a follower, not a trailblazer.

In June 1580, Savorgnan thanks Pinelli for sending him metal wheels, a comedy, and a treatise addressing the borders of the Ottoman empire.[66] The cumulative effect of these items may be jarring to our eyes, although not any more than the assortment of topics in John Locke's journal,[67] just to cite a master note-taker of the early modern period, but it is indicative of what these mathematicians and practical men judged as relevant. This letter was written in Osoppo, or twenty miles north-west of Udine, where Savorgnan housed an armoury and a collection of mechanical devices; at the time, he was interested in shipbuilding and had been responsible for dozens of bastions across the Venetian territory. The exchange makes it clears that Pinelli did not take an idle or scholarly pleasure in pseudo-Aristotelian mechanics, but that he supplied heavy metal pieces and thus investigated the field by means of machines.[68] If we zoom out within the MS that preserves this bundle of letters, covering the period from the spring of 1578 to the end of 1580,[69] we also notice the reciprocal integration between the development of rigorous mechanical terminology and translation efforts.

The key historical actors, seen in context, were Guidobaldo dal Monte (1545–1607) and the traveller and polymath Filippo Pigafetta (1533–1604), who produced a vernacular version of dal Monte.[70] The *editio princeps* of the *Mechanicorum liber* appeared in 1577 in folio, but the format changed to quarto four years later for the Italian translation, which was openly constructed for a diverse and less wealthy audience.[71] Indeed, Pigafetta's dedicatory epistle in the 1581 edition highlights Savorgnan's

[64] Matteo Valleriani, *Galileo Engineer*, Dordrecht: Springer 2010.

[65] Antonio Favaro, *Galileo Galilei a Padova: Ricerche e scoperte*, Padua: Antenore 1968, p. 60, is incredibly dismissive of Pinelli.

[66] BAM, MS R 121 sup., f. 11^{r}.

[67] Yeo, *Notebooks, English Virtuosi*, p. 197.

[68] Bertoloni Meli, *Thinking with Objects*, p. 32, makes the useful point that the mechanical *esperienze* of this cohesive intellectual group are quite alien to our way of thinking.

[69] BAM, MS R 121 sup., ff. 4^{r}–24^{r}. See Antonio Becchi, Domenico Bertoloni Meli, and Enrico Gamba (eds), *Guidobaldo del Monte (1545–1607): Theory and Practice of the Mathematical Disciplines from Urbino to Europe*, Berlin: Edition Open Access 2013, and Mary Henninger-Voss, 'Working Machines and Noble Mechanics: Guidobaldo del Monte and the Translation of Knowledge', in: *Isis* 91 (2000), pp. 233–59.

[70] For more on Pinelli and Pigafetta, see Andrea Savio, *Tra spezie e spie: Filippo Pigafetta nel Mediterraneo del Cinquecento*, Rome: Viella 2020, pp. 15–67. There would be a lot to make of Pigafetta's travelogues as a counterpart of mathematical exchages in Pinelli's papers, especially the 1591 *Relatione del reame del Congo*, which essentially develops a Marco Polo device (a merchant dictating to an informer) to criticize the moral contradictions of Spanish colonialism from a viewpoint favourable to the global reach of papal Rome.

[71] A first list of professionals is in Guidobaldo dal Monte, *Le mechaniche [...] tradotte in volgare dal Sig. Filippo Pigafetta*, Venice: Francesco de' Franceschi 1581, f. 91^{r}: 'di capitani di guerra, d'ingegneri, et di qualsi voglia artefice', later incremented to 'architetti, scultori, marinari, muratori, maestri di legname, funditori d'atra bigliare, fabbri, et altri tali artefici, et ad ogni persona ingegnosa' (f. 92^{r}).

residence as a suitable and admirable framework for the tools and epistemic research carried out from dal Monte's viewpoint:

> I was also delighted to realize that his household compared to a shop floor, where all the weapons are neatly displayed in their stalls, as well as to a warehouse of military machines and pulleys, since [Savorgnan] must have designed out of his ingenuity a dozen of different shapes, made in part to roll and in part to lift heavy loads with minimal friction, like the one which only has a barbed wheel and at its full strength can pull five cannons [...] and including that other, which just with an individual push on the turning handle can move twelve thousand pounds.[72]

If Pigafetta saw Pinelli's papers on Savorgnan in MS R 121 sup., to which he may have been privy, he would have found there a drawing of a pulley,[73] which is the perfect companion to his preface. We can presume that the readers of the time had a similar perspective on Pinelli's own household despite it being refigured or fetishized as an exclusively patrician site of knowledge in our scholarship.[74] To further illustrate this point, I wish to draw attention to the fact that Pinelli's collection of mathematical plates, instructional broadsheets or lessons about an instrument were frequently more valuable than the object itself.[75] In the case of Fabrizio Mordente's reduction compass, designed to divide circles into equal parts and allow scaled representations, its instructions sold for 120 lire as against 35 for the instrument alone. Perhaps, the different price was an economical ruse to keep the cost of tuition high; this is what happened to Galileo's compass in 1606, when a pirated copy of the accompanying manual forced him to publish.[76] The expert Contarini, who was enthusiastic about Mordente's device, criticized a similar design by Moleto by speculating that soldiers were unlikely to master its intricacies.[77] This demonstration took place in Pinelli's

[72] Dal Monte, *Le mechaniche*, ff. b1ᵛ–2ʳ: 'presi anco diletto in vedere la sua abitatione essere a guisa d'una bottega d'arme politamente a suoi luoghi serbate, e un magazino di machine bellicose et da muover pesi, avendone [il Savorgnan] fabricate di sua industria forse dodici di maniere differenti, parte da strascinare et parte da alzare, con pochissima forza, smisurati pesi, come quella che ha una sola rota co' denti et all'erta tira cinque de' suoi canoni [...] et quell'altra la quale con un'oncia di forza sola, posta nel manico che la volge, da il moto a dodicimila libbre di peso'.

[73] BAM, MS R 121 sup., f. 17ʳ. On machine books, see Marcus Popplow, 'Why Draw Pictures of Machines? The Social Contexts of Early Modern Machine Drawings', in: *Picturing Machines, 1400–1700*, ed. W. Lefèvre, Cambridge, MA: MIT Press 2004, pp. 17–52.

[74] Renaissance collectors are tricky to historicize, often exposed to the failure of ignoring the other side of elite practices. Pinelli is no exception. He commissioned a Greek-style binding for some of his books: Anna Gialdini, 'Antiquarianism and Self-Fashioning in a Group of Bookbindings for Gian Vincenzo Pinelli', in: *Journal of the History of Collections* 29 (2017), pp. 19–31; but he was also immersed in the craft of making books. In a letter of July 1572, Pinelli sends Mocenigo his recipe to produce glue from flour, used by his artisan Prospero to bind books according to their format (BAM, MS G 272 inf., f. 24ʳ).

[75] In what follows, I follow the valuable account of Henninger-Voss, 'Comets and Cannonballs', pp. 18–21, but I am interested in the making of scientific knowledge as a form of paperwork, instead of comparing paper (library) with the external world (objects); a similar take is in Matthew Eddy, 'The Interactive Notebook: How Students Learned to Keep Notes During the Scottish Enlightenment', in: *Book History* 19 (2016), pp. 86–131, and Renée Raphael, *Reading Galileo: Scribal Technologies and the Two New Sciences*, Baltimore: Johns Hopkins University Press 2017.

[76] Mario Biagioli, 'Replication or Monopoly? The Economies of Invention and Discovery in Galileo's Observations of 1610', in: *Science in Context* 13 (2000), pp. 547–92.

[77] BAM, MS A 71 inf., ff. 23ʳ–28ᵛ (opinions of Contarini and Moleto).

library; and the owner looks more like someone who presides over a competition, rather than a witness. Indeed, the ubiquitous Moleto asked Pinelli many times if he could buy instruments on his behalf,[78] alluding or wishing to fulfil legal eligibility, not a collector's wishes. It is natural that Barozzi knew how these mathematical practitioners read their texts collectively, apart from the distribution via intermediaries, which was in fact common at the time. It is noteworthy that the official *imprimatur* for his Latin translation of Proclus' commentary on Euclid came from Ettore Ausonio, a mathematician in the orbit of the Accademia della Fama and strongly associated with Pinelli.[79] This is another piece of the puzzle that instills comprehensible scepticism about the recurrent view of this circle acting merely as a mediating infrastructure and not, for example, a panel with the recognized ability to settle and arbitrate scientific matters.

One of the surprising decisions taken by Pinelli regarding Barozzi was that he produced a personal transcription of his Greek juridical collection but apparently ignored the entries pertaining to rhetoric, Patristic, and of course mathematics.[80] The next two files that document contacts between the two, and Pinelli's emblematic respect for Barozzi, pertain to the same type of media—epistolary exchange—and could be treated as remaining testimonies of the maturity of the late sixteenth-century discussion about Sacrobosco's geometry and its shortcomings.

The first of these files is a set of three letters between Barozzi in Venice and an illustrious Roman Jesuit, Christopher Clavius (1537–1612), whose *Commentary on Sacrobosco's 'Sphere'*, first published in 1570, was regularly reprinted and was the first to directly name and confront Copernicus on the earth's cosmic centrality.[81] Pinelli requested a remarkably neat copy and placed the letters as an independent bundle marked with a title.[82] In the opening letter Clavius addresses Barozzi with careful admiration and suggests to him that the reputation of his book would have been greater in Rome, if he had restrained from criticizing Sacrobosco for minor 'quibbles'.[83] The response of Barozzi, while thoroughly courteous, is a long restatement of how those quibbles are really major astronomical mistakes, repeated out of tradition, and ends with the formulation of doubts regarding how the natural movement of Mercury affects the ecliptic.[84] At this point, Clavius declines further answers and sends a brief message via intermediaries.[85] Similar, in tone and thematic, is another

[78] BAM, MS S 105 sup., ff. 48^{r} and 53^{r}.

[79] Venice, State Archive, Riformatori dello Studio di Padova, busta n. 259; cited in Ventrice, 'Ettore Ausonio matematico', p. 1146.

[80] BAM, MS R 110 sup., ff. 232^{r}–236^{r}.

[81] Edward Grant, 'The Partial Transformation of Medieval Cosmology by Jesuits in the Sixteenth and Seventeenth Centuries', in: *Jesuit Science and the Republic of Letters*, ed. Mordechai Feingold, Cambridge, MA: MIT Press 2003, pp. 127–55.

[82] BAM, MS S 81 sup., ff. 255^{r}–260^{v}; Pinelli's title on f. 254^{r}.

[83] BAM, MS S 81 sup., f. 255^{v}: 'sarebbe in maggior reputatione il suo libro, senza queste bagatelle di Sacrobosco'.

[84] BAM, MS S 81 sup., f. 260^{r}, for Mercury's orbit; it is interesting to note that Barozzi, complying with an unwritten code of civility, informs Clavius that as soon as he received his printed edition, the *Cosmographia*, he sent it to the binder, as an acceptable preliminary to become an object of study, and also that Barozzi refers to his numbering of Sacrobosco's mistakes, which in his view or perhaps in his personal copy of the text are identified by the mark '+', as a visual aid to memory.

[85] BAM, MS S 81 sup., f. 260^{v}.

letter written by Barozzi in 1585 in response to one of Pinelli's cosmological doubts. This second paper is now preserved in another manuscript (Figure 10.3),[86] and it amounts to an erudite excursus on Greek authorities, perhaps to please Pinelli's taste. On the back of the original envelope, Barozzi dates his piece and shows an interesting awareness of the archival logic followed by Pinelli because attributing the label 'De circulo semper apparente apud Aristotelem' has the double effect of inscribing the discussion within the Peripatetic field and also the epistemic effect of making it sound like an effort in textual exegesis.

For Pinelli, it was natural to be contacted for the retrieval and perusal of documents that could be of public interest and even contain heterodox beliefs. This occurred, for example, during Giordano Bruno's trial.[87] In this respect, he enjoyed a notable concession by the inquisitor in Venice, who allowed Pinelli to read several prohibited copies that would be otherwise hard to obtain.[88] It should come as no surprise, then, if Pinelli kept a copy of the 1587 official document with which the Inquisition charged Barozzi with apostasy, heresy, and engagement in occult sciences.[89] What seems exceptionally intriguing is what Pinelli added immediately after Barozzi's condemnation. He assembled a transcription of Ptolemy's *Liber de Analemmate*, presumably inspired by the famous humanist and mathematician in Urbino, Federico Commandino (1509–1575). In 1562, Commandino had revised a fragmentary translation by William of Moerbeke by adding a new Latin version with a commentary.[90] As if to capture together the stereographic and inquisitorial angle of a leading, late Renaissance mathematical career, Pinelli's evidence on Barozzi presents in itself the

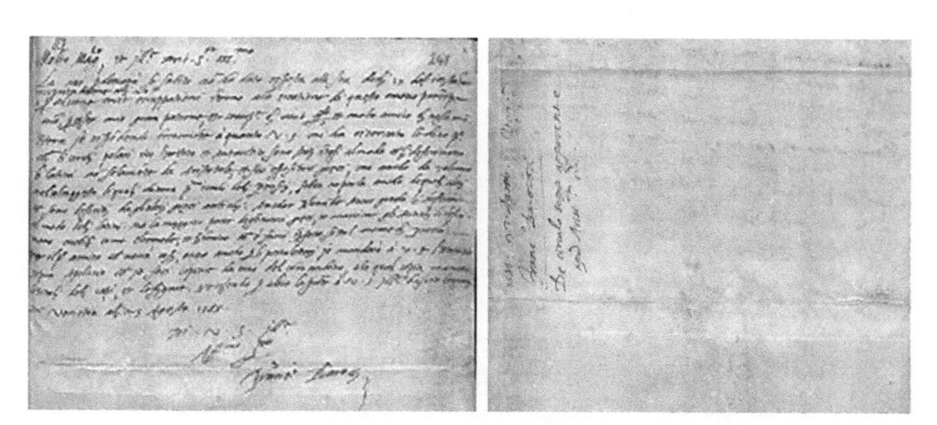

Figure 10.3 Original letter of Barozzi to Pinelli in 1585, with the cover.

[86] BAM, MS S 105 sup., f. 241^{r-v}.

[87] F. A. Yates, 'Giordano Bruno: Some New Documents', in: *Revue Internationale de Philosophie* 16 (1951), pp. 174–99, was certainly the first to observe that Pinelli was following, thanks to his correspondents, Bruno's difficult European career; see G. Giglioni, 'The "Horror" of Bruno's Magic: Frances Yates Gives a Lecture at the Warburg Institute (1952)', in: *Bruniana & Campanelliana* 20 (2014), pp. 477–97.

[88] Hannah Marcus, *Forbidden Knowledge: Medicine, Science, and Censorship in Early Modern Italy*, Chicago: University of Chicago Press 2020, p. 143.

[89] BAM, MS R 109 sup., ff. 282^{r}–287^{r}.

[90] BAM, MS R 109 sup., ff. 288^{r}–292^{v}.

clear example of an allocation of excerpts that might not follow 'heads' or keywords, as in the Latin tradition of *loci communes*, but remains thoughtful and was intended to assist memorization and recall.

Mathematical scholarship and politics

With this background in mind, we are better placed to reassess Pinelli's handling of mathematical practice, as well as the sometimes forceful and jarring filing habits that came with it. This *adjacency* should be seen as a shared, sustaining mechanism of disciplinary intersection and not just as a hazard of cultural transmission. By aligning mathematics with matters of state, but also calendrical or mercantile transactions, Pinelli is feeding his digressive curiosity.[91] He is, on the other hand, acting as a steward for a range of private users—consumers and producers alike—who pushed beyond the recognizable category of university lecturers both in terms of scholarly interests and social roots. A sharper distinction may be drawn here between those who idealized an individual investigator, often based on leisure and a wavering philosophical distrust in appearances, in some quarters, and those practitioners who valued and promoted a collective, cooperative approach. Both styles can be found across Pinelli's papers. In this context, however, there was considerable overlap between the fields.

To be sure, within the Renaissance tradition of the *cultura animi*, being a virtuoso is a remedy to different failures of self-government.[92] And yet, at the core of Pinelli's performance as a virtuoso the main preoccupation is with archival practices. Like antiquarians or natural philosophers, mathematicians were undoubtedly eager to communicate their passions to students and peers. Pinelli's collection was designed to facilitate such an exchange. Indeed, part of his papers suggest that 'mathematization of nature' was for him synonymous with a type of miscellaneous filing that could greatly benefit scientific validation. In this light, his filing system can be constructed as a corrective to dogmatism and the imperfections of the mind's distemper. Even in a transparently non-Baconian fashion, the act of making lists fights the defectiveness of human nature.

Beyond this ethical configuration, the affinity between record-keeping and instruments certainly bolstered the function of Pinelli's library as the central 'clearing house' for early modern science in the Veneto. As messy and hopelessly cumulative as a first view of the scope of learning within Pinelli's archive might be—which includes the mathematical sciences—it does not necessarily follow that historical actors were fully bewildered.

As observed earlier, mathematicians or engineers gathered at Pinelli's house with the utility of their knowledge in mind. Often, tangible results of their discussions ensued or were preserved in the shape of broadsheets, drawings or plates. Replicating was necessary for Pinelli. It also enlarged the geographical compass of his collection.

[91] Christian Coppens, 'Curiositas or Common Places: Private Libraries in the Sixteenth Century', in: *Biblioteche private in età moderna e contemporanea. Atti del convegno internazionale di Udine, 18–20 ottobre 2004*, ed. Angela Nuovo, Milan: Bonnard 2005, pp. 33–42.

[92] Sorana Corneanu, *Regimens of the Mind: Boyle, Locke, and the Early Modern Cultura Animi Tradition*, Chicago: University of Chicago Press 2011, pp. 79–113.

The court of Urbino held him in great consideration, as Pinelli's biographer, Paolo Gualdo, reports; this esteem was not empty, but generated a concrete audience in the Paduan circle for Guidobaldo dal Monte's mechanical studies. Another example is Bernardino Telesio. Despite the anti-Aristotelian polemics, Pinelli was known as an admirer of Telesian philosophy, as we know from an insert in the *Varii de naturalibus rebus libelli* published in 1590.[93] When we study the archival traces left by this solid network of intellectual relationships, we notice how scientific dialogue was stimulated by technical illustration and we also appreciate how difficult it is, for historians, to shift from the traditional template in which an important patron visits Pinelli's library and takes full advantage of its resources, to then publish something else that was informed by them, to a mode in which it is Pinelli's agency that provides the inception to the epistemic movement. Given the importance of Aristotelian meteorology in Pinelli's collection, which boasted several unpublished treatments of the subjects by established university teachers as well as free intellectuals, it is not entirely surprising to see a protracted discussion of sea tides in Pinelli's MS R 99 sup., which includes sheets of mathematical calculations and culminates with a diagrammatic image to mark the lowest reachable level during the water decrease in the lagoon (Figure 10.4). The exact same image is present in a holographic manuscript by the

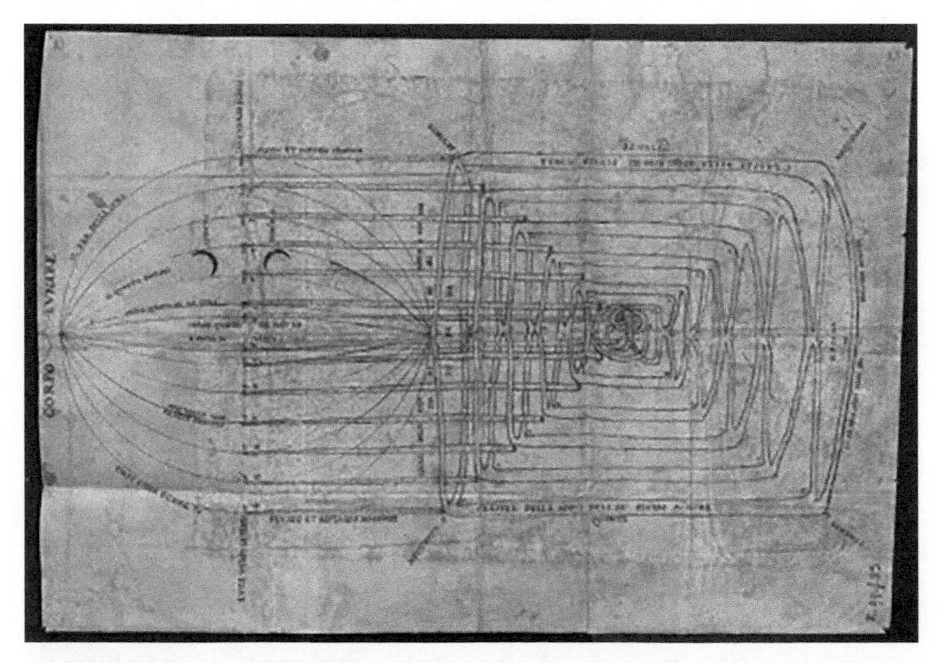

Figure 10.4 Mathematical diagram of the phenomenon called *bassa da mar*, related to sea tides.

[93] Oreste Trabucco, 'Telesian Controversies on the Winds and Meteorology', in: *Bernardino Telesio and the Natural Sciences in the Renaissance*, ed. P. D. Omodeo, Leiden: Brill 2019, pp. 96–115.

state engineer Cristoforo Sabbadino (1489–1560), which is housed in Venice.[94] This pattern of 'replication and dissemination' persisted into the Galilean period, since in the late 1590s Galileo himself copied a paper in Pinelli's collection on the optical properties of mirrors, presumably for personal use, taken from Ettore Ausonio's treatise *Theorica speculi concavi sphaerici* (Figure 10.5).[95]

In some ways, given the Aristotelian bricolage in Pinelli's circle, these practitioners were increasingly fostered (and constrained) by distinct sets of communicative practices. They looked at scribal production as service or favours that investigators could perform for each other, as well as a safe alternative to publication in printed books. In particular, mathematicians approached their work as always unfinished, always under construction. This absence of completion generated stocks of papers which came to be valued as repositories of knowledge. Indeed, Pinelli's archive is suitable to observe the recommendation of paper-based search engines across the Venetian communities of natural science and the spread of confidence on the principle that shared interests were safeguarding knowledge.

Another discernible feature of mathematical adjacency has to do with Pinelli's obvious belief that to be a good mathematician one needed to be grounded in many other disciplines. For inasmuch as information could flow through Pinelli's hands it simultaneously reminds us of the familiar image of a late Renaissance polymath and also displays multiple points of reference, as if one were browsing in a genuine storefront. In doing so, Venetian mathematical scholarship emerged as a type of complementary site of practice. In Ambrosian manuscripts such as D 173 inf. or D 178 inf. one

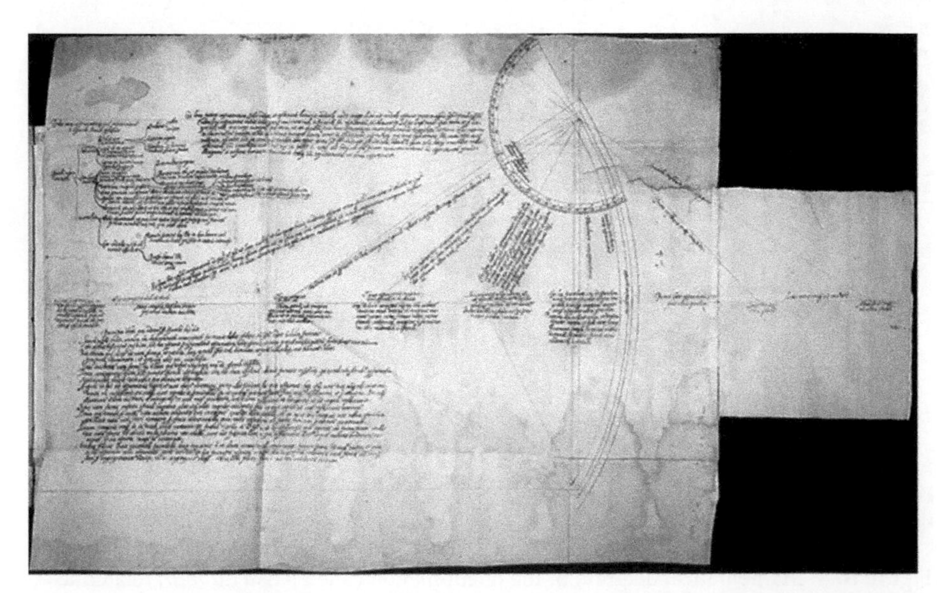

Figure 10.5 Galileo Galilei's copy of Ettore Ausonio's *Theorica speculi concavi sphaerici.*

[94] Venice, Marciana Library, MS It IV 51.

[95] Sven Dupré, 'Ausonio's Mirrors and Galileo's Lenses: The Telescope and Sixteenth-Century Practical Optical Knowledge', in: *Galilaeana* 2 (2005), pp. 145–80.

could verify the remarkable amount of geometrical problems, issues of trigonometry, and eccentric mathematical annotations which Pinelli's papers reveal. Some of these materials openly declare an academic origin and show themselves to be a transcription of 'honeste et utili speculationi', originally aired in a pedagogical context.[96] The majority of them, however, tend to emphasize that they were meant for a private motive, as with Alessandro Piccolomini's description of a mechanical design for a night watch.[97]

The most sustained combination between mathematics and other bits of collecting activity in Pinelli's archive, one may argue, concerns politics and has gone almost unnoticed until recently. In this field, Pinelli's own recollection contradicts the often-repeated assumption that, out of a shy temperament or his virtuoso's modesty, he never transcended the format of passive note-taking. Evidence of original *mésalliances* between politics and mathematics is abundant in the Ambrosian archive, but the manuscript that I first offered for scrutiny in this chapter is, in fact, a classic example of a miscellany pivoted on politics and would repay further analysis.[98]

The last four units of Pinelli's MS R 122 sup. relate to the theme of the *ragion di stato*.[99] First, we read Leonardo Salviati's 'Discorso' on Tacitus (ff. 404^{r}–410^{r}), which presents marginal glosses, and is probably copied from a printed edition. Secondly, an anonymous and brief *Consideratione di quanta importanza sia ad un Principe o ad altro Ministro Pubblico l'havere scritture di stato* (ff. 412^{r}–412^{v}), seemingly copied by Pinelli himself, or by his habitual scribe, cements the general perception that the Renaissance state has every advantage in maintaining a functional and centralized body of intelligence in writing. The next two items in this bundle are thorny as a philological document but exceptionally well suited to offer access to the reflections of Pinelli on the subject. On f. 413^{r}, we encounter a page that is developed horizontally, like a synoptic table but also like the technical drawings that conclude the hydraulic annotations on Venice's waters. To this summary, Pinelli adds a systematic title: 'to appropriately rule a state, it is necessary to gain a true cognition of its inner schemes' (*notitia vera de' i disegni*). It was a shared belief, among political neophytes, that a proper study of a state's mechanism required mathematical aptitude.

Right next to this last page, we find another summary in Pinelli's own hand (ff. 414^{r}–416^{r}), which is entitled *Alcuni capi per un trattato di ragioni di stato* and appears to be a reduction, divided in four points, of a printed treatise—possibly by Giovanni Botero whose name is nonetheless absent. Sandwiched between these points, on f. 415^{r}, there is another file that is unknown to the Ambrosian catalogues: Pinelli's unusual transcription of an independent letter, which was evidently written to answer an explicit solicitation by the collector and contains a skeptical, unflattering

[96] BAM, MS D 173 inf., fol. 14^{r}.

[97] BAM, MS D 178 inf., f. 54^{r}.

[98] Filippo De Vivo, 'How to Read Venetian *Relazioni*', in: *Renaissance and Reformation* 32 (2011), pp. 25–59, considers Pinelli as an example of the elite readership of these reports, but does not extend his valuable analysis to the logic of how they were stored; see also A. Nuovo, 'Manuscript Writings on Politics and Current Affairs in the Collection of Gian Vincenzo Pinelli (1535–1601)', in: *Italian Studies* 66 (2011), pp. 193–205.

[99] On Pinelli and Botero, see Blythe Alice Raviola, *Giovanni Botero: Un profilo fra storia e storiografia*, Milano: Bruno Mondadori 2020, pp. 45–50.

opinion. Botero, this anonymous writer says, delivers much less than he promises on face value—perhaps, to Pinelli's own dismay, if he agreed.

All the texts assembled by Pinelli at the end of R 122 sup. insist on the idea of 'penetrating' into the arcana of power as well as on the court seen as a 'notebook of all state machinations'. Likewise, some anonymous advice on f. 412r suggests that a ruler should keep agents or ambassadors, not for obedience's sake, but to 'penetrate' each other's mind and thus enjoy a wiser self-government. Prudence, the text continues, could aptly derive from the abundance of historical examples (*per abbundare d'essempii*), and it is plain to see how in Pinelli's reckoning modern business transactions are the result of the same epistolary practice, instructional sheets, reports and foreign negotiations that intersect the world of scientific exploration. Politics and mathematics, in other words, are reaping a peculiar reward from a scribal proliferation (*tante scritture*, f. 412v).[100] Whatever was the origin of this passage, copied or invented by Pinelli himself in reaction to literature on the topic, and regardless of his personal opinions on Botero, the accentuation is clear: it is our modern prerogative to have opened the way to these writings, and a 'diplomatic' revolution has preceded a more traditional, 'scientific' one. Pinelli's synoptic table, on f. 413r, essentially articulates the same principles, by dividing matters into two areas: in one bottom branch we witness Machiavellian ideas on how a state is fortified by land and sea, money, army, and so on; the upper division pertains to a typology of sources within Renaissance intelligence.[101] By design, politics and mathematic find their meaning through 'reports and discourses'. Put differently, the archival turn unfolds the mathematization of nature.

Communicating mathematically and polymathically

In short, Pinelli's methodizing is self-consciously 'modern'. It takes full advantage of new techniques of note-taking, some of which coalesced as tools in natural history, while others developed as a side-effect of Ramist and tabular interests in the printing world. Pinelli's archive represents as well the richest estuary of learned academies inside a Venetian territory. This serendipitous confluence of interests and methods left a mark in the language of Pinelli himself, who referred to his research as a necessity to operate in an in-depth and endlessly discursive fashion.[102]

Along with its empirical bent and copious commentary it is ultimately this nervous apprehension of the ethical imperative to amass particulars, and not, as commonly stated, a reciprocation of well-placed friends who procured *specimina*, that

[100] BAM, MS R 122 sup., f. 412r–v: 'ma è necessario penetrare nelli Archivii de' Prencipi et vedere i loro negotiati, le lettere, le instruttioni et l'altre cose che in esse si conservano et in questa parte da pochi anni in qua si è molto aperta la via a persone di negotii perché sono usciti in luce tanti negotiati di cose diverse di Prencipi, tante relationi d'Ambasciatori forestieri, tante scritture sopra leghe, sopra Concilii, sopra conclavi'.

[101] BAM, MS R 122 sup., f. 413r: 'per reporti/ per discorsi/ per spie et dispotion di questo et quello/ per via d'Ambasciatori et questi trattando con i grandi hanno a ponderar diligentemente li costume dei grandi, il valor loro, il consiglio, et le maniere di tutti, e del Prencipe stesso'.

[102] BAM, MS G 272 inf., f. 53r: 'raggionare di continuo'.

significantly distances Pinelli's library from the foremost humanist-collectors of the previous era—and first among them, Pietro Bembo.[103]

On the basis of this assessment, one must confront now a Janus-faced image of Pinelli that has been transmitted in our sources and in our literature. At one end of the spectrum, we encounter Girolamo Mercuriale, himself neglected in recent years, who at the peak of his career was respected as physician and savant. In a letter of May 1604, reflecting on Pinelli soon after his death, Mercuriale claimed that he did not possess the skills of a courtier.[104] Possibly, Mercuriale's unceremonious dismissal rehearses the circumstance that in the late 1570s Pinelli planned to move to Rome, and join the city's global ambitions, but a relocation never occurred. This view of Pinelli as an anti-courtesan clashes with the celebration of his prudent minimalism and of his irenic disposition, which allowed him to entertain a fruitful dialogue with the likes of Dudith, Scaliger, and many others, despite their Protestant or heterodox views.

Both poles of interpretation are two-dimensional, in the end. They aim at binding an actor, Pinelli, according to categories that reflect poorly the innovative engagements with his objects of knowledge. Unlike the creator of a *Kunstkammer*, he never collected pure curiosities; his mathematical papers also did not undergo the transformation into purely collectible items, at least not in the beginning, when they were still imbricated in post-Aristotelian agendas of natural inquiry. A better rationale would entail, therefore, to carefully re-evaluate Pinelli's communicative agency inside the collection, and to treat the library like a proper experimental workshop. Everything was meant to be 'taken out'—and eventually realigned or further spliced onto conjoined binders of files: diagrams, individual print runs, or Tycho Brahe's instrumental album. To understand the knowledgeable Pinelli is to recalibrate both objects and their holder as if inside a toolbox, or more concretely like playing along with an astronomical volvelle dial in Peter Apian's *Cosmographia* (first published in 1524). With 'stand-alone', paper instruments such as those, hand and print move along fluid boundaries; and it becomes easier for us to accept a history of early modern media as experimental substrates in the owner's (or maker's) atelier.

From a historical perspective, the unintentional ambiguity of Pinelli's agency—whether he remains a passive note-taker or an active inventor, instigating research questions in a collegial environment—seems largely to coincide with what has been called 'applied knowledge' as opposed to 'pure knowledge'.[105] Definitive answers would require a lengthy strategy but by sampling the interaction between Kepler and Galileo in the 1590s, I wish to show what a sensitive premium of explanation we could extract from a correct disentangling of the 'polymathic' and 'mathematical' sides of the scientific conversations orchestrated *by* or *through* Pinelli's circle.

Soon after Kepler had published his ground-breaking treatise in 1596, the *Mysterium cosmographicum*, which heralded a 'neo-Platonic' reform of its entire disciplinary field, his problem had been to secure readers of suitable disposition and

[103] Massimo Danzi, *La biblioteca del Cardinal Pietro Bembo*, Geneva: Droz 2005.

[104] Nuovo, 'Pinelli's Collection of Catalogues', p. 137.

[105] S. Roux, 'Forms of Mathematization (14th–17th centuries)', in: *Early Science and Medicine* 15 (2010), pp. 319–37.

appropriate scholarly calibre. First, he looked to the South.[106] Despite sending gift copies of the imprint via intermediaries Kepler's attempt to initiate a conversation, and perhaps even orchestrate a prototypical 'peer review' of the monograph's key themes, with Italian astronomers proved initially unsuccessful. Then Galileo intervened.

It is not entirely clear who facilitated Kepler's reaching of him.[107] In a highly sensitive reading,[108] Massimo Bucciantini ascribes this mediation to the activity of Edmund Bruce (*fl.* 1585–1606), a notable alumnus of the *natio Anglica* at the University of Padua, who along with other diplomats and scholars, including Henry Savile, regularly attended Pinelli's house. Indeed, as we now know, Bruce did more than passing on mathematical intelligence back to Central Europe. His key research interests oscillated between astronomy and antiquarian historiography, and he was instrumental to Pinelli's desire to assemble a conspicuous ethico-political folder about England and Scotland.[109] Like other interlocutors of Kepler, including Johannes Wacker von Wackenfels (1550–1619), the German dedicatee of the *De nive sexangula*, a short and brilliant essay on crystallography first printed in 1611, and presumably like Pinelli himself, Bruce was enthusiastic about Giordano Bruno's cosmology.

Moving forward, every line of communication that presented itself to Kepler was almost a refraction in an archipelago of courtly performances and eccentric scholarly pursuits. Galileo's response to his *Mysterium*, as well, was at once personal and cross-fertilized by the Paduan sociability of the time. Yet, anxiety and theoretical rivalry notwithstanding, the two cosmologists adopted a pose of mutual courtesy. Still, they had different agendas. Galileo tried to adapt Kepler's astronomical data to his science of motion, at a particular stage of its development. As one learns from Jochen Büttner's meticulous analysis, Galileo drew a diagram to capture the relations between 'mechanical' and astronomical thinking (Figure 10.6), responding to ch. 20 in the original monograph.[110] This effort was largely outside Kepler's remit, and would likely have disappointed him.

By comparing self-assurance and internal contradictions of this scribal draft with unmatched skill,[111] Büttner was able to date Galileo's reaction more precisely to *c.*1604, and to clarify that the solution's shortcomings required engaging in new proportionalities between the degrees of speed and the timing of the fall. From my point of view, the remarkable feature is that the articulated cross-section of the cosmos made in Tübingen and known to historians as Tabula III, one of the most famous

[106] To economize space, I can only point to a fuller treatment of *Mysterium* in my article 'Kepler's snow: the epistemic playfulness of geometry in seventeenth-century Europe', in British Journal for the History of Mathematics 37/2 (2022), pp. 117–37.

[107] R. Biancarelli Martinelli, 'Paul Homberger: Il primo intermediario tra Galileo e Keplero', in: *Galilaeana* 1 (2004), pp. 171–81, maintains that the *Mysterium* was given to Galileo by the Austrian composer Homberger, who studied in Padua, but does not speculate on what to make of this.

[108] M. Bucciantini, *Galileo e Keplero: Filosofia, cosmologia e teologia nell'Età della Controriforma*, Turin: Einaudi 2000, pp. 93–116.

[109] See Robert Westman, *The Copernican Question: Prognostication, Skepticism and Celestial Order*, Berkeley: University of California Press 2011, p. 365, and Philo, 'English and Scottish Scholars', pp. 52–7.

[110] Jochen Büttner, 'Galileo's Cosmogony', in: *Largo campo di filosofare: Eurosymposium Galileo 2001*, ed. J. Montesinos and C. Solís, La Orotava: Fundación Canaria Orotava 2001, pp. 391–402.

[111] Büttner, 'Galileo's Cosmogony', p. 400.

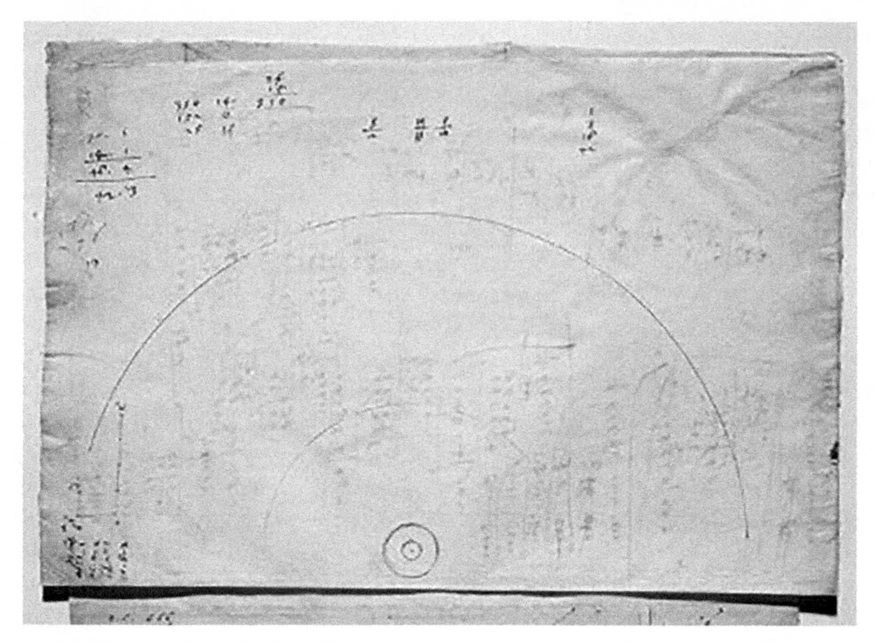

Figure 10.6 Galileo Galilei's diagram, made in response to his reading of Kepler, *c.*1604.

illustrations in the entire history of science, was effectively undone by calculations that were scribbled in the margins of a folio page turned upside down.[112]

Of course, Mästlin, Kepler and their accompanying printing entourage knew that to arrive at the woodcut showing the planets like nested solids in the wake of those German polyhedral representations, often carved in ivory, they needed to rely on a myriad of numbers. Yet, a decision was taken (by whom it is unclear) just to emphasize the final artistic product—no doubt, to raise its marketability as a side-effect. Once we move all this to Northern Italy, following Kepler's desire to break into an international community, we can observe how the algebraic foundations that the makers of *Mysterium* sought to 'blackbox' inside the cosmological broadsheet reclaimed their position in the centre stage. Galileo, one of the treatise's intended readers, temporarily reverted the printed thickness of Kepler's edition to a purely computational impulse. Why did he do so or what do we stand to gain? Where did this epistemic movement take place? The second question is the easier to answer. Although the draft was produced in or around 1604, it is fair to say that Galileo encountered Kepler in Pinelli's library, in the long shadow of mathematical intelligence that was cast from inside that household. But then, was Pinelli an enabling polymath, in possession of Transalpine documents that he saw thanks to itinerant acquaintances? Did he serve primarily as a postal drop box or did he actively contribute to Galileo's mathematical reshaping of

[112] Reiterating two points made by Raphael, *Reading Galileo*: (i) experimental traditions did not replace, but went hand in hand with scribal and humanistic technologies; (ii) astronomy was often reduced to mixed mathematics.

Keplerian astronomy? A proper solution to these queries cannot be undertaken here, because it would require a comparative look at other research folders of the period where one is similarly puzzled about Pinelli's role: Paolo Sarpi on magnetism,[113] and Giordano Bruno's atomist geometry,[114] to name but two.

Unless the issue is posed too narrowly or abruptly, however, I wish to entertain the notion that Galileo's reaction to Kepler reactivated a type of Aristotelian abstractionism. If this is true, Pinelli's circle becomes ideal as an intellectual infrastructure, which is something that can be said even in the case in which the Pisan scientist simply followed a paper trail left by Kepler on the strength of the library's cosmopolitan network, but equally compelling as a starting point for an investigation which, admittedly, sits quite comfortably within the Ambrosian papers documenting a season of long debates on mathematics, astronomy, and their relation to logic.

From Pinelli's viewpoint, if we choose to highlight it instead of using it as a setting in the background, a small but veritable army of practitioners and artisans—including scribes, notaries, or similar professionals—arose to meet the demands of what I identified as the 'problematic' assembling of science of this cultural group. Unauthorized copies of Paduan lectures and treatises disseminated scribally coexisted with medicinal recipes and gossip from barbers or patricians, as well as with the compiling of Jesuit *litterae annuae*. This rich proliferation is another symptom of the creative expansion that was initiated, not only archived, by this library. Likewise, given the growing impulse of mathematical performance and discussion, it is obvious that Pinelli expected the young Galilei in the 1590s to follow the path already trodden by his trusted collaborator Moleto.[115] Naturally, his judgement was wrong or reductive, historically speaking, but Pinelli was not off the mark in presuming that Moleto's outline of mathematical practice constituted, in Venice and elsewhere in the Republic, the correct, metropolitan brand for the discipline. For a while, in fact, before the path of Moleto and Galileo diverged, there was an intersection. This juncture remains consequently the most promising tool to explain how in Pinelli's world Aristotelianism and mathematization cooperated and how in turn their blending facilitated the reception of Kepler in Italy.

Tantalizing or bulging as they may seem, Ambrosian miscellanies are our best source to delineate this early modern culture of record-keeping, especially when we can ascertain that we are indeed working on Pinelli's original folders. This is the case of MS R 104 sup., which contains a giant 'R' on f. 212^{r} as a topical index, even though the rest of the manuscript was rebound at Borromeo's request and is now in poor state of repair.

This miscellaneous codex is of special interest, albeit a puzzle in terms of its contents. It assembles cosmographical excerpts, political dispatches, and philosophical

[113] C. Sander, 'Early-Modern Magnetism: Uncovering New Textual Links between Leonardo Garzoni SJ (1543–1592), Paolo Sarpi OSM (1552–1623), Giambattista Della Porta (1535–1615), and the Accademia dei Lincei', in: *Archivum Historicum Societatis Iesu* 85 (2016), pp. 303–63.

[114] P. Rossini, 'New Theories for New Instruments: Fabrizio Mordente's Proportional Compass and the Genesis of Giordano Bruno's Atomist Geometry', in: *Studies in History and Philosophy of Science* 76 (2019), pp. 60–8.

[115] For a deeper contextualization, which includes also Daniele Barbaro, see Manfredo Tafuri, *Venice and the Renaissance*, Cambridge, MA: MIT Press 1989, p. 247.

booklets, some of which reveal great originality in this context.[116] Throughout its different and independent articulations, MS R 104 sup. pursues a vernacular theory of 'signs', related to the Averroistic interpretation of Aristotelian physics. In a fundamental way, the method employed responds to a problem-solving attitude and it implements the idiosyncratic natural philosophy that was practiced in Pinelli's circle. As in other examples discussed above, mathematical practice is coextensive with reflections on the conglomerate nature of the early European states. The later fascicles of this manuscript, which are philologically separate,[117] contain a letter written by Galileo to Alvise Mocenigo, of 1594, in which he evaluates an invention of Hero of Alexandria, providing a mechanical drawing in the end.[118] This transcription is quite calligraphic; the letter's dedicatee, moreover, is indexed on the margins. From these small details and obviously given the date and Mocenigo's importance for this group, we can safely assume a strong interest by Pinelli, who supplemented the technical opinion of Galileo with a small anthology of quotes taken from the illustrious Peripatetic thinker Jacopo Zabarella (1533–1589). Pinelli took these extracts from Zabarella's published work on *De anima*; this explains their epistemological orientation. Yet, in a novel scribal environment, the pieces were organized in such a way as to culminate with the exposition of two geometrical problems, moving from theory to practice.[119]

This multifarious evidence is a very relevant embodiment of the ideal Aristotelian mathematization assembled by Pinelli during Galileo's early Paduan years. To add additional strength to this historical reconstruction I would finally like to briefly emphasize another section of MS R 104 sup. in which Pinelli makes two successive copies of a letter written to him by Edmund Bruce, the English astronomer and virtuoso who in Bucciantini's reading of this episode provided Galileo with a copy of Kepler's tract.[120] The interest of this source lies in Bruce's careful description of the Tabula Peutingeriana, a thirteenth-century map in parchment showing the state-run road network of the Roman Empire. In Bruce's lexicon we can see an application of Pinelli's collection to social geography and through named interlocutors such as Alvise Saraceno and Onofrio Panvinio we may also tease out a local network of antiquarians which demonstrates a collective reconnaissance, similar to the 'Britannia' of the first Baconian circles.[121]

Conclusion: note-taking and note-sharing

Pinelli's papers were neither private nor public. They might appear to us messy and fragmentary, and indeed the majority of the individual pieces stand alone; they are

[116] In particular, a letter of Johannes Caselius of Helmstedt (BAM, MS R 104 sup., ff. 260r–262r), an opinion by Ulisse Aldrovandi on a passage of the *Historia animalium* (ff. 342r–v), and Pinelli's letter to Guido Panciroli on Livy, with his reply (ff. 349r–351r).

[117] There are clear stitches visible on BAM, MS R 104 sup., f. 378v, and the watermark analysis confirms this independence.

[118] BAM, MS R 104 sup., f. 376r.

[119] BAM, MS R 104 sup., ff. 378r–379r.

[120] BAM, MS R 104 sup., f. 237r–v and ff. 238r–239v, for the respective transcriptions.

[121] For this tradition of topographical writing, from Camden's *Britannia* of 1586 to Childrey's *Britannia Baconica* of 1660, see Angus Vine, *In Defiance of Time: Antiquarian Writing in Early Modern England*, Oxford: Oxford University Press 2010, and Yale, *Sociable Knowledge*, pp. 5–8.

'fair copies' or the production of scribes specifically engaged in the act of copying. However, when they are sufficiently close to the interlocking shapes that Pinelli originally designed for them, these papers often make a strong case for a hybrid cognitive style of doing and communicating science. Within such a style, mathematical interests seem interspersed with, and partially overwhelmed by, a concern for ethnography and politics—which sometimes falls under the guidance of a local epistemic genre, the Venetian reports or *relazioni*. Pinelli asserts an archival voice, which I proposed to call his scribal self, discreetly, mostly by supplying transcriptions as a basis for comparison and by suggesting ingenious templates in which scientific information should be recollected and shared. To place this last point in a concrete setting, we might never know if Pinelli actively reshaped a dialogue between Kepler and Galileo, or if it was just a function of the commodification of courtly culture. But it adds considerable nuances to appreciating how, in the same library, this Paduan collector was making Zabarella's theory 'react' with geometrical problems of the kind entertained by engineers in the Arsenal.

In the end, what we learn about Pinelli's habits of thought suggests that it is increasingly necessary to study him as a pivotal agency in the many-headed construction of Venetian natural philosophy, instead of gazing at him as a scrivener or as a heroic, distracted virtuoso in whose prestigious chests historians can go treasure-hunting. Pinelli's scribal self, moreover, is simultaneously operating in and beyond academia, and across learned and practical traditions. His papers offer interesting traces of vernacular culture which refute an often too clear-cut distinction between this lower production and the Latin used by Aristotelians. In fact, Pinelli asked his questions in a manner that clearly testifies to the participation in a cycle of knowledge expansion. For example, in a manuscript dominated by the figure of the alchemist and mathematician Ettore Ausonio, Pinelli added a marginal note to query 'how is optics necessary to astronomy?'.[122]

In another manuscript, organized by Pinelli around Moleto's lectures in the 1580s, with such diverse topics as mechanics, Sacrobosco, artillery, wine-making in Vicenza, and a treatise on the Nile, Pinelli underlines two aspects of Moleto's treatments. First, he remarks, when Aristotle used the word *mechanicus* he really meant an engineer; second, Euclid should be seen as a 'follower of the Platonic doctrine', that is, as the text insists, a theoretician who shows how one could produce nested solids.[123] Our customary view of late Renaissance cultures of collecting helps us for sure to situate these episodes. One could resort, in this case, to Pinelli's list of Moleto's mathematical instruments, which he acquired after the latter's death and was drafted in several copies. At the end of one inventory, we see a jarring mixture of books and objects: meteorological treatises on sea tides and other cosmological ones on the *Sphaera*, together with a playful album of mathematical sketches, an astrolabe and a mechanical clock.[124] But to fully make sense of this evidence, one must admit that a baseline of information on Pinelli's note-taking and assembling is required. For a query on

[122] BAM, MS G 119 inf., f. 12^{r}.

[123] BAM, MS S 100 sup., f. 213^{r}: 'sectator Platonicae doctrinae'. Incidentally, this is how a historical actor familiar with Pinelli's library would have understood Kepler's cosmic design in the 1596 treatise, which was disseminated, via gift-giving, across the Alps.

[124] BAM, MS S 94 sup., f. 171^{r}.

optics or Plato is intelligible as a fractured statement, but opens an entirely more corroborating canvas if it is made to reflect the scribal apprenticeship and exceptional care with which themes were 'bound'.

By 1600, in Pinelli's circle, there were many doubts about Pliny and the erudition that played such an integral part in that tradition. When these co-investigators progressively moved to paper technology, their efforts could be seen as an epistemic effect of Aristotelian *historia*. What Brian Ogilvie called the 'science of describing' continued to instil a deep respect and to inspire a condensation of observation and empirical experience.[125] In itself, and for seventeenth-century naturalists, this is not at all surprising; after all, Richard Yeo has told us how even a prodigious memory, like the one that was sometimes attributed to missionaries and religious men, gave way at the turn of the century to notes, aids, or paper compendia.[126] What is remarkable, in Pinelli, is that the building blocks of his cultural activity were 'problem texts', or an inquisitive technique that allowed him to juxtapose, as if in a question-and-answer format applied to archival science, letters on one side, and opinions, diagrams, sketches, and *adversaria* on the other. There may not be an easy name for this data. Yet, it was so pervasive and indispensable to Pinelli that he reiterated it for humbler practitioners as well as in his 'Savilian' manuscripts.[127] In a significant way, Pinelli undertook a collection of doubts and he probably understood doubts as problems, too.[128] However chaotic the surface of his media may seem, the reward lies in these structural similarities.

[125] Ogilvie, *Science of Describing*, pp. 262–4.
[126] Yeo, *Notebooks, English Virtuosi*, p. 231.
[127] See, for example, BAM, MS D 243 inf. and MS P 227 sup.
[128] BAM, MS D 142 inf., f. 442^{r}: 'problema est propositio dubia'.

11

Mathematical Businesses

Seventeenth-Century Practitioners and their Academic Friends

Philip Beeley

Flight from contagion

As is well known, during the months following the outbreak of plague in April 1665,[1] all those who could afford to do so, materially and professionally, left London to see out the epidemic in safer quarters, and only returned to the metropolis when it was considered safe to do so, that is to say, at the beginning of the following year.[2] Most active members of the Royal Society, including William Petty (1623–1687), Paul Neile (1613–1686), and Robert Moray (1608–1673), decamped to Oxford, the university city in which the Society had been born and where two of the leading figures, the Savilian professor of geometry John Wallis (1616–1703) and the natural philosopher Robert Boyle (1627–1691) lived and worked.[3] It would be another three years before Boyle returned to London. When, in September, the meetings of the Royal Society would otherwise have resumed following the summer recess, Boyle offered his Oxford rooms for this purpose instead:

> I did not know why we might not, though not as a society yet as a company of Virtuosi renew our meetings & being put upon nameing the day & place I proposed Wednesday as an auspicious day, being as you know that of our former assemblys, & for the place till they could be better accommodated I offered them my Lodging, where over a dish of Fruite we had a great deale of pleasing Discourse, & some Experiments that I showed them [...].[4]

Among those who had taken refuge in Oxford with his wife and children was the accountant and mathematical author John Collins (1626–1683), who knew the city

[1] Unless otherwise indicated, all dates are given old style, i.e. according to the Julian calendar employed in England until 1752, with New Year's Day falling on 25 March. For ease of reference to standard editions of correspondence, dates of letters are given in footnotes according to both the Julian and Gregorian calendars.

[2] See A. Lloyd Moote and Dorothy C. Moote, *The Great Plague: The Story of London's Most Deadly Year*, Baltimore: Johns Hopkins University Press 2004, pp. 71, 79–82.

[3] See *The History of the Royal Society of London, for Improving of Natural Knowledge, from its first rise*, ed. Thomas Birch, 4 vols, London: for A. Millar 1756–7, vol. II, pp. 61–2; Robert E. W. Maddison, *The Life of the Honourable Robert Boyle F.R.S.*, London: Taylor & Francis 1969, pp. 122–3.

[4] Robert Boyle to Henry Oldenburg, 30 September/[10 October] 1665, in *The Correspondence of Henry Oldenburg*, ed. A. Rupert Hall and Marie Boas Hall, 13 vols, Madison, WI and London: University of Wisconsin Press et seq. 1965–86, vol. II, pp. 535–7 at p. 537.

well, having been born nearby at Water Eaton, and having spent a year, in 1641, working as apprentice to the local bookseller Thomas Allam (17th century).[5] William Austin, the father of Collins's wife Bellinda (?–1680) had been brought to Oxford to work in the kitchen at Wadham College during the Interregnum.[6] Under circumstances that are unclear, and likely to remain so, on 12 August 1665, Collins met with Wallis and was shown some of the treasures of the Savilian Library, the library exclusively for the use of the two Savilian professors, including a manuscript introduction to Galileo Galilei's (1564–1642) mechanics.[7] Collins, who in earlier years had produced a number of popular books on topics such as merchants' accounts, dialling, or the use of the quadrant, presented Wallis with a stereometric problem, to calculate the volume of liquor contained in part of an elongated spheroid (Figure 11.1):[8]

> A Probleme proposed to mee by Mr Joh. Collins Aug. 12. 1665
>
> Suppose *EεpεEP* a wine or beer Vessel, (part of the Spheroide *PBpB*,) whose semidiameter at the Middle is *CP*, at the Heads *δE*: Half the length *CP*, *Kφ*, or *HE*: and filled (more than half but not so high as *E*:) as high as the plain surface *γγγγ* (the vessel lying Horizontally:) It is demanded, How much is the liquor contained: or, the magnitude of the Portion *γγεpεγγ*.

From the record which Wallis kept of this meeting, and which he incorporated years later in his major work on the mathematical sciences, we learn that already in the presence of Collins, whom he describes as 'a talented man' and one 'knowledgeable in mathematical matters', he set out the key points of the problem's solution and its demonstration.[9] After they had parted, Wallis set about producing an extensive, five-page solution to the problem his visitor had posed, and sent it to him a few days

[5] See Anthony Wood, *Athenae Oxonienses: An exact history of all the writers and bishops who have had their education in the most ancient and famous University of Oxford*, 2 vols, London: for Thomas Bennet 1691–2, vol. II, p. 794: 'At 16 years of age Joh. Collins was put an Apprentice to a Bookseller (one Tho. Allam) living without the Turl-gate of Oxon, but Troubles soon after following he left that Trade'. Allam is recorded as having collaborated with the Oxford printer Leonard Lichfield on various publications between 1636 and 1655.

[6] John Collins to John Beale, 20/[30] August 1672, Cambridge University Library, MS Add. 9597/13/5, f. 83^r–$85a^v$; *Correspondence of Scientific Men of the Seventeenth Century*, ed. Stephen Jordan Rigaud, 2 vols, Oxford: at the University Press 1841, vol. I, pp. 195–203 at p. 202. See Barbara J. Shapiro, *John Wilkins 1614–1672: An Intellectual Biography*, Berkeley: University of California Press 1969, p. 122: 'Wilkins had somehow managed to get as Wadham's master cook Mr William Austin, who had served as Prince Charles's cook when he was Prince of Wales'. Shapiro points out that after the Restoration Austin returned to the royal kitchen. On Collins's marriage see William Letwin, *The Origins of Scientific Economics. English Economic Thought 1660–1776*, London: Methuen 1963, p. 102.

[7] Collins mistakenly thought afterwards that the manuscript had been written by Galileo. See John Wallis to John Collins, 7/[17] August 1666, *Correspondence of John Wallis (1616–1703)*, 4 vols (to date), ed. Philip Beeley and Christoph J. Scriba, Oxford: Oxford University Press, 2003ff, vol. II, pp. 278–81 at pp. 279–80: 'The manuscript of Galilee, which you saw with mee; is not his Mechanick Problems; but rather an Introduction to them'.

[8] Oxford, Bodleian Library, MS Don. d. 45, f. 279^v–277^v (foliation reversed). Further autograph copies of Collins's problem with Wallis's solution are found in MS Don. e. 12, f. 7^r–10^r and f. 12^r–16^r. The former copy includes a synopsis of his calculation.

[9] John Wallis, *Mechanica: sive, de motu, tractatus geometricus*. 3 parts, London: William Godbid for Moses Pitt 1670–1, vol. II, pp. 477–8; John Wallis, *Opera mathematica*, 3 vols and suppl., Oxford: at the University Press 1693–99, I, pp. 870–1: 'Libet hic subjungere, Problematis Solutionem quod ante aliquot annos mihi solvendum proposuit Vir ingeniosus, & rerum Mathematicarum peritus, D. Joh.

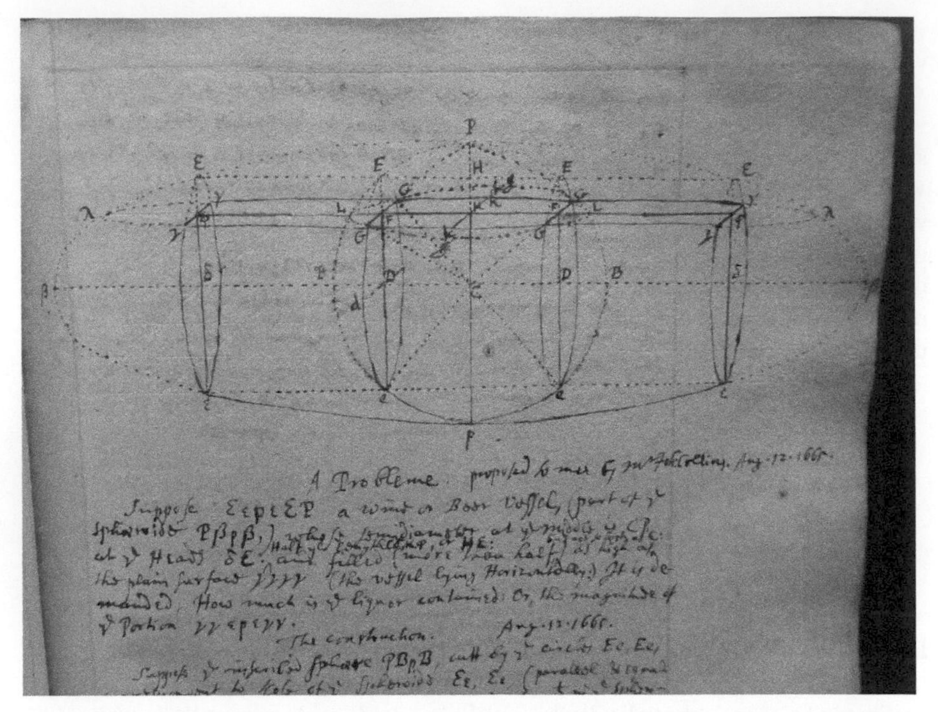

Figure 11.1 Wallis, Diagram of problem proposed to him by Collins, 12 August 1665.

later.[10] It would appear from Wallis's remarks that he showed him the method he intended to use before their meeting concluded.

There seems to have been no further contact between the two men until the beginning of the following year, by which time Collins had long since returned to London and taken up again his prime occupation as accountant in the Excise Office.[11] Writing at the beginning of January, he thanks Wallis for the 'unexpected favours' he had received when at Oxford:

> The unexpected favours I received from you when at Oxford cannot but make me mindfull of those obligations of thankes I ought to returne, as a Testimony whereof vouchsafe to accept of a few bookes I herewith send viz. Mengoli Geometria

Collins, (Augusti 12° 1665.) tanquam rem maxime desideratam; & quotidiani usus, in mensurandis vasis vinariis & cervisiariis, partim depletes. Cujus ego Problematis Solutionem, ejusque Demonstrationem, eodem statim die, summatim ei exhibui; (&, quantum scio, omnium primus absolve) ad hunc sensum'. There are two Latin copies of the problem proposed by Collins and Wallis's solution among the Savilian professor's papers. See Oxford, Bodleian Library, MS Don. e. 12, f. 7^r–10^v (versos blank) and f. 12^r–16^v (versos blank).

[10] Wallis, *Mechanica*, p. 478; *Opera mathematica*, vol. I, p. 871: 'Quod ipsum paucis post diebus, latius aliquanto explicatum, ad illum etiam misi'.

[11] See John Collins to John Wallis, 2/[12] January 1665/6, *Correspondence of Wallis*, vol. II, pp. 188–90 at p. 190; See also Letwin, *Origins of Scientific Economics*, pp. 101–2.

Speciosa, Antimi Farbii Opusculum Geometricum et Caravagii Mediolanensis Applicationum Doctrina which handles the Limits of such Aequations as have but two Nomes.[12]

Typically employing the 'do ut des' principle of the Republic of Letters, he repays Wallis for his kindness in Oxford and particularly his efforts in solving the stereometric problem by sending him a number of mathematical books he had succeeded in procuring from Italy, including Pietro Mengoli's (1626–1686) *Geometriae speciosae elementa* (Bologna 1659) and Honoré Fabri's (1607–1688) *Opusculum geometricum* (Rome 1659).[13] When Wallis later sends him papers on the nature of the meridian line contained in Gerard Mercator's (1512–1594) sea chart, Collins repays the Savilian professor for his kindness with an exemplar of Henry Sutton's (d. 1665) printed quadrant he had varnished and at the same time encloses a copy of what he describes as a Dutch tract on perfect numbers by the Polish mathematician Jan Brożek, otherwise known under his Latinized name Johannes Broscius (1585–1652).[14]

Collins also uses his first letter to Wallis following his return to London to supply him with latest mathematical news, most notably of the deaths in London during the recent outbreak of the plague of Anthony Thompson (d.1665) and Sutton, whom he describes as 'two of the best Mathematicall Instrument Makers'. In view of Sutton's importance in mathematical circles, Collins takes care to point out that John Marke (*fl.* 1665–1679), one of Sutton's former workmen, had taken over the business, and that he would personally give him instruction wherever necessary.[15] Furthermore, by way of showing just how well apprised he was of contemporary developments in the mathematical sciences, Collins reports on new books by Blaise Pascal (1623–1662) and Gilles Personne de Roberval (1602–1675) that were expected anytime from France, and on deliberations concerning the meridian line based on the sea charts of Edward Wright (1558–1615) and Gerard Mercator. He rounds off by seeking to establish the foundations of an intellectual friendship:

And if any thing come to my knowledge about these matters I should be willing to communicate it, and humbly crave the favour when your occasions draw you to

[12] John Collins to John Wallis, 2/[12] January 1665/6, *Correspondence of Wallis*, vol. II, pp. 188–90 at p. 190.

[13] On this principle see Saskia Stegeman, *Patronage and Services in the Republic of Letters: The Networks of Theodorus Janssonius van Almeloveen (1657–1712)*, Amsterdam: APA-Holland University Press 2005, pp. 169–73; Nora Gädeke, 'Gottfried Wilhelm Leibniz', in: *Les grands intermédiaires culturels de la République des lettres du xve au xviiie siècles*, ed. Christiane Berkvens-Stevelinck, Hans Bots, and Jens Häseler, Paris: Honoré Champion 2005, pp. 257–306; Françoise Waquet and Hans Bots, *La république des lettres*, Paris: Belin; Brussels: De Boeck 1997.

[14] John Collins to John Wallis, 28 February/[10 March] 1665/6, *Correspondence of Wallis*, vol. II, pp. 191–4. The tract was Jan Brożek, *De numeris perfectis disceptatio*, Amsterdam: Blaeu 1637. It had first appeared in a year earlier in Krakow, printed by Antoni Wośinski.

[15] John Collins to John Wallis, 2/[12] January 1665/6, *Correspondence of Wallis*, vol. II, pp. 188–90 at p. 189: 'the quondam Servant of Mr Sutton, John Marke, being now returned is about taking his Masters house; he desires the presentment of his most humble service. We hope he may prove as good a Workeman as his deceased Master, and that mite of knowledge I have attained to, I shall most willingly serve him with'.

> London, to affoard me the cognisance thereof that I may have some further opportunity administred of expressing my Gratitude. In the interim if you shall vouchsafe a Line or two in returne it needs no other direction but to me as an Accomptant at the excise office in Bartholomew Lane London [...].[16]

Collins could scarcely have been more explicit in his desire that he and Wallis from then on engage in a mutually beneficial scientific exchange whether in person or through correspondence. By using his address at the Excise Office, he was able to remind the Savilian professor of his own position as a government accountant without any noticeable implication of inferiority. As we shall see, Collins more often than not adopted a contrary position. There were good grounds for his suggesting a possible meeting in London, for he was no doubt aware that Wallis often travelled there on University business on account of his being the institution's archivist. At other times, Wallis would pass through the metropolis on his way to visit relatives in his native Kent.[17]

Collins and his scientific milieu

It had not been necessary for Collins to inform Wallis of the tragedy that had befallen Thompson and Sutton. Such was the standing of these two men within London's scientific circles, and not just within the mercantile, seafaring, printing, and practitioner milieus Collins frequented, that the secretary of the Royal Society, Henry Oldenburg (1618?–1677), who remained in London throughout the Great Plague, had immediately communicated the news to those members who were already in Oxford or had fled there to escape the contagion.[18] The reaction of Robert Moray, one of the most influential early fellows and an important patron of the mathematical sciences, speaks to the widely felt impact and is captured in just a few poignant words: 'wee all here are much troubled with the loss of poor Thomson & Sutton'.[19]

Collins's own connection with Sutton went back a long way.[20] After having spent seven years serving on board on English merchantman, most of the time engaged by the Republic of Venice in its ongoing war against the Turkish fleet in the Eastern Mediterranean, Collins returned to London in 1649 during the heady days of the Revolution.[21] Making use of the skills he had acquired during his spare time at sea or

[16] Ibid.

[17] See for example John Wallis to John Collins, 7/[17] August 1666, *Correspondence of Wallis*, vol. II, pp. 278–81 at p. 280: 'I passed through London not long since, both as I went into Kent, & as I came back again'.

[18] See Marie Boas Hall, *Henry Oldenburg: Shaping the Royal Society*, Oxford: Oxford University Press 2002, pp. 95–101.

[19] Robert Moray to Henry Oldenburg, 10/[20] October 1665, *Correspondence of Oldenburg*, vol. II, pp. 539–62 at p. 561.

[20] Frances Willmoth, *Sir Jonas Moore: Practical Mathematics and Restoration Science*, Woodbridge: Boydell Press 1993, p. 164.

[21] See Collins's account of his life in the preface to his *Introduction to Merchants-Accompts: containing seven distinct questions or accompts*, London: William Godbid for Robert Horne 1674, sig. B1^r–B2^r.

during the year beforehand spent as junior clerk to John Marr (*fl.* 1614–1647) in the household of the then Prince of Wales, he taught elementary arithmetic, accountancy, and the art of handwriting. We know also that he attended the regular public lectures on astronomy and geometry given at Gresham College, for on one occasion he asked Samuel Foster (*c.*1600–1652), the astronomy professor and author of a number of practical books, for advice on mathematical publications he should read.[22] Foster was a practitioner with an academic background, having studied at Emmanuel College, Cambridge. In response to Collins's request he cited a list of predominantly Continental publications by the likes of Johannes Scheubel (1494–1570), Michael Stifel (1487–1567), Christoph Clavius (1537–1612), and Marino Ghetaldi (1568–1626), while the only English author he cited was Thomas Harriot (1560–1621):

> Now as to the booke it selfe, Dr Croone and Mr Colwall can attest that the late Mr Foster of Gresham College seldome heard it mentioned but tooke occasion to utter his Dislike of it [...]. By reason whereof in anno 1649 I asked Mr Foster what Authors he would advise unto: he replyed that the Algebra of Scheubelius (out of which Mr Bunning hath taken some of his Notes), Stifelius, Clavius, Dybuadius, Stevin did fully handle the Surds and Euclids irrationall lines, that Harriott, Herigon, Deschartes & Ghetaldus sufficiently the Specious and Exegetick part, not mentioning Vieta or Mr Oughtred [...].[23]

Noticeably absent in this list is any reference to William Oughtred's (1575–1660) *Clavis mathematicae*. That book had first been published in 1631, and was still a popular text at the English universities at the time, while being increasingly looked down upon in London's mathematical circles.[24]

Foster represents an important early link between Collins and Wallis. In 1646 or 1647 (Wallis was unable to be precise about the year) he, too, had a significant encounter with the man who was then enjoying his second stint as holder of the post of Gresham professor of astronomy.[25] Wallis was a minister of the cloth in London at the time and no more than an aspiring mathematician, taking part in weekly scientific meetings which would later be seen by him as the true forerunner of the Royal

[22] See John Collins to John Wallis, *c.* 10/[20] February 1666/7, *Correspondence of Wallis*, vol. II, pp. 311–18 at pp. 313–14. Among Foster's publications on topics in the realm of practical mathematics we find his *The Uses of a Quadrant* (London 1652) and *The Art of Dialling* (London 1638). On contributions by Gresham professors to practical learning see Mordechai Feingold, 'Gresham College and London Practitioners: The Nature of the English Mathematical Community', in: *Sir Thomas Gresham and Gresham College: Studies in the Intellectual History of London in the Sixteenth and Seventeenth Centuries*, ed. Francis Ames-Lewis, Aldershot: Ashgate 1999, pp. 174–88 at p. 179.

[23] John Collins to John Wallis, *c.* 10/[20] February 1666/7, *Correspondence of Wallis*, vol. II, pp. 311–18 at pp. 313–14. William Croone (1633–84) was professor of rhetoric at Gresham College; Daniel Colwall (d.1691) was a patron of the sciences and Fellow of the Royal Society.

[24] See John Collins to John Wallis, 2/[12] February 1666/7, *Correspondence of Wallis*, vol. II, pp. 301–7 at p. 304, and John Collins to Francis Vernon, mid-December 1671; Cambridge University Library MS Add. 9597/13/5, ff. 68^r–69^v; *Correspondence of Scientific Men*, vol. I, pp. 151–6 at p.152.

[25] Foster was first appointed astronomy professor following the death of Henry Gellibrand (1597–1637), but resigned after eight months. He was reappointed in May 1641 and remained in post until his death in July 1652. See John Ward, *The Lives of the Professors of Gresham College: to which is prefixed the life of the founder, Sir Thomas Gresham*, London: John Moore for the author 1741, pp. 85–8, 90.

Society.[26] Foster, along with men such as Theodore Haak (1605–1690), Jonathan Goddard (1616–1675), and John Wilkins (1614–1672), was a regular attendee at those meetings and it was probably in that context that he submitted to Wallis a theorem on the spherical triangle. Wallis published this theorem together with its demonstration in part two of *Mechanica* (London 1670–71), noting explicitly that they were 'by our countryman' Samuel Foster, not very long ago professor of astronomy in Gresham College, London.[27]

When this work was reedited for publication in his *Opera mathematica* (Oxford 1693–9), Wallis added the reference to possible dates and a suggestion that the theorem had perhaps derived from Thomas Harriot, whose scientific heritage he was by now particularly keen to promote:

> Who [*sc.* Samuel Foster] showed me a theorem on spherical triangle around the year 1646 or 1647, which had already been established by our countrymen a while ago, and which had been first discovered, if I remember correctly, by our Harriot.[28]

Mathematical authorship

Within a year of his appointment as accountant to the Excise Office, in 1652, Collins had published his first two books, effectively comprising material he had earlier used in teaching mathematics. There was a small book on decimal arithmetic, which also covered simple and compound interest as well as annuities, which was later reprinted after his death, when the editor summed up its contents so (Figure 11.2):

> This Book is a fit Companion for all Gentlemen, Merchants, Scriveners, and other Trades-men, that deal much in lending of Money upon Interest, Mortgages, buying of Estates either in Fee, Copy, or Lease, holding Annuities, Rent Charges, Forbearance of Money, Discompt, or any other way concerning Interest, &c.[29]

[26] See Philip Beeley, 'Eine Geschichte zweier Städte: Der Streit um die wahren Ursprünge der Royal Society', in: *Acta Historica Leopoldina* 49 (2008), pp. 135–62, and Mordechai Feingold, 'The Origins of the Royal Society Revisited', in: *The Practice of Reform in Health, Medicine, and Science, 1500–2000*, ed. Margaret Pelling and Scott Mandelbrote, Aldershot: Ashgate 2005, pp. 167–83.

[27] Wallis, *Mechanica*, vol. II, p. 475: 'Quae est etiam D. Samuelis Fosteri nostratis emonstration, in Collegio Greshamensi Londini Astronomiae non ita pridem Professoris'. The wording in the second edition of this work is slightly different. See Wallis, *Opera mathematica*, vol. I, p. 869.

[28] Wallis, *Opera mathematica*, vol. I, p. 869: 'Qui hoc mihi, de Triangulo Sphaerico, Theorema indicavit, Anno circiter 1646 aut 1647; ut apud nostros tum dudum receptum; & ab Harrioto nostro, si satis memini, primo repertum'. On Wallis's engagement for Harriot's scientific legacy see Jacqueline A. Stedall, 'Rob'd of Glories: The Misfortunes of Thomas Harriot and his Algebra', in: *Archive for History of Exact Sciences* 54 (2000), pp. 455–97; Jacqueline A. Stedall, *A Discourse Concerning Algebra: English Algebra to 1685*, Oxford: Oxford University Press 2007; Philip Beeley, '"Our Learned Countryman": Thomas Harriot and the Emergence of Mathematical Community in Seventeenth-Century England', in: *Thomas Harriot: Science and Discovery in the English Renaissance*, ed. Robert Fox, London: Routledge 2023, pp. 72–102.

[29] John Collins, *The Doctrine of Decimal Arithmetick*, ed. J. D. London: R. Holt for Nathanial Ponder 1685. The original publication was entitled *The Doctrine of Decimal Arithmetick, Simple Interest, &c., abridged*, London 1665. As the later publication makes clear, it was abridged for portability in a letter case.

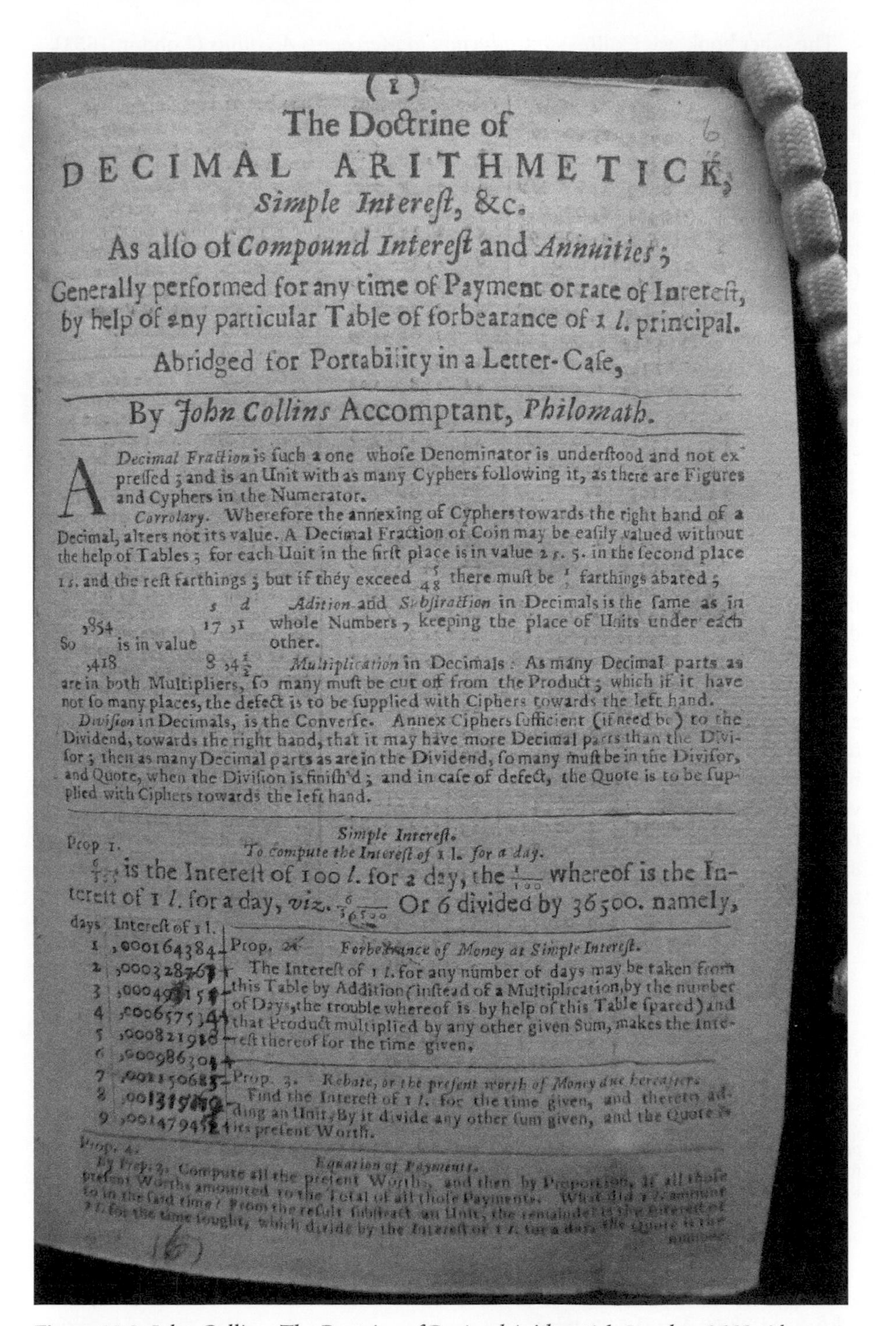

(1)

The Doctrine of

DECIMAL ARITHMETICK,

Simple Interest, &c.

As also of *Compound Interest* and *Annuities*;

Generally performed for any time of Payment or rate of Interest, by help of any particular Table of forbearance of 1 *l*. principal.

Abridged for Portability in a Letter-Case,

By *John Collins* Accomptant, *Philomath*.

A *Decimal Fraction* is such a one whose Denominator is understood and not expressed; and is an Unit with as many Cyphers following it, as there are Figures and Cyphers in the Numerator.

Corrolary. Wherefore the annexing of Cyphers towards the right hand of a Decimal, alters not its value. A Decimal Fraction of Coin may be easily valued without the help of Tables; for each Unit in the first place is in value 2 *s*. 5. in the second place 1 *s*. and the rest farthings; but if they exceed $\frac{5}{48}$ there must be $\frac{1}{}$ farthings abated;

		s	*d*
	,854	17	,1
So	is in value		
	,418	8	,4½

Adition and *Subsiraction* in Decimals is the same as in whole Numbers, keeping the place of Units under each other.

Multiplication in Decimals: As many Decimal parts as are in both Multipliers, so many must be cut off from the Product; which if it have not so many places, the defect is to be supplied with Ciphers towards the left hand.

Division in Decimals, is the Converse. Annex Ciphers sufficient (if need be) to the Dividend, towards the right hand, that it may have more Decimal parts than the Divisor; then as many Decimal parts as are in the Dividend, so many must be in the Divisor, and Quote, when the Division is finish'd; and in case of defect, the Quote is to be supplied with Ciphers towards the left hand.

Simple Interest.

Prop 1. *To compute the Interest of 1 l. for a day.*

$\frac{6}{365}$ is the Interest of 100 *l*. for a day, the $\frac{1}{100}$ whereof is the Interest of 1 *l*. for a day, *viz*. $\frac{6}{36500}$ Or 6 divided by 36500. namely,

days	Interest of 1 l.
1	,000164384
2	,000328767
3	,00049[illegible]15[illegible]
4	,000657534
5	,000821918
6	,000986301
7	,001150685
8	,00131[illegible]
9	,001479452

Prop. 2. *Forbearance of Money at Simple Interest.*

The Interest of 1 *l*. for any number of days may be taken from this Table by Addition (instead of a Multiplication, by the number of Days, the trouble whereof is by help of this Table spared) and that Product multiplied by any other given Sum, makes the Interest thereof for the time given.

Prop. 3. *Rebate, or the present worth of Money due hereafter.*

Find the Interest of 1 *l*. for the time given, and thereto adding an Unit, By it divide any other sum given, and the Quote is its present Worth.

Prop. 4. *Equation of Payments.*

By Prop. 3. Compute all the present Worths, and then by Proportion, If all those present Worths amounted to the Total of all those Payments. What did 1 *l*. amount to in the said time? From the result subtract an Unit, the remainder is the Interest of 1 *l*. for the time sought, which divide by the Interest of 1 *l*. for a day, the Quote is the [illegible]

Figure 11.2 John Collins, *The Doctrine of Decimal Arithmetick*, London 1665, title page.

The other book was Collins's *Introduction to Merchants Accounts* (London 1653), which covered questions such as the production of ledgers, factorage, and how to keep ships' accounts.[30] It went through several editions up to 1697, after first appearing as an appendix to Gerard de Malynes's (*fl.* 1585–1627) *Consuetudo, vel, Lex Mercatoria*, a widely read handbook for merchants which had first been published in 1622.[31] It appeared there as part of a compendium alongside Richard Dafforne's (*fl.* 1635) *Merchant's Mirrour*, which the Northampton-born author produced after returning from many years' residence in Amsterdam[32], and the London notary John Marius's (*fl.* 1651) *Advice concerning bils of exchange* (London 1655).[33] Dafforne's book comprised an introduction to book-keeping along with several journals and ledgers, and Collins confidently suggested that his method of presentation was more advantageous for the beginner:

> As to my Booke of Accompts it was first published in 1652 as an Appendix to Lex Mercatoria and Dafforne's booke of Accompts called the Merchants Mirrour wherein there are 3 severall Journalls and Leidgers, and he that understands any of them might be presumed to understand mine also.[34]

Before the end of the decade, immersing himself deeper into London's multi-faceted mathematical circles, Collins succeeded in producing three more books, likewise directed at the practical market, but this time drawing primarily on other publications, rather than on personal teaching experience. The *Sector on a Quadrant* (London 1659) was a work specially commissioned by Sutton, and aimed at providing a market for his paper and card instruments, whose durability Collins had been able to enhance through the application of varnish, alongside the instruments in wood or brass he otherwise produced. The quadrant itself had been designed by Thomas Harvey (*fl.* 1657–1663), whom Collins describes in the epistle to the reader as 'my loving friend'.[35] Harvey having insufficient leisure to write an explanatory booklet, Sutton asked Collins to take on the task instead. It was a division of labour, as he explained to Wallis, on the occasion of sending him an exemplar of the quadrant itself:

> At the request of Mr Sutton I wrote a despicable treatise of quadrants. His designe was to demonstrate himselfe to be a good workeman in cutting the Prints of those quadrants and thereby to obtaine Customers, mine to Improove the Prints by Vernish,

[30] John Collins, *An introduction to merchants accounts, containing five distinct questions of accounts*, London: James Flesher for Nicholas Bourn 1653.

[31] Gerard de Malynes, *Consuetudo, vel, lex mercatoria, or the ancient law-merchant. Divided into three parts: according to the essential parts of trafficke*, London: Adam Islip 1622. Malynes was an early advocate for free trade. See Joyce Oldham Appleby, *Economic Thought and Ideology in Seventeenth-Century England*, Princeton, NJ: Princeton University Press 1978, pp. 41–8, 50–1.

[32] Richard Dafforne, *Merchant's Mirrour: or, directions for the perfect ordering and keeping of his accounts*, London: R. Young for N. Bourn 1635.

[33] John Marius, *Advice concerning bils of exchange*, 2nd edn, London: William Hunt 1655.

[34] John Collins to James Gregory, 19/[29] October 1675; Cambridge University Library, MS Add. 9597/13/6, f. 157^{r}–158^{v}; *James Gregory Tercentenary Volume*, ed. Herbert Westren Turnbull, London: G. Bell & Sons 1939, pp. 337–43 at pp. 342–3.

[35] John Collins, *The Sector on a Quadrant, or a treatise containing the description and use of four several quadrants*, London: J. M. for George Hurlock 1659, sig. $A2^{r-v}$.

> which I was certaine I could accomplish 12 yeares since, to a better lustre then this I herewith send (together with a sheete of my Booke) which I now send and the which, Commaculated with Dirt or Inke, will be washed away without dammage.[36]

There is an abundance of evidence of Collins collaborating closely with Sutton, on the eve of the Restoration, but these remarks made to Wallis uniquely reveal him to have been practically engaged in the making of instruments, albeit in the process of finishing. It is quite possible that at that time there was a stronger artisanal component to his professional activity than otherwise appears from his letters and publications.

The second book that came out in 1659, *Geometricall Dyalling*, is a case in point (Figure 11.3). As Collins had in earlier years received instruction from one of the foremost constructors of sundials, John Marr, and had additionally produced this publication, likely he was himself involved in the construction, although no evidence of this has come down to us. Ostensibly an explication and demonstration of 'divers difficulties in the Works of the Learned Mr. Samuel Foster', notably his *Art of Dialling* (London 1638) and *Uses of the Quadrant* (London 1652),[37] Collins's *Geometricall Dyalling* seeks to show how methods for constructing particular cases can be derived from the Gresham professor's general scheme for drawing and 'inscribing the Requisites [*sc.* dials] in oblique leaning plains'.[38] According to the narrative Collins presents in his epistle to the reader, Foster had conveyed this general scheme to the mathematical practitioner Thomas Rice (*fl.* 1640–1660), employed as one of the gunners at the Tower of London, in 1640. Himself a pupil of Foster, Rice enjoyed a good reputation as a diallist, being 'much exercised in making of Dyalls in many eminent places in the City', and was according to Collins concerned that the scheme should be made public. Alongside Sutton, he is named by Collins as having promoted the publication.[39]

Not unusually for that period, fear of a significant discovery being lost to mathematics serves to provide the motivation for another man's publication, although much of what *Geometricall Dyalling* contains undoubtedly was the result of Collins's own efforts. But it was not just a question of furthering the interests of mathematics itself. Following Foster's death as well as that of his friend and executor Edmund Wingate

[36] John Collins to John Wallis, 28 February/[10 March] 1665/6, *Correspondence of Wallis*, vol. II, pp. 191–4 at pp. 193–4. See Letwin, *Origins of Scientific Economics*, pp. 108–9. Collins's knowledge of varnish is revealed also in a letter written to Oldenburg probably in December 1668: London, Royal Society, Classified Papers 24, No, 19; *Correspondence of Oldenburg*, vol. V, pp. 211–13.

[37] Samuel Foster, *The Art of Dialling, by a new, easie, and most speedy way. Shewing, how to describe the houre-lines upon all sorts of plaines, howsoever, or in which latitude soever scituated*, London: John Dawson for Francis Eglesfield 1638; Samuel Foster, *The Uses of a Quadrant fitted for daily practise*, ed. A[nthony] T[hompson], London: for Francis Eglesfield 1652. In later years of his life Foster was severely infirm, and friends such as Anthony Thompson and John Twysden assisted him in publishing his work.

[38] John Collins, *Geometricall Dyalling: or, Dyalling performed by a line of chords onely, or by the plain scale*, London: Thomas Johnson for Francis Cossinet 1659, title page.

[39] Ibid., epistle to the reader: 'Moreover, Mr. Rice added, that in regard of the death of the Author, and since of his Executor, who had the care and inspection of his Papers, I should do well to Study out the Demonstration of the former Scheme, and make it publick, the rather, because it hath been neglected. The manner of Inscribing the hour-lines being already published in a Treatise of the Authors, Intituled, Posthumi Fosteri, this desire of his, which was also furthered by Mr. Sutton and others, I am confident is fully effected in the following Treatise'.

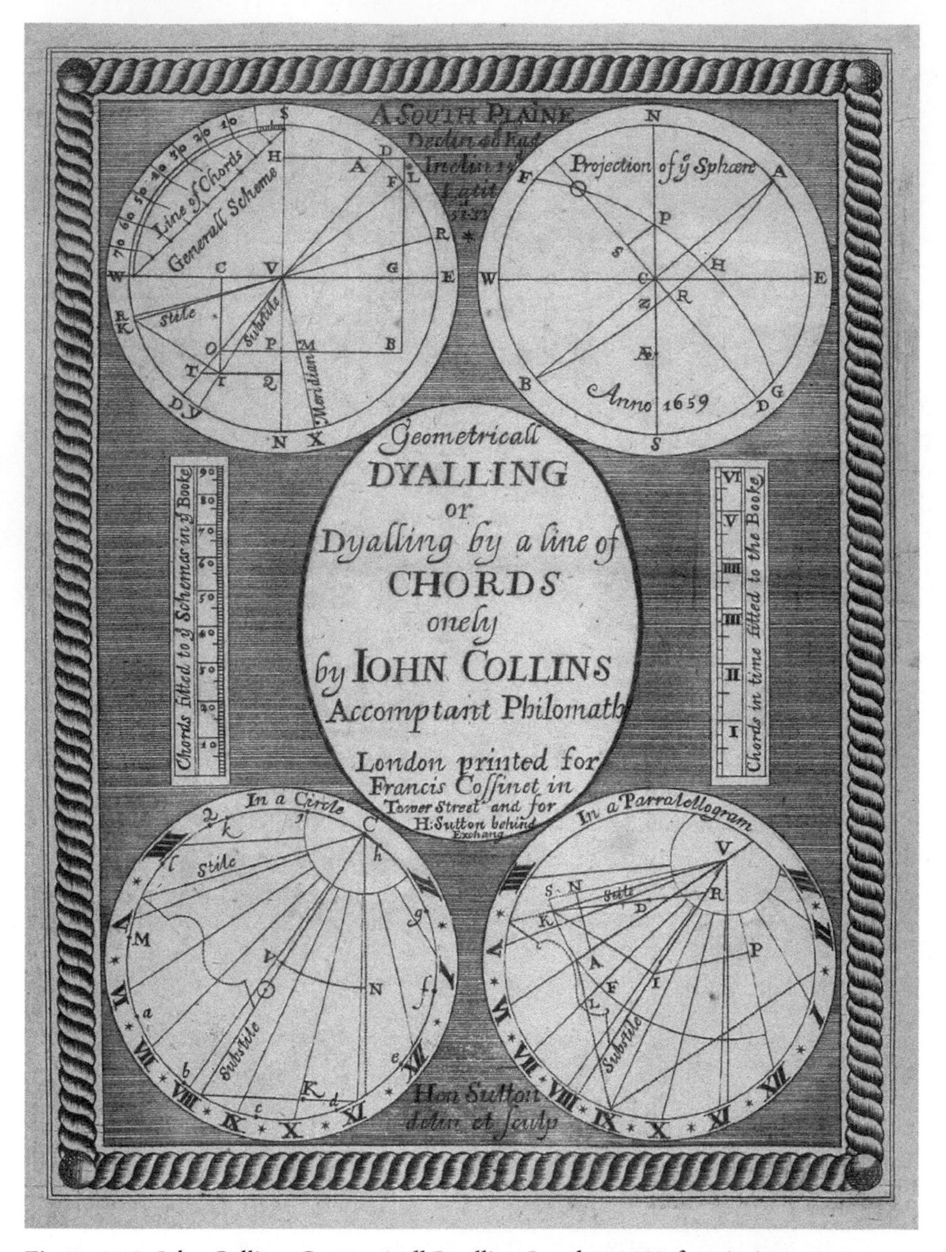

Figure 11.3 John Collins, *Geometricall Dyalling*, London 1659, frontispiece.

(1596–1656), the Gresham professor's good name had evidently been sullied through the appearance of numerous illegal and imperfect copies of various works.[40]

[40] See the epistle to the reader in his posthumous edition of a previously unpublished tract by Foster, *Posthuma Fosteri: The description of a ruler, upon which is inscribed divers scales: and the Uses thereof:*

The frontispiece of *Geometricall Dyalling* comprises an elaborate set of engravings by Sutton which illustrate some of the methods of projection of the sphere and of inscribing hours employed by Collins. They replicate some of the many illustrations throughout the book for which Sutton was responsible. Here, the close links between mathematical authorship, bookselling, and instrument making in the metropolis are eminently visible and reaffirm particularly Collins's close association with Sutton at the time. An advertisement strategically placed after the list of contents informs readers that various scales fitted to any radius along with all manner of other mathematical instruments 'either for Sea or Land' were to be had, 'exactly made in Brass or Wood' by Henry Sutton in Threadneedle Street. There were reciprocal benefits for the two men and for the bookseller, Francis Cossinet (*fl.* 1659), through this association: Collins was able to illustrate the value of Sutton's instruments to the book's readership, notably mariners and diallers, and therefore attract customers to Sutton's workshop in Threadneedle Street, while the good names of both Sutton and Collins in London's practical milieus would provide the necessary guarantee to Cossinet that *Geometricall Dyalling* was a viable enterprise. At the same time, Cossinet could guarantee interest in the book to the author and the illustrator since his shop in Tower Street specialized in publications for seamen and conveniently also sold maps and instruments, too.

The mathematician and the astrologer

Notable among Cossinet's other publications was *Natura prodigiorum* by the astrologer John Gadbury (1627–1704), who drawing on another branch of mathematical tradition styled himself not dissimilar to other practitioners as 'philomathematicus'.[41] Gadbury was a friend of Collins and the early biographies of the two men are remarkably similar: they were born around the same time in Oxfordshire villages just a few miles apart and as young men took up apprenticeships in the university city for a short time, before seeking to improve their chances in London. Having learnt the trade under the tutelage of the astrologer and medical practitioner Nicholas Fiske (1579–1659), Gadbury began publishing annual almanacs and ephemerides, while also venturing into fields such as navigation, astronomy, and exploration. From the early 1650s onwards, he together with Joshua Childrey (1625–1670) and John Goad (1616–1689) embarked upon an ambitious programme of reforming astrology along the lines of natural philosophy and later succeeded in garnering the support in this endeavour of numerous fellows of the Royal Society such as John Aubrey (1626–1697), John Beale (1608–1683), and Jonas Moore (1617–1679). Collins, who

invented and written by Mr. Samuel Foster, late Professor of Astronomie in Gresham-Colledg, London: Robert and William Leybourn for Nicholas Bourn 1652, sig. A3^{r-v}, where he points out that there were not only illegal copies of the work in hand in circulation: 'We thought fit farther to advertise thee, that there are abroad in particular hands, imperfect Copies of some other Treatises of the same Author: Namely, An easie Geometricall way of Dialling. Another most easie way to project houre-lines upon all kinde of Superficies, without respect had to their standing, either in respect of Declination or Inclination. A Quadrant fitted with lines for the solution of most Questions of the Sphere: with some other things of the like nature.'

[41] John Gadbury, *Natura prodigiorum: or, a discourse touching the nature of prodigies*, London: for Francis Cossinet 1665.

was elected to the Royal Society in 1667 on the proposal of the former Savilian professor of astronomy and then Bishop of Salisbury Seth Ward (1617–1689), also gave encouragement to this programme.[42] In his collection of 150 nativities, published in 1662, Gadbury included that of his friend Collins, based on his precise place and date of birth, and describing him thus:

> This native is a Student in the Mathematicks, and hath published many excellent Pieces therein; which declare his Judgement and Fancie to be sharp and active enough: for you must know, it is impossible for any Man, without an excellent Understanding, to be a good Mathematician; the Study is so large and various.[43]

In his description of Collins, Gadbury echoes the self-perception of the London mathematician, who claimed that he was elected to the Royal Society for his mathematics, on recommendation, although he was 'a meane person'.[44] The image painted is one of a man who was mathematically skilled, but lacked the social graces and manners of speech of learned gentlemen, and who nonetheless despite little in the way of formal education succeeded in producing considerable contributions to his science:

> That his Elocution is but indifferent, (although the matter of which he speaks be always substantial) I know to be true, the Native being my Friend and Country-man: and it is no wonder at all; (neither is it a disgrace of shame, for him, or any Man, to wear the Livery of the Heavens!) [...] It is to be noted, this Native applyed himself to the Study of the Mathematicks, without much (if any) Assistance therein; and onely by his own Labour and Industry, hath (for the advantage of his Country) published, (1.) His Sector upon a Quadrant: (2.) Geometrical Dyalling, or Dyalling performed by a Line of Chords only, (3.) The Marriner's plain Scale, new plain'd. All which, by the ablest Mathematicians, are, for their Facility and Plainness, held the best of their kind extant.[45]

[42] See Bernard Capp, *Astrology and the Popular Press: English Almanacs 1500–1800*, London: Faber and Faber 1979, pp. 185–90. Seth Ward, it should be noted, was a vehement opponent of astrology. See his anonymously authored *Vindiciae Academiarum containing, some briefe Animadversions upon Mr. Websters Book, stiled, The Examination of Academies*, Oxford: Leonard Lichfield for Thomas Robinson 1654, p. 30: 'But the mischiefe is, we are not given to Astrology [...] Nay call it that ridiculous cheat, made up of nonsence and contradictions, founded only upon the dishonesty of Impostors, and the frivolous curiosity of silly people'.

[43] John Gadbury, *Collectio geniturarum: or, a collection of nativities, in CL genitures*, London: John Cottrel 1662, p. 186. Gadbury also published a nativity of his 'very good Friend' William Leybourn, printer, with his brother Robert, of *Posthuma Fosteri* and himself author of numerous mathematical text books; pp. 187–8.

[44] John Collins, 'A Narrative of the Case of John Collins Accomptant in relation to Employments about his Majesties Affaires', Kew, The National Archives, SP 29/398, f. 261^{r}–$261a^{v}$ at f. $261a^{r}$: 'The said Collins though a meane person was in 1667 admitted a Member of the Royall Society, upon an account they had received that the said Collins might be usefull in printing & promoting Mathematicks, which he presumeth he hath accordingly been'. See also Michael Hunter, *Science and Society in Restoration England*, Cambridge: Cambridge University Press 1981, pp. 72–3.

[45] Gadbury, *Collectio geniturarum*, pp. 186–7.

A sense of community

No opportunity was missed by mathematical practitioners to cite the work of respected friends and contemporaries, but also their own publications. Collins concludes his epistle to the reader in *Geometricall Dyalling* by noting that anyone arriving without sufficient knowledge of procedures such as delineating chords or drawing parallels would find them dealt with in his new treatise *Navigation by the Mariners Plain Scale new plain'd*, which was then still at the press.

Collins considered the last of his three books to come out in 1659 to be an introductory work suitable in particular for prospective seafarers, with part of the first chapter explicitly conveying 'geometrical rudiments'. In the short epistle to the reader, he cites his own time at sea and the lack of such a publication 'suitable to the meanest Capacity, and performed by a Scale of small bulk and price' as his chief motivations, noting that the tables contained such as those of tangents and sines could be confirmed, if necessary, solely by the use of a bare ruler and compasses.[46] The only other instruments required are printed in the frontispiece: a full-sized print of an azimuth compass and plain scale, engraved by Sutton, and suitable for cutting out and creating paper or wooden instruments.[47] But *Mariners Plain Scale* is also remarkable in a number of other ways, too. Collins conceived the work as promoting the public good which he here identifies with the interests of the East India Company, to whom it is dedicated as promising and presaging 'no less then the future felicity of this Nation'.[48] With the Company once again on a sound commercial footing after years of Cromwell's opposition to its monopoly, Collins would have known that English mariners and merchants had to confront and deal with the rivalry with the Dutch, recognized to be the finest and most efficient seafarers in the world. Collins saw his book as potentially furthering the aims of the Company through improved training of seafarers in particular. His impressive dedication is buttressed by a poem dedicated to Collins himself by his friend the London-born arms-painter and land-surveyor Sylvanus Morgan (1620–1693), celebrating the three works published that year.[49] Among Morgan's numerous publications is also be to found a book on dialling.[50] What perhaps is most remarkable is a particular concession to the commercial interests of Cossinet, and indeed Sutton as well, in that *Mariners Plain Scale* is split up into three parts each of which was to be offered for sale separately, this being, as Collins writes, 'for the ease of the Buyer, and advantage of the Stationer, who finde small Bulks more convenient for Sale than great'.[51] *Mariners Plain Scale* and Collins's other

[46] John Collins, *Navigation by the Mariners Plain Scale new plain'd: or, a treatise of geometrical and arithmetical navigation*, London: Thomas Johnson for Francis Cossinet 1659, First Book, sig. A2^{v}.

[47] Ibid. Collins provides explanations of the construction and uses of these instruments in the course of the first part of the book: pp. 13–14 ('The Description of the Scale in the Frontispiece'), pp. 37–8 ('The Azimuth Compass in the Frontispiece described').

[48] Ibid., sig. A1^{v}. See Arnold A. Sherman, 'Pressure from Leadenhall: The East India Company Lobby, 1660–1678', in: *Business History Review* 50 (1976), pp. 329–55; Appleby, *Economic Thought*, pp. 102–3, 166–8.

[49] Collins, *Mariners Plain Scale*, First Book, sig. A3^{r}.

[50] Sylvanus Morgan, *Horologiographia optica. Dialling Universall and Particular*, London: R & W Leybourn for Andrew Kemb and Robert Boydell 1652.

[51] Collins, *Mariners Plain Scale*, Second Book, sig. A2^{r}. The final division of the publication into parts appears to have been decided after printing had begun and does not correspond entirely to the principal

books were intended for a ready market of diallers, seafarers, merchants, and tradesmen in England's bustling metropolis. They were relatively inexpensive to produce, and were offered at an affordable price. Above all, there was no shortage of buyers.[52]

Collins and Pell: promoting algebra and practical mathematics

In many ways the trajectory of Collins's career at the end of the Protectorate would have appeared little different to that of countless other mathematical practitioners in London who made their living through teaching, book production, or instrument making. However, he soon sought actively to tap into the more theoretical discussions of academically minded contemporaries, most notably through dealings with the mathematician and political agent John Pell (1611–1685), who having been sent by Oliver Cromwell (1599–1658) on a diplomatic mission to Switzerland in 1654 had only returned to England four years later, shortly before the Lord Protector's death. Pell had long cultivated contacts with members of London's community of practical mathematicians, including John Leake (*fl.* 1640–1686), Henry Bond (*c.*1600–1678), and John Marr. Among his papers there is a note of a geometrical problem proposed to him by Leake on 17 February 1641,[53] while in a letter to Marin Mersenne (1588–1648) the previous year he described an encounter with Marr at Richmond in 1638.[54] It is highly likely that Marr, from whom Collins received his earliest mathematical knowledge, would have mentioned this encounter to his junior clerk. More recently, he would have had professional commerce both with Leake and Bond.

The earliest recorded meeting between Collins and Pell was on 21 August 1662. In a memorandum made the following day, Pell notes that 'Mr John Collins of the Excise Office told me that he extremely desired the Demonstrations of two truths confirmed by experience'.[55] In archetypal fashion the practitioner sought the assistance of the theoretician in providing a rigorously derived proof for what hitherto had been successfully practised only. The truths concerned were directly related to Collins's practical work on questions of navigation and victualling, and publications resulting from these. They included a gauging theorem which Pell considered Collins himself to have devised:

title page where no mention of separate books is made. The first book comprises 'Geometricall Rudiments' (pp. 1–56) and a 'Treatise of Navigation' with a separate title page which does refer to the three books but not to its purported contents (pp. 1–128). The Second Book, is devoted to 'The Drawing and Delineating of the Analemma' (pp. 1–68), while the Third Book shows 'the Uses of a Line of Chords, in resolving all the Cases of Spherical Triangles Stereographically' (pp. 1–36).

[52] See Philip Beeley, 'Practical Mathematicians and Mathematical Practice in later Seventeenth-Century London', in: *British Journal for the History of Science* 52 (2019), pp. 225–48.

[53] London, British Library, MS Add. 4407, f. 143^r. See Noel Malcolm and Jacqueline Stedall, *John Pell (1611–1685) and his Correspondence with Sir Charles Cavendish*, Oxford: Oxford University Press 2005, p. 93.

[54] See John Pell to Marin Mersenne, [14]/24 January 1640, *Correspondance du P. Marin Mersenne, Religieux Minime*, ed. Cornelis de Waard, René Pintard, Bernard Rochot, et al., 17 vols, Paris: PUF and CNRS 1933–88, vol. IX, pp. 57–68 at p. 60. In this letter Pell also mentions seeing Bond.

[55] London, British Library, MS Add. 4425, f. 274^r.

1. The equality of the Divisions of the Mercatorian line with the divisions of the Logarithmicall-Tangent line.
2. The Theoreme (which I thinke hee invented) by which men may Gauge vessels partly drawen out.[56]

Nor was this request unique, as a further example from two years later confirms.[57] In both recorded cases, Pell duly produced his solutions within a matter of days and no doubt communicated them to Collins by letter, although only his own records survive.

Collins was clearly not satisfied by the response he received on this first occasion, for he raised almost identical questions at his meeting with Wallis in Oxford in August 1665 and in correspondence with the Savilian professor shortly after that meeting. Already in his first letter, Collins pondered whether the meridian line of Gheert Cremer, better known as Gerard Mercator, be the same or in the same ratio as Edmund Gunter's (1581–1626) logarithmic scale of tangents, but this does not appear to have provoked a satisfactory response,[58] for he took up the topic again in the following letter:

> Moreover that the Meridian Line of Mercators Chart should seeme (as it doth) to be the same with the Logarithme Tangents (viz. that the adding of natural Secants should Constitute a Logarithme tangent, though an unwonted ratio) is Mysterium aliquod grande proposed long since to Mr Briggs and Mr Gunter, but not approoved or disprooved.[59]

The statement of the mathematical theory underlying the rectangular orthomorphic projection of Mercator's sea chart was first given by Edward Wright (1561–1615) in his nautical tables, where he introduced a division of parts of the meridian on that chart between the Equator and parallels of latitude compiled by the addition of secants at intervals of a minute of arc at the Equator.[60] Wright, whose name Collins felt would more appropriately be given to the sea chart, thereby unknowingly came to present in his improved tables successive values that were later seen to correspond to those of a logarithmic tangent line.[61]

[56] Ibid.

[57] Problem of 4 March 1664, London, British Library, MS Add. 4422, f. 159r–162r.

[58] John Collins to John Wallis, 2/[12] January 1665/6, *Correspondence of Wallis*, vol. II, pp.188–90. Wallis's reply, in which he set out considerations on Edward Wright's and Mercator's sea charts, has not survived.

[59] John Collins to John Wallis, 28 February/[10 March] 1665/6, *Correspondence of Wallis*, vol. II, pp. 191–4 at pp. 192–3.

[60] E[dward] W[right], *Certaine Errors in Navigation*, London: Valentine Sims 1599, sig. D1r–D1v. See Jim Bennett, 'Mathematics, Instruments and Navigation, 1600–1800', in: *Mathematics and the Historian's Craft: The Kenneth O. May Lectures*, ed. Glen van Brummelen and Michael Kinyon, New York: Springer 2000, pp. 43–56 at pp. 47–8; Margaret E. Schotte, *Sailing School: Navigating Science and Skill, 1550–1800*, Baltimore: Johns Hopkins University Press 2019, pp. 100–3.

[61] See Collins, *Mariners Plain Scale*, First Book, Treatise of Navigation, pp. 34–5: 'till our late famous Countrey-man Mr. Edward Wright invented that excellent Chart, called Mercators Charts, but ought more properly to be called Wrights Chart'.

Around the same time as the exchanges between Collins and Wallis on the topic, an article by Niklaus Kauffman (1620?–1687), better known as Nicolaus Mercator, appeared in the *Philosophical Transactions*, in which the German-born mathematician and astronomer sought to clarify the theory of map projection of his namesake: 'The line of Artificial Tangents, or the Logarithmical Tangent-line, beginning at 45 deg. and taking every half degree for a whole one, is found to agree pretty near with the Meridian-line of the Sea-Charte'.[62] Nicolaus Mercator was keen to know whether the similarity between nautical and logarithmic tables was purely coincidental or reflected a fundamental relationship and issued a challenge to his contemporaries to tackle the problem alongside him:

> And seeing all these things do depend on the solution of this Question, Whether the Artificial Tangent-line be the true Meridian-line? It is therefore, that I undertake, by God's assistance, to resolve the said Question. And to let the world know the readiness and confidence, I have to make good this undertaking, I am willing to lay a Wager against any one or more persons that have a mind to engage, for so much as another Invention of mine (which is of less subtlety, but of far greater benefit to the publick) may be worth to the Inventor.[63]

Nicolaus Mercator was known to be working on a demonstration, but Collins evidently was doing likewise, for he would was soon able to send one to Pell, as he reported to Thomas Brancker (1633–1676):

> In lieu of that from Mr Mercator I gave him my Demonstration that the Meridian line of the Sea Chart, called Mercators, is a Logarithme Tangent—he affirming he had another of his own.[64]

However, neither Collins, Mercator, nor Pell succeeded in explaining why the table of latitudes for a rhumb line in a nautical chart should behave like a table of logarithms. This honour fell instead to the Scottish mathematician James Gregory (1638–1675), a friend of Collins, who in an appendix to his *Exercitationes geometricae*, published in 1668, showed that the meridian line is analogous to a scale of logarithmic tangents of the half complements of the latitude.[65]

[62] Nicolaus Mercator, 'Certain Problems Touching some Points of Navigation', in: *Philosophical Transactions* 13 (4 June 1666), pp. 215–18. See Joseph Ehrenfried Hofmann, *Nicolaus Mercator (Kauffman): Sein Leben und Wirken vorzugsweise als Mathematiker*, Abh. Akad. Wiss. Lit. Mainz, Math.-Naturw. Kl., Jahrgang 1950, No. 3, Wiesbaden: Steiner Verlag 1950, pp. 9–11.

[63] Mercator, 'Certain Problems', p. 217.

[64] John Collins to Thomas Brancker, 19/[29] November 1667, London, British Library, Add. MS. 4416, f. 211^{r-v}. See also Collins to Henry Oldenburg, *c.* 30 September/[10 October] 1676, *Correspondence of Oldenburg*, vol. XIII, pp. 83–92 at p. 89; Rigaud, *Correspondence of Scientific Men*, vol. I, pp. 211–20 at p. 219.

[65] James Gregory, *Exercitationes geometricae*, London: William Godbid and Moses Pitt 1668, pp. 14–24 esp. p. 19: 'Ex praedictis manifeste patet Lineam Merdianam Planisphaeri Nautici esse scalam tangentium antificialium arcuum, qui sunt semisses complementorum latitudinum, posito radio loco unitatis, quoniam (ut patet ex Trigonometria) praedictae differentiae sunt eaedem cum dictis tangentibus.'

Collins and Dary: the mathematics and politics of gauging

Questions relating to the calculation of volumes of vessels were a focal point of contemporary correspondence between Collins and his friend the tobacco-cutter and dialler Michael Dary (1613–1679), for whom he helped obtain employment, in 1663, as gauger of the excise in Bristol. A mathematician of some competence, Dary eked out a living through a modest combination of low-ranking government occupations and the sale of publications for the practical market such as *Dary's Diarie*, an account of the construction and use of a quadrant which came out in 1650 (Figure 11.4).[66] Like Collins, he maintained close ties to Henry Sutton and he referred the readers of the *Diarie* to the instrument maker in Threadneedle Street, where the quadrant described 'or any other Mathematicall Instrument, either for Sea or Land, is exactly made in Brasse or Wood'.[67] But there was a deeper goal to Dary's mathematical engagement. As gauger, he sought to refine the process of determining the capacity of the variously shaped vats, hogsheads, and barrels in which inn-keepers and vintners kept

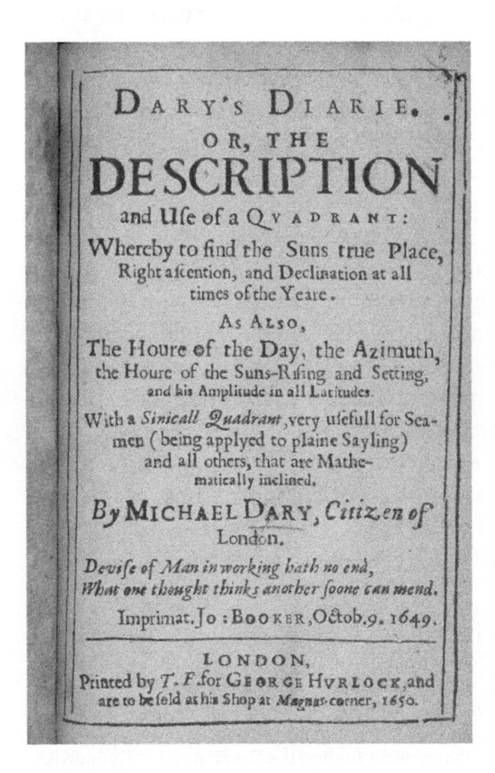
DARY'S DIARIE.
OR, THE
DESCRIPTION
and Uſe of a QVADRANT:
Whereby to find the Suns true Place, Right aſcention, and Declination at all times of the Yeare.
AS ALSO,
The Houre of the Day, the Azimuth, the Houre of the Suns-Riſing and Setting, and his Amplitude in all Latitudes.
With a *Sinicall Quadrant*, very uſefull for Seamen (being applyed to plaine Sayling) and all others, that are Mathematically inclined.
By MICHAEL DARY, *Citizen of* London.
Deviſe of Man in working hath no end,
What one thought thinks another ſoone can mend.
Imprimat. Jo: BOOKER, Octob. 9. 1649.
LONDON,
Printed by *T. F.* for GEORGE HURLOCK, and are to be ſold at his Shop at *Magnus-corner*, 1650.

Figure 11.4 Michael Dary, *Dary's Diarie or, the Description and Use of a Quadrant*, London: T. F. for George Hurlock 1650, title page.

[66] Michael Dary, *Dary's Diarie: Or, the description and use of a quadrant*, London: T.F. for George Hurlock 1650.

[67] Ibid., sig. A3^{v}.

their wine by determining the capacity from the linear measures of the vessel. Old-fashioned empirical methods such as emptying the contents of a vessel into containers of known capacity were to be replaced by mathematically elegant ones.[68]

Following his arrival in Bristol, Dary wrote to Collins to thank him for the assistance provided in obtaining the position, but then turned rapidly to the topic of 'those tuns or vessels which brewers do use', for he was able to announce the discovery of 'an excellent general theorem' which had given him great mental satisfaction and above all 'an incredible facility and celerity in calculating, the contents of those vessels from inch to inch'.[69] He did not go into detail other than to say that he did it all 'by addition of the primes (or heads) of the first, second, and third differences', in other words by means of figurate numbers. The complete explanation and corresponding table were eventually published six years later in *Dary's Miscellanies*, a copy of which he immediately sent to Isaac Newton (1642–1726/7).[70]

Dary's early letter to Collins is important not just on account of the mathematical news it transmits but also because of the strong sense of community it conveys among practitioners in London. Having left the metropolis behind him, Dary asked his correspondent to pass on his heartiest greetings to John Leake, Thomas Harvey, Henry Sutton, John Rowley (1630–after 1665), 'and all the rest of our brethren as if I had named them in particular'.[71]

Special respects are accorded senior members of the community, namely John Reynolds (*fl.* 1599–1663), sometime assistant master of the Mint, and Henry Bond, who in later years had become reader in navigation in the Royal Dockyard at Chatham, and was additionally father-in-law to Dary.

Like so many of his employments, that in Bristol eventually failed, too, and Dary again hit hard times. As in the past, Collins again stepped in and succeeded in procuring him employment as gauger of the excise in Newcastle.[72] By now acting as Dary's patron, Collins sought, some two years later, to obtain for him the position of master of the newly founded Mathematical School at Christ's Hospital, for which purpose he wrote a glowing testimonial to John Frederick (1601–1685), president of the Hospital and for many years member of Parliament for London. Alongside praising Dary's diverse contributions to the mathematical sciences, actual and prospective, Collins points out that he had recently been given a gunner's place in the Tower of London with the support of the mathematician Jonas Moore (1617–1679) and his friend the poet and clerk of the ordnance Edward Sherburne (1618–1702).[73]. Further support was provided by Newton, who was by now in regular correspondence

[68] See Letwin, *Origins of Scientific Economics*, pp. 107–8; Hunter, *Science and Society*, p. 96.

[69] Michael Dary to John Collins, 23 April/[3 May] 1663, Cambridge University Library, MS Add. 9597/13/6, f. 22^{r-v}; *Correspondence of Scientific Men*, vol. I, pp. 99–101.

[70] Michael Dary, *Dary's Miscellanies: Being, for the most part, a brief collection of mathematical theorems*, London: W[illiam] G[odbid] 1669, pp. 29–43. The 'Table of Figurate Numbers for the Speedy Calculation and exact Correction of the first, second and third Differences, in finding the Content of the Cylindroid or Prismoid' is on pp. 40–1. See also John Collins to James Gregory, 24 December 1670/[3 January 1671], University of St Andrews Library, ms31009, ff. 16^{r}–17^{v}; Turnbull, *James Gregory*, pp. 153–9 at p. 154.

[71] Michael Dary to John Collins, 23 April/[3 May] 1663, Cambridge University Library, MS Add. 9597/13/6, f. 22^{r-v}; *Correspondence of Scientific Men*, vol. I, pp. 99–101 at p. 100.

[72] John Collins to James Gregory, *c.* November 1671, University of St Andrews Library, ms31009, f. 29^{r}–32^{v}; Turnbull, *James Gregory*, pp. 193–204 at p. 196.

[73] John Collins to John Frederick, 24 June/[4 July] 1673, Cambridge University Library, MS Add. 9597/13/5, f. 82^{r}–82a^{v}; *Correspondence of Scientific Men*, vol. I, pp. 204–6. See also *The Diary of Robert*

with Dary on mathematical topics, the latter typically sending questions to or asking the opinion of the former.[74]

Even a testimonial from the Cambridge mathematician was unable to swing the issue. The post at Christ's Hospital went to John Leake instead, Dary having been considered to be too old to teach the school's forty pupils. It is not known how long Dary stayed at the Tower, but he certainly continued his activities at a gauger, for in December 1674 he published a fourpenny pamphlet explaining divergent measurements of one and the same cask or tun found in Greenwich, to which he gave the humorous title *A Tale of a Tub*.[75]

However, Dary increasingly turned his attention to algebraic equations, posing problems on this topic both to Gregory and to Newton, and sometimes indirectly to these two men via Collins.[76] Gregory, it must be said, was not overly impressed by what he saw of Dary's work in algebra, but Collins sought to reassure the Scottish mathematician that through reading Kinckhuysen's work on the topic there had been a change for the better:

> You will see by the inclosed that Mr Dary hath improoved himselfe in Algebra, which he did cheifly by having leisure, and reading Kinckhuysen at the farthing Office, where I gott him to attend, during my absence in the mornings, and where I shall desire his further attendance, and he having promised to undergoe the Labour, I doubt not but to find others that will contribute to his encouragement.[77]

It is noticeable that Collins makes no attempt to downplay his own role in Dary's improvement. Not only was it he that created the circumstances under which Dary was able to find the necessary leisure for his mathematical studies, but also it is clear that Collins lent him the book that made this improvement possible. Collins had long held Gerard Kinckhuysen's (1625–1666) *Algebra ofte stel-konst* (Haarlem 1661) in high esteem and had been the prime mover in efforts to get a Latin translation of the work published in England.[78]

Hooke, 1672–1680, ed. Henry W. Robinson and Walter Adams, London: Taylor & Francis 1935, pp. 39, 48; Willmoth, *Sir Jonas Moore*, pp. 147, 196.

[74] Isaac Newton to John Collins, 17/[27] September 1673, Cambridge University Library, MS Add. 9597/2/18, f. 36^{r-v}; *The Correspondence of Isaac Newton*, ed. Herbert Westren Turnbull, Joseph Frederick Scott, A. Rupert Hall and Laura Tilling, 7 vols, Cambridge: Cambridge University Press 1967–81, vol. I, pp. 307–8.

[75] Michael Dary, *A Tale of a Tub, or the Greenwich Problem*, London: for William Shrowsbury 1674. In the section devoted to stereometry in John Mayne's *Socius mercatoris: or the Merchant's Companion*, London: William Godbid for Nathaniel Crouch 1674, p. 149, the author describes his indebtedness to Dary's work on gauging, especially as set out in *Dary's Miscellanies.*

[76] See for example John Collins to James Gregory, 23 November/[3 December] 1674, University of St Andrews Library, ms31009, f. 66^{r-v}; Turnbull, *James Gregory*, pp. 290–2 at p. 291; John Collins to Isaac Newton, July? 1675, Cambridge University Library, MS Add. 3964, pp. 59–62; *Correspondence of Newton*, vol. I, pp. 346–7.

[77] John Collins to James Gregory, 4/[14] September 1675, University of St Andrews Library, ms31009, f. 79^{r-v}; Turnbull, *James Gregory*, pp. 327–8 at p. 327. For an example of Gregory's critical comments on Dary's work on equations see James Gregory to John Collins, 11/[21] September 1675, Cambridge, Cambridge University Library, MS Add. 9597/13/6, f. 154^{r-v}; Turnbull, *James Gregory*, pp. 328–9.

[78] See for example John Collins to James Gregory, 14/[24] March 1671/2, University of St Andrews Library, ms31009, f. 41^{r}–42^{v}; Turnbull, *James Gregory*, pp. 224–5 at p. 225, and Christoph J. Scriba, 'Mercator's Kinckhuysen-Translation in the Bodleian Library at Oxford', in: *British Journal for the History of Science* 2 (1964), pp. 45–58.

In September and October 1674, Dary exchanged letters with both Gregory and Newton on ways of handling equations of the form $z^n + bz^p + R = 0$ and could scarcely hide his enthusiasm; when sending copies of his latest results he announced to the Cambridge mathematician 'I Cannot refrain from sending you this: I have had fresh Thoughts this morning about those two sorts of Equations which wee have latly Bandied about, and I have attained an Universall series for any Equation of two Cossique notes'.[79] Older friends were sometimes also called upon to provide assistance when Dary came across problems with algebraic equations he was unable to solve by his own efforts. Thus, when in 1675 he tried unsuccessfully to find a method for solving the quartic equation $+ yyyy + 8yyy - 24yy + 104y - 676 = 0$, which he called 'this soure crabb', he sent the task to Collins who in turn passed it on to Pell.[80]

Fittingly, Dary's last publication was devoted to the topic of adfected equations and appeared in 1678. It was rather different to his first book on equations, a modest introductory work he had brought out some fourteen years earlier after reading a tract by the Cartesian Florimond De Beaune (1601–52).[81] As with *A Tale of a Tub*, it took the form of a fourpenny pamphlet, and was 'submitted to the censure' of four very different friends, two of whom, Jonas Moore and Edward Sherburne, had earlier been credited with helping him obtain his post as gunner at the Tower. The other two were James Hoare, junior (d.1679), chief clerk at the Mint, and the astrologer George Wharton (1617–1681), for whom recently the lucrative post of treasurer and paymaster of the ordnance had been created, and whose illustrious circle included the mathematicians William Brouncker (*c.* 1620–1684) and Charles Scarburgh (1615–1694).[82] Dary's pamphlet, comprising three concise chapters, was anything but illustrious, the words at the end signifying the sole means by which its survival as printed document might be secured: 'I was desired by some to Print this in this Form, and leave a pretty wide Margent on the left side, that so they might bind it up with any of my Books, or with any other Book what they please'.[83]

Collins and Wallis: the politics of adequate scientific exchange

Collins was similarly inclined to excessive modesty when it came to his own publications. For some time following his meeting with Wallis in Oxford and his subsequent return to London he had expected to see the Savilian professor during a visit to the metropolis. Such an occasion would have provided him with an ideal opportunity to show Wallis his gratitude by personally giving him some of his mathematical

[79] Michael Dary to Isaac Newton, 15/[25] October 1674, Cambridge University Library, MS Add. 9597/2/18, f. 41^{r-v}; *Correspondence of Newton* I, p. 326.

[80] London, British Library, MS Add. 4425, f. 57. See Malcolm and Stedall, *John Pell*, p. 278.

[81] Michael Dary, *The General Doctrine of Equation reduced into brief precepts*, London: for Nathaniel Brook 1664.

[82] See Willmoth, *Jonas Moore*, pp. 146–7; Capp, *Astrology and the Popular Press*, pp. 41, 44–5, 194–6; Michael Hunter, *John Aubrey and the Realm of Learning*, London: Duckworth 1975, p. 47; Christopher Edgar Challis, *A New History of the Royal Mint*, Cambridge: Cambridge University Press 1992, pp. 356–7. James Hoare junior was elected fellow of the Royal Society in 1669, having been proposed by the president, William Brouncker.

[83] Michael Dary, *The Doctrine of Adfected Equations Epitomized*, London: M. Clark for the author 1678.

publications. When the anticipated visit failed to materialize, Collins instead sent a copy of his *Introduction to Merchants Accounts*, but included the comment that the book on account of its subject matter was scarcely worthy of acceptance. He stopped short of also sending his *Mariners Plain Scale new plain'd*, describing that work as being 'altogether unworthy the view of a Geometer', most likely because of that book's rather chaotic composition. On the other hand, he was able to indicate that through his support and instruction the standard of Marke's workmanship had improved:

> Upon the expectation of your arrivall here I have hitherto forborne to send you my booke of accompts, which is so excentrique to your studies as I thought unworthy of your acceptance, but notwithstanding have sent it herewith. That pasted quadrant was of Mr. Markes fixing, but now he is more carefull. As to my Navigation it is altogether unworthy the view of a Geometer or that Character you are pleased to affoard it.[84]

Over time, the exchanges between Collins and Wallis become increasingly collegial, and increasingly concerned with topics which were of interest and importance to both of them. Thus, when Collins, in the summer of 1666, requested that Wallis on his next visit to London bring with him his copy of Luca Valerio's (1553–1618) *Subtilium indagationum liber primus* (Rome 1632), one of a number of contemporary works aimed at improving the Archimedean method of exhaustion, he did so expressly with the purpose of assessing its approach for the purposes of gauging. Wallis gladly agreed: 'That of Valerius, (if I forget it not,) I intend to bring with mee next time I come to London, that you may see it as you desire'.[85]

In this setting, their roles were clearly defined. Collins would send news from the mathematical circles he moved in or of books which had arrived from abroad. Sometimes, he would pose practical questions on topics such as on Edmund Gunter's logarithmically divided scale or on problems in trigonometry that would be of particular interest to gaugers. In his replies, Wallis would patiently set out his response to the various points raised. If their correspondence was initially in this sense somewhat one-sided, they soon found areas of common interest, particularly on questions of algebra, where Collins proved to be a more than an adequate interlocutor. An important vehicle of this change of focus was their involvement in the publication of John Pell's *Algebra*, a book whose genesis was unique even by seventeenth-century English standards.

Originally intended simply to be a translation of the *Teutsche Algebra* written by the Swiss mathematician and administrator Johann Heinrich Rahn (1622–1676), and as such registered with the Stationers' Company in London, things only changed after the translator Thomas Brancker discovered that Rahn's former teacher had been John Pell, and that he happened then to be in London. The two men decided that Pell would not only review ongoing work on the translation, but also was free to make

[84] John Collins to John Wallis, 2/[12] August 1666, *Correspondence of Wallis*, vol. II, pp. 275–8 at p. 275.

[85] John Wallis to John Collins, 7/[17] August 1666, *Correspondence of Wallis*, vol. I, pp. 278–81 at p. 280. Interestingly, Henry Oldenburg in his letter to the English Resident in Florence, John Finch, of 10/[20] April 1666 passed on a request for the procurement of a copy of this work. See *Correspondence of Oldenburg*, vol. III, pp. 86–7. Unfortunately, there is no indication from whom this request originated.

suitable amendments and additions. Ultimately, the translated work was considerably expanded through the efforts of Pell, particularly in the latter sections.[86] Collins's role in the production was crucial, particularly after first Pell and then Brancker moved to Cheshire: Pell, to take up residence with his patron William Brereton in Brereton Hall, and Brancker having been presented by Brereton to the rectory at nearby Tilston. Collins was the man in London largely responsible for checking the proofs and seeing that everything reached the printer, while Brancker and John Pell completed the editorial work in Cheshire.[87] Wallis's task was more focused: he was in charge of checking painstakingly the long table of incomposite or prime numbers that made up the final section of the book.[88]

The publishing agent

Collins's good connections in the London book market had already led to a productive correspondence with another theoretical mathematician some three years before he made Wallis's acquaintance. He was at the time aware of the commercial success of Isaac Barrow's (1630–1677) Latin edition of Euclid's *Elements of Geometry*, first published in 1655,[89] followed five years later by an English translation,[90] and he knew that the man recently elected to the new Lucasian chair in mathematics at Cambridge had prepared, but not published, Latin editions of Apollonius, Archimedes, and Theodosius. As with the Latin *Elements*, these translations were faithful to the original Greek texts, but with the contents concisely presented using symbolic notation derived from Oughtred's *Clavis*. Collins soon sought to persuade Barrow to publish these texts in London.

There was a strong transactional nature to the two men's early correspondence. Collins sent Barrow details of books from abroad he was able to procure, while occasionally the Cambridge mathematician would cite the titles of other publications he desired. By way of repayment, Barrow sent Collins copies of mathematical books from his own extensive library, and eventually supplied him a catalogue, with prices of the various volumes noted, so that something like equivalence in value could be achieved; Collins proceeded likewise.[91] From Barrow's earliest surviving letter we

[86] On the background to Pell's edition see Christoph J. Scriba, 'John Pell's English Edition of J. H. Rahn's *Teutsche Algebra*', in: *For Dirk Struik: Scientific, Historical and Political Essays in Honor of Dirk. J. Struik*, ed. Robert S. Cohen, John J. Stachel, and Marx W. Wartofsky, Dordrecht: D. Reidel 1974, pp. 261–74; Malcolm and Stedall, *John Pell*, pp. 199–208, 306–12.

[87] Malcolm and Stedall, *John Pell*, p. 203. See also Scriba, 'John Pell', p. 271.

[88] See John Collins to John Pell, 18/[28] July 1668 and this letter's enclosure, *Correspondence of Wallis*, vol. II, pp. 326–9.

[89] *Euclidis elementorum libri XV. breviter demonstrati*, ed. Isaac Barrow, Cambridge: University Press for William Nealand 1655. See Mordechai Feingold, 'Isaac Barrow: Divine, Scholar, Mathematician', in: *Before Newton: The Life and Times of Isaac Barrow*, ed. Mordechai Feingold, Cambridge: Cambridge University Press 1990, pp. 1–104 at p. 43.

[90] *Euclide's Elements; the whole fifteen books. Compendiously demonstrated*, ed. Isaac Barrow, London: R. Daniel for William Nealand in Cambridge 1660.

[91] See Isaac Barrow to John Collins, 12/[22] November 1664, Cambridge University Library MS Add. 9597/13/6, f. 6^{r-v}; *Correspondence of Scientific Men*, vol. II, pp. 40–1 at p. 41; Isaac Barrow to John

learn that he had recently received from Collins a number of books, including the *Apiaria universae philosophiae mathematicae* (1642) of Mario Bettini (1582–1657) and probably Willebrord Snell's (1580–1626) *Cyclometricus, de circuli dimensione* (1621), together with one of his own.[92]

Nor did Collins neglect his own needs as a practical mathematician. We know, for example, that he sought Barrow's advice on a series of trigonometrical questions that were of interest to gaugers. However, the Cambridge mathematician was at pains to point out his own limitations, as a theoretical mathematician, in assessing the quality of more practically-orientated books; indeed, in one of his early letters he implores Collins in future to adopt a more abstract style of writing since this corresponded more to his own way of thinking:

> I received the booke of Snellius, & that of your owne composition, which you were pleased to bestow upon me, & for which I thanke you, though I must confesse it doth somewhat surpasse my capacity, who have little acquainted my selfe with that kind of Practicks, & indeed hardly with any; that little study I have employed upon Mathematical busynesses being never designed to any other use, then the bare knowledge of the generall reasons of things, as a scholar, & no further, so that if you propound any thing to me, I pray please to doe it in the most generall & abstract termes, as neere to the Geometricall style, as you can; otherwise I shall hardly understand the questions.[93]

There were evident limits to the transactional model the two men adopted, for Barrow did not reciprocate Collins's wide interests but instead tailored his requirements to the topics on which he was actually working. Thus, he specifically asked Collins to procure a copy of Bonaventura Cavalieri's *Trigonometria*, but only rather grudgingly accepted Collins's offer of Michael Stifel's (1487–1567) *Algebra* (1545) or Pietro Mengoli's (1626–1686) *Novae quadraturae arithmeticae, seu de additione fractionum* (1650), citing his lack of leisure for reading books not central to his academic concerns:

> As for Mengolus (in whom I never read any thing) & Stifelius his Algebra, & what ever other bookes, I referr it wholy to your discretion, & shall be glad to have what you shall thinke good, & though my employments allow me but little leisure to peruse them thoroughly, yet I shall so farr looke into them as to give you my judgment of them; if you so require.[94]

Mengoli's book on arithmetical quadratures had long held a fascination for Collins. Already in his *Doctrine of Decimal Arithmetick* he presumed, without having seen the work, that it contained a useful treatment of the addition of fractions in arithmetical

Collins, 29 November/[9 December] 1664, Cambridge University Library MS Add. 9597/13/6, f. 7^{r-v}; *Correspondence of Scientific Men*, vol. II, pp. 42–3 at p. 42.

[92] See Isaac Barrow to John Collins, end of 1663, Cambridge University Library, MS Add. 9597/13/6, f. 2^{r-v}; *Correspondence of Scientific Men*, vol. II, pp. 32–5.

[93] Ibid.

[94] Ibid.

progressions relevant to the calculation of interest.[95] When the copy of Mengoli eventually arrived, towards the end of the following year, Barrow deferred giving an assessment for lack of time, although he did not neglect to remark on the adverse impression the work had made on him even under cursory perusal.[96] Although he returned briefly to the question of Mengoli at the beginning of the following year, he did little more than reaffirm his earlier opinion.[97]

Soon their discourse turned to a topic of mutual interest, the properties of the hyperbola, where the Cambridge mathematician conveniently was able to draw on extensive teaching notes in Latin which he replicated largely verbatim in his letters. He also set out a number of theorems that he believed to be new concerning the surfaces of conoids and spheroids.[98] The pattern was now one of epistolary instruction. Collins would indicate where he had difficulty in understanding what Barrow had written, and Barrow would duly use his next letter to address those points causing problems.[99] The exchange of books all but disappeared into the background by this time, but we know that Collins did send Barrow copies of various parts of Pell's *Algebra* as these came off the press.[100]

In his self-appointed role as publishing agent, Collins sought early on in his correspondence with Barrow to persuade the Cambridge mathematician to commit to a new book project. Evidently, he had found the London printer William Godbid (d.1678) willing to take on Barrow's editions of Apollonius and Archimedes, briefly demonstrated using the algebraic notation of Oughtred.[101] All that was required was for Barrow to review and then release his manuscripts of the texts concerned. Barrow

[95] John Collins, *The Doctrine of Decimal Arithmetick*, p. 7: 'If you are to Aequate an Annuity, at Simple Interest, I presume a Compendium may be found in Mengolus his Arithmetical Quadratures (a Book I never saw) who it's probable by a Compendium gets the fact of an Arithmetical Progression, and adds Fractions that have a constant Numerator, and an Arithmetical Progression for their Denominators'. In a letter to James Gregory written two years later he expressed the desire to see Mengoli's Geometria speciose 'that is not here to be had'. See John Collins to James Gregory, early March 1668, Cambridge University Library, MS Add. 9597/13/6, f. 92^{r}-93^{v}; Turnbull, *James Gregory*, pp. 45–8 at p. 46. Interestingly, Oldenburg sought to obtain a copy of the work from Finch in 1666. See Henry Oldenburg to John Finch, 10/[20] April 1666, *Correspondence of Oldenburg*, vol. III, pp. 86–7. Over three years later, he asked Sluse to obtain a copy at the Frankfurt book fair. See Henry Oldenburg to René François de Sluse, 14/[24] September 1669, *Correspondence of Oldenburg*, vol. VI, pp. 232–5/235–6. It is not recorded for whom the copies were intended.

[96] See Isaac Barrow to John Collins, 12/[22] November 1664, Cambridge University Library, MS Add. 9597/13/6, f. 6^{r-v}; *Correspondence of Scientific Men*, vol. II, pp. 40–1: 'For I perceive he [*sc.* Mengoli] doth affect to use abundance of new definitions & uncouth terms, so that one must as it were learne new languages to attaine to his meaning, though it may be only somewhat ordinary is couched under them'.

[97] See Isaac Barrow to John Collins, 1/[11] February 1666/7, Cambridge University Library, MS Add. 9597/13/6, f. 9^{r-v}; *Correspondence of Scientific Men*, vol. II, pp. 46–7.

[98] See Isaac Barrow to John Collins, 6/[16] March 1667/8, Cambridge University Library, MS Add. 9597/13/6, f. 10^{r-v}; *Correspondence of Scientific Men*, vol. II, pp. 48–56 at p. 54.

[99] See for example Isaac Barrow to John Collins, 28 March/[7 April] 1668, Cambridge University Library, MS Add. 9597/13/6, f. 11^{r}–12^{v}; *Correspondence of Scientific Men*, vol. II, pp. 56–64.

[100] See Isaac Barrow to John Collins, 14/[24] May 1668, Cambridge University Library, MS Add. 9597/13/6, f. 13^{r-v}; *Correspondence of Scientific Men*, vol. II, pp. 65–7 at p. 65.

[101] It is important to note that Barrow only ascribed algebra the role of a useful notational technique; he denied that it could be an independent mathematical discipline. On this topic see Helena M. Pycior, *Symbols, Impossible Numbers, and Geometric Entanglements: British Algebra through the Commentaries on Newton's Universal Arithmetick*, Cambridge: Cambridge University Press 1997, pp. 162–5; Michael S. Mahoney, 'Barrow's Mathematics: Between Ancients and Moderns', in: *Before Newton: The Life and Times of Isaac Barrow*, ed. Mordechai Feingold, Cambridge: Cambridge University Press 1990, pp. 179–249 at pp. 189, 195–6, 200–2.

was, however, reluctant to engage with a commercial stationer, unless a good offer were made, having been cheated out of payments due to him by the publisher of his Euclid edition, William Nealand (*fl.* 1649–1662). Perhaps as a way of improving the terms of an agreement, Barrow claimed that a friend had already offered to cover the costs of printing the two classical authors if Nealand's heirs agreed to release the rights to the Euclid edition so that this alongside his rendition of Theodosius's *Spherics* could be included, too.[102]

Nothing, however, came out of these efforts and Collins remained silent on the topic until some three years later, when he sought to persuade Barrow to publish his Cambridge lectures on geometry along with his lectures on optics. Although he repeatedly implored Barrow to release the lecture notes, the Cambridge mathematician was obdurate. Collins, by now a member of the Royal Society, decided to adopt more drastic measures and in November or December 1668 wrote to the newly elected vice-chancellor of Cambridge, Edmund Boldero (1609–1679). Boldly purporting to express the wish of members of the Royal Society, Collins openly sought Boldero's assistance in achieving his long-held aim of seeing Barrow's works into print. Specifically, he entreated the vice-chancellor to use the power conferred on him by his office to

> prevaile with the said Mr Barrow to publish some other good Bookes by him intended, as his Comment on Archimedes, on the Sphaericks, his owne Perspective, Projections, Elements of Plaine Geometry, but those abovementioned [*sc.* the *Treatise of Opticks* and Apollonius's *Conics*] seeme at present to be in the greatest forwardnesse and in other respects the sooner to be desired, and seeing England affoards persons no whit inferiour to the best of Forreigners, we may presume wee doe not trespasse in urging a common benefit to Learning.[103]

It is not clear how far this approach was officially sanctioned, if at all. Collins's strategy might thus have seemed risky, but in the event it proved successful, and shortly afterwards he was able to announce to James Gregory that 'Mr Barrow doth hereafter intend a Treatise of Opticks for the Presse but is not yet ready'.[104] Nor was this announcement presumptuous, for the Cambridge lectures were published already in the following year, under the imprint of Collins's chosen printer Godbid. In the preface, Barrow expressly thanked those who helped him most in the undertaking. His successor on the Lucasian chair, and confrère at Trinity College, Isaac Newton, is praised for having reviewed and corrected the text of the lectures, while Collins is thanked for his skill and effort in bringing the volume to press.[105] He says he

[102] Isaac Barrow to John Collins, 3/[13] March 1664/5, Cambridge University Library MS Add. 9597/13/6, f. 8^{r-v}; *Correspondence of Scientific Men*, vol. II, pp. 44–6 at p. 45. See Feingold, 'Isaac Barrow', pp. 42, 44.

[103] John Collins to Edmund Boldero, 1668, Cambridge University Library, MS Add. 9597/13/6, f. 1^{r-v}; *Correspondence of Scientific Men*, vol. I, pp. 137–8. See Feingold, 'Isaac Barrow', p. 70.

[104] John Collins to James Gregory, 30 December 1668/[9 January 1669], University of St Andrews Library, ms31009, f. 4^{r-v}; Turnbull, *James Gregory*, pp. 54–8 at p. 55.

[105] Isaac Barrow, *Lectiones XVIII, Cantabrigiae in scholis publicis habitae; in quibus opticorum phaenomenon genuinae rationes investigantur, ac exponuntur. Annexae sunt lectiones aliquot geometricae*, London: William Godbid for John Dunmore and Octavian Pulleyn 1669, To the reader, 'quorum

will call Collins 'not undeservedly the Mersenne of our community' because of his commitment in realizing both his own works and those of others. Contemporary scholars soon adopted this appellation.

The three classical authors had to wait longer. It was not until 1675 that Barrow's editions of Archimedes's works, the first four books of Apollonius's *Conics*, and Theodosius's *Sphaerics* left the press, the publisher this time being Robert Scott, the most assiduous London bookseller to the Continent, whom Collins held in high regard.[106] No mention is made of Collins in the preface to these works, despite his having played a decisive role in their publication. But Collins's efforts did not stop there. Constantly aware of the need to ensure that mathematical books were successfully marketed, he soon sent Oldenburg a set of notes for him to use in writing his account of the editions for the *Philosophical Transactions*. In these, he drew attention not only to Barrow's rationale in using algebraic notation, but also to future plans on the part of the publisher:

> Sir you may take what followes for your Account of Dr Barrows edition of Archimedes Apollonius Theodosius [...]
>
> 4. That all the three Bookes viz Archimedes Apollonius and Theodosius are delivered in a breife Symbolicall method of Expression, pursuant to the Sense Propositions and demonstrations of the Ancients unlesse where the Author thought fitt to enlarge and otherwise demonstrate some of the Propositions from more easy Principles of his owne herein pursuing his owne former method in which long since he published an entire Euclid in Octavo.
>
> 5. That the Stationer Mr Scot intends to reprint the said Euclid as a part of another Volume of the Ancients in the said method, the residue of which Volume may be an abridgement of Pappus, Serenus, the 3 latter Bookes of Apollonius with such other small Tracts of the Ancients as have either been recovered, or restored by the Industry of this and the last age, but these Tracts being not as yet all in a readinesse for the Presse, the Learned that have any Lucubrations thereon or have prepared any of them for publick view would much oblige the Republic of Literature in communicating the same.

unus [...] D. Isaacus Newtonus, college noster [...] exemplar revisit, aliqua corrigenda monens, sed & de suo nonnulla penu suggerens, quae nostris alicubi cum laude innexa cernes, alter (quem nostrae gentis haud immerito Mersennum dixero, cum sua tum aliorum opera provehendis hisce literis natum) D. Joh. Collinsius, ingente suo cum labore editionem procuravit'. See Feingold, 'Isaac Barrow', pp. 69–70.

[106] *Archimedis opera: Apollonii Pergaei conicorum libri IIII. Theodosii sphaerica: method nova illustrata, & succincte demonstrata*, ed. Isaac Barrow, 3 parts, London: William Godbid for Robert Scott 1675. On the background to this edition see Feingold, 'Isaac Barrow', pp. 44–5, 75–7. On Scott see John Collins to Thomas Baker, 24 April/[4 May] 1677, Cambridge University Library, MS Add. 9597/13/5, f. 29^{r-v}; *Correspondence of Scientific Men*, vol. II, pp. 20–2 at p. 22: 'Mr Scot a bookseller in Little Britain, who drives a foreign trade'. In his letter to Isaac Newton dated 30 April/[10 May] 1672, Cambridge University Library, MS Add. 9597/2/18, f. 17^{r-v}; *Correspondence of Newton*, vol. I, pp. 146–8 at p. 147, he points out that 'our Latin Booksellers here are averse to the Printing of Mathematicall Bookes there being scarce any of them that have a forreigne Correspondence for Vent, and so when such a Coppy is offered, in stead of rewarding the Author they rather expect a Dowry with the Treatise'. Scott was an exception, being involved also in procuring scholarly manuscripts from the continent. See the letter addressed to him from Christopher Anderson in Rome, dated 22 November 1673, Cambridge University Library, MS Add. 9597/13/5, f. 5^{r-v}; *Correspondence of Scientific Men*, vol. I, pp. 210–11.

> 6. The Stationer likewise sells the former treatises of the Reverend Author being Opticall and Geometricall Lectures to the latter whereof there are now made some elegant additions de Maximis et Minimis.[107]

After only slightly rearranging and polishing these notes, Oldenburg duly printed them as the review in his journal.[108] Nothing came of Scott's intention to publish a collection of ancient mathematical texts despite Collins's call for contributions, while the 'elegant additions' he referred to had appeared the previous year in the latest edition of Barrow's *Lectiones geometricae*.[109] On the other hand, Barrow's Latin commentary on Archimedes's *De Sphaera et cylindro* and a new corrected impression of his edition of the *Elements* were published in 1678, a year following his death.[110]

Collins, Newton, and the Kinckhuysen project

It was Barrow who first introduced Collins to Newton, or rather to the work of Newton, for he did not reveal the identity of the author for some time. Initially, Barrow informed Collins simply of a friend of his in Cambridge who had recently brought him some papers 'wherein he hath sett downe methods of calculating the dimensions of magnitudes like that of Mr Mercator concerning the hyperbola, but very generall; as also of resolving aequations'.[111] Eleven days later, at the end of July 1669, he proceeded to send Collins these papers by post. They comprised none other than the treatise 'De Analysi per aequationes numero terminorum infinitas', in which Newton presents among other things his method for applying infinite series to the resolution of problems such as determining the lengths and areas of curves.[112] This was the first instance of Newton's mathematical writings being circulated within a limited circle of scientific friends and Barrow was keen to ensure that nothing went amiss, requesting that Collins acknowledge their receipt and return them to him after perusal.[113] In fact, Collins not only read 'De Analysi' but also made a copy of the work for himself. From this he made, around 1677, a further copy for Wallis who used it in preparing his *Treatise of Algebra*. Only in his next letter to Collins, written some three weeks later, did Barrow reveal the identity of the author, drawing attention to Newton's youth and

[107] London, Royal Society, Classified Papers, vol. XXIV, No. 33; *Correspondence of Oldenburg*, vol. XI, pp. 315–16.

[108] *Philosophical Transactions* 114 (24 May 1675), pp. 314–15.

[109] Isaac Barrow, *Lectiones opticae & geometricae: in quibus phaenomenon opticorum genuinae rationes investigantur, ac exponuntur: et generalia curvarum linearum symptomata declarantur*, 2 parts, London: William Godbid for Robert Scott 1674, vol. II, pp. 149–51.

[110] Lectio reverendi et doctissimi viri D. Isaaci Barrow [...] in qua Theoremata Archimedis de Sphaera & Cylindro, per methodum Indivisibilium investigate, ac breviter demonstrata exhibentur, London: John Redmayne for J. Williams, 1678; Euclidis Elementorum libri XV. Breviter demonstrati, opera Is. Barrow [...] et prioribus mendis typographicis nunc demum purgati, London: John Redmayne for J. Williams 1678.

[111] Isaac Barrow to John Collins, 20/[30] July 1669, London, Royal Society, MS 81, No. 1; *Correspondence of Newton* I, pp. 13–14 (excerpt).

[112] London, Royal Society MS 81, No. 4; *The Mathematical Papers of Isaac Newton*, ed. Derek Thomas Whiteside, 8 vols, Cambridge: Cambridge University Press 1967–81, vol. II, pp. 206–46/207–47.

[113] Isaac Barrow to John Collins, 31 July/[10 August] 1669, London, Royal Society, MS 81, No. 2; *Correspondence of Newton*, vol. I, p. 12 (excerpt).

describing him as being of 'an extraordinary genius & proficiency in these things'.[114] Contrary to his earlier instruction, he now invited Collins to impart the papers to William Brouncker, president of the Royal Society, if he so desired.[115]

Collins went on to introduce Newton's name into the wider Republic of Letters, by citing it alongside the full title of 'De Analysi', while also sketching out the work's significance, in notes he prepared for Oldenburg so that the secretary of the Royal Society could include them in his next letter to the Flemish mathematician René François de Sluse (1622–1685).[116] That letter was duly sent a few days later and also included a long list of mathematical books Sluse was requested to procure. This list had originated from Collins, but Oldenburg indicated instead that the books were for the Royal Society, presumably to avoid a personal demand appearing too prominent. Only in the postscript did he mention two requests explicitly from Collins, 'a man most studious in mathematics'.[117] These were, first, for a geometrical construction relating to the axes of conic sections and, second, for Christoph Grienberger's (1561–1636) catalogue of fixed stars.

Unsurprisingly, Collins passed on the news about Newton to his friend James Gregory, too, but a good two months later. This delay is significant, because apart from reminding Gregory of Barrow's warm praise for Newton in the preface to his lectures on optics, Collins suggests that Newton had invented his general series method of quadratures before Mercator had published his method of squaring the hyperbola in *Logarithmotechnia* (1668):

> Mr Barrow hath resigned his Lecturers place to one Mr Newton of Cambridge whome he mentioneth in his Optick Praeface as a very ingenious person, one who (before Mercators Logarithmotechnia was extant) invented the same method and applyed it generally to all Curves, and diverse wayes to the Circle, which possibly he may send up to be annexed to Mr Barrowes Lectures.[118]

The assertion that Newton had invented his method before reading Mercator must have originated from Newton himself and clearly anticipates his later claims to priority over Gottfried Wilhelm Leibniz (1646–1716) in the discovery of the calculus. In all likelihood it was sown in Collins's mind through Barrow, for records suggest that Newton and Collins only became directly acquainted with one another after this

[114] Isaac Barrow to John Collins, 20/[30] August 1669, London, Royal Society MS 81, No. 3; *Correspondence of Newton*, vol. I, pp. 14–15 (extract).

[115] Ibid.

[116] John Collins to Henry Oldenburg, *c.* 12/[22] September 1669; *Correspondence of Oldenburg*, vol. VI, pp. 226–9.

[117] Henry Oldenburg to René François de Sluse, 14/[14] September 1669, *Correspondence of Oldenburg*, vol. VI, pp. 232–5/235–6: 'Desiderat Dominus Collins, Vir Mathematum studiosus, ut, si commode fieri possit, ipsi mittas Griembergeri Catalogum fixarum stellarum'. See also Henry Oldenburg to René François de Sluse, 2/[12] April 1669, *Correspondence of Oldenburg*, vol. V, pp. 468–70/470–1, where Oldenburg introduces Collins as 'vir Mathematum amantissimus, et Scriptorum Mathematicorum helluo incomporabilis'.

[118] John Collins to James Gregory, 25 November/[5 December] 1669, University of St Andrews Library, ms31009, f. 7^{r-v}; Turnbull, *James Gregory*, pp. 73–4.

letter to Gregory was written.[119] In a letter written to Gregory on Christmas Eve the following year, Collins gave an account of his first meeting with Newton in London, at which he asked his help with summing a harmonic progression:

> I never saw Mr Newton (who is younger then your selfe) but twice viz somewhat late upon a Saturday night at his Inne, I then proposed to him the adding of a Musicall Progression, the which he promised to consider and send up, I told him I had done something in it, and would send him what Considerations I had about it, but his came up (before I sent him mine) without any Indication of his method. And againe I saw him the next day having invited him to Dinner, in that little discourse we had about Mathematicks, I asked him what he would make the Subject of his first Lectures, he said Opticks proceeding where Mr Barrow left [...] having no more acquaintance with him, I did not thinke it becomming to urge him to communicate any thing, the rather in regard Mr Barrow told me the Mathematick Lecturer there is obliged either to print or put 9 Lectures yearly in Manuscript into the publick Library, whence Coppies of them might be transcribed.[120]

Collins first wrote to Newton at the end of 1669, following their meeting in London. As with Wallis and Barrow, he initiated the correspondence by sending Newton a number of questions, in this case on series as well as one on the resolving of equations by means of tables. Newton expended an inordinate amount of time particularly on the first question, his response taking up almost all of his reply. As a sweetener, Collins sent the new Lucasian professor through Barrow's hands part of Wallis's *Mechanica*, which he was then seeing through the press in London. He also sent him by the same path a copy of Kinckhuysen's *Algebra ofte Stel-konst* (1661), with Nicolaus Mercator's handwritten Latin translation interleaved, with the expressed desire that Newton would identify defects in the work and suggest how they might be rectified (Figure 11.5).[121] Collins's intention thereby was that the elaborated translation would be an up-to-date introductory text attractive to a potential bookseller and capable of serving the needs of 'young Students of Algebra'.[122] Newton took his time with his review, but by July the following year he had returned the copy of Kinckhuysen to Collins with his marginal notes inserted, or, as he wrote, 'intermixed with the Authors discourse'.[123] Newton's additions, among other things on irrational roots and

[119] According to Trinity College records, Newton travelled to London on 26 November 1669 and returned to Cambridge on 8 December. See Whiteside's introduction to *Mathematical Papers of Newton*, vol. II, pp. x–xi. There are justifiable grounds for suspecting the authenticity of this record, given that Collins wrote to Gregory precisely the day before Newton appears to have arrived in London and two days before the most probable date of their first meeting, on 27 November. Of course, if the meeting occurred before Collins wrote to Gregory there would be no need to postulate Barrow acting as intermediary.

[120] John Collins to James Gregory, 24 December 1670/[3 January 1671], University of St Andrews Library, ms31009, f. 16^{r}–17^{v}; Turnbull, *James Gregory*, pp. 153–9 at p. 154.

[121] The surviving original manuscript, contained in the copy of Kinckhuysen, was discovered by Christoph J. Scriba nearly fifty years ago. See his 'Mercator's Kinckhuysen-Translation'.

[122] John Collins to Isaac Newton, 13/[23] July 1670, Cambridge University Library, MS Add. 9597/2/18, f. 72^{r}–73^{v}; *Correspondence of Newton*, vol. I, pp. 32–3.

[123] Isaac Newton to John Collins, 11/[21] July 1670, Cambridge University Library, MS Add. 9597/2/18, f. 6^{r-v}; *Correspondence of Newton*, vol. I, pp. 30–1. Earlier, in his letter of 19/[29] January, Newton had written 'Your Kinck-Huysons Algebra I have made some notes upon. I suppose you are not so much in hast of

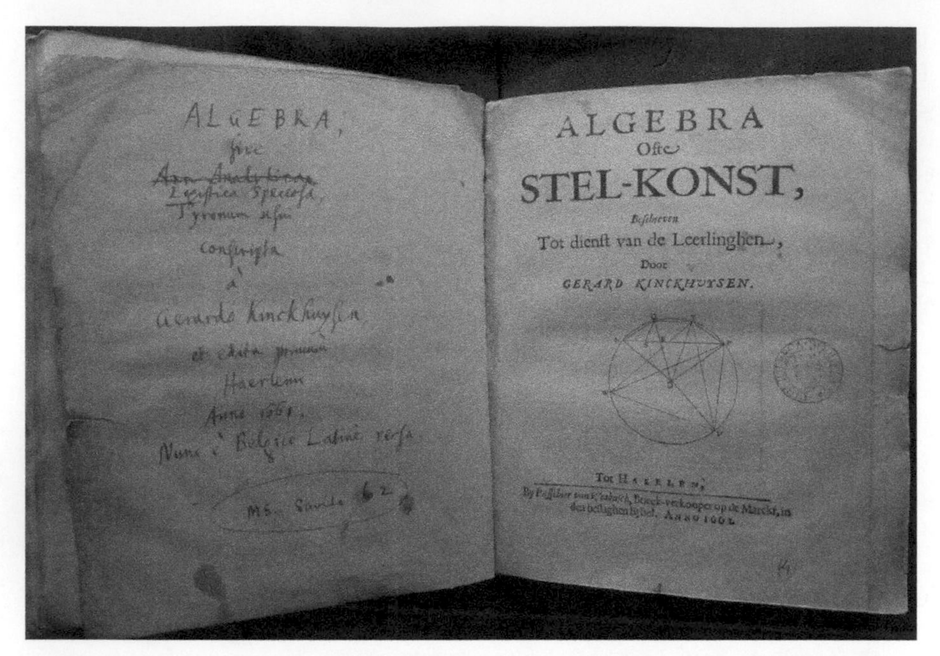

Figure 11.5 Gerard Kinckhuysen, *Algebra ofte Stel-Konst, Haarlem*: Wesbusch 1661 with Mercator's Latin translation and autograph notes by Newton interweaved.

the extraction of cubic roots, the latter drawn from Ferguson's *Labyrinthus algebrae* (1667), found Collins's approval, but the tirelessly active philomath proposed that further additions might be drawn from a number of other useful Dutch and German publications to which readers would scarcely have access and to this end promptly sent Barrow's successor copies of Scheubel, Van Ceulen, and Hume.[124]

Unfortunately, the aims of the two men did not entirely coincide. While Collins was keen to devise with Newton's help an attractive publication package, Newton

it, which makes me doe that onely at my leisure'. See Cambridge University Library, MS Add. 9597/2/18, f. 1^{r}–$1a^{v}$ at f. 1^{v}; *Correspondence of Newton*, vol. I, pp. 16–20 at p. 20. In February, he reiterated his willingness to write notes on rectifying mistakes or defects of Kinckhuysen's work, but suggested that it did not warrant a formal commentary as Collins had evidently recently proposed. See Isaac Newton to John Collins, 6/[16] February 1669/70, Cambridge University Library, MS Add. 9597/2/18, f. 2^{r-v}; *Correspondence of Newton*, vol. I, pp. 23–5 at p. 24. In his first of two letters written to Collins in July, he confessed that he was not quite sure what was required of him. See Isaac Newton to John Collins, 11/[21] July 1670, Cambridge University Library, MS Add.9597/2/18, f. 6^{r-v}; *Correspondence of Newton*, vol. I, pp. 30–1 at p. 30: 'I know not whither I have hit your meaning or noe'.

[124] See John Collins to Isaac Newton, 13/[23] July 1670, Cambridge University Library, MS Add., f. 7^{r-v}; *Correspondence of Newton*, vol. I, pp. 32–3. The books concerned were Johann Scheubel, *Algebrae compendiosa facilisque descriptio*, Paris: Cauellat 1551; Ludolph van Ceulen, *Fundamenta arithmetica et geometrica*, Leiden: Colster 1615; James Hume, *Traité de la trigonométrie*, Paris: De la Coste 1636. Collins also mentioned Frans van der Huyps and Sieur le Tenneur as being authors of note on the topic of irrationals. He had earlier lent Newton his copy of Johan Jacob Ferguson, *Labyrinthus algebrae*, The Hague: Johannes Rammazeyn 1667. Francis Vernon tried unsuccessfully to obtain for Collins a copy of Jacques Alexandre le Tenneur's anonymously published *Traité des quantitez incommenssurables*, Paris: I. Dedin 1640.

was primarily concerned to provide Collins the scientific expertise he required. It was not Newton's desire to get into print, and certainly not under his own name.[125] On the other hand, he felt it would be wrong to publish the book simply with Kinckhuysen's name on the title page when the original text had been substantially altered, and suggested instead appending words to the effect that it had been 'supplemented by another author'.[126] Newton thus made clear that he did not want his name to appear anywhere in the publication, something Collins found hard to understand.[127]

It is important to note that Newton's declared aversion to being named did not signify a change of heart towards Collins's project itself. On the contrary, having received the books that had been sent him, Newton asked Collins to return the copy of Kinckhuysen so that he could expand further his existing notes and comments.[128] Not unexpectedly, Newton's willingness to work further on the project produced an enthusiastic response on the part of its initiator. Not only did Collins immediately grant the Cambridge mathematician more time, but he also suggested that the Kinckhuysen edition would stand him in good stead in his future scholarly endeavours, not least with the Royal Society, to which he would be elected a year and a half later:[129]

> Perceiving by your last that you are willing to take some more paines at present with Kinckhuysen I remand the same, but doe not presse your selfe in time, your paines herein will be acceptable to some very eminent Grandees of the R Societie who must be made acquainted therewith, and forasmuch as Algebra may receive a further Advancement from your future endeavours and that you are more likely than any man I know, herein to oblige the Republick of Learning [...].[130]

Collins used the opportunity of this letter to send Newton an arithmetical study of equations he had drawn up in response to Pell's suggestion that it might be possible to solve all equations by means of tables. He evidently hoped in this way to bring his own deliberations on the topic more strongly into their discussion.[131]

[125] Isaac Newton to John Collins, 11/[21] July 1670, Cambridge University Library, MS Add. 9597/2/18, f. 6r–v; *Correspondence of Newton*, vol. I, pp. 30–1: 'For I assure you I writ what I send you not so much with a designe that they should bee printed as that your desires should bee satisfied to have me revise the booke'.

[126] Ibid.: 'There remains but one thing more & that's about the Title page if you print these alterations which I have made in the Author: For it may bee esteemed unhandsom & injurious to Kinck huysen to father a booke wholly upon him which is soe much alter'd from what hee had made it. But I think all will bee safe if after the words [nunc e Belgico Latine versa,] bee added [et ab alio Authore locupletata] or some other such note.'

[127] John Collins to Isaac Newton, 13/[23] July 1670, Cambridge University Library, MS Add. 9597/2/18, f. 6r–v; *Correspondence of Newton*, vol. I, pp. 30–1: 'Lastly why you should desire to have your Name unmentioned I see not, but if it be your will and command so to have it, it shall be observed'.

[128] Isaac Newton to John Collins, 16/[26] July 1670, Cambridge University Library, MS Add. 9597/2/18, f. 8r–v; *Correspondence of Newton*, vol. I, pp. 36–8.

[129] Newton was elected to the Royal Society on the proposal of Seth Ward on 11/[21] January 1671/2. See Birch, *History of the Royal Society*, vol. III, pp. 1–2.

[130] John Collins to Isaac Newton, 19/[29] July 1670, Cambridge University Library, MS Add. 9597/13/6, f. 9r–v; *Correspondence of Newton*, vol. I, pp. 36–8 at pp. 36–7.

[131] The enclosure has not survived. Collins later sent a similar paper called a 'Narrative about Aequations' to Gregory. See John Collins to James Gregory, 1/[11] November 1670, University of St Andrews Library, ms31009, f. 18r–v, f. 37r–38v; Turnbull, *James Gregory*, pp. 109–12, 113–17.

Newton did not hurry to respond. His reply, written over two months later, was unfortunately only able to offer Collins a short and less than encouraging assessment of his paper's contents.[132] More regrettable still will have been Newton's late change of heart following the summer break on the rationale of his contributing to the Kinckhuysen project. Now firmly focused on developing further his own analytical methods, Newton had come to see that his additions to the *Algebra ofte Stel-konst* would be increasingly at odds with the original text methodologically. Thus, in working through the fourth part of the book, he had found that most of the problems were solved

> not by any generall Analyticall method, but by particular & contingent inventions, which though many times more concise than a generall method would allow, yet in my judgment, are lesse propper to instruct a learner [. . .].[133]

Although Newton considered the possibility of composing an introductory work of his own, he ultimately rejected this option not only because of the plethora of such works already available, but more tellingly because he was averse to seeking the praise of others, even if they be valued members of the Royal Society:

> But considering that by reason of severall divertisements I should bee so long in doing it as to tire your patience with expectation, & also that there being severall Introductions to Algebra already published I might thereby gain the esteeme of one ambitious among the crowd to have my scribbles printed, I have chosen rather to let it passe without much altering what I sent you before.[134]

Having brought his participation in the Kinckhuysen project effectively to an end, Newton returned the books Collins had lent him along with his only slightly revised notes and comments. It would appear that there was mutual acceptance of the project's demise. It was not spoken of again. While the two men's correspondence continued to flourish, it settled down forthwith into a productive exchange of mathematical news. Collins nonetheless still held out hope that something of the combined efforts of Mercator and Newton could be saved. In December 1670, he announced to Gregory that 'Kinckhuysens Introduction to Algebra with notes thereon and additions thereto made by the learned Mr Isaac Newton of Cambridge (at the request of Dr Barrow) is ready for the Presse'.[135]

[132] Isaac Newton to John Collins, 27 September/[7 October] 1670, Cambridge University Library, MS Add. 9597/2/18, f. 10^{r-v}; *Correspondence of Newton*, vol. I, pp. 42–4 at p. 43: 'though the speculation bee pretty I much suspect it will never becom usefull for the solving of aequations'.

[133] Ibid., pp. 43–4.

[134] Ibid., p. 44.

[135] John Collins to James Gregory, 24 December/[3 January] 1670/1; University of St Andrews Library, ms31009, f. 16^{r}–17^{v}; Turnbull, *James Gregory*, pp. 153–9 at pp. 156–7.

Collins and Gregory: from scientific instruments to quadratures

By this time, Gregory was already Regius Professor of Mathematics at St Andrews, a post to which he had been appointed with the help of his patron, the statesman and natural philosopher Robert Moray (1608/9–1673). The two men had first met under quite different circumstances, when, in 1663, the Aberdeen-educated mathematician spent time in London while his first book, *Optica promota*, was being printed.[136] In this work Gregory had not only provided a masterful account of the optical properties of mirrors and lenses, but also the earliest description of a reflecting telescope which he had designed and hoped then to get constructed in the metropolis. After their chance meeting, Collins put Gregory in touch with the skilled glass-grinder and instrument maker Richard Reeves (d.1666) and his associate, Christopher Cocks (or Cox, *fl.* 1660–1696). Reeves was however unable to make a parabolic mirror of the necessary curvature and so Gregory, who was about to leave for Padua to pursue his mathematical studies further under Stefano degli Angeli (1623–1697), abandoned the project.[137] There was no further contact until some four years later when Collins recalled his earlier meeting with Gregory after being lent a copy of the Scottish mathematician's *De quadratura circuli et hyperbolae* (1667) by the bookseller Samuel Thompson (d.1668). Thompson, whose premises were in Little Britain, was a friend of Gregory's and had already sent this or another copy of *De quadratura circuli* to Wallis, probably at the request of Moray.[138] Gregory, seemingly unaware of the poor state of the mathematical book market in England, had hoped that Thompson would get *De quadratura circuli* reprinted. This was not to be, but at least Thompson was able to inform Collins of Gregory's whereabouts in Italy.[139]

This was an opportunity not to be missed. Apart from renewing his acquaintance with Gregory, Collins saw an ideal path opening up to get hold of some of the Italian mathematical books he craved. In the letter he subsequently wrote, he cited specifically Mengoli's *Geometria speciosa* (1659), but other authors were also named. Furthermore, he enclosed a carefully prepared list of rare titles, evidently hoping

[136] James Gregory, *Optica promota, seu abdita radiorum reflexorum & refractorum mysteria, geometrice enucleata*, London: F. Hayes for S. Thomson 1663. Gregory's description of the reflecting telescope is on pp. 92–5.

[137] See James Gregory to John Collins, 23 September/[3 October] 1672, Cambridge University Library, MS Add. 9597/13/6, f. 138^{r-v}; Turnbull, *James Gregory*, pp. 241–3. Newton, keen to promote his own design of a reflecting telescope, called part of Gregory's account into question. See James Gregory to John Collins, 7/[17] March 1673, Cambridge University Library, MS Add. 9597/13/6, f. 139^{r}–140^{v}; Turnbull, *James Gregory*, pp. 258–62 at p. 259; Isaac Newton to John Collins, 9/[19] April 1673, Cambridge University Library, MS Add. 9597/2/18, f. 33^{r}–34^{v}; *Correspondence of Newton*, vol. I, pp. 269–71.

[138] See John Wallis to John Collins, 27 February/[8 March] 1667/8, *Correspondence of Wallis*, vol. II, pp. 429–31, and John Wallis to Robert Moray?, ? February 1667/8, *Correspondence of Wallis*, vol. II, p. 433.

[139] See John Collins to John Pell, 11/[21] February 1667/8, London, British Library Add. MS 4278, f. 332^{r-v}: 'Mr Gregorie [the Scot] now at Padua, hath sent over a small Booke published by himself, to Mr Thompson to reprint, wherein he hath squared the Circle and Hyperbola, by a new geometrico-Algebraick method of converging Polygons, and that in most vast Numbers'. See also John Collins to James Gregory, 30 December 1668/[9 January 1669], University of St Andrews Library, ms31009, f. 4^{r-v}: 'Mr Samuell Thompson is dead [...] I thinke <you> might doe well to let your Tract de Vera Circuli et Hyperbolae mensura be <printed> with your intended Opticks and Astronomy'.

to find favour among virtuosi in London by further enhancing the Royal Society's library collection. First of all, though, he was keen to remind Gregory of their earlier meeting and present his credentials:

> Sir, it was once my good hap to meete with you in an Alehouse or in Sion Colledge, and though I have not been educated at Universities and so my attainments are meane, yet I have an ardent love to these Studies, and endeavouring to raise a Catalogue of Mathematicall Bookes and to procure scarce ones for the use of the Royall Society and my owne delight, I crave your assistance in procuring what I mention in this letter, or the inclosed Paper.[140]

The self-deprecating remarks Collins uses here are replicated in other letters to academically trained men. But they are counterbalanced by his newly won status as fellow of the Royal Society. Gregory himself would not be admitted for another eight months, and when he was elected, on 11 June 1668, it was on the proposal of Collins—the one and only time that he proposed a new member. But Collins did not use membership of the Royal Society simply to validate the potentially expensive request to Gregory. He was also able to point out that he, too, was a good friend of Moray, and that Gregory's assistance in book procurement would be 'very acceptable' to the Scottish statesman.

There was a scientific pretext to Collins's letter, too: he wanted to take up the topic of Gregory's book, though not so much on account of the controversial analytical procedure he had employed.[141] Instead, he sought to convey his good scientific connections and awareness of contemporary mathematical discussion by citing the work of other contemporary authors such as Walter Warner (1563–1643), William Brouncker, and Wallis, taking the opportunity specifically to cite at length a letter he had recently received from Barrow on the properties of the hyperbola.[142] Thereby facilitating the exchange of mathematical intelligence, he was also able to point out the significance of the quadrature of the hyperbola to practical mathematicians, notably to the methods employed by gaugers:

> the Quadrature of the Hyperbola is a Proposition very necessary in Guaging and consequently of great use in relation to the Kings revenue, for many Brewers tons are like Silver tankards trunci conici circulares divided into two Partitions with a plaine erect to the Base, to hold Liquors of different strengths, and also stand stooping, and the quadrature of the Hyperbola doth capacitate us to cube any Segment of a Cone.[143]

[140] John Collins to James Gregory, early March? 1668, Cambridge University Library, MS Add. 9597/13/6, f. 92^{r}–93^{v}, at f. 92^{v}; Turnbull, *James Gregory*, pp. 45–48, at p. 47.

[141] See Davide Crippa, *The Impossibility of Squaring the Circle in the 17th Century. A Debate among Gregory, Huygens and Wallis*, Cham: Birkhäuser 2019, especially pp. 35–73, and Christoph J. Scriba, *James Gregorys frühe Schriften zur Infinitesimalrechnung*, Gießen: Selbstverlag des Mathematischen Seminars 1957, especially pp. 13–20.

[142] Isaac Barrow to John Collins, 6/[16] March 1667/8, Cambridge University Library, MS 9597/13/6, f. 10^{r-v}; Rigaud, *Correspondence of Scientific Men*, vol. II, pp. 48–56.

[143] John Collins to James Gregory, early March? 1668, Cambridge University Library, MS Add. 9597/13/6, f. 92^{r}–93 at f. 92^{r}; Turnbull, *James Gregory*, pp. 45–8 at p. 45.

Nor does he neglect to mention the ideas of his friend Michael Dary, who had asserted 'a connection between the quadrature of the hyperbola and that of a segment of a circle', before proceeding to explain Dary's approach in more detail.

Later on, the two men would go on to conduct a productive epistolary commerce after Gregory had returned to Scotland. In this context, Collins effectively served as intermediary between Gregory and Newton, skilfully avoiding potential flashpoints in areas where the two professional mathematicians achieved remarkably similar breakthroughs, the reflecting telescope and the binomial theorem. As with his other illustrious friends, Collins was the indefatigable purveyor of mathematical news, the conscientious supplier of books, as well as being publishing agent par excellence. It was to Collins that Gregory turned when seeking assistance in obtaining mathematical instruments for study or instruction,[144] and astronomical instruments for his planned observatories, first at St Andrews, and later at Edinburgh.[145] But for all his effort, Collins's opening letter to Gregory was not a resounding success. Not only was Gregory about to leave Italy by the time the letter reached him in Padua so that he was unable to procure the books Collins wanted, but also, without any pretence of gentlemanly politeness, he used his reply to tear into all the suggested alternative approaches to quadrature Collins had cited.[146]

Collins as seen by himself and others

There was a deeply rooted tension in Collins's character between his humble origins and his scientific aspirations. While he occasionally spoke of himself as being of 'meane character',[147] he succeeded in gaining a solid reputation as a man of exceptional talent at promoting the mathematical sciences, whether through the communication of scientific intelligence across his network, the procurement of rare books from abroad for the Royal Society, or simply the lending to others of books

[144] See for example John Collins to James Gregory, 2/[12] February 1668/9, University of St Andrews Library, ms31009, f. 1^{r-v}; Turnbull, *James Gregory*, pp. 65–7; John Collins to James Gregory, 15/[25] March 1668/9, University of St Andrews Library; Turnbull, *James Gregory*, pp. 70–2.

[145] Gregory's now missing letter to Collins of 23 May 1672 evidently contained a request for the purchase of instruments. See also Gregory's subsequent letter, dated 6/[16] August 1672; Cambridge University Library, MS Add. 9597/13/6, f. 137^{r-v}; Turnbull, *James Gregory*, pp. 240–1. There are numerous further examples, such as John Collins to James Gregory, 23 November/[3 December] 1674, 31 December 1674/[10 January 1675], and 20/[30] April 1675; respectively University of St Andrews Library ms31009, ff. 66^{r-v}, 68^{r-v}, and 69^{r-v} and Turnbull, *James Gregory*, pp. 290–2, pp. 294–5, and pp. 296–7. Gregory also sought and received advice from the Astronomer Royal. See James Gregory to John Flamsteed, 19/[29] July 1673 and John Flamsteed to James Gregory, 2/[12] August 1673, Cambridge University Library, RGO 1/43, ff. 70^{r}–71^{v} and RGO 1/36, ff. 24^{r}–25^{v}; *The Correspondence of John Flamsteed, First Astronomer Royal*, ed. Eric G. Forbes, Lesley Murdin, and Frances Willmoth, 3 vols, Bristol: Institute of Physics Publishing 1995–2002, vol. I, pp. 225–56 at 233–6.

[146] James Gregory to John Collins, 26 March/[5 April] 1668, Cambridge University Library, MS Add. 9597/13/6, f. 107^{r-v}; Turnbull, *James Gregory*, pp. 49–51.

[147] See for example John Collins to Edward Bernard, 16/[26] March 1670/1, *Correspondence of Wallis*, vol. III, pp. 431–5 at p. 431: 'I will not goe about to detaine you with a Discourse to intimate how happy it is for a Man inferioris Subsellii, and a Non-Academick to have the honour of the Acquaintance with the learned, such as you are'; John Collins, 'A Narrative of the Case of John Collins Accomptant in relation to Employment about his Majesties affaires', Kew, The National Archives, SP29/398, f. 261^{r}–261a^{r} at f. 261a^{r}: 'The said Collins though a meane person was in 1667 admitted a Member of the Royall Society'.

from his personal collection to further their own investigations. For a time, he hoped to set up a stationer's business of his own as a way to having more direct influence on the mathematical publishing market, but this project failed for want of sufficient financial resources.[148] As a facilitator of the growth of knowledge with a clear sense of purpose he was fittingly described by contemporaries as 'England's Mersenne'.[149]

Driven by his desire to see the indigenous mathematical culture flourish, Collins devised marketable book projects, hurried along reluctant authors, and where necessary adopted a hands-on approach accompanying publications through the printing process. Not without reason did he describe himself as the midwife to John Pell's *Algebra* or to John Wallis's *Mechanica*.[150] At the same time, he shared the fear of his academic friend Wallis that mathematicians in England and Scotland, through their disinclination to publish, were hindering their chances of being seen to compete adequately with the prodigious achievements of contemporaries on the Continent. Barrow, Pell, and above all Newton were all entreated repeatedly by him to be more forthcoming. When seeking to persuade Gregory to publish his treatise on dioptrics in English, Collins suggested that only in this way would he be able to outshine the Italian Jesuit mathematician Francesco Eschinardi (1623–1703): 'and herein I have no other concernment, but that of one that desires the advancement of knowledge, and rather that your name and abilities should be in esteeme than that of Eschinardus'.[151]

It is a measure of the esteem that Collins himself came to enjoy as a 'Non-Academick' that he willingly and confidently offered advice to his academic friends Barrow and Newton in preparing their works for publication. His opinion was clearly valued both as a mathematician and as one seeking to promote mathematical learning. He gave similar support to Wallis, too. Around the end of the 1670s, the Savilian professor sent him a new draft of what would later become the *Treatise of Algebra* in order to canvas his views on its suitability for publication.[152] At the time, Collins was putting together a proposal for a collection of works on algebra and Wallis's tract was evidently envisaged to be a major part of this.[153] The report Collins returned to

[148] See John Collins to John Beale, 20/[30] August 1672, Cambridge University Library, MS Add. 9597/13/5, f. 83^r–85a^v at f. 84^r; *Correspondence of Scientific Men*, vol. I, pp. 195–204 at p. 200: 'I am about to turne Stationer my selfe, and as I have been, so I beleive I shall continue, as eager as any man to get good Bookes printed'. Shortly afterwards in the same letter he wrote, f. 84^v/p. 201: 'I intend god willing to set up a Stationers Trade [...] and afterwards hope to fall into the Printing of Bookes'.

[149] See for example Barrow, *Lectiones XVIII*, Epistle to the reader; Edward Bernard to John Collins, 3/[13] April 1671, MS Add. 9597/13/5, f. 35^r–36^v; *Correspondence of Scientific Men*, vol. I, pp. 158–60 at p. 159: 'yourself, the very Mersennus and intelligence of this age'. See also Thomas Strode, 'Apollonius analiticus', Oxford, Bodleian Library, MS Savile 43, f. 4^v: 'I was desired by our Mersennus having seene an essay of mine on this subiect [*sc.* conics], to undertake this part'. In the margin Strode has written 'Mr John Collins'.

[150] See John Collins to Thomas Baker, 10/[20] February 1676/7, Cambridge University Library; MS Add. 9597/13/6, f. 176^v–177^r; *Correspondence of Scientific Men*, vol. II, pp. 14–16 at p. 15, and John Collins to Thomas Baker, 23 May/[2 June] 1677, Cambridge University Library, MS Add. 9597/13/5, ff. 30^r–31^v; *Correspondence of Scientific Men*, vol. II, pp. 23–6 at p. 24.

[151] John Collins to James Gregory, 29 September/[9 October] 1670, University of St Andrews Library, ms31009, f. 52^r–53^v; Turnbull, *James Gregory*, pp. 105–7 at p. 107.

[152] The first draft of the Treatise had been sent to Collins in 1676. See Birch, *History of the Royal Society*, vol. IV, pp. 166–7.

[153] The proposal drawn up by Collins was outlined on his behalf by William Croone at the meeting of Council of the Royal Society on 21 January 1679/80. It was noted thereby that Collins was 'ready to print

Wallis is important for at least three reasons.[154] First, he was remarkably forthright in requesting the correction of what he sees as significant omissions in the text, such as the absence of an account of Barrow's 'Optick and Geometrick Lectures'. Collins suggested that by rectifying this omission Wallis would not only please Newton and the University of Cambridge, but also possibly revive the sale 'of a booke that is slow'.[155] Second, Collins ventured to correct Wallis's account of Newton's work on series on the basis of the tract *De analysi per aequationes infinitas*, a copy of which had been in his possession since 1669.[156] And third, he responded to Wallis's apology for having written the work in the vernacular by pointing to the paucity of local booksellers who traded Latin works abroad, and reminding him at the same time of the aim of the Royal Society 'to promote and encourage the printing of Mathematicks and other bookes of Art in our owne tounge'.[157]

Despite Collins's efforts to speed up the process, Wallis's work on producing a final draft of the *Treatise of Algebra* dragged on and it was not until near the end of 1682 that he was able to draw up a proposal to print the work and have this presented to the Royal Society. Although it did not receive the imprimatur of the Society, members were encouraged to take out subscriptions.[158] Earlier that same year, Collins's proposal for printing another tract on algebra, the *Geometrical Key* of Thomas Baker (1625?–1689), in which the author devised a geometrical construction different from that of Descartes for solving biquadratic equations, was likewise presented and approved.[159] Following publication of the work in 1684, Baker was elected member of the Royal Society, having been proposed by the former mathematics master at Christ's Hospital Mathematical School, Edward Paget (1652–1703).[160]

two volumes of algebra, written by Dr. Wallis, Mr. Baker, Mr. Newton, &c. provided the society would engage to take off 60 copies after the rate of 1d. ½ a sheet'. See Birch, *History of the Royal Society*, vol. IV, p. 4.

[154] John Collins for John Wallis, ?1678, Cambridge University Library MS Add. 3977, No. 13; MS Add. 4007 (B), f. 745^r–749^r; *Correspondence of Newton*, vol. II, pp. 241–4.

[155] Ibid. Collins suggested further that in this way Wallis would 'encourage Stationers in future undertakings'.

[156] Ibid. 'Sir I am loath to incurre your displeasure, but yet must take liberty to tell you some things concerning your intended Explanation of Mr Newtons Series If I had been so minded I could about 9 yeares since namely at the beginning of 1669 have imparted to you a full treatise of his of that Argument but did not'. On this topic see also the editor's remarks in Newton, *Mathematical Papers*, vol. II, pp. 206–7.

[157] John Collins for John Wallis, ?1678, Cambridge University Library, MS Add. 3977, No. 13; *Correspondence of Newton*, vol. II, pp. 241–4. There is another copy of this text: MS Add. 4007 (B), f. 745^r–749^r.

[158] Abraham Hill presented the proposal to the Council of the Royal Society on Collins's behalf at the meeting on 22 November 1682, while Robert Hooke read out the contents of the *Treatise* to members at their meeting later that day. See Birch, *History of the Royal Society*, vol. IV, pp. 166–7. It was not until May 1685 that Wallis was finally able to present a copy of the complete printed text to the Royal Society. See Charles A. Rivington, 'Early Printers to the Royal Society 1663–1708', in: *Notes and Records of the Royal Society* 39 (1984), pp. 1–27 at p. 16, and Birch, *History of the Royal Society*, vol. IV, p. 396.

[159] Thomas Baker, *Clavis geometrica catholica: sive janua aequationum reserata/The Geometrical Key: or the gate of equations unlock'd*, London: J. Playford 1684. As indicated by the title, this work is in Latin with facing page English translation. On the presentation of Collins's proposal to the Royal Society and the recommendation that it be printed, see Birch, *History of the Royal Society*, vol. IV, pp. 156, 162.

[160] Paget made the proposal on 12 November 1684; Baker was duly elected a week later. See Birch, *History of the Royal Society*, vol. IV, pp. 328, 332.

Interestingly, Baker, who had studied for a time at Magdalen Hall, Oxford, did not publicly acknowledge the support he received from Collins. This was unusual and can be contrasted with the effusive praise expressed by the mathematical teacher John Kersey (1616–1677) in the preface of his two-volume work on algebra, the *Elements of that Mathematical Art, commonly called Algebra* (1673):

> I cannot but declare to the World, that my old and much respected Friend, Mr. John Collins, a person well known to be both singularly skilfull in, and an industrious Promoter of the Mathematicks in general, hath been a principal Instrument in bringing this Work to light, as well by animating me to Compile it, as by endeavouring to procure it to be well Printed.[161]

Right up to the end, Collins did not neglect his practical interests. After attending the Act at Oxford in July 1683, he travelled on horseback to Malmesbury in order to inspect the ground where it was proposed to construct a canal linking the Thames to the Bristol Avon, a project then being promoted by a Dorset gentleman by the name of Francis Matthews (*fl.* 1655–1684).[162] Severe illness contracted during that outing led to his death in November of that year.[163]

When the antiquarian John Aubrey, who was concerned in the Avon canal project, informed Wallis of the seriousness of Collins's condition, the Savilian professor responded by noting that 'the progress of Mathematick Learning' owed much to the industry of their mutual friend.[164] Later, both Wallis and Aubrey mourned Collins's death, but the enormity of the loss was felt particularly among the London mathematical community where he had been such a dominant and influential figure. The instrument maker and mathematics teacher Euclid Speidell (1631–1702), son of the mathematician John Speidell (1577–1649), exemplified his centrality to that community in terms that would have served as a fitting epitaph, remarking that

> he was not only excellent in Mathematical Arts and Sciences, but of a very good, affable and frank Nature to Communicate any thing he knew to any Lover and Enquirer of those things, and hath left behind him those Mathematical Works which will continue his Fame amongst the Lovers and Students therein. He also in his Life-time, promoted the Publishing of other Men's Mathematical Works; as the Elaborate *Algebra* of the Learned John Kersy, who was my Father's Disciple about 1645: And also of the Learned Baker's *Algebra*, and several others. He was a Man of great correspondence with Mathematical Persons in foreign Parts, and thereby could give

[161] John Kersey, *Elements of that Mathematical Art, commonly called Algebra, expounded in four books*, 2 vols, London: William Godbid for Thomas Passinger 1673, vol. I, sig. b3^{r}.

[162] See James Long to John Aubrey, 5/[15] March 1683, Oxford, Bodleian Library, MS Aubrey 12, f. 279^{r-v}. The Oxford-educated 2nd Baronet James Long (*c.*1617–1692) was formerly MP for Malmesbury and a close friend of Aubrey. On Aubrey's support for the scheme to link the Thames with the Avon see Hunter, *John Aubrey*, pp. 66–7.

[163] See the obituary printed in Birch, *History of the Royal Society*, vol. IV, pp. 232–4.

[164] John Wallis to John Aubrey, 17/[27] September 1683, Oxford, Bodleian Library, MS Aubrey 13, f. 343^{r-v}.

Information of any New or Old Mathematical Book; and till my Acquaintance with him, I was ignorant of Foreign Authors; being but young when my Father dyed, and not then having taken any Pains in these Studies: So that by the said Collins's Information and Means, I have heard of, and seen, some Foreign Mathematical Authors of Note and Esteem.[165]

[165] Euclid Speidell, *Logarithmotechnia: or, the Making of Numbers called Logarithms to Twentyfive Places*, London: Henry Clark for the author 1688, p. 3. See also Collins's assessment of the consequences of non-payment for his government services in his 'Narrative of the Case of John Collins', f. 261a^r, where he puts his admittance to the Royal Society down to 'an account they had received that the said Collins might be usefull in printing & promoting Mathematicks, which he presumeth he hath accordingly been, commonly spending much time and paines in correspondence with others, drawing of Schemes and correcting at the Presse, to his damage about 20^l p. annum, all which will cease if Collins be not paid, to the great preiudice of Algebraick and other Mathematicall Learning, and of the affaires of the RS'.

12

'All of this was born on Paper'

The Mathematics of Tunnelling in Eighteenth-Century Metallic Mines

Thomas Morel

Introduction

The eighteenth century has rightly been described as a pivotal period for mathematical practices during which a 'quantifying spirit', or '*esprit géométrique*', developed across a broad spectrum of disciplines. The evolution was noticeable in the academic world, where it ranged from political economy to celestial mechanics, and was felt among practitioners as well, in areas as diverse as chemistry and forestry.[1] While it is clear that an ever-increasing number of areas then came to use numbers and geometry in a more extensive fashion (i.e. simply using more quantified information), the thesis of a more intensive mathematization of nature needs to be clarified. After all, a silent rise of practical mathematics had already been observed in several economic and technical domains from the Renaissance onward.[2] Moreover, the concept of 'quantification', as useful as it may be, is in a sense too broad: In the eighteenth century, there was for instance little in common between the ready reckoners or rules of thumb used by unlearned foresters in their daily business and the 'application of mathematical methods' deductively obtained by scholars.[3] The development of theories and instruments of mathematical computation and representation was still highly uneven and, crucially, mostly uncoordinated from one field of human activity to another. Modern technical schools were created mostly in the aftermath of the violent conflicts that raged later in the century, beginning after the Seven Years War and then, more vigorously, after the Napoleonic Wars.

This chapter presents a case study about the use of geometry in metallic mining, and more specifically in the planning and development of long-distance tunnels for the drainage of water. Mining is especially relevant given its importance for the general development of sciences and technology in the eighteenth century. The social and political context of the Holy Roman Empire, from where most mining experts

[1] Tore Frängsmyr, J. L. Heilbron, and Robin E. Rider, *The Quantifying Spirit in the 18th Century*, Berkeley: University of California Press 1990.

[2] Thomas Morel, 'Mathematics and Technological Change: The Silent Rise of Practical Mathematics', in *Bloomsbury Cultural History of Mathematics*, vol. 3, 2024, pp. 179–206; H. Floris Cohen, 'The "Mathematization of Nature": The Making of a Concept, and How It Has Fared in Later Years', in: *Historiography of Mathematics in the 19th and 20th Centuries*, ed. Volker R. Remmert, Martina R. Schneider, and Henrik Kragh Sorensen, Cham: Springer 2016, pp. 143–60.

[3] Frängsmyr, Heilbron, and Ryder, *The Quantifying Spirit in the 18th Century*, p. 2.

came, played an important role, too. As cameralism and mercantilism promoted massive investments from the various states, projectors were under pressure to improve their ability to monitor large projects by presenting more detailed and quantified plans.[4] Could mathematics, however, fully capture such complex enterprises, or did they only offer the misleading reassurance and pseudo-scientificity of geometrical figures and numbers? Can we trust the engaging maps, precise time frames, and the balanced budget conserved in archives? To what extent exactly was geometry involved in the planning process, and later during excavation operations? Which kind of mathematical practitioners were involved, and how did they work?

If the particularities of each domain forbid general and all-encompassing assertions, this example will nevertheless be useful to understand how engineers and administrators gradually came, in their daily practices, to rely almost naturally on mathematics. Far from being a static set of rules, using geometry and arithmetic in mining was a process of constant adaptation to growing technical challenges. The digging of drainage tunnels in large mining districts shows how quantification and surveys became extraordinarily efficient in the second part of the eighteenth century, when methods to compute excavations and counter-excavations became widespread.[5] The central case study of this chapter is set in the Harz Mountains of central Germany, as this region, which belonged to Hanover, had close ties to Great Britain. I will describe how a long-standing tradition of subterranean geometers, comparatively efficient institutions, and highly skilled administrators worked together from 1771 to 1777 to plan the Deep George Tunnel (*Tiefer-Georg Stollen*), and later monitored its digging until the final breakthrough in 1799.[6]

This success was undoubtedly the most resounding of its kind on the European scene. It was not, however, an isolated or surprising event: engineering projects were at the time widely discussed in journals and in academies. The techniques employed on the digging site represent the culmination of a series of incremental changes in the way mathematics was used in metallic mines. Minor alterations in the way surveys were performed, then represented on mining maps, and finally used by mining councils during their deliberations, triggered at length a dramatic evolution in the status of mathematics. The success of the operation proved that it was possible to conceive major operations from scratch by relying solely on geometry.

[4] While focused on the French case, Chandra Mukerji, *Impossible Engineering: Technology and Territoriality on the Canal Du Midi*, Princeton, NJ: Princeton University Press 2009, offers general reflections about the relationship between political power and early modern engineering. Ursula Klein, *Technoscience in History: Prussia, 1750–1850*, Cambridge, MA: MIT Press 2020, offers a compelling set of case studies on Prussia, from which the mathematical disciplines, however, are largely absent.

[5] This development, it should be noted, largely predates the Industrial Revolution. Major underground canals were almost simultaneously designed in England, most famously the Worsley Navigable Levels on the initiative of Francis Egerton, 3rd Duke of Bridgewater (1736–1803), but in the context of coal mining where technical challenges are hardly comparable. See Francis Henry Egerton, *Description du plan incliné souterrain exécuté entre les deux biefs des canaux souterrains dans les Houillières de Walkden-Moor, en Angleterre, par le Duc de Bridgewater*, Paris: Bureau des annales des arts et manufactures 1812.

[6] The present chapter builds on work carried out in a previous study, Thomas Morel, *Underground Mathematics: Craft Culture and Knowledge Production in the Holy Roman Empire*, Cambridge: Cambridge University Press 2022, and summarizes several general arguments of that book.

In the first section, I will outline the scope of geometry in mining in the early eighteenth century, underlining its legal and technical role. Surveys and maps were routinely used in a piecemeal fashion, for instance to plan the installation of waterwheels or in holing from one gallery to the next. Improvement in precision gradually led to the idea of planning tunnels in straight lines over long distances, using a new setting-out technique. This is detailed in the second and third sections with the example of the Deep George Tunnel. I describe successively how mathematics was used during the planning phase, to inform the decisions of the mining council, and later to monitor the digging operations. While the mathematics used by surveyors Rausch and Länge was not particularly advanced for its time, its systematic use and the close cooperation with the rest of the administration resulted in an astonishing efficiency, as can be read in administrative reports, travel diaries, or in publications made by natural scientists. The last section analyses to what extent this practical geometry was taught in universities or mining academies, and presents several contemporary textbooks written by mathematicians or mining experts.

Subterranean geometry and the digging of mine tunnels

From the Middle Ages onward, the importance of metallic mining was threefold. First, the extraction of silver was the main source of metal for coinage. Second, the raw silver ore contained lead and several other useful metals, such as cobalt, quicksilver, or bismuth, which could be used in many arts and crafts. Third, the concentration of capital and the technical hurdles faced in mining districts acted as a powerful catalyst for technical and scientific progress.[7] When one thinks of science and technology in mining, however, mathematics rarely comes to mind despite fulfilling numerous functions within the mining administration: reckoning and distribution of profits, drawing and computation of machines, setting of concession limits, and so on.[8]

In the early decades of the mining boom, at the turn of the sixteenth century, the main role of geometry was a legal one. In crowded fields, such as depicted on the famous mining altar of St. Anne's Church in Annaberg (1522), setting concession limits underground was crucial.[9] Surveys were performed either by the

[7] On early modern mining, the general reference work is Klaus Tenfelde, Christoph Bartels, and Rainer Slotaa (eds), *Geschichte des deutschen Bergbaus. Der alteuropäische Bergbau: von den Anfängen bis zur Mitte des 18. Jahrhunderts*, vol. 1, Münster: Aschendorff Verlag 2012. On the specific importance of mining to master technical and scientific processes, see Franz Mathis, *Die deutsche Wirtschaft im 16. Jahrhundert*, Enzyklopädie deutscher Geschichte 11, Munich: Oldenbourg Verlag 1992, pp. 53–85; Eric H. Ash, 'German Miners, English Mistrust, and the Importance of Being "Expert"', in: Franz Mathis, *Power, Knowledge, and Expertise in Elizabethan England*, Baltimore: Johns Hopkins University Press 2004, pp. 19–54.

[8] The most important reckoning master (*Rechenmeister*) of the sixteenth century, Adam Ries, went on to work for the Saxon mining administration: see Walter Schellhas, *Der Rechenmeister Adam Ries (1492 bis 1559) und der Bergbau*, Freiberg: Bergakademie 1977. On the general importance of mathematics in mining, see Morel, *Underground Mathematics*.

[9] St Anne's mining altar was painted by Hans Hesse (*c.*1470–1539), and is now listed on the UNESCO World Heritage List as part of the *Erzgebirge* region (https://whc.unesco.org/en/list/1478).

mining masters themselves, or by jurors and local surveyors; practices could vary widely from one region to another. Geometry was ubiquitous, and yet there was little time for theoretical considerations. Most technical duties of the time were tackled by a clever use of the mining compass and measuring ropes.[10] Drainage tunnels, for instance, were neither planned on paper nor dug in straight lines. From the mouth of a gallery, miners simply followed the sinuous ore veins, as can be seen in Figure 12.1, and occasionally dug ventilation shafts. In any case, they would avoid the much harder 'deaf stone' surrounding the metallic veins, for it was much costlier and slower to dig there. Districts were made of numerous small private enterprises; most pits had a limited length, and conversely a limited depth, limiting the need for comprehensive or large-scale surveying operations.

The mining boom soon faded in the Holy Roman Empire, as its provinces were flooded with silver and gold from Potosi and other mines in the New World. The Thirty Years War (1618–48) further derailed the economy of mining regions. In a legal framework in which the investors' *Kuxen* (mining shares) not only entitled them to large dividends in good times, but forced shareholders to make up for losses; this

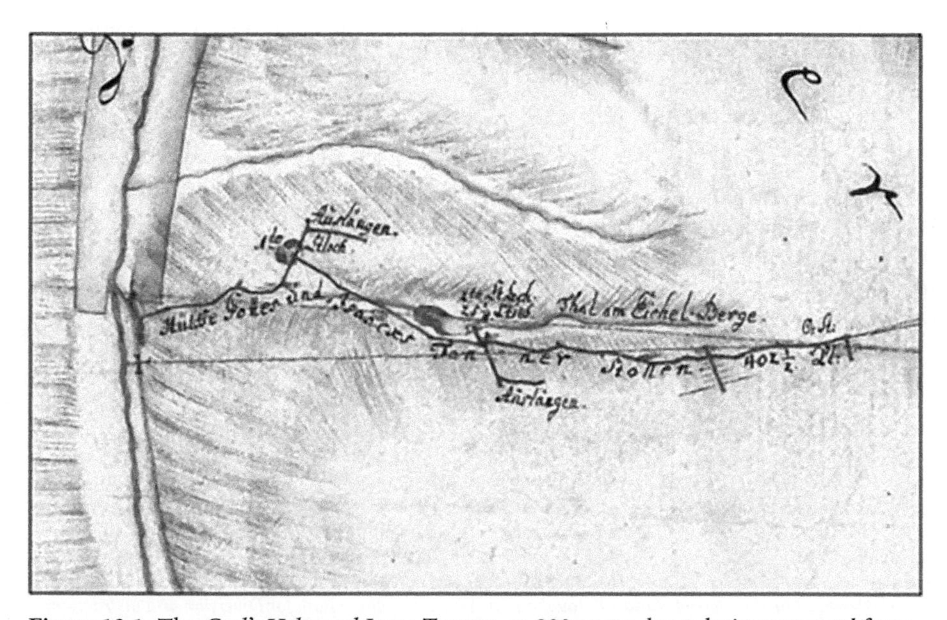

Figure 12.1 The *God's Help and Isaac Tanner*, an 800-metre-long drainage tunnel from the sixteenth century. Observe how the tunnel was dug sinuously following the silver ore vein, with prospecting branches [*Auslängen*] and close, shallow ventilation shafts. The thin straight line represents the route of the modern Deep George Tunnel (see Figure 12.2).

[10] See Morel, *Underground Mathematics*, chapter 2, 'A Mathematical Culture: The Art of Setting Limits'.

led to the abandonment of many districts.[11] Local rulers stepped in, both to preserve a long-term source of bullion and because the organization of mining districts required a great stability over time. In most places, the mining administration—once a mere regulatory body—now assumed the technical direction of all mines and the supervision of future development. This new cameralist policy was labelled '*Direktionsprinzip*'.[12]

This new political context had a direct influence on the way geometry was implemented in the mines. In most regions, all concessions were gradually connected and merged into large districts supervised by a local surveyor. The once isolated pits, tunnels, and galleries were now considered as the roads and buildings of underground towns: travel literature, for instance, mentions mines 'which, for the Number of Workmen and Passages, looks like a subterraneous City'.[13] In that context, the drainage tunnels became even more important, as a whole region might in some cases depend on one or a couple of these *Wasserstollen*. In 1684, an 'Administration of Drainage Tunnels and Trenches' was founded in Saxony on the initiative of Abraham von Schönberg (1640–1711), captain-general of the mining administration (*Oberberghauptman*). 'The *Stollen*', Schönberg wrote, 'are the heart and the key of mountains, and give the most continuation to mining, just as they require great costs'.[14]

At the end of the seventeenth century, it thus became common to draw mining maps. The bigger charts, which could be several metres long, were used to depict entire districts and to plan the development of extraction. More precise maps were also drawn in specific cases to assist the decisions of the mining council. Using these plans (and later combined plans and elevations), surveyors found out how to ascertain on paper the direction and find out the difference of height between various tunnels or inaccessible points.[15] The underground surveyors were now in charge of interlinking the countless mining pits to the existing drainage tunnels. Connecting two galleries that were not located on the same vein, however, was a difficult operation:

[11] General background about the economy of mining districts and how the *Kuxen* (mining shares) system worked can be found in Tina Asmussen, 'The *Kux* as a Site of Mediation: Economic Practices and Material Desires in the Early Modern German Mining Industry', in: Susanna Burghartz, Lucas Burkart, and Christine Göttler (eds), *Sites of Mediation: Connected Histories of Places, Processes, and Objects in Europe and Beyond, 1450–1650*, Intersections, vol. 47, Leiden: Brill 2016, pp. 159–82.

[12] On the relationship between mining and cameralism, see Friedrich P. Springer, 'Über Kameralismus und Bergbau', in: *Der Anschnitt* 62 (5/6) (2010), pp. 230–41.

[13] G. Nelson, *The Wonders of Nature Throughout the World Display'd, Both for Diversion and Instruction*, London: Watts 1740, p. 110. The author's discourse is here based on Edward Brown's *Travels*.

[14] Abraham Schönberg, *Ausführliche Berg-Information*, Leipzig und Zwickau: Fleischer und Büschel 1693, p. 190: 'die Erb-Stollen das Hertz und Schüssel der Gebürge sind | und den Bergwerck | die meiste Fortsetzung geben | auch grosse Kosten erfordern'. On Schönberg and the new administration, see Wolfgang Jobst and Walter Schellhas, 'Abraham von Schönberg, Leben und Werk: Die Wiederbelebung des Erzgebirgischen Bergbaus nach dem dreißigjährigen Krieg durch Oberghauptmann Abraham von Schönberg', in: *Freiberger Forschungshefte, Kultur und Technik*, vol. D 198, Leipzig and Stuttgart: Verlag für Grundstoffindustrie 1994, pp. 74–133.

[15] Hans Joachim Alberti, *Entwickelung des bergmännischen Rißwesens*, Diplom-Arbeit im Markscheide-Institut der Bergakademie Freiberg, 1927; Frank Kirnbauer, 'Die Entwicklung des Grubenrißwesens in Österreich', in: *Blätter für Technikgeschichte* 24 (1962), pp. 60–129.

> When such excavations are driven toward each other in deaf [hard] stone, without an ore vein or a cleft showing a sure sign of the direction to be taken; then the hour [of the compass] which the *Markscheider* gives for the direction should be very carefully observed and kept. And one can hardly let the surveyor survey often enough to prevent mistakes, or to correct in due time the mistakes already made.[16]

This operation, named *Querschlag* or *Durchschlag* in the dialect of miners (and *holing* or *cross-sections* in English), was all the more difficult given the sinuousness of galleries and the impossibility of using the methods of triangulation common in open-air surveying. Johann Gottfried Jugel (1707–1786), surveying and mining director in Prussia, underlined how important precision was during these operations:

> From this, one can then see what actually hinges on an accurate mine survey at the surface [...] a lot of money is often used in order to produce such holing in the mountains, oftentimes quite a few hundred fathoms until one gets where one needs to be [...] It frequently requires an amazing amount of money, especially when the breakthrough isn't reached, which is when the trouble really starts, since one often does not even know where one should go after such a drift.[17]

All the existing sources, be they printed textbooks, handwritten instructions, or even legal documents, emphasize how crucial the accuracy of measurements was. When dealing with holing, or more generally with the digging of ventilation shafts or drainage tunnels, 'various [people] knowledgeable of mines, together with the *Markscheider*, should survey the mine painstakingly and accurately'.[18] Countless testimonies show that mistakes regularly happened in the first half of the eighteenth century. In Ilmenau, for instance, archive documents recall how mining shareholders started legal proceedings against their local mining master in 1720: 'the aforementioned mining director Keller has made enormous tunnel-related mistakes [*enormiter Stollen-Fehler*]'. Investors complained that the intended 'breakthrough had been missed, not only by 4 fathoms in the height, but also by 3 fathoms on the side', making the tunnel all but useless. They further asked that Keller be condemned 'to

[16] Friedrich Wilhelm von Oppel and Johann Gottlieb Kern, *Bericht von Bergbau*, Leipzig: Crusius 1772, p. 109: 'Werden dergleichen Oerter in Queergestein gegen einander getrieben, ohne daß ein Gang oder Kluft den sichern Wegweiser der zu nehmenden Richtung giebt; so ist die Stunde, welche der Markscheider zur Richtung abgiebt, sehr sorgsam wahrzunehmen und beyzubehalten. Und man kan den Markscheider kaum öfters genung nachziehen lassen, um vor Fehlern gewarnt zu seyn, oder begangene Fehler zeitig ausbessern zu können'.

[17] Johann Gottfried Jugel, *Gründlicher und deutlicher Begriff von dem gantzen Berg-Bau-Schmeltz-Wesen und Marckscheiden, in Drey Haupt-Theite eingentheilet*, Berlin: Rüdiger 1744, p. 225: 'Hieraus kann man nun sehen, was auf einem accuraten Gruben-Zug beym Tageziehen eigentlich ankömmt [...] oft vieles Geld angewendet wird, um solche Durchschlag durch die Erden, ja öfters wohl etliche hundert Lachter, zu machen, ehe man dahin gelangt, wo mal solchen benöthiget ist [. . .] Dazu wird oft ein erstaunliches Geld erfordert, zumalen oft wenn er durchgesetzet, nicht getroffen ist, sodann geht erst die Noth recht an, weil man oft nicht einmal weiß, wo es nun nach solcher Strecke zu gehen soll'.

[18] Jugel, *Gründlicher und deutlicher Begriff*, p. 225.

reimburse the expenses incurred in this case' and even be arrested.[19] If such extreme cases were rare, deviations were very common and had a major impact on local economies. A holing meant to be achieved in a few years could take decades until the final breakthrough, and costs often ballooned accordingly. The age-old custom of praying every Sunday for the completion of important galleries was obviously more than an outdated tradition.[20]

Let us take the Electorate of Saxony as an example of the traditional way in which geometry could be used to monitor the digging of a drainage tunnel in the mid-eighteenth century. In Freiberg, the largest district of the Ore Mountains, the main tunnel used for the drainage of water—the *Alter Tiefer Stollen*—dated in part from the fourteenth century and was reaching full capacity. In the 1720s and 1730s the local surveyor had attempted in vain to connect this tunnel to various neighbouring galleries, hoping to divert part of the water flow. The repeated failures were costly, and demonstrated that maps (when they had not been lost) were often inaccurate or inadequate.[21]

In 1747, the mining administration finally ceased to consider patchy solutions or temporary expedients, and decided on a major overhaul. The whole process was documented in a report written by Carl Eugenius Pabst von Ohain (1718–1784), the captain-general who completed the project a quarter of a century later. The first action at the time was not, however, to order a general survey of the district, but rather to organize 'renewed visitations of the mine buildings, frequent inspections, and examinations above ground and thorough comparisons and assessments of the few and hardly usable reports still available'.[22] After five years of inconclusive considerations, the captain-general finally convinced the higher administration in Dresden, the state's capital, to pay the surveyors Richter and Müller for a full survey, with an emphasis on measuring the relative depths of galleries. The inertia of mining administrations is understandable if one considers both the high costs of these operations and the risks entailed.

On the basis on the surveyors' 1752 report, the mining administration decided simply to connect several existing galleries, avoiding long cross-sections that were considered too risky. The project progressed slowly for the next fifteen years, until a new technique was introduced in 1767. Thanks to the oral instructions of Bohemian mine technicians, the Saxons learned how to dig tunnels from both ends by driving mines in opposition to one another, which required careful computations and a great care in ascertaining directions (see Figure 12.2). This was a major decision: in a

[19] NLA HA, BaCl Hann. 84a, Nr. 7266, unpaginated, in the *Species Facti*: 'ersagter berg *Director* Keller *enormiter* Stollen-Fehler begangen'. Letter from the investors, 11 December 1720: 'angegebenen Durchschage, nicht nur in der Höhe 4 Lachter, sondern auch auf der Seite 3 Lachter gefehlet, zu Ersetzung derer dißfalls verhorten Unkosten *condemnie*ret, auch darüber gar in *Arrest* gebracht'.

[20] See also Wilfried Liessmann, *Historischer Bergbau im Harz*, Heidelberg: Springer 2010, pp. 56–7.

[21] See Carl Eugen Pabst von Oheim, *Geschichte des 'Thurmhofer Hilfsstollens' bei Freiberg mit dem Bericht von C.E. Pabst von Ohain aus dem Jahr 1772*, ed. Jensund Berichte vom sächsischen Bergbau 4, Kleinvoigtsberg: Kugler Verlag 1998, pp. 14–18. The corresponding archive is Sächsisches Staatsarchiv, Bergarchiv Freiberg, 40010 Bergamt Freiberg, Nr. 2790.

[22] Pabst von Oheim, *Geschichte des 'Thurmhofer Hilfsstollens'*, p. 22: 'Wiederholte Befahrungen der Berggebäude, häufige Besichtigungen und Untersuchungen übertage, sorgfältige Vergleichung und Beurtheilung der noch vorhandenen wenigen und kaum zu gebrauchenden Nachrichten'.

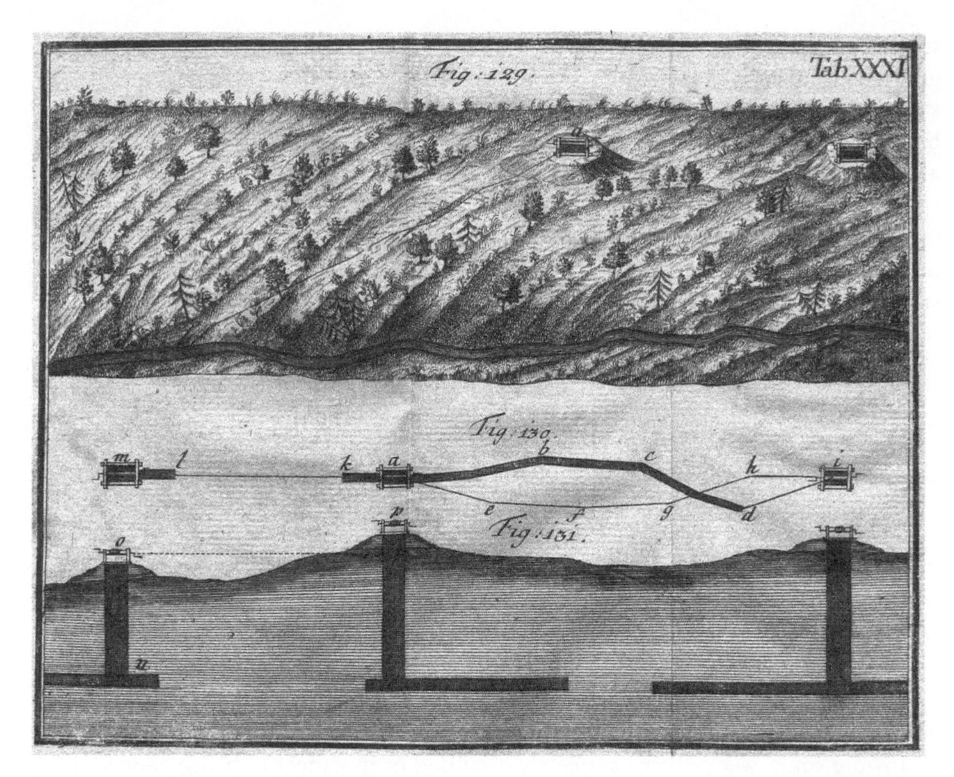

Figure 12.2 Example of a long-distance connection between two galleries, as depicted in Franz Ludwig von Cancrin, *First Principles of the Science of Mining and Salt Mining* [*Erste Gründe der Berg und Salzwerkskunde*], Frankfurt-am-Main 1776.

way, switching from small-scale connections to tunnelling over long distances was a change akin to the evolution from coastal navigation to open sea journeys based on careful measurements. Risky as it was, the gamble paid off: 'The success surpassed even the highest hopes', wrote Pabst von Ohain: after two decades of uncertain labour, the new method enabled the tunnel to be completed within five years. The captain-general's description of this 25-year-long project clearly indicates that there was not yet a full confidence in the ability of surveyors to plan long-distance tunnelling with sufficient precision.[23]

As the previous examples have shown, an undeniable trust in the power of geometry already existed in the early eighteenth-century mining world. However, this appreciation should not be confused with later conceptions of a largely mathematized engineering. Surveys and maps were merely seen as useful tools, which could be used along the way to respond to the circumstances that miners faced. Even by the mid-century, most mining administrations thought it wise to divide a large project in

[23] Pabst von Oheim, *Geschichte des 'Thurmhofer Hilfsstollens'*, p. 30: 'Der Erfolg übertraf noch die davon gefaßte Hoffnung'.

smaller portions, confident that short-distance errors would have less impact. Digging a gallery from both ends or with an intermediary point, while maintaining a straight line over long distances, was not considered a realistic option. On the other hand, the way mathematics was used gives us insight into how science, technology, and planning formed a coherent system. While geometry and quantification were rightfully seen as important, they were not yet systematically used from the beginning to plan large enterprises, because this very possibility simply felt unachievable.[24] The next section will document the pivotal moment when the accuracy was found to be sufficient and comprehensive planning was introduced.

'All of this was born on paper': the planning phase (1771–77)

Let us now turn to another mining region, the Harz mountains, where the first fully mathematized planning of a drainage tunnel was realized in the last quarter of the eighteenth century. In spite of political intricacies, the twin cities of Clausthal and Zellerfeld had been extracting ore from the same web of rich ore veins for centuries.[25] An elaborated system of water ponds known as the *Wasserwirtschaft*, together with sophisticated pumping machines, had kept the extraction going.[26] Still, as the Harz mining pits reached unknown depths, they came to encounter the same problems as the other mining regions: As the local mines 'went deeper by the day', captain-general von Reden noted, the quantity of ground water had substantially increased. 'Many ore veins' were either completely drowned or in danger of being flooded. All the existing solutions were reaching their limits, becoming 'so expensive' to operate as to endanger the economy of the whole region.[27]

The decision to build a new drainage tunnel from scratch was not a light one, and the planning phase lasted for seven long years. In 1770, a first report was written by Georg Andreas Steltzner (1725–1802), who had climbed all the career rungs from ore crusher to mining master. Steltzner had greatly improved the water supply for pumps and water-wheels in the previous decade, but 'the machines, construction of ponds, water pipes and other installations' were not sufficient any more to keep 'the deepest mine workings' from flooding.[28] Captain-general Friedrich Wilhelm von Reden

[24] Pabst von Oheim, *Geschichte des 'Thurmhofer Hilfsstollens'*, p. 8 describes this well: 'mehrfach änderte man die Richtung'.

[25] The next sections will focus on the policy of the Hanoverian state, to which Clausthal belonged. Zellerfeld was governed by the Braunschweig-Lüneburg authorities: see Christoph Bartels, *Vom frühneuzeitlichen Montangewerbe zur Bergbauindustrie: Erzbergbau im Oberharz 1635–1866*, Bochum: Deutsches Bergbau-Museum 1992, pp. 46–8.

[26] Concerning the *Wasserwirtschaft*, the technical water system of the Harz, see Martin Schmidt, *Die Wasserwirtschaft des Oberharzer Bergbaus*, Neuwied: Neuwieder Verlagsgesellschaft 1989.

[27] Claus Friedrich von Reden, *Rede bei dem feyerlichen Anfange des tiefen Georg-Stollen-Baues unweit der Bergstadt Grund*, Clausthal: Wendeborn 1777, pp. 3–5. On the situation in the early 1770s, see Liessmann, *Historischer Bergbau im Harz*, pp. 170–4.

[28] NLA HA, BaCl Hann. 84a, Nr. 9803, doc. 1a, *pro memoria*, 10 November 1770: 'die Machinen, Teichbau Waßerleitungen und andere Anlagen' [...] 'denen tieffen Gruben'. On Steltzner, see Christoph Bartels, 'Der Harzer Oberbergmeister Georg Andreas Steltzner (1725–1802) und die Montanwissenschaften in der zweiten Hälfte des 18. und am Beginn des 19. Jahrhunderts', in: *Staat, Bergbau und Bergakademie: Montanexperten um 18. und frühen 19. Jahrhundert*, Vierteljahrschrift für Sozial- und Wirtschaftsgeschichte

(1752–1815) agreed to consider a new drainage tunnel while pondering the pros and cons. 'A brand-new work' would be 'very lengthy and costly', he acknowledged, compared to a simple repurposing of existing tunnels by means of new connections.[29] On the other hand, a new tunnel dug in a straight line might offer long-term relief while crossing promising prospects. The council prudently ordered a report:

> Since it is necessary for such an important matter to be taken into careful consideration, and that all its circumstances be considered thoroughly: You will authorize that all records about the galleries [of the town of] Grund be inspected, and the surveyor Rausch will be charged with collecting all the information that could figure on local mining maps.[30]

From then on, the mining council would be in weekly, if not daily, contact with the two surveyors working on the project. Samuel Gottlieb Rausch (d.1778) had been appointed in Clausthal in 1742 and had subsequently trained Johann Christian Heinrich Länge (d.1803), who now worked for the twin city of Zellerfeld.[31] As we saw in the first section of this chapter, the first move of the mining administration was usually not to order new surveys, which were costly operations, but to gather the existing information. In this case, the existence of precise maps and reports allowed for a geometrical engineering expertise that the old surveyor Rausch could produce from his working cabinet. The crucial question to be answered was the following: 'whether the local mines could benefit so much from the construction of a deep tunnel as to warrant the gamble of such high costs?'[32]

The vice captain-general August von Veltheim (1741–1801) was not convinced by the audacious proposal, and in fact he suggested a completely different, and more traditional, solution. Von Veltheim's idea was to repurpose existing galleries instead of building a new tunnel, a perspective that might not sound fashionable, he readily admitted, but required 'incomparably less costs and time than this [other] proposal.'[33]

223, ed. Hartmut Schleiff and Peter Konečný, Stuttgart: Franz Steiner 2015, pp. 279–80; Herbert Dennert, *Bergbau und Hüttenwesen im Harz: vom 16.–19. Jh. dargest. in Lebensbildern führender Persönlichkeiten*, Clausthal-Zellerfeld: Pieper 1986.

[29] Johann Christian Gotthard, *Authentische Beschreibung von dem merkwürdigen Bau des Tiefen Georg Stollens am Oberharze*, Wernigerode: Carl Samuel Struck 1801, p. 75; NLA HA, BaCl Hann. 84a, Nr. 9803, doc. 1b, meeting report, unpaginated: 'langwierig und viele Kosten erfodern'.

[30] NLA HA, BaCl Hann. 84a, Nr. 9803, doc. 1b, meeting report, unpaginated: 'Wie aber nothwendig, daß diese, so *importante* Sache in genügsame Überlegung genommen, und nach allen ihren Umstände gründlich erwogen werde: so wolten Sie veranlaßen, daß in der *Communion* die sämtlichen Acten von denen Gründerschen Stollens aufgesuchte würden, auch dem Marckscheidern *Rausch* hiemit aufgeben, das jenige, was davon etwa bey den hiesigen Rißen vorhanden seyn mögte, bey die Hand zu kriegen'.

[31] It is possible to gain detailed insights into their knowledge, for Rausch's handwritten textbook has been preserved. Written in 1740, as he was training with his father Samuel Rausch (d.1742), it is now conserved at the Mining Academy in Freiberg (TU BAF–UB XVII 13, *Anleitung zur Marckscheide-Kunst*). The unusual importance given to draining tunnels and water ponds reflects their local importance in the Harz region.

[32] NLA HA, BaCl Hann. 84a, Nr. 9803, doc. 1b, meeting report, unpaginated: 'ob denen hiesigen Gruben durch die Herführung eines tiefen Stollens so viel Nutzen zu wachsen können, daß dagegen die großen anzuwendende Kosten zu *hazardir*en wären'.

[33] NLA HA, BaCl Hann. 84a, Nr. 9803, doc. 3, *pro memoria* of 18 December 1772: 'ungleich weniger Kosten und Zeit, als jener Vorschlag erfordert'.

As mining administrations had been doing for generations, the seasoned official proposed to judiciously connect existing tunnels and to improve their efficiency using modern digging methods. The risks associated with the long-distance planning of tunnels would disappear and be replaced by several shorter cross-sections. For this alternative project to be viable, however, precise computations about the depth of existing mine workings had to be made by the surveyors. While initial observations seemed to favour von Veltheim's plan, surveyor Rausch soon discarded the possibility. 'He has presented to the mining office a map [...] on which the relation of one [gallery] to the other is presented', summed up Steltzner in his report, 'and the mining office has forwarded the map to me'.[34] The mining master then used all the available evidence to assess the two alternative proposals, using several mining maps and his own 'visual inspections' (*Augenschein*) of the most important spots. At the beginning of 1774, mining master Steltzner finally came to the conclusion that the idea of a deeper gallery, despite its novelty, was the most promising one.

During this first planning phase, most of the work had been carried out on paper, using ancient reports, existing field books and, most importantly, the growing stock of precise mining maps. The two surveyors relied on the vast trove of data gathered over the last century, which were punctually complemented by new surveys of a few limited areas for which a greater accuracy was critical. This crucial remark shows the transformation that subterranean geometry had recently undergone. In this region where mining had been going on for centuries, an extensive knowledge of the environment already existed. Going on site was expensive and, more importantly, direct observation could do little to decide between competing proposals, when these were so glaringly different and involved distant places.

Now that a consensus had been reached about the necessity of a new, then-unnamed drainage tunnel, the second task was to ascertain its main direction, the *Stollenlinie*. This was not a mere technical decision but entailed far-reaching economic and political considerations.[35] An important question was to decide where exactly the entrance of the gallery—its *Mund* (mouth)—would be located. This obviously conditioned the depth of the future gallery, but also influenced the direction it would follow. The lowest points were farthest away from the mines; they would thus require more time and costly investments in aeration shafts, machines, and workers. Some proposed routes crossed several promising spots for further prospecting, each of which had to be carefully assessed. It was also crucial to decide which mines of the vast existing complex would be connected first, while trying to make the most of the existing infrastructure in order to lower the costs. In the end, the easiest part of the project was the naming of the gallery: since King George III of Great Britain and Hanover was open to the idea of funding the project, it was named after him the Deep George Tunnel (*Tiefer-Georg Stollen*).[36]

[34] NLA HA, BaCl Hann. 84a, Nr. 9803, doc. 9, *pro memoria*, 16 January 1774: 'Dieser hat Nr. 10. Trin. 1773 davon denn Riß in BergAmte vorgezeiget, in welchen die verhältniß eines gegen das andern bestimt ist, mir ist von BergAmt das Riß Zugestellet'.

[35] A technical description of the mathematical operations entailed in this process can be found for instance in Bennett Hooper Brough, *A Treatise on Mine-Surveying*, London: Griffin 1913, pp. 252–60.

[36] See NLA HA, BaCl Hann. 84a, Nr. 9803, doc. 12, report of 26 May 1774 to the royal government, and Gotthard, *Authentische Beschreibung*, p. 30.

In order to make a proper comparison between several possible pathways, a full survey was then ordered in the summer of 1774.[37] These measurements were recorded by Rausch and Länge in dozens of field books (*Observationsbücher*), allowing us to follow precisely the actions of the subterranean surveyors until the final completion of the project.[38] This operation perfectly illustrates how mine surveying was performed in the late eighteenth century. On 9 August, Länge began his work by taking a reference point low in the valley, close to the mouth of an abandoned gallery in the city of Grund. Using his semicircle, compass, and surveying chain, the surveyor went through the hills and woods and arrived at the entrance of the *Old-Blessing* mine on 13 September. The pit was located in the main district to be drained, and from there Rausch went down the mine. It took him six days to reach the deepest point, where the planned tunnel was to be connected. After five weeks of continuous work, the surveyor was finally able to ascertain accurately the relative height and position of the two points.[39] In order to understand how surveys were made, we can here quote the Swiss natural scientist Jean-André Deluc. Deluc visited the Harz mines three times while the digging of the Deep George tunnel was under way. As a *Reader* to Queen Charlotte, he would then write (and later publish) long *Lettres physiques et morales sur l'histoire de la terre et de l'homme* describing his scientific travels with a wealth of details. 'The geometrical measures' of mine surveyors, Deluc wrote in an article published in the *Philosophical Transactions of the Royal Society of London*, 'be made in so very singular a manner, that one does stand in need of this experience, to be persuaded of their exactness', and continued:

> A twisted brass wire five *toises* [c.10m] long, two puncheons, a semi-circle, and a compass, are all the instruments made use of by the subterraneous Geometer. By means of his two puncheons, he extends his wire in the direction of the way which he is measuring: and by practice he acquires a habit of always stretching it to the same degree. His semi-circle, which is very light, being suspended at the middle of the wire, shews him its inclination. By this means he has a right-angled triangle, of which the hypothenuse and angle at the base are known. He has consequently the vertical height and horizontal distance gone over. After this he suspends his compass to the wire, in order to find out its declination, and consequently the direction of his horizontal line. It is in this manner that he draws the plan and section of these subterraneous labyrinths. It is likewise by this means that he goes over hills and vales, in order to determine points corresponding to his pits and galleries.[40]

Once this initial survey was completed, surveyors Rausch and Länge then linked their initial reference point in Grund to all other envisaged entrances of the future tunnel by means of new measurements. They drew their results on a map, to which a lengthy

[37] NLA HA, BaCl Hann. 84a, Nr. 9803, doc. 16 and 17.

[38] In the *Bergarchiv* Clausthal, the section NLA HA Dep. 150 contains some forty field books related to the digging of the Deep George Tunnel.

[39] NLA HA Dep. 150 Acc. 2018/700 Nr. 252, p. 1 for the beginning, p. 31 for the entrance into the *Old Blessing* pit, p. 36 for the last part of this first survey.

[40] Jean André Deluc, 'Barometrical Observations on the Depth of the Mines in the Hartz', in: *Philosophical Transactions of the Royal Society of London* 67 (1777), pp. 422–4.

report was added. Four possible entrances and pathways were listed, together with estimates for several parameters, including lengths, costs and—most importantly—the depth that each would add compared to existing draining tunnels.[41] The result was checked multiple times, most thoroughly by mining master Steltzner, who laconically added in the margins 'perused and found correct'.[42]

The large-scale map, and most importantly the underlying data, offered solid quantitative evidence on the basis of which a decision could then be taken. At the end of that year, a one-week conference was organized between the political authorities of Hanover and the mining administration.[43] The aim of captain-general von Reden was to secure political support to finance the costly project. Trained in cameral sciences at the University of Göttingen, he ordered a new set of maps, this time less technical and more figurative, as well as synthetic tables gathering the main characteristics of the proposals.[44] This material was then brought to London and presented to King George III.[45] The project went through a couple of years of hesitations about the specifics of the financing plan (a process which we will not relate here, but in which maps again played a crucial role).[46] Meanwhile, surveyors continued their work, providing data on demand to feed the ongoing debates and considering numerous alternative solutions.

In 1777, the planning phase for a new, deeper gallery to drain ground water was coming to an end. At that point, the Deep George tunnel was the longest tunnel ever projected. When completed, it would be 284 metres deep and, unlike its predecessors, would not follow either the sinuous ore veins or existing galleries. The new galleries would be bored through hard stone following mostly straight lines, in order to connect the mines with the lowest exit point in the region. This fully mathematized project was radically new in its use of quantification, but it was not the mere application of abstract academic theories. The precise plans for its realization had been established by practitioners who took into account the technical and local specificities of the Harz mountains. Their proposal was practicable, which means that it included a credible cost estimate and a timetable. In contrast to the piecemeal approaches that had prevailed in previous generations, the geometry of the mine was the ubiquitous instrument that informed virtually all the decisions taken during the elaboration of the project. It was indeed innovative to consider mathematics as offering both a methodological framework in which projecting could take place, and as the most convenient tool to assess and choose between several technical options. From a

[41] Other parameters included the number of aeration shafts needed and the duration of the work, as well as the most promising prospecting areas. One copy of the map is conserved in NLA HA, BaCl Hann. 84a, Nr. 9803. See doc. 24 for the report, dated 24 October.

[42] NLA HA Dep. 150 Acc. 2018/700 Nr. 252, p. 102.

[43] See Bartels, *Vom frühneuzeitlichen Montangewerbe zur Bergbauindustrie*, pp. 388–92.

[44] See, for example, NLA HA, BaCl Hann. 84a, Nr. 9803, unpaginated, and NLA HA, BaCl Hann. 84a, Nr. 9805, unpaginated (reports of 8 August and 13 August 1776).

[45] One copy of this map is conserved in Rissarchiv des LBEG, Clausthal, B832. Von Reden also produced estimates about the costs and duration of the project, with its cash burn detailed by semesters. These estimates, conserved in the archives (see the last page of 9805) turned out to be mostly, although not entirely, accurate.

[46] George III's private councillor and the Hanoverian administration had to negotiate with the Principality of Brunswick, which shared the political authority over parts of the Harz mines: see NLA HA, BaCl Hann. 84a, Nr. 9805.

modern point of view, this dual use of mathematics hardly seems remarkable, but it certainly impressed contemporaries. The ability of surveyors to 'see through stones' using geometry amazed Jean-André Deluc, who underlined for other peers at the Royal Society the complexity of the surveying process:

> The path that this tunnel will take is neither horizontal, nor in a straight line: & it is there again that the miner's skill can be witnessed. The gallery must be regularly inclined, so that the water runs off; & it must be inclined as little as possible, so as not to lose depth. [...] Besides, it should follow certain contours, either to pass under valleys where shafts can be drilled, or to avoid parts that are too hard and take too long to drill, or parts that are too soft and require stamping. Sometimes again, if the surveyor sees near his path some gallery already drilled for other purposes, he detours in order to take advantage of the work already done.[47]

Monitoring through geometry (1777–99)

The first phase of the project was thus completed by mid-1777. On 1 July, a general visitation was organized for the whole mining administration. The proposed route was followed from the existing mine workings—where the tunnel would eventually arrive to drain underground water—to its future mouth some 10 kilometres way, in the small town of Grund. The last outstanding decision was whether it was desirable to use the existing mouth of the old *God's Help and Isaac Tanner* tunnel close to Grund and reuse the 130 fathoms (*c.* 250 metres) of existing galleries. 'During the bespoken visitation', a report stated, 'the surveyor Mr. Länge presented a map', and 'the attendees were convinced' that the existing infrastructure had been dug with too 'many slants', as can be seen in Figure 12.1 where the two proposals are compared.[48] This is an undeniable indication that confidence in the accuracy of geometric methods was growing: the traditional method of using existing winding tunnels to lower the costs was now considered hazardous. After a last round of double-checking, Länge confirmed that the mining master, Steltzner, 'stands by the correctness of his measurements up to a difference of a few inches'.[49] On 26 July, a solemn ceremony was organized with cannon salutes and a miners' parade, and the first pickaxes were handed out.

[47] Jean André Deluc, *Lettres physiques et morales sur l'histoire de la terre et de l'homme: Addressees a la reine de la Grande Bretagne, par J. A. de Luc*, La Haye et Paris: De Tune et Duchesne 1779, vol. 4, pp. 623–4: 'Cette route que devra tenir la *Galerie*, n'est ni horizontale, ni en droite ligne: & c'est la encore que se montre l'habileté des Mineurs. La galerie doit être régulièrement inclinée, pour que les eaux s'écoulent; & elle doit l'être le moins possible, pour ne pas perdre de la profondeur [...] Il faut de plus qu'elle suive certains contours; soit pour passer sous des Vallées où l'on puisse percer des puits; soit pour éviter des parties trop dures qui prendroient beaucoup trop de temps à percer, ou des parties trop molles qui exigeraient des étampages. Quelque fois encore, si le Géomètre voit près de sa route quelque *Galerie* toute percée pour d'autres usages, il se détourne, afin de profiter de ce travail déjà fait'.

[48] Gotthard, *Authentische Beschreibung*, p. 73: 'Bey der gedachten Befahrung producirte der damalige Communion Markscheider Herr Länge einen Special-Riß [...] überzeugten sich die Anwesenden [...] wodurch dasselbe viele Krümmungen erhalten hatte'.

[49] NLA HA BaCl Hann. 84a Acc. 8 Nr. 2465/1, Extracts from Protocol, 20 July 1777: 'daß er bis auf eine Differenz von eine paar Zollen vor die Richtigkeit seines Zuges einstehe'.

The work of surveyors Rausch and Länge was, however, far from over, as the site of the Deep George Tunnel required constant attention and fine-tuning. While the general direction of the operation was known with certainty, the operation itself was not simply a matter of digging from the mouth in the valley up to the existing mine workings. In the late eighteenth century, even accounting for the use of blasting powder, boring speed averaged less than one metre per week. At this pace, drilling a ten-kilometre tunnel in one direction only would have taken about two centuries. The mining administration thus decided to rely on the new technology—mentioned in the previous section—of pursuing digging operations from both ends, a method that had already been successfully introduced in Lower-Hungary and Saxony. Still, starting from both sides would only cut the duration of the digging operation in half. The decision was therefore taken to use fifteen intermediary points. From each of those points, a ventilation shaft would be dug with great care to the exact depth the tunnel was supposed to have; excavations and counter-excavations would then have to follow the gallery's slope in order to meet precisely. 'It had been roughly calculated' that, by using this method, 'this tunnel, could be cleared in about 20 years, if it were to be worked vigorously'.[50]

These operations were described in some detail by Jean-André Deluc. On 13 July 1778, he arrived in Clausthal for a second trip, the purpose of which was again to conduct barometric experiments. With reference to captain-general von Reden and other mining officials who accompanied him Deluc recalled, 'we visited a mining city, close to the place where the new drainage tunnel they are working on will be coming out. We saw on our way several of the shafts being dug to speed up the works'.[51] One year after the start of digging operations, 'six shafts are being dug, or have been, on the pathway of the tunnel. Once sunk to the necessary depth, two pairs of Miners will leave from the bottom of each one, heading in opposite directions on this planned route'.[52] Both the robustness of these geometric methods and the economy of means amazed Deluc: 'This is thus certainly of one of *geometry*'s nicest applications; and when the miner is proud of his art, this will not surprise me'.[53]

Concretely, the two mine surveyors had various tasks to perform to ensure that the project was heading in the right direction. They were especially careful, of course, when it came to the connection of various sections of the tunnel. As soon as routine visitations indicated that an excavation was about to meet a counter-excavation, a local survey was performed to monitor the exact relative position of the two groups

[50] Gotthard, *Authentische Beschreibung*, p. 29: 'Man hatte ohngefähr berechnet, daß dieser Stollen, sofern er recht schwunghaft betrieben werden würde, in ohngefähr 20 Jahren durchschlägig gemacht werden könnte'.

[51] Deluc, *Lettres physiques et morales*, vol. 4, p. 621: 'nous allâmes visiter [...] l'une des *Villes de Mines*, & près du lieu où sortira la nouvelle *Galerie d'ecoulement* à laquelle on travaille. Nous vîmes sur notre chemin plusieurs des Puits qu'on perce pour accélérer l'ouvrage'.

[52] Deluc, *Lettres physiques et morales*, vol. 4, p. 623: 'Six puits se percent actuellement, ou sont percés, sur la route de la Galerie; & lorsqu'ils se seront enfoncés à la profondeur nécessaire, il partira deux couples de Mineurs du fond de chacun, qui se dirigeront en sens contraire sur cette route prévue'.

[53] Deluc, *Lettres physiques et morales*, vol. 4, p. 625: 'Voilà donc certainement une des plus belles applications de la *Géométrie*; & quand le Mineur est glorieux de son Art, je ne faurois m'en étonner.'

of miners, to ensure that both galleries had the same direction (*Streichen*) and inclination (*Fallen*). Indeed, there were always some discrepancies between the projections and the real course of a tunnel: in one report, surveyor Länge complained about the 'incorrigible imperfection of the instruments and the deviation of the magnetic needle'.[54] Moreover, even the best miners could not dig in perfectly straight lines, meaning that regular inspections were needed. As the years went by, even the slow drying-up of the paper on which the original large-scale maps had been drawn made them unreliable to ascertain directions with the desired precision. It was thus necessary, after several years, to proceed to new observations 'and give the latest instructions about the direction of the connection based on them'.[55]

Let us follow the first section of the work: in July 1777, a first group of miners began working from the deepest point in the valley, digging the mouth of the tunnel (the *Orifice* on Figure 12.3). A second group was supposed to lead a counter-excavation in the direction of the mouth from the sixth ventilation shaft (the *Puits d'airage* or *Lichtloch*, also on Figure 12.3). This shaft already existed from sixteenth-century works, and the surveyors had thought it could simply be repurposed. As Steltzner inspected the field-books of surveyors Rausch and Länge, however, he found out that the shaft's depth had been wrongly taken to be sufficient. The mining master suddenly alerted the mining council in early October: 'according to a [new] survey, surveyor Länge reports that the point in the mouth of the Deep George tunnel would be 1 fathom, 7 eighths, and 7 inches [*c.* 3.5 m] deeper than the highest point in the 6th ventilation shaft where the counter excavation should be set'.[56] Taking into account the (very low) slope between the two locations, computations showed that the existing shaft was precisely 1 fathom, 5 eighths and, 2 inches [*c.* 3m] too high relative to the projected path of the Deep George Tunnel. Minor discrepancies could always have been corrected during the smoothing phase (*Nachhauen*), but such a 'considerable divergence' was not tolerable.[57] This alert led to a complete review of all the surveyor's calculations about depth, which fortunately showed that the error was localized, and did not affect the general path of the tunnel. The ventilation shaft was simply dug slightly deeper, and the digging work resumed. Almost two years later, 'on the night of 23 to 24 July 1779', Steltzner reported that the first connection had been made between an excavation and a counter-excavation, for a total length of some 450 metres (see Figure 12.3, point R^1). The mining master soberly expressed his hopes that 'the other connections may also happen without major obstacles', and immediately redeployed the miners to the next sections.[58]

[54] NLA HA, BaCl Hann. 84a, Nr. 9825, unpaginated, report of 10 January 1783: 'incorrigibelen Unvollkommenheit der Instrumente und der Abweichung der Magnet-Nadel'.

[55] NLA HA, BaCl Hann. 84a, Nr. 9825, unpaginated, report of 10 January 1783: 'den Zug von neuen verrichten, und nach denselben die letzten Anweise Linien zum Durchschlag angeben'.

[56] NLA HA BaCl Hann. 84a Acc. 8 Nr. 2465/1, Extract Protocolli 2 Quartals Luciae 1777 recto: 'nach gemachter *Observation* gebe der Markscheider Länge an, daß der Punct des tiefen Georg Stollen im Mundloch 1 Ltr 7 achtel 7 Zoll tiefer stünde als der Punct bey dem 6ten Lichtloche im hangenden wo die Gegen-Örter [...] angesetzet werden solten'.

[57] The discrepancy would not, strictly speaking, have endangered the whole project, but the loss of depth would have affected the next sections and made the tunnel slightly less useful.

[58] NLA HA BaCl Hann. 84a Acc. 8 Nr. 2465/1, report from 10 July 1796, first page: 'beträchtliche Abweichung'; report from 25 June 1779, first page: 'die übrigen Durchschläge auch ohne große Hinderniß [...] erfolgen möge'.

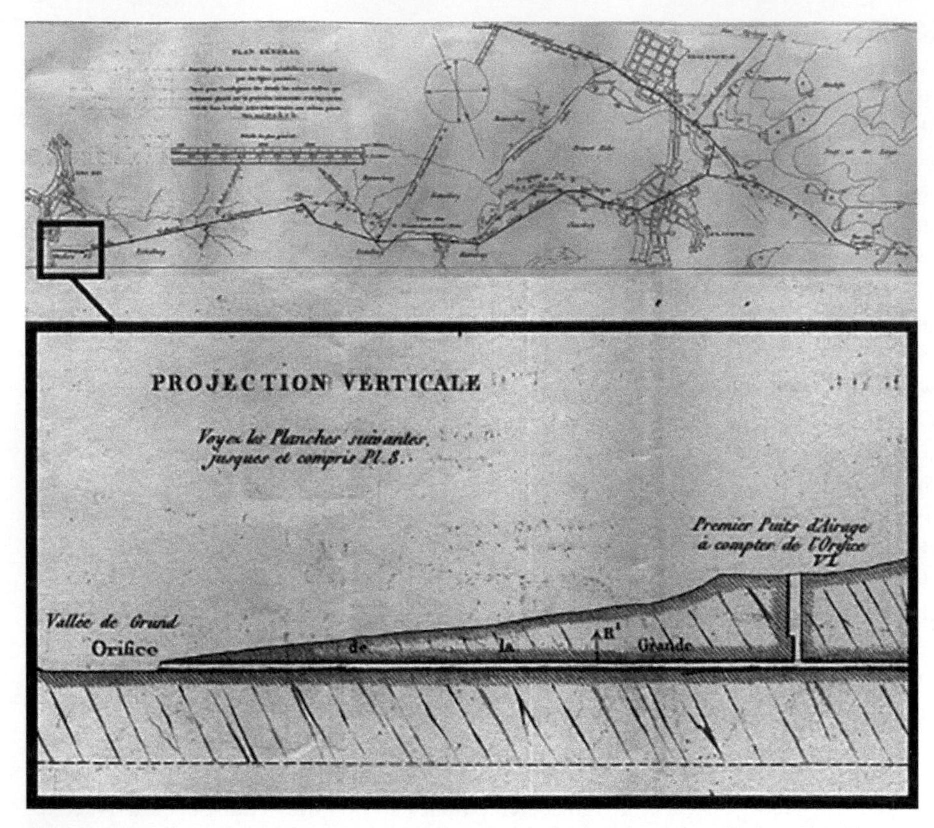

Figure 12.3 Elevation of the Deep George Tunnel, from Héron de Villefosse's *De la richesse minérale* (1810). This small portion of the work, some 450 metres long, was the first section to be completed in 1779, when the excavation dug from the mouth of the gallery [*Orifice*] met the counter-excavation led from the sixth ventilation point [*Puits d'airage*] at the point R^1.

The Deep George Tunnel offers valuable lessons about the scope and nature of practical mathematics in the late eighteenth century. The mathematical tools tirelessly used by Rausch and Länge, while involving a large volume of data, were all things considered rather elementary. So were the methods used by mining master Steltzner to check their results and the corrections that were ultimately implemented. There were no theoretical quandaries or higher analysis, but all this does not mean that surveying was an easy task. The real difficulty was to use stable procedures, ensure the accuracy and stability of measurements over the years, and finally scale-up these seemingly easy methods.[59] The suspended compass, the semicircle, and the surveying

[59] One century earlier, mining master Balthasar Rösler (1605–1673) had already noted the problem of ensuring that methods working over short distances could be scaled up. In his *Mirror of Mining*, he wrote that subterranean geometry 'seems to be a minor art, and the one who manages to correctly measure 6 or 10 angles thinks—and might well convince himself—that he is already a master. But if he is asked to perform

chain produced impressive results thanks in good part to the stable technical and cultural system in which they were operating. The culture of systematic surveying had developed over generations and was in its very nature cumulative. A vast trove of data, regularly updated and improved by subterranean surveyors, was recorded in registers and maps (Figure 12.4). Measurements had a codified place in planning operations and informed the decision-making process. Quantitative data was presented to the mining council, ensuring that technical choices would be based on collective deliberations. A comparable project for an underground canal in France (*Canal de Picardie*) was pursued by the *mécanicien* Pierre-Joseph Laurent from 1764 onwards, that is, almost at the same time as the Deep George Tunnel. In France, however, the lack of a stable culture of engineering and decision-making led to important surveying mistakes, cost overruns, and ultimately the abandonment of the project in 1775.[60]

By the end of the eighteenth century, surveying methods in the German-speaking mining states had reached such a level of certainty that captain-general von Reden and mining master Steltzner barely exceeded their budget, something that had rarely been the case in previous times. Numerous field books and bills for the payment of surveys indeed reveal that the work went smoothly, with negligible deviations.[61] Digging the Deep George Tunnel was supposed to take 'about 20 years', and in fact the inauguration took place in 1799, after twenty-two years of work. Until that date, surveyors constantly assumed considerable responsibilities.

Tunnelling in mathematical and mining textbooks

This chapter has hitherto been concerned mainly with practising surveyors working in real-world settings. This last section will highlight how subterranean geometry briefly attracted a wider audience, or more precisely a wider readership, during the last third of the eighteenth century. Not only were several academic textbooks published specifically on the topic, but major sites were frequently visited by a diverse audience, and reported upon by journals. The most famous university professor of mathematics of the time, Abraham Gotthelf Kästner (1719–1800), published in 1775 his *Notes on Subterranean Geometry*—precisely at the same time as surveyors Länge and Rausch were doing their planning work on the Deep George Tunnel.

a measurement of a hundred or more angles, it will never work' (Balthasar Rösler, *Speculum metallurgiae politissimum, oder, Hell-polierter Berg-Bau-Spiegel*, Dresden: Winckler 1700, p. 87).

[60] See Louis Thbaut, *Le mécanicien anobli Pierre-Joseph Laurent, 1713–1773: des mines d'Anzin au Canal de Saint-Quentin*, thèse de 3e cycle, Lille III 1974, pp. 265–372.

[61] I do not mean to say that there were no problems at all, which given the scale of the project would be surprising, only that none of these problems were related to measurements or calculations. For instance, when the miners were close to the *Wilhelm* mining pit, the gallery floor turned out to be too crackled (*zerklüftet*) and thus let the water go down the underlying mine. A 78-fathom-long cast iron conduit (*Gefluder*) was made as a bridge, and the operations continued. See Antoine-Marie Héron de Villefosse, *Über den Mineral-Reichtum: Betractungen über die Berg-, Hütten- und Salzwerke verschiedener Staaten, sowohl hinsichtlich ihrer Production und Verwaltung, als auch des jetzigen Zustandes der Bergbau- und Hüttenkunde, Zweyter Band, Des technischen Theils erste und zweyte* Abtheilung, Sondershausen: Bernhard Friedrich Voigt 1822, pp. 174–5.

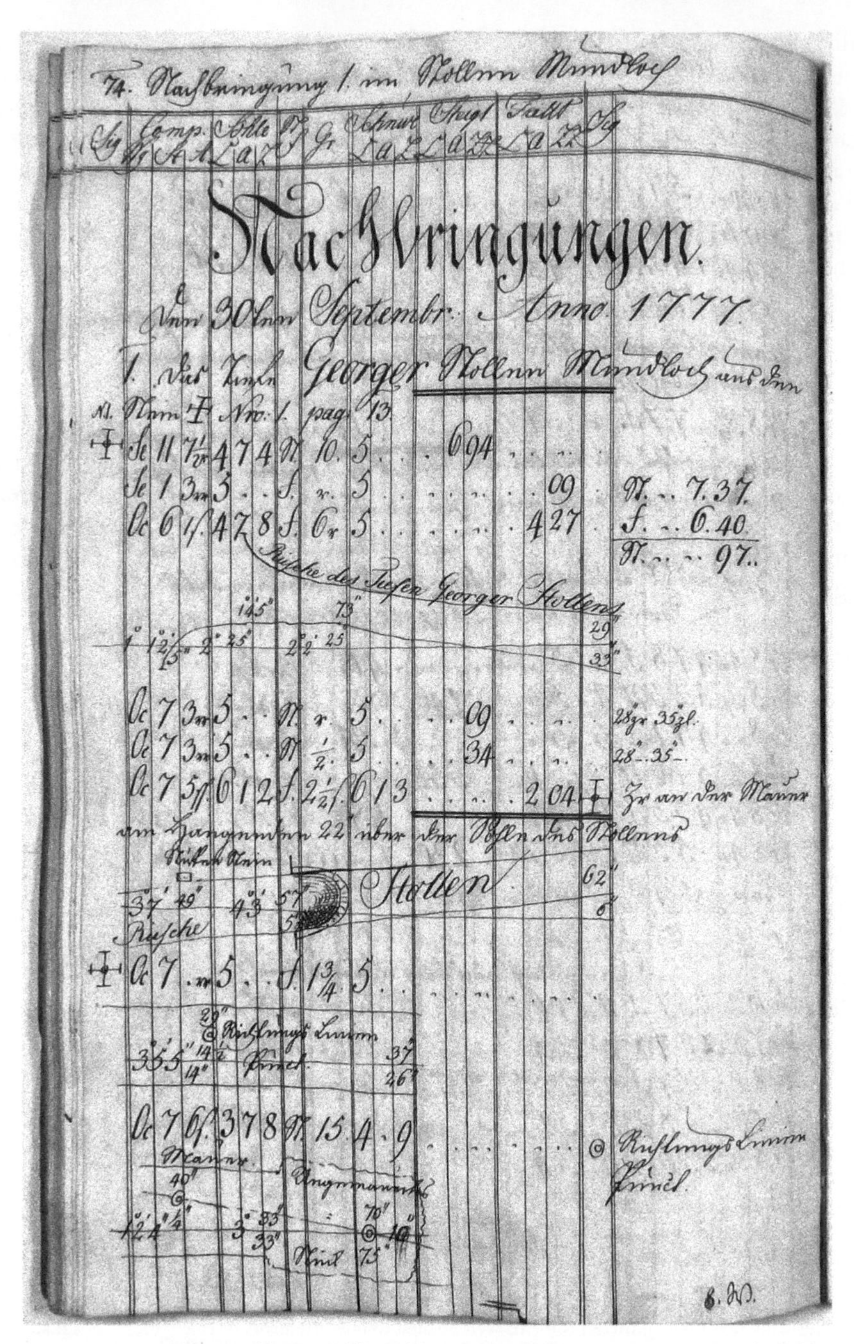

Figure 12.4 Additional 'reports' [*Nachbringungen*] from surveyor Länge's field book, 30 September 1777. Surveying from fixed points represented by symbols carved in stones (here the '╪'), underground surveyors could later refine their measurements. Each line describes the measurement from one point to another; the data tables are supplemented with sketches to remove ambiguities.

Kästner was professor of mathematics and physics at the University of Göttingen, and member of the privy council (*Hofrat*) of the Hanoverian state. The mathematician decided to lecture on the topic at his university, and seems to have had a genuine, if academic, interest in the discipline. In the introduction of the book, Kästner recalled visiting Saxon mines almost thirty years earlier. At that time, the modern tunnelling methods used to dig the Deep George Tunnel (as described above) did not even exist. The fact that Kästner's experience was outdated is not really relevant here, however, given that the content of his *Notes* was not based on actual practices, but on his wide academic review of the existing literature. In the end, he chose to use a fairly outdated textbook by Johann Friedrich Weidler (1691–1755), originally written in Latin in 1726, to which he added remarks and corrections; the result was then published as a stand-alone book.[62] While practitioners had written more recent and comprehensive works on the topic, the university professor considered that these 'more complete instructions [...] were not suitable for academic lectures'.[63]

The example of Kästner might seem caricatural, as an extreme form of academic pedantry. It is in fact quite representative of his time. Eighteenth-century academic textbooks used in German universities were, in most cases, not meant to offer an adequate hands-on training, even when they dealt with practical mathematics. These works were meant to be used as compendiums, and brought together material to be taught in one semester to a heterogeneous body of students. In that context, it is not surprising that Kästner's *Notes on Subterranean Geometry* did not mention the mathematical operations of tunnelling, or even mine tunnels. The goal of a university course was not to offer an operative, useful knowledge but to provide a theoretical overview on a broad range of disciplines.[64]

A contemporary French textbook, emphasizing how this discipline was 'little known among us', concluded that 'it is clear, however, that there is not a single problem of subterranean geometry which does not immediately belong to what we call practical geometry'.[65] While this was in a certain sense true concerning the theoretical aspects of mine surveying, the French textbooks did not include real-life cases or concrete data. While such works could certainly give an interested reader an idea of what the discipline was about, their content was not sufficient to learn how to work effectively as a surveyor. Nonetheless, the fact that many standard introductions to applied mathematics came to include sections on subterranean geometry reveals that the public awareness of the topic had been raised.

[62] Abraham Gotthelf Kästner, *Anmerkungen über die Markscheidekunst, nebst einer Abhandlung von Höhenmessungen durch das Barometer*, Göttingen: Vandenhoeck 1775, introduction.

[63] Kästner, *Anmerkungen*, introduction: 'vollständigere Anleitungen [...] sind nicht für akademische Vorlesungen'.

[64] On the teaching of mathematics in German universities, see Heidi Kühn, '*Die Mathematik im deutschen Hochschulwesen des 18. Jahrhunderts (unter besonderer Berücksichtigung der Verhältnisse an der Leipziger Universität)*', unpublished dissertation, Karl-Marx-Universität Leipzig 1988. On Kästner's activity as professor, see Desirée Kröger, '*Abraham Gotthelf Kästner als Lehrbuchautor, unter Berücksichtigung weiterer deutschsprachiger mathematischer Lehrbücher für den universitären Unterricht*', unpublished dissertation, Bergische Universität Wuppertal 2014.

[65] Antoine de Genssane, *La géométrie souterraine: ou traité de géométrie-pratique, appliqué a l'usage des travaux des mines*, Montpellier: Rigaud, Pons 1776, introduction: 'peu connu parmi nous [...] Il est cependant constant qu'il n'y a pas un Problême dans la Géométrie Souterraine, qui n'appartienne immédiatement à ce que nous appellons Géométrie-Pratique'.

In the eighteenth century, mining sites were becoming a fashionable place for natural scientists, and for a much broader audience as well. The University of Göttingen, with its large student population, was only a few hours' ride from the Deep George Tunnel and the rest of the Harz mines. Christoph Wilhelm Jakob Gatterer (1759–1838), *Privatdozent* for mineralogy and natural history and a colleague of Kästner, thus published in 1785 a first volume of *Instructions for Touring the Harz and Other Mines with Benefits*, indicating that the book was meant to 'serve as a handbook for lectures', but could 'be used as handbook for travellers' too.[66] Despite his frequent mentions of the Deep George Tunnel, and several promises of a 'complete description' of the digging operations 'in one of the following volumes of this work', Gatterer never managed to complete the undertaking.[67]

Beyond the genre of university textbook, which given its intended audience could hardly present in detail the mathematics of tunnelling, there is a second kind of academic literature that dealt with the *Markscheidekunst*: the works written by mining school professors, which were used to teach in those institutions. In the last third of the eighteenth century, several *Bergakademien* (mining academies) had been founded, mainly in German-speaking states, the most famous ones being located in Schemnitz (1762, now Banská Štiavnica, Slovakia) and Freiberg (1765).[68] The most prominent textbook on the subject was written by Johann Friedrich Lempe in 1782 and entitled *Gründliche Anleitung zur Markscheidekunst* (*Detailed Instruction on Subterranean Geometry*). J. F. Lempe (1757–1801) was a former miner and student of the Freiberg mining academy, who had then completed his mathematical studies at the University of Leipzig with the analyst Carl Friedrich Hindenburg (1741–1808).

Lempe's textbook is an 800-page work that shows a deep understanding of the more recent developments of the discipline and yet, in this compendious tome, tunnelling methods per se are only dealt with in passing.[69] This is all the more surprising given that drainage tunnels were indeed a crucial element of every mining district, at least until the introduction of electric pumping machines in the late nineteenth century. In the corresponding passage in Lempe's textbook, one does not even find the concrete step-by-step process that we have seen in action with surveyors Rausch and Länge. The focus of this textbook section is on trigonometric formulae with which the direction of the main tunnel line can be calculated directly, or the determination of relative positions of exterior points known from one or the other side. In other words,

[66] Christoph Wilhelm Jakob Gatterer, *Anleitung den Harz und andere Bergwerke mit Nuzen zu bereisen*, Göttingen: Vandenhoeck 1785, introduction, p. vii: 'Einmal soll es als Handbuch zu Vorlesungen dienen [...] Zweytens, kann dasselbe als Handbuch von Reisenden gebraucht werden'.

[67] The Deep George tunnel is mentioned three times in vol. 2, a complete description is promised in vols. 3 and 4, and finally in vol. 5, pp. 445–6, from which the quote is taken: 'Eine ausführliche Beschreibung [...] kommt [...] in einem der folgenden Bände dieses Werkes'.

[68] The wider context in which the education of mine engineers was reformed is presented in detail in Donata Brianta, 'Education and Training in the Mining Industry, 1750–1860: European Models and the Italian Case', in: *Annals of Science* 57/3 (2000), pp. 267–300.

[69] Johann Friedrich Lempe, *Gründliche Anleitung zur Markscheidekunst*, Leipzig: Crusius 1782, pp. 474–7.

it presents not concrete surveying procedures but an opinion about their underlying theoretical viewpoint.

To sum up the present analysis of eighteenth-century mathematical textbooks, none of these works were sufficient to learn the mathematics of tunnelling. There seems to be a gap between the best available printed works and the actual methods used by contemporary surveyors. In order to understand why, one has to clarify the context in which these books were used.[70] At the Freiberg mining academy, for instance, Lempe acted as professor of mathematics, and only taught the 'theoretical subterranean geometry' to a handful of gifted students. In contrast, practical instructions were delivered directly in the mining pits by surveyor Richter (who, incidentally, had supervised the last years of the *Alter Tiefer Stollen* studied in the first section of the present chapter).[71] There was no meaningful dispute or opposition between the professor and the surveyor. Richter and Lempe had both been trained in the mining pits of Saxony, and both were heavily influenced by mining master Scheidhauer (d.1784), who wrote extensively on subterranean geometry, although his manuscripts were never published.[72] The discrepancy between Lempe's textbook and Richter's teaching simply reflects the difficulty of articulating theory and practice in the late eighteenth century. The age-old companionship system, which had proved its worth, was still largely in place. In fact, it is only in the mid-nineteenth century that mining academies were reformed to offer a full-scale mathematization, under the influence of Julius Weisbach (1806–1871).[73]

Conclusion

> Experience on many mountains has already shown that connections for light or ventilation [shafts] do not always succeed; on the contrary, these are rarely well executed; for one often has missed by many fathoms, and nothing has been gained after much labour and sour toil.[74]
>
> Johann Gottfried Jugel, *Geometria Subterranea*, 1773, p. 490

[70] In fact, the handful of textbooks in which the procedures of excavations and counter-excavations (*Ort und Gegenort*) are presented in a useful fashion are textbooks about mining technology, such as Franz Ludwig von Cancrin, *Erste Gründe der Berg und Salzwerkskunde*, vol. 6.2, Frankfurt am Main: Andrea 1776, pp. 288–9.

[71] Pabst von Oheim, *Geschichte des 'Thurmhofer Hilfsstollens'*, p. 33.

[72] On Scheidhauer, see Kerrin Klinger and Thomas Morel, 'Was ist praktisch am mathematischen Wissen? Die Positionen des Bergmeisters J. A. Scheidhauer und des Baumeisters C. F. Steiner in der Zeit um 1800', in: *NTM Zeitschrift für Geschichte der Wissenschaften, Technik und Medizin* 26/3 (2018), pp. 267–99.

[73] See, most notably, Julius Weisbach, *Die neue Markscheidekunst und ihre Anwendung auf bergmännische Anlage. Erste Abtheilung: die trigonometrischen und Nivellir-Arbeiten über Tage, sowie die Anwendung auf die Anlage des Rothschönberger Stollns*, Braunschweig: Vieweg 1851.

[74] Johann Gottfried Jugel, *Geometria subterranea, oder Unterirdische Messkunst der Berg- und Grubengebäude, insgemein die Markscheidekunst genannt, etc.*, Leipzig: Kraus 1773, p. 490: 'Auf vielen Gebürgen hat schon die Erfahrung bewiesen, daß dergleichen Durchschläge auf Licht und Wetter nicht allezeit zugetroffen, sondern nur selten wohl gerathen sind; denn öfters hat man viele Lachter damit gefehlet, und mit aller sauren Mühe und Arbeit nichts gewonnen'.

Underground canals, shafts, and drainage tunnels were among the largest projects of the eighteenth century, and the Deep George Tunnel became the most significant one for contemporaries. The length of these subterranean works required multiple excavations and counter-excavations, while the mountainous landscape and the depth of mining pits made surveying operations difficult. The nature of these enterprises thus offers a good case study to observe how practitioners gradually achieved an efficient mathematization of nature in the eighteenth century. Johann Carl Freiesleben (1774–1846), who studied at the Freiberg mining academy with Alexander von Humboldt (1769–1859), toured the site of the Deep George Tunnel, praising 'the value of subterranean geometry, which appeared in such a bright light during the execution of this project'.[75]

The mathematics of tunnelling highlights an often-overlooked consequence of the new quantifying spirit that developed at the time: elementary arithmetic and geometry, when put into practice systematically by experienced artisans, could hope to achieve impressive results. More often than not, major technical evolutions were sparked by a highly efficient and integrated use of well-known tools and methods, not by higher mathematics or new theories. In this kind of practical mathematics, innovation was not deduced from theories or applied by *savants*. It was slowly incorporated by engineers themselves, who relied on data and experience gathered by previous generations, within an existing cultural and administrative system.

The idea of fully mathematized planning had previously been unthinkable. Although Renaissance surveyors used and valued geometry, it was considered as a useful tool—on some occasions, at least—and not as a set of general laws of nature. The traditional way of using surveys and mining maps in a piecemeal fashion made sense in the technical environment of the time, even if we now know it to be fundamentally more costly and risky. It was the combination of new (and cheaper) technical methods, the slow accumulation of data (field books, maps) and finally a growing need to quantify and obtain reliable planning that led to a full mathematization.

In the world of eighteenth-century practitioners, geometry was an essential cog in large, multifaceted technical systems, rather than an abstract solution to well-defined problems.[76] This episode can teach us one last important fact about the dynamics of professional mathematics at the time. As our survey of mathematics textbooks has shown, there were important discrepancies between the textbooks used at universities and the methods actually used in the mines. The training of mine surveyors was largely unconnected with the university courses of the time, or indeed with what could be derived from most printed books. The actual methods have been mostly preserved in messy unpublished notes and field books, in project reports and

[75] Johann Carl Freiesleben, *Bemerkungen über den Harz. Erster Theil: Bergmännische Bemerkungen*, Leipzig: Schäfer 1795, p. 296: 'über den Werth der Markscheidekunst, dey bey der Ausführung dieses Unternehmens in so hellem Lichte erschien'.

[76] In this sense, the historical dynamics of practical mathematics may resemble more strongly what can be observed in technological systems than the evolution of abstract sciences.

mining maps. Practical geometry was sustained by an efficient system of companionship in the robust institutional frameworks of the local mining administrations.[77] The large number of printed works about mining sciences, the theory of machines and surveying methods by local engineers, on the other hand, indicates that this technical field was rapidly developing. This movement culminated with the foundation of mining academies, in which original syntheses between theory and practice were explored.

[77] See Rainer Sennewald, 'Die Stipendiatenausbildung von 1702 bis zur Gründung der Bergakademie Freiberg 1765/66', in: *Technische Akademie Freiberg, Festgabe zur 300. Jahrestag der Gründung der Stipendienkasse für die akademische Ausbildung im berg- und Hüttenfach zu Freiberg in Sachsen*, Freiberg: TU Bergakademie 2002, pp. 407–29; Thomas Morel, 'Le microcosme de la géométrie souterraine: échanges et transmissions en mathématiques pratiques', in: *Philosophia Scientiae* 19/2 (2015), pp. 17–36.

PART IV
THE PRACTICE AND TEACHING OF MATHEMATICS

13
Climbing the Social Ladder
Johannes Faulhaber's Path from Schoolmaster to Fortification Engineer

Ivo Schneider

Two portraits of Johannes Faulhaber are preserved in the form of broadsides (*Einblattdrucke*) (see Figures 13.1 and 13.2). They show Faulhaber aged 35 and 50 respectively. The main purpose of these two prints was to inform potential clients about the knowledge and skills Faulhaber could acquaint them with. Even if the two artists who made the engravings claimed absolute authenticity for these portraits, they were obviously more concerned to document the social status of the man portrayed than with his individual characteristics. Thus, the 35-year-old Faulhaber displays a receding hairline in accordance with the contemporary image of somebody who, in contrast to a craftsman, depends less on his hands for his work. Fifteen years later, Faulhaber has regained a full head of hair, presumably because the image of an architect and fortification engineer did not afford a high forehead.

Both portraits display Faulhaber's right hand holding a pair of compasses, indicating his familiarity with the solution of geometrical problems. Whereas Faulhaber in 1630 is called a well-known engineer and a man well versed in every part of mathematics, in 1615 he figures as a man engaged in the city of Ulm as a teacher of (elementary) mathematics. He displays in the inornate habit of a German schoolmaster his social affiliation with the professional group of teachers of reading, writing, and calculation skills. As such, he appears as one occupying a rather modest but respectable position in the society of his day. That Faulhaber was able to familiarize interested clients over and above these elementary skills with more demanding mathematical knowledge and with the mysteries of the biblical numbers is suggested by the list of six publications in German, cited below, which for the most part, as is stated, were also available in Latin. These publications serve to indicate Faulhaber's ability to teach things which went far beyond the reach of an ordinary schoolmaster and so gave him an advantage over his peers in Ulm and in neighbouring free cities like Nuremberg or Augsburg.

Faulhaber's peers were so-called *Rechenmeister* who taught their subject in German, in contrast to the members of the faculty of arts in universities who taught in Latin. In order to qualify as a *Rechenmeister* one had to pass a period of apprenticeship of five to six years in the school of an established *Rechenmeister*. At least in the second half of the sixteenth century the apprenticeship ended in several German cities with a final examination in which over and above proving his ability to master the four basic arithmetic operations the candidate needed to solve quadratic equations in one unknown in the so-called *Coss*—a term derived from the Italian

Figure 13.1 Johannes Faulhaber (aged 35). From a broadside advertisement for the Rechenmeister Johannes Faulhaber, Augsburg 1615.

Figure 13.2 Johannes Faulhaber (aged 50), line engraving by Sebastian Furck, n.p. 1630.

term 'cosa' which was used by the *maestri d'abaco* in Italy to signify the unknown in equations. A German *Rechenmeister* had also to teach how to determine areas and volumes in connection with the solution of commercial and financial problems.

Apprenticeship and first years as a schoolmaster

Faulhaber, who was born as the youngest son of a weaver in the Freie Reichsstadt (Imperial Free City) of Ulm in 1580, worked for some years as an unskilled weaver in his father's workshop following the latter's early death before he started a four-year apprenticeship in Ulm with David Selzlin, a 'Schreib- und Rechenmeister'. Thereafter, he worked for a year and a half as a so-called Provisor in the school of Johann Krafft, another 'Schreib- und Rechenmeister' in that city. Surviving sources contain no information as to whether Faulhaber had to pass an examination after his apprenticeship, as was the case elsewhere. After he had heard, in 1600, that one of the six (originally four) positions as a schoolmaster in Ulm had become vacant, he applied for permission to open his own school there. Despite vehement objections on the part of Krafft, local officials granted the permission. With a loan of 200 florins from the city he was able to purchase a house for his school. He paid off the loan during the following four years. Shortly afterwards, in September 1600, he married Ursula Esslinger.

Being married was a prerequisite for a future schoolmaster because most of the pupils and students he would teach lived and were provided for in his household. For their son's boarding and upkeep, the parents had to pay a certain fee, the amount of which was not fixed but mainly determined by the prevailing market. We learn this from Faulhaber's correspondence with Sebastian Kurz who operated a school in Nuremberg.

Faulhaber's first obligation in this position, for which he was paid a modest salary by the city, was to teach children how to read, write, and calculate. When he advertised his skills alongside his portrait, in 1615, he hoped for a competitive advantage by hinting to the publications he could claim so far. The list reflects the main subjects which kept Faulhaber occupied even beyond that year. These are cossic algebra, the use of new scientific instruments he had designed himself, and his enduring enthusiasm for the interpretation of the biblical numbers.

In that same year, besides publishing his portrait Faulhaber publicly proposed a series of five problems concerning sums of powers of integers to 'all philosophers and mathematicians in Europe'. However, with the exception of Faulhaber's friend and former student Johann Remmelin nobody accepted the challenge. Indeed, one of Faulhaber's opponents sought to vilify him three years later as somebody who, as the lowly son of a weaver, suffered from an overestimation of his own status and abilities.

Even if the occupation of a weaver was considered as hardly less respectable than that of a *Rechenmeister* Faulhaber was eager to transcend his former social status in two ways. The first was to be acknowledged as a prophet sent in order to communicate God's will to the world. The second was more down to earth: to become known to those governing and ruling and eventually to become their counsellor. Faulhaber tried to bring about and justify such a rise in his standing through his portrait of 1630. In this portrait he not only claims the title of an engineer, but also wears a distinguished habit with a gold medal he had received from Prince Mauritz of Orange, thereby hinting at historic sources which confirmed the participation of men named Faulhaber in chivalric tournaments 400 years earlier.

No wonder that he divided people into, on the one hand, admirers and supporters and, on the other, bitter opponents who criticized his various claims as unjustified. Such conflicts influenced his career for the whole of his life and brought Faulhaber again and again into conflict with the city authorities and the Protestant church.

Already during his apprenticeship, Faulhaber had started to develop an impressive network of correspondents. Many of them he had met in the arena of mathematical problems.

Faulhaber's *Lustgarten* and the reaction to it by Peter Roth from Nuremberg

One fruit of this concern with problems constitutes his first publication, the first on the list of his portrait from 1615. It pertains to his *Arithmetischer Cubicossischer Lustgarten* ('Arithmetical cubicossical pleasure garden') from 1604, a booklet of less than one hundred pages, which contained 160 mostly cubic problems.[1] In it he gave one—sometimes two, or all—of the solutions of the problems, but did not disclose the way he had found them. Nonetheless, this publication sufficed to convince his readers that he was able to solve these problems. But anyone wanting to grasp the way by which he had found the solutions had to go to Faulhaber himself, who was prepared to instruct interested clients privately for payment. In this way, Faulhaber used his *Lustgarten* as a kind of sales catalogue at the market for mathematical goods, to which belonged in this case polygonal and pyramidal numbers, their sums and related problems, as well as the solution of cubic equations.

As far as cubic equations are concerned, the *Lustgarten* is mainly informed by Cardano's *Artis magnae, sive de regulis algebraicis liber unus*, but also by Michael Stifel's *Die Coss Christoff Rudolffs* from 1554 and Johann Jung's *Rechenbuch* from 1578, in the title of which Jung promises to give the solutions of cubic equations according to the rules of algebra.[2]

Three examples may illustrate Faulhaber's form of presenting his problems, some of which do not lead to cubic but in the end to quadratic equations, the solution of

[1] Johannes Faulhaber, *Arithmetischer Cubicossischer Lustgarten. Darinnen Hundert vnd Sechtzig Blümlein/ das ist/ außerlesner schöner künstlicher Exempel mit Newen* Inventionibus *gepflantzet werden. Welche theils auß Hieronymo Cardano/ vnnd andern Lateinischen Scribenten versetzt vnnd gezogen: Theils aber insonderheit die liebliche Polygonalische Röslin/ von newen zum Lust erzogen worden*, Tübingen: Erhard Cellius 1604.

[2] Johann Jung, *Rechenbuch auff den Ziffern vnd Linien/ darinne allerley Kauffmans handlung/ nach art der Regel de Tri vnd Welschen Practica/ sampt der Regel Falsi/ dardurch die Exempla der acht Regeln Coß auffgelöset werden/ Neben ausziehung der wurtzeln Arithmetischer Progression/ So wol den Regeln Algebre/ vnd andern aufflösungen Cubicossischer vergleichungen/ so vor niemals am Tage gewesen/ alles ordentlich gestelt durch Johann Jungen Rechenmeister zu Lübeck*, Johann Balhorn d.J. [1578].

which seems to be presupposed by him. Problem LIII is an example of this: A owns 12 florins more than B. If you subtract the third power of B's fortune from the third power of A's fortune you get 1161 florins. Let A's fortune be x and B's y, then you have

$$x = y + 12 \text{ and } x^3 - y^3 = 1161.$$

This leads to the quadratic equation

$$36y^2 + 432y + 1728 = 1161,$$

with the two solutions $y = -1.5$ ($x = 10.5$) and $y = -10.5$ ($x = 1.5$), the second of which is not mentioned by Faulhaber. Obviously Faulhaber saw no problem in confronting his readers with negative fortunes.

The following problem LIV really is cubic. It reads: I own 12 florins more than my fellow. The square of my fortune surpasses the cube of my fellow's fortune by 128 florins. Faulhaber, by answering that I own 8 and my fellow −4 florins, leaves out the solutions

$$x = \frac{1}{2}(29 \pm \sqrt{41}) \text{ and } y = \frac{1}{2}(5 \pm \sqrt{41}).$$

In Problem X, Faulhaber asks whether the solution of the equation

$$x^{12} + 36x^3 + 7 = 28x^6,$$

which he represented in cossic signs, can be found by 'regulierte Rechnung' (regulated calculation). His answer is 'yes', and he gives as a result the two roots

$$x = \sqrt[3]{3 \pm \sqrt{2}}$$

without any further comment. The substitution $y = x^3$ leads to a biquadratic equation. If one splits the corresponding polynomial into two quadratic factors, equivalent to the solution of the biquadratic equation, one can see that Faulhaber considered only one of the two resulting quadratic equations. Typical for this time, the given solutions obviously appeared acceptable to Faulhaber's readers. The solution of problem X can be understood as being indicative of Faulhaber's knowledge of how to solve the general quartic already in 1604.

Thirty-four problems in the *Lustgarten* have to do with figurate numbers, especially pyramidal numbers. Readers who wanted to solve problems with pyramidal numbers had first to learn from Faulhaber how to determine polygonal and pyramidal numbers and then apply a formula for solving the corresponding type of cubic equation (see Figure 13.3).

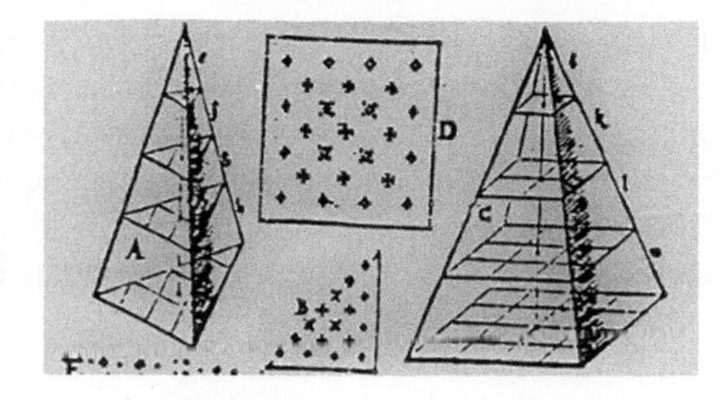

Figure 13.3 Geometrical structure of cubic, quadratic, and pyramidal numbers, after Johannes Remmelin, Mysterium arithmeticum, n.p. 1615, sig. B4[r].

If x is the (natural) number of points (pebbles) on a side of the polygon which constitutes the polygonal number, and if the x-th term of the arithmetical series with the first term 1 and the difference d, and

$$1 + (x - 1)d$$

is called the polygonal root of x, then the polygonal number of x is the sum of the polygonal roots from 1 to x : $\sum_{i=1}^{x} [1 + (i - 1)d] = x + \binom{x}{2} . d.$

For each polygon, d equals the number of corners minus 2, that is: 1 for triangular numbers, 2 for quadrangular numbers, etc.

Accordingly, the pyramidal number of x is the sum of the polygonal numbers from 1 to x:

$$\sum_{r=1}^{x} \left[r + \binom{r}{2} . d \right] = \binom{x+1}{2} + \binom{x+1}{3} . d = \frac{dx^3 + 3x^2 + (3 - d)x}{6}.$$

By signifying a natural number n in the formulae for figurate numbers by x, the unknown *res* or *cosa* in the cossic form of equations, Faulhaber integrated this part of number theory into his understanding of the *Coss* or algebra.

The market reacted immediately to Faulhaber's *Lustgarten* with the request that he publish the methods by which he had solved his problems, which, however, Faulhaber refused. One of his competitors, Peter Roth, who worked as a *Rechenmeister* in Nuremberg, had found out that Faulhaber was actually quite serious in his refusal to publish his way of solving the 160 problems.

Like the Italian *maestri d'abaco*, German *Rechenmeister* often used to set problems of varying complexity for one another; when challenged in this way a *Rechenmeister* would be expected to find a solution to the problem and, in turn, to challenge the original contender with a new problem. By this system of mutual challenges a kind of ranking amongst the *Rechenmeister* was established, which allowed interested clients to select the best for their instruction.

Over time, Faulhaber became the centre of a network of people with whom he corresponded and exchanged results, who in part translated for him, for example from Latin, and who informed him about new publications and provided him with these or vice versa. Generally, the German authors of arithmetic and cossic textbooks did not constitute a homogeneous social group.[3] It consisted of *Schreib- und Rechenmeister* who taught in privately organized schools, like Faulhaber and Adam Ries, whose most popular publication on arithmetic ran into more than ninety editions. The incredible success of some German arithmetical books—Ries's was not the only successful one—was due to the fast-growing demand for the ability to calculate with Indo-Arabic numerals in written form, especially in the light of constantly expanding trade. This form of calculating still competed in the sixteenth century with the older so-called 'Linienrechnen' ('line reckoning'), which was carried out in a way similar to calculations with the abacus on lines engraved in a table, the 'Rechentisch' ('reckoning table'). Since the 'Linienrechnen' did not allow any later control of the correctness of the calculations performed and was slower than written calculations with the Indo-Arabic numerals on paper, it disappeared eventually from the teachings of the *Rechenmeister* and in ordinary arithmetic.

Alongside university teachers such as Michael Stifel or Henricus Grammateus (Heinrich Schreyber), the authors of German arithmetic books were members of other occupations, like the gauger Jakob Köbel, who, however, had studied at the University of Heidelberg, where he gained the degree of *baccalaureus* of both laws. Other authors were amateur mathematicians, so-called philomaths, most of whom had enjoyed a higher education and accordingly could claim a higher social standing, or belonged to the socially very diffuse group of mathematical practitioners stretching from instrument makers to architects and fortification engineers.

To the group of mathematicians—as the term was understood at the time—belonged also astronomers like Johannes Kepler. This was less because the term used to describe a theoretical astronomer was 'mathematicus' than because such astronomers were well-trained mathematicians able to contribute to contemporary mathematical discussion and production. Kepler, who certainly influenced the young Descartes, was sceptical about the auspicious claims of algebraists of his time and considered algebra in the form of the *Coss* as inferior to the Euclidean *Elements* as far as rigour goes.[4] Over and above simple calculating skills, the *Coss* offered the possibility of solving problems outside the reach of ordinary arithmetic. In the second half of the sixteenth century, knowledge of how to solve quadratic equations had spread so far through various publications that it had lost its former status of secret knowledge. The publication of Geronimo Cardano's *Ars magna*, which appeared in Nuremberg in 1545, opened for the *Coss* the domain of cubic problems and the solution of the pertinent cubic equations.

The first German mathematician who reacted to the *Ars magna* was Michael Stifel. He had earlier turned to Luther's Protestantism and served as a parish priest in

[3] Ivo Schneider, 'Ausbildung und fachliche Kontrolle der deutschen Rechenmeister vor dem Hintergrund ihrer Herkunft und ihres sozialen Status', in: *Verfasser und Herausgeber mathematischer Texte der frühen Neuzeit*, ed. Rainer Gebhardt, Schriften des Adam-Ries-Bundes Annaberg-Buchholz, vol. 14, Annaberg-Buchholz 2002, pp. 1–22.

[4] Ivo Schneider, 'Trends in German mathematics at the time of Descartes' stay in southern Germany', in: *Mathématiciens français du XVIIe siècle: Pascal, Descartes, Fermat*, ed. M. Serfati and D. Descotes, Clermont-Ferrand: Presses Universitaires Blaise Pascal 2008, pp. 45–67, refers to parts of Kepler's *Harmonices Mundi* (Frankfurt 1619).

different places. In his Latin *Arithmetica integra* of 1544 he outlined the arithmetical and cossic knowledge available in the Germany of his day. His *Coss* included a method for solving quadratic equations, for which he introduced the mnemonic AMASIAS (with A, M, A&S, I, A&S representing the work steps). In 1545, there appeared Stifel's *Deutsche Arithmetica* and in the following year his *Rechenbuch von der Welschen und Deutschen Practick*. Five years before he became a teacher of mathematics at the University of Jena, in 1559, he had re-edited the *Coss* of Christoff Rudolff to which he had added among other things his form of the solution of cubic equations.[5] Christoff Rudolff was a student of Henricus Grammateus at the University of Vienna; his German *Coss* had already been published in 1525 by the publisher Cephaleus in Argentorati (Strasbourg),[6] and when in the 1550s demand for this book could not be satisfied any more, it was re-edited by Stifel. By 1553, Stifel was certainly familiar with Cardano's *Ars magna*, which had already been available for several years. Seemingly, Stifel considered Cardano's solutions of cubic equations as admissible only if they were rational or belonged to the realm of Euclidian irrationalities, that is to say, if they could be constructed as a line segment according to the tenth book of the *Elements*. At least we know that Johann Junge, a *Rechenmeister* who was trained in Nuremberg and had become a schoolmaster in Lübeck, interpreted Stifel in this way.

Independently of the problem of accepting the extension of the value range (*Wertebereich*) of the solutions of cubic equations given by Cardano half a century earlier, and about 25 years after Junge's *Rechenbuch*, which was published only in a small edition,[7] Stifel's re-edition of Rudolf's *Coss* caused an increasing demand for relevant instruction. Johannes Faulhaber reacted to this situation with his *Lustgarten* of 1604 and in this way increased the already existing demand.

One can distinguish two different strategies in the sixteenth century for advertising textbooks on arithmetic and the *Coss* in the book market in Germany—the first as accompaniments to oral instruction by the *Rechenmeister* and the second as volumes self-sufficient for autodidacts. Instructions for autodidactic learning were usually given in algorithmic form together with many problems and their solutions as a means to control the effect of learning; this form was generally restricted to easily accessible 'common' knowledge. Material accompanying oral instruction in the form of problems and the results of their solution later degenerated into a mere sales catalogue for putative or very new knowledge. It seems to be clear, for obvious economic reasons, that new results and new mathematical skills were only rarely included (if at all) in the material published for autodidactic learning. New skills and results were normally communicated only in private instruction, as special secrets, and their transformation into more or less common knowledge was therefore effected only by a slow process of diffusion.

Economic interests made the *Rechenmeister* keep their clients dependent on their authority for as long as possible. Later, Descartes taught his readers to use their

[5] Michael Stifel, *Die Coß Christoffs Rudolffs. Die schönen Exempeln der Coß Durch Michael Stifel Gebessert vnd sehr vermehrt*, Königsberg: Alexander Lutomyslensis 1553.

[6] Christoff Rudolff, *Behend und Hubsch Rechnung durch die kunstreichen regeln Algebre, so gemeincklich die Coß geneñt werden*, Strasbourg: Cephaleus 1525.

[7] For a long time no extant copy could be found in any library; only in the 1990s was a copy found in the library of the Germanisches Museum in Nuremberg.

own minds, effectively freeing them from any other authority such as that of the *Rechenmeister*. He summarized equivalent modes of solution by abstraction, and gave explanations and occasionally demonstrations of mathematical theorems by reducing them to accepted premises. In contrast, the *Rechenmeister* used to claim new different-looking forms of known solutions to a problem as inventions of their own, and aimed to find as many as possible, since by the standards of the time whoever could offer more different solutions for the same problem (or at least solutions that looked different) was considered as better.

Most of the clients of the *Rechenmeister* were for a long time content to receive solutions to concrete problems in the form of algorithms without any proof or even hint of an understanding of the algorithmic procedure. In addition, they often received rules in the form of cues indicating which algorithm to apply to what kind of problem. This form of communication stemmed from a tradition of oral teaching. As long as the clients were content with such a form of instruction and did not ask for explanations and proofs, there was no way to find out that many of the solutions were equivalent. The mentality behind this behaviour is paralleled by a common language, in which clearness and definiteness were obviously not understood as virtues. This is illustrated on an elementary level by the lack of any real orthography: so, for example, three different spellings for the conjunction *und* (corresponding to the English 'and') can be found on one page; in an important official document a family name is spelt in different ways on the same page; and writers took delight in using a great variety of different expressions and words for the same thing.

The payments well-to-do clients were willing to offer for instruction in new mathematical methods were responsible for competition among the *Rechenmeister* for quick and, if possible, exclusive information about such methods. Prices depended of course on the supply of new (sometimes only allegedly new) methods. The safest way to secure exclusive information was to create it oneself. It seems clear that the system offered sufficient reward to stimulate creativity on the one hand while, on the other, truly creative solutions were scarce. One reason for this state of affairs was that by aiming at the 'invention' of as many different modes of solution as possible for the same problem—as such, a relatively easy endeavour—*Rechenmeister* were distracted from looking for solutions for really challenging new problems.

Descartes, who served as their main critic, had unmasked the procedures of the *Rechenmeister* as mere tricks, showing that most of the modes of solution and fields of applications advertised as new and different by one or other *Rechenmeister* had a common mathematical structure. This can readily be seen by comparing the notation for the powers of the unknown used by the *Rechenmeister* and by Descartes. The cossic signs for the different powers of the unknown were generally not based on each other. By this means the *Rechenmeister* suggested that for every new power a new 'theory' was required in order to deal with it, whereas Descartes's notation for the powers of the unknown, which we still use today, was easy to understand and did not burden the memory (Figure 13.4).

The story of the meeting between Faulhaber and Descartes in the winter of 1619/20, which dates back to an account of Descartes's life and work from Daniel Lipstorp published in 1653, is a metaphor for the confrontation of two different

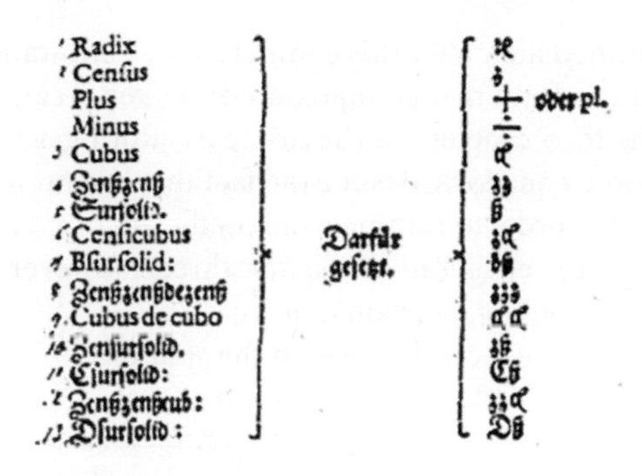

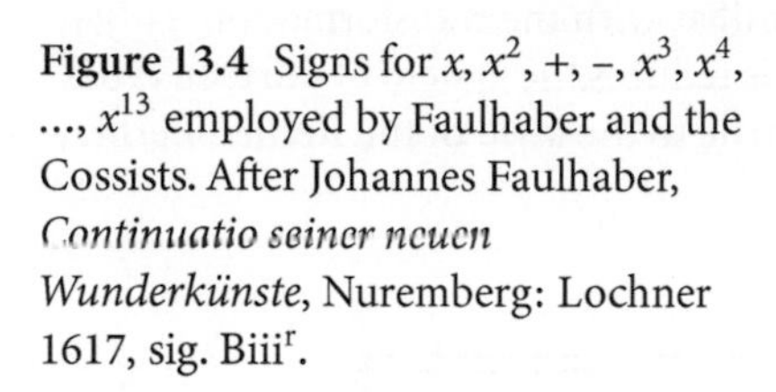

Figure 13.4 Signs for x, x^2, $+$, $-$, x^3, x^4, ..., x^{13} employed by Faulhaber and the Cossists. After Johannes Faulhaber, *Continuatio seiner neuen Wunderkünste*, Nuremberg: Lochner 1617, sig. Biii^r^.

mathematical styles. In his story, Lipstorp leaves no doubt as to his conviction regarding the superiority of Descartes's style, which rendered the style of the *Rechenmeister* as represented by Faulhaber outdated.[8]

In 1605, Peter Roth announced that he was going to publish a book which would not only contain all of Cardano's rules for the solution of cubic equations but also the way to find the solutions of all the problems contained in Faulhaber's *Lustgarten*. Like Faulhaber, Roth took advantage of the fact that published information about Cardano's solutions of cubic equations had hardly been available since the *Ars magna* had been published sixty years earlier. Therefore, Roth considered the time ripe for the publication of Cardano's rules in German in a self-explanatory way and well illustrated by their application to Faulhaber's problems. Faulhaber, who doubted that Roth was capable of achieving this aim, was deeply shocked when he learned that the first part of Roth's *Arithmetica Philosophica* of 1608 included the rules for the solution of cubic equations according to Cardano. Like Cardano, Roth differentiated between thirteen types of cubic equations, where the coefficients a, b, c may not be negative:

1. $x^3 + bx = c$
2. $x^3 + bx = c$
3. $x^3 + c = bx$
4. $x^3 = ax^2 + c$
5. $x^3 + ax^2 = c$
6. $x^3 + c = ax^2$
7. $x^3 + ax^2 + bx = c$
8. $x^3 + bx = ax^2 + c$
9. $x^3 + ax^2 = bx + c$
10. $x^3 = ax^2 + bx + c$
11. $x^3 + c = ax^2 + bx$
12. $x^3 + bx + c = ax^2$
13. $x^3 + ax^2 + c = bx$

[8] Kurt Hawlitschek in his *Johann Faulhaber (1580–1635) and René Descartes (1596–1650): auf dem Weg zur modernen Wissenschaft*, Ulm: Stadtbibliothek 2006, considers Lipstorp as a reliable witness for a personal meeting between Faulhaber and Descartes in Ulm.

Immediately after this enumeration, Roth stated the fundamental theorem of algebra in the form that an equation of degree n can have at most n roots. In the following thirteen chapters of the first part, Roth gave the solutions for the thirteen types of cubic equations, despite the fact that he was familiar with the transformation $x = y - a/3$ in order to eliminate the coefficient of x^2: this reduces the system of thirteen types of cubic equations to the first three. However, true to the style of the *Rechenmeister*, Roth did not mention this.

For the second case with the solution

$$x = \sqrt[3]{\frac{c}{2} + \sqrt{\left(\frac{c}{2}\right)^2 - \left(\frac{b}{3}\right)^3}} + \sqrt[3]{\frac{c}{2} - \sqrt{\left(\frac{c}{2}\right)^2 - \left(\frac{b}{3}\right)^3}}$$

he remarked that there are three real roots, one positive and two negative, if the radicand of the square root is negative. The direct determination of these three real roots by the trisection of an angle was unknown to Roth, as it still was to Cardano.

The second part of the *Arithmetica Philosophica* contained the solutions of all the problems in Faulhaber's *Lustgarten*, except for one, with all necessary explanations.[9] In the final third part Roth posed 102 problems which afforded the solutions of equations of degree four, five, six, and seven. This part of the *Arithmetica Philosophica* functioned in the same way as the *Lustgarten* as a catalogue of the new mathematical skills that Roth would teach only privately for payment.

Generally, it can be said that Faulhaber, and even more so Roth, ceased to have recourse to geometrical or other intuition concerning the form of the accepted roots. They dealt with negative numbers, which could not be represented in Euclidean geometry before the introduction of the distinction of a positive from a negative direction. Like Cardano, both Faulhaber and Roth admitted cube roots, which cannot be reduced to Euclidean irrationalities.

Roth, however, did even better than Faulhaber. His formal approach, totally detached from any real situation, allowed him to give complex values for amounts of money, as in the solution of his problem 24 for a quartic equation, where the fortunes of two persons are represented by

$$x = -21 \pm \sqrt{-144} \text{ and } y = 3d\sqrt{-3}.$$

In order to impress their clients with seemingly new skills beyond modifications of known methods, the *Rechenmeister* surprised their readers with new 'disciplines', to which they added an appropriate new terminology. Thus, Faulhaber developed for figurate numbers such as pyramidal numbers, the origins of which date back to Greek antiquity, a fancy new nomenclature based on Greek names of numbers. By this, a

[9] Faulhaber noticed on p. 86 of his *Newer Arithmetischer Wegweyser. Zu der Hochnutzlichen freyen Rechenkunst, mit Newen Inventionibus geziert*, Ulm: Johann Meder 1614 (2nd edn 1617), that Roth had failed to solve problem 149 of his *Lustgarten*. See Kurt Hawlitscheck, *Johann Faulhaber 1580–1635: eine Blütezeit der mathematischen Wissenschaften in Ulm*, Ulm: Stadtbibliothek 1995, p. 159.

Rechenmeister like Faulhaber could make their clients believe that they had created an important new theory consisting of a multitude of complicated formulas, knowledge of which they would sell to those interested.

Another example of the style of the *Rechenmeister* when they displayed their skills is given by Roth's solution of the last and most complicated problem in Faulhaber's *Lustgarten*. Roth was not content with just one or only even solutions. Instead, he claimed to have twenty-eight different modes of solving the problem. According to the value system acknowledged by the *Rechenmeister*, someone who knew twenty-eight ways of solving a problem was considered to be far better than someone who knew only one or two.

The mainly financial interests, and with them part of the value system of the professional group of *Rechenmeister*, also make understandable certain aspects of the typical style in which they presented their mathematical materials. As long as mathematical amateurs paid for their instruction by professional mathematicians they became co-owners of what had been revealed to them. This meant not only a disadvantage for the professional *Rechenmeister* and their peers but also provided an incentive for amateurs, who were ostensibly not interested in making money with their newly acquired mathematical skills, to become the sole owners of such knowledge if they managed to create it themselves.

With the availability of printed editions of the works of the great Greek mathematicians in the original language, or in Latin translation, from the second half of the sixteenth century onwards, clients became acquainted with mathematical proofs and so correspondingly the demand for explanations and proofs in mathematics increased. The *Rechenmeister* tried to satisfy this demand. If Faulhaber boasted of giving an 'infallible demonstration' of the correctness of a formula, he did not deduce his result from acknowledged premises. Instead, he showed the effectiveness of the formula, say, of the sum of the first n natural numbers raised to the power 13 in a concrete case like $n = 7$. The sceptical customer could compare the result of his addition of seven terms with the result delivered by the formula in the case of $n = 7$. Identical results would be sufficient to persuade the most sceptical client of the effectiveness, if not of the correctness, of the applied formula. However, a satisfactory mathematical proof would have been afforded in this case by complete induction, which came into general mathematical usage only after the publication of Jakob Bernoulli's *Ars conjectandi* in 1713.

The mathematically interested and talented reader eventually learned, often with the help of private instruction, to create new results of his own. It was only a question of time before the most gifted clients of the *Rechenmeister* asked for a presentation of their results according to the standards they had learned from the extant texts of the great Greek mathematicians. This created a real dilemma for the *Rechenmeister*: if they changed the form of their presentation into one oriented towards the texts of Greek mathematics, every other *Rechenmeister* could eventually buy or read the book and so include its content in his own teaching. In this way, authors would lose their advantage in information and so also their economic competitiveness. Alternatively, they could simply satisfy the expectations of their better clients only as long as their teachings were not published or no one else made a better offer. However, a decisive change came about first with the publications of François Viète and later, in 1637, with the publication of the *Géométrie* of the 'amateur' Descartes. The most talented former clients of the *Rechenmeister* turned eventually to Descartes and his followers.

The development of mathematics in the seventeenth century testifies that the professional group of *Rechenmeister* and mathematical practitioners could no longer satisfy the demands of well-informed and mathematically talented amateurs and that they eventually ceased to belong to the producers of new mathematics. Signs of this transitional process, in which mathematical production moved away from the 'professionals' to economically disinterested amateurs, can be seen quite early in Italy in the quarrels between Cardano and Tartaglia in the middle of the sixteenth century. The intellectual duel on the battlefield of mathematics between the attorney for professional mathematics van Roomen and the mathematical 'amateur' Viète at the end of the sixteenth century offers another example. Mathematical challenges functioned as a kind of intellectual duel to decide the ranking among mathematicians.

Interested clients could now find all the explanations in Roth's book that Faulhaber originally was only willing to give orally, for payment. Deprived of the financial rewards he had hoped for, Faulhaber planned to pay Roth back for the damage he had caused by the publication of the solutions to his cubic problems by threatening to publish the solutions of all the problems contained in the third part of the *Arithmetica Philosophica.* Since Roth was not impressed by this threat, Faulhaber offered to buy the rest of the edition of his work. Only after nearly the whole edition had been sold, in 1609, and despite its extraordinarily high price, was Roth ready to concede to Faulhaber the remaining eight copies against the promise not to publish the solutions of his or Faulhaber's problems within the next two years.

In this competition with Roth, who died in 1617, Faulhaber appeared to most of his contemporaries to have been the loser, and perhaps correctly so: he had after all failed to publish the solutions of the problems contained in the third part of Roth's *Arithmetica Philosophica* while the author was still alive. But after the two years' publishing moratorium had elapsed, Faulhaber began to work again in this field. He later published the reduction of the general quartic to a cubic equation in verbal form, long before Descartes published his version, but never returned to the problems of Roth. One reason, and perhaps the most effective motive for Faulhaber's reservation, was quite simply that he later occupied himself more with other fields, evidently more rewarding to him, such as surveying or fortification.

Faulhaber's interpretation of biblical numbers

Besides a couple of smaller publications concerning scientific instruments, the list accompanying Faulhaber's portrait from 1615 contained two papers which derived from his dedication to the interpretation of biblical numbers. So far, nothing has suggested the religious raptures and heterodox views of the apparently very ambitious Faulhaber that would later involve him in numerous conflicts with the municipality and the representatives of the Protestant church in Ulm. Such a change only became visible in 1606 when he began to associate himself with the landlord of his courtyard, the sanctimonious baker Noah Kolb. On some occasion, Kolb suggested to

Faulhaber that he possessed a form of enlightenment deriving directly from God, and the mathematician was strengthened in this conviction by his confessor, Johann Bartholome, who was a preacher at the Ulm minster. According to a contemporary report, Bartholome 'ordained and initiated' Faulhaber as one of the latter-day prophets. Soon afterwards, Faulhaber announced in Ulm and other cities that doomsday was imminent. As a result, the municipal authorities of Ulm decided, late in 1606, to imprison him. Soon after, he was released on account of his wife's pregnancy, but for some time his freedom of movement (*Bewegungsfreiheit*) as well as his contacts with others were severely restricted. In 1611, Holy Communion was denied to Faulhaber because of a suspicion of sorcery, among other things. Since he continued to remain in contact with Kolb, Faulhaber was publicly reprimanded two years later and all further dealings with the baker forbidden. When Kolb was put on trial and subsequently executed in 1615, accused of fornication, Faulhaber was one of the witnesses questioned. As this episode reveals, Faulhaber repeatedly displayed disobedience towards the municipal authorities of the city and the church despite the risks such recalcitrant behaviour entailed, even if his transgressions were quite different from those of Kolb.

The inadmissible speculations that Faulhaber was accused of concerned first and foremost his interpretations of the biblical numbers, which were based on the formulae that he had adopted and developed for the determination of polygonal and pyramidal numbers.

To obtain these formulae he applied various methods, among them a difference calculus.[10] Invariably the seven biblical numbers 2300, 1290, 1335, 666, 1260, 1600, and 1000, repeatedly and figuratively interpreted, played a special role. The fact that the great mathematicians and philosophers of antiquity did not know anything about the biblical numbers was understood by Faulhaber as an indication of God's intention to hide such information until his time.[11] His claims to be able to reveal the intentions of God by means of the interpretation of the biblical numbers became explicit and visible when he published his *Newer Mathematischer Kunstspiegel* ('New artistic mirror of mathematics'). In addition to the descriptions of an instrument for surveying and of a special set of compasses, this book contains speculations about the biblical numbers. A Latin translation by Johannes Remmelin appeared in the same year.[12] Both publications prompted representatives of the church authorities to object to the number speculations contained in them, as had happened on earlier occasions. The Ulm printer of both treatises was subsequently urged not to print any more works of Faulhaber or Remmelin without permission of the council. Apparently, Faulhaber was only moderately impressed by the admonitions of the clerical and secular authorities in Ulm. In 1613, he published *Andeutung Einer vnerhörten newen Wunderkunst* ('Indication of an unheard new miraculous art') which was printed in Nuremberg and contained an interpretation of the biblical numbers as pyramidal numbers.

[10] See Ivo Schneider, *Johannes Faulhaber (1580–1635): Rechenmeister in einer Zeit des Umbruchs*, Basel: Birkhäuser 1993, chapters 5 and 7.

[11] Johannes Faulhaber, *Miracula Arithmetica, zu der Continuation seines Arithmetischen Wegweisers gehörig*, Augsburg: David Francken 1622, p. 30.

[12] Johannes Faulhaber, *Speculum Polytechnum Mathematicum nouum*, translated by Johannes Remmelin, Ulm: Typis Ioannis Mederi 1612.

In this book Faulhaber delivered an extremely concise explanation of the sense of such number speculations. According to him the biblical numbers are of the form

$$\binom{n+1}{3}d + \binom{n+1}{2},$$

where n and d are natural numbers with $n > 1$. According to Faulhaber, the biblical numbers are characterized by being indicated by God. Since most of the biblical numbers treated by Faulhaber are divisible by 10 the choice of $n = 4$ yields the equation $10d + 10 = b$ with $d = \frac{b-10}{10}$, which gives for each $b > 10$ divisible by 10 a natural number d. This does not hold for 666, for which there exists only one non-trivial representation, with $n = 3$ and $d = 165$.

In 1613, another book of Faulhaber's appeared, namely his *Himlische gehaime Magia Oder Newe Cabalistische Kunst/ vnd Wunderrechnung/ Vom Gog vnd Magog* ('Heavenly secret magic or new Cabbalistic art and miraculous calculus about Gog and Magog'), published by Remmelin in Ulm and printed in Nuremberg. Faulhaber dedicated the book to the Holy Roman Emperor Matthias and sent a copy to him just before the beginning of the Reichstag in Regensburg. It is apparent from the dedication that Faulhaber expected protection from the Emperor against his opponents in the city of his birth. Three testimonies concerning technical inventions that precede the actual text were intended to demonstrate to the Emperor and other powerful men that Faulhaber was an accomplished inventor and at the same time someone capable of interpreting the biblical numbers. They show us that Faulhaber considered his Cabbalistic interpretations as comparable in their usefulness and applicability to his technical inventions.

The Emperor's protection would have served Faulhaber well, because representatives of the clerical authorities in Ulm soon became unhappy about his move into the realm of rather daring number speculations. They accused him of departing from his true field of competence as a mathematician and instead of composing in an unacceptable manner a prophecy 'of letters, numbers, and sealed words'. In consequence, he was ordered to appear at the Dombauhütte in Ulm to be questioned by clergymen about the meaning of his statements and his justification of the use of notions like 'magic'.

In the preface and in the conclusion of his *Himlische gehaime Magia* ('Heavenly secret magic') Faulhaber claims that, unlike those who believed they could decipher such divine secrets by common sense, he had learned something about the secrets hidden in the 'heavenly numbers' from God in person. Here is a key to understanding the conflict between Faulhaber and his many critics. Until the end, Faulhaber emphasized the intentionally esoteric character of the 'biblical numbers'. If God had wanted man to be able to discover the secrets by means of common sense, he would have made it possible to decipher the numbers in the Book of Revelation much earlier. This however, at least in Faulhaber's view, was not God's intention at all. Instead, God had chosen certain individuals to decipher the secret meaning of the biblical numbers. Such individuals had to be enlightened by God and, obviously, the result of the enlightenment lay outside the realm of human faculties. Faulhaber further maintained that he himself belonged to those enlightened by God, who had elevated him (in his own estimation) beyond all necessity to a position where he could supply explanations for the knowledge to which he had gained access. To most of his contemporaries such a claim was palpably unacceptable, because they could not discern in Faulhaber much (or, indeed, anything at all) to justify it, and moreover they were themselves quite capable of giving insightful explanations into the origin of Faulhaber's very vague utterances about the meaning of the biblical numbers. Faulhaber was thus depicted openly as a charlatan or a misguided religious fanatic.

In the text of the *Himlische gehaime Magia*, a booklet of ten pages, Faulhaber referred to the unsubstantiated necessity of assigning by means of a 'general key' to every 'heavenly miraculous number its philosophical algebraic weight' and by doing so to fix certain measures of time from which predictions of the occurrence of important events could be deduced. He also provided a 'word calculus' (*Wortrechnung*), based on not just one but four alphabets, which he used to decipher a biblical saying that he had 'observed through God's grace' in different texts of both the Old and New Testaments. Only in 1619, the year in which the Emperor Matthias died, did Remmelin publish this biblical saying together with the requisite word calculus in an apology for Faulhaber, entitled the *Sphyngis Victor*. The saying was 'Gog and Magog, a high regent comes from the offspring of Japhet'. According to the biblical book of Ezekiel, Gog from the land of Magog leads at the end of time an army of nations against the nation of Israel. After initial successes Gog is defeated together with his followers by God himself. In many of his Cabbalistic works, Faulhaber refers to Gog, who, in the context of a very old Christian tradition, had assumed the traits of the Antichrist.

Faulhaber and his contemporaries often used a word calculus in which the letters of an alphabet are given numerical values and vice versa, part of a long tradition going back at least to the secret Jewish teachings of the Cabbala. One goal of this word calculus, which had been revived in the sixteenth and seventeenth centuries, was the interpretation of certain sayings by means of the assignment of numbers to the letters in them, and conversely the hiding of names and sayings by means of numbers. A prominent precursor of Faulhaber in Germany was Michael Stifel, who in his anonymously published *Rechen Büchlein Vom End Christ* of 1532 had used word calculus to identify the Beast of the Book of Revelation[13] with Leo X, pope at the time that Luther nailed his 95 theses to the door of the Castle Church in Wittenberg. Stifel also used a text taken from the Bible in order to predict the end of the world on 18 October 1533. This prediction, which was obviously false, ended in disaster for all who had believed in it. It was only through being rescued by Luther that Stifel escaped from those who demanded compensation for losses suffered by the prediction.

Among the signs given most attention by the followers of Faulhaber were the celestial phenomena, because they belonged to God's sphere of immediate influence. To a certain extent the events seen in the sky were placed in direct relation to events on earth in the sense of the macrocosm–microcosm relationship of Paracelsus. Not only did many ordinary people see themselves as more or less helplessly

[13] Revelation 13:18.

dependent on the events visible in the sky, but also political actors such as the Imperial military commander Wallenstein reached their decisions in accordance with celestial phenomena.

The Rosicrucian movement

The opposition between the followers of such interpreters of tokens like Faulhaber and their enemies was intensified by the so-called Rosicrucian movement, triggered by the circulation of the two first Rosicrucian manifestos, *Fama Fraternitatis* and *Confessio Fraternitatis*. These appeared first in the form of handwritten versions and then, after 1614, in printed form.[14] It is not known who wrote the *Fama* and the *Confessio*, although it is generally assumed that the author or authors were to be found in a circle of Protestant theologians in Tübingen.

In both texts a brotherhood of the Rosicrucian cross is mentioned. Furthermore, in the *Fama* the history of the brotherhood is accompanied by references to its teachings. The basic tenor is a criticism of established authorities like Aristotle in the field of philosophy and Galen in the field of medicine. In order to read the only book that possessed authority, that is to say, the book of nature written by God, the Rosicrucians based their teachings on neo-Pythagorean, neo-Platonic, and hermetic ideas in addition to the microcosm–macrocosm correspondence in the harmony between man and nature, and the writings of Paracelsus and the Cabbala. Numbers and their properties were in their view attributed with extraordinary explanatory power. It was supposedly by means of God's grace that members of the original Rosicrucian brotherhood acquired knowledge about the book of nature. The new generation of the brothers, who had rediscovered the grave of the founder of the order, foresaw a new reformation of mankind on the basis of the knowledge found in the book of nature, and called upon the readers of the *Fama*, which they had written, to express themselves with respect to this first communication of the brotherhood.

The conglomerate of expectations contained in the *Fama* was so lengthy that over the period between 1614 and 1625 it led to more than four hundred known Rosicrucian publications. The tremendous impact of the *Fama* and the two next Rosicrucian manifestos was not caused by the novelty of the ideas contained in them. After all, many of the ideas, such as the Cabbalistic or neo-Platonic body of thought, the expectation of a more comprehensive reformation based on a balance between theology and science, the critical discussion of the rigid forms of scholastic science, and the idea of the creation of new forms of science that would lead to the actual solution of

[14] Anon, *Fama Fraternitatis, Oder Brüderschafft/ des Hochlöblichen Ordens des R. C. An die Häupter/ Stände und Gelehrten Europae*, in: *Allgemeine vnd General Reformation der gantzen weiten Welt. Beneben der Fama Fraternitatis, Deß Löblichen Ordens des Rosenkreutzes/ an alle Gelehrte und Häupter Europae geschrieben: Auch einer kurtzen Responsion, von dem Herrn Haselmeyer gestellet/ welcher deßwegen von den Jesuitern ist gefänglich eingezogen/ und auff eine Galleren geschmiedet: Itzo öffentlich in Druck verfertiget/ und allen trewen Hertzen communiciret worden*, Cassel: Wilhelm Wessel 1614, pp. 91–128 and *Confessio Fraternitatis, Oder Bekanntnuß der löblichen Bruderschafft deß hochgeehrten Rosen-Creutzes/ an die Gelehrten Europae geschrieben*, in: ibid., pp. 54–82.

real practical problems, could all be found in texts that were published before and independently of the Rosicrucian manifestos.

Rather, the success of the *Fama* was caused by the clever grouping of expectations that were fostered by completely different circles. These were united only in their dissatisfaction with the after-effects of all forms of dogmatism. Correspondingly, the reactions to the Roiscrucian manifestos were also multifaceted. Not surprisingly, the main body of opponents consisted of representatives of orthodox Protestantism. Quite sensational legal proceedings initiated by representatives of the Protestant church in Württemberg show that by around 1620 the Protestant orthodoxy had succeeded in marginalizing the Rosicrucian movement and at the same time declaring as irrational the presuppositions for using Cabbalistic, number-mystical, and alchemistic elements in their beliefs and predictions.

However, this appraisal hardly influenced the self-conception of those who were attacked by the orthodoxy, as two examples will show. Among the adherents of the Rosicrucian movement, the fairly large number of mystics who followed a Cabbalistic tradition were in general mathematically trained and, as far as their mathematics went, they stood above any accusation of irrationality. According to Daniel Mögling, an at times fervent follower of the movement, the Rosicrucians abandoned a literal understanding of Holy Scripture in order to read the 'true book of life' with the 'eyes of the mind' and interpret it in harmony with the Bible. The realm of everything that is accessible to human reason could only be transcended with God's help. However, the means needed in order to assure oneself of God's help were no longer rational. Moreover, to be able to effect God's help required His special grace, bestowed not on everybody but only on a few. If the symbolic language of the Holy Scripture was interpreted in the right way, they argued, it would not contradict what the eyes of reason see in the book of nature or what is on the pages that can only be interpreted by means of God's grace.

The followers of the Rosicrucian movement and their opponents can be viewed as representatives of two mentalities typical of this time. The orthodox Protestants, on the side of the opponents, represent the mentality of independence that is reached socially by means of personal responsibility and, in the area of knowledge, by restricting the means of acquiring knowledge to the activity of the human mind. From their point of view, God had created the world in accordance with a plan accessible to human understanding: it seemed agreeable to God for man to study the plan of the Creation and, in the sense of that plan, to explain as natural phenomena the many events seen by others as miracles brought about by God. This mentality corresponded to a high state of preparedness for competition and conflict.

In contrast, the mentality of the followers of the Rosicrucian movement was characterized by a longing for harmony in a world without conflicts, that would be guaranteed by unselfish labour for God and mankind in combination with access to new transcendent knowledge in the possession of a group of the chosen; these exceptionally gifted individuals would take over the responsibility that according to the orthodoxy everyone had to bear individually.

From the very beginning, the texts of the Rosicrucians appealed to Faulhaber and they confirmed his ideas in many respects. Like many other followers of the Rosicrucian movement, he tried for years in vain to get in touch with the legendary brotherhood of Rosicrucians.[15] An anonymous Latin text published in 1615, of which Johannes Remmelin, Faulhaber's friend, later claimed to be the author, was one such futile attempt.[16] This text was like a birdcall directed explicitly to the 'above all enlightened and highly laudable men of the Fama of the Rosicrucian brotherhood'. This did not protect Faulhaber from being seen as a member of the brotherhood by the authorities during the most intense disputes about the meaning of the comet of 1618. His booklet on the interpretation of that comet, *Fama siderea nova*, in which he again manipulated the number 666, and of which the title began with the word 'Fama', like the first of the Rosicrucian texts, was associated in Ulm with suspiciously observed meetings of the Rosicrucian movement in which Faulhaber, at least from the point of view of the church authorities, played a significant part.[17] The authorities discovered, for example, that Faulhaber secretly met up to seventy people, the *Rosenkreutz Brueder* ('Rosicrucian brothers'), who apart from such 'Conventicula' communicated with each other in writing.

The special attention paid to Faulhaber by the clerical and municipal authorities was sparked off by his interpretation of the comet of 1618 that created big waves. In 1618, Johannes Kepler had observed three comets that were predominantly identified as one and the same heavenly body.[18] The first was only faintly visible from the end of August to the end of September and was apparently ignored by most people, as was the second, albeit not by Faulhaber. A third comet, easily visible to everyone, was only seen in November 1618. From both the Catholic and Protestant pulpits sermons about its significance were delivered; and, in a flood of leaflets and treatises, astronomers and self-proclaimed experts on comets attempted to satisfy the curiosity of an intensely interested public.

The authors of the many texts on comets can be classified into two groups on the basis of their views: the first group, to which also Faulhaber and most followers of the Rosicrucian movement belonged, viewed the comet as a 'preacher of penance' (*Bussprediger*) put by God on the 'heavenly pulpit' (*Kanzel des Himmels*) to announce God's wrath and punishment if men did not abandon their sinful ways. The other group saw its particular task as appeasing the population, plagued in any case with numerous problems, needs, and fears. Its representatives were concerned with explaining the comet as a natural phenomenon—for example as something atmospheric—without any further significance. The critics of Faulhaber (who belonged to the second group) accused him of claiming, two months after the event and without justification, the arrival of the comet as confirmation of his

[15] Letter to Rudolf von Bunau of 21/31 January 1618, Stadtarchiv Ulm, J1 Autographen, Nr. L20; Protokoll des Pfarrkirchenbaupflegamts, 27 July 1619, Stadtarchiv Ulm, A Rep. 13-A 6876.

[16] [Johannes Remmelin], *Mysterium Arithmeticum Sive, Cabalistica & Philosophica Inventio, nova admiranda & ardua, qua Numeri Ratione et Methodo computentur, Mortalibus à Mundi Primordio Abdita, et ad Finem non sine singulari omnipotentis Dei provisione revelata. Cum Illuminatissimis laudatissimisque Fraternitatis Roseae crucis Famae Viris humiliter & syncerè dicata*, [n.p.] 1615.

[17] Protokoll des Pfarrkirchenbaupflegamts, 27 July 1619, Stadtarchiv Ulm, A Rep. 13-A 6876.

[18] Johannes Kepler, *De Cometis Libelli Tres*, Augsburg: Typis Andreæ Apergeri 1619; see Johannes Kepler, *Gesammelte Werke*, Vol. VIII, Munich: C. H. Beck'sche Verlagsbuchhandlung 1963, pp. 129–262 esp. p. 177; cf. Werner Landgraf, 'Über die Bahn des zweiten Kometen von 1618', *Sterne* 61 (1985), pp. 351–3.

prediction. One of Faulhaber's friends objected that besides Faulhaber other 'credible learned people' (*glaubwürdige gelehrte Leute*) had also seen a comet, in September 1618 (although it had not been very visible), and that God had 'ascribed an irresistible power' (*Kraft erteilt wurde*) to Faulhaber's prediction, made well in advance in a calendar for the year 1618, that a comet would appear on 1 September.

The Protestant Imperial free city of Ulm had ordered this one-page calendar for the year 1618 to be produced with the intention of giving it to the city's civil servants. Among the special references and predictions made in the calendar was the entry 'comet' on 1 September. However, the small size of the entry, on a single page covering all the days of the year, along with the limited distribution of the calendar, did not make it suitable to inform the public at large of Faulhaber's prediction. In a letter dated 26 August 1618, Faulhaber himself had informed a friend of his, Matthäus Beger, in the Imperial free city of Reutlingen, of his observations of the comet he had seen in August, with the intention that they would be conveyed to Professor Michael Maestlin in nearby Tübingen. Beforehand Beger had only been informed by the university[19] that in Tübingen and surroundings the comet would not be seen before November. This meant that Faulhaber was the first to claim that he had discovered the comet, in observations already carried out in August.

Faulhaber saw no limit to the validity of his prediction through the fact that the comet was observed later in Tübingen. In his *Fama Siderea Nova*, published by Daniel Mögling in 1619 under the pseudonym Julius Gerhardinus Goldtbeeg from Jena, he interpreted the observation without any restrictions as confirmation of his special God-given faculties to interpret the comet as a divinely sent token. In the heated discussions following the appearance of the *Fama*, the main opponents of Faulhaber were the principal of the grammar school in Ulm, Johann Baptist Hebenstreit, and one of his colleagues, the 'Praeceptor' Zimbertus Wehe, who tried to hide behind a pseudonym of his own in his two pamphlets written against Faulhaber. Hebenstreit and Wehe were acquaintances of Faulhaber who, until they began their criticism of the *Fama*, had occasionally visited him at his house. Hebenstreit's about-face came as a particular surprise to Faulhaber, who had given him instruction in relation to the observation of the comet(s) of 1618, and Hebenstreit had also helped correct the text of the *Fama*. Through his subsequently published booklet *Cometen Fragstuck* (The question of the comet), Hebenstreit sought to exploit the general interest in the appearances of the comet(s) as quickly as possible. However, a mistake he made in the booklet, confusing Mars with Arcturus, was apparently quickly discovered by one of his competitors in the flourishing market of texts on comets and used as a basis for a destructive criticism of his text. Indeed, Hebenstreit's views of the essence and the location of the comet provoked fairly lively disagreement at the time.

In his *Cometen Fragstuck*, Hebenstreit had already asserted, without mentioning Faulhaber, that his eyes were 'too foolish' (*zu blöd*) to see a comet on 1 September that

[19] Mathæus Beger, *Problema astronomicum: Die Situs der Stermen, Planetarum oder Cometarum zu observirn ohne Instrumenta, allein mit einem geraden Lineal oder Faden*, [n.p.] 1619, sig. D1^{v}.

would only be visible much later.[20] In his second Latin booklet of 1619,[21] Hebenstreit extensively attacked the possibility of predicting the appearance of comets on the basis of Cabbalistic number speculations. At the same time, he criticized the contents of *Fama siderea nova*, again without mentioning Faulhaber. Moreover, Hebenstreit and Wehe both suggested that Faulhaber had taken the prediction of the comet from a publication of the Imperial mathematician and astronomer Johannes Kepler. Kepler had in his *Prognosticon* for the year 1618, in a section about diseases, granted the possibility of the appearance of a comet, because no comet had been observed since 1607.[22] Although at the time it was not yet known that comets return after a period characteristic for their orbit, Kepler obviously assumed, on the basis of the astronomical observations he was familiar with, that comets would appear more or less regularly. Kepler's argument was based on experience and was reasonable, although he could not explain the observed regularity. It was presented so as to make comets appear as something natural and not as something wonderfully supernatural.

Faulhaber was familiar with Kepler's *Ephemeris* for 1618, in which for September the ecliptical longitude of Mars—calculated with reference to the meridian through Uraniborg, Tycho Brahe's observatory on the island of Hven—and the ecliptical latitude of the moon as well were fixed at 3° 33′.[23] Wehe followed Hebenstreit[24] and in his reconstruction of the background for Faulhaber's prediction of a comet in 1618 connected the possibility of the appearance of a comet that Kepler had mentioned, without a more precise indication of the time, with the fact that the longitude of Mars and the latitude of the moon had the same value in the *Ephemeris* for 1618 on the first of September based on the Julian calendar.[25] According to Wehe, Faulhaber had against all reason and every scientific rule interpreted the double occurrence of this value 3° 33′ as the double occurrence of the number 333, that is, as the Number of the Beast 666, and had construed this as a divine indication of the occurrence of a special event.

The testimonies of Hebenstreit and Wehe about the way in which Faulhaber predicted the comet are completely in accordance with the style of the Cabbalistic speculations Faulhaber engaged in elsewhere. Neither Faulhaber nor any of his defenders contradicted this point raised by the two critics. This is true both of *Vorläufer einer*

[20] Johann Baptist Hebenstreit, *Cometen Fragstuck/ auß der reinen Philosophia, Bey Anschawung, deß in diesem 1618. Jahr/ in dem Obern Lufft schwebenden Cometen, erläutert/ vnd auff etlicher Gelehrten vnd Vngelehrten Gegehren/ an Tag gegeben.* Ulm: Johann Meder 1618.

[21] Johann Baptist Hebenstreit, *De Cabala Log-Arithmo-Geometro-Mantica, variis nuper artibus spargi coepta, & Orbi Europaeo obtrusa, dissertatiuncula*, Ulm: Johann Meder 1619.

[22] Quoted from Justus Cornelius, *Vindiciarvm Favlhaberianarvm Prodromus*, Ulm 1619, p. 17. The original source was Johannes Kepler, *New vnnd Alter Schreib Calender sambt dem Lauff vnd Aspecten der Planeten auff das Jahr Christi M. DC. XVIII. Prognosticum Astrologicum auff das Jahr MDCXVIII. Von natürlicher Influentz der Sternen in diese Nidere Welt*, Linz: Johann Blancken 1618.

[23] Johannes Kepler, *Ephemeris nova Motuum Coelestium ad annum vulgaris aerae M D C XVIII. Ex obseruationibus potissimum Tychonis Brahei, Hypothesibus Physicis, & Tabulis Rvdolphinis; Nova etiam formâ disposita, ut Calendarii Scriptorii usum praebere possit. Ad Meridianum Vranopyrgicum in freto Cimbrico, quem proximè circumstant Pragensis, Lincensis, Venetus, Romanus*, Linz [no year]; in: Kepler, *Gesammelte Werke*, Vol. XI, 1, Munich: C.H. Beck'sche Verlagsbuchhandlung 1983, pp. 75–94, in particular p. 91.

[24] Hebenstreit, *De Cabala*, p. 26.

[25] [Zimbertus Wehe] alias Hisaias sub cruce, *Expolitio famae sidereae novae Faulhaberianae*, Ulm: Parnasische Truckerey 1619, p. 23.

Rechtfertigung Faulhabers ('Precursors of a justification of Faulhaber'), a text written under the pseudonym of Justus Cornelius in defence of Faulhaber,[26] and of the *Fortsetzung der Rechtfertigung Faulhabers* ('Continuation of the Justification of Faulhaber'), which appeared subsequently, written by an author who used the pseudonym C. Euthymius de Brusca.[27] Both texts refer explicitly to the specifications of the positions of Mars and the moon in Kepler's *Ephemeris* for 1618.[28]

C. Euthymius de Brusca first of all asserts that Faulhaber saw Kepler's calendar for 1618, shown to him by Hebenstreit, in which a comet was only mentioned as appearing in December 1618, long after the actual appearance.[29] A few pages later, after admitting the correctness of Wehe's description, he attempts to paper over the cracks with the statement that Faulhaber owed the discovery that 666 is a 'tessaracondexagonal number' with the square root 6, and at the same time a prismatic number on the basis of a nonagon (a polygon with nine sides) with the same square root, to the 'speculation and consideration' (*Speculation vnd Betrachtung*) of the longitude and latitude of, respectively, Mars and the Moon for 1 September 1618.[30] As a matter of fact, 666 can be represented as a polygonal number of square root 6 corresponding to a 46-gon, that is as the sixth term of an arithmetic sequence of the second order with first term 1 and difference 44, and also as a prismatic number of square root 6 corresponding to a nonagon, that is as six times the sixth term of an arithmetic sequence of the second order with first term 1 and difference 7.

However, the two representations offer no connection to Kepler's value of 3° 33′ of the longitude of Mars, and the latitude of the Moon on 1 September 1618. After all, it is very probable that Hebenstreit and Wehe were correct in their account of the way in which Faulhaber found his prediction of a comet; it even looks as if Hebenstreit and Wehe, both of whom before the conflict about the comet of 1618 had a friendly relationship to Faulhaber, did not even have to speculate about Faulhaber's method, but instead learned about it either directly or indirectly through intermediaries.

It was only once the clerical establishment felt that its authority could be eroded by 'prophets' like Faulhaber that Hebenstreit in particular began his attacks on Faulhaber's two texts on comets and at the same time induced the church to start an investigation of his theses. Thus, in a colloquium in the autumn of 1619, which was not open to the public, an attempt was made to answer the main question: whether Faulhaber's prediction of a comet was the result of a divine inspiration or of his own speculations. Faulhaber's testimony that he owed his knowledge about biblical numbers only to his zeal when studying arithmetic, and to prayer, saved him from further sanctions.

Faulhaber's prediction of a comet, and the subsequent debates about the seriousness and reliability of his means to achieve it, took place during the first years of the

[26] Cornelius, *Vindiciarvm*.

[27] C. Euthymius de Brusca, *Vindiciarum Faulhaberianarum. Continuatio. Das ist Rechtmessige Rettung/ Herrn Johann Faulhabers Mathematici zu Ulm Famae Sidereae, Wider Die Ehrenrüge Teutsche DiffamationSchrifften/ Expolitio Famae sidereae, &c. und Postulatum aequitatis plenissimum, &c. genant/ Welche M. Zimpertus Wehe Lateinischer Schulen Collaborator zu Ulm. Under dem falschen Namen Hisaiae sub Cruce als durch offentlichen Truck spargirt hat*, Moltzheim: Stephan Bidermann 1620.

[28] Cornelius, *Vindiciarvm*, p. 13, and de Brusca, *Vindiciarvm*, p. 24.

[29] De Brusca, *Vindiciarvm*, p. 18.

[30] Ibid., p. 24.

Thirty Years' War and also in the interval between the creation of his two portraits. The events of the following years helped him after several setbacks to improve his social position considerably.

When it had become known that, contrary to a promise given at the beginning of 1621, Faulhaber had spoken again about Gog and Magog, absolution after confession was denied to him by the church. However, he then obtained absolution from another confessor whom he was able to mislead. He also took Holy Communion in spite of repeated admonitions by Dr Dieterich not to participate; the result was that he was excluded from Holy Communion by the clerical authorities in Ulm.[31] Things escalated until the end of the year 1621. During this time the accusation of Faulhaber's conscious disrespect for and deception of the authorities played a decisive role. In the same year, a text appeared anonymously[32] in which the disciplinary actions by the municipal authorities against Faulhaber were vehemently attacked.[33] The authorities thereupon saw themselves prompted to proceed more strongly against Faulhaber, who denied all knowledge of the text and its author. However, two intercepted letters from Faulhaber to his friend the physician Dr Verbezius, and the testimonies of the nobleman Hans Ludwig Schad, who had had dealings with Faulhaber and Verbezius,[34] fuelled the distrust and suspicion against Faulhaber so much that it was considered justified to put him in prison.[35] The intention was to question Faulhaber about the divine enlightenments that he had repeatedly claimed to have had and also about his participation in the brotherhood of the Rosicrucians.[36] However, Faulhaber absconded from prison just before Christmas Eve that same year and fled to Augsburg.[37] There he was also excluded from the Holy Communion by the municipal authorities. After Johann Fugger the Elder and others from Augsburg had spoken up for him in Ulm and the authorities there had promised not to imprison Faulhaber again,[38] he returned to the city of his birth in March 1622. Just three months later he fled to Tübingen after having not obeyed another invitation by the authorities in Ulm to exculpate himself at the Dombauhütte.[39] While he was in the neighbouring city, theologians of the university told him that the Greek text of the New Testament, where the biblical number 666 occurs, was corrupt.[40] This may have tipped the balance, persuading Faulhaber after much deliberation that he could return to Ulm. This he did at the beginning of 1624. After a discussion and reconciliation with representatives of the church he signed a profession of faith that was acceptable to the local clerical authorities.[41] Nonetheless, Faulhaber never ceased to claim the status

[31] Protokolle des Pfarrkirchenbaupflegamts, 20 and 23 March 1621, Stadtarchiv Ulm, A Rep. 13-A 6876.

[32] Anon, *Gründliche Warhaffte Erzehlung Was in den Etlich Jahr wehrenden aber noch nit zu End gebrachten Stritten zwischen Johann Faulhaber und Gegentheil sich verloffen, von einer eifrigen Christlichen Persohn getreulich an Tag geben*, [n.p.] 1621; one suspects that David Verbez was the author of this text.

[33] Jakob Neubronner, manuscript of a biography of Faulhaber preserved in Stadtarchiv Ulm, p. 22f.

[34] Rats-Protokoll, Nr. 71, 21 November 1621, Stadtarchiv Ulm, A Rep. 5-A 3530.

[35] Rats-Protokoll, Nr. 71, 17, 19, and 20 December 1621, Stadtarchiv Ulm, A Rep. 5-A 3530.

[36] Rats-Protokoll, Nr. 71, 6 April 1621, Stadtarchiv Ulm, A Rep. 5-A 3530.

[37] Rats-Protokoll, Nr. 71, 24 December 1621, Stadtarchiv Ulm, A Rep. 5-A 3530.

[38] Rats-Protokoll, Nr. 72, 20 March 1622, Stadtarchiv Ulm A Rep. 5-A 3530.

[39] Protokoll des Pfarrkirchenbaupflegamts, 15 May 1622, Stadtarchiv Ulm, A Rep. 13-A 6848.

[40] Johann Faulhaber to Sebastian Kurz, 20 February 1623, Paris, Bibliothèque Nationale de France, Fonds Allemand 219 (formerly nouv. acq. 4419), f. 289^{r-v}.

[41] Protokoll des Pfarrkirchenbaupflegamts, 9 February 1624, Stadtarchiv Ulm, A Rep. 13-A 6848.

of a person selected and endowed by God with the special faculty of being able to interpret the biblical numbers, even if such a claim did not improve his standing in Ulm.

By this time, Faulhaber had already spent several years extending his expertise in more practical fields. Ever since the beginning of the war in 1618 he had turned his hand to surveying and fortification in Ulm and contributed importantly to building up the city's defences. In the autumn of 1622 he went to Basel, where he was engaged as a fortification engineer until January 1624. During his stay in Basel Faulhaber visited the Netherlands in 1623 for a short time in order to meet Prince Moritz, who after he had passed the relevant examination offered him a permanent position as engineer with a high salary. Although Moritz gave him his half-length portrait in gold as a present Faulhaber refused the offer.[42] His highly favourable appraisal in the Netherlands induced the burgomaster of Ulm to engage Faulhaber officially as an engineer for the next three years. From that time onwards, Faulhaber appended his signature on letters with 'Ingenieur' (engineer) in various forms—sometimes abbreviated to 'Ing'. or extended to 'bestellter Ingenieur' (appointed engineer). After achieving this title he had reached the social status of the man in the portrait of 1630.

His salary in Ulm, although much lower than that offered by Prince Moritz and also lower than his salary in Basel, was ten times his original salary as a *Rechenmeister*. After the end of the first contract with the city of Ulm, in 1628, Faulhaber received much less from Ulm for new engagements. Thus, his salary in 1631 for the following two years was only 200 florins. Of course, Faulhaber was back then free to accept offers from other cities, provided that the city of Ulm had agreed to it. So he worked temporarily for cities like Schaffhausen, Nikolsburg in Mähren, Frankfurt, Memmingen, and Lauingen. The expert opinions he delivered alternated with interpretations, many of which were meaningless commonplaces garnished with quotations from the Bible. In particular, his claims to have invented incredible new technical gadgets did not withstand critical scrutiny, as a group of contemporary experts in Augsburg revealed in 1620 and 1621. A good many of the 'secrets and sciences' that Faulhaber had claimed for himself were already to be found in the technical literature of the time. One suspects it is likely that the same holds true of some of Faulhaber's claims concerning his mathematical achievements. Obviously, Faulhaber was far more reliable and capable in this field, even if we do not know all the sources he might have used.

In the *Miracula Arithmetica* of 1622 and in the *Academia Algebrae* published in 1631,[43] Faulhaber continued his earlier work on arithmetic progressions of higher order and their sums, in particular sums of powers of natural numbers, which he

[42] Johann Faulhaber to Sebastian Kurz, 30 April 1624, Paris, Bibliothèque Nationale de France, Fonds Allemand 219 (formerly nouv. acq. 4419), f. 286^{r-v}.

[43] Johannes Faulhaber, *Academia Algebrae. Darinnen die miraculossiche Inventiones, zu den höchsten Cossen weiters continuiert vnd profitiert werden. Dergleichen zwar vor 15. Jahren den Gelehrten auff allen Vniversiteten in gantzem Europa proponiert, darauff continuiert, auch allen Mathematicis inn der gantzen weiten Welt dediciert, aber bißhero, noch nie so hoch, biß auff die regulierte, Zensicubiccubic Coß, durch offnen Truck publiciert worden. Welcher vorgesetzet ein kurtz Bedencken, Was einer für Authores nach ordnung gebrauchen sole, welcher die Coß fruchtbarlich, bald, auch fundamentaliter lehrnen vnd ergreiffen will*, Augsburg: J. Remmelin 1631.

now extended up to the exponent 17. For these he provided general formulae, which according to the title of the second of these booklets belonged for him to algebra.

The proofs of his formulae lacked the mathematical strength of nineteenth-century mathematicians. For odd $r = 2s + 1$, $s = 1, \ldots, 8$, he found the sums in the form

$$\sum n^{2s+1} = \left(\sum n\right)^2 \cdot \left[\sum_{i=1}^{s} ai\left(\sum n\right)^{s-i}\right].$$

For even $r = 2s$, $s = 2, \ldots, 8$, he found the sums in the form

$$\sum n^{2s} = \sum n^2 \cdot \left[\sum_{i=1}^{s} bi\left(\sum n\right)^{s-i}\right].$$

Faulhaber did not in all cases calculate the concrete values of the coefficients a_i and b_i, correctly.[44] For the determination of higher sums with sum exponents up to $t = 10$, he used the relation:

$$^{t+1}\sum n^r = \frac{1}{t}\left[(n+t).^t \sum n^r - {}^t\sum n^{r+1}\right].$$

It can be seen as a tragedy that Faulhaber, who not only in the *Miracula Arithmetica* but also in the *Academia Algebrae* had displayed his ability as a creative mathematician, received little or no reaction from outside his immediate vicinity during his lifetime. The long title of his *Academia Algebrae* reflects his efforts over a period of many years to arouse the interest of the 'mathematicians of the world' in his findings, especially those concerning sums of powers of integers.

If the *Academia Algebrae* constitutes the peak of Faulhaber's development as a mathematician it does not begin to exhaust his mathematical merits. Faulhaber had found, presumably independently of others, a three-dimensional form of the Pythagorean theorem. His compendium of everything that concerns applications of mathematics especially to fortification, which appeared from 1630 until 1633 in four parts under the title *Ingenieurs-Schul* ('School of engineers'), deals for the first time in German with the then new calculation device, the decimal logarithms of Briggs. In contrast to his former less responsible treatment of the accomplishments of others, Faulhaber was very careful in the *Ingenieurs-Schul* to quote his sources and to set out what he had learned from them. In this way, he raised the level of mathematical knowledge in a manner that made it difficult and soon impossible for the *Rechenmeister* to contribute to the further development of mathematics.

[44] Donald E. Knuth found that the sums up to the exponent 23 are correct but for the following exponents, 24 and 25, they are not; see Donald E. Knuth, 'Johann Faulhaber and sums of powers', in: *Mathematics of Computation* 61 (1993), pp. 277–94.

14

The Difficult Relation of Surveyors with Algebra

The Hundred Mathematical Questions of Cardinael

Albrecht Heeffer

Introduction

At the beginning of the seventeenth century there was a growing scepticism concerning the use of algebra for solving geometrical problems. One might expect the resistance against algebra to have come mostly from lay practitioners of mathematics such as merchants, instrument-makers, surveyors, reckoning masters, gauging experts, and engineers. But, in fact, this was not the case. The resistance came also from scholarly mathematicians and philosophers of science, while many lay practitioners embraced the new algebra. Notable adversaries of algebra within the scholarly culture included Johannes Kepler and Claude Mydorge. Kepler, who had knowledge of algebra through his friend Jobst Bürgi, a clock- and instrument-maker, looked down on its use with disdain, considering it a tool for merchants unworthy to tackle problems within the art of geometry. Concerning construction problems with mean proportionals he wrote: 'We conclude that these algebraic analyses make no contribution to our present concerns, nor do they set up any degree of knowledge that can be compared with what we discussed earlier'—with 'earlier' obviously referring to his use of classical Greek rules for geometrical construction.[1]

Another mathematical scholar, Claude Mydorge, a good friend of Descartes and Mersenne, was equally opposed to the use of algebra for construction problems in geometry, especially for conic sections. Mydorge's main and most important work is *Prodromi catoptricorum et dioptricorum* in which he completely ignores algebraic methods.[2] Mydorge was equally opposed to the use of instruments for tracing ellipses and parabolae: his second book of the *Prodromi catoptricorum et dioptricorum* concerns the geometrical description of such conic sections in plane geometry by points, a subject not treated by Apollonius. His insistence on respecting the rules of classical construction was of course in direct opposition to the new programme of his close friend Descartes, who considered algebra as the analytical tool of choice for solving

[1] Quoted from *The Harmony of the World* by Henk Bos, *Redefining Geometrical Exactness: Descartes' Transformation of the Early Modern Concept of Construction*, New York: Springer 2001, p. 192, who has a dedicated section on Kepler's objections to algebraic methods (pp. 189–93).

[2] Claude Mydorge, *Claudii Mydorgii [...] Prodromi catoptricorum et dioptricorum: sive conicorum operis ad abdita radii reexi et refracti mysteria praeuij & facem praeferentis*. Paris: Ex typographia I. Dedin 1639.

all geometrical problems. Interesting in this regard is a quote attributed to Descartes by John Pell in a letter of 1657 to Hartlib (which was never sent):

> the advantage of those new calculations in geometry, is most easily understood by comparing Midorge with Roberval. Mons. Midorge hath head ten times fitter for geometry than Mons. Robervals, I say ten times. And yet Roberval shall doe ten yea a hundred times more in geometry than Midorge. Because Roberval is more exercised in that logisticall way both for inquisition & demonstration & he hath more patience to pursue a tedious calculation when it is not his good fortune to take up the shortest way at the first setting out. Where-as Midorge keeps nearer to the old fashion; which oftentimes brings a man into a labyrinth but gives him no clew to lead him out of it.[3]

So, Kepler and Mydorge were representatives of the purists within the scholarly tradition who resisted the use of algebra for geometrical problem-solving. But what were the objections to algebra within the lay culture? On the Continent, sons of merchants, surveyors, gaugers, and artisans were taught the basics of arithmetic and geometry at private schools. During the fourteenth and fifteenth centuries the centres of education were primarily the *abbaco* schools in the commercial cities of Northern Italy.[4] With the advent of book printing during the sixteenth century, knowledge of arithmetic spread over Europe and led to more reckoning schools, in Germany and the Low Countries. Antwerp became an important centre for such schools during the second half of the sixteenth century.[5] Algebra was not part of the curriculum at these schools, but the teachers were well acquainted with the subject and many authored manuscript treatises or published books on algebra. There survive about two hundred manuscripts from between 1390 and 1500 by Italian *abbaco* masters containing problems solved by algebra.[6] During the sixteenth century arithmetic teachers such as Adam Ries and Grammateus in Germany wrote treatises on algebra while in the Low Countries we find the authors Simon Stevin, Nicolas Petri, Jan Stampioen,[7] Eduard Mellema, Michiel Coignet, and Valentin Mennher.

With the fall of Antwerp in 1585 many teachers, engineers, and practitioners of practical geometry—Stevin and van Lansberge being the most important—fled north to Holland, which became the new centre for engineers. Prince Mauritz asked Stevin to found a new school of engineering at Leiden to teach sons of artisans and surveyors in the Dutch language rather than in Latin. The school was established in 1600 with Ludolf van Ceulen and Simon Fransz van Merwen as the first teachers, to be

[3] Noel Malcolm and Jacqueline A. Stedall, John Pell *(1611–1685) and His Correspondence with Sir Charles Cavendish: The Mental World of an Early Modern Mathematician*, Oxford: Oxford University Press 2005, pp. 168–9.

[4] Warren van Egmond, '*The Commercial Revolution and the Beginnings of Western Mathematics in Renaissance Florence, 1300–1500*', PhD dissertation, Indiana University, Ann Arbor, MI 1976.

[5] Ad Meskens, *Practical Mathematics in a Commercial Metropolis: Mathematical Life in Late 16th Century Antwerp*, Berlin: Springer 2015.

[6] Warren van Egmond, *Practical Mathematics in the Italian Renaissance: A Catalogue of Italian Abbacus Manuscripts and Printed Books to 1600*, Florence: Istituto e Museo di Storia della Scienza 1980.

[7] Nicolaus Petri, *Practicque, om te leeren rekenen, cijpheren ende boeckhouwen, met die regel coss ende geometrie seer profijtelijcken voor alle coopluyden*, Amsterdam: Cornelis Claesz 1583; Jan Stampioen, *Algebra, ofte Nieuwe stel-regel, waer door alles ghevonden wordt inde wiskonst, wat vindtbaer is*, 's Gravenhage: ghedruckt ten huyse vanden autheur 1639.

followed by father and son van Schooten. Stevin's book *Wisconstige gedachtenissen* (Mathematical thoughts) was used for teaching and together with the other vernacular books of van Ceulen and Stampioen it provides a representative account of the curriculum: geometry, surveying, trigonometry, arithmetic, perspective, cosmology, and bookkeeping.[8] Although Stevin devoted a significant part of his *L'Arithmetique* to algebra, the Rule of Coss, as it was called, was not taught to engineers or surveyors at Leiden.[9]

Also used as a textbook throughout the seventeenth century was Johan Sems and Jan Pieterszoon Dou's *Practijck des Landmetens: Leerende alle rechte ende cromzydige Landen, Bosschen, Boomgaerden, ende ander velden meten, soo wel met behulp des quadrants, als sonder het selve* (The practice of surveying: teaching land surveying with rectangular and irregular sides, forests, orchards, and other plots of land, both using the quadrant or without), printed at Leiden.

This short survey suffices to demonstrate that the simple explanation for the resistance to algebra by practitioners of mathematics in the lay tradition was a lack of knowledge and that the claim that algebra was mostly practised by the scholarly tradition is easily contradicted by many counterexamples. Obviously many lay surveyors, gaugers, or engineers lacked the skills to solve complex geometrical problems by algebra, but authors of books on engineering or surveying who advertised on the frontispiece that all problems in the book are solved without the aid of algebra did so not because they lacked knowledge of algebra but because they were convinced it was the right way. In the rest of this chapter we will investigate their motivations via a prime example of the early seventeenth century, a popular work on surveyors' geometry by Sybrandt Hanszoon Cardinael. We will demonstrate that while Cardinael's problem-solving techniques are devoid of any kind of symbolism, they are based on a long tradition of geometrical algebra going back to 1800 BC in Mesopotamia.

Cardinael's background

Sybrandt Hanszoon Cardinael was born in 1578 in Harlingen, a small fishing and shipping town in the shadow of the university city of Franeker, in the northern Dutch province of Friesland.[10] His parents, Hans Sybrandt—hence the name Hanszoon

[8] Simon Stevin, *Wisconstige gedachtenissen, Inhoudende t'ghene daer hem in gheoeffent heeft den [...] vorst ende heere, Mavrits prince van Oraengien*, Leyden: Inde Druckerye van Ian Bouvvensz 1605–8; Ludolf van Ceulen, *De arithmetische en geometrische fondamenten*, Leyden: Ioost van Colster, ende Iacob Marcus 1615; also: *Fundamenta arithmetica et geometrica*, Leiden: Colster 1615; Jan Stampioen, *Resolutie ende ontbindinghe der twee vraegh-stucken, in den jare 1632*, Rotterdam: op de weduwe van Matthiis Bastiaensz 1634.

[9] Simon Stevin, *L'arithmetique de Simon Stevin de Bruges: contenant les computations des nombres Arithmetiques ou vulgaires: Aussi l'Algebre, auec les equations de cinc quantitez: Ensemble les quatre premiers liures d'Algebre de Diophante d'Alexandrie, maintenant premierement traduicts en François*, Leyden: Christophle Plantin 1685.

[10] The genealogy of the Cardinael family and several related family members has been studied by R. M. A. De Jong, *Genealogie van de familie De Gavere*, genealogysk jierboek, Ljouwert: Fryske Akademie 1996. The description in this chapter of Cardinael's life and his book closely follows the introductory chapter of Matthijs H. Sitters and Albrecht Heeffer, *The Hundred Geometrical Questions by Sybrandt Hansz Cardinae: Early-modern Dutch Geometry from the Surveyor's Tradition*, Boston: Docent Press 2016.

(son of Hans)—belonged to the Mennonite church of the Netherlands (in Dutch *doopsgzinden* or 'baptism-minded'), named after Menno Simons (1496–1561), a Roman Catholic priest from Friesland who converted to Anabaptism around 1536. Several mathematicians and engineers of that time were Mennonites, such as Willem Janszoon Blaeu (1571–1638), Cornelis Drebbel (1572–1633), Jacob Metius (1575–1650), Abraham de Graaf (1635–1713), and Gerard Kinckhuysen (1625–1706).

After father Hans died (probably in 1600), Sybrandt Hanszoon moved to Amsterdam in 1605. There, he married Levijntje Panten, the daughter of a reckoning master, Lieven Panten, in 1607. The Panten family moved from Brughes in Flanders to Harlingen because of the religious persecutions at the end of the sixteenth century. Levijntje gave birth to six children between 1610 and 1625. Sybrandt took over the reckoning school of his father-in-law, probably soon after arriving in Amsterdam, since Lieven Panten died in 1603. He taught geometry, arithmetic, surveying, astronomy, navigation, and bookkeeping. The first proof of his mathematical activities in Amsterdam stems from a recreational problem he publicly announced in 1606 about the shadows of three sticks of unequal length.[11] In the year 1606 he also published a rebate book, with tables of discounts useful for merchants. From 1608 he was active as a surveyor, as several maps are extant from his hand. In 1613 he was commissioned for the mapping of the third enlargement of Amsterdam.[12]

Not much is known about Sybrandt Hansz. until his first book on geometry, *Hondert Geometrische questien met hare solutien* (Hundred geometric questions with their solutions), was published in 1612. After the establishment of *Costers' Nederduytsche Academie* in 1617 he became one of the teachers. This school taught in Dutch for a growing class of merchants and artisans in Amsterdam, but did not last long because of political and religious turmoil. In 1620 Cardinael published a book on arithmetic (*Arithmetica ofte Reecken-konst*), the same year that a new edition of the *Hondert Geometrische questien* (*HGQ*) also appeared. Apart from his surveying work, he might also have been active as a wine gauger at that time since his name appears in an official document.[13] In 1635, he published a book in Dutch on cosmology rejecting Copernican astronomy. His main argument against Copernicanism, as expounded in the second and third chapter, stems from Aristotelian notions of natural and forced motion. From 1639 onwards several editions appear of his earlier work on arithmetic, edited for their use in schools. He died in 1647 and was buried on 26 November in the family tomb which he had constructed the year before.

While Cardinael currently is not recognized in the history of mathematics at the level of his Dutch contemporaries, such as Willebrord Snellius, Frans van Schooten, or Christiaan Huygens, he certainly was well regarded in his time. As a witness of his recognition we can mention several lesser-known historical facts.[14] To resolve a

[11] Published as appendix 4 in: Gerard Kinckhuysen, *Algebra, ofte Stel-konst, beschreven tot dienst van de leerlinghen*, Haerlem: Passchier van Wesbusch 1661, pp. 42–5, figure on p. 36.

[12] L. Jansen, 'De Derde Vergroting van Amsterdam', in: *52nd Yearbook of the Amstelodamum Historical Society*, Amsterdam: J. H. de Bussy 1959, pp. 42–89 at p. 51.

[13] Reported by J. G. van Dillen, *Bronnen tot de geschiedenis van het bedrijfsleven en het gildewezen van Amsterdam*, The Hague: Nijhoff 1929–74, and discussed by Matthijs H. Sitters, *Sybrandt Hansz. Cardinael 1578–1647: Rekenmeester en wiskundige: zijn leven en zijn werk*, Hilversum: Verloren 2008, p. 51.

[14] For further references, see Sitters, *Cardinael* (2008), pp. 59–62.

polemic on the theory of navigation, he was invited by the Admiralty of Amsterdam to take a seat in a commission together with Plancius, Blaeu, Stevin, Snellius, and Dou during the years 1612–14, which clearly places him at the centre of the scientific elite. Also there is the curious story, recorded by Dirck Rembrandtszoon van Nierop, that René Descartes took the trouble to visit Cardinael, and tried to convince him to accept the Copernican cosmological model. While Descartes's effort was to no avail, it shows that Cardinael was a person to be reckoned with. Jan Stampioen de Jonge (1610–1653), who was commissioned by Constantijn Huygens to tutor his son, advised the young Christiaan Huygens to study Cardinael's *HGQ* for an understanding of geometry. And further, as discussed below, he was also internationally recognized through the translations of his work into German and English. In a study published in *Isis*,[15] Rudd's adaptation of Cardinael's work was considered as one of the most influential geometry books of the seventeenth century.

The book and its editions

Cardinael's *Hondert Geometrische Questien* most often appeared bound together with later editions of the *Practijck des landmetens* (see Figure 14.1) and a book on surveyor's instruments also by Sems and Dou (see Figure 14.2), the bound edition noting on its frontispiece: 'Expanded by Hundred geometrical questions, by Sybrandt Hansz. Reckoning master at Amsterdam'. There appear to be at least two series of editions, one printed by Willem Janszoon, 'op het water in de vergulde sonnewyser' at Amsterdam, and a second one by Jan Janszoon, also in Amsterdam 'op het water in de Pas-Caert'; both printers are from the Blaeu family. It is certainly the case that at least two editions of the Willem Janszoon printings of Cardinael's book can be identified: one with and one without a dedication to Prince Mauritz and Duke Willem Lodewijk. The one without the dedication appears to be the later one.

The later editions of Cardinael's book were printed in Amsterdam by Jan Janszoon, 'op het water in de Pas-Caert', as was the *Practijck des landmetens*. The typesetting differs from the Willem Janszoon editions but the woodcuts and copper plates appear to be identical. Two different editions printed by Jan Janszoon can be identified, the earlier having 127 pages (16 octavo quires of 8 pages each, numbered A1 to H8) and the later 118 numbered pages. The later edition is mostly found bound together with the *Practijck des landmetens*, though separate copies also exist. Of three copies owned by Ghent University library—one printed by Willem Janszoon and two by Jan Janszoon—only one (by Jan) is bound together with the *Practijck des Landmetens*.[16]

None of the editions of *HGQ* lists a printing date, which makes it difficult to identify and distinguish the different print runs. Several historians have suggested dating based on secondary sources. The pioneer of Dutch history of mathematics, Bierens de Haan, included biographical data on Cardinael in his *Bouwstoffen* at the end of the nineteenth century, and an incomplete bibliography in his *Bibliographie néerlandaise*

[15] L. Karpinski and F. Kokomoor, 'The Teaching of Elementary Geometry in the Seventeenth Century' in: *Isis* 10 (1928), pp. 21–32.

[16] The reproductions in our text are taken from a copy of this latter edition owned by A. Heeffer.

Practijck des Lantme-
tens: Leerende alle rechte ende krom-
zydige Landen / Boſſchen / Boomgaer-
den/ ende andere velden meten/ ſo wel
met behulp des Quadrants / als
ſonder het ſelve.

Mitſgaders alle Landen deelen in gelijcke ende on-
gelijcke deelen op verſcheyden manieren, met eenige nieuwe gecalculeerde Tafelen daer toe dienende.

Gecomponeert door Johan Sems, ende Jan Pieterſz.
Dou, geadmitteerde Landtmeters.

*Vermeerdert met hondert Geometriſche Queſtien met
haer Solutien. Door* S Y B R A N T H A N S Z.
Rekenmeeſter tot Amſterdam.

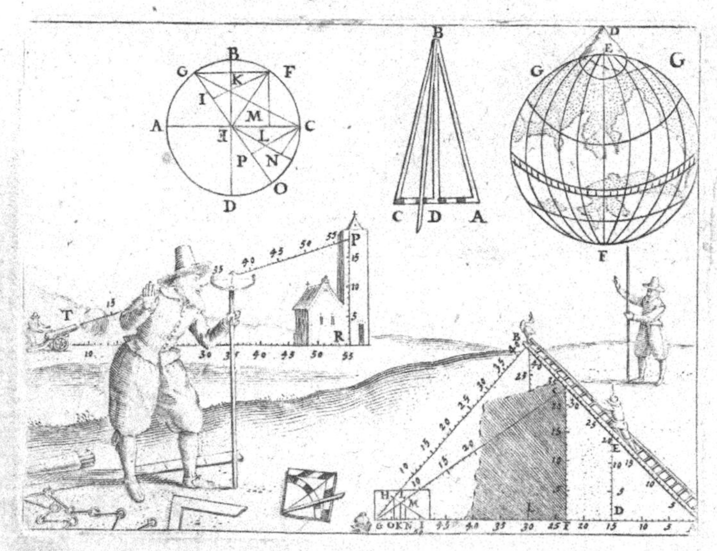

Gedruckt tot Amſterdam by Jan Janſz. op het
Water / in de Pas-Caert.

Figure 14.1 Frontispiece of the book on surveying by Sems and Dou (copy owned by A. Heeffer).

Van het gebruyck der
Geometriſche inſtrumenten.
Leerende alle ongenaeckelijcke lenghten breetten/ wijtten/hooghten ende diepten/met behulp van ſom-mighe Geometriſche inſtrumenten af meten/ ſoo wel ſonder calculatie / als met behulp der ſelvighen.

Deſghelijcks Caerten maecken, ſoo wel van eenige Landtſchappen met hare behoorlijcke Steden, Dorpen, Caſte len ende Slooten, als van eenighe particuliere Velden, ende hoemen een gantſche Provincie,mitſgaders de Middellinie ende ommeloop des Aerdtbodems ſal afmeten, ende een Stadt, Sterckte ofte Caſteel inde grondt legghen, met meer andere konſtighe ſtucken der Geometrie belanghende.

Door *Iohan Sems* ende *Ian Pieterſz. Dou.*

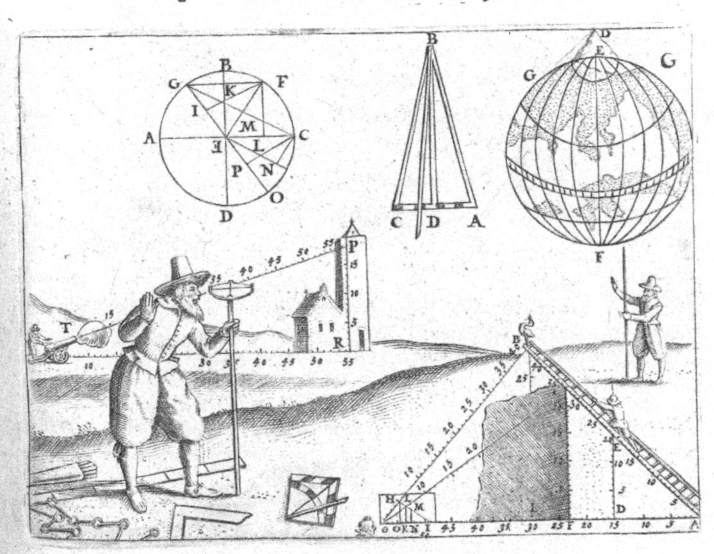

Ghedruckt tot Amſterdam by Jan Janſz, op het Water/ inde Pas-Caert.

Figure 14.2 Frontispiece of the book on surveying instruments by Sems and Dou (copy owned by A. Heeffer).

historique-scientifique.[17] Jan van Maanen in his PhD dissertation, as well as in a later article devoted to Cardinael, dates the first edition to or before 1612.[18] In an article of 2003, Matthijs Sitters follows the date of 1612 for the first edition but in his PhD dissertation on Cardinael he consistently uses 1614 as the publication date of the book, which appeared bound together with the second edition of *Practijck des landmetens* published in 1614.[19]

A bibliographical study by Klaas Hoogendoorn provides us with the most reliable data on the publication history.[20] Hoogendoorn determined an upper bound for the first edition through a reference to his book in the dedication to *Het gulden zegel der grote zeevaart* (The golden seal of great navigation) by Jan Hendrik Jarichs van der Leys, dated 1614.[21] The publication date of 1612, proposed by Bierens de Haan and van Maanen, is therefore most likely correct. The second edition by Willem Janszoon is dated around 1620 because from 1623 Willem started adding Blaeu to his name. However, because the German edition by Willem Janszoon appeared in 1617, likely the Dutch version was published that year or earlier. The two editions by Jan Janszoon are dated by Hoogendoorn much later than previously assumed, based on book catalogues of that era—around 1635 and 1645—though this contradicts the publication date by Jan Janszoon of the German editions of *Practijck des Landmetens* in 1617.

Translations

Soon after the publication of the first Dutch edition of the *HGQ* the book was translated into German by Sebastian Kurz (1576–1659), known in Latin as Curtius (see Figure 14.3), a reckoning master from Nuremberg who had earlier published several works on merchant arithmetic, geometry, and algebra (see Figure 14.4).[22] He was a

[17] David Bierens de Haan, *Bouwstoffen voor de geschiedenis der wis—en natuurkundige wetenschappen in Nederland*, 2 vols, Amsterdam 1878–87, pp. 13–72; D. Bierens de Haan, *Bibliographie néerlandaise historique-scientifique des ouvrages importants dont les auteurs sont nés aux 16e, 17e et 18e siècles, sur les sciences mathématiques et physiques avec leurs applications*, Rome: Imprimerie des sciences mathématiques et physiques 1883, p. 52.

[18] Jan van Maanen, 'Facets of seventeenth-century mathematics in The Netherlands', PhD dissertation, Utrecht University 1987; Jan van Maanen, 'Cardinael in de geschiedenis van de wiskunde', in: *Nieuw archief voor wiskunde*, series 5, 4/1 (2003), pp. 51–5.

[19] Matthijs H. Sitters, 'Sybrandt Hansz. Cardinael (1578–1647) Meester in de meetkunde', in: *Niew Archief voor wiskunde*, series 5, 4/4 (2003), pp. 309–16; Matthijs H. Sitters, 'Sybrandt Hansz. Cardinael 1578–1647: Rekenmeester en wiskundige: zijn leven en zijn werk', PhD dissertation, Groningen 2007.

[20] Klaas Hoogendoorn, 'De werken van de meetkundige rekenmeester Sybrandt Hansz. Cardinael (1578–1647): Proeve van een bibliografie', in: *De boekenwereld* 23 (2006–7), pp. 276–90.

[21] Hoogendoorn, 'De werken'.

[22] Sebastian Curtius, *Ein newes wolgegründtes Rechenbuch: nach rechter Art der Practic, mit manchem schöonen geschwinden Vortheil, neben gründlicher Aussführung der gantzen und gebrochenen: sampt den Proportionen Speciebus [...] Dessgleichen zum Beschluss eine Aufföosung mancherley küunstlicher cossischer unnd polygonalischer Exempla, etlicher fürnemer und kunstreicher Rechenmeistern, so zu End irer Rechenbucher one Facit gesetzt seyn: dergleichen vorhin in Druck nie aussgangen, nun aber der lieben Jugend zu gedeylicher Wolfart mit Fleiss zusammen colligirt und jetzt erstmals in Druck gegeben*, Nuremberg: In Verlegung Conrad Baurn 1604; Sebastian Curtius, *Compendium Arithmeticae, Das ist: Ein Neues, Kurtzes und Wolgegründtes SchulRechenbüchlein, von allerley Hauß—und Kau_manns rechnungen [...]: Meinen lieben Discipulis*, Nuremberg: Fuhrmann 1610.

Figure 14.3 Sebastian Kurz.

friend of Johan Faulhaber (1580–1635), the surveyor of the city of Ulm who also published on algebra, and they exchanged hundreds of letters. Kurz also translated Dou's books on mathematical instruments and surveying.[23] These last two were published in Amsterdam and may have been commissioned by the printer.

There appears to have been only one edition of the German translation of *HGQ*, printed in Amsterdam by Willem Janszoon with dedication dated 1 January 1617. An introduction has been added on operations with surds, referring to the works of Christof Rudolff and Michael Stifel, and at the end of the book twelve geometrical and arithmetical problems have also been added, acknowledging Johann Heer and Paulo Schmidt (possibly students of Kurz). Otherwise, the translation is very faithful to the original, using the same woodcuts and exactly the same pagination.

Another translation of the *HGQ* was into English by Thomas Rudd (1583?–1656) (see Figure 14.5). Rudd was an English military engineer who learned Dutch through his service in the Low Countries from around 1612 to 1627.[24] In 1627, King Charles I appointed him as his first chief military engineer in charge of the fortifications in Wales. In 1659 he wrote a supplement to a handbook for military engineers, first

[23] Sebastian Curtius and Jan Pieterszoon Dou, *Tractat vom machen und Gebrauch eines Neugeordneten mathematischen Instruments: inn welchem underschiedliche Künstliche stuck die Geometriae betreffende verfasset und begriffen seind*, Amsterdam: Bey Wilhelm Janss 1616; Curtius and Dou, *Practica des Landmessens: darinn gelehrt wird, wie man alle recht-, und krumseitige Land, Wälder, Baumgärten [...] sowoh mit dem Quadranten, als ohne demselben messen soll*, Amsterdam: Jansson 1616.

[24] Historical documents on the city of Breda mention Rudd as an official surveyor and cartographer in 1618: A. J. van der Aa, *Geschiedkundige beschrijving van de stad Breda en hare omstreken*, Gorinchem: Noorduyn en Zoon 1845, p. 14. A PhD dissertation by Schaefer lists Rudd as official surveyor at the province of Brabant as early as 1612: Christoph Schaefer, '"Krygsvernuftelingen": Militäringenieure und Fortifikation in den Vereinigten Niederlanden', PhD dissertation, University of Giessen 2001.

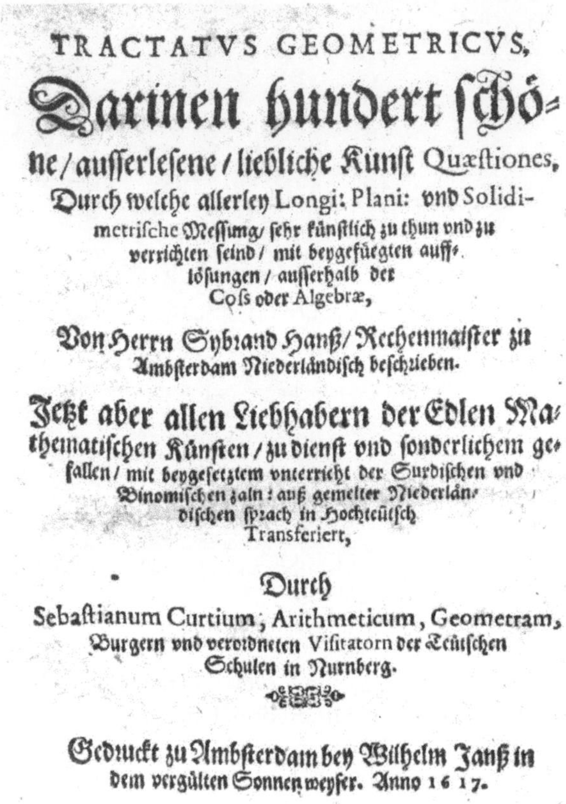

TRACTATVS GEOMETRICVS,
Darinen hundert schö-
ne/ausserlesene/liebliche Kunst Quæstiones,
Durch welche allerley Longi: Plani: vnd Solidi-
metrische Messung/ sehr künstlich zu thun vnd zu
verrichten seind/ mit beygefüegten auff-
lösungen/ ausserhalb der
Coss oder Algebræ,

Von Herrn Sybrand Hanß/ Rechenmaister zu
Ambsterdam Niederländisch beschrieben.

Jetzt aber allen Liebhabern der Edlen Mathematischen Künsten/ zu dienst vnd sonderlichem ge-
fallen/ mit beygesetztem vnterricht der Surdischen vnd
Binomischen zaln: auß gemelter Niederlän-
dischen sprach in Hochteütsch
Transferiert,

Durch
Sebastianum Curtium, Arithmeticum, Geometram,
Burgern vnd verordneten Visitatorn der Teütschen
Schulen in Nurnberg.

Gedruckt zu Ambsterdam bey Wilhelm Janß in
dem vergülten Sonnenweyser. Anno 1617.

Figure 14.4 The frontispiece of the German edition of *HGQ*.

published in 1650, Elton's *Compleat Body of the Art Military*. According to Frances Yates, Thomas Rudd was an enlightened Rosicrucian and possibly identical to Dr Rudd, the author of occult works such as a Treatise on *Angel Magic*.[25] In 1651, he added a version of the text of John Dee's well-known *Mathematical preface* to his edition of the first six books of Euclid's *Elements*.

Rudd's adaptation of the *HGQ* was published in 1650 by Robert Leybourn in London. The book commences with an introduction of over 50 pages explaining techniques typical for the surveyor. For the second part he borrowed heavily from Cardinael, translating 79 problems and adding some others of his own.[26] The translation is close to the original but adapted to the English units of measures. Rudd also occasionally changed the context. Cardinael's question 48 concerns digging a ditch in farmland which becomes a fish pond in Rudd's question 87. However, Rudd did not take much effort to conceal his source. Most of the drawings and the values used for the questions are identical to those of Cardinael.

[25] Frances A. Yates, *The Rosicrucian Enlightenment*, London: Routledge and Kegan Paul 1972, p. 258.
[26] See Sitters and Heeffer, *Hundred Geometrical Questions*, Appendix 5.2 for a cross table.

A
HUNDRED
GEOMETRICAL
QVESTIONS,

WITH

Their Solutions and Demonſtrations:
Moſt of them being reſolved both Arithmetically and Geometrically, by Lines and Numbers, &c.

By Captain THOMAS RUDD,
Chiefe Engineer to His late Majeſty.

LONDON,
Printed by ROBERT LEYBOURN,
for ROBERT BOYDEL, at the Bulwark neere Tower-Hill, *M DC L.*

Figure 14.5 Frontispiece of Rudd's adaptation from Cardinael.

The context

Practical geometry

Almost all of the questions posed by Cardinael are set within a practical context and many concern surveying in a direct or indirect way. Direct references to surveying pertain to the measurements of heights with the help of instruments (Q9, 12, 17, 100).[27] Indirect references to surveying are the questions in which certain sides of a regular or irregular piece of land have been measured, and in which the length of unknown sides have to be established (as in Q3).[28] The whole of subgroup 10, transformation of geometrical shapes with conservation of area, are classic problems from the surveyors' tradition. Also subgroup 11, partitioning a convex polygon into equal parts or parts with prescribed ratios, belongs to the oldest practices of land administration. As an illustration, Figure 14.6 shows a Babylonian tablet from the Ur-III period (*c.*2100–2000 BCE).

Although this precedes Cardinael by more than three millennia, it concerns problems which undoubtedly were also part of early modern practices. The tablet is a

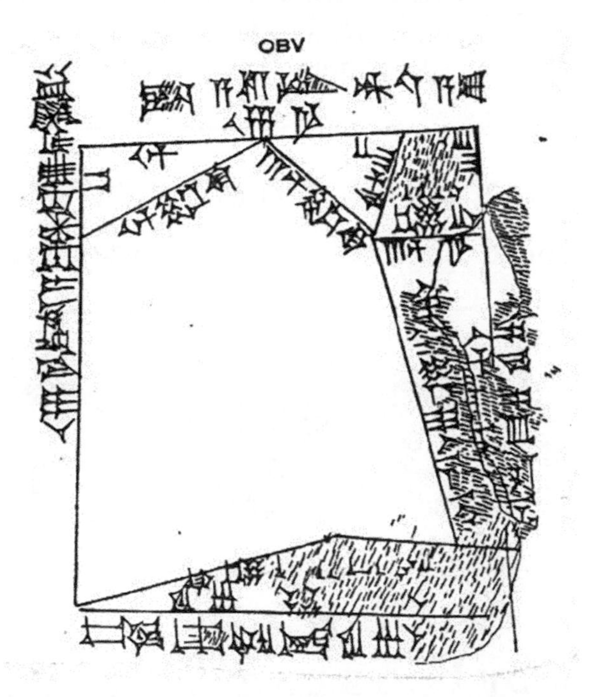

Figure 14.6 A legal document describing the partitioning of an irregular piece of land into five parts of equal area (c. 2100–2000 BCE).

[27] These Q-numbers refer to the questions by Cardinael as found in the critical edition by Sitters and Heeffer, *Hundred Geometrical Questions*, on pp. 39, 42, 47, and 158.

[28] See Sitters and Heeffer, *Hundred Geometrical Questions*, §3.2.1 for a classification of similar problems into groups and subgroups.

juridical document describing how an irregular piece of land needs to be partitioned into five parallel strips of equal area. It is discussed in detail by Jöran Friberg.[29] Given the area of the outer quadrangle, the area of a piece of land can be determined by subtracting the calculated area of the neighbouring triangles and trapezoids. Hence, with the area known and length of 16 n given (see Figure 14.7) the problem is reduced to determining the widths of the five pieces s_1 to s_5. Though the extant document does not specify how the land can be divided into pieces of equal area, Friberg convincingly argues that doing so requires knowledge of solving quadratic problems. Since the document lists some of the values s_1 to s_5, we can assume that someone was able to do the calculation. Friberg shows in detail how the calculation can be performed by geometrical algebra. The techniques needed for solving the division problem are shown to be consistent with lay surveyors' methods from around 1800 BCE.

The discovery that Old Babylonian surveyors could solve quadratic problems goes back to the 1930s in Göttingen. It was Otto Neugebauer who first realized that quadratic problems were solved on tablet YBC 6967, of around 1800 BCE. The problem can be summarized as follows. Given that the product of two unknown values equals 1 (in sexagesimal, or 60 in decimal) and that one unknown (igibūm) is 7 greater than the other (igūm), what are these values?

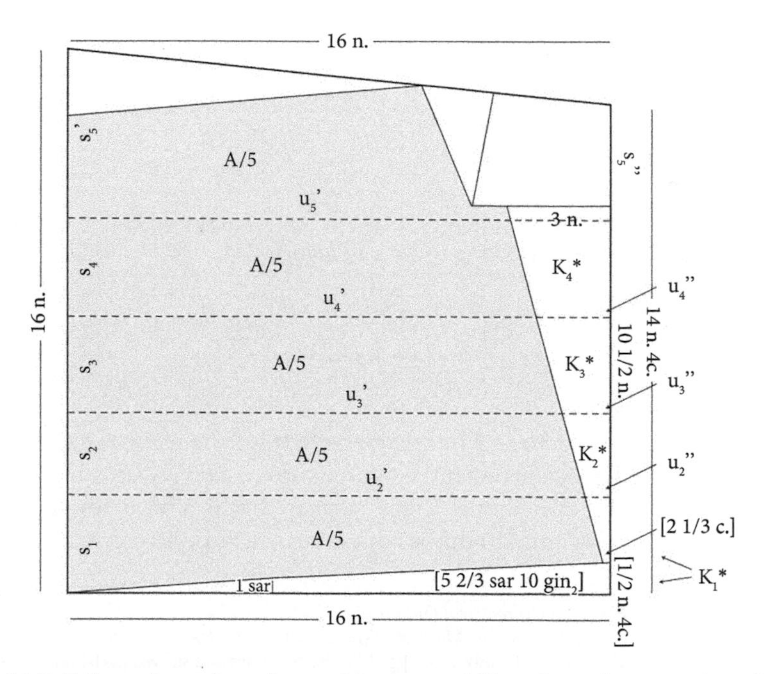

Figure 14.7 Friberg shows how the partitioning could have been done consistently with prevailing methods (Friberg, 'A Geometric Algorithm').

[29] Jöran Friberg, 'A Geometric Algorithm with Solutions to Quadratic Equations in a Sumerian Juridical Document from Ur III Umma', in: *Cuneiform Digital Library Journal* 3 (2009), pp. 1–27.

This problem corresponds to case 12 of the eighteen cases distinguished by Sitters and Heeffer, of which we will show an example below.[30] It concerns a rectangular area where the product and the difference between the larger and smaller sides are given.

In his book with Abraham Sachs, Neugebauer writes that '[t]he problem treated here belongs to a well known class of quadratic equations' and gives a purely algebraic/arithmetical interpretation of the tablet, as is shown in his translation:[31]

Obverse

1. [The *igib*]*ūm* exceeded the *igūm* by 7.
2. What are [the *igūm* and] the *igibūm*?

3–5. As for you—halve 7, by which the *igibūm* exceeded the *igūm*, and (the result is) 3;30.

6–7. Multiply together 3;30 with 3;30 and (the result is) 12;15.

8. To 12;15, which resulted for you,
9. add [1,0, the produ]ct, and (the result is) 1,12;15.
10. What is [the square root of 1],12;15? (Answer:) 8;30.
11. Lay down [8;30 and] 8;30, its equal, and then

Reverse

1–2. subtract 3;30, the *takilūm*, from the one,

3. add (it) to the other.
4. One is 12, the other 5.
5. 12 is the *igibūm*, 5 the *igūm*.

Neugebauer believes that the operations in the text correspond to the solution of a system of two equations, amounting to the calculation

$$\left.\begin{matrix} x \\ y \end{matrix}\right\} = \sqrt{\left(\frac{7}{2}\right)^2 + 1,0} \pm \frac{7}{2}.$$

When comparing Neugebauer's translation with this formula one can indeed see the relation. The difference between the two unknowns being 7, it is halved (lines 3–5), which is 3; 30 (or 3½ decimal). This is squared, which becomes 12;15 (lines 6–7) or $(3\frac{1}{2})^2 = 12\frac{1}{4}$ decimal. To this is added 1,0, which is the product of the two

[30] Sitters and Heeffer, *Hundred Geometrical Questions*.

[31] Otto Neugebauer and Abraham Sachs, *Mathematical Cuneiform Texts*, American Oriental Series 29, New Haven, CT: American Oriental Society 1945, pp. 129–30. Numbers are shown in the base 60 system. Where we normally use the decimal point, in the sexagesimal system it is customary to use a semicolon. Higher-level multiples and lower-level fractions are separated by a comma. The sexagesimal number 3;30 thus corresponds to 3 units and 30/60 fractions of a unit, or in decimal 3½. The square brackets fill in text missing due to damage to the tablets. The front text on the tablet is numbered 1–11, the lines on the back from 1 to 4.

unknowns. In line 10 the root of this sum is taken, being 8;30 (or 8½ decimal). The two unknowns then are 8;30 ± 3;30, being 12 and 5.

From the end of the 1970s, Jens Høyrup published several articles on Old Babylonian mathematics, advancing the thesis that their algebraic methods depended on lay surveyors' practices.[32] Through a thorough semantic analysis of the Akkadian text, he determined that certain verbs, interpreted by Neugebauer as arithmetical operations, instead refer to concrete operations. What Neugebauer reads as 'to subtract', Høyrup understands as 'to tear off'; 'to halve' becomes 'to break in two'; and 'adding together' becomes 'pasting together'. In this context, 'lay down' now makes sense as a concrete action. Rather than solving quadratic equations, Old Babylonian surveyors were using a cut-and-paste geometry to solve problems, very much as Cardinael was doing. Eleanor Robson provides a slightly different translation but does adopt the interpretation of concrete cut-and-paste operations such as 'break in half' and 'put down' for this and similar Old Babylonian tablets.[33]

The two unknown values *igūm* and *igibūm* can be considered as the sides of a rectangle with area 1 (or 60 decimal). In a first step the area exceeding the square with side *igūm* is cut in half (see Figure 14.8). In the following step, half of the rectangle which was cut is pasted onto the bottom of the original rectangle (see Figure 14.9), which gives meaning to the term 'laying down' which puzzled Neugebauer. The area of the original rectangle remains unchanged (the white area in Figure 14.9). The open space then is a square of side 3½. So, the shaded square in Figure 14.9 has an area of 12¼. The area of the completed square is therefore known: 72¼. From this, the side can be derived, which is 8½. It now becomes evident that the two unknown values are 8½ ± 3½ or 12 and 5.

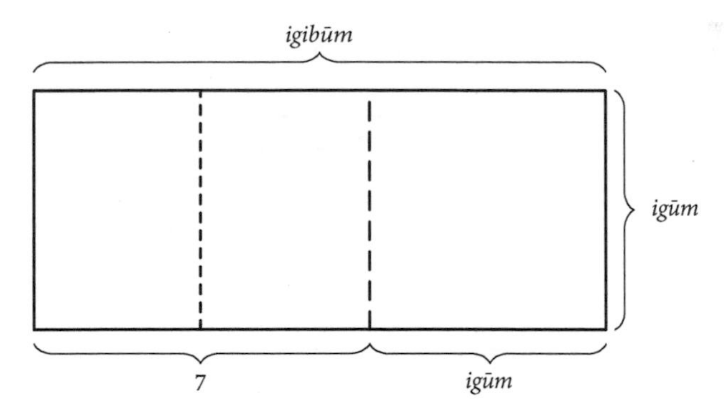

Figure 14.8 The geometrical interpretation of YBC 6967 by Høyrup, *Lengths, Widths, Surfaces*, pp. 55–8.

[32] Culminating in: Jens Høyrup, *Lengths, Widths, Surfaces: A Portrait of Old Babylonian Algebra and Its Kin*, Studies and Sources in the History of Mathematics and Physical Sciences, New York: Springer 2002.

[33] Eleanor Robson, *Mathematics in Ancient Iraq: A Social History*, Princeton, NJ: Princeton University Press 2008, pp. 113–14.

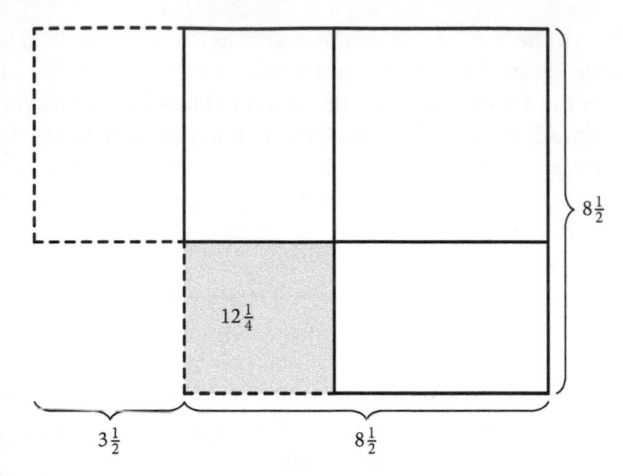

Figure 14.9 The cut-and-paste solution to YBC 6967 from Høyrup, Lengths, Widths, Surfaces, pp. 55–8.

This new view makes Neugebauer's symbolic interpretation, as a problem belonging to 'a class of quadratic equations', inconsistent with Høyrup's lexical analysis. When substituting $y = x + 7$ into $xy = 60$, one arrives at the quadratic equation $x^2 + 7x = 60$. The roots for the quadratic equation $ax^2 + bx + c = 0$ are

$$\left.\begin{matrix} x_2 \\ x_1 \end{matrix}\right\} = -\frac{\mathrm{b}}{2\mathrm{a}} \pm \sqrt{\frac{b^2 - 4ac}{4a^2}}$$

with $a = 1$, $b = 7$ and $c = -60$; this amounts to

$$x_1 = \frac{-7}{2} + \sqrt{\frac{(-7)^2 + 240}{4}} = 5$$

and

$$x_2 = \frac{-7}{2} - \sqrt{\frac{(-7)^2 + 240}{4}} = -12.$$

The negative solution only arises in the context of symbolic algebra and is avoided in the Old Babylonian cut-and-paste geometry as well as in Cardinael's techniques for solving problems. Curiously, this geometrical solution to the problem of tablet YBC 6967 is identical to Sitters and Heeffer's case 12, with the accompanying figure closely resembling our Figure 14.9.[34] Cardinael applied this method in Q51 which we quote fully in English translation:

[34] Sitters and Heeffer, *Hundred Geometrical Questions*, §3.1.5.

Question 51.
When the number of 567 musketeers were known to me and that they encircle the number of lancers aforementioned (being 624) four men thick, and that the proportion of the breadth AB against the length *AD* were now unknown to me, then to find AB and AD [see Figure 14.10].

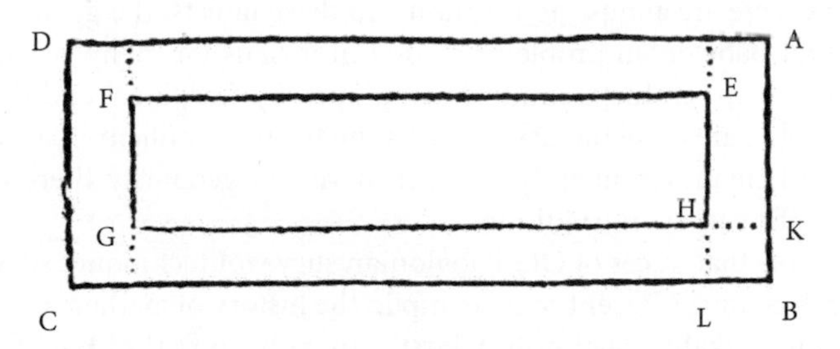

Figure 14.10 Cardinael's Question 51: A battalion is arranged in a rectangular form with the lancers on the outside, four rows thick, covered by the musketeers on the inside.

So, cut off of the quadrangle *ABCD* as much as GDCN and paste it to *BK* which makes *BKIL* [see Figure 14.11]. So, then *GALM* is a square, making *AG* or *AL* 40, as *BAD* is equal to *LAG*, the square *AGML* then is 1600. Now, add the number of musketeers (being 576) to 624, the number of lancers, the sum making 1200. This subtracted from 1600 there remains 400 for the square *KNMI*. The √ꝫ [square root] from this is 20 for *IK*, which is equal to BL. So *BL* makes 20, this subtracted from *LA* 40, there remains 20 for *AB*. So, *AD* must be 60, as *GD* is equal to *BL*.

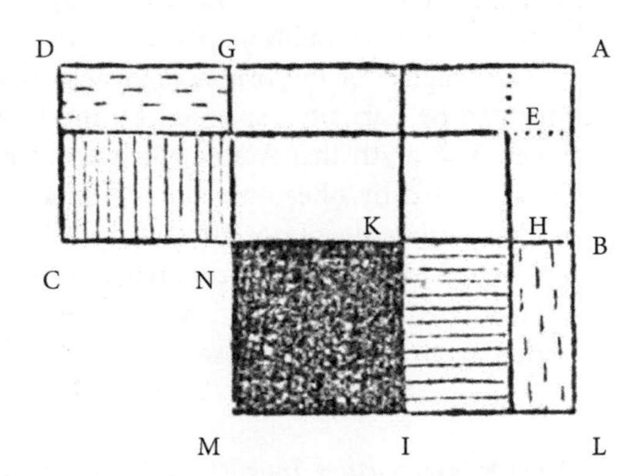

Figure 14.11 Cardinael's Question 51: solution.

Though the problem of Q51 is more complex than tablet YBC 6967, Cardinael's solution is based on the same technique: cutting off an area *DGNC* from the given rectangle, pasting it at the bottom, completing the square *GALM*, determining its side *AG*, which allows the calculation of the solution. We note that the first two sentences of the corresponding explanation by Cardinael show how his terminology is remarkably similar to the procedure using cut-and-paste operations known to Old Babylonian surveyors.

Equally noteworthy is that the Old Babylonian problem also appears with the same values as Cardinael's problem Q14, with a rectangle of area 60 and sides 5 and 12, though the solution method is slightly different. The choice of values may seem arbitrary in the context of early modern geometry but not for Old Babylonian mathematics where they are each others' reciprocals. When 60 is a unit, then 1/5 equals 12 and 1/12 equals 5.

However, as there are almost no diagrams on these tablets, the geometrical interpretation of Old Babylonian problems eluded historians for many years. The close correspondence between the techniques employed by Cardinael and those of Old Babylonian algebra are quite remarkable. Despite the three millennia separating Old Babylonian mathematics from early modern surveyors' geometry, there seems to be a continuity in the surveyors' tradition.

So, is it possible that traces of Old Babylonian surveyor techniques survived up to the seventeenth century? Recent scholarship in the history of mathematics gives support to such a remarkable relationship. Firstly, there is Book II of Euclid's *Elements*. Following the geometrical interpretation of Old Babylonian algebra, Friberg goes quite far in claiming a close relationship.[35] He meticulously shows how propositions II.2–3, II.4,7, II.5–6, II.8, II.9–10, II.11,14, and II.12–13 correspond to surveyor techniques found on Old Babylonian tablets. In his view, these propositions by Euclid are a reformulation of Old Babylonian geometrical models but in abstract terms, without the use of numbers. Cardinael's methods and techniques are obviously related to Euclid's *Elements*,[36] but in spirit they may be even closer to surveyor formulations of Old Babylonian mathematics, including the use of concrete numbers and the reference to tangible operations. Take a typical example such as Q35, where Cardinael uses the terms 'to halve' (*halveer*), 'to cut off' (*neemt weg*) and 'to paste' (*addeer*).

Apart from an influence through the scholarly tradition, passed on through authoritative texts such as from Euclid, there may have been an important but neglected influence by the oral tradition, passed down from master surveyors to their apprentices. The enigma of the origin of Arabic algebra at the beginning of the ninth century can be partially explained by the Old Babylonian heritage of surveyors' practices. The myth that Arabic algebra descended from the *Arithmetica* of Diophantus, created by fifteenth-century humanists, is currently unsustainable.[37] The claim that Arabic algebra originated from India, as did the Arabic numerals, was already refuted more than a century ago by Léon Rodet.[38] With Ancient Greece and India crossed off the list, few options remained, until the Dutch historian Hubertus Busard noted a close correspondence between a Mediaeval Latin text

[35] Jöran Friberg, *Amazing Traces of a Babylonian Origin in Greek Mathematics*, Singapore: World Scientific Press 2007.

[36] As shown by Sitters and Heeffer, *Hundred Geometrical Questions*, §3.1.5.

[37] Albrecht Heeffer, 'On the Nature and Origin of Algebraic Symbolism', in: *New Perspectives on Mathematical Practices: Essays in Philosophy and History of Mathematics*, ed. B. Van Kerkhove, Singapore: World Scientific Publishing 2009, pp. 1–27.

[38] Léon Rodet, 'L'algèbre d'Al-Khārizmi et les méthodes indienne et grecque', in: *Journal asiatique* 11 (1878), pp. 5–98.

and Old Bablyonian methods.[39] The text he edited is the *Liber mensurationem*, attributed to Abū Bakr, of which the Arabic original has been lost. The first part is on *misāḥa* (surveying), and bears a close resemblance to Old Babylonian problems. Jens Høyrup further developed the theory that Old Babylonian surveyor techniques survived the decline of the Sumerian empire and were passed on through the oral tradition of lay culture. Its influence is also notable in the tradition of mediaeval works on practical geometry, of which Fibonacci's *Practica Geometriae* is the main proponent.[40]

From Fibonacci's *Practica Geometriae* to Cardinael, the line of influence becomes more direct. Practical geometry expanded rapidly during the Renaissance. While during the Middle Ages the discipline depended mostly on surveying, it soon expanded to several other areas: fortification, perspective, ballistics, gauging, navigation, cartography, hydrostatics. From the fifteenth century onwards, in Italy the traditional empirical engineers such as Taccola (Mariano di Jacopo, 1382–*c*.1453) were transformed into mathematically literate military engineers such as Buonaittu Lorini and Vittorio Zonca (1568–1603). In a well-informed study by Mario Biagioli it is argued that these engineers not only migrated upwards in social status but they also significantly advanced the role and recognition of practical mathematics in Renaissance society.[41] Also in France, military engineers firmly established themselves in the service of the dukes by the end of the sixteenth century, while publishing works on fortification, surveying, mathematical instruments, and often also their own editions of Euclid's *Elements*. Some representative figures are Joseph Boillot, Claude Flamand, Jean Errard, and Salomon de Caus. In the Low Countries the engineering tradition first flourished in Flanders through Simon Stevin (1548–1620) and Philippe van Lansberge (1561–1632).

It thus comes as no surprise that Cardinael took the pragmatic choice of solving his hundred geometrical problems without the use of the Rule of Coss. It was pedagogically and commercially sound to do so given his audience of surveyors and engineers and it fitted his personal preference for ancient Greek ruler-and-compass constructions. While Cardinael composed his book during the Twelve Years' Truce (1609–1621) of the Eighty Years' War, he even accounted for the military context by including problems on military formation.[42]

Geometrical algebra

Since we have used the term 'geometrical algebra' for many of the techniques employed by Cardinael we should point to a long-standing dispute about the use of

[39] Hubertus L. L. Busard, 'L'algèbre au Moyen Âge: Le 'Liber mensurationum' d'Abū Bekr', in: *Journal des Savants* (1868), pp. 65–125.

[40] Barnabas Hughes, *Fibonacci's De practica geometrie*, New York: Springer 2008.

[41] Mario Biagioli, 'The Social Status of Italian Mathematicians, 1450–1600', in: *History of Science* 27 (1989), pp. 1–75.

[42] As in subgroup 3 (Q36–39, Q49–53) of Sitters and Heeffer, *Hundred Geometrical Questions*.

this term, as a disclaimer to the reader. Let us start with a notable quote from Michael S. Mahoney in his book on Fermat:

> Any historian of mathematics conscious of the perils and pitfalls of Whig history quickly discovers that the translation of past mathematics into modern symbolism and terminology represents the greatest danger of all. The symbols and terms of modern mathematics are the bearers of its concepts and methods. Their application to historical material always involves the risk of imposing on that material, a content it does not in fact possess.[43]

We are aware that our use of modern symbolism in the mathematical commentaries poses the risk of obtruding an algebraic content into the geometrical techniques employed by Cardinael. However, we believe it is beneficial to the reader to convince him- or herself of the validity of Cardinael's geometrical reasoning by means of modern symbolism. Starting from the literal translation of the Dutch text provided, a modern rendering of the mathematical reasoning, within the context of the systematic treatment of techniques, is a legitimate exercise. When using AB^2 in a mathematical commentary, this corresponds to Cardinael's somewhat misleading phrasing 'the square AB', where AB is actually the side of a square, a choice of words which again is remarkable as 'Old Babylonian mathematics names squares and their sides identically'.[44] Cut-and-paste operations on, for example, two squares can be represented by operators as in $AB^2 - CD^2$ for 'subtracting the square of *CD* from the square of *AB*' or $AB^2 + CD^2$ for 'adding the squares of *AB* and *CD* together'.

As for our use of the term 'geometrical algebra' it may be appropriate to give readers some background on an important controversy in the history of mathematics which started some forty years ago.

In 1975, Sabetai Unguru started the polemic in a long article in the prestigious journal *Archive for History of Exact Sciences*. Unguru is a Romanian classicist and historian of mathematics who migrated to Israel, and is best known for his edition of the *Conics* by Apollonius. Starting with a critical assessment of several claims by van der Waerden in his *Science Awakening*, he considers geometrical algebra 'a monstrous, hybrid creature, a contradiction in terms, a logical impossibility' and argues that there is no algebra in Euclid's *Elements*.[45]

During the following years, three respected mathematicians with a longtime interest in the history of mathematics responded furiously to Unguru's article. The first, B. L. van der Waerden, responded by demonstrating that mathematicians can easily see and interpret an algebraic basis in Euclid's *Elements*.[46]

Obviously, this is a fact Unguru would not deny. His objection was that such interpretation is historically not correct. By defining algebra as 'the art of handling

[43] Michael S. Mahoney, *The Mathematical Career of Pierre de Fermat (1601–1665)*, Princeton, NJ: Princeton University Press 1973, pp. xii–xiii.

[44] Robson, *Mathematics in Ancient Iraq*, p. 115.

[45] Sabetai Unguru, 'On the Need to Rewrite the History of Greek Mathematics', in: *Archive for History of Exact Sciences* 15 (1975), pp. 67–114 at p. 77.

[46] B. L. van der Waerden, 'Defense of a "Shocking" Point of View', in: *Archive for History of Exact Sciences* 15 (1976), pp. 199–210.

algebraic expressions like $(a + b)^2$ and solving equations like $x^2 + ax = b$', van der Waerden did not serve his case very well. What mathematicians see as equations are often not symbolic equations at all, as illustrated by the example of Old Babylonian algebra in the previous section.

Hans Freudenthal also responded in the same journal, claiming that '*Elements* V is algebra and nothing else'.[47] The third mathematician to respond was André Weil, in a short note asking: 'let us stop the disease before it gets fatal'.[48] Unguru did not succeed in getting a response to his critics published in the same journal, but it was accepted by *Isis*.[49] By the end of the seventies, it became widely accepted that an algebraic interpretation of Euclid, which started during the Arabic period and was pursued by Renaissance mathematicians, is historically unsound. Historians of mathematics should especially be careful when using symbolic representations of geometrical operations from the *Elements*. Let us discuss one subtle example.

Euclid's *Elements*, Book IX, Proposition 21 states that 'if any multitude whatsoever of even numbers is added together, then the whole is even'. The proof is very simple and does not depend on any other propositions, only on the definition of number. A number is even when it has a half part. Given numbers *AB*, *BC*, *CD*, and *DE* (shown by Euclid as line segments) being even, the whole *AE* (the concatenation of line segments) must also be even since it has a half part, being the sum of the half parts of the original numbers. A symbolic rendering of the reasoning could be: take an arbitrarily long finite sum of even numbers $2a + 2b + 2c + ... + 2n$. This can be rewritten as $2(a + b + c + ... + n)$ and is obviously even. However convincing the mathematical reasoning here is, the proof differs from Euclid's original. In a second attempt, we could employ the definition of an even number in our symbolic representation and write an even number a as $c + c$, thus exemplifying that it has a half part c. Suppose we have two even numbers $a = c + c$ and $b = d + d$, whose sum is $a + b = (c + c) + (d + d)$. This we can rewrite as $a + b = (c + d) + (c + d)$, showing that the sum consists of two equal parts, which makes it even. Again, this proof is different from Euclid's original. In the two examples we have either used the commutativity or the associativity of addition, while Euclid's proof depends only on the definition of even numbers. Thus by representing the original proof in a symbolic way we have made it a different proof, sound but depending on valid truths other than the original one.

After this historiographical contextualization we can now address the question of what geometrical algebra could entail. The criticism of Unguru was directed towards the unjustified interpretation of geometrical concepts, operations, and proofs in ancient Greek mathematics as algebra and more specifically symbolic algebra. However, if we understand algebra as an analytic problem-solving method for arithmetical or geometrical problems which uses some abstract representation of unknown values to reason, then geometrical algebra makes sense. Symbolic algebra depends

[47] Hans Freudenthal, 'What Is Algebra and What Has Been Its History?', in: *Archive for History of Exact Sciences* 16 (1977), pp. 189–200.

[48] André Weil, 'Who Betrayed Euclid?', in: *Archive for History of Exact Sciences* 19 (1978), pp. 91–93.

[49] Sabetai Unguru, 'History of Ancient Mathematics: Some Reflections on the State of the Art', in: *Isis* 70 (1979), pp. 555–65

on a symbolic model for representations.[50] Symbolism is a script, which delegates the cognitive load of problem solving to an external environment, thus facilitating our reasoning. An alternative model is the geometrical model. Difficult arithmetical problems can be solved algebraically by using a cut-and-paste geometry as in Old Babylonian algebra and as Cardinael also did. As traced in a recent publication by Jens Høyrup, the concept of geometrical algebra was introduced by Hans Georg Zeuthen, not Paul Tannery as often assumed.[51] Høyrup points out that Zeuthen's original conception of geometrical algebra was that of 'a tool for visualization of these as well as other operations in the form of the geometrical representation and handling of general magnitudes and the operations to be undertaken with them'. This original conceptualization of geometrical algebra is close to what we have described as a geometrical model for algebra, parallel to the symbolic model. So, Unguru's criticism should be narrowed down to symbolic translations of ancient Greek mathematics. There was a geometrical algebra practised by Old Babylonian scribes and it apparently survived in the surveyors' tradition to the time of Cardinael.

[50] Albrecht Heeffer, 'The Symbolic Model for Algebra: Functions and Mechanisms', in: *Model-Based Reasoning in Science and Technology, Abduction, Logic, and Computational Discovery*, ed. Lorenzo Magnani, Walter Carnielli, and Claudio Pizzi, Studies in Computational Intelligence 314, Heidelberg: Springer 2010, pp. 519–32.

[51] Jens Høyrup, 'What is "geometric algebra," and what has it been in historiography?', Contribution to the session Histoire de l'historiographie de l'algébre, Séminaire Histoire et Philosophie des Mathématiques, CNRS, SPHERE, & projet ERC SAW, Université Paris Diderot, 11 January 2016, in: *AIMS Mathematics* 2/1 (2017), pp. 128–60.

15
The Life Mathematick
John and Euclid Speidell, and the Centrality of Instruments in Seventeenth-Century Pedagogy

Boris Jardine

Introduction

In one of his briefer 'Lives',[1] John Aubrey noted that 'Mr [John] Speidell [...] taught Mathematiques in London, and published a booke in 4to, named Speidel's Geometrical Extraction. London. 163[...] which made young men have a love to Geometrie'.[2] Our own knowledge of John Speidell—and his aptly named son Euclid—extends only a little way beyond these few lines. Father and son both published enough to provide us with some prefatory gleanings, and a handful of other sources give us a sense of the Speidells' connections and interests. John Speidell wrote on arithmetic, geometry and logarithms, and is known to have taught mathematics in London—privately on Drury Lane and Queen Street, and also at Sir Francis Kynaston's short-lived academy, the 'Musaeum Minervae'.[3] Euclid Speidell edited and republished some of his father's work, taught mathematics at his house on Threadneedle Street, and spent his later years as a customs officer. Yet as with so many of the early modern mathematical teachers who operated 'beyond the learned academy'—especially in London—we know virtually nothing of the circumstances, motivations, and personalities of this mathematical family.

Recently, however, an extensive biographical treatment of both John and Euclid Speidell, written by the latter in 1689, has been identified in the Monson Papers at Lincolnshire Archives (Figure 15.1).[4] This bears the descriptive title 'The Life of John

[1] On the naming of Aubrey's project see John Aubrey, *Brief Lives, with An Apparatus for the Lives of our English Mathematical Writers*, ed. Kate Bennett, 2 vols, Oxford: Oxford University Press 2015, vol. 1, p. xci.

[2] Ibid., p. 752.

[3] Both men are given brief treatments in E. G. R. Taylor, *The Mathematical Practitioners of Tudor and Stuart England*, Cambridge: Cambridge University Press 1970 (first edn 1954), pp. 195 (John Speidell), 250 (Euclid Speidell). A detailed treatment of John Speidell's contribution to the development of logarithms is to be found in Florian Cajori, *A History of Elementary Mathematics, with Hints on Methods of Teaching*, New York: Macmillan 1950 (first edn 1896), pp. 164–5.

[4] Lincoln, Lincolnshire Archives, Monson Papers, classmark: MON 7/21. Hereafter 'The Life of John Speidell [...] and of his Son Euclid'. The manuscript is cited, though is not discussed and was apparently not consulted, in Kate Bennett's magisterial edition of Aubrey's *Brief Lives*, vol. 2, p. 1716. I first chanced upon the MS in 2014 while using the National Archives search engine; since then Benjamin Wardhaugh has made excellent use of the manuscript in his chapter on Euclid Speidell in *The Book of Wonders: The Many Lives of Euclid's Elements*, Glasgow: HarperCollins 2020, pp. 225–9. I discuss the provenance and possible circumstances of composition in my conclusion below.

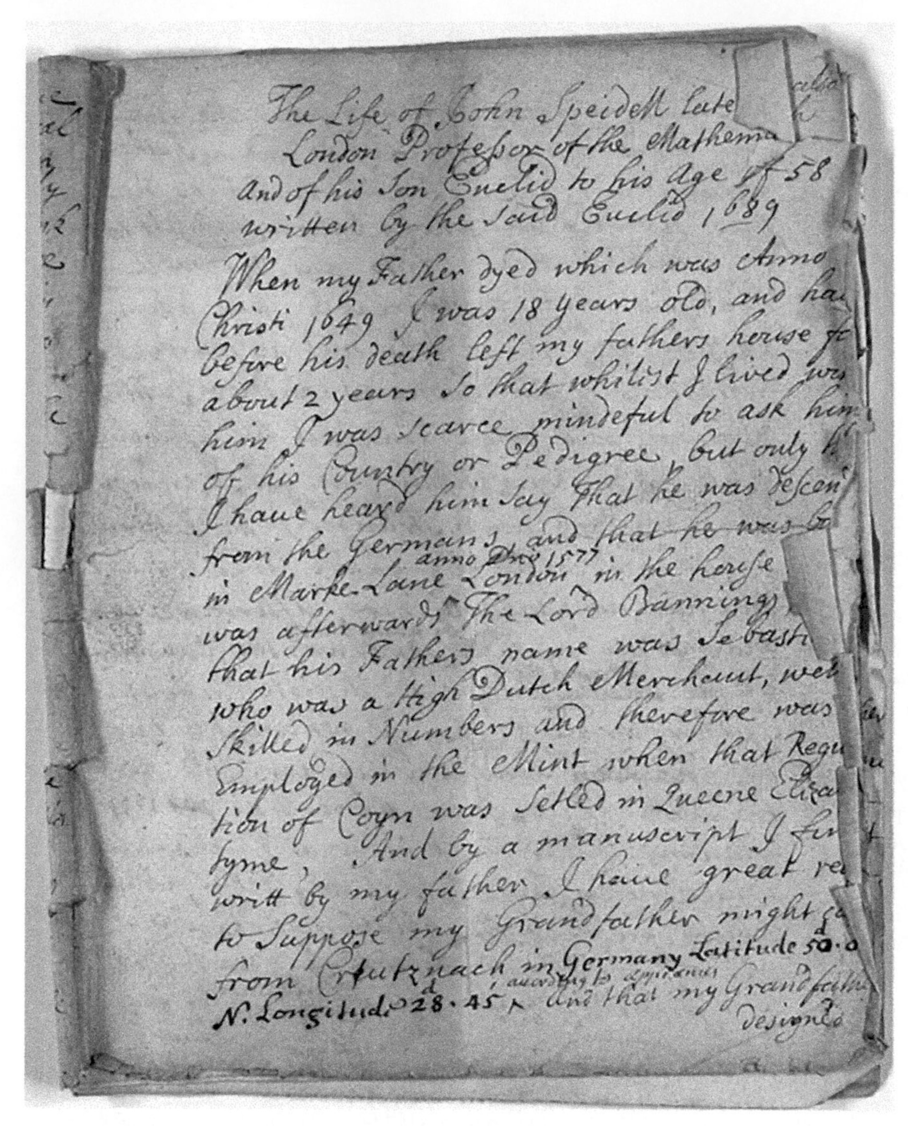
The Life of John Speidell late
London Professor of the Mathema
And of his Son Euclid to his Age of 58
written by the said Euclid 1689
When my Father dyed which was Anno
Christi 1649 I was 18 years old, and ha
before his death left my fathers house fo
about 2 years So that whilst I lived wi
him I was scarce mindeful to ask him
of his Country or Pedigree, but only
I haue heard him Say That he was descen
from the Germans, and that he was bo
anno Dni 1577
in Marke-Lane London in the house
was afterwards The Lord Bannings
that his Fathers name was Sebasti
who was a High Dutch Merchant, wel
Skilled in Numbers and therefore was
Employed in the Mint when that Regu
tion of Coyn was Setled in Queene Eliza
tyme, And by a manuscript I fin
writt by my father I haue great re
to Suppose my Grandfather might c
from Creutznach in Germany Latitude 50.0
N. Longitude 28.45 according to Appianus And that my Grandfath
designed

Figure 15.1 First page of Euclid Speidell's memoir.

Speidell late[...] / London Professor of the Mathema[ticks] / and of his Son Euclid to his Age of 58 / written by the said Euclid 1689', and consists in 16 pages of text, though the narrative is evidently unfinished. Although obscure in places, this is a highly revealing document, covering an enormous span of time: light is shed on three generations of the Speidell family, and much information is supplied about the nature of teaching and learning mathematics for almost a century: John Speidell was born in

1577 and was active from 1607; Euclid died in 1702, and edited his last mathematical work in 1698.

In Euclid Speidell's manuscript (hereafter 'The Life'), his father John Speidell is revealed to have been a pious, unworldly character, chiefly concerned with music and mathematics; Euclid was more cosmopolitan and adventurous, spending his best years at sea. In addition to the plentiful biographical data that the manuscript furnishes, its major contribution is the texture it lends to our understanding of early modern mathematical careers. As Hester Higton comments, private teachers are 'one of the most elusive subsets of the mathematical culture'.[5] More generally, questions of 'higher' or technical education in this period have been hotly debated ever since Lawrence Stone's 1964 article positing an 'educational revolution in England, 1560–1640'.[6] The case of the Speidells makes clear that a large amount of mathematical instruction was concerned with instruments. This apparently straightforward statement has not, I argue, been fully understood: instruments were, for perhaps the majority of teachers and their pupils, the beginning and end of mathematical instruction—a way into the subject and also its conclusive application to practical problems.[7] Historical and contemporary debates over the 'correct' order of mathematical instruction have obscured the absolute centrality of instruments to instruction in mathematics, as well as to the shaping of mathematical careers and uses of learning. This naturally goes beyond the limited case of John and Euclid Speidell. But then 'The Life' is not limited to these two men. Incidental biographical information also calls into question the priority of two of the most important mathematical instruments of the era: the 'Gunter' sector and the plain scale. This takes us back to the beginning of public mathematical instruction in England, in the closing years of Elizabeth's reign, and also forward to Euclid Speidell's collaborations in the 1650s and '60s. In all of the examples considered here—John Goodwyn and the sector, John Speidell and the plain scale, Euclid Speidell and his connection to Henry Sutton—instruments and instrumental practice are at the heart of mathematical teaching.

[5] Hester Higton, 'Elias Allen and the Role of Instruments in Shaping the Mathematical Culture of Seventeenth-Century England', unpublished PhD thesis, University of Cambridge 1996, p. 25.

[6] Lawrence Stone, 'The Educational Revolution in England, 1560–1640', in: *Past & Present* 28 (1964), pp. 41–80. For mathematical teaching, focusing on London and mathematical practice, see Stephen Johnston, 'Mathematical Practitioners and Instruments in Elizabethan England', in: *Annals of Science* 48 (1991), pp. 319–44; Deborah Harkness, *The Jewel House: Elizabethan London and the Scientific Revolution*, New Haven, CT: Yale University Press 2007, ch. 3; Philip Beeley, 'Practical Mathematicians and Mathematical Practice in Later Seventeenth-Century London', in: *British Journal for the History of Science* 52 (2019), pp. 225–48; Lesley B. Cormack, 'The Commerce of Utility: Teaching Mathematical Geography in Early Modern England', in: *Science & Education* 15 (2006), pp. 305–22. For mathematics at the universities see Mordechai Feingold, *The Mathematicians' Apprenticeship: Science, Universities and Society in England, 1560–1640*, Cambridge: Cambridge University Press 1984. For numeracy more generally compare Keith Thomas, 'Numeracy in Early Modern England: The Prothero Lecture', in: *Transactions of the Royal Historical Society* 37 (1987), pp. 103–32, and Jessica Otis, '"Set Them to the Cyphering Schoole": Reading, Writing, and Arithmetical Education, circa 1540–1700', in: *Journal of British Studies* 56 (2017), pp. 453–82.

[7] The case has been made, for one group, in Anthony Turner, 'Mathematical Instruments and the Education of Gentlemen', in: *Annals of Science* 30 (1973), pp. 51–88.

The Speidells: an immigrant family in Elizabethan London

The primary value of a source such as this is the biographical data it offers, and the light shed on intellectual and practical communities. Ever since the pioneering work of E. G. R. Taylor in the 1940s and '50s, we have had a sense of the mathematical networks that were formative in the development of the mathematical arts, *c.*1550–1700.[8] This is now a very well-known community of creative and technical achievement, including scholars, craftsmen, practitioners (gunners, navigators, surveyors, etc.), teachers, courtiers, gentlemen, and even nobility.[9] With Euclid Speidell's autobiography in hand we now have an excellent resource with which to analyse the claims of historians who have used and critiqued the notion of that archetypal extra-academic persona, the 'mathematical practitioner', and in particular the relationship between teaching and instrumentation.[10]

Early in 'The Life', Euclid reveals that he has 'heard [his father] say':

> that he was descend[ed] from the Germans and that he was b[orn] in Marke Lane London anno D[omi]ni 1577. [...] that his Fathers name was Sebast[ian] who was a High Dutch Merchant, we[ll] skilled in Numbers and therefore was Employed in the Mint when that Regu[la]tion of Coyn was setled in Queen Eliza[beth's] tyme[11]

This confirms what we might have guessed before: John Speidell's father was Sebastian 'Speydell' or 'Spydell', a German merchant who settled in London around 1561 and died in 1597. Sebastian Speydell was indeed involved in the recoinage of 1561, and a little later was a founder member (with Thomas Thurland) of the Company of Mines Royal.[12]

[8] Taylor, *Mathematical Practitioners*; for networks see Jacqueline Stedall, 'Tracing Mathematical Networks in Seventeenth-Century England', in: *The Oxford Handbook of the History of Mathematics*, ed. Eleanor Robson and Jacqueline Stedall, Oxford: Oxford University Press 2009, pp. 133–52; also Philip Beeley, 'Practical Mathematicians'.

[9] See, for instance, Eric H. Ash, *Power, Knowledge, and Expertise in Elizabethan England*, Baltimore, MD: Johns Hopkins University Press 2004; Jim Bennett, 'The Challenge of Practical Mathematics', in: *Science, Culture and Popular Belief in Renaissance Europe*, ed. Stephen Pumfrey, Paolo L. Rossi, and Maurice Slawinski, Manchester: Manchester University Press 1991, pp. 176–90; Stephen Johnston, 'Making Mathematical Practice: Gentlemen, Practitioners and Artisans in Elizabethan England', unpublished PhD thesis, University of Cambridge 1994; Larry Stewart, 'Other Centres of Calculation, or, Where the Royal Society Didn't Count: Commerce, Coffee-Houses and Natural Philosophy in Early Modern London', in: *British Journal for the History of Science* 32 (1999), pp. 133–53.

[10] On this relationship see Turner, 'Mathematical Instruments and the Education of Gentlemen'; Katherine Hill, '"Juglers or Schollers?" Negotiating the Role of a Mathematical Practitioner', in: *British Journal for the History of Science* 31 (1998), pp. 253–74; Hester Higton, 'Does Using an Instrument Make You Mathematical? Mathematical Practitioners of the 17th Century', in: *Endeavour* 25 (2001), pp. 18–22; Adam Mosley, 'Objects, Texts and Images in the History of Science', in: *Studies in History and Philosophy of Science* 38 (2007), pp. 289–302; Margaret E. Schotte, 'Nautical Manuals and Ships' Iinstruments, 1550–1800: Lessons in Two and Three Dimensions', in: *The Routledge Companion to Marine and Maritime Worlds 1400–1800*, ed. Claire Jowitt, Craig Lambert, and Steve Mentz, Abingdon: Routledge 2020, pp. 273–97.

[11] 'The Life of John Speidell [...] and of his Son Euclid', p. 1.

[12] For John Speydell at the Mint see M. B. Donald, *Elizabethan Monopolies: The History of the Company of Mineral and Battery Works from 1565 to 1604*, Edinburgh: Oliver and Boyd 1961, p. 27; for Speydell and the Company of Mines Royal see ibid., p. 10, and M. B. Donald, *Elizabethan Copper: The History of the Company of Mines Royal 1568–1605*, Whitehaven: Michael Moon 1989 (first edn 1955), p. 49.

Euclid writes that he has seen 'a manuscript [...] writt by my father', which reveals Sebastian had come from 'from Creutznach in Germany Latitude 50°.0 N. Longitude 028°.45', to which Euclid adds by caret 'according to Appianus'.[13] With this information we can trace the Speidells back to the Palatinate, and the town now known as Bad Kreuznach. This adds to our store of knowledge about Sebastian Speydell, who was reputed to have had a vast income from his trading activities. Euclid's manuscript reveals, however, that in spite of this the family experienced fluctuating fortunes:

> my Grandfather designed my Father to have beene a Merchant but what with great Losses and my Fathers [...] Extraordinary Love to Musick & Mathematicks he was diverted and wholly betooke himself to the Mathematical Arts[14]

It is not clear from Euclid's phrasing when the family's 'great Losses' occurred. He might mean that his grandfather Sebastian died in reduced circumstances, but there is a strong suggestion that it was his father John who had lost the family fortune.[15]

The next piece of biographical information supplied by Euclid about his father is that

> about 28 years of his age (he told me) [John Speidell] marryed with one John Goodwins widdow, which Goodwin had beene Eminent in Educating youth in making them fitt for Merchants & Trades[16]

This apparently incidental detail in fact has significance for our consideration not only of the Speidells but of mathematical culture more generally, because it reveals, for the first time, that the teacher John Goodwyn died in or before 1605.[17] It is possible that John Speidell had been a pupil of Goodwyn's—either way the period following Sebastian's death in 1597 must have been occupied with mathematical study, because we know that John Speidell was teaching in the City of London from 1607, around the time he also invented a novel kind of divided scale, to which I return below.[18]

Whether or not Goodwyn was Speidell's master, it is worth taking a brief diversion to consider the significance of Goodwyn's date of death, which was previously unknown and bears on an important question of priority in the invention of the instrument that came to be known as the 'Gunter sector'. This also sets the stage chronologically for the establishment of John Speidell as a 'professor of Mathematicks' in early seventeenth-century London.

[13] 'The Life of John Speidell [...] and of his Son Euclid', p. 1.

[14] Ibid., p. 2.

[15] Sebastian Speydell's will survives but only specifies the division of his estate, not the specific amounts. See TNA PROB/11/89/131.

[16] 'The Life of John Speidell [...] and of his Son Euclid', p. 2.

[17] Recalling that John Speidell was born in 1577, we can see that 'about 28 years of his age' must date John's marriage to Goodwyn's widow to *c.*1605. Goodwyn's biography, as we will see, requires emendation, but see Taylor, *Mathematical Practitioners*, p. 194, and Gerard L'E. Turner, *Elizabethan Instrument Makers: The Origins of the London Trade in Precision Instrument Making*, Oxford: Oxford University Press 2000, p. 72.

[18] For the 1607 origin of Speidell's teaching see John Speidell, *A Geometricall Extraction, or a compendious collection of the chiefe and choyse problemes*, London: Edward Allde 1617, sig. A4r–v.

The 'Goodwyn sector'?

Goodwyn is first heard of in 1597, when he was recommended by William Barlow in the latter's *Navigator's Supply*. Barlow tells us that Goodwyn is

> A man unskilfull in the Lattin tongue, yet having proper knowledge in Arithmetike, and Land-measuring, in the use of the Globe, and sundry other Instruments: And hath obteined, partly by his own industrie, and by reading of English Writers (whereof there are many very good) and partly with conference with learned men, (of which hee is passing desirous) such ready knowledge and dexteritie of teaching and practising the groundes of those Artes, as (giving him but his due) I have not been acquainted with his like.[19]

There is plenty here to consider. Barlow's book was an attempt to promote mathematical techniques and the use of instruments in navigation; as he says, by the end of the sixteenth century there were some (if not perhaps 'many') books in English that could furnish an education in this sort, but only one teacher known to him, namely Goodwyn. He goes on to lament the lapse of the London mathematical lectureship previously held by Thomas Hood ('a learned man of sufficiencie answerable of very honest and courteous behaviour, affable to resolve beginners of their doubts'); and in this way the *Navigator's Supply* can be placed alongside contemporary proposals for institutional support for the mathematical arts, and in particular their application to navigation.[20] Only a year later, Gresham College itself was to be established in London: the 1590s were therefore a decisive decade in the provision of education in the mathematical arts.[21]

As for Goodwyn himself, the use of instruments is prominent in his qualifications, as befits a book illustrated by the premier instrument maker of the 1590s, Charles Whitwell.[22] An advertisement on the title-page is even more explicit on the matter:

[19] William Barlow, *The Navigators Supply: conteining many things of principall importance belonging to Navigation, with the description and use of diverse Instruments framed chiefly for that purpose*, London: G. Bishop, R. Newbery, and R. Barker 1597, sig. K2.

[20] In 1588 Thomas Hood had been appointed Mathematical Lecturer of the City of London, and a year after Barlow's book, in 1598, Gresham College was established. Prior to both, Humfrey Gilbert had proposed a Royal Academy, in which two professors were to teach arithmetic and geometry, with their applications in astronomy, navigation, and military matters. For Hood see Francis R. Johnson, 'Thomas Hood's Inaugural Address as Mathematical Lecturer of the City of London (1588)', in: *Journal of the History of Ideas* 3 (1942), pp. 94–106, and Johnston, 'Mathematical Practitioners and Instruments in Elizabethan England', esp. pp. 330ff. For Gilbert and his proposals see Henry Ellis, 'Copy of a Plan proposed to Queen Elizabeth by Sir Humphry Gilbert, for instituting a London Academy', in: *Archaeologia* 21 (1827), pp. 506–20; David W. Waters, *The Art of Navigation in England in Elizabethan and Early Stuart Times*, London: Hollis and Carter 1958, p. 243.

[21] For Gresham College (in relation to mathematics and instrumentation) see Allan Chapman, 'Gresham College: Scientific Instruments and the Advancement of Useful Knowledge in Seventeenth-Century England', in: *Bulletin of the Scientific Instrument Society* 56 (1998), pp. 6–13; Allan Chapman, 'No Small Force: Natural Philosophy and Mathematics in Thomas Gresham's London', in: *Sir Thomas Gresham and Gresham College: Studies in the Intellectual History of London in the Sixteenth and Seventeenth Centuries*, ed. Francis Ames-Lewis, Aldershot: Ashgate 1999, pp. 146–73.

[22] On Whitwell (and his engravings for Barlow's book) see Turner, *Elizabethan Instrument Makers*, pp. 29–31 and 40–1.

> If any man desire more ample instructions concerninge the use of these instruments, hee may repayre unto Jhon [*sic*] Goodwin dwellinge in Bucklerburye teacher of the growndes of these artes.

Here we have evidence that—in addition to Thomas Hood's well known reliance on instruments in his teaching—John Goodwyn built at least part of his reputation around instruction in the devices illustrated by Whitwell in Barlow's book. The writer on surveying Arthur Hopton thought that Goodwyn had been the inventor of the circumferentor, or surveyors compass—an attribution that cannot be correct but suggests that Goodwyn designed or improved instruments in addition to teaching their use.[23] And Goodwyn may well have been tutor to Hopton and Aaron Rathborne; he was certainly tutor to the mathematical practitioner John Reynolds.[24] All of these men were involved in the development and promotion of instruments in their respective domains.

Somewhat surprisingly, learning the date of Goodwyn's death sheds light on his relationship with the instrument trade, and in particular the development of the calculating instrument called the sector. This is an ingenious device, simultaneously developed by a number of scholars and craftsmen in the last years of the sixteenth century. The sector consists in a pair of identical scales or sets of scales, hinged together at one end, and it uses the principle of similar triangles to turn complex operations into a simple matter of transposition using a pair of dividers. The basic design of the sector, using just one doubled scale, was formulated around the same time by Thomas Hood, Galileo Galilei, and, possibly, Michel Coignet.[25] But the addition of numerous other scales, extending the instrument into a wide range of disciplines, is usually attributed to Edmund Gunter, who published his form of the instrument in 1623 but claimed to have invented it in 1607.[26]

The transition from Hood's sector to 'Gunter's' is as much about the use of instruments as it is about their underlying principles. In Hood's original treatise on the instrument almost all of the 'uses' concern the proportion of lines, angles, and areas—which is to say that the sector is for use within geometry, and its practical applications are to the drawing of plans, shapes, and diagrams, primarily in land surveying. Gunter's treatment is entirely different: his aim is to use the sector to find numerical values for proportions. The first 'use' of the instrument, for example, is given as follows:

> let there be given three lines A, B, C, to which I am to find a fourth proportionall. [L]et A, measured in the line of *lines* be 40, B 50, and C 60, and suppose the question be this. If 40 *Monthes* give 50 *pounds*, what shal 60?[27]

[23] See Taylor, *Mathematical Practitioners*, p. 98.

[24] See Boris Jardine, 'Henry Sutton's Collaboration with John Reynolds (Gauger, Assayer and Clerk at the Royal Mint)', in: *Bulletin of the Scientific Instrument Society* 130 (2016), pp. 4–7 at p. 5; Norman Biggs, 'John Reynolds of the Mint: A Mathematician in the Service of King and Commonwealth', in: *Historia Mathematica* 48 (2019), pp. 1–28.

[25] See Turner, *Elizabethan Instrument Makers*, pp. 70–2 for a discussion of the claimants and their cases.

[26] Edmund Gunter, *The Description and Use of the Sector*, London: William Jones 1623, p. 143.

[27] Ibid., p. 11.

In addition to this change in emphasis—from proportion to number—Gunter added specific scales that allowed numerical calculations in a range of disciplines, most notably navigation. Naturally, as the principle on which the instrument is based is the same, many uses and propositions overlap between Hood and Gunter. Yet the shift in emphasis is clear: Hood's instrument emerges from geometrical practice; Gunter's is designed to solve problems in arithmetic.[28]

This is an immensely significant development. Mathematics in this period was, outside of the old universities, almost entirely a practical matter. As Higton has shown, for example, the vast majority of mathematical books published in the seventeenth century were concerned with practical applications, especially in astronomy, navigation, and surveying.[29] The role of instruments in these subjects is largely self-evident. Hood's sector had effectively made geometry into an instrumental practice. Now, with Gunter's version of the sector, arithmetic could follow.

Recall that Gunter's design dates (on his own testimony) from 1607, and that Goodwyn was dead by 1605. How, then, can it be possible that amongst the surviving corpus of early sectors there are three made to Gunter's design inscribed 'John Goodwin Scu[lpsit]'?[30] Until now these (Figure 15.2) have seemed merely a curiosity. They were at one point thought to have been made by Goodwyn himself, but Gerard Turner has correctly identified the engraving style as that of Elias Allen, who was

Figure 15.2 Brass sector by Elias Allen, 1623, made to design of Edmund Gunter.

[28] On the relationship between geometry and number in this period see the insightful comments in Higton, *Elias Allen and the Role of Instruments*, pp. 125–7.

[29] Ibid., p. 12 *et seq.*

[30] These are described in Turner, *Elizabethan Instrument Makers*, pp. 280–1.

apprentice and successor to Charles Whitwell, and was made free of the Worshipful Company of Grocers in 1612.[31] Turner supposed that these three sectors must have been made *for* Goodwyn.[32] But Euclid Speidell's manuscript reveals that Goodwyn was dead long before this was even a possibility.

Now, the 'Goodwyn' sectors are in fact *identical* with the design published by Gunter in 1623, and made by Allen from that point on.[33] But as we have seen Goodwyn was already deceased before Gunter's initial design of the instrument (*c.*1607). So what we appear to have is the posthumous attribution of Gunter's design to Goodwyn, by Elias Allen—the very artisan charged with making the instrument by Gunter. This tangled web of teachers, scholars, and artisans must also include Charles Whitwell, Allen's master, who collaborated with Goodwyn and made a pair of 'transitional' sectors that are further developed than Hood's, less well advanced than those put into print by Gunter in 1623.[34]

In light of previous interest in questions of priority and the sector, it is tempting to argue that Goodwyn first applied the sector to the wide range of uses published by Gunter in 1623, and that we should rightly call the 'Gunter sector' the 'Goodwyn sector'. However, this is not my aim here. Instead, I will make the weaker but more broadly significant argument that Goodwyn also taught the use of the sector, and adapted Hood's design in the same way as Gunter, i.e. to shift emphasis from geometry to arithmetic, and from theory to practice. From Gunter we already know that the sector was being used in this way from 1607, because he tells us that his manuscript was in demand from that date. But the connection to Goodwyn takes us back a little further, and also extends the audience significantly. The circulation of Gunter's text was presumably limited by personal contact and hearsay; it was also written in Latin; Goodwyn was offering his services as a teacher, in the vernacular, to anyone who could visit him at his house in the City of London, at Bucklersbury.

This brings us back to the lives and careers of John and Euclid Speidell, because like Goodwyn and Gunter they made instruments central to their teaching—specifically instruments designed to aid calculation and therefore address the era's well-known lack of general numeracy.[35]

An 'Extraordinary Love to Musick & Mathematicks'

This diversion into the life and times of John Goodwyn is also useful as it sets the stage for John Speidell's debut as a mathematical teacher in London, as part of the generation immediately following Hood and Goodwyn. In the 1590s, these men were

[31] For Allen see Higton, *Elias Allen and the Role of Instruments.*

[32] Turner, *Elizabethan Instrument Makers*, p. 72.

[33] Most of Allen's sectors date from around this time—probably including the 'Goodwyn' sectors.

[34] See Turner, *Elizabethan Instrument Makers*, pp. 222–3 (the Whitwell sectors described here are held at the Science Museum, London). See also Higton, *Elias Allen and the Role of Instruments*, pp. 195–7, for a 'transitional' Allen sector at the National Museum of Scotland; also Whipple Museum inv. nos. 6643 and 6654 (these instruments are unpublished but can be seen at, respectively, https://collections.whipplemuseum.cam.ac.uk/objects/15396 and https://collections.whipplemuseum.cam.ac.uk/objects/15407).

[35] See Thomas, *Numeracy in Early Modern England.*

virtually alone in offering tuition; but by the second decade of the seventeenth century the astronomical author G. Gilden could write that

> if any Noble spirite bee desirous to adorne his understanding with the rich ornament of Mathematicall knowledge [...] never were there better or nearer helpes to attaine thereto, then at present, in this Cittie, by the Methodicall instructions of very many learned Professours[36]

From the preface to John Speidell's 1617 book *A Geometricall Extraction* we know that he began teaching mathematics in the City of London in 1607,[37] so he was clearly part of the change that had occurred between the 1590s and 1610s. Around the same time Speidell also invented a calculating scale. Unfortunately we lack a description of the scales on this instrument, and no example survives. All that we know of it comes from the same 1617 preface, in which Speidell writes that

> not only these Problems contained in this booke [...] but much more viz. in Arithmeticke, Geometrie, Astronomie, Navigation, Surveighing, fortification, architecture, taking of heights & distances, and all other parts of the Mathematicks, &c. may be performed by a Mathematicall Scale[,] now newly (this present yeare) by me invented, farre beyond my former scale made in Anno. 1607[,] the which with all other Mathematicall instruments, are made by my loving friends Mr Elias Allen, over against St. Clements Church in the Strand, in Brasse; and Mr. John Tomson in Hosier lane by Smithfield (in Wood) and may also both in Wood and Brass be had, with the instructions thereof by me at my house[38]

Just like the inscription 'John Goodwin Scu[lpsit]' on sectors made after Goodwyn's death, this statement is hard to interpret, and raises questions of priority. It has led to the belief that Speidell should be called the inventor of the navigator's 'plain scale', first described in print in English in 1624 by John Apsley.[39] Yet as we can see navigation is only one of the many applications of Speidell's instrument, which should properly be termed a 'universal' instrument, following period usage.[40] From what we know about the development of multi-function rules with engraved scales in this period, it would be more appropriate to invoke the name of Edmund Gunter again, this time as the supposed inventor of the 'Gunter scale', a rule capable of precisely the range of functions Speidell claimed for his own.[41]

[36] G. Gilden, *A Prognostication, Published for this yeere of our Lord and Saviour Jesus Christ. 1616*, London: [The Worshipful Company of Stationers] 1616, sig. B2r.

[37] John Speidell, *A Geometricall Extraction*, sig. A4r–v.

[38] Ibid.

[39] See, for example, Waters, *The Art of Navigation*, pp. 445–6.

[40] See Boris Jardine, 'On Being Compendious: Universal Instruments and the Mathematical Arts', in Fabian Kraemer and Itay Sapir (eds), *Coping with Copia: Epistemological Excess in Early Modern Art and Science*, Amsterdam: Amsterdam University Press, forthcoming.

[41] See Otto van Poelje, 'Gunter Rules in Navigation', in: *Journal of the Oughtred Society* 13 (2004), pp. 11–22.

Further evidence for this comes a little further on in Euclid's memoir, when he writes that his father invented

> a Rule for the Easy & speedy measuring of Board Timber & Stone, which was long before either Partridge or Lybourn published theirs[42]

This must refer to Partridge's important 1661 book *The Description and Use of an Instrument, Called the Double Scale of Proportion*—one of the earliest printed descriptions of the instrument that came to be known as the slide-rule—and to Leybourn's 1667 book on the 'line of proportion', i.e. the 'Gunter Scale'.[43]

Here we can draw the same conclusion as we did with Goodwyn and the sector. Speidell can confidently be claimed as one of those who innovated in the application of numerical functions to engraved scales. But neither he nor Gunter can really be called the 'inventor' of this technique, which was known in the late sixteenth century. Two general antecedents for divided scales of the kind advertised by Speidell are the printed tables of trigonometric functions and the workman's engraved scales (for example for measures of timber or board) that both became common in the sixteenth century.[44] By the turn of the seventeenth century the notion of representing numerical functions in non-linear engraved scales appears to have been common. A particularly important instrument in this connection is the 'reigle platte', developed at Antwerp by Michel Coignet:[45] certainly the reigle platte can perform the functions Speidell claims of his own rule—and only Gunter's later application of a logarithmic line takes us beyond these early instruments. Given Speidell's interest in logarithms it is likely that he, too, applied logarithmic scales to his own instruments soon after Napier's publication of the technique in 1614.

As we have seen with Goodwyn and the sector, questions about the development of instruments in this period are complicated by a number of factors. Old instruments were described as 'new' inventions; existing forms of instruments could be augmented or adapted, sometimes resulting in entirely new devices. The history of instruments with divided scales is especially complex, because the instruments themselves often circulated either with manuscript descriptions or the promise of verbal explanation; moreover, many of these instruments were made of wood or even paper, and survival rates are low.[46]

[42] 'The Life of John Speidell [...] and of his Son Euclid', p. 4.

[43] Seth Partridge, *The Description and Use of an Instrument, Called the Double Scale of Proportion*, London: William Leybourn and William Wright 1661; William Leybourn, *The Line of Proportion, Commonly Called Gunter's Line, Made Easie*, London: J.S. and G. Sawbridge 1667.

[44] See Otto van Poelje, 'Gunter Rules in Navigation'; Stephen Johnston, 'The Carpenter's Rule: Instruments, Practitioners and Artisans in 16th-Century England', in: *Proceedings of the Eleventh International Scientific Instrument Symposium*, ed. G. Dragoni, A. McConnell, and G. L'E. Turner, Bologna: Grafis Edizione 1994, pp. 39–45.

[45] Ad Meskens, 'Michiel Coignet's Contribution to the Development of the Sector', in: *Annals of Science* 54 (1997), pp. 143–60.

[46] For paper instruments see David. J. Bryden, 'The Instrument Maker and the Printer: Paper Instruments Made in Seventeenth Century London', in: *Bulletin of the Scientific Instrument Society* 55 (1997), pp. 3–15.

Another factor is clearly a reticence that some teachers felt about publishing descriptions of instruments, the use of which they could explain to paying pupils.[47] John Speidell, for all of his interest in instruments, never published a single description of any of them, and in fact in one of his advertisements includes 'other secret inventions to be taught in private', alongside 'Geometry and Astronomy, by practice and demonstration, with the use of severall instruments for surveying of Lands' etc.[48] William Oughtred, for very different reasons, withheld information on the instruments he had designed, until prompted to publish by pupils, instrument makers, or in response to rival teachers.[49] By no means did all teachers conduct themselves in this way—but we must acknowledge that secrecy was just as much a marketing tool as was openness, especially for those without institutional support.

Speculations about priority and invention aside, we can see that designing and teaching the use of instruments was central to mathematical pedagogy in the early decades of the seventeenth century. Typically the question about the role of instruments in teaching has been understood in light of the controversy of the early 1630s between Richard Delamain and William Oughtred, with Delamain arguing for instruments as an easy way in to harder subjects, and Oughtred responding that instruments could distract from theoretical grounding.[50] But the evidence of the lives of men like Goodwyn and Speidell is that this is a false opposition—or at least that Oughtred's was an extreme position, not held by many of his contemporaries. Certainly there could be a prejudice against instruments amongst scholarly communities—witness the episode reported in Aubrey's *Lives* when Gunter himself was supposedly denied the Savilian Professorship in geometry for his over-reliance on instruments[51] (though here one suspects that this was a later and rhetorical comment on class and status, rather than a faithful record of Gunter's 'interview').[52] There were also, it is true, some teachers who seem not to have incorporated instruments into their lessons, in particular where these concerned arithmetic and its applications, especially in accounting.[53] But if any generalization is possible about teaching mathematics in early modern England, it is that instruments were fully and unproblematically part of the 'informal syllabus' on offer from private teachers and in the few lectures open to the public.

Evidence for this is plentiful in Euclid Speidell's memoir of himself and his father. After recounting his father's important work on logarithms—well documented elsewhere[54]—Euclid describes two more of John Speidell's instruments:

[47] For secrecy amongst mathematical practitioners see Johnston, 'Mathematical Practitioners', pp. 328ff.

[48] John Speidell, *A Briefe Treatise for the Measuring of Glass, Board, Timber, or Stone, Square or Round [...]*, London: printed by Thomas Harper, p. 164—[date incomplete], leaf following title-page.

[49] See Hill, '"Juglers or Schollers?"', esp. pp. 257–8.

[50] See Hill, '"Juglers or Schollers?"'; A. J. Turner, 'William Oughtred, Richard Delamain and the Horizontal Instrument in Seventeenth Century England', in: *Annali dell'Istituto e Museo di Storia della Scienza di Firenze* (1981) 6, pp. 99–125; Frances Willmoth, *Sir Jonas Moore: Practical Mathematics and Restoration Science*, Woodbridge: Boydell Press 1993, Chapter 2.

[51] Aubrey, *Brief Lives*, vol. 1, pp. 264–5.

[52] Willmoth, *Sir Jonas Moore*, p. 3.

[53] See Thomas, 'Numeracy in Early Modern England'.

[54] See in particular the numerous works of Florian Cajori on this topic, e.g. *A History of Elementary Mathematics*, pp. 164ff. for John Speidell.

> he also contrived a Guaging Rodd which was used like a Sector standing at a certain Angle to finde or cast up the content of a Cask, of whose use I have yet found any printed Copy[55]

Again, this instrument must remain obscure owing to the lack of printed descriptions. It may have been something like the gauging instrument Richard Collins claimed to have invented around 1660, as this was indeed used 'like a Sector'—in other words it was a large, hinged rule which could be opened to take the measure of the ends, width, and circumference of a cask, with the volume then read off the scales by means of another divided scale.[56]

That instruments were important in John Speidell's teaching is clear not just from his own testimony and that of his son, but also in the reception of his works. As Aubrey says, the *Geometricall Extraction* 'made young men have a love to Geometrie'.[57] But there is also evidence that it was understood in the context of mathematical practice and instrumentation. For example, in 1648 the otherwise unknown John Darker created an elaborate illustrated manuscript which transforms John Speidell's *Geometricall Extraction* into a full course in both abstract and applied geometry. The title page gives a full description of this new hybrid work, and visually combines John Speidell's text with Edmund Gunter's *Works*, from which the illustrations are copied (Figure 15.3).

From Euclid's comments we can glean that his father was moderately successful as a teacher and author through the first three decades of the seventeenth century. From Aubrey we know that one of John Speidell's pupils was the politician Edmund Wylde (1618–1695): Aubrey was keen to learn the method of division that Speidell had taught his friend Wylde.[58] But this is the extent of our knowledge of Speidell's private clientele.

More can be said about what must have been the pinnacle of John Speidell's career, namely his appointment as Professor of Geometry at Sir Francis Kynaston's 'Musaeum Minervae', which opened on Bedford Street, Covent Garden, in 1635.[59] Here Speidell taught geometry—including its practical applications—as part of a course of modern gentlemanly training.[60] The full seven-year course offered at the Musaeum covered an extraordinary range of subjects, including fencing, music, languages, heraldry, the study of antiquities and coins, the procedures of the common law, husbandry, dancing and deportment, riding, sculpture, and writing. The sciences were represented in both their theoretical and practical aspects: medicine

[55] 'The Life of John Speidell [...] and of his Son Euclid', p. 4.

[56] Richard Collins, *The Countrey Gauger's Vade Mecvm, or Pocket-Companion: Being Decimal Tables for the Speedy Gauging of Small Brewing Vessels [...] And Also for the Gauging of Cask in Ale or Wine Measure [...] Also the Description and Use of an Instrument for the Gauging of Small Brewing Vessels*, London: W. Godbid, M. Pitt and Anthony Owen 1677.

[57] Aubrey, *Brief Lives*, vol. 1, p. 752.

[58] Aubrey, *Brief Lives*, vol. 2, p. 1652.

[59] On the Musaeum Minervae see Cesare Cuttica, 'Sir Francis Kynaston: The Importance of the "Nation" for a 17th-century English Royalist', in: *History of European Ideas* 32 (2006), pp. 139–61; G. H. Turnbull, 'Samuel Hartlib's Connection with Sir Francis Kynaston's "Musaeum Minervae"', in: *Notes and Queries* 97 (1952), pp. 33–7. The key primary source is [Sir Francis Kynaston], *The Constitutions of the Musaeum Minervae*, London: T.P. and Thomas Spencer 1636.

[60] This account is largely derived from Cuttica, 'Sir Francis Kynaston', pp. 142ff.

Figure 15.3 Title-page of John Darker's 1648 version of John Speidell's *Geometricall Extraction*, expanded into a full treatise on geometry and instrumentation.

(physiology and anatomy), astronomy, navigation, optics, fortification, and of course geometry rounded out the course of study. It seems likely that John Speidell was well suited to the job owing to his cosmopolitan learning. Unlike many of his contemporaries who could profess knowledge of geometry and practical mathematics, John was also (according to Euclid) conversant with a number of the languages on offer at the Musaeum. The learned, pious, and practical son of a wealthy German merchant was an ideal candidate for the professorship at the nascent and Continentally oriented academy. A more personal connection between John Speidell and Francis Kynaston may have been forged through their shared love of music—I have already quoted Euclid's comment that his father had an 'Extraordinary Love to Musick', and from Hartlib we know that Kynaston had a remarkable collection of musical manuscripts and was himself 'a great lover of Music'.[61]

[61] Turnbull, 'Samuel Hartlib's Connection', p. 34, quoting Hartlib's *Ephemerides* for 1635.

The period of success, however, was to be short-lived. The Musaeum Minervae did not outlive its master, Sir Francis Kynaston, who died in 1642.[62] Given that the Musaeum is not mentioned by Euclid, it is possible that he was not aware of the details of this episode in his father's life—but the general circumstances of learning in the period are made clear in his comment that

> before the |Civil *add.*| wars broke out his Practice was chiefly among the Courtiers |For in those days The Mathematical Arts were had in more Esteeme by the Gentry *add.*| but when they left London he was forct to apply himself other ways which did not encourage him so much or prove so beneficial[63]

The 1640s are indeed murky years for our understanding of John Speidell's career. His only publication from this decade is *A Briefe Treatise for the Measuring of Glass, Board, Timber, or Stone, Square or Round*, published by Thomas Harper—though unfortunately the final number has been omitted from the date. This book marks a distinctly practical turn in Speidell's output, which may reflect a somewhat urgent need of funds—a notion that is naturally speculative but which is borne out by the prominent and rather pleading advertisement offering instruction in arithmetic for as little as 'halfe an houre'. Speidell's publisher, meanwhile, provides useful context for the struggles caused by the Civil Wars: Harper had been one of the more successful stationers of the 1630s—publishing around 17 books per year—but his output plummeted to fewer than 5 per year in the 1640s.[64] Another significant factor may have been the demise of Speidell's previous publisher Elizabeth Allde in 1640; Allde was the widow of Edward Allde (d.1628) and between them husband and wife had published all of Speidell's works up to then, beginning with his 1609 collection *Certaine Verie Necessarie and Profitable Tables*.[65]

The remainder of Euclid's comments on his father give a sense of the elder Speidell's character:

> He lived to a good old Age, and was very healthful of constitution not having shared of any bed sickness from his Infancy to his Death bed, and then he dyed of a Dropsie, which might be occasioned not only from age, but from a studious sedentary life and if I may say too much denying himself of comon society & enjoyment, for |my mother would say she *add.*| never knew him to drinck in Tavern or the Alehouse but when he could not reach home never going to morning or Evening Clubbs. He was very Religious in observing the Lords day, and though there happned in his life tyme great alterations in Religious Worship, yet he held it of absolute necessity to do the worke of the Lords day. he was not a litle Charitable to his Ability And

[62] See ibid., for an account of the demise of the Musaeum (the actual date of which is uncertain).

[63] 'The Life of John Speidell [...] and of his Son Euclid', p. 5.

[64] Matteo A. Pangallo, 'Correction to Plomer's biography of Thomas Harper', in: *Notes and Queries* 254 (2009), pp. 203–5.

[65] On Edward and Elizabeth Allde see *A Dictionary of Printers and Booksellers in England, Scotland and Ireland, and of Foreign Printers of Englis Books 1556–1640*, ed. R. B. McKerrow, London: Bibliographical Society 1910, pp. 5–6.

> though he foresaw he could not leave much for a wife & six children, yet he believed with David the Lord would provide for them and so he did Praised be his Name for it[66]

One senses in these comments that the younger man was neither so strict in his religious observance nor so sedentary in his ways, and this is borne out by Euclid's description of his own education and career at sea.

The education of Euclid Speidell (1631–1702)

Given his father's evident talents as a teacher, it is hardly surprising that Euclid's mathematical education was reasonably extensive. Early in life (up to age 13) we learn that 'my Father did [teach?] me arithmetick as far as Trade might req[uire]';[67] and Euclid adds that he was also at this point in a position to instruct the young children of the Puritan parliamentarian Sir Robert Harley (called Harlow by Euclid) in arithmetic.[68] The proximity of Harley's house to Westminster School was fortuitous for Euclid, who was, he says, spotted by the Master, Dr Richard Busby (1606–1695), a noted grammarian, linguist, and disciplinarian who also taught Wren, Hooke, and Locke.[69] Euclid's account of his meeting with Busby is intriguing:

> going through the abbey |bare headed *add.*| the then & now Master the Rever[end] Doctor Busby |being in the abbey *add.*| called me to him, and after having interrogated me of my Parents & Education, and if I then went to the Latin School, and understanding by me, I did not then, bidd me tell my father, if he would |have me proceed father in Grammar learning and *add.*| let me come to his school he would give me my Learning for nothing, which my Father accepted[70]

Busby's charity is perhaps explained by the political circumstances already mentioned: Busby was a Royalist holding on to his liberty, and, as we have seen, John Speidell's own fortunes were failing owing to the demise of the Royalist-backed Musaeum Minervae and dwindling clientele at court. Another apparently idle comment by Euclid gestures to the temper of the time: he mentions in passing that the Charing Cross was on his route to school, and that he remembers its destruction, which took place in 1643. This event must have been dramatic for Euclid, as it was personally overseen by none other than Robert Harley, whose children he was then teaching.

After leaving Westminster School in 1646, Euclid spent two years continuing his mathematical education with his father:

[66] 'The Life of John Speidell [...] and of his Son Euclid', p. 4.
[67] Ibid., p. 7.
[68] Ibid.
[69] See G. F. Russell, *Memoir of Richard Busby D.D. (1606–1695) with Some Account of Westminster School in the Seventeenth Century*, London: Lawrence and Bullen 1895.
[70] 'The Life of John Speidell [...] and of his Son Euclid', p. 7.

> when I desisted going to School my father sett me to study farther in arithmetick, and to have ready hand in Practical Geometry, espetially in his Geometrical Extraction, and to have some common uses of the Plane scale & sector and before I might leave my fathers house might to in arithmetick as far as Doubl Position, and also beene in the Field with him several tymes to take the survey of several fields, and in anno 1648, we tooke the survey of Woods at Eltham in Kent. I might learn also some thing of Instrumental Dyalling, but as to Triggonometry by Calculation either plane or Spherical I was a stranger[71]

The sequence here is instructive. Euclid had already had tuition in arithmetic, fitting him for teaching or for pursuing a trade, and now he moved on to advanced arithmetic and basic geometry. Instruments for calculation and measurement were to follow (the plane scale and sector, discussed above), along with some surveying and dialling. In 1627 John Speidell had published his *Breefe Treatise on Sphaericall Triangles*, so it is somewhat surprising to learn that Euclid left his father's teaching ignorant of trigonometry. But we should also pause to consider that having basic arithmetic and geometry and knowing the use of various instruments may have been considered sufficient education for a wide range of trades. Too often modern educational priorities are projected back onto the past, especially when it comes to the use of instruments. We are rarely willing, it seems, to accept that the use of instruments may have been an end in itself: the conclusion of a mathematical education rather than a diversion or way-point to 'deeper' learning. Note also the separation of instrumental and theoretical understanding: by using the plain scale and sector, and constructing sundials, Speidell would have been *using* trigonometry, perhaps even quite advanced spherical trigonometry—yet he considers himself a 'stranger' to the subject. This would seem to confirm the suspicion, often articulated in relation to the Oughtred/Delamain controversy, that the use of instruments was separate to the acquisition of theoretical knowledge.

The instrumental turn was taken further after Euclid began his career as a clerk:

> I was placed with a Lawyer of the Inner Temple in whose Lodgings I mett with an old Clarke very expert in those affairs but naturally having a Genious to Mathematical Arts we betooke our selves to more Demonstrable studies and he being a pretty Manuist was hugely delighted in making himself Mathematical Instruments for his own studyes & diversion[72]

But here Euclid's mathematical narrative drops off suddenly, and by 1652 (having 'a great minde to goe to sea') we board ship and set sail for what was evidently a period of great adventure for our chronicler.

This nautical section is hard to paraphrase, and also to interpret. It occupies almost the entirety of Euclid's personal narrative, and is very lively and full of incident; however it has no bearing on Euclid's mathematical career, and ends abruptly with Euclid in pursuit of yet another Dutch fleet.

[71] Ibid., pp. 8–9.
[72] Ibid., p. 9.

We now therefore leave the manuscript altogether and must complete Euclid's biography from other sources. Following the pattern established by Goodwyn and John Speidell, the next piece of evidence we have for Euclid's career is in the form of an ambiguous instrument. This is a calculating scale, perhaps not dissimilar to the one designed by his father, made by the craftsman Henry Sutton and inscribed '*In usum Euclidis Speidell Angli*' (Figure 15.4). This rule is undated, but was made before 1665, when Henry Sutton died.[73] Most likely it dates from between 1657, when the first of a group of scales with applications in dialling were made, and 1662, when Euclid Speidell finally followed in his father's footsteps and set up as a teacher of mathematics in Threadneedle Street, where he was lodger to a Virginal Maker, almost certainly Stephen Keene.[74]

Threadneedle Street was also the address of Henry Sutton, so here we have quite a concentrated craft, teaching, and entrepreneurial community. We can extend this through Speidell's other connections to form one part of the complex mathematical network that is so characteristic of seventeenth century London:[75] in the 1670s Speidell was, as he says, 'in Company' with the practitioner Michael Dary,[76] and he was also around this time close to John Collins.[77] This was good company. All three men were talented mathematicians from humble backgrounds: Euclid's we know well; Dary was by trade a tobacco-cutter, who used his self-taught mathematics to obtain patronage and the positions of exciseman and gunner in the tower;[78] Collins was, like Euclid, a clerk-turned-practitioner, who (like Euclid and Dary) was involved in excise (Accountant to the Excise Office).[79] All three men struggled to

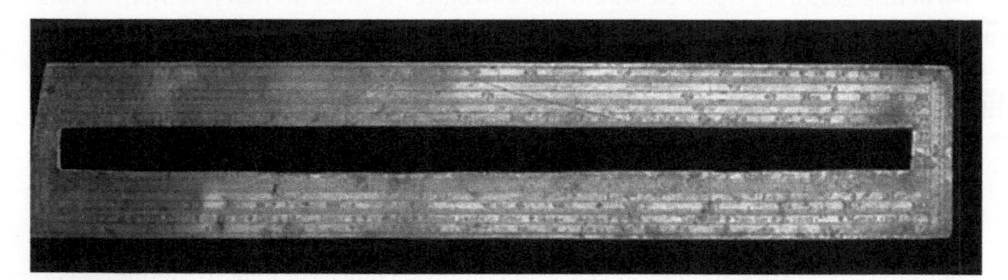

Figure 15.4 Dialling scale by Henry Sutton, inscribed '*In usum Euclidis Speidell Angli*'.

[73] For Sutton see Catherine Eagleton and Boris Jardine, 'Collections and Projections: Henry Sutton's Paper Instruments', in: *Journal of the History of Collections* 17 (2005), pp. 1–17; Jardine, 'Henry Sutton's Collaboration with John Reynolds'; Jim Bennett, '"That Incomparable Instrument Maker": The Reputation of Henry Sutton', in: *The Whipple Museum of the History of Science: Objects and Investigations*, ed. J. Nall, L. Taub, and F. Willmoth, Cambridge: Cambridge University Press 2019, pp. 83–100.

[74] On Keene see Darryl Martin, 'The English Virginal', unpublished PhD thesis, University of Edinburgh 2003.

[75] Stedall, 'Tracing Mathematical Networks'; Beeley, 'Practical Mathematicians'.

[76] Euclid Speidell, *Logarithmotechnia: Or, The Making of Numbers Called Logarithms: To Twenty Five Places, from a Geometrical Figure, with Speed, Ease and Certainty [...]*, London: Henry Clark, Euclid Speidell and Philip Lea 1688, p. 1.

[77] For Collins and Speidell see Philip Beeley, '"To the Publike Advancement": John Collins and the Promotion of Mathematical Knowledge in Restoration England', in: *BSHM Bulletin: Journal of the British Society for the History of Mathematics* 32 (2017), pp. 61–74 at pp. 73–4.

[78] See Taylor, *Mathematical Practitioners*, p. 217.

[79] Beeley, 'To the Publike Advancement'.

live from mathematics, designed and promoted instruments, and profited to varying degrees from the increasing significance of customs and excise in the latter half of the seventeenth century.

The inscription on Sutton's rule ('*In usum ...*') is presumably a form of advertisement for Euclid's services as a teacher. As with his father, inference and suggestion are necessary in assessing his teaching practice. The earliest known advertisement for his services is to be found in John Brown's 1662 *The Triangular Quadrant: Or the Quadrant on a Sector*, which introduced a multi-function astronomical instrument with particular applications to navigation. The advert for Euclid is terse but does suggest that his particular expertise was in navigational matters—and this of course would fit with his recent experiences on board ship.[80] The Sutton instrument has scales for the construction of sundials as well as for navigation, and this fits with the range of subjects covered in Brown's book. Euclid also followed his father more closely in offering tuition in arithmetic, specifically as taught in John Speidell's *Arithmetical Extraction*, which Euclid edited for a second edition in 1686. By this point, however, Euclid had been employed as a customs officer for more than a decade, and was using blank space in his book to advertise the tuition offered by Reeve Williams; we must therefore suppose that at some point in the 1670s or early 1680s he stopped offering instruction.

In many ways Euclid's career in Restoration London mirrors that of his father in the first half of the century. Both collaborated with and promoted instrument-makers; both published opportunistically; both taught a wide range of subjects and specialized in instrumentation. Perhaps the main difference between the two is precisely in the unworldly/cosmopolitan divide identified at the outset. John Spiedell was not nearly so well connected as his son Euclid was to be. The one really significant connection he made, to Sir Francis Kynaston, evidently did not result in a successful career, and by the 1640s his fortunes were on the wane. Elias Allen was an erstwhile collaborator, but more support was needed. Euclid seems to have spent far less of his career attempting to profess or even pursue mathematics, yet he ended up apparently comfortably employed as a customs officer, republishing his father's works and—if Hooke is to be believed—plagiarizing more talented mathematicians. Certainly, by the end of the century there were more opportunities for those with even a slight mathematical training, and the networks of publishers, artisans, practitioners, and scholars established at the beginning of the century were far more complex and better developed by its end.

Conclusion

Owing to its fragmentary and incomplete nature it is difficult to build strong arguments on Euclid's narrative. Rather, it offers glimpses of many aspects of mathematical teaching over an exceptionally long period. An incidental detail—the date of John Goodwyn's death—allows us to look anew at a group of enigmatic sectors, and to

[80] John Brown, *The Triangular Quadrant: Or the Quadrant on a Sector. Being a general Inftrument For Land or Sea Observations. Performing all the uses of the ordinary Sea Inftruments [...]*, London: [John Brown and Henry Sutton] 1662, p. 24.

add Goodwyn tentatively to the list of those involved with the development of the instrument in the crucial decade 1595–1605. Whatever the specifics of Goodwyn's involvement, we have more evidence that instruments were central to the teaching of mathematics, and that later writers could claim credit for inventions, while craftsmen like Allen might (for whatever reason) attempt to set the record straight. The use of instruments in teaching arithmetic is an important and under-studied development; in many ways it 'completes' the range of topics that could be instrumentalized in early modern mathematics, at least until the full significance of algebra was recognized in the latter part of the seventeenth century.

For John Speidell we have more information on an exemplary career, beginning around the time of a boom in mathematics teaching in London, and again based on the design of and instruction in mathematical instruments. And we know *a little* more about that most enigmatic of institutions, the Musaeum Minervae. For Euclid Speidell, of course, we now have direct and personal evidence about many aspects of his career—though partly because of the lack of a conclusion to the manuscript these do not resolve into a very clear picture. Again, instruments were part of his instruction, and these as we know from other sources were essential to the third of his careers, as a mathematical teacher, beginning in the early 1660s.

One question remains: why did Euclid decide to write the story of his life and parentage? There is nothing in the manuscript itself to help us answer this question. The provenance is intriguing but also sheds no light.[81] Owing to the date it is tempting to link it to John Aubrey's 'Brief Lives' project; in the year the manuscript was composed, 1689, Robert Hooke tantalizingly 'told Aubrey of *Speidell* etc', which may refer to the manuscript *or* to Hooke's claim, made at the same time, that Euclid Speidell had plagiarized Nicholas Mercator.[82] We know from Aubrey's writings on the education of young gentlemen that he was interested in the Speidells, father and son, recommending John's 'Rules for Measuring Timber' for the 'Mathematical Instruments' section of the library of the ideal school, and Euclid Speidell's Arithmetical Common Questions for teaching mathematics.[83]

Certainly Euclid's manuscript meets Aubrey's requirements, not just for the antiquarian and infinitely digressive *Brief Lives* but for the more tightly focused project that he called 'Lives of Our English Mathematical Writers'. As his most recent editor Kate Bennett puts it, Aubrey sought to provide a model for 'a new intellectual community in which obscure as well as luminous persons would be recognized for their contribution to the advancement of learning, and in which support networks of family,

[81] The manuscript, as mentioned, is in the Monson Papers at Lincolnshire Archives. The provenance appears to be as follows: Rev. Thomas Speidell (d.1836) is known to have left his books to his cousin Capt. George Eilers (1777–1842), and amongst these were various works of his ancestors, including John Speidell; Eilers in turn left these books to William John Monson, 6th Baron Monson of Burton (1796–1862). As there are many other papers relating to Thomas Speidell, it is a reasonable conjecture that the present MS followed the same route, thence to the Monson papers, which have been at Lincolnshire Archives since 1951. Information on the Speidell–Eilers–Monson provenance comes from the flyleaf of a sammelband of John Speidell's works, offered as Lot 601 in the Sotheby's sale of 18–19 September 2018, 'The Erwin Tomash Library on the History of Computing'.

[82] Aubrey, *Brief Lives*, vol. 2, p. 1716.

[83] J. E. Stephens (ed.), *Aubrey on Education*, London: Routledge & Kegan Paul, 1972, pp. 85, 103–4, 166 n. 7.

patronage, intellectual mentoring, and education would become apparent'.[84] Euclid's manuscript reveals that for all the precariousness of a mathematical life in early modern England, there was a remarkable persistence about the 'philomaths' who clung to their subject, could never really escape it and managed, somehow to build their careers from instruments, invention, teaching and mathematical practice.[85]

[84] Kate Bennett, 'John Aubrey and the "Lives of Our English Mathematical Writers"', in: *The Oxford Handbook of the History of Mathematics*, ed. Eleanor Robson and Jacqueline Stedall, Oxford: Oxford University Press, pp. 329–52.

[85] For Euclid Speidell as 'philomath' see the title page to his *Logarithmotechnia*.

16

James Thomson Senior and Mathematics at the Belfast Academical Institution, 1814–1832

Mark McCartney

Thomson's early life and education

When the Ulster Historical Society first unveiled a commemorative blue plaque in Belfast on the site of the birthplace of Lord Kelvin, the plaque had a somewhat embarrassing spelling error: Kelvin's name was given as William Thom*p*son—with a redundant p. The plaque was later replaced, at not inconsiderable expense, with one correctly stating that William Thomson had been born on the site. In fairness to whoever performed the final check on the text, it is a common enough error, propagated in more than one book on physics and its history. Indeed, arguably it is not as obvious or embarrassing a mistake as may appear at first sight. Lord Kelvin's grandfather, would doubtless have affirmed the spelling of Thompson as correct, for so it appeared on the family Bible and on other family documents. It seems that Kelvin's father, James, unilaterally changed the spelling of his surname at Glasgow University on discovering that the Scottish spelling typically omitted the *p*.[1] It was, perhaps, the act of a young man who was eager to fit in to, and progress within, the lofty world of higher learning. For progress, betterment, and the value of education were key parts of James Thomson Snr's life.

James Thomson was born on 13 November 1786 at the family farm just outside Ballynahinch in County Down. He was the second youngest of six children.[2] His older sisters taught him to read, and, according to his own eldest daughter, Elizabeth, 'He taught himself arithmetic from a dilapidated copy of Bonycastle [*sic*] which he was lucky enough to find; not only mastering its contents, but supplying many pages that were wanting'.[3] The 'Bonycastle' in question was John Bonnycastle's *The Scholar's Guide to Arithmetic* which was first published in 1780. As a child of about eleven he managed to construct both sundials and a night dial (or nocturnal): 'He used to tell his children how he had puzzled over an old book on navigation which contained a

[1] Silvanus P. Thompson, *The Life of William Thomson, Baron Kelvin of Largs*, London: Macmillan & Co. 1910, p. 3; W. Innes Addison, *The Matriculation Albums of the University of Glasgow: From 1728 to 1858*, Glasgow: James Maclehose and Sons 1913, p. 247.

[2] While Elizabeth King, *Lord Kelvin's Early Home*, London: Macmillan & Co. 1909, 1, mentions only the four children older than James, Thomas Hamilton in his brief sketch of Thomson in *Belfast Literary Society 1801–1901: Historical Sketch, with Memoirs of some Distinguished Members*, Belfast: M'Caw, Stevenson & Orr 1902, p. 60, lists six, in order of descending age: Robert, Kitty, Mary, John, James, and Elizabeth.

[3] King, *Lord Kelvin's Early Home*, p. 1.

chapter on dialling, and felt disheartened because he could not understand it; and how he subsequently found that his difficulties had arisen from the fact that the book was all wrong. One night sitting up with his father and elder brother, Robert, to guard the orchard, while he watched the stars slowly revolving, he thought out for himself the true principle of dialling, and made a dial for his father's farm which told the time correctly.'[4] The dials, though simply made with lines scraped on to slate, were proud enough family possessions to be still kept by one of his grandchildren in the early 20th century.[5]

There are two further tales regarding Thomson's precocity. The first comes from a short booklet addressed to 'Sabbath School Scholars' telling them of three boys from Ballynahinch who went on to do great things. Of James Thomson (who, it must be said gets the shortest coverage, of just over one page in a twenty-eight page booklet) it states: 'It is yet told in the neighbourhood, that one morning his father sent him out to clean the byre, and passing soon after, and finding the place not cleaned, and James making diagrams on the floor, he gave him a good scolding, and said he would never be fit for anything'.[6]

On its own it is tempting to dismiss this as local folklore, perhaps based on the myth of Archimedes' death. However, a second story may lend some credence to it. A footnote in the biography of Henry Montgomery, who in Thomson's adult life was his colleague at the Belfast Institution, states that, while looking after his father's cattle, the young James

> was frequently observed drawing with a piece of chalk, or sharp-pointed stone, on an ordinary roofing slate, strange hieroglyphical figures, over which he would pore in silent abstraction; but which, to his father and the family, were quite unintelligible. On one occasion, however, the late Rev. Dr. Armstrong of Dublin, then a young man, assistant to Dr. Bruce in the Belfast Academy, happening to call at Thomson's place, was shown some of these drawings, and, to his surprise, discovered them to be abstruse mathematical problems, carefully worked out. He at once examined him on the subject, found the stuff he was made of, and urged his father to send him to school, and, if possible, to college; telling him that, in all probability, he would arrive at high distinction.[7]

Even allowing for some exaggeration, James was clearly a very able and inquisitive child. It seems that his father took Armstrong's advice. Near to the Thomson's farm the Rev. Dr Samuel Edgar 'Having a large family to provide for, and also a taste for teaching' had opened a 'Classical school [...] in one of the office-houses of his farm, fitted up for the purpose'.[8] James Thomson attended and advanced quickly through

[4] Joseph Larmor and James Thomson (eds), *Collected Papers in Physics and Engineering by James Thomson*, Cambridge: Cambridge University Press 1912, p. xiv.

[5] Larmor and Thomson, *Collected Papers*, p. xv.

[6] William J. Patton, *Three Ballynahinch Boys: A New Year's Address to the Young for 1880*, Belfast: Archer & Sons 1880, p. 4. A 'byre' is a cow shed.

[7] John A. Crozier, *The Life of the Rev. Henry Montgomery, LL.D.*, vol. 1, London: Simpkin, Marshall & Co. 1875, p. 162. Henry Montgomery was Head-master of English at the Belfast Academical Institution from 1817 to 1839.

[8] Patton, *Three Ballynahinch Boys*, p. 3.

the ranks of pupils, eventually becoming Edgar's assistant. Writing a letter of recommendation in 1813 for Thomson's appointment at the Belfast Academical Institution, Edgar stated

> Mr Thompson [*sic*] has lived in my family for these seven years past, excepting the time of his attendance at College, and acted as first Usher in my school. In that time he has taught the Classics, Geography, Use of the Globes, Algebra, Fluxions and other parts of school education, with credit and success. His proficiency in those branches, in Arithmetic, in Mathematics, & other departments of science you already know, or will know, from the testimony of those who are able to judge: any further observations on this particular are therefore unnecessary. His aptness to teach is everyway equal to his literary attainments. He was for some years a scholar in my school, and on being advanced to the rank of teacher had to gain amongst those who had been his class fellows that authority so essentially necessary to a master. This he had the art to acquire, and has had the dexterity to maintain.[9]

Writing to his son William, in 1845, Thomson recalls that around the age of 21, 'I was teaching eight hours a day at Dr Edgar's, and during the extra hours—often fagged and comparatively listless—I was reading Greek and Latin to prepare me for entering College, which I did not do till nearly two years after'.[10] Thus, it was not until he was almost 24 that James entered Glasgow University, in the autumn of 1810.

While at Glasgow, Thomson was clearly a hardworking and able student. He won a number of prizes, each awarded at the Annual Distribution of Prizes 'in the presence of a numerous meeting of the University, and of many Reverend and respectable gentlemen of the City and neighbourhood'[11] on the first of May each year. In 1810–11 he was listed amongst the eight prize winners in the Junior Mathematical Class. A certain 'Ja. Thompson, Ireland' was also listed as one of the seven students awarded prizes 'for the best Translations from Tacitus, Plautus, Livy, Ovid, and Caesar, into English'. Given the change in spelling that James seems to have introduced to his surname while in Glasgow, it is reasonable to expect that this is also our Thomson from Ballynahinch. In the 1811–12 session he won, along with one Robert Park (also listed as being from Ireland), the Senior Mathematical Class prize, and was one of the nine students awarded a prize for Natural Philosophy for students who showed 'propriety of conduct, exemplary diligence, and display of eminent abilities during the session'. In the 1811–12 session he was also one of six students awarded a prize for Latin verses, and again won a prize for Latin translation, a prize which he picked up for third time at the end of the 1812–13 session.[12]

James graduated *Artium Magister* in 1812, which after only two years study appears at first sight a rather short course. However, in the eighteenth century, Francis Hutcheson, professor of moral philosophy at Glasgow from 1729 to 1746, and Ulsterman, had promoted the idea that Glasgow should allow Irish students who had been taught

[9] Public Records Office of Northern Ireland, SCH/524/7B/7/58.

[10] Crosbie Smith and M. Norton Wise, *Energy & Empire: A Biographical Study of Lord Kelvin*, Cambridge: Cambridge University Press 1989, p. 8.

[11] W. Innes Addison, *Prize Lists of the University of Glasgow from session 1777–78 to session 1832–33*, Glasgow: Carter & Pratt 1902, p. 137.

[12] Addison, *Prize Lists of the University of Glasgow*, pp. 138, 141, 144, 147, 154.

at private teaching academies to proceed to a Glasgow MA after only two years. While this was doubtless popular with students making the then arduous journey across the Irish Sea, it was controversial in other quarters. In 1772, the Bishop of Down and Connor complained that 'he had incurred much disapprobation for admitting into orders persons who had obtained degrees after two years' attendance at Glasgow, and that, if degrees continued to be granted on such terms, he would be obliged to reject applicants founding upon them'.[13] The University seems to have failed to take decisive action on the matter, but at least in the case of James Thomson no injustice was likely to have been done either to the degree, or the student, as the prizes he won clearly indicate a strong and motived undergraduate. After gaining his MA, Thomson returned to Glasgow for a further two years and according to his daughter 'attended most of the medical classes, and went through the complete theological course with a view to entering the ministry'.[14] It seems however that Thomson was simultaneously considering a teaching career, as in 1812 he put himself forward 'as a candidate for the department of teaching of Mathematicks or Natural philosophy, or both if they should be taught by the same person'.[15] The post applied for was in the not yet opened Belfast Academical Institution.

Teaching at the Belfast Academical Institution

The aim of the Belfast Academical Institution (or Inst. as it is universally known in Belfast today) was to be both a school and a college combined, with the college part aiming to provide students with an education equivalent to the Scottish universities, to thus save young men from Ulster the expense and inconvenience of crossing the Irish Sea. While on a good day the journey from Belfast to Greenock, some 25 miles west of Glasgow, could take ten hours, in adverse conditions, and with a less favourable boat, it could take three to four days.[16] On at least one occasion while a student James Thomson and some of his fellow travellers to Glasgow asked to be put ashore on the Ayrshire coast, as they were confident that it would be faster to walk the remaining journey than to sail it![17]

The Belfast Academical Institution was not the first school in Belfast. In 1785 the Belfast Academy was founded. Its initial intention was also to be both a school and college combined, but the desire for a college part of the Academy was never fulfilled. Hence, when the plans for the Belfast Academical Institution were being put forward there was one dissenting voice in the town: Rev. William Bruce, Principal of the Belfast Academy. Not only was the Academical Institution being founded by some of his political enemies, but, more pragmatically, he foresaw a drain on potential student numbers to his own school. Although Rev. Bruce went to the length of writing

[13] James Coutts, *A History of the University of Glasgow: From Its Foundation in 1451 to 1909*, Glasgow: James Maclehose and Sons 1909, pp. 308–9.

[14] King, *Lord Kelvin's Early Home*, p. 10.

[15] Smith and Wise, *Energy & Empire*, p. 10. Note that while there is no reason to doubt the accuracy of this quote, the reference given by Smith and Wise to the Public Records Office of Northern Ireland file (SCH/524/3C/1) does not correspond to the letter quoted, and I have been unable to identify it in the Records Office electronic catalogue.

[16] Thompson, *Life of William Thomson*, p. 4.

[17] King, *Lord Kelvin's Early Home*, p. 7.

an anonymous article in the *Belfast News Letter* seeking to discredit the project, his objections made no difference. The foundation stone was laid in 1810, and the school, which remains as one of the most impressive pieces of architecture in Belfast, opened on 1 February 1814.[18] Though today it is in the centre of a large city, when it was opened the Academical Institution sat on the very edge of a Belfast which in total occupied only a square mile. Belfast, however, was then a town which was growing rapidly. Between 1813 and 1831, which almost exactly covers Thomson's time in the town, the population nearly doubled, going from 27,832 to 53,737.

Advertisements for masters in the school section of the Academical Institution were placed in November 1813. There were four applicants for the post of teacher of mathematics and arithmetic, and James was unanimously appointed to the job on 4 January 1814. No sooner had James arrived back in Belfast from Glasgow than it was suggested that he also be appointed as professor of mathematics in the college department. At a meeting, on 8 February 1814, of the joint boards of managers of the Institution 'held specially in consequence of the arrival of Mr. Thomson the Mathematician' it was moved 'that Mr Thompson [*sic*] is our Professor of Mathematics and that the Professor of Mathematics be endowed with a salary of one hundred pounds per annum for three years'. The decision was adjourned until the next week, but was defeated by 14 to 4.[19] Although the professorship of mathematics did not fall into James's hands immediately, by the end of 1815 the job was his, and thus he found himself with two jobs within the Institution, one in each of the school and college departments. The description of Thomson simply as 'the Mathematician' seems to have been a common one in Belfast, and he was occasionally referred to as such in the minutes of meetings within the Institution, and beyond.

Early correspondence between James and the Board of Management of the Academical Institution reveals him to be a determined man. One of the historians of the school, John Jamieson, puts Thomson's occasional feistiness down to his feeling he had been snubbed early on in his time there by being left off a committee which organized weekly assemblies within the school. It was an oversight which prompted 'a long but dignified yet insistent letter to the Board' in which 'he protested not only against this shabby treatment but against other instances of incivility and a lack of respect he claimed not as a favour but as a right'.[20] It is also possible, however, that such a letter was simply an early example of the assertive *modus operandi* of an ambitious and capable young man.

As an example of Thomson's assertiveness it is hard to resist reproducing part of a letter, dated 7 February 1815, to Robert Simms, Assistant Secretary to the Board of Managers:

> Dear Sir, Be so good as to desire the Board to take into their consideration the state of my class-room. They are well aware that even before the late vacation, it was too small to accommodate the pupils attending [...] I may also observe that I have the largest classes & the smallest room. The room is scarcely capable of <u>accommodating</u> sixty

[18] John Jamieson, *The History of the Royal Belfast Academical Institution: 1810–1960*, Belfast: William Martin & Son 1959, p. 6.

[19] Minutes of Joint Board of Managers and Visitors 1807–1814, pp. 299, 306, 311, 312. Public Records Office of Northern Ireland, SCH/524/3A/1/1.

[20] Jamieson, *History of the Royal Belfast Academical Institution*, p. 16.

> boys at once; but, at present, at one period of the day, it contains no fewer than ninety six, at another about ninety [...] A highly unpleasant and unwholesome heat is thus generated in the room; and it is of consequence uncomfortable and unwholesome for the teachers and the pupils. Besides it is almost impossible to preserve proper order. When the pupils are crowded together, idle and mischievous boys have an opportunity of annoying their neighbours [...] It may be asked what remedy I would propose. I know of none but to take seven or eight feet in length from Mr. Spence's room. It is totally impossible to wait for new buildings...the credit of the teacher, of the department & of the Institution is at stake [...]. I need scarcely mention that in case of the room's being enlarged, additional desks and forms will be needed.

Thomson then adds as a postscript, least the Assistant Secretary be in any doubt that he expects a speedy resolution, 'It will oblige one to be informed what determination the Board may come to on this matter'.[21]

At the next meeting of the Joint Board of Managers it was noted that a member of staff had agreed to swap rooms, even though in the letter of the 7th Thomson, with the thoroughness of a man who has thought through all the possibilities, had explicitly ruled it out as an option.[22] What makes the letter of 7 February all the more amusing is that these complaints about 'unpleasant and unwholesome heat' came only a few weeks after a letter signed jointly by James Thomson, Thomas Spence (writing master), and James Knowles (English master) complaining 'that our rooms are at present very badly supplied with fires [...] It is scarcely necessary to mention that in the Arithmetical and Writing school it is almost impossible to proceed with business when the fingers of the boys are cramped with cold'.[23] A member of staff who complained that his room was too cold in January, but too hot in February, must have left the Joint Board of Managers at least a little bemused.

While building a career in the Academical Institution James Thomson was also building his life in other directions. In the summer of 1816, Margaret Gardiner, 'a young, bright and lively girl',[24] came to Belfast from Glasgow to visit her cousin Dr William Cairns, who was professor of logic and belles letters in the college department of Inst. James and Margaret were introduced, and in a letter to her sister she records what seems to have been their first meeting:

> I had the honour of the Mathematician as my walking companion. His first appearance is about as awkward as can be; he looks as if he were thinking of a problem and so modest he can scarcely speak, but when tête-à-tête he improves amazingly in the way of speaking. On our forenoon walk we had a most edifying and feeling discussion on sea-sickness and the best mode of preventing it. But in the evening we were much more sublime. I suppose the moon rising in great beauty, and Jupiter shining with uncommon lustre, called forth the Professor's energies, and I got a very instructive and amusing lecture upon astronomy.'[25]

[21] Public Records Office of Northern Ireland, SCH/524/7B/9/15.

[22] Minutes of Joint Board of Managers and Visitors 1814–1821, 14 February 1815. Public Records Office of Northern Ireland, SCH/524/3A/1/2.

[23] Public Records Office of Northern Ireland, SCH/524/7B/9/10.

[24] Larmor and Thomson, *Collected Papers*, p. xv.

[25] Larmor and Thomson, *Collected Papers*, p. xvi.

Evidently the power of discussions on sea-sickness and astronomy should not be underestimated as a method of wooing a wife, as the next summer Thomson travelled to Glasgow and the pair were married. After honeymooning in the Highlands of Scotland they returned to Belfast. Thomson built two houses opposite the Institution on College Square East; one to raise a family in, and the other to rent out. The couple had seven children in Belfast; Elizabeth (1818–1896), Anna (1820–1857), James (1822–1892), William (1824–1907), John (1826–1847), Margaret (1827–1831), and Robert (1829–1905).

The childhood memories of the eldest daughter, Elizabeth, are gathered together in *Lord Kelvin's Early Home*. She paints a picture of a happy upbringing with a father who was both hard working, and eager to ensure his children benefited from a good education. It was an education which he and Margaret in large part provided. A letter from Thomson's brother Robert, who still lived in Ballynahinch, to his sister, who now lived in America, gives a perspective on family life in Belfast: 'Brother James and his family are well. Have five very fine children, two girls and three boys. They are handsome and kind, and fond of their friends; and what is more, are wonderfully apt in learning; but I don't think it strange, as both father and mother are drilling them. Their mother has a good education, and Elizabeth speaks French well.'[26] The word 'drilling' sets a somewhat harsh note, which is out of key with the overall tone of Elizabeth's warm evocation of her upbringing. Her recollections are certainly filled with examples of the education of her and her siblings, but it seems to have been a process which, while doubtless well regimented, was enthusiastically and fully entered into by both parents and children alike.

School books at the Belfast Institution

While at home Thomson was not only raising and educating his children, he was also writing books (see Table 16.1). He rose at 4 a.m., with coffee, cream and lamp, to work on these before heading across the road to take 8 a.m. and 11 a.m. classes in the college department, and afternoon classes in the school department.

The books Thomson published give an indication of the range of both his teaching and his prodigious work ethic while in Belfast. The texts which went through the most editions, and were published outside the author's own lifetime, were predominantly volumes targeted at a school audience: *A Treatise on Arithmetic*, *An Introduction to Modern Geography*, *The First Six and Eleventh and Twelfth Books of Euclid*, and *An Elementary Treatise on Algebra*. The books on arithmetic, Euclid, and algebra were all adopted as textbooks by the National Board of Education in Ireland, and certainly by the mid-1840s Thomson was reaping considerable income from them.[27]

Note also that although Thomson's edition of Euclid was not published until 1834, by which time he was professor of mathematics in Glasgow, he seems to have begun work on it while in Belfast. The 1831 first edition of *An Introduction to Differential and*

[26] King, *Lord Kelvin's Early Home*, pp. 40–1, footnote.
[27] Smith and Wise, *Energy & Empire*, pp. 33–4, note that in 1845 alone the profits from his textbooks totalled £378 14s 8d.

Table 16.1 Books authored by James Thomson.

Title	Year of publication	Publisher of first edition	Last known edition	Year of last known edition
A Treatise on Arithmetic in Theory and Practice	1819	Joseph Smith (Belfast)	72	1880
Key to Thomson's Treatise on Arithmetic	1825	Simms & McIntyre (Belfast)	edition not listed	1881
Elements of Plane and Spherical Trigonometry, with the First Principles of Analytic Geometry	1825	Joseph Smith (Belfast)	4	1844
An Introduction to Modern Geography, with an appendix containing an outline of astronomy and the use of globes	1827	Simms & McIntyre (Belfast)	27	1857
Remarks on the Phenomena of the Heavens as they appear from Different Bodies of the Solar System	1827	A Mackay (Belfast)	1	1827
An Atlas adapted to a Treatise on Modern Geography	1828	Simms & McIntyre (Belfast)	Listed as 'new edition'	1844
An Introduction to the Differential and Integral Calculus, with an appendix, illustrative of the theory of curves	1831	Simms & McIntyre (Belfast)	2	1849
The First Six and the Eleventh and Twelfth Books of Euclid's Elements; with notes and illustrations, and an appendix in five books	1834	Adam & Charles Black (Edinburgh)	15	1860
An Elementary Treatise on Algebra, Theoretical and Practical	1844	Longmans (London)	14	1863
Key to Thomson's Treatise on Algebra	1847	Longmans (London)	1	1847

Sources: The 'keys' to the books on arithmetic and algebra are texts which give worked solutions to the questions in those books. The only volume not a book related to Thomson's teaching is *Remarks on the Phenomena of the Heavens* which is a sixteen-page booklet republishing a set of three articles which were 'drawn up hastily, to be read before a Literary Society' and 'afterwards given to fill a corner in a newspaper'. Although the *Atlas* is listed as authored by Thomson in both the catalogues of the Bodleian Library, Oxford and the Linen Hall Library, Belfast, it is more accurate to say that it is authored by the cartographer R. Pattison who drew the maps 'under the direction and superintendance [*sic*] of Mr. Thomson'.

Information on last known editions was sourced via the catalogues of the Linen Hall Library, Belfast, the British Library, London, the Bodleian Library, Oxford, and the online repository of books at www.archive.org.

Integral Calculus contains an advertisement (see Figure 16.1) stating that Thomson's edition of Euclid is 'preparing for publication'. Figure 16.1 also gives an indication, via number of editions, of the success of the other books which Thomson authored during his time in Belfast.

Thomson's first book, *A Treatise on Arithmetic in Theory and Practice*, was well received and when the second edition was produced in 1825 the publisher proudly added quotations from reviews in an advertisement at the front of the book. The *Belfast News Letter* praised the book as 'the most useful treatise on the subject we have ever seen', while the *London Literary Gazette* noted 'its adaptation to the use of Irish pupils. All the variations of English and Irish Currency, and Measures are given, and their proportions displayed in a very ingenious way'. Finally the *London Monthly Review* commented that '[t]he questions, moreover, are so contrived that while they serve all the purposes of Arithmetical lessons to the Student, they give him information of many important facts in Commerce, Geography, Astronomy, Chronology, Chemistry, and other branches of knowledge'.[28] With regard to these 'applied' questions Thomson states at the beginning of the first edition that '[i]n the formation of questions of this latter class, recourse has been made to a great number of works and authentic documents, which must be unknown to the generality of pupils; and it is hoped that what is thus presented may tend to excite in the learner a desire to acquire further information on the subjects which will thus be introduced to his notice'.[29]

In publishing his *Arithmetic* Thomson was entering an already crowded market, as Augustus De Morgan's published list of *Arithmetical books from the invention of the printing press to the present time* (the present time in question being 1847), shows.[30] Previous to using his own book, the texts used by students at the Belfast Institution had been John Bonnycastle's *The Scholar's Guide to Arithmetic* (first published 1780) and Gough's *Practical Arithmetic* (first published 1758).[31] According to De Morgan, Gough's book 'had such extensive currency in Ireland [...] that the name of the author became almost synonymous with arithmetic; in so much that when Professor Thomson's Arithmetic was first published in that country, it went by the name of "Thomson's Gough"'.[32]

All three books have chapters covering material which ranges from basic techniques in arithmetic to application chapters on topics such as profit and loss, barter, interest, and annuities, and of course these applied topics naturally lead to practical questions for the pupils. However, as noted in the quotation from the *London Monthly Review* above, it is the addition of extra factual knowledge in many of the questions that sets Thomson apart from Gough and Bonnycastle. As an example, consider Thomson's section on simple division. After a page of 'drill' questions on

[28] James Thomson, *A Treatise on Arithmetic in Theory and Practice*, Belfast: Simms & McIntyre 1819, p. 2.

[29] James Thomson, *A Treatise on Arithmetic in Theory and Practice*, 2nd edn, Belfast: Simms & McIntyre 1825, p. ii.

[30] Augustus De Morgan, *Arithmetical books from the invention of the printing press to the present time*, London: Taylor and Walton 1847.

[31] Note from James Thomson to Robert Simms, 1815. Public Records Office of Northern Ireland, SCH/524/7B/9/83.

[32] De Morgan, *Arithmetical Books*, pp. 79–80.

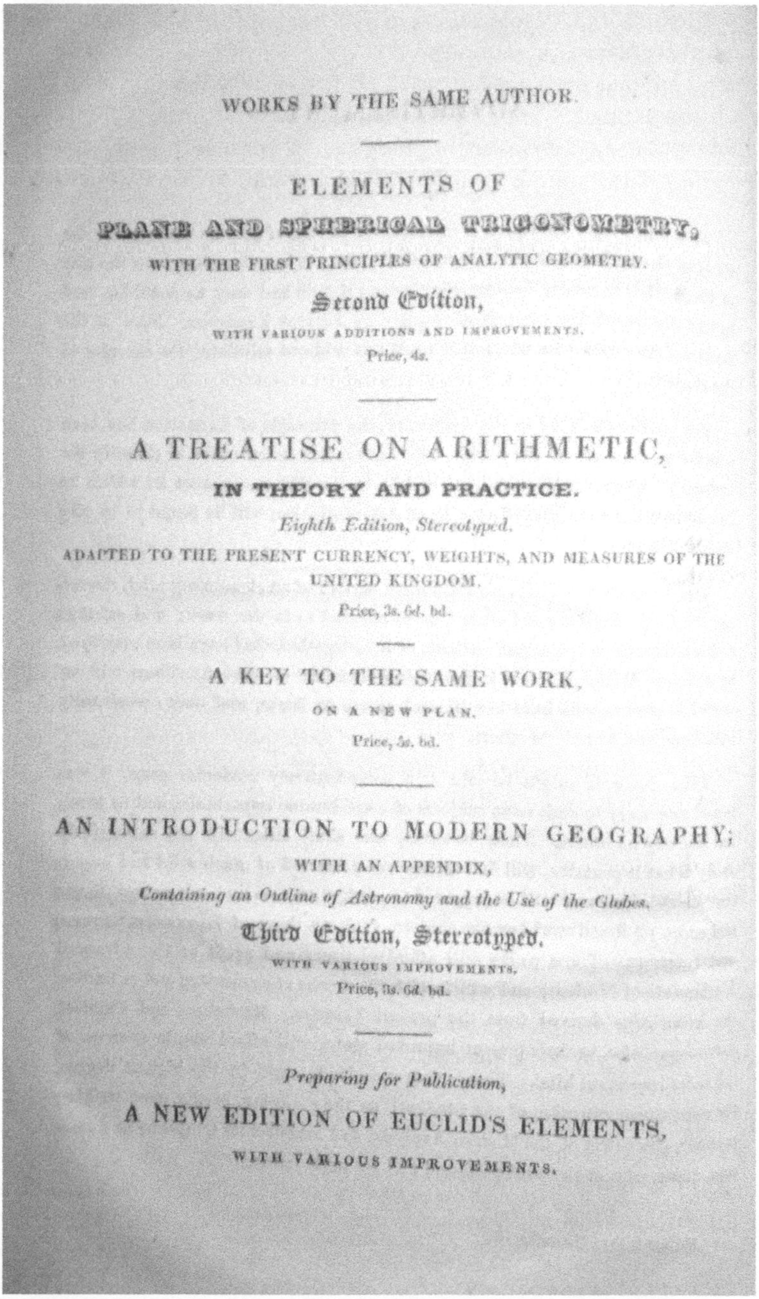

WORKS BY THE SAME AUTHOR.

ELEMENTS OF

PLANE AND SPHERICAL TRIGONOMETRY,

WITH THE FIRST PRINCIPLES OF ANALYTIC GEOMETRY.

Second Edition,

WITH VARIOUS ADDITIONS AND IMPROVEMENTS.

Price, 4s.

A TREATISE ON ARITHMETIC,

IN THEORY AND PRACTICE.

Eighth Edition, Stereotyped.

ADAPTED TO THE PRESENT CURRENCY, WEIGHTS, AND MEASURES OF THE UNITED KINGDOM.

Price, 3s. 6d. bd.

A KEY TO THE SAME WORK,

ON A NEW PLAN.

Price, 5s. bd.

AN INTRODUCTION TO MODERN GEOGRAPHY;

WITH AN APPENDIX,

Containing an Outline of Astronomy and the Use of the Globes.

Third Edition, Stereotyped,

WITH VARIOUS IMPROVEMENTS.

Price, 3s. 6d. bd.

Preparing for Publication,

A NEW EDITION OF EUCLID'S ELEMENTS,

WITH VARIOUS IMPROVEMENTS.

Figure 16.1 Advertisement from the first edition of Thomson's *An Introduction to the Differential and Integral Calculus*, published 1831. Note that Thomson's edition of Euclid, which was not to appear until 1834, is listed as 'preparing for publication'. Note, too, that *A Treatise on Arithmetic* has gone through eight editions since 1819; *An Introduction to Modern Geography*, three editions since 1828; and *Elements of Plane and Spherical Trigonometry*, two editions since 1825.

division of the form '1457924651÷1204', he has a set of further questions applied to more practical problems, for example:[33]

The linen exported from Ireland in 1809, was 43,904,382 yards: of how many pieces, each containing 25 yards, did this quantity consist? Answ. 1,756,175 $\frac{7}{25}$.

Note that that these questions are 'simple division' because they involve only numbers of one denomination. An example of a compound division would be £73 18s 2d ÷ 14.

A comparable question to the above from Gough's 1815 edition of his *Practical Arithmetic*[34] reads:

> A shop-keeper bought a piece of cloth containing 42 yards for £22 10s. of which he sells 27 yards for £15 15s. How many yards has he left, and what is their cost? Answ. 15 yards, which cost £6 15s.

Both questions show how basic arithmetical techniques are applied in daily business, but whereas Gough gives reasonable figures, Thomson gives actual figures. Assuming they are indeed drawn from 'authentic documents', gathering such data would have come at considerable effort to the author, and one hopes would indeed 'excite in the learner' a desire to learn more.

In his job as master in mathematics and arithmetic in the school department, Thomson also taught geography. Indeed, one of his first questions on being offered the post in 1814 was to inquire 'whether the teaching of Geography belongs to my department'.[35] This was a perfectly reasonable question in an era when geography commonly fell within the established disciplines of mathematics or natural philosophy. Thus, for example, when Belfast's first school, the Belfast Academy, was opened, geography, arithmetic, algebra, astronomy, navigation, and bookkeeping were all listed under the subject of 'mathematicks'.[36] Thomson, eager to extend his fiefdom, happily took the subject, and in 1827 produced another textbook, *An Introduction to Modern Geography*. In presenting to the public, in 1828, the atlas which formed a companion to the book, the publishers stated that the text was so up to date that 'none of the School Atlases already published are adapted to it' and noted that Thomson's geography had sold more than 2,000 copies in less than six months.[37]

The text on geography is the book by Thomson which is almost certainly the most charming for the modern reader. The main body of the book takes each county or region and discusses it under the headings of ports, lakes, rivers, towns, climate, population, education, religion, and so on. To this are added extensive footnotes, which in some cases take up over 90% of the printed page, and are provided to 'enliven the study' of the subject with 'lighter and more entertaining material'. Under the sections discussing the character of the peoples in various countries, Thomson delivers what,

[33] Thomson, *Treatise on Arithmetic*, 1st edn, p. 32.

[34] John Gough, *Practical Arithmetic in Four Books*, Belfast: Simms & McIntyre 1815, p. 52.

[35] Letter from James Thomson to Joseph Stevenson, 8 January 1814. Public Records Office of Northern Ireland, SCH/524/7B/8/1.

[36] A. T. Q. Stewart, *Belfast Royal Academy: The First Century 1785–1885*, Antrim: Greystone Press 1985, p. 2.

[37] *Atlas adapted to Thomson's Modern Geography*, Belfast: Simms & McIntyre 1828, p. i.

with the separation of time, can be seen amusing stereotypes. Thus, 'the French are in general spritely, and fond of amusement; ingenious and polite; and strongly influenced by a love of distinction and glory', whereas, 'notwithstanding all the exertions of the Russian sovereigns, the great mass of the people are but half civilised. They are superstitious, and are in general very fond of spirituous liquors'.[38]

When it came to the description of his own country, Thomson states that 'the lower Irish are considered a lively, shrewd people, and warm in their attachments and antipathies. In many instances however, particularly in the south and west of the kingdom, they commit acts of turbulence and cruelty, arising from bad education and habits'. After briefly noting that there is but one University in Ireland (Trinity College), he then adds, in a substantial footnote, what amounts to a glowing recommendation for his own establishment:

> The Institution of Belfast was established by public subscription; and affords extensive courses of lectures, on Latin, Greek, Hebrew, Logic and Belles Lettres, Moral Philosophy and Metaphysics, Mathematics, Natural Philosophy, Anatomy, and other subjects. This seminary is open to persons of all religious denominations, and is adopted in particular by the Presbyterians of Ireland as a place of education for their clergy. It has also a series of schools for the primary branches of education, and for languages ancient and modern. It was opened for teaching in 1814.[39]

Before moving on to discuss the books by Thomson which were targeted at the college side of the Belfast Institution, it is worth mentioning the one book, or more accurately booklet, in Table 16.1 which is not a teaching text: *Remarks on the Phenomena of the Heavens, as they Appear from Different Bodies in the Solar System*, published in 1827. This is a sixteen-page booklet republishing a set of three articles which were 'drawn up hastily, to be read before a Literary Society' and 'being of a light nature, they were afterwards given to fill a corner in a newspaper, at the request of the Editor'. The Society in question was the Belfast Literary Society, and the newspaper *The Belfast News Letter*. Thomson had joined the Belfast Literary Society in 1818, acted as its president for 1821–22, and over the period from 1819 to 1831 read a total of nine papers, nearly all on mathematics, astronomy and geography.[40]

Despite Thomson's self-deprecating comments, *Remarks on the Phenomena of the Heavens* is an enjoyable piece of popular astronomy, imagining how an inhabitant of the sun, moon, Jupiter and its moons, Saturn and its rings, or a comet would view the rest of the solar system. Thus, says Thomson, for a solar astronomer the planets 'would never appear to *wander* in the heavens; and hence the name by which he would designate them would have a very different import from the original

[38] James Thomson, *An Introduction to Modern Geography*, Belfast: Simms and McIntyre 1827, pp. 43, 103.

[39] Thomson, *Introduction to Modern Geography*, pp. 34–5.

[40] Hamilton, *Belfast Literary Society*, p. 187 lists them as: 'On the tides' (1819); 'A view of the progress of mathematics among the Saracens' (1821); 'A sketch of the progress of mathematical science among the Greeks' (1822); 'Essay on the opinions that have been formed respecting the nature and phenomena of the fixed stars' (1824); 'On rivers' (1826); 'On the celestial phenomena, as seen from other bodies, in the solar system' (1827); 'Currents at sea' (1829); 'Two unpublished letters of Doctor Thomas Reid' (1830); 'Remarkable instances of hereditary talent among men of science' (1831).

signification of our term *planet*'. On the moon the earth would always be at nearly the same place in the sky. A comet dweller would have an unprecedented range of views of the sun, from along its highly elliptic orbit, but 'with the purposes which they [comets] are intended to serve in the economy of the universe we are totally unacquainted'. Thomson ends the piece by noting that astronomy gives 'evidence of endless and astonishing variety in the works of the Creator [...] that do so truly and so strikingly declare the glory of God',[41] with this last phrase alluding clearly to Psalm 19, verse 1.[42]

College books at the Belfast Institution

By the late 1820s the college department of the Belfast Academical Institution had an average of 193 students enrolled over the academic sessions 1826–27, 1827–28, and 1828–29 (see Table 16.2). While these numbers are small compared to the average number of students enrolled at well-established Scottish institutions such as the University of St Andrews (which over the period 1826–36 was 273[43]) or Glasgow University (where over the years 1826–36 the average enrolment in the Faculty of Arts was 556[44]), they are creditable for such a new and relatively small college.

In 1827, Thomson found himself writing, at the behest of a church committee, to Thomas Chalmers (who at this point in his career was professor of moral philosophy at St Andrews), to persuade him to preach at the opening of the new Fisherwick Presbyterian church which was being completed just a few metres away from the Thomson's home. As part of the correspondence Thomson sent Chalmers a copy of his *Elements of Plane and Spherical Trigonometry* and also described aspects of the teaching in the college classes at the Belfast Academical Institution:

> We have not the power of giving degrees; but at the end of our course of literature and science we give, after examination, what we call a General Certificate signed by all the members of Faculty; and this is treated with the same respect as a degree in arts from the universities by the religious bodies whose students we instruct. Our course for obtaining this is at least three years; and tho' from the additional expense to the students and the practice already established, we have not ventured to render a longer term imperative, we are using our influence and advice to cause the students to add another year; and the better order of them are in many instances complying with our wishes. In the formation of our course we have chiefly followed the Glasgow system, as the Presbyterian clergy here had almost all been educated there and were partial

[41] James Thomson, *Remarks on the phenomena of the heavens, as they appear from different bodies in the solar system*, Belfast: A. Mackay 1827, pp. 2, 4, 14, 16.

[42] For a detailed assessment of James Thomson's religious beliefs see Andrew R. Holmes, 'James Thomson Sr. and Lord Kelvin: Religion, Science, and Liberal Unionism in Ulster and Scotland', in: *Journal of British Studies* 50 (2011), pp. 100–24.

[43] James Maitland Anderson, *The Matriculation Roll of the University of St Andrews 1747–1897*, Edinburgh: William Blackwood and Sons 1905, pp. xxiv and lxxxvii–lxxxviii. This average assumes that no student who matriculated in a given year dropped out of their studies, and that the duration of study was four years.

[44] Coutts, *History of the University of Glasgow*, p. 343.

Table 16.2 Enrolments on the college section of the Belfast Academical Institution for the academic sessions 1826–27 to 1829–30.

Subject	1826–27	1827–28	1828–29	1829–30
Logic and Belles Lettres	70	70	76	77
Moral Philosophy and Metaphysics	32	48	53	49
Natural Philosophy	33	27	32	43
Mathematics	89	100	97	126
Latin	35	31	16	26
Greek	55	47	48	44
Hebrew	39	39	33	37
Anatomical Class	10	16	26	9
Divinity (Synod of Ulster)	28	23	15	18
Divinity (Seceding Synod)	34	30	32	33
Total students enrolled	181	193	205	–

Sources: minutes of the joint board of managers and visitors, December 1828–July 1836, April 1829, p. 22; Public Records Office of Northern Ireland, SCH/524/3A/1/4; note by James Thomson, dated 8 December 1829 on number of students attending college classes. Public Records Office of Northern Ireland, SCH/524/7B/23/46.
Note that across all three years mathematics is the largest class. No total number is given in the original data for the session 1829–30. The two divinity classes reflect the fact that Presbyterianism in Ulster was at this time made up of more than one group, and that there were two divinity professors. The first divinity professor of the Seceding Synod was James Thomson's teacher in Ballynahinch, Samuel Edgar, with his son John Edgar succeeding him in the role in 1826.

to the system. On this we have engrafted a system of strict examinations somewhat after the plan in Dublin and in the English Universities. The student on entering for the first time is examined on Latin and Greek. Owing to the state of the schools over the country, we are often obliged to admit students who are but indifferently acquainted with these languages. This is a great evil, and is but partially remedied by their subsequent attendance on the Latin and Greek classes. We have been doing as much, however, as circumstances will permit to obviate it; and our exertions are much aided by the religious bodies, who examine their students before they authorise them to attend college, and are gradually becoming more strict, and requiring a higher degree of preparation. By these means the schools are gradually improving, and we are now beginning to get several well educated lads who have been instructed by our own students. At the commencement of each succeeding session, the student is examined on the courses studied during the previous one, to keep up his attention during the summer. What we consider of much more importance, however, is that all regular students, that is, who mean to take General Certificates, are required to submit to a public examination at the end of the session on the business of the classes attended.[45]

[45] James Thomson to Thomas Chalmers CHA4.86.33 (9 June 1827), New College Library Archive, Edinburgh.

In the same letter, he goes on to emphasize that the examinations are 'long and strict' and that the students are 'with few exceptions remarkable for their diligence and good conduct'. Further, after noting that the General Certificate course in mathematics is made up of the first six books of Euclid, plane trigonometry, algebra, and mensuration, he notes that those taking a more extensive 'medal course' also study spherical trigonometry, conic sections and 'a considerable part of the Differential and Integral Calculus'. These two courses were called the junior and senior mathematical courses respectively. A relatively small number of students went on to take the senior course, with Thomson recording in a note in December 1829 that he had 100 students in the junior class and 26 in the senior[46] with comparable numbers of 93 in the junior class and 24 in the senior being recorded for the 1831–32 session.[47] The *Belfast Magazine and Literary Journal* for May 1825 contains not only a report on the examinations that had taken place at the end of the previous month at the Institution, but also lists a selection of the questions which were set.[48] These questions are reproduced in the Appendix to this chapter. Although more complete lists of questions exist in manuscript form[49] for examinations over the academic years 1824–5 to 1831–2, the selection published in May 1825 give a representative sample of both the level of question and the topics examined. Examination questions frequently give a better feel for the standard and content of a course of study than a published curriculum, and thus while Thomson claimed to Chalmers that the senior course covered 'a considerable part' of the calculus, there is little evidence of this being the case from the 1825 examination questions, with questions on calculus making up a small minority of those listed. In fairness to Thomson, to be in contention for the medal associated with the 'medal course' students had to take a further examination which did contain more calculus. However, typically this examination was taken by only one or two students. For example, in the April 1828 medal paper, which was sat by one student (one Robert Wilson), of the 19 questions listed, 13 were on calculus.[50]

In addition to the examination questions taken by all members of the class, Thomson also set students three individual questions to be completed, one assumes, one student at a time in Thomson's presence. Each year Thomson recorded a summary of the questions and his assessment of the answers in a neat table. As an example, in the April 1828 examinations a typical set of three questions given to a senior student (in this case Henry Kelso) are noted by Thomson as 'Euclid III 31', 'Area of polygon', and '$\frac{x-1}{7} - \frac{x-3}{9} = 2$', with his assessments of Kelso's answers being 'V.W.' (which from the context is probably 'very well') for the first two and 'A slight mistake' for the third.[51]

[46] Note by James Thomson, dated 8 December 1829 on number of students attending college classes. Public Records Office of Northern Ireland, SCH/524/7B/23/46.

[47] Minute Book of the Joint Board of Managers and Visitors 1828–1836, July 1832, p.186. Public Records Office of Northern Ireland, SCH/524/3A/1/4.

[48] *Belfast Magazine and Literary Journal* 1/4 (May 1825), pp. 384–6.

[49] Public examination of the Junior Mathematical Class, 26 April 1825; MS13/N/2, Royal Belfast Academical Institution, MS13/N/3; Belfast Institution, College Department. Mathematical Classes 1828–29, MS13/N/4, Queen's University Belfast Library Special Collections.

[50] Royal Belfast Academical Institution, MS13/N/3, Queen's University Belfast Library Special Collections. Although Thomson lists 19 questions, he notes that the last three were 'not proposed', meaning that Mr Wilson had an examination where 13 out of 16 questions were on calculus.

[51] Royal Belfast Academical Institution, MS13/N/3, Queen's University Belfast Library Special Collections.

Thomson's assertion to Chalmers that the General Certificate 'is treated with the same respect as a degree in arts from the universities' is borne out by the fact that the level and style of questions he used in his Glasgow A.M. examinations in the 1830s was the same level and style used by him in Belfast in the 1820s.[52]

The prominence of trigonometry in the curriculum at the Institution is reflected in the fact that in 1825 Thomson published *Elements of Plane & Spherical Trigonometry*, which stated on the title page that the book was 'the substance of the first part of the senior course of mathematics taught at the Belfast Institution' (Figure 16.2). At sixty-eight pages, the 1825 edition was Thomson's shortest book, but it still covered substantial territory. Compound and multiple angle formulae, sine and cosine rules, spherical trigonometry with practical application to right ascension and declination, longitude and latitude, and dialling are dealt with, all finished off with ten pages at the end on analytic geometry. In the section on multiple angle formulae Thomson shows how these can be derived via the 'curious imaginary formulas' based on the manipulation of the expression $\cos A + \sin A\sqrt{-1}$. In the second, much expanded, edition of the book Thomson states that 'The first edition of this work was intended chiefly as a Text-Book for the use of the students in the BELFAST INSTITUTION; and it was therefore written, not as a regular and complete treatise on Trigonometry, but as an outline to be filled up, and illustrated orally in the Lectures'. The phrase 'to be filled up' is to be taken literally. Of the four copies of the first and second editions of the book held by the Linen Hall Library in Belfast one of the second editions and both of the first editions are bound with alternate blank and printed pages, with some of the blank pages containing notes taken by students. A review of the second edition in the *Dublin Literary Gazette* states that 'Though this is a second edition, it is, in fact, the first offered to the public; the former having been intended chiefly as a text-book for Dr. Thomson's pupils'[53] which suggests that the first edition was not distributed beyond Belfast.

The other college level book which Thomson published while in Belfast was his *Introduction to the Differential and Integral Calculus*. Coming at the end of his time at the Academical Institution it is unlikely to have had much classroom use, and indeed if the 1825 examination questions reflect the general state of the curriculum, the volume, which includes topics such as functions of two variables, finite differences, and the calculus of variations, contains much more than was taught. At over two hundred and forty pages it was substantially longer, and covered a wider range of material, than Baden Powell's 1829 *Short Treatise on the Principles of the Differential Calculus* which was targeted at students at the University of Oxford. It is hard to resist the conclusion that in publishing his book on calculus, Thomson was not so much producing a treatise for use by his students as advertising his own scholarly credentials. One Dublin writer, who commenced his review of the book by quipping that he was old enough to remember the days when students at Trinity College Dublin would have responded to a book on calculus with the words *Graecum est legi non potest* (i.e. it's all Greek to me), praised not only the text and its author, but also the fact the

[52] Record of Students Attendance, Junior Mathematical Class, 1832–33 MS13/N/5, Queen's University Belfast Library Special Collections.

[53] *The Dublin Literary Gazette*, no. 7 (13 February 1830), p. 108.

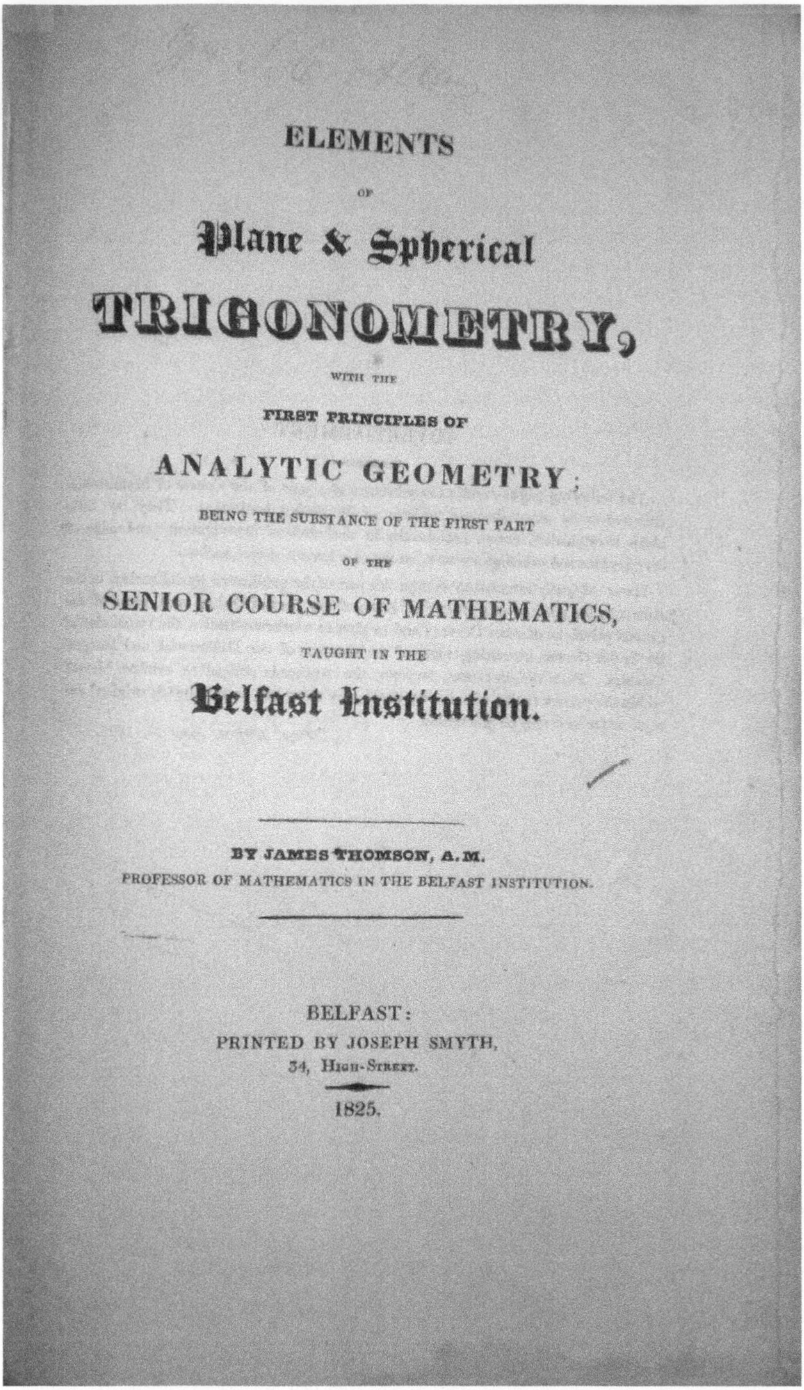

ELEMENTS

OF

Plane & Spherical

TRIGONOMETRY,

WITH THE

FIRST PRINCIPLES OF

ANALYTIC GEOMETRY;

BEING THE SUBSTANCE OF THE FIRST PART

OF THE

SENIOR COURSE OF MATHEMATICS,

TAUGHT IN THE

Belfast Institution.

BY JAMES THOMSON, A.M.

PROFESSOR OF MATHEMATICS IN THE BELFAST INSTITUTION.

BELFAST:

PRINTED BY JOSEPH SMYTH,

34, High-Street.

1825.

Figure 16.2 Title page of the first edition of Thomson's *Elements of Plane and Spherical Trigonometry*, published 1825.

such higher learning was no longer confined in Ireland to the University of Dublin. The reviewer's one qualm was that he felt that Thomson's use of Lagrange's method as a basis of calculus was unhelpful.[54] Thomson, it seems, agreed and he rewrote the second edition using the method of limits.

Leaving Belfast

One of the young Lord Kelvin's earliest memories was the excitement in the family home in Belfast when it was announced that his father was to be awarded an honorary LLD by Glasgow University. But in the same year, although they were not aware of them, storm clouds were gathering over the family.

At the conclusion of a letter dated 10 April 1829 to Thomas Chalmers, who by this stage was professor of divinity at Edinburgh University, Thomson stated that 'Mrs Thomson joins me in the kindest regards. She is very slowly recovering from a premature confinement'.[55] The couple's youngest son, Robert, had been born in February. Both mother and baby were initially unwell, and although Robert recovered and lived a long life, Margaret did not. She died the next April, to be followed to the same grave by their youngest daughter, also named Margaret, the following year. James Thomson endeavoured to fill the huge void that was left. The cots of the youngest children were moved into his bedroom,[56] and, according to his eldest daughter Elizabeth, in the evenings '[h]e gathered us about him, and in every way strove to supply the place of our lovely mother. He was indeed both father and mother to us, and watched over us continually'.[57]

Whether his wife's death was an important motivator for James Thomson to leave Belfast or not is unclear, but by 7 December 1830 he is again writing to Chalmers, this time about applying for the soon to be vacant chair of mathematics at Glasgow and asking if Chalmers might be willing to contact some of the electors on his behalf, or write a testimonial in support of his application. Hearing no response, he wrote again two weeks later pleading that 'any thing from you would be of the greatest service'.[58] It seems that Chalmers was not forthcoming with a testimonial, as nothing from him is included in the eleven testimonials which James Thomson gathered and bound at the end of 1830.[59] Of the eleven references, nine note Thomson's books, and the only publication which is explicitly mentioned (by three) is *Introduction to the Differential and Integral Calculus*. 'As to Dr. THOMSON'S acquirements as a Mathematician, I shall speak very briefly; for his publications show that he has directed his attention to the highest branches of analytical science—the Integral Calculus, the Calculus of Differences, the Calculus of Variations', wrote S. J. McClean, a Fellow of Trinity College

[54] *The National Magazine* 2/1 (1831), pp. 90–4.

[55] James Thomson to Thomas Chalmers CHA4.129.13 (1829), New College Library Archive, Edinburgh.

[56] Hamilton, *Belfast Literary Society*, p. 63.

[57] King, *Lord Kelvin's Early Home*, p. 87.

[58] James Thomson to Thomas Chalmers CHA4.149.31, CHA4.149.33 (1830), New College Library Archive, Edinburgh.

[59] Testimonials in favour of James Thomson. MS Gen 1752/1/2/2, Special Collections, Glasgow University.

Dublin. Thomson's latest book had come along at just the right time to help boost his chances. In fact, it had been published just in time. The publication date of *Introduction to the Differential and Integral Calculus* is 1831, with Thomson's preface being dated November 1830. Given that all three testimonials which explicitly mention the calculus text are dated either 15 or 16 December 1830, clearly Thomson had sent copies of the book hot off the presses to the respective referees, all three of whom were in Dublin.

Although Thomson thought that elections to the chair would take place in January 1831, there was considerable delay and on 24 November 1831 he wrote to Chalmers once again, asking if he would contact the newly appointed Lord Rector of Glasgow in his favour stating, perhaps to Chalmers' relief, 'I feel quite ashamed again to trouble you on this subject. But this will, I believe, be the last time. At present my prospect of success is rather encouraging. An additional vote however, may be of great consequence'.[60] Thomson's views on his prospects were accurate. On Friday 16 December 1831, he was elected to the Glasgow chair. It was a position which he occupied until his death in 1849. By the end of the next week following his election Thomson had made it clear that he was willing to teach for the remainder of the academic session at the Belfast Institution,[61] and thus give time for his successor to be sought. Although the Institution advertised the two positions, of professor of mathematics in the college department and teacher of mathematics in the school, in such a way that they could be filled separately, initially they were filled by one man: Rev. William Mulligan. Mulligan, however, died by drowning in August 1833. Thereafter two members of staff were employed; John Radford Young to teach mathematics in the college department, and Rev. Isaiah Steen in the school department. Steen it seems published only one textbook during his career: *A Treatise on Mental Arithmetic in Theory and Practice* in 1846,[62] but Young, though almost entirely self-taught, published a large number of books, both before and after his arrival in Belfast.[63]

On Wednesday 1 August 1832, Belfast gave Thomson a proud farewell. In the afternoon he was presented with silver plate in the Common Hall of the Institution, and then at 6 p.m. guests gathered at the Royal Hotel, Donegall Place for dinner, speeches, and toasts. Perhaps the most effusive speech came from one of Thomson's ex-pupils, Joseph Napier, a Dublin barrister who, although only 27 years old, claimed that

> Before [Thomson's] arrival amongst us, mathematical and physical science was in its dusky morning twilight. I can remember when its name was but little honoured, and its various branches but sparingly cultivated. Now I can testify with an honest pride, that many a humble cottager, and many a humble peasant in Ulster, can turn from the

[60] James Thomson to Thomas Chalmers CHA4.169.35 (1831), New College Library Archive, Edinburgh.

[61] Minute Book of the Faculty of the Belfast Academical Institution, 1818–1832, 22 December 1831, p. 259. Public Records Office of Northern Ireland, SCH/524/3C/1

[62] Isaiah Steen, *A treatise on mental arithmetic in theory and practice*, Belfast: Henderson 1846. Second and third editions were published in 1848 and 1857.

[63] E. I. Carlyle, rev. Alan Yoshioka, 'Young, John Radford (1799–1885)', in: *Oxford Dictionary of National Biography*, Oxford: Oxford University Press 2004, https://doi.org/10.1093/ref:odnb/30274.

> labours of his daily toil to scan, with philosophic eye, the mysterious wonders of the heavens; to strengthen his intellect with the deductions of reason and the certainty of demonstration, and refresh and delight his mind with the pleasures and pursuits of science.[64]

Napier's words, for all their rhetorical flourish, point to the fact that in James Thomson the rapidly growing town of Belfast, and the young and ambitious Belfast Academical Institution, had an enthusiastic and respected mathematical educator whose books were bringing him financial success and recognition well beyond the confines of the province of Ulster. In his letter of resignation, he stated that he had spent 'a large proportion of the best, and, I have no doubt happiest days of my life'[65] at what at the end of 1831 had become the *Royal* Belfast Academical Institution. But the opportunity of a chair at one of Scotland's universities was always going to be impossible for Thomson to resist, and so after a delay to avoid the worst of a cholera outbreak in Glasgow, the family moved across the Irish Sea in October 1832.[66]

During his time at Glasgow, Thomson continued to be a successful educator, producing new editions of some of his books, and also publishing three more: an edition of Euclid; a book on algebra; and a 'key' giving solutions to problems in the algebra text (Table 16.1). Although he published a small amount of research, none of it was of any great import.[67] He also modernized the mathematics curriculum and was part of the movement to reform aspects of the university. However, the most important thing that he ultimately bequeathed to Glasgow was his son William, later Lord Kelvin. William Thomson occupied the chair of natural philosophy at Glasgow from 1846 until 1899, and his huge fame as *the* natural philosopher in the Victorian era brought corresponding glory and renown to the university. It is not for nothing that when James Coutts's *A History of the University of Glasgow* was published in 1909, with over 500 years of history to choose from, the image picked for the frontispiece was that of Lord Kelvin.

Although James had escaped the outbreak of cholera in 1832, an outbreak of 1848–49 ended his life. He died, delirious, on Friday 12 January 1849. His son William recorded that shortly before his death

> he burst out rather faintly into a very incoherent set of expressions of numbers in all varieties of arithmetical denominations, hurrying rapidly from one to another, and giving the answer or saying "That's right! Now what is seven hundred and eighty-six inches equal to?" and so on for several minutes. His mind wandering back to the school times 20 or 30 years ago![68]

[64] *Belfast News Letter*, Friday 3 August 1832, p. 2.

[65] Jamieson, *History of the Royal Belfast Academical Institution*, p. 16.

[66] King, *Lord Kelvin's Early Home*, p. 97.

[67] James Thomson, 'On the true and extended interpretations of formulae in spherical trigonometry', in: *The London and Edinburgh Philosophical Magazine and Journal of Science* 10 (1837), pp. 18–24; James Thomson, 'A geometrical proposition', in: ibid. 15 (1839), pp. 41–2; James Thomson, 'Investigation of a new series for the computation of logarithms: With a new investigation of a series for the rectification of the circle', in: *Transactions of the Royal Society of Edinburgh* 14/1 (1840), pp. 217–23.

[68] King, *Lord Kelvin's Early Home*, p. 241.

However, rather than end on such a sorrowful note, it is perhaps better to give the last words to John Nichol, a childhood friend of Thomson's children in Scotland, and a man whose father, J. P. Nichol, had been James Thomson's colleague as professor of astronomy at Glasgow:

> Old Dr. James was one of the best of Irishmen, a good mathematician, an enthusiastic and successful teacher, the author of several valuable school books, a friend of my father's, and himself the father of a large family, the members of which have been prosperous in the world ... Good-hearted, he was shrewdly alive to his interest without being selfish, and would put himself to some trouble, and even expense, to assist his friends. He was a stern disciplinarian, and did not relax his discipline when he applied it to his children, and yet the aim of his life was their advancement.[69]

Appendix: Junior and Senior Mathematical Course Questions

The Belfast Magazine and Literary Journal lists the following questions as being asked at the April 1825 examinations for the Junior and Senior Mathematical Courses of the Belfast Academical Institution. The list of questions is prefaced with 'At the examinations on MATHEMATICS, the following questions, out of many others, were all answered by the Students'; however it seems reasonable to assume that they provide a representative subset of the questions asked.

Junior Mathematical Course

1. Given the perimeter and altitude of an isosceles triangle; to construct it.
2. Given the diagonals and the ratio of the sides of a parallelogram; to construct it.
3. If a tangent drawn through the vertex of a triangle inscribed in a circle, meet the base produced. The line bisecting their angle of intersection cuts the triangle in such a manner, that each of the segments next the vertex, is a mean proportional between the segments next the base. Required a proof.
4. If a vertical angle of a triangle be double of one of the angles at the base, the rectangle under the sides is equal to the rectangle under the base and the line bisecting the vertical angle.
5. Given the base, the difference of the sides, and the difference of the angles at the base of a plane triangle; to construct it.
6. In a given circle, to inscribe a rectangle, having its sides in a given ratio.
7. Prove the fourth proposition of the second book of Euclid by proportion.
8. On a given base to construct a triangle having its other sides in a given ratio, and its area a maximum.
9. Prove that the segments of the base of a triangle made by a perpendicular, are proportional to the cotangents of the adjacent angles at the base; and show from this a method of resolving a triangle, when two sides and the contained angle are given.
10. Prove the third proposition of the sixth book of Euclid by trigonometry.
11. Prove that the base of a plain triangle is to the sum of the sides, as the cosine of half the vertical angle is to the sine of half the difference of the angles at the base.

[69] William A. Knight, *Memoir of John Nichol*, Glasgow: James MacLehose & Sons 1896, pp. 19–20.

12. Prove that the base of a plain triangle is to the difference of the sides, as the cosine of half the vertical angle is to the sine of half the difference of the angles at the base.

Resolve the following equations:-

13. $\frac{3x^2 - 8x+2}{x-3} = \frac{6x^2 + 13x - 4}{2x}$.
14. $x = a + \sqrt{3x^2 - 4a^2}$.
15. $x - y = a$ and $x^3 - y^3 = b$.
16. $\frac{3x^2 - 11x + 2}{x - 2} = \frac{6x^2 - 24x}{2x - 5}$.

Senior Mathematical Course

1. Prove that $\cos A = \frac{1}{2}\sqrt{1 + \sin 2A} + \frac{1}{2}\sqrt{1 - \sin 2A}$.
2. Prove that $\frac{\sin A + \sin B}{\cos A + \cos B} = \tan(\frac{1}{2}(A + B))$.
3. Prove that $4 \cos A^3 = \cos 3A + 3 \cos A$.
4. In a plain triangle, prove from the formula $\cos A = \frac{b^2 + c^2 - a^2}{2bc}$, that the area is equal to $\sqrt{s(s - a)(s - b)(s - c)}$, where $s = \frac{1}{2}(a + b + c)$.
5. In a spherical triangle, prove that the sines of the sides are proportional to the sines of the opposite angles, from the fundamental formula, $\cos a = \cos A \sin b \sin c + \cos b \cos c$.
6. Given the sun's declination at a given hour, on a given day, to find the latitude.
7. What two places on the tropic of Cancer are each 5,000 miles distant from Belfast?
8. Trace the mutations in the signs of the chord of a variable circular arc.
9. If two sides of a spherical triangle be each 45°, the cosine of the remaining side is equal to the square of the cosine of half the opposite angle.
10. If two angles of a spherical triangle be each 45°, the square of the cosine of half the remaining angle is equal to half the square of the cosine of half its opposite side.
11. In a right angled, isosceles, spherical triangle, the sine of one of the equal angles is to the sine of 45°, as the radius is to cosine of half the hypotenuse.
12. To investigate the mode of finding the latitude and longitude of any of the heavenly bodies from its right ascension and declination; and conversely.
13. Given the latitude and longitude, or the right ascension and declination of two stars, to find their distance asunder.
14. Given the distances of a comet from two known stars; to find its latitude, longitude, right ascension, and declination.
15. Find the differential of $(a + x)^x$.
16. Find the integral of $dx.\frac{x^4 + a^4}{x + a}$.
17. Find the equation of the parabola, the focus being the origin.
18. Find the subtangent of the hyperbola.
19. Find the subtangent of the curve whose equation is $x^2y = a^3$.
20. Find the area of the same curve.

Bibliography

Aa, A. J. van der, *Geschiedkundige beschrijving van de stad Breda en hare omstreken*, Gorinchem: Noorduyn en Zoon 1845.

Abram, William Alexander, 'Memorial of the Late T.T. Wilkinson, F.R.A.S, of Burnley', in: *Transactions of the Historic Society of Lancashire and Cheshire* 4 (3rd ser.) (1875), pp. 77–94.

Ackerberg-Hastings, Amy, 'John Playfair on British Decline in Mathematics', in: *BSHM Bulletin: Journal of the British Journal for the History of Mathematics* 23 (2008), pp. 81–95.

Adams, Andrew and Richard Woodman, *Light Upon the Waters: The History of Trinity House 1514–2014*, London: The Corporation of Trinity House 2013.

Adams, John, *The Young Sea-officer's Assistant*, London: Lockyer Davis 1773.

Adamson, Ian R., 'The Administration of Gresham College and its Fluctuating Fortunes in the Seventeenth Century', in: *History of Education* 9 (1980), pp. 13–25.

Addison, W. Innes, *Prize Lists of the University of Glasgow from Session 1777–78 to Session 1832–33*, Glasgow: Carter & Pratt 1902.

Addison, W. Innes, *The Matriculation Albums of the University of Glasgow from 1728–1858*, Glasgow: James Maclehose and Sons 1913.

Adrain, Robert, 'A View of the Diophantine Analysis', in: *Liverpool Apollonius*, no. 2 (1824), pp. 86–91.

Akveld, L. M. and W. J. van Hoboken (eds), *Maritieme geschiedenis der Nederlanden: Zeventiende eeuw, van 1585 tot ca 1680*, 4 vols, Bussum: De Boer Maritiem 1977.

Alberti, Hans Joachim, *Entwickelung des bergmännischen Rißwesens*, Diplom-Arbeit im Markscheide-Institut der Bergakademie Freiberg, 1927.

Alberti, Samuel J. M. M., 'Amateurs and Professionals in One County: Biology and Natural History in Late Victorian Yorkshire', in: *Journal of the History of Biology* 34 (2001), pp. 115–47.

Alberti, Samuel J. M. M., 'Natural History and the Philosophical Societies of Late Victorian Yorkshire', in: *Archives of Natural History* 30/2 (2003), pp. 342–58.

Alberti, Samuel J. M. M., 'Conversaziones and the Experience of Science in Victorian England', in: *Journal of Victorian Culture* 8/2 (2003), pp. 208–30.

Alborn, Timothy, 'A Calculating Profession: Victorian Actuaries among the Statisticians', in: *Science in Context* 7 (1994), pp. 433–68.

Alborn, Timothy, 'Quill-Driving: British Life-Insurance Clerks and Occupational Mobility, 1800–1914', in: *Business History Review* 82 (2008), pp. 31–58.

Albree, Joe and Scott H. Brown, '"A Valuable Monument of Mathematical Genius": *The Ladies' Diary* (1704–1840)', in: *Historia Mathematica* 36 (2009), pp. 10–47.

Almeida, André Ferrand de, '"Arrumar as terras, os rios e os montes": os jesuitas matemáticos e os mapas do Brasil meridional, 1720–1748', in: *Manoel de Azevedo Fortes (1660–1749): Cartografia, Cultura e Urbanismo*, ed. Mário Gonçalves Fernandes, Porto: Gedes 2006, pp. 99–122.

Amicable Society, *The Charter of the Corporation of the Amicable Society for a Perpetual Assurance-office; Together with the By-laws thereunto Belonging*, London: George Sawbridge 1710.

Anderson, James Maitland, *The Matriculation Roll of the University of St Andrews 1747–1897*, Edinburgh: William Blackwood and Sons 1905.

Anon., *Fama Fraternitatis: Oder Entdeckung der Bruderschaft des Hochlöblichen Ordens des R.C. An die Häupter, Stände und Gelehrten in Europa*, in: *Allgemeine vnd General Reformation der gantzen weiten Welt. Beneben der Fama Fraternitatis, Deß Löblichen Ordens des Rosenkreutzes, an alle Gelehrte und Häupter Europae geschrieben: Auch einer kurtzen Responsion, von dem Herrn Haselmeyer gestellet, welcher deßwegen von den Jesuitern ist gefänglich eingezogen, und auff eine Galleren geschmiedet: Itzo öffentlich in Druck verfertiget, und allen trewen Hertzen communiciret worden*, Cassel: Wilhelm Wessel 1614.

Anon., *Chymische Hochzeit: Christiani Rosencreutz. Anno 1459. Arcana publicata vilescunt; et gratiam prophanata amittunt. Ergo: ne Margaritas obijce porcis, seu Asino substerne rosas*, Strasbourg 1616.

Anon., *Gründliche Warhaffte Erzehlung Was in den Etlich Jahr wehrenden aber noch nit zu End gebrachten Stritten zwischen Johann Faulhaber und Gegentheil sich verloffen, von einer eifrigen Christlichen Persohn getreulich an Tag geben*, o. O. 1621.

Anon., *A Catalogue of the Curious Mathematical, &c Books of the Late Mr. Edw. Rollinson* [London 1775].

Anon., [Obituary of Charlotte Hutton], in: *The Gentleman's Magazine* (October 1794), pp. 960–1.

Anon., [Review of Hutton, *Dictionary*], in: *English Review* 28 (July 1796), pp. 14–19.

Anon., [Review of Hutton, *Dictionary*], in: *Critical Review* (November 1796), pp. 302–5.

Anon., 'Lineal Sections', in: *The Student*, no. 1 (1797), p. 38.

Anon., [Review of Hutton, *Dictionary*], in: *The Monthly Review* (1798), p. 185.

Anon., 'Charles Hutton', in: *Public Characters*, 10 vols, London 1799–1809, vol. 2, pp. 97–123.

Anon., [Review of Hutton *et. al* (eds), *Abridgement*, vol. 1], in: *The British Critic* (1803), p. 540.

Anon., 'Mr. Woodhouse on the rectification of the hyperbola', in: *The Gentleman's Magazine* 85 (1815), pp. 18–22.

Anon., 'Lagrange's "A Treatise upon Analytical Mechanics ..."', in: *The Monthly Review or Literary Journal* 78 (October 1815), pp. 211–13.

Anon., *A Catalogue of the Entire, Extensive and Very Rare Mathematical Library of Charles Hutton, L.L.D.*, [London 1816].

Anon., *Tribute of Respect to Charles Hutton, LL.D. F.R.S. &c. &c.*, [London 1822].

Anon., [Obituary of Charles Hutton], in: *Monthly Magazine* 55 (March 1823), pp. 137–42.

Anon., [Obituary of Charles Hutton], in: *The Edinburgh Annual Register* 16 (December 1823), pp. 328–31.

Anon., [Obituary of Charles Hutton], in: *The Mathematical Repository* (N.S.) 5 (1830), pp. 187–96.

Anon., *Catalogue of a Miscellaneous Collection of Books: Being the Valuable and Scientific Library of the Late Dr. Olinthus Gregory [...] Which Will Be Sold by Auction by Messrs. Southgate and Son ... on Thursday, March the 17th, 1842 and following day* [London 1842].

Anon., 'Mathematics', in: *The Preston Chronicle*, issue 1703, Saturday 19 April 1845.

Anon., 'Mathematics', in: *The Preston Chronicle*, issue 1713, Saturday 28 June 1845.

Anon., 'University Honours', in: *The Preston Chronicle*, issue 2266, Saturday 2 February 1856, p. 4.

Anon., [Obituary of Charles Ansell], in: *Proceedings of the Royal Society of London* 34 (1882), pp. vii–viii.

Anon., 'Rambling Remarks', in: *British Weekly: A Journal of Social and Christian Progress*, 1 July 1897, p. 184.

Ansell, Charles, *Tables Exhibiting the Law of Mortality: Deduced from the Combined Experience of Seventeen Life Assurance Offices*, London: J. King 1843.

Appleby, Joyce Oldham, *Economic Thought and Ideology in Seventeenth-Century England*, Princeton, NJ: Princeton University Press 1978.

Apt, A. J., 'Wright, Edward (bap. 1561, d. 1615), Mathematician and Cartographer', in: *Oxford Dictionary of National Biography*, 2004, https://doi.org/10.1093/ref:odnb/30029.

Araújo, Renata de, 'Manoel de Azevedo Fortes e o estatuto dos Engenheiros Portugueses', in: *Manoel de Azevedo Fortes (1660–1749): Cartografia, Cultura e Urbanismo*, ed. Mário Gonçalves Fernandes, Porto: Gedes 2006, pp. 15–34.

Archibald, Raymond, 'Notes on Some Minor English Mathematical Serials', in: *The Mathematical Gazette* 14 (1929), pp. 379–400.

Ash, Eric H., *Power, Knowledge, and Expertise in Elizabethan England*, Baltimore: Johns Hopkins University Press 2004.

Ashworth, William J., 'Baily, Francis (1774–1844)', in: *Oxford Dictionary of National Biography*, Oxford: Oxford University Press 2007, https://doi.org/10.1093/ref:odnb/1077.

Asmussen, Tina, 'The *Kux* as a Site of Mediation: Economic Practices and Material Desires in the Early Modern German Mining Industry', in: *Sites of Mediation: Connected Histories of Places, Processes, and Objects in Europe and Beyond, 1450–1650*, ed. Susanna Burghartz, Lucas Burkart, and Christine Göttler, Intersections 47, Leiden: Brill 2016, pp. 159–82.

Atlas adapted to Thomson's Modern Geography, Belfast: Simms & McIntyre 1828.

Attar, Karen, 'Augustus De Morgan (1806–1871), his Reading, and his Library', in: *The Edinburgh History of Reading: Modern Readers*, ed. Mary Hammond, Edinburgh: Edinburgh University Press 2020.

Aubrey, John, *Brief Lives, with An Apparatus for the Lives of our English Mathematical Writers*, ed. Kate Bennett, 2 vols, Oxford: Oxford University Press 2015.

Ayres, John, *The trades-mans Copy book, or, Apprentices Companion*, London: by the author 1688.

Ayres, John, *The Penmans daily Practise*, London: by the author 1690.

Azevedo Fortes, Manoel de, *Tratado do Modo o mais facil, e o mais exacto de Fazer as Cartas Geograficas*, Western Lisbon: Pascoal da Sylva 1722; https://purl.pt/16976.

Azevedo Fortes, Manoel de, *O Engenheiro Portuguez*, 2 vols, Western Lisbon: Manoel Fernandes da Costa 1728–29, https://purl.pt/14547.

Azevedo Fortes, Manoel de, *Oraçaõ Academica, que pronunciou Manoel de Azevedo Fortes, na presença de Suas Magestadas, hindo a Academia ao Paço em 22 de Outubro de 1739*, no place: no publisher 1739, https://dspace.uevora.pt/ri/handle/123456789/266.

Azevedo Fortes, Manoel de, *Logica Racional, Geometrica, e Analitica*, Lisbon: Jozé Antonio Plates 1744, https://digital.bbm.usp.br/handle/bbm/2264

[Azevedo Fortes, Manoel de], *Elementos das Mathematicas, ou Tractado da grandeza em geral*, Biblioteca Nacional de Portugal, cód. 1861; cód. 5194//1; cód. 6205//17 (incomplete).

Azevedo Fortes, Manoel de, *Geometria Espiculativa; Trigonometria Espherica; Modo de riscar e dar aguadas nas plantas melitares*, Biblioteca Pública de Évora, cód. Manizola 258, https://dspace.uevora.pt/ri/handle/123456789/281

Babbage, Charles, *Reflections on the Decline of Science in England, and on Some of its Causes*, London: Fellowes 1830.

Baily, Francis, *An Account of the Several Life-assurance Companies Established in London: Containing a View of their Respective Merits and Advantages*, London: Richardson 1810.

Baily, Francis, *The Doctrine of Life-annuities and Assurances: Analytically Investigated and Explained, Together with Several Useful Tables Connected with the Subject and a Variety of Practical Rules for the Illustration of the Same*, London: J. Richardson 1810.

Baily, Francis, *An Account of the Revd. John Flamsteed, the First Astronomer Royal*, London: printed by order of the Lords Commissioners of the Admiralty 1835.

Baker, Alexi, '"Humble Servants," "Loving Friends" and Nevil Maskelyne's Invention of the Board of Longitude', in: *Maskelyne: Astronomer Royal*, ed. Rebekah Higgitt, London: Hale Books 2014, pp. 203–28.

Baker, Thomas, *Clavis geometrica catholica: sive janua aequationum reserata/The Geometrical Key: or the gate of equations unlock'd*, London: J. Playford 1684.
Baldwin, R. C. D., 'Borough, William (bap. 1536, d. 1598), Explorer and Naval Administrator', in: *Oxford Dictionary of National Biography*, 2008, https://doi.org/10.1093/ref:odnb/2915.
Ball, Walter William Rouse, *A History of the Study of Mathematics at Cambridge*, Cambridge: at the University Press 1889.
Barbin, Évelyne, 'On French heritage of Cartesian geometry in Elements from Arnauld, Lamy and Lacroix', in: *"Dig Where You Stand" 5: Proceedings of the Fifth International Conference on the History of Mathematics Education*, ed. Kristín Bjarnadóttir et al., Utrecht: Freudenthal Institute 2015, pp. 11–28.
Bardi, Alberto, 'Scientific Interactions in Colonial, Multilinguistic and Interreligious Contexts: Venetian Crete and the Manuscript Marcianus latinus VIII.31 (2614)', in: *Centaurus* 63 (2021), pp. 339–52.
Barlow, William, *The Nauigators Supply: Conteining Many Things of Principall Importance Belonging to Nauigation, with the Description and Vse of Diuerse Instruments Framed Chiefly for that Purpose; but Seruing also for Sundry Other of Cosmography in Generall*, London: G. Bishop, R. Newbery and R. Barker 1597.
Barnard, Toby, 'The Hartlib Circle and the Cult and Culture of Improvement in Ireland', in: *Samuel Hartlib and Universal Reformation. Studies in Intellectual Communication*, ed. Mark Greengrass, Michael Leslie, and Timothy Raylor, Cambridge: Cambridge University Press 1994, pp. 281–97.
Barnes, C. L., *The Manchester Literary & Philosophical Society*, Manchester: [*s.n.*] 1938.
Barrett, C. R. B., *The Trinity House of Deptford Strond*, London: Lawrence & Bullen 1893.
Barrett, Katy, 'The Wanton Line: Hogarth and the Public Life of Longitude', PhD dissertation, University of Cambridge 2014.
Barrow, Isaac (ed.), *Euclidis elementorum libri XV. breviter demonstrati*, Cambridge: Cambridge University Press for William Nealand 1655.
Barrow, Isaac (ed.), *Euclide's Elements; the Whole Fifteen Books. Compendiously Demonstrated*, London: R. Daniel for William Nealand in Cambridge 1660.
Barrow, Isaac, *Lectiones XVIII, Cantabrigiae in scholis publicis habitae; in quibus opticorum phaenomenon genuinae rationes investigantur, ac exponuntur. Annexae sunt lectiones aliquot geometricae*, London: William Godbid for John Dunmore and Octavian Pulleyn 1669.
Barrow, Isaac, *Lectiones opticae & geometricae: in quibus phaenomenon opticorum genuinae rationes investigantur, ac exponuntur: et generalia curvarum linearum symptomata declarantur*, 2 parts, London: William Godbid for Robert Scott 1674.
Barrow, Isaac (ed.), *Archimedis opera: Apollonii Pergaei conicorum libri IIII. Theodosii sphaerica: method nova illustrata, & succincte demonstrata*, 3 parts, London: William Godbid for Robert Scott 1675.
Barrow, Isaac, *Lectio reverendi et doctissimi viri D. Isaaci Barrow [...] in qua Theoremata Archimedis de Sphaera & Cylindro, per methodum Indivisibilium investigate, ac breviter demonstrata exhibentur*, London: John Redmayne for J. Williams 1678.
Barrow, Isaac, *Euclidis Elementorum libri XV. breviter demonstrati, opera Is. Barrow [...] et prioribus mendis typographicis nunc demum purgati*, London: John Redmayne for J. Williams 1678.
Barrow–Green, June, '"A Senior Wrangler among Senior Wranglers": Ellis's Mathematical Education' in: *A Prodigy of Universal Genius: Robert Leslie Ellis, 1817–1859*, ed. Lukas M. Verburgt, Springer 2022.
Bartels, Christoph, *Vom frühneuzeitlichen Montangewerbe zur Bergbauindustrie: Erzbergbau im Oberharz 1635–1866*, Bochum: Deutsches Bergbau-Museum 1992.

Bartels, Christoph, 'Der Harzer Oberbergmeister Georg Andreas Steltzner (1725–1802) und die Montanwissenschaften in der zweiten Hälfte des 18. und am Beginn des 19. Jahrhunderts', in: *Staat, Bergbau und Bergakademie. Montanexperten um 18. und frühen 19. Jahrhundert*, Vierteljahrschrift für Sozial- und Wirtschaftsgeschichte 223, ed. Hartmut Schleiff and Peter Konečný, Stuttgart: Franz Steiner 2015, pp. 279–80.

Becchi, Antonio, Domenico Bertoloni Meli, and Enrico Gamba (eds), *Guidobaldo del Monte (1545–1607): Theory and Practice of the Mathematical Disciplines from Urbino to Europe*, Berlin: Edition Open Access 2013.

Becher, Harvey, 'Radicals, Whigs, and Conservatives: The Middle and Lower Classes in the Analytical Revolution at Cambridge in the Age of Aristocracy', in: *British Journal for the History of Science* 28 (1995), pp. 405–26.

Beeley, Philip, 'A Philosophical Apprenticeship: Leibniz's Correspondence with the Secretary of the Royal Society, Henry Oldenburg', in: *Leibniz and His Correspondents*, ed. Paul Lodge, Cambridge: Cambridge University Press 2004, pp. 47–73.

Beeley, Philip, 'Eine Geschichte zweier Städte: Der Streit um die wahren Ursprünge der Royal Society', in: *Acta Historica Leopoldina* 49 (2008), pp. 135–62.

Beeley, Philip, '"To the Publike Advancement": John Collins and the Promotion of Mathematical Knowledge in Restoration England', in: *BSHM Bulletin: Journal of the British Society for the History of Mathematics* 32 (2017), pp. 61–74.

Beeley, Philip, 'Practical Mathematicians and Mathematical Practice in later Seventeenth-Century London', in: *British Journal for the History of Science* 52/2 (2019), pp. 225–48.

Beeley, Philip, 'Leibniz and the Royal Society Revisited', in: *Leibniz's Legacy and Impact*, ed. Julia Weckend and Lloyd Strickland, New York: Routledge 2020, pp. 23–52.

Beeley, Philip, '"A designe Inchoate": Edward Bernard's Planned Edition of Euclid and its Scholarly Afterlife in Late Seventeenth-Century Oxford', in: *Reading Mathematics in Early Modern Europe: Studies in the Production, Collection, and Use of Mathematical Books*, ed. Philip Beeley, Benjamin Wardhaugh, and Yelda Nasifoglu, London: Routledge 2021, pp. 192–229.

Beeley, Philip, '"Our Learned Countryman": Thomas Harriot and the Emergence of Mathematical Community in Seventeenth-Century England', in: *Thomas Harriot: Science and Discovery in the English Renaissance*, ed. Robert Fox, London: Routledge 2023, pp. 72–102.

Beeley, Philip and Christoph J. Scriba (eds), *Correspondence of John Wallis (1616–1703)*, 4 vols (to date), Oxford: Oxford University Press, 2003–.

Beger, Mathæus, *Problema astronomicum: Die Situs der Sternen, Planetarum oder Cometarum zu observirn ohne Instrumenta, allein mit einem geraden Lineal oder Faden*, [n.p.] 1619.

Bell, D., *An Experiment in Education: The History of Worcester College for the Blind, 1866–1966*, London: Hutchinson 1967.

Bell, H. T. M. and Maria Panteki, 'Harley, Robert (1828–1910), Mathematician and Congregational Minister', in: *Oxford Dictionary of National Biography*, 2008, https://doi.org/10.1093/ref:odnb/33715.

Bellhouse, David R., 'A New Look at Halley's Life Table', in: *Journal of the Royal Statistical Society: Series A (Statistics in Society)* 174 (2011), pp. 823–32.

Bellhouse, David R., *Leases for Lives: Life Contingent Contracts and the Emergence of Actuarial Science in Eighteenth-Century England*, Cambridge: Cambridge University Press 2017.

Bennett, Jim, 'The Challenge of Practical Mathematics', in: *Science, Culture and Popular Belief in Renaissance Europe*, ed. Stephen Pumfrey, Paolo L. Rossi, and Maurice Slawinski, Manchester: Manchester University Press 1991, pp. 176–90.

Bennett, Jim, 'Instruments and Practical Mathematics in the Commonwealth of Richard Hakluyt', in: *Hakluyt and Oxford*, ed. Anthony Payne, London: The Hakluyt Society 2017, pp. 35–52.

Bennett, Jim, '"That Incomparable Instrument Maker": The Reputation of Henry Sutton', in: *The Whipple Museum of the History of Science: Objects and Investigations*, ed. J. Nall, L. Taub and F. Willmoth, Cambridge: Cambridge University Press 2019, pp. 83–100.

Bennett, Jim, 'Mathematics, Instruments and Navigation, 1600–1800', in: *Mathematics and the Historian's Craft: The Kenneth O. May Lectures*, ed. Glen van Brummelen and Michael Kinyon, New York: Springer 2000, pp. 43–56.

Bennett, J. M., *Sir James Cockle: First Chief Justice of Queensland, 1863–1879*, Sydney: Federation Press 2003.

Bennett, Kate, 'John Aubrey and the "Lives of our English mathematical writers"', in: *The Oxford Handbook of the History of Mathematics*, ed. Eleanor Robson and Jacqueline Stedall, Oxford: Oxford University Press, 2009, pp. 329–52.

Bergen, Amanda, 'A Philosophical Experiment: The Wilberforce Memorial School for the Blind *c.*1833–1870', in: *European Review of History: Revue européenne d'histoire* 14/2 (2007), pp. 147–64.

Bernardo, Luís Manuel A. V., *O Projecto Cultural de Manuel de Azevedo Fortes*, Lisbon: Imprensa Nacional-Casa da Moeda 2005.

Bernoulli, Nikolaus, *Dissertatio Inauguralis Mathematico-juridica de Usu Artis Conjectandi in Jure*, Basil: Johannis Conradi 1709; English translation by Thomas Drucker in: Haberman and Sibbett, *History of Actuarial* Science, vol. I, pp. 187–96.

Biagioli, Mario, 'The Social Status of Italian Mathematicians, 1450–1600', in: *History of Science* 27 (1989), pp. 1–75.

Biagioli, Mario, 'Replication or Monopoly? The Economies of Invention and Discovery in Galileo's Observations of 1610', in: *Science in Context* 13 (2000), pp. 547–92.

Bierbrier, Morris L., *Who Was Who in Egyptology*, 4th rev. edn, London: Egypt Exploration Society 2012.

Bierens de Haan, David, *Bouwstoffen voor de geschiedenis der wis- en natuurkundige wetenschappen in de Nederlanden*, 2 vols, Amsterdam, 1878–87.

Bierens de Haan, David, *Bibliographie néerlandais historique scientifique*, Rome: Imprimerie des Sciences Mathématique et Phisique 1883.

Biggs, Norman, 'T. P. Kirkman, Mathematician', in: *Bulletin of the London Mathematical Society* 13 (1981), pp. 97–120.

Biggs, Norman, 'John Reynolds of the Mint: A Mathematician in the Service of King and Commonwealth', in: *Historia Mathematica* 48 (2019), pp. 1–28.

Birch, Thomas (ed.), *The History of the Royal Society of London, for Improving of Natural Knowledge, from its first rise*, 4 vols, London: for A. Millar 1756–57.

Black, M. A., *Chronological List & Statistical Chart of the Life Assurance Associations Established in the United Kingdom from 1706 to 1863 Showing Where They Are, When & How They Disappeared*, London: Edward Stanford 1864.

Blair, Ann, 'Note-Taking as an Art of Transmission', in: *Critical Inquiry* 31 (2004), pp. 85–107.

Blumhardt, J. F., *Catalogue of the Library of the India Office*, vol. 2, part 5: *Marathi and Gujarati Books*, London 1908.

Blundeville, Thomas, *M. Blundevile his Exercises*, London: Iohn Windet 1594.

Boltz, C. L., *Seventy Five Years of Popular Science: Record of the Hampstead Scientific Society, 1899–1974*, London: Hampstead Scientific Society 1974.

Booth, Philip M., '"Freedom with Publicity"—The Actuarial Profession and United Kingdom Insurance Regulation from 1844 to 1945', in: *Annals of Actuarial Science* 2 (2007), pp. 115–45.

Bos, Henk, *Redefining Geometrical Exactness: Descartes' Transformation of the Early Modern Concept of Construction*, New York: Springer 2001.

Boteler, Nathaniel, *Six Dialogues about Sea-Services between an High-Admiral and a Captain at Sea*, London: for Moses Pitt 1685.
Bottomley, James, 'Notes on the Early History of the Literary and Philosophical Society', in: *Proceedings of the Manchester Literary and Philosophical Society* 25 (1885–86), pp. 3–7.
Boulenger, Jean, *La Geometrie Pratique*, new edition with notes and a 'Traité de l'Arithmetique par Geometrie' by Jacques Ozanam, Paris: Michel David 1691.
Bourne, William, *A Regiment for the Sea: Conteyning Most Profitable Rules, Mathematical Experiences, and Perfect Knovvledge of Nauigation, for All Coastes and Countreys: Most Needefull and Necessarie for All Seafaring Men and Trauellers, as Pilotes, Mariners, Marchants.* [*et*] *c. Exactly Deuised and Made by VVilliam Bourne*, Imprinted at London: By [Henry Bynneman for] Thomas Hacket, and are to be solde at his shop in the Royall Exchaunge, at the signe of the Greene Dragon 1574.
Bourne, William, *A Regiment for the Sea, and Other Writings on Navigation*, ed. E. G. R. Taylor, Cambridge: Cambridge University Press for the Hakluyt Society 1963.
Bowyer, T. H., 'James, Sir William, First Baronet (1722–1783), Naval Officer and Director of the East India Company', in: *Oxford Dictionary of National Biography*, 2004, https://doi.org/10.1093/ref:odnb/14626.
Boyer, Marjorie Nice, 'Pappus Alexandrinus', in: *Catalogus Translationum et Commentariorum*, ed. P. O. Kristeller and F. E. Cranz, Washington DC: Catholic University of America Press 1969.
Bragagnolo, Manuela, 'Geografia e politica nel Cinquecento: La descrizione di città nelle carte di Gian Vincenzo Pinelli', in: *Laboratoire Italien* 8 (2008), pp. 163–93.
Brears, Peter, *Of Curiosities & Rare Things: The Story of Leeds City Museums*, Leeds: Friends of Leeds City Museums 1989.
Breimer, D. D., J. C. M. Damen, J. S. Freedman, M. Hofstede, J. Katgert, T. Noordermeer, and O. Weijers, *Hora Est! On Dissertations*, Leiden: Universiteitsbibliotheek Leiden 2005, https://openaccess.leidenuniv.nl/handle/1887/17795.
Brianta, Donata, 'Education and Training in the Mining Industry, 1750–1860: European Models and the Italian Case', in: *Annals of Science* 57/3 (2000), pp. 267–300.
Broucke, Jan van den, *Instructie der Zeevaert*, Rotterdam: Abraham Migoen 1609/10.
Brough, Bennett Hooper, *A Treatise on Mine-Surveying*, London: Griffin 1913.
Brown, John, *The Triangular Quadrant, or, The Quadrant on a Sector: Being a General Instrument for Land or Sea Observations: Performing All the Uses of the Ordinary Sea Instruments, as Davis Quadrant, Forestaff, Crosstaff, Bow, with More Ease, Profitableness, and Conveniency, and as Much Exactness as Any or All of Them: Moreover, It May Be Made a Particular and a General Quadrant for All Latitudes, and Have the Sector Lines Also: To Which Is Added a Rectifying Table to Find the Suns True Declination to a Minute or Two, Any Day or Hour of the 4 Years: Whereby to Find the Latitude of a Place by Meridian, or Any Two Other Altitudes of the Sun or Stars*, [London] 1662.
Brown, W., *Reports of Cases Argued and Determined in the High Court of Chancery: During the Time of Lord Chancellor Thurlow, of the Several Lords Commissioners of the Great Seal, and of Lord Chancellor Loughborough, from 1778 to 1794*, London: W. Clarke and Sons 1819.
Browne, Horace B., *Chapters of Whitby History, 1823–1946. The Story of Whitby Literary and Philosophical Society and of Whitby Museum*, Hull: A. Brown & Sons 1946.
Brożek, Jan, *De numeris perfectis disceptatio*, Amsterdam: Blaeu 1637.
Bruce, John, *A Memoir of Charles Hutton*, Newcastle 1823.
Bruijn, Jaap R., *Schippers van de VOC in de achttiende eeuw aan de wal en op zee*, Amsterdam: Bataafsche Leeuw 2008.
Bruijn, Jaap R., F. S. Gaastra, I. Schöffer, with E. S. van Eyck van Heslinga, *Dutch–Asiatic Shipping in the 17th and 18th centuries*, http://resources.huygens.knaw.nl/retroboeken/das.

Brusca, C. Euthymius de, *Vindiciarum Faulhaberianarum. Continuatio. Das ist Rechtmessige Rettung/ Herrn Johann Faulhabers Mathematici zu Ulm Famae Sidereae, Wider Die Ehrenrüge Teutsche DiffamationSchrifften/ Expolitio Famae sidereae, &c. und Postulatum aequitatis plenissimum, &c. genant/ Welche M. Zimpertus Wehe Lateinischer Schulen Collaborator zu Ulm. Under dem falschen Namen Hisaiae sub Cruce als durch offentlichen Truck spargirt hat*, Moltzheim: Stephan Bidermann 1620.

Bryden, David. J., 'The Instrument Maker and the Printer: Paper Instruments Made in Seventeenth Century London', in: *Bulletin of the Scientific Instrument Society* 55 (1997), pp. 3 15.

Bucciantini, M., *Galileo e Keplero: Filosofia, cosmologia e teologia nell'Età della Controriforma*, Turin: Einaudi 2000.

Buckland, William, 'Account of an Assemblage of Fossil Teeth and Bones of Elephant, Rhinoceros, Hippopotamus, Bear, Tiger, and Hyæna, and Sixteen Other Animals; Discovered in a Cave at Kirkdale, Yorkshire, in the Year 1821: With a Comparative View of Five Similar Caverns in Various Parts of England, and Others on the Continent', in: *Philosophical Transactions of the Royal Society of London* 112 (1822), pp. 171–236.

Burke, Peter, 'The Renaissance Dialogue', in: *Renaissance Studies* 3/1 (1989), pp. 1–12.

Burton, Anthony, 'Lit and Phil Museums in the North-West', in: *Manchester Memoirs: being the Memoirs and Proceedings of the Manchester Literary and Philosophical Society* 155 (2016–2017), pp. 101–8.

Busard, Hubertus L. L., 'L'algèbre au Moyen Âge: Le 'Liber mensurationum' d'Abū Bekr', in: *Journal des Savants* (1868), pp. 65–125.

Butters, Suzanne B., *The Triumph of Vulcan: Sculptors' Tools, Porphyry, and the Prince in Ducal Florence*, Florence: Olschki 1996, vol. 2, pp. 454–9.

Büttner, Jochen, 'Galileo's Cosmogony', in: *Largo campo di filosofare: Eurosymposium Galileo 2001*, ed. J. Montesinos and C. Solís, La Orotava: Fundación Canaria Orotava 2001, pp. 391–402.

Cajori, Florian, 'Discussion of Fluxions: from Berkeley to Woodhouse', in: *The American Mathematical Monthly* 24 (1917), pp. 145–54.

Cajori, Florian, *A History of Elementary Mathematics, with Hints on Methods of Teaching*, New York: Macmillan, 1950 (1st edn 1896).

Campbell-Kelly, Martin, Mary Croarken, Raymond Flood, and Eleanor Robson, eds, *The History of Mathematical Tables: From Sumer to Spreadsheets*, Oxford: Oxford University Press 2003.

Cancrin, Franz Ludwig von, *Erste Gründe der Berg und Salzwerkskunde*, vol. 6.2, Frankfurt am Main: Andrea 1776.

Capp, Bernard, *Astrology and the Popular Press: English Almanacs 1500–1800*, London: Faber and Faber 1979.

Capp, Bernard, 'Coelson [Colson], Lancelot (1627–1687?), Astrologer and Medical Practitioner', in: *Oxford Dictionary of National Biography*, 2004, https://doi.org/10.1093/ref:odnb/5995.

Cardano, Girolamo (Geronimo), *Artis magnae, sive de regulis algebraicis liber unus*, Nürnberg 1545.

Carlyle, E. I., rev. Alan Yoshioka, 'Young, John Radford (1799–1885)', in: *Oxford Dictionary of National Biography*, Oxford: Oxford University Press 2004, https://doi.org/10.1093/ref:odnb/30274.

Carneiro, Ana, Ana Simões and Maria Paula Diogo, 'Enlightenment Science in Portugal: The *Estrangeirados* and their Communication Networks', in: *Social Studies of Science* 30/4 (2000), pp. 591–619.

Carugo, Adriano, 'L'insegnamento della matematica all'Università di Padova prima e dopo Galileo', in: *Storia della cultura veneta*, ed. Girolamo Arnaldi and Manlio Pastore Stocchi, Vicenza: Neri Pozza 1984, vol. 4/II, pp. 151–99.

Cash, James, *Where There's a Will There's a Way. Or, Science in the Cottage: An Account of the Labours of Naturalists in Humble Life*, London: Robert Hardwicke 1873.

Cassels, J. W. S., 'The Spitalfields Mathematical Society', in: *Bulletin of the London Mathematical Society* 11 (1979), pp. 241–58.

Cassinet, R., 'L'aventure de l'édition des Éléments d'Euclide en arabe par la Société Typographique Médicis vers 1594', in: *Revue française d'histoire du livre* 62 (1993), pp. 5–51.

Ceulen, Ludolf van, *De arithmetische en geometrische fondamenten*, Leiden: Ioost van Colster, ende Iacob Marcus 1615.

Ceulen, Ludolf van, *Fundamenta arithmetica et geometrica*, Leiden: Colster 1615.

Challis, Christopher Edgar, *A New History of the Royal Mint*, Cambridge: Cambridge University Press 1992.

Chaplain, William, 'William Mountaine, F.R.S., Mathematician', in: *The American Neptune* 60 (1960), pp. 185–90.

Chapman, Allan, 'Gresham College: Scientific Instruments and the Advancement of Useful Knowledge in Seventeenth-Century England', in: *Bulletin of the Scientific Instrument Society* 56 (1998), pp. 6–13.

Chapman, Allan, 'No Small Force: Natural Philosophy and Mathematics in Thomas Gresham's London', in: *Sir Thomas Gresham and Gresham College: Studies in the Intellectual History of London in the Sixteenth and Seventeenth Centuries*, ed. Francis Ames-Lewis, Aldershot: Ashgate 1999, pp. 146–73.

Charmantier, Isabelle and Staffan Müller-Wille, 'Worlds of Paper: An Introduction', in: *Early Science and Medicine* 19 (2014), pp. 379–97.

Claessens, G., 'Imagination as Self-Knowledge: Kepler on Proclus' *Commentary on the First Book of Euclid's Elements*', in: *Early Science and Medicine* 16 (2011), pp. 182–3.

Clark, E. Kitson, *The History of 100 Years of Life of the Leeds Philosophical and Literary Society*, Leeds 1924.

Clark, Peter, *British Clubs and Societies 1580–1800: The Origins of an Associational World*, Oxford: Clarendon Press 2000.

Cleeland, J. and S. Burt, 'Charles Turner Thackrah: A Pioneer in the Field of Occupational Health', in: *Occupational Medicine* 45/6 (1995), pp. 285–297.

Cleirac, Estienne, *Us et costumes de la mer, divisées en 3 parties*, Bordeaux: G. Millanges 1647.

Coelson, Lancelot, *The Poor Man's Physician and Chyrugion*, London: printed by A.M. for Simon Miller at the Starre in St Pauls Churchyard 1656; 2nd ed., 1663.

Coelson, Lancelot, *Speculum Perspicuum Uranicum, or, An Almanac* [...], London: Company of Stationers 1676/77.

Coggeshall, Henry, *The Art of Practical Measuring, Easily Perform'd, by a Two-Foot Rule*, London: printed for Richard King 1729.

Cohen, H. Floris, 'The "Mathematization of Nature": The Making of a Concept, and How It Has Fared in Later Years', in: *Historiography of Mathematics in the 19th and 20th Centuries*, ed. Volker R. Remmert, Martina R. Schneider, and Henrik Kragh Sorensen, Cham: Springer 2016, pp. 143–60.

Coke, Roger, *A Discourse of Trade. In two Parts. The first treats of the Reason of the Decay of the Strength, Wealth, and Trade of England. The latter, of the Growth and Increase of the Dutch Trade above the English*, London: for H. Brome and R. Horne 1670.

Collier, John, *Compendium Artis Nauticæ. Being the Daily Practice of the Whole Art of Navigation*, London: sold by J. Harbin, B. Motte, F. Simons, W. Meadows, S. Goodwin, S. Fitzer, C. Digby and E. Baldwin 1729.

Collier, John, *A Letter to the Practisers, Promoters, and Learners of Navigation*, London: printed by W. Pearson for the author 1730.

Collins, John, *An Introduction to Merchants Accounts, Containing Five Distinct Questions of Accounts*, London: James Flesher for Nicholas Bourn 1653.

Collins, John, *The Sector on a Quadrant, or a Treatise Containing the Description and Use of Four Several Quadrants*, London: J. M. for George Hurlock 1659.

Collins, John, *Geometricall Dyalling: or, Dyalling Performed by a Line of Chords Onely, or by the Plain Scale*, London: Thomas Johnson for Francis Cossinet 1659.

Collins, John, *Navigation by the Mariners Plain Scale new plain'd: or, a Treatise of Geometrical and Arithmetical Navigation*, London: Thomas Johnson for Francis Cossinet 1659.

Collins, John, *The Doctrine of Decimal Arithmetick, Simple Interest, &c., Abridged*, London 1665.

Collins, John, *An introduction to Merchants-Accompts: Containing Seven Distinct Questions or Accompts*, London: William Godbid for Robert Horne 1674.

Collins, John, *The Doctrine of Decimal Arithmetick*, ed. J. D. London: R. Holt for Nathanial Ponder 1685.

Collins, Richard, *The Countrey Gauger's Vade Mecvm, or, Pocket-Companion: Being Decimal Tables for the Speedy Gauging of Small Brewing Vessels either of a Circular, Elliptical, or Rectilineal Base: and Also for the Gauging of Cask in Ale or Wine Measure, either Full or Part Empty: Also the Description and Use of an Instrument for the Gauging of Small Brewing Vessels*, London: W. Godbid, M. Pitt and Anthony Owen 1677.

Collis, Robert, *The Petrine Instauration: Religion, Esotericism and Science at the Court of Peter the Great, 1689–1725*, Leiden: Brill 2012.

Colson, Nathaniel, *The Mariners New Kalendar*, London: printed by J. Darby for William Fisher, at the Postern-Gate near Tower-Hill; Robert Boulter, at the Turks-Head; and Ralph Smith at the Bible in Corn-Hill, near the Royal Exchange 1677.

Colson, Nathaniel, *The Mariners New Kalendar*, J. Darby for William Fisher at the Postern-Gate near Tower-Hill, Thomas Passenger, at the Three Bibles on London-Bridg, and Eliz. Smith, at the Bible in Corn-hill, near the Royal Exchange 1688.

Compagnie royale d'assurance, *Prospectus de l'établissement des assurances sur la vie*, Paris: Lottin l'aîné et Lottin 1788.

Conceição, Margarida Tavares da, 'A teoria nos textos portugueses sobre engenharia militar: o *Engenheiro Portuguez* e os tratados de fortificação', in: *Manoel de Azevedo Fortes (1660–1749): Cartografia, Cultura e Urbanismo*, ed. Mário Gonçalves Fernandes, Porto: Gedes 2006, pp. 35–55.

Conde, Antónia Fialho, 'Alentejo (Portugal) and the Scientific Expertise in Fortification in the Modern Period: the Circulation of Masters and Ideas', in: *The Circulation of Science and Technology: Proceedings of the 4th International Conference of the ESHS*, ed. Antoni Roca-Rosell, Barcelona: Societat Catalana d'Història de la Ciència i de la Tècnica 2012, pp. 246–52.

Conde, Antónia Fialho and M. Rosa Massa-Esteve, 'Teaching Engineers in the Seventeenth Century: European Influences in Portugal', in: *Engineering Studies* 10/2–3 (2018), pp. 115–32.

Coolidge, Julian L., 'Robert Adrain and the Beginnings of American Mathematics', in: *The American Mathematical Monthly* 33 (1926), pp. 61–76.

Coppens, Christian, 'Curiositas or Common Places: Private Libraries in the Sixteenth Century', in: *Biblioteche private in età moderna e contemporanea. Atti del convegno internazionale di Udine, 18–20 ottobre 2004*, ed. Angela Nuovo, Milan: Bonnard 2005, pp. 33–42.

Cordier, Samson Le, *Instruction des pilotes*, Havre de Grâce: J. Gruchet 1683; Veuve de G. Gruchet & Pierre Faure 1748; Chez P. J. D. G. Faure 1754.

Corley, T. A. B. and A. J. Crilly, 'Cockle, Sir James (1819–1895), Lawyer in Australia and Mathematician', in: *Oxford Dictionary of National Biography*, 2004, https://doi.org/10.1093/ref:odnb/5788.

Cormack, Lesley B., 'The Commerce of Utility: Teaching Mathematical Geography in Early Modern England', in: *Science & Education* 15 (2006), pp. 305–22.

Corneanu, Sorana, *Regimens of the Mind: Boyle, Locke, and the Early Modern Cultura Animi Tradition*, Chicago: Chicago University Press 2011.

Cornelius, Justus, *Vindiciarvm Favlhaberianarvm Prodromus*, Ulm: Johann Blancken 1619.

Cortés de Albacar, Martín, *Breve compendio de la sphera y de la arte de navegar*, Seville: A. Alvarez 1551.

Cortés, Martín, *The Arte of Navigation. Conteynyng a Compendious Description of the Sphere, with the Makyng of Certen Instrumentes and Rules for Navigations* [...], London: Richard Jugge, 1561, 1572; tr. Richard Eden, ed. David W. Waters. Delmar, NY: Scholars' Facsimiles 1992.

Costa, Shelley, 'The *Ladies' Diary*: Society, Gender and Mathematics in England, 1704–1754', PhD thesis, Cornell University 2000.

Costa, Shelley, 'The *Ladies' Diary*: Gender, Mathematics, and Civil Society in Early-Eighteenth-Century England', in: *Osiris* 17 (2002), pp. 49–73.

Cotton, Joseph, *Memoir on the Origin and Incorporation of the Trinity House of Deptford Strond*, London: J. Darling 1818.

Coutts, James, *A History of the University of Glasgow: From its Foundation in 1451 to 1909*, Glasgow: James Maclehose and Sons 1909.

Craik, Alex D. D., *Mr Hopkins' Men: Cambridge Reform and British Mathematics in the Nineteenth Century*, London: Springer-Verlag 2007.

Craik, Alex D. D., 'Mathematical Analysis and Physical Astronomy in Great Britain and Ireland, 1790–1831: Some New Light on the French Connection', in: *Revue d'histoire des mathématiques* 22 (2016), pp. 223–94.

Crilly, Tony, *Arthur Cayley: Mathematician Laureate of the Victorian Age*, Baltimore: Johns Hopkins University Press 2006.

Crilly, Tony, Steven H. Weintraub and Paul R. Wolfson, 'Arthur Cayley, Robert Harley and the Quintic Equation: Newly Discovered Letters 1859–1863', in: *Historia Mathematica* 44/2 (2017), pp. 150–69.

Crippa, Davide, *The Impossibility of Squaring the Circle in the 17th Century: A Debate among Gregory, Huygens and Wallis*, Cham: Birkhäuser 2019.

Croarken, Mary, 'Providing Longitude for All: The Eighteenth Century Computers of the Nautical Almanac', in: *Journal of Maritime Research* 4 (2002), pp. 106–26.

Croarken, Mary, 'Astronomical Labourers: Maskelyne's Assistants at the Royal Observatory, Greenwich, 1765–1811', in: *Notes and Records of the Royal Society of London* 57 (2003), pp. 285–98.

Croarken, Mary, 'Tabulating the Heavens: Computing the Nautical Almanac in 18th-Century England', in: *IEEE Annals of the History of Computing* 25/3 (2003), pp. 48–61.

Crone, Ernst, 'Pieter Holm en zijn Zeevaartschool', in: *De Zee: Zeevaartkundig Tijdschrift* 52 (1930), pp. 136–44, 185–95, 270–80, 352–62, 416–24, 489–97, 560–8, 642–51, 704–16.

Crone, Ernst, *Cornelis Douwes, 1712–1773: Zijn Leven en Zijn Werk; Met Inleidende Hoofdstukken over Navigatie en Zeevaart-Onderwijs in de 17de en 18de eeuw*, Haarlem: H. D. Tjeenk Willink & zoon 1941.

Crosland, Maurice and Crosbie Smith, 'The Transmission of Physics from France to Britain: 1800–1840', in: *Historical Studies in the Physical Sciences* 9 (1978), pp. 1–61.

Crozier, John A., *The Life of the Rev. Henry Montgomery, LL.D.*, vol. 1, London: Simpkin, Marshall & Co. 1875.

Cruz, Jozé Gomes da, *Elogio Funebre de Manoel de Azevedo Fortes*, Lisboa: Jozé da Silva da Natividade 1754.

Curth, Louise Hill, 'Medical Advertising in the Popular Press' in: *From Physick to Pharmacology: Five Hundred Years of British Drug Retailing*, ed. Louise Hill Curth, Aldershot: Ashgate 2006, pp. 29–48.

Curtius, Sebastian, *Ein newes wolgegründtes Rechenbuch: nach rechter Art der Practic, mit manchem schönen geschwinden Vortheil, neben gründlicher Aussführung der gantzen und gebrochenen: sampt den Proportionen Speciebus [...] Dessgleichen zum Beschluss eine Aufflösung mancherley künstlicher cossischer unnd polygonalischer Exempla, etlicher fürnemer und kunstreicher Rechenmeistern, so zu End irer Rechenbucher one Facit gesetzt seyn: dergleichen vorhin in Druck nie aussgangen, nun aber der lieben Jugend zu gedeylicher Wolfart mit Fleiss zusammen colligirt und jetzt erstmals in Druck gegeben*, Nurnberg: In Verlegung Conrad Baurn 1604.

Curtius, Sebastian, *Compendium Arithmeticae, Das ist: Ein Neues, Kurtzes und Wolgegründtes SchulRechenbüchlein, von allerley Hauß- und Kauffmanns rechnungen [...]: Meinen lieben Discipulis*, Nurnberg: Fuhrmann 1610.

Curtius, Sebastian and Jan Pieterszoon Dou, *Tractat vom machen und Gebrauch eines Neugeordneten mathematischen Instruments: inn welchem underschiedliche Künstliche stuck die Geometriae betreffende verfasset und begriffen seind*, Amsterdam: Bey Wilhelm Janss 1616.

Curtius, Sebastian and Jan Pieterszoon Dou, *Practica des Landmessens: darinn gelehrt wird, wie man alle recht-, und krumseitige Land, Wälder, Baumgärten [...] sowoh mit dem Quadranten, als ohne demselben messen soll*, Amsterdam: Jansson 1616.

Cuttica, Cesare, 'Sir Francis Kynaston: The Importance of the "Nation" for a 17th-Century English Royalist', in: *History of European Ideas* 32 (2006), pp. 139–61.

Dafforne, Richard, *Merchant's Mirrour: or, Directions for the Perfect Ordering and Keeping of His Accounts*, London: R. Young for N. Bourn 1635.

Dale, William, *Calculations Deduced from First Principles: In the Most Familiar Manner by Plain Arithmetic, for the Use of the Societies Instituted for the Benefit of Old Age*. London: J. Ridley 1772.

Danzi, Massimo, *La biblioteca del Cardinal Pietro Bembo*, Geneva: Droz 2005.

Dary, Michael, *Dary's Diarie. Or, the Description and Use of a Quadrant*, London: T.F. for George Hurlock 1650.

Dary, Michael, *The General Doctrine of Equation Reduced into Brief Precepts*, London: for Nathaniel Brook 1664.

Dary, Michael, *Dary's Miscellanies: Being, for the Most Part, a Brief Collection of Mathematical Theorems*, London: W[illiam] G[odbid] 1669.

Dary, Michael, *A Tale of a Tub, or the Greenwich Problem*, London: for William Shrowsbury 1674.

Dary, Michael, *The Doctrine of Adfected Equations Epitomized*, London: M. Clark for the author 1678.

Daston, L. (ed.), *Science in the Archives: Pasts, Presents, Futures*, Chicago: University of Chicago Press 2017.

Davids, C. A., 'Het zeevaartkundig onderwijs voor de koopvaardij in Nederland tussen 1795 en 1875. De rol van het Rijk, de lagere overheid en het particuliere initiatief', in: *Tijdschrift voor Zeegeschiedenis* 4 (1985), pp. 164–90.

Davids, C. A., *Zeewezen en Wetenschap: De Wetenschap en de Ontwikkeling van de Navigatietechniek in Nederland Tussen 1585 en 1815*, Amsterdam: De Bataafsche Leeuw 1986.

Davids, C. A., 'Het navigatieonderwijs aan personeel van de VOC', in: *De VOC in de kaart gekeken: Cartografie en navigatie van de Verenigde Oostindische Compagnie 1602–1799*, ed. Patrick van Mil and Mieke Scharloo, The Hague: SDU 1988, pp. 65–74.

Davies, Griffith, *Tables of Life Contingences*, London: Longman and Co. 1825.
Davies, J. D., 'Warren, Sir William (bap. 1627, d. 1695), Naval Contractor', in: *Oxford Dictionary of National Biography*, 2004, https://doi.org/10.1093/ref:odnb/58160.
Pen-and-Ink [T. S. Davies], 'On the Cultivation of Geometry in Lancashire', in: *Notes and Queries* 57 (1850), pp. 436–8.
Davies, Thomas Stephens, 'XXIX. Geometry and Geometers No. II', in: *London, Edinburgh, and Dublin Philosophical Magazine and Journal of Science*, ser. 3, 33/221 (1848), pp. 201–6.
Davies, Thomas Stephens, 'LXXIX. Geometry and Geometers No. VII', in: *London, Edinburgh, and Dublin Philosophical Magazine and Journal of Science*, ser. 4, 1/7 (1851), pp. 536–44.
Davis, John, *The Seamans Secrets*, London: Thomas Dawson 1595.
Dechalles, Claude François Millet, *Les Elemens d'Euclide*, new edition, Paris: Estienne Michallet 1690.
Dee, John, 'Mathematicall Praeface', in: Euclid, *The Elements of Geometrie*, tr. Henry Billingsley, London: Iohn Daye 1570.
Deluc, Jean André, 'Barometrical Observations on the Depth of the Mines in the Hartz', in: *Philosophical Transactions of the Royal Society of London* 67 (1777), pp. 422–4.
Deluc, Jean André, *Lettres physiques et morales sur l'histoire de la terre et de l'homme. Addressees a la reine de la Grande Bretagne, par J. A. de Luc*, 5 vols, La Haye et Paris: De Tune et Duchesne 1779–80.
De Moivre, Abraham, *Annuities upon Lives, or, The Valuation of Annuities upon any Number of Lives, as also, of Reversion*, London: W. Pearson 1725.
De Moivre, Abraham, *The Doctrine of Chances: Or, a Method of Calculating the Probability of the Events in Play*, 2nd edn, London: Woodfall 1738; 3rd edn, London: Millar 1756.
De Morgan, Augustus, *Arithmetical Books from the Invention of the Printing Press to the Present Time*, London: Taylor and Walton 1847.
Dennert, Herbert, *Bergbau und Hüttenwesen im Harz: vom 16.–19. Jh. dargest. in Lebensbildern führender Persönlichkeiten*, Clausthal-Zellerfeld: Pieper 1986.
Denniss, John, *Figuring it Out: Children's Arithmetical Manuscripts, 1680–1880*, Oxford: Huxley Scientific Press 2012.
Deparcieux, Antoine, *Essai sur les probabilités de la durée de la vie humaine*, Paris: Guerin 1746.
Despeaux, Sloan Evans, 'The Development of a Publication Community: Nineteenth-Century Mathematics in British Scientific Journals', PhD thesis, University of Virginia 2002.
Despeaux, Sloan Evans, 'A Voice for Mathematics: Victorian Mathematical Journals and Societies', Chapter 7 in: *Mathematics in Victorian Britain*, ed. Raymond Flood, Adrian Rice, and Robin Wilson, Oxford: Oxford University Press 2011, pp. 155–74.
Despeaux, Sloan Evans, 'Mathematical Questions: A Convergence of Mathematical Practices in British Journals of the Eighteenth and Nineteenth Centuries', in: *Revue d'histoire des mathématiques* 20/1 (2014), pp. 5–71.
Despeaux, Sloan Evans, 'Connected by Questions and Answers: The Milieu of Mathematical Editors of English Commercial Journals, 1775–1854', in: *Circulation des mathématiques dans et par les journaux: Histoire, territoires, publics*, ed. Hélène Gispert, Jeanne Peiffer, and Philippe Nabonnand, forthcoming.
Dickinson, Harry W., *Educating the Royal Navy: Eighteenth-and Nineteenth-Century Education for Officers*, London and New York: Routledge 2007.
Dijksterhuis, Fokko Jan, 'Duytsche Mathematique and the Building of a New Society: Pursuits of Mathematics in the Seventeenth-Century Dutch Republic', in: *Mathematical Practitioners and the Transformation of Natural Knowledge in Early Modern Europe*, ed. Lesley B. Cormack, Steven A. Walton, and John A. Schuster, Cham: Springer 2017, pp. 167–81.

Dillen, J. G. van, *Bronnen tot de geschiedenis van het bedrijfsleven en het gildewezen van Amsterdam*, The Hague: Nijhoff 1929–74.

Dodson, James, *The Mathematical Repository*, vol. 3, London: John Nourse 1755.

Donald, M. B., *Elizabethan Monopolies: The History of the Company of Mineral and Battery Works from 1565 to 1604*, Edinburgh: Oliver and Boyd 1961.

Donald, M. B., *Elizabethan Copper: The History of the Company of Mines Royal 1568–1605*, Whitehaven: Michael Moon 1989; first edn 1955.

Drew, B., *The London Assurance: A Chronicle*, London: The London Assurance 1928.

Dudley, Robert, *Dell'Arcano del Mare*, Florence 1646, 1647.

Dupré, Sven, 'Ausonio's Mirrors and Galileo's Lenses: The Telescope and Sixteenth-Century Practical Optical Knowledge', in: *Galilaeana* 2 (2005), pp. 145–80.

Dürer, Albrecht, *Etliche underricht zu Befestigung der Stett, Schloss und Flecken*, Nüremberg 1527.

Eagleton, Catherine and Boris Jardine, 'Collections and Projections: Henry Sutton's Paper Instruments', in: *Journal of the History of Collections* 17 (2005), pp. 1–17.

Eddy, Matthew, 'Tools for Reordering: Commonplacing and the Space of Words in Linnaeus' Philosophia Botanica', in: *Intellectual History Review* 20 (2010), pp. 227–52.

Eddy, Matthew, 'The Interactive Notebook: How Students Learned to Keep Notes During the Scottish Enlightenment', in: *Book History* 19 (2016), pp. 86–131.

Egerton, Francis Henry, *Description du Plan incliné souterrain exécuté entre les deux biefs des canaux souterrains dans les Houillières de Walkden-Moor, en Angleterre, par le Duc de Bridgewater*, Paris: Bureau des annales des arts et manufactures 1812.

Egmond, Warren van, 'The Commercial Revolution and the Beginnings of Western Mathematics in Renaissance Florence, 1300–1500', PhD dissertation, Indiana University, Ann Arbor, MI 1976.

Egmond, Warren van, *Practical Mathematics in the Italian Renaissance: A Catalogue of Italian Abbacus Manuscripts and Printed Books to 1600*, Florence: Istituto e Museo di Storia della Scienza 1980.

Ehrhardt, Caroline, 'Tactics: In Search of a Long-Term Mathematical Project (1844–1896)', in: *Historia Mathematica* 42/4 (2015), pp. 436–67.

Ellerton, Nerida and M. A. Clements, *Rewriting the History of School Mathematics in North America 1607–1861: The Central Role of Cyphering Books*, Dordrecht: Springer 2012.

Ellerton, Nerida and M. A. Clements, *Samuel Pepys, Isaac Newton, James Hodgson, and the Beginnings of Secondary School Mathematics: A History of the Royal Mathematical School Within Christ's Hospital, London 1673–1868*, Cham: Springer 2017.

Elliott, Chris, *Egypt in England*, Swindon: English Heritage 2012.

E[lliott], E. B., 'Obituary Notice: Robert Harley', in: *Proceedings of the London Mathematical Society, Second Series* 9 (1911), pp. xii–xv.

Elliott, Paul, 'The Origins of the "Creative Class": Provincial Urban Society, Scientific Culture and Socio-Political Marginality in Britain in the Eighteenth and Nineteenth Centuries', in: *Social History* 28/3 (2003), pp. 361–87.

Elliott, Paul, 'Towards a Geography of English Scientific Culture: Provincial Identity and Literary and Philosophical Culture in the English County Town, 1750–1850', in: *Urban History* 32/3 (2005), pp. 391–412.

Elliott, Paul, *The Derby Philosophers: Science and Culture in British Urban Society, 1700–1850*, Manchester: Manchester University Press 2009.

Ellis, Henry, 'Copy of a Plan Proposed to Queen Elizabeth by Sir Humphry Gilbert, for Instituting a London Academy', in: *Archaeologia* 21 (1827), pp. 506–20.

Emerson, William, *The Doctrine of Fluxions: Not Only Explaining the Elements Thereof, but Also its Application and Use in the Several Parts of Mathematics and Natural Philosophy*, London: Printed for J. Richardson 1757.

Falconer, Isobel 'Rotheram, John (*c*.1750–1804), Natural Philosopher', in: *Oxford Dictionary of National Biography*, 2004, https://doi.org/10.1093/ref:odnb/24153.

Faulhaber, Johannes, *Arithmetischer Cubicossischer Lustgarten. Darinnen Hundert vnd Sechtzig Blümlein, das ist, außerlesner schöner künstlicher Exempel mit Newen Inventionibus gepflantzet werden. Welche theils auß Hieronymo Cardano, vnnd andern Lateinischen Scribenten versetzt vnnd gezogen: Theils aber insonderheit die liebliche Polygonalische Röslin, von newen zum Lust erzogen worden*, Tübingen: Erhard Cellius 1604.

Faulhaber, Johannes, *Speculum Polytechnum Mathematicum nouum*, trans. Johannes Remmelin, Ulm: Typis Ioannis Mederi 1612.

Faulhaber, Johannes, *Himlische gehaime Magia Oder Newe Cabalistische Kunst, vnd Wunderrechnung, Vom Gog vnd Mogog. Darauß die Weisen, Verständigen vnd Gelerten, so diser Göttlichen Kunst genugsam erfahren, heimlich observieren vnd fleissig außrechnen mögen, die Beschaffenheit deß grossen Christenfeindts Gog vnd Magogs. Auß Teutschem, Lateinischem, Griechischem vnnd Hebraischem, Kunst vnd wunder Alphabeth, in verborgene Retzel eingewickelt, vnd in den Truck gegeben*, Nürnberg 1613.

Faulhaber, Johannes, *Newer Arithmetischer Wegweyser. Zu der Hochnutzlichen freyen Rechenkunst, mit Newen Inventionibus geziert*, Ulm: Johann Meder 1614; 2nd edn 1617.

Faulhaber, Johannes, *Gemein offen Auß-Schreiben, Deß Ehrnvösten, Weitberümbten vnd Sinnreichen Herren Johann: Faulhabers, Burgers vnd bestellten Mathematici in Vlm, etc. Vor disem Schrifftlich beschehen: An alle Philosophos, Mathematicos, sonderlich Arithmeticos vnd Künstler, so auff allen Vniversiteten vnd Schulen, oder anderer Orthen in Europa sein möchten*, ed. Friedrich Schwedler, Augsburg 1615.

Faulhaber, Johannes, *Miracula Arithmetica, zu der Continuation seines Arithmetischen Wegweisers gehörig*, Augsburg: David Francken 1622.

Faulhaber, Johannes, *Ingenieurs-Schul, Erster Theyl: Darinnen durch den Canonem Logarithmicvm alle Planische Triangel zur fortification, oder Architectura Militari, Optica, Geodaesia, Geometria, etc. gar leichtlich und behänd zu solviren, gelährt wird, darneben die Doctrina Triangulorum Sphaericorum zur Geographia, Gnomonica Astronomia gehörig auch zu sehen. Auß Adriano Vlacq, Henrico Briggio, Nepero, Pitisco, Berneckhero vnd andern hochberümbten Authorn gezogen, vnd als den besten Safft vnd Kerrn in ein kurtz Compendium gebracht. Mit angehenckten Miraculosischen Kunst Quaestionen, dergleichen hiebevor nie gesehen*, Frankfurt am Main 1630.

Faulhaber, Johannes, *Academia Algebrae. Darinnen die miraculossiche Inventiones, zu den höchsten Cossen weiters continuiert vnd profitiert werden. Dergleichen zwar vor 15. Jahren den Gelehrten auff allen Vniversiteten in gantzem Europa proponiert, darauff continuiert, auch allen Mathematicis inn der gantzen weiten Welt dediciert, aber bißhero, noch nie so hoch, biß auff die regulierte, Zensicubiccubic* Coß, *durch offnen Truck publiciert worden. Welcher vorgesetzet ein kurtz Bedencken, Was einer für Authores nach ordnung gebrauchen sole, welcher die Coß fruchtbarlich, bald, auch fundamentaliter lehrnen vnd ergreiffen will*, Augsburg: J. Remmelin 1631.

Faulhaber, Johannes, *Anderer Theil der Ingenieurs Schul. Darinnen die Regular Fortification, sampt den Aussenwercken, durch vnd ohne Rechnung mit newen Inventionibus gelehrt werden. Welches nicht allein durch den Canonem Logarithmicum, sondern auch durch new Inventierte Instrument, so deutlich vnd klar für die Augen gestelt wird, daß auch einer gleichsamb ohne Mundlichen Bericht, solche Fortifications Kunst leichtlich begreiffen kan*, Ulm 1633.

Faulhaber, Johannes, *Ingenieurs Schul Dritter Theil. Darinnen Die Irregular Figuren zu Fortificirn, durch vnnd ohne Rechnung mit vnderschidlichen Newen Inventionibus gelehrt werden. Insonderheit etliche Secreta, wie theils durch die Algebram die subtileste Questiones, so in der Fortification für fallen mögen, zu solviren, vnd theils durch new Inventierte Instrument vnderschidliche Irregulares Figuras Royal zubezaychnen, auch wie man sonsten im Bawen grossen Vnkosten ersparen, vnd dannocht alles zum bestand richten könde. Auß aygner Erfahrung, in grosser berümpter Stätten Fortifications Gebäwen practicirt, vnd jetzo vff vielfaltiges begeren an den Tag gegeben*, Ulm 1633.

Faulhaber, Johannes, *Ingenieurs Schul, Vierdter Teil. Von Fortificatione Practica Offensiva et Defensiva. Da Gelehrt würdt, wie deß Feindes Vöstungen zu belägern, sich darvor Zubeschantzen, solche zu Vmbzinglen, Zubeschiessen, Zubestürmen vnnd Zuerobern: Entgegen wie ein jede Statt/ wann sie vom Feind Belägert, mit vnderschidlichen Fortifications Wercken, vnd gegenwehren, wider allen Gewalle, mit Göttlicher Hülff defendiert vnd Beschirmbt werden möchte. Mit etlichen Newen Mechanischen Inventionibus vnd Stratagematibus fürgestelt/ dergleichen bißhero im Truck nit gesehen*, Ulm 1633.

Fauque, Danielle, 'Les Écoles d'hydrographie en Bretagne au XVIIIe siècle', in: *Mémoires de la Société d'histoire et d'archéologie de Bretagne* 78 (2000), pp. 369–400.

Favaro, Antonio, 'Amici e corrispondenti di Galileo Galilei: Giuseppe Moletti', in: *Atti del Real Istituto Veneto di Scienze Lettere ed Arti* 77 (1918), pp. 47–118.

Favaro, Antonio, *Galileo Galilei a Padova: Ricerche e scoperte*, Padua: Antenore 1968.

Feingold, Mordechai, *The Mathematicians' Apprenticeship: Science, Universities and Society in England, 1560–1640*, Cambridge: Cambridge University Press, 1984.

Feingold, Mordechai, 'Isaac Barrow: Divine, Scholar, Mathematician', in: *Before Newton: The Life and Times of Isaac Barrow*, ed. Mordechai Feingold, Cambridge: Cambridge University Press 1990, pp. 1–104.

Feingold, Mordechai, 'Gresham College and London Practitioners: The Nature of the English Mathematical Community', in: *Sir Thomas Gresham and Gresham College. Studies in the Intellectual History of London in the Sixteenth and Seventeenth Centuries*, ed. Francis Ames-Lewis, Aldershot: Ashgate 1999, pp. 174–88.

Feingold, Mordechai, 'The Origins of the Royal Society Revisited', in: *The Practice of Reform in Health, Medicine, and Science, 1500–2000*, ed. Margaret Pelling and Scott Mandelbrote, Aldershot: Ashgate 2005, pp. 167–83.

Ferguson, Johan Jacob, *Labyrinthus algebrae*, The Hague: Johannes Rammazeyn 1667.

Ferreira, Maria Elisabete Barbosa, 'Teoria(s) de Proporções em Portugal na Primeira Metade do Século XVIII', master's dissertation, Braga: Escola de Ciências da Universidade do Minho 2013, http://hdl.handle.net/1822/29384.

Ferreira, Nuno Alexandre Martins, 'Luís Serrão Pimentel (1613–1679): Cosmógrafo Mor e Engenheiro Mor de Portugal', master's dissertation, Lisbon: Faculdade de Letras da Universidade de Lisboa 2009, http://hdl.handle.net/10451/467.

Fiocca, Alessandra, 'Giuseppe Moleto (1531–1588), matematico al servizio dei Gonzaga e della Repubblica di Venezia', in: *Contributi di scienziati mantovani allo sviluppo della matematica e della fisica*, ed. Fabio Mercanti e Luca Tallini, Cremona: Monotipia Cremonese 2001, pp. 111–29.

Fleure, H. J., 'The Manchester Literary and Philosophical Society', in: *Endeavour* 6 (1947), pp. 147–51.

Forbes, Eric G., 'Collections II: The Crawford Collection of Books and Manuscripts on the History of Astronomy, Mathematics, etc., at the Royal Observatory, Edinburgh', in: *British Journal for the History of Science* 6 (1973), pp. 459–61.

Forbes, Eric G., Lesley Murdin, and Frances Willmoth (eds), *The Correspondence of John Flamsteed, the First Astronomer Royal*, 3 vols, Bristol: Institute of Physics Publishing, 1995–2002.

Forsyth, A. R., 'James Cockle [obituary]', in: *Proceedings of the London Mathematical Society* 26 (1895), pp. 551–4.

Foster, Samuel, *The Art of Dialling; by a New, Easie and Most Speedy Way. Shewing, How to Describe the Houre-Lines upon All Sorts of Plaines, Howsoever, or in Which Latitude Soever Scituated*, London: John Dawson for Francis Eglesfield 1638.

Foster, Samuel, *The Uses of a Quadrant Fitted for Daily Practice*, ed. A[nthony] T[hompson], London: for Francis Eglesfield 1652.

Foster, Samuel, *Posthuma Fosteri: The Description of a Ruler, upon Which is Inscribed Divers Scales: and the Uses Thereof: Invented and Written by Mr. Samuel Foster, Late Professor of Astronomie in Gresham-Colledg*, London: Robert and William Leybourn for Nicholas Bourn 1652.

Frängsmyr, Tore, J. L. Heilbron, and Robin E. Rider, *The Quantifying Spirit in the 18th Century*, Berkeley: University of California Press 1990.

Freiesleben, Johann Carl, *Bemerkungen über den Harz. Erster Theil: Bergmännische Bemerkungen*, Leipzig: Schäfer 1795.

Hans Freudenthal, 'What Is Algebra and What Has Been its History?', in: *Archive for History of Exact Sciences* 16 (1977), pp. 189–200.

Friberg, Jöran, 'A Geometric Algorithm with Solutions to Quadratic Equations in a Sumerian Juridical Document from Ur III Umma', in: *Cuneiform Digital Library Journal* 3 (2009), pp. 1–27.

Friberg, Jöran, *Amazing Traces of a Babylonian Origin in Greek Mathematics*, Singapore: World Scientific Press 2007.

Fuller, Mary C., 'Arctics of Empire: the North in Principal Navigations (1598–1600)', in: *The Quest for the Northwest Passage: Knowledge, Nation and Empire, 1576–1806*, ed. Frédéric Regard, London: Routledge 2016, pp. 15–29.

Furlotti, Barbara, 'Connecting People, Connecting Places: Antiquarians as Mediators in Sixteenth-Century Rome', in: *Urban History* 37 (2010), pp. 386–98.

Gadbury, John, *Collectio geniturarum: or, A Collection of Nativities, in CL Genitures*, London: John Cottrel 1662.

Gadbury, John, *Natura prodigiorum: or, A Discourse Touching the Nature of Prodigies*, London: for Francis Cossinet 1665.

Gädeke, Nora, 'Gottfried Wilhelm Leibniz', in: *Les grands intermédiaires culturels de la République des lettres du xve au xviiie siècles*, ed. Christiane Berkvens-Stevelinck, Hans Bots, and Jens Häseler, Paris: Honoré Champion 2005, pp. 257–306.

Galilei, Galileo, *Discorsi e dimostrazioni matematiche, intorno a due nuove scienze attenenti alla mecanica & i movimenti locali*, Leiden 1638.

Garber, Elizabeth, *The Language of Physics: The Calculus and the Development of Theoretical Physics in Europe, 1750–1914*, Boston: Birkhäuser 1999.

Garcia, João Carlos, 'Manoel de Azevedo Fortes e os mapas da Academia Real da História Portuguesa, 1720–1736', in: *Manoel de Azevedo Fortes (1660–1749): Cartografia, Cultura e Urbanismo*, ed. Mário Gonçalves Fernandes, Porto: Gedes 2006, pp. 141–73.

Garcia de Palacio, Diego, *Instrvción Návtica*, Mexico: Pedro Ocharte 1587.

Garcie, Pierre, *Le grand routier et pilotage et enseignement pour ancrer tant ès ports, havres qu'autres lieux de la mer*, Rouen: chez Jehan Burges 1521.

Gardner, A. D. and E. W. Jenkins, 'The English Mechanics' Institutes: The Case of Leeds 1824–42', in: *History of Education* 13/2 (1984), pp. 139–52.

Gascoigne, John, *Cambridge in the Age of the Enlightenment: Science, Religion and Politics from the Restoration to the French Revolution*, Cambridge: Cambridge University Press 1989.

Gascoigne, John, 'Sensible Newtonians: Nicholas Saunderson and John Colson', in: *From Newton to Hawking: A History of Cambridge University's Lucasian Professors of Mathematics*, ed. Kevin C. Knox and Richard Noakes, Cambridge: Cambridge University Press 2003, pp. 171–204.

Gaskell, Elizabeth Cleghorn, *Mary Barton: A Tale of Manchester Life*, vol 1, London: Chapman and Hall 1848.

Gatterer, Christoph Wilhelm Jakob, *Anleitung den Harz und andere Bergwerke mit Nuzen zu bereisen*, Göttingen: Vandenhoeck 1785.

Gavine, David, 'Mackay Andrew, (1758–1809), Teacher of Navigation', in: *Oxford Dictionary of National Biography*, 2004, https://doi.org/10.1093/ref:odnb/17552.

Genssane, Antoine de, *Le géométrie souterraine: ou traité de géométrie-pratique, appliqué a l'usage des travaux des mines*, Montpellier: Rigaud, Pons 1776.

G. H. R., 'The Late W. S. B. Woolhouse', in: *Journal of the Institute of Actuaries* 31 (1894), pp. 362–5.

Gialdini, Anna, 'Fonti codicoogiche e archivistiche per la ricostruzione della biblioteca di Michael Sophianos', in *Miscellanea Graecolatina II*, ed. Federico Gallo and Lisa Benedetti, Rome: Bulzoni 2014, pp. 287–323.

Gialdini, Anna, 'Antiquarianism and Self-Fashioning in a Group of Bookbindings for Gian Vincenzo Pinelli', in: *Journal of the History of Collections* 29 (2017), pp. 19–31.

Gibelin, Jacques (ed.), *Abrégé des Transactions Philosophiques de la Société Royale de Londres*, Paris 1787–91.

Gietermaker, Claas Hendricksz, *'t Vergulde Licht der Zeevaert*, Amsterdam: Doncker 1677; Amsterdam: J. van Keulen 1710.

Gialdini, Anna, *Driehoex-rekening bestaende in de verklaringe en ontbindinge der platte driehoecken [...] Met een discours tusschen een schipper en stuurman, aengaende de zeevaert*, [Amsterdam]: gedruckt voor den Autheur 1665.

Gialdini, Anna, *Le Flambeau reluisant, ou proprement Thresor de la navigation*, tr. J. Viret, Amsterdam: Henri Donker 1667.

Giglioni, G., 'The "Horror" of Bruno's Magic: Frances Yates Gives a Lecture at the Warburg Institute (1952)', in: *Bruniana & Campanelliana* 20 (2014), pp. 477–97.

Gilden, G., *A Prognostication, Published for this Yeere of our Lord and Saviour Jesus Christ. 1616*, London 1616.

Girbal, François, *Bernard Lamy (1640–1715): étude biographique et bibliographique*, Paris: Presses universitaires de France 1964.

Gisborne, Thomas, 'On the Benefits and Duties Resulting from the Institution of Societies for the Advancement of Literature and Philosophy', in: *Memoirs of the Literary and Philosophical Society of Manchester* 5/1 (1798), pp. 70–88.

Glaisyer, Natasha, *The Culture of Commerce in England, 1660–1720*, Royal Historical Society Studies in History, Woodbridge: The Boydell Press 2011.

Gompertz, Benjamin, 'On the Nature of the Function Expressive of the Law of Human Mortality, and on a New Mode of Determining the Value of Life Contingencies', in: *Philosophical Transactions of the Royal Society of London* 115 (1825), pp. 513–83.

Gotthard, Johann Christian, *Authentische Beschreibung von dem merkwürdigen Bau des Tiefen Georg Stollens am Oberharze*, Wernigerode: Carl Samuel Struck 1801.

Gough, John, *Practical Arithmetic in Four Books*, Belfast: Simms & McIntyre 1815.

Goulding, Robert, 'Numbers and Paths: Henry Savile's Manuscript Treatises on the Euclidean Theory of Proportion', in: *Reading Mathematics in Early Modern Europe: Studies in the*

Production, Collection, and Use of Mathematical Books, ed. P. Beeley, Y. Nasifoglu, and B. Wardhaugh, London: Routledge 2021, pp. 33–61.

Gowing, Margaret, 'Science, Technology and Education in England in 1870', in: *Oxford Review of Education* 4 (1978), pp. 3–17.

Graaf, Abraham de, *De kleene Schatkamer*, Amsterdam: voor den Autheur by d'Erfg. van Paulus Matthysz. 1688.

Grant, Edward, 'Celestial Motions in the Late Middle Ages', in: *Early Science and Medicine* 2/2 (1997), pp. 129–48.

Grant, Edward, 'The Partial Transformation of Medieval Cosmology by Jesuits in the Sixteenth and Seventeenth Centuries', in: *Jesuit Science and the Republic of Letters*, ed. Mordechai Feingold, Cambridge, MA: MIT Press 2003, pp. 127–55.

Grattan-Guinness, Ivor, *Convolutions in French Mathematics, 1800–1840: From the Calculus and Mechanics to Mathematical Analysis and Mathematical Physics*, Basel: Birkhäuser 1990.

Gray, G. J., rev. Anita McConnell, 'Dodson, James (*c.*1705–1757), Mathematician and Actuary', in: *Oxford Dictionary of National Biography*, 2008, https://doi.org/10.1093/ref:odnb/7756.

Gray, Jeremy, 'Overstating Their Case? Reflections on British Mathematics in the Nineteenth Century', in: *BSHM Bulletin: Journal of the British Society for the History of Mathematics* 21 (2006), pp. 178–85.

Gregory, James, *Optica promota, seu abdita radiorum reflexorum & refractorum mysteria, geometrice enucleata*, London: F. Hayes for S. Thomson 1663.

Gregory, James, *Exercitationes geometricae*, London: William Godbid and Moses Pitt 1668.

Gregory, Olinthus, *A Treatise of Mechanics*, London 1806.

[Gregory, Olinthus], 'A Review of Some Leading Points in the Official Character and Proceedings of the Late President of the Royal Society', in: *Philosophical Magazine*, series 1, no. 56 (1820), pp. 161–74, 241–57.

Gregory, Olinthus, 'Brief Memoir of the Life and Writings of Charles Hutton', in: *Imperial Magazine* 5 (March 1823 [obituary dated 1 February]), pp. 201–27.

Grendler, Marcella, 'A Greek Collection in Padua: The Library of Gian Vincenzo Pinelli (1535–1601)', in: *Renaissance Quarterly* 33 (1980), pp. 386–416.

Grendler, Paul F., *The Roman Inquisition and the Venetian Press*, 1540–1605, Princeton, NJ: Princeton University Press 1975.

Grier, Jason, 'Navigation, Commercial Exchange and the Problem of Long-Distance Control in England and the English East India Company, 1673–1755', PhD dissertation, York University 2018.

Gualdo, Paolo, *Vita Ioannis Vincentii Pinelli, Patricii Genuensis*, Augsburg: Markus Welser 1607.

Guicciardini, Niccolò, *The Development of Newtonian Calculus in Britain 1700–1800*, Cambridge: Cambridge University Press 1989.

Guicciardini, Niccolò, 'Dot-Age: Newton's Mathematical Legacy in the Eighteenth Century', in: *Early Science and Medicine* 9/3 (2004), pp. 218–56.

Gulizia, Stefano, 'The Philosophy of Mathematics in Gian Vincenzo Pinelli's Papers', in: *Bruniana & Campanelliana* 25 (2019), pp. 459–74.

Gulizia, Stefano, 'Francesco Patrizi da Cherso and the Anti-Aristotelian Tradition: Interpreting the *Discussiones Peripateticae* (1581)', in: *Intellectual History Review* 29 (2019), pp. 561–73.

Gulizia, Stefano, 'Kepler's Snow: The Epistemic Playfulness of Geometry in Seventeenth-Century Europe', in: *British Journal for the History of Mathematics* 37/2 (2022), pp. 117–37.

Gunter, Edmund, *The Description and Use of the Sector*, London: William Jones 1623.

G[wyther], R. F., 'Arthur Cayley [obituary]', in: *Memoirs and Proceedings of the Manchester Literary and Philosophical Society, Fourth Series* 9 (1895), pp. 235–7.

Haberman, Shelby and Trevor A. Sibbett, *History of Actuarial Science*, 10 vols, London: W. Pickering & Chatto 1995.

Hakluyt, Richard, *The Principall Nauigations, Voiages and Discoueries of the English Nation*, London: George Bishop and Ralph Newberie 1589; 2nd edn in 3 vols, London: George Bishop, Ralph Newberie, and Robert Barker 1599–1600.

Hall, A. Rupert and Marie Boas Hall (eds), *The Correspondence of Henry Oldenburg*, 13 vols, Madison, WI: University of Wisconsin Press 1965–86.

Hall, Marie Boas, *Henry Oldenburg: Shaping the Royal Society*, Oxford: Oxford University Press 2002.

Halley, Edmond, 'An Estimate of the Degrees of the Mortality of Mankind, Drawn from Curious Tables of the Births and Funerals at the City of Breslaw; With an Attempt to Ascertain the Price of Annuities upon Lives', in: *Philosophical Transactions of the Royal Society of London* 17 (1693), pp. 596–610.

Hamilton, Thomas, *Belfast Literary Society 1801–1901: Historical Sketch, with Memoirs of some Distinguished Members*, Belfast: M'Caw, Stevenson & Orr 1902.

Harkness, Deborah E., *The Jewel House: Elizabethan London and the Scientific Revolution*, New Haven, CT: Yale University Press 2007.

Harley, R., 'On the Rev. T. P. Kirkman's Method of Resolving Algebraic Equations', in: *Proceedings of the Literary and Philosophical Society of Manchester* 8 (1868–69), pp. 4–20.

Harley, R., 'Sir James Cockle [obituary]', in: *Memoirs and Proceedings of the Manchester Literary and Philosophical Society, Fourth Series* 9 (1895), pp. 215–28.

Harley, R., 'James Cockle [obituary]', in: *Proceedings of the Royal Society of London* 59 (1896), pp. xxx–xxxix.

Harley, R., 'Robert Rawson [obituary]', in: *Proceedings of the London Mathematical Society* 4 (1907), pp. xv–xvii.

Harley, R., *Brief Biographical Sketch of Robert Rawson*, London: James Harley 1910.

Harris, G. G., *The Trinity House of Deptford, 1514–1660*, London: Athlone Press 1969.

Harvey, George, 'Extract of Letter', in 'Scientific Transactions of the Meeting: Tuesday Morning' in: *Report of the First and Second Meetings of the British Association for the Advancement of Science*, London: John Murray 1833, pp. 58–9.

Haselden, Thomas, *The Description and Use of that Most Excellent Invention Commonly Call'd Mercator's Chart*, London: printed for the author 1722.

Haselden, Thomas, *The Seaman's Daily Assistant*, London: printed for W. and J. Mount, T. Page, and Son 1761.

Hawlitschek, Kurt, *Johann Faulhaber 1580–1635: eine Blütezeit der mathematischen Wissenschaften in Ulm*, Ulm: Stadtbibliothek 1995.

Hawlitschek, Kurt, *Johann Faulhaber (1580–1635) and René Descartes (1596–1650): auf dem Weg zur modernen Wissenschaft*, Ulm: Stadtbibliothek 2006.

Heal, Ambrose, *The English Writing-Masters and Their Copy-Books, 1570–1800*, Cambridge: at the University Press 1931.

Heard, John, *From Servant to Queen: A Journey through Victorian Mathematics*, Cambridge: Cambridge University Press 2019.

Heath, T. L., *The Thirteen Books of Euclid's Elements*, 3 vols, Cambridge: Cambridge University Press 1908.

Hebenstreit, Johann Baptist, *Cometen Fragstuck, auß der reinen Philosophia, Bey Anschawung, deß in diesem 1618. Jahr, in dem Obern Lufft schwebenden Cometen, erläutert, vnd auff etlicher Gelehrten vnd Vngelehrten Gegehren, an Tag gegeben*, Ulm: Johann Meder 1618.

Hebenstreit, Johann Baptist, *De Cabala Log-Arithmo-Geometro-Mantica, variis nuper artibus spargi coepta, & Orbi Europaeo obtrusa, dissertatiuncula*, Ulm: Johann Meder 1619.

Heeffer, Albrecht, 'On the Nature and Origin of Algebraic Symbolism', in: *New Perspectives on Mathematical Practices. Essays in Philosophy and History of Mathematics*, ed. B. Van Kerkhove, Singapore: World Scientific Publishing 2009, pp. 1–27.

Heeffer, Albrecht, 'The Symbolic Model for Algebra: Functions and Mechanisms', in: *Model-Based Reasoning in Science and Technology, Abduction, Logic, and Computational Discovery*, ed. Lorenzo Magnani, Walter Carnielli, and Claudio Pizzi, Studies in Computational Intelligence 314, Heidelberg: Springer 2010, pp. 519–32.

Heesen, Anke te, 'The Notebook: A Paper Technology', in: *Making Things Public: Atmospheres of Democracy*, ed. B. Latour and P. Weibel, Cambridge, MA: Harvard University Press 2005, pp. 582–9.

Hendriks, F., 'The Case Book of John Rowe, of London and Exeter, from 1775 to 1790. Edited from the original MS., with an Introductory Notice' in: *The Assurance Magazine and Journal of the Institute of Actuaries* 7 (1857), pp. 136–48.

Hendriks, F., 'Contributions to the History of Insurance, and of the Theory of Life Contingencies, with a Restoration of the Grand Pensionary De Wit's Treatise on Life Annuities', in: *The Assurance Magazine* 2 (1852), pp. 222–58.

Henninger-Voss, Mary, 'Working Machines and Noble Mechanics: Guidobaldo del Monte and the Translation of Knowledge', in: *Isis* 91 (2000), pp. 233–59.

Henninger-Voss, Mary, 'Comets and Cannonballs: Reading Technology in a Sixteenth-Century Library', in: *The Mindful Hand: Inquiry and Invention from the Late Renaissance to Early Industrialization*, ed. L. Roberts, S. Schaffer, and P. Dear, Amsterdam: Royal Netherlands Academy of Arts and Sciences 2007, pp. 11–33.

Heren XVII, *Instructie Van de Eygenschap der Winden, In het vaerwater Tusschen Nederlandt en Java*, Amsterdam: Paulus Mathysz. Oost-Indische Compagnie 1671.

Herschel, J. F. W., *Memoir of Francis Baily, Esq*., London 1845.

Hewins, W. A. S. and Robert Brown, 'Milne, Joshua (1776–1851)', in: *Oxford Dictionary of National Biography*, Oxford: Oxford University Press 2008, https://doi.org/10.1093/ref:odnb/18784.

Higgitt, Rebekah, 'Why I Don't FRS My Tail: Augustus De Morgan and the Royal Society', in: *Notes and Records of the Royal Society* 60/3 (2006), pp. 253–9.

Higgitt, Rebekah, 'Equipping Expeditionary Astronomers: Nevil Maskelyne and the Development of "Precision Exploration"', in: *Geography, Technology and Instruments of Exploration*, ed. Fraser MacDonald and C. W. J. Withers, Farnham: Ashgate 2015 pp. 15–36.

Higgitt, Rebekah, '"Greenwich near London": The Royal Observatory and Its London Networks in the Seventeenth and Eighteenth Centuries', in: *British Journal for the History of Science* 52/2 (2019), pp. 297–322.

Higgitt, Rebekah, Jasmine Kilburn-Toppin, and Noah Moxham, '*Science and the City*: The Spaces and Geographies of Metropolitan Science', in: *Science Museum Group Journal* 15 (2021), https://dx.doi.org/10.15180/211506/001.

Higgitt,Rebekah, Jasmine Kilburn-Toppin, and Noah Moxham, *Metropolitan Science: London Sites and Cultures of Knowledge and Practice, 1600–1800*, Bloomsbury, forthcoming.

Higton, Hester, 'Elias Allen and the Role of Instruments in Shaping the Mathematical Culture of Seventeenth-Century England', PhD thesis, University of Cambridge 1996.

Higton, Hester, 'Does Using an Instrument Make You Mathematical? Mathematical Practitioners of the 17th Century', in: *Endeavour* 25 (2001), pp. 18–22.

Hill, Katherine, 'Mathematics as a Tool of Social Change: Educational Reform in Seventeenth-Century England', in: *The Seventeenth Century* 12 (1997), pp. 23–36.

Hill, Katherine, '"Juglers or Schollers?" Negotiating the Role of a Mathematical Practitioner', in: *British Journal for the History of Science* 31 (1998), pp. 253–74.

[Hilton, W.], [Editor's Note], in: *The Student*, no. 3 (1799), p. 72.

Hobson, Anthony, 'A Sale by Candle in 1608', in: *The Library* 26 (1971), pp. 215–33.

Hodgson, James, *A System of Mathematics*, 2 vols, London: Printed for Thomas Page, William and Fisher Mount, at the Postern on Tower-Hill 1723.

Hoffmann, Christoph, 'The Pocket-Schedule: Note-Taking as a Research Technique', in: *Reworking the Bench: Research Notebooks in the History of Science*, ed. F. L. Holmes, J. Renn, and H.-J. Rheinberger, Dordrecht: Springer 2003, pp. 183–202.

Hofmann, Joseph Ehrenfried, *Nicolaus Mercator (Kauffman): Sein Leben und Wirken vorzugsweise als Mathematiker*, Abhandlungen der Mathematisch-Naturwissenschaftlichen Klasse, Akademie der Wissenschaften und der Literatur in Mainz, Jahrgang 1950.

Hogarth, Peter J. and Ewan W. Anderson, *'The Most Fortunate Situation': The Story of York's Museum Gardens*, York: Yorkshire Philosophical Society 2018.

Holm, Pieter, *Mond-Exame Voor de Stierlieden van Het Schip Recht door Zee*, [Amsterdam]: the Author 1759.

Holmes, Andrew R., 'James Thomson Sr. and Lord Kelvin: Religion, Science, and Liberal Unionism in Ulster and Scotland', in: *Journal of British Studies* 50 (2011), pp. 100–24.

Holton, David (ed.), *Literature and Society in Renaissance Crete*, Cambridge: Cambridge University Press 1991.

Hoogendoorn, Klaas, 'De werken van de meetkundige rekenmeester Sybrandt Hansz. Cardinael (1578–1647). Proeve van een bibliografie', in: *De boekenwereld* 23 (2006–7), pp. 276–90.

Horst, F. van der, 'Daar donderd het canon van Vinckel aan de Theems!: Hendrik August, baron van Kinckel (1747–1821)', in: *Marinekapiteins uit de achttiende eeuw: Een Zeeuws elftal*, ed. J. R. Bruijn, A. C. Meijer and A. P. van Vliet, Den Haag/Middelburg: Koninklijk Zeeuwsch Genootschap der Wetenschappen 2000, pp. 167–79.

Howarth, O. J. R., *The British Association for the Advancement of Science: A Retrospect 1831–1921*, London: British Association for the Advancement of Science 1922.

Howson, A. G., *A History of Mathematics Education in England*, Cambridge: Cambridge University Press 1982.

Høyrup, Jens, *Lengths, Widths, Surfaces: A Portrait of Old Babylonian Algebra and Its Kin*, Studies and Sources in the History of Mathematics and Physical Sciences, New York: Springer 2002.

Høyrup, Jens, 'What Is "Geometric Algebra," and What Has It Been in Historiography? Contribution to the Session Histoire de l'historiographie de l'algèbre, Séminaire Histoire et Philosophie des Mathématiques, CNRS, SPHERE, & projet ERC SAW, Université Paris Diderot, 11 January 2016', in: *AIMS Mathematics* 2/1 (2017), pp. 128–60.

Hudson, J. W., *The History of Adult Education, in Which is Comprised A Full and Complete History of the Mechanics' and Literary Institutions, Athenæums, Philosophical, Mental and Christian Improvement Societies, Literary Unions, Schools of Design, etc., of Great Britain, Ireland, America, etc.*, London: Longman, Brown, Green & Longmans 1851.

Hues, Robert, *Tractatus de globis et eorum vsu*, 2nd edn, London: Thomas Dawson 1594.

Hues, Robert, ed. Clements R. Markham, *Tractatus de globis et eorum usu: a treatise descriptive of the globes constructed by Emery Molyneux and published in 1592*, London: The Hakluyt Society 1889.

Hughes, Barnabas, *Fibonacci's* De practica geometrie, New York: Springer 2008.

Hume, James, *Traité de la trigonométrie*, Paris: De la Coste 1636.

Hunter, Michael, *John Aubrey and the Realm of Learning*, London: Duckworth 1975.

Hunter, Michael, *Science and Society in Restoration England*, Cambridge: Cambridge University Press 1981.

Hunter, Michael, *Establishing the New Science: The Experience of the early Royal Society*, Woodbridge: The Boydell Press 1989.

Hunter, Michael, *The Royal Society and its Fellows, 1660–1700*, Oxford: The Alden Press for The British Society for the History of Science, 2nd edn 1994.

Hunter, Michael, *Archives of the Scientific Revolution: The Formation and Exchange of Ideas in Seventeenth-Century Europe*, Woodbridge: The Boydell Press 1998.

Hutton, Catherine, *Reminiscences of a Gentlewoman of the Last Century*, Birmingham 1891.

Hutton, Charles, *The School-master's Guide: or, A Complete System of Practical Arithmetic, Adapted to the Use of Schools*, Newcastle 1764.

Hutton, Charles, *A Treatise on Mensuration, Both in Theory and Practice*, Newcastle and London 1770.

Hutton, Charles (ed.), *The Diarian Miscellany*, 5 vols, London 1775.

Hutton, Charles, 'The Force of Fired Gun-Powder, and the Initial Velocities of Cannon Balls, Determined by Experiments', in: *Philosophical Transactions* 68 (1778), pp. 50–85.

Hutton, Charles, 'An Account of the Calculations Made from the Survey and Measures Taken at Schehallien, in Order to Ascertain the Mean Density of the Earth', in: *Philosophical Transactions* 68 (1778), pp. 689–788.

Hutton, Charles, *Mathematical Tables*, London 1785.

Hutton, Charles, *The Compendious Measurer: Being a Brief, Yet Comprehensive, Treatise on Mensuration and Practical Geometry*, London 1786.

Hutton, Charles, *Tracts, Mathematical and Philosophical*, London 1786.

Hutton, Charles, *Elements of Conic Sections: with Select Exercises in Various Branches of Mathematics and Philosophy. For the Use of the Royal Military Academy at Woolwich*, London: Printed for J. Davis 1787.

Hutton, Charles, 'Experiments on the Expansive Force of Freezing Water, Made by Major Edward Williams of the Royal Artillery, at Quebec in Canada [. . .] Communicated in a Letter from Charles Hutton [...] to Professor John Robison, General Secretary of the Royal Society of Edinburgh', in: *Transactions of the Royal Society of Edinburgh* 2 (1790), pp. 23–8; reprinted in: *The Literary Magazine and British Review* 6 (January 1791), pp. 20–2.

Hutton, Charles, *A Mathematical and Philosophical Dictionary*, London 1795–96.

Hutton, Charles, *A Course of Mathematics*, 2 vols, London 1798.

Hutton, Charles, 'On the Calculations for Ascertaining the Mean Density of the Earth', in: *Philosophical Magazine* 38 (1811), pp. 112–16.

Hutton, Charles, *Tracts on Mathematical and Philosophical Subjects*, 3 vols, London 1812.

Hutton, Charles, 'On the Mean Density of the Earth', in: *Philosophical Transactions* 111 (1821), pp. 276–92; reprinted in: *Philosophical Magazine* 58/279 (1821), pp. 3–13.

Hutton, Charles, George Shaw, and Richard Pearson (eds), *The Philosophical Transactions of the Royal Society of London, from Their Commencement, in 1665, to the Year 1800: Abridged, with Notes and Biographical Illustrations*, 18 vols, London 1803–9.

Iliffe, Rob, 'Mathematical Characters: Flamsteed and Christ's Hospital Royal Mathematical School', in: *Flamsteed's Stars: New Perspectives on the Life and Work of the First Astronomer Royal, 1646–1719*, ed. Frances Willmoth, Woodbridge: The Boydell Press with the National Maritime Museum 1997, pp. 115–44.

Iordanou, I., 'The Professionalization of Cryptology in Sixteenth-Century Venice', in: *Enterprise & Society* 19 (2018), pp. 979–1013.

Irigoin, Jean, 'Les ambassadeurs à Venise et le commerce des manuscrits grecs dans les annees 1540–1550', in: *Venezia centro di mediazione tra Oriente e Occidente (secoli XV–XVI). Aspetti e problemi*, ed. H. G. Beck, M. Manoussacas, and A. Pertusi, Florence: Olschki 1977, pp. 399–415.

Irvine, Wm. Fergusson, *A Short History of the Township of Rivington in the County of Lancaster with Some Account of the Church and Grammar School*, Edinburgh: Ballantyne Press, 1904.

Israel, Jonathan, *The Dutch Republic: Its Rise, Greatness, and Fall, 1477–1806*, Oxford: Clarendon Press 1995.

Ives, Andrea, *Admiral Long's Foundation & Burnt Yates School: 250 Years of History*, Harrogate: Printed for the author 2014.

Jalobeanu, Dana, 'Core Experiments, Natural Histories and the Art of *Experientia Literata*: The Meaning of Baconian Experimentation', in: *Society and Politics* 5 (2011), pp. 88–103.

Jamieson, John, *The History of the Royal Belfast Academical Institution: 1810–1960*, Belfast: William Martin & Son 1959.

Jansen, L., 'De Derde Vergroting van Amsterdam', in: *52nd Yearbook of the Amstelodamum Historical Society*, Amsterdam: J. H. de Bussy 1959, pp. 42–89.

Jardine, Boris, 'Henry Sutton's Collaboration with John Reynolds (Gauger, Assayer and Clerk at the Royal Mint)', in: *Bulletin of the Scientific Instrument Society* 130 (2016), pp. 4–7.

Jardine, Boris, 'On Being Compendious: "Universal" Instruments and the Mathematical Arts', in: *Coping with Copia: Epistemological Excess in Early Modern Art and Science*, ed. Fabian Kraemer and Itay Sapir, Amsterdam: Amsterdam University Press, forthcoming.

Jobst, Wolfgang and Walter Schellhas, 'Abraham von Schönberg, Leben und Werk: Die Wiederbelebung des Erzgebirgischen Bergbaus nach dem dreißigjährigen Krieg durch Oberghauptmann Abraham von Schönberg', in: *Freiberger Forschungshefte, Kultur und Technik*, vol. D 198, Leipzig: Verlag für Grundstoffindustrie 1994, pp. 74–133.

Johns, Adrian, *The Nature of the Book: Print and Knowledge in the Making*, Chicago: University of Chicago Press 1998.

Johnson, Francis R., 'Thomas Hood's Inaugural Address as Mathematical Lecturer of the City of London (1588)', in: *Journal of the History of Ideas* 3 (1942), pp. 94–106.

Johnson, W., 'Contributors to Improving the Teaching of Calculus in Early 19th-Century England', in: *Notes and Records of the Royal Society of London* 49 (1995), pp. 93–103.

Johnston, Stephen, 'Mathematical Practitioners and Instruments in Elizabethan England', in: *Annals of Science* 48 (1991), pp. 319–44.

Johnston, Stephen, 'Making Mathematical Practice: Gentlemen, Practitioners and Artisans in Elizabethan England', PhD thesis, University of Cambridge 1994.

Johnston, Stephen, 'The Carpenter's Rule: Instruments, Practitioners and Artisans in 16th-Century England', in: *Proceedings of the Eleventh International Scientific Instrument Symposium*, ed. G. Dragoni, A. McConnell, and G. L'E. Turner, Bologna 1994, pp. 39–45.

Johnston, Stephen, 'The Identity of the Mathematical Practitioner in 16th-Century England', www.mhs.ox.ac.uk/staff/saj/texts/mathematicus.htm.

Jones, Clifford, *The Sea & the Sky: The History of the Royal Mathematical School of Christ's Hospital*, Horsham: Christ's Hospital 2015.

Jones, W. D., *Records of the Royal Military Academy*, 2nd edn, Woolwich: Royal Artillery Institution 1895.

Jong, R. M. A. De, *Genealogie van de familie De Gavere*, genealogysk jierboek, Ljouwert: Fryske Akademie 1996.

Jonkers, A. R. T., *Earth's Magnetism in the Age of Sail*, Baltimore: Johns Hopkins University Press 2003.

Jugel, Johann Gottfried, *Gründlicher und deutlicher Begriff von dem gantzen Berg-Bau-Schmeltz-Wesen und Marckscheiden, in Drey Haupt-Theite eingentheilet*, Berlin: Rüdiger 1744.

Jugel, Johann Gottfried, *Geometria subterranea, oder Unterirdische Messkunst der Berg- und Grubengebäude, insgemein die Markscheidekunst genannt, etc.*, Leipzig: Kraus 1773.

Jung, Johann, *Rechenbuch auff den Ziffern vnd Linien, darinne allerley Kauffmans handlung, nach art der Regel de Tri vnd Welschen Practica, sampt der Regel Falsi, dardurch die Exempla der acht Regeln Coß auffgelöset warden, Neben ausziehung der wurtzeln Arithmetischer Progression, So wol den Regeln Algebre, vnd andern aufflösungen Cubicossischer vergleichungen, so vor niemals am Tage gewesen, alles ordentlich gestelt durch Johann Jungen Rechenmeister zu Lübeck*, Lübeck: Johann Balhorn d.J. [1578].

Jungnickel, Christa and Russell McCormmach, *Cavendish: The Experimental Life*, 2nd edn, n.p., 1999.

Kargon, Robert, *Science in Victorian Manchester: Enterprise and Expertise*, Manchester: Manchester University Press 1977.

Karpinski, L. and F. Kokomoor, 'The Teaching of Elementary Geometry in the Seventeenth Century' in: *Isis* 10 (1928), pp. 21–32.

Kästner, Abraham Gotthelf, *Anmerkungen über die Markscheidekunst, nebst einer Abhandlung von Höhenmessungen durch das Barometer*, Göttingen: Vandenhoeck 1775.

Keefer, H., 'Johannes Faulhaber, der bedeutendste Ulmer Mathematiker und Festungsbaumeister', in: *Württembergische Schulwarte* 4 (1928), pp. 1–12.

Keiding, Niels, 'The Method of Expected Number of Deaths, 1786–1886–1986', in: *International Statistical Review* 55(1987), pp. 1–20.

Keller, V., A. M. Roos, and E. Yale (eds), *Archival Afterlives: Life, Death, and Knowledge-Making in Early Modern British Scientific and Medical Archives*, Leiden: Brill 2018.

Kennefick, Ciara, 'The Contribution of Contemporary Mathematics to Contractual Fairness in Equity, 1751–1867', in: *Journal of Legal History* 39/3 (2018), pp. 307–39.

Kenrick, J., 'A Retrospect of the Early History of the Yorkshire Philosophical Society', in: *Annual Report of the Council of the Yorkshire Philosophical Society for MDCCCLXXIII*, York 1874, pp. 34–44.

Kepler, Johannes, *New vnnd Alter Schreib Calender sambt dem Lauff vnd Aspecten der Planeten auff das Jahr Christi M. DC. XVIII. Prognosticum Astrologicum auff das Jahr MDCXVIII. Von natürlicher Influentz der Sternen in diese Nidere Welt*, Linz: Johann Blancken 1618.

Kepler, Johannes, *Ephemeris nova Motuum Coelestium ad annum vulgaris aerae M D C XVIII. Ex obseruationibus potissimum TYCHONIS BRAHEI, Hypothesibus Physicis, & Tabulis RVDOLPHINIS; Nova etiam formâ disposita, ut Calendarii Scriptorii usum praebere possit. Ad Meridianum Vranopyrgicum in freto Cimbrico, quem proximè circumstant Pragensis, Lincensis, Venetus, Romanus*, Linz [no year].

Kepler, Johannes, *De Cometis Libelli Tres*, Augsburg: Typis Andreæ Apergeri 1619.

Kepler, Johannes, *Gesammelte Werke*, 26 vols, Munich: C. H. Beck 1937–2017.

Kersey, John, *The Elements of That Mathematical Art, Commonly Called Algebra, Expounded in Four Books*, 2 vols, London: William Godbid for Thomas Passinger and Benjamin Hurlock 1673.

Kersey, John, *The Third & Fourth Books of the Elements of Algebra*, London: Printed by William Godbid for Thomas Passinger 1674.

Kinckhuysen, Gerard, *Algebra, ofte Stel-konst, beschreven tot dienst van de leerlinghen*, Haerlem: Passchier van Wesbusch 1661.

King, Elizabeth, *Lord Kelvin's Early Home*, London: Macmillan & Co. 1909.

Kirchvogel, Paul A., 'Faulhaber, Johann', in: *Complete Dictionary of Scientific Biography*, vol 4, Charles Scribner's Sons, 2008, pp. 549–53.

Kirkman, Thomas P., 'On the Perfect Partitions of $r^2 - r + 1$', in: *Transactions of the Historic Society of Lancashire and Cheshire* 9 (1856–57), pp. 127–42.

Kirkman, Thomas P., 'On the Theory of Groups and Many-Valued Functions', in: *Memoirs of the Literary and Philosophical Society of Manchester*, 3rd ser. 1 (1862), pp. 274–398.

Kirkman, Thomas P., 'The Complete Theory of Groups, Being the Solution of the Mathematical Prize Question of the French Academy for 1860', in: *Proceedings of the Literary and Philosophical Society of Manchester* 3 (1863–64/1864–65), pp. 133–152.

Kirkman, Thomas P., *Philosophy without Assumptions*, London: Longmans, Green, and Co. 1876.

K[irkman], W. W., 'Thomas Penyngton Kirkman [obituary]', in: *Memoirs and Proceedings of the Manchester Literary and Philosophical Society*, 4th ser. 9 (1895), pp. 238–43.

Kirnbauer, Frank, 'Die Entwicklung des Grubenrißwesens in Österreich', in: *Blätter für Technikgeschichte* 24 (1962), pp. 60–129.

Klein, Ursula, *Technoscience in History: Prussia, 1750–1850*, Cambridge, MA: MIT Press 2020.

Klinger, Kerrin and Thomas Morel, 'Was ist praktisch am mathematischen Wissen? Die Positionen des Bergmeisters J. A. Scheidhauer und des Baumeisters C. F. Steiner in der Zeit um 1800', in: *NTM Zeitschrift für Geschichte der Wissenschaften, Technik und Medizin* 26/3 (2018), pp. 267–99.

Knight, William A., *Memoir of John Nichol*, Glasgow: James MacLehose & Sons 1896.

[Knowles, John], 'Preface' in: *The Student*, no. 1 (1797).

Knuth, Donald E., 'Johann Faulhaber and Sums of Powers', in: *Mathematics of Computation* 61 (1993), pp. 277–94.

Kool, Marjolein, *Die conste vanden getale: Een studie over Nederlandstalige rekenboeken uit de vijftiende en zestiende eeuw, met een glossarium van rekenkundige termen*, Hilversum: Verloren 1999.

Kröger, Desirée, 'Abraham Gotthelf Kästner als Lehrbuchautor, unter Berücksichtigung weiterer deutschsprachiger mathematischer Lehrbücher für den universitären Unterricht', dissertation, Bergische Universität Wuppertal 2014.

Kühn, Heidi, 'Die Mathematik im deutschen Hochschulwesen des 18. Jahrhunderts (unter besonderer Berücksichtigung der Verhältnisse an der Leipziger Universität)', dissertation, Leipzig: Karl-Marx-Universität 1988.

[Kynaston, Sir Francis], *The Constitutions of the Musaeum Minervae*, London 1636.

Laird, W. R., 'The Scope of Renaissance Mechanics', in: *Osiris* 2 (1986), pp. 43–68.

Laird, W. R., *The Unfinished Mechanics of Giuseppe Moletti: An Edition and English Translation of His Dialogue on Mechanics, 1576*, Toronto: University of Toronto Press 2000.

Lamy, Bernard, *Traité de la Grandeur en general*, Paris: André Palard 1680; later editions with the title *Elemens des Mathematiques ou Traité de la Grandeur en general*, 2nd edn, Paris: André Palard 1689; 3rd edn, Florentin Delaulne 1704; 4th edn, Paris: Nicolas Pepie 1715.

Landgraf, Werner, 'Über die Bahn des zweiten Kometen von 1618', *Sterne* 61 (1985), pp. 351–3.

Larmor, Joseph and James Thomson (eds), *Collected Papers in Physics and Engineering by James Thomson*, Cambridge: Cambridge University Press 1912.

Lastman, Cornelis Jansz, *Schat-kamer des grooten seevaerts-kunst*, Amsterdam 1621.

Lastman, Cornelis Jansz, *Beschrijvinge van de Kunst der Stuer-luyden* (ed. princ. 1642), Amsterdam: Symon Cornelisz. Lastman 1657.

Lawn, Brian, *The Rise and Decline of the Scholastic 'Quaestio Disputata', With Special Emphasis on Its Use in the Teaching of Medicine and Science*, Leiden: Brill 1993.

Leitão, Henrique, 'Jesuit Mathematical Practice in Portugal, 1540–1759', in: *The New Science and Jesuit Science: Seventeenth Century Perspectives*, ed. Mordechai Feingold, Dordrecht: Springer 2003, pp. 229–47.

Lempe, Johann Friedrich, *Gründliche Anleitung zur Markscheidekunst*, Leipzig: Crusius 1782.

Letwin, William, *The Origins of Scientific Economics. English Economic Thought 1660–1776*, London: Methuen 1963.

Levy-Eichel, Mordechai, '"Suitable to the Meanest Capacity": Mathematics, Navigation and Self-Education in the Early Modern British Atlantic', in: *The Mariner's Mirror* 103/4 (2017), pp. 450–65.

Lewin, Christopher G., *Pensions and Insurance before 1800: A Social History*. East Linton: Tuckwell 2002.

Leybourne, Thomas, 'Preface', in: *The Mathematical Repository* 2 (1801), pp. v–vi.
Leybourn, William, *The Compleat Surveyor: containing the whole Art of Surveying of Land, by the Plain Table, Theodolite, Circumferentor, and Peractor*, London: R. & W. Leybourn for E. Brewster and G. Sawbridge 1653.
Leybourn, William, *Arithmetick, Vulgar, Decimal, and Instrumental. In three parts*, London: R. and W. Leybourn for G. Sawbridge 1657.
Leybourn, William, *Arithmetick: vulgar, decimal, instrumental, algebraical. In four parts.* 3rd edition, London: S. Streater for George Sawbridge 1668.
Lied, Liv Ingeborg and Marilena Maniaci (eds), *Bible as Notepad: Tracing Annotations and Annotations Practices in Late Antique and Medieval Biblical Manuscripts*, Berlin: De Gruyter 2018.
Liessmann, Wilfried, *Historischer Bergbau im Harz*, Heidelberg: Springer 2010.
Linschoten, Jan Huygen van, *Reys-gheschrift van de navigatien der Portugaloysers in Orienten*, Amsterdam: Cornelis Claesz 1595.
Linschoten, Jan Huygen van, *His Discours of Voyages into ye Easte & West Indies Deuided into Foure Bookes.* London: J. Windet for J. Wolfe 1598.
Linschoten, Jan Huygen van, *Itinerario, voyage ofte schipvaert naer Oost ofte Portugaels Indien. Deel 4 en 5*, ed. J. C. M. Warnsinck, The Hague: Martinus Nijhoff 1939.
Lonsdale, Henry and Joshua Milne, *The Life of John Heysham, M.D. and His Correspondence with Mr. Joshua Milne Relative to the Carlisle Bills of Mortality*, London: Longmans, Green, and Co. 1870.
Lott, Frederick Barnes, *The Centenary Book of the Leicester Literary and Philosophical Society*, Leicester: [*s.n.*] 1935.
Loughney, Claire, 'Colonialism and the Development of the English Provincial Museum, 1823–1914', PhD thesis, University of Newcastle 2006.
Lowth, Robert, *A Short Introduction to English Grammar*, 2nd edn, London 1763; 3rd edn, 1769.
Lubenow, William C., *'Only Connect': Learned Societies in Nineteenth Century Britain*, Woodbridge: The Boydell Press 2015.
Lucchi, Piero, 'Un trattato di crittografia del Cinquecento: le *Zifre* di Agostino Amadi fra cultura umanistica e cultura dell'abaco', in: *Matematica e cultura*, ed. Michele Emmer, Milan: Springer Italia 2004, pp. 39–50.
Lutun, Bernard, 'Des Ecoles de Marine et Principalement des Ecoles d'Hydrographie (1629–1789)', in: *Sciences et Techniques en Perspective* 34 (1995), pp. 3–30.
Maanen, Jan van, 'Facets of Seventeenth-Century Mathematics in The Netherlands', PhD dissertation, Utrecht University 1987.
Maanen, Jan van, 'Cardinael in de geschiedenis van de wiskunde', in: *Nieuw archief voor wiskunde*, ser. 5, 4/1 (2003), pp. 51–5.
Macfarlane, Alexander, *Lectures on Ten British Mathematicians of the Nineteenth Century*, Mathematical Monographs 17, New York: John Wiley & Sons 1916.
Machado, Diogo Barbosa, *Bibliotheca Lusitana*, vol III, Lisbon: Ignacio Rodrigues 1752.
MacHale, Desmond, *George Boole: His Life and Work*, Dublin: Boole Press 1985.
Mackay, Andrew, *The Theory and Practice of Finding the Longitude at Sea or Land*, 2nd edn, 2 vols, Aberdeen: Printed for the Author by J. Chalmers & Co. 1801.
MacLean, Gerald M., *The Rise of Oriental Travel: English Visitors to the Ottoman Empire, 1580–1720*, Basingstoke: Palgrave Macmillan 2004.
MacLean, I., *Episodes in the Life of the Early Modern Learned Book*, Leiden: Brill 2020.

MacLeod, R. M., J. R. Friday, and C. Gregor, *The Corresponding Societies of the British Association for the Advancement of Science 1883–1929: A Survey of Historical Records, Archives and Publications*, London: Mansell 1975.

Mackenzie, Eneas, *A Descriptive and Historical Account of Newcastle-upon-Tyne*, Newcastle 1827.

Maclaurin, Colin, *A Treatise of Fluxions: In Two Books*, Edinburgh 1742.

Maclaurin, Colin, *The Collected Letters of Colin MacLaurin*, ed. Stella Mills, Nantwich: Shiva 1982.

M[acMahon], P. A., 'Robert Harley, 1828–1910', in: *Proceedings of the Royal Society* 91 (1915), pp. i–v.

Maddison, Robert E. W., *The Life of the Honourable Robert Boyle F.R.S.*, London: Taylor & Francis 1969.

Madelaine, J., [Review of Hutton, *Nouvelles expériences d'artillerie*], in: *Journal des sciences militaires* 5 (1826), pp. 350–79.

Mahoney, Michael S., *The Mathematical Career of Pierre de Fermat (1601–1665)*, Princeton, NJ: Princeton University Press 1973.

Mahoney, Michael S., 'Barrow's Mathematics: Between Ancients and Moderns', in: *Before Newton: The Life and Times of Isaac Barrow*, ed. Mordechai Feingold, Cambridge: Cambridge University Press 1990, pp. 179–249.

Makepeace, Chris E., *Science and Technology in Manchester: Two Hundred Years of the Lit. and Phil*, Manchester: Manchester Literary & Philosophical Publications 1984.

Makreel, Dirk, *Lichtende leydt-starre der groote zee-vaert*, Amsterdam: H. Doncker 1671.

Malcolm, Noel and Jacqueline A. Stedall, *John Pell (1611–1685) and His Correspondence with Sir Charles Cavendish: The Mental World of an Early Modern Mathematician*, Oxford: Oxford University Press 2005.

Maltby, H. J. M. and W. P. Winter, *Fifty Years of Local Science, 1875–1925: A Record of Fifty Years of Work Done by Members of the Bradford Natural History and Microscopical Society and the Bradford Scientific Association*, Bradford: [*s.n.*] 1925.

Malynes, Gerard de, *Consuetudo, vel, lex mercatoria, or the ancient law-merchant. Divided into three parts: according to the essential parts of trafficke*, London: Adam Islip 1622.

Maltby, H. J. M. and W. P. Winter, *Consuetudo; vel, Lex Mercatoria; or, the Ancient Law-merchant, in Three Parts*, London: Basset 1686.

Mancosu, Paolo, 'Aristotelian Logic and Euclidean Mathematics: Seventeenth-Century Developments of the *Quaestio de certitudine mathematicarum*', in: *Studies in the History and Philosophy of Science* 23 (1992), pp. 241–65.

Manno, A., 'Giulio Savorgnan: *machinatio* e *ars fortificatoria* a Venezia', in: *Cultura scienze e tecniche nella Venezia del Cinquecento: Atti del convegno internazionale di studi Giovanni Battista Benedetti e il suo tempo*, ed. Antonio Manno, Venice: Istituto veneto di scienze, lettere e arti 1987, pp. 227–45.

Manzo, Silvia, 'Probability, Certainty and Facts in Francis Bacon's Natural Histories', in: *Skepticism in the Modern Age*, ed. José R. Maia Neto, Gianni Paganini, and John Christian Laursen, Leiden: Brill 2009, pp. 123–38.

Maranta, B., *Methodi cognoscendorum simplicium libri tres*, Venice: Valgrisi 1559.

Marcus, Hannah, *Forbidden Knowledge: Medicine, Science, and Censorship in Early Modern Italy*, Chicago: University of Chicago Press 2020.

Marcus, Hannah and Paula Findlen, 'Deciphering Galileo: Communication and Secrecy before and after the Trial', in: *Renaissance Quarterly* 72 (2019), pp. 953–95.

Marius, John, *Advice Concerning Bils of Exchange*, 2nd edn, London: William Hunt 1655.

Markham, Albert Hastings (ed.), *The Voyages and Works of John Davis the Navigator*, London: The Hakluyt Society 1880.

Markham, Clements R., *A Life of John Davis, the Navigator, 1550–1605: Discoverer of Davis Straits*, London: George Philip & Son 1889.
Martin, Darryl, 'The English Virginal', PhD thesis, University of Edinburgh 2003.
Martindale, Adam, *The Country-Survey-Book: or Land-Meters Vade-Mecum. Wherein the Principles and practical Rules for Surveying of Land, are so plainly (though briefly) delivered, that any one of ordinary parts (understanding how to add, subtract, multiply and divide,) may by the help of this small Treatise alone, and a few cheap Instruments easy to be procured, Measure a parcel of Land, and with judgment and expedition Plot it, and give up the Content thereof*, London: for R. Clavel and T. Sawbridge 1692.
Martin, Darryl, *The Country Almanack for the Year, 1675. Suted to the several Capacities, Humours, and Occasions of Gentlemen, Scholars, Travellers, and Husband-men, &c.*, London: F. L. for the Company of Stationers 1675.
Martinelli, R. Biancarelli, 'Paul Homberger: Il primo intermediario tra Galileo e Keplero', in: *Galilaeana* 1 (2004), pp. 171–81.
Maskell, Joseph, *Collections in Illustration of the Parochial History and Antiquities of the Ancient Parish of Allhallows Barking, in the City of London*, London: Bryan Corcoran & Co. 1864.
Maskelyne, Nevil, 'A Proposal for Measuring the Attraction of Some Hill in this Kingdom by Astronomical Observations', in: *Philosophical Transactions* 65 (1775), pp. 495–99.
Maskelyne, Nevil, 'An Account of Observations Made on the Mountain Schehallien for Finding its Attraction', in: *Philosophical Transactions* 65 (1775), pp. 500–42.
Mathis, Franz, *Die deutsche Wirtschaft im 16. Jahrhundert*, Enzyklopädie deutscher Geschichte 11, München: Oldenbourg Verlag 1992.
Maxwell, Susan M., 'Hues, Robert (1553–1632), Mathematician and Geographer', in: *Oxford Dictionary of National Biography*, 2008, https://doi.org/10.1093/ref:odnb/14045.
Maxwell, Susan M., 'Molyneux, Emery (d.1598), Maker of Globes and Ordnance', in: *Oxford Dictionary of National Biography*, 2008, https://doi.org/10.1093/ref:odnb/50911.
Mayhew, Robert (ed.), *The Aristotelian* Problemata Physica*: Philosophical and Scientific Investigations*, Leiden: Brill 2015.
Mayne, John, *Socius mercatoris: or the Merchant's Companion*, London: William Godbid for Nathaniel Crouch 1674.
McConnell, Anita, 'Davis, John (d.1621), Sailor', in: *Oxford Dictionary of National Biography*, 2008, https://doi.org/10.1093/ref:odnb/7284.
McKerrow, R. B. (ed.), *A Dictionary of Printers and Booksellers in England, Scotland and Ireland, and of Foreign Printers of English Books 1556–1640*, London: The Bibliographical Society 1910.
Medina, Pedro de, *Arte de navegar*, Valladolid: F. Fernández de Córdoba 1545.
Pedro de Medina, *Svma de cosmographía* (1561), ed. Juan Fernández Jiménez, [Valencia]: Albatros 1980.
Mee, Jon, 'The Transpennine Enlightenment, 1780–1840', in: *Annual Report of the Yorkshire Philosophical Society* (2016), pp. 65–71.
Mee, Jon and Jennifer Wilkes, 'Transpennine Enlightenment: The Literary and Philosophical Societies and Knowledge Networks in the North, 1781–1830', in: *Journal of Eighteenth-Century Studies* 38/4 (2015), pp. 599–612.
Meli, Domenico Bertoloni, *Thinking with Objects: The Transformation of Mechanics in the Seventeenth Century*, Baltimore: Johns Hopkins University Press 2006.
Melmore, Sidney, 'Some Letters from Charles Hutton to Robert Harrison', in: *The Mathematical Gazette* 30 (1946), pp. 71–81.
Mercator, Nicolaus, 'Certain Problems Touching Some Points of Navigation', in: *Philosophical Transactions* 13 (4 June 1666), pp. 215–18.

Meschini, Anna, *Michele Sofianòs*, Padua: Liviana 1981.
Meskens, Ad, 'Michiel Coignet's Contribution to the Development of the Sector', in: *Annals of Science* 54 (1997), pp. 143–60.
Meskens, Ad, *Practical Mathematics in a Commercial Metropolis: Mathematical Life in Late 16th Century Antwerp*, Berlin: Springer 2015.
Meyer, G. F., 'Devices in Mathematics', in: *American Association of Instructors of the Blind: Twenty-Seventh Biennial Convention, held at Watertown, Massachusetts, June 23 to 27, 1924*, pp. 202–8.
Meyer, W. R., 'Le Mesurier, Paul (1755–1805), Merchant and Politician', in: *Oxford Dictionary of National Biography*, 2008, https://doi.org/10.1093/ref:odnb/16428.
Michel, Nicolas and Ivahn Smadja, 'Mathematics in the Archives: Deconstructive Historiography and the Shaping of Modern Geometry (1837–1852)' in: *British Journal for the History of Science* 54/4 (2021), pp. 423–41.
Mikkeli, Heikki, 'The cultural programmes of Alessandro Piccolomini and Sperone Speroni at the Paduan Accademia degli Infiammati in the 1540s', in: *Philosophy in the Sixteenth and Seventeenth Centuries: Conversations with Aristotle*, ed. Constance Blackwell and Sachiko Kusukawa, London: Routledge 1999, pp. 76–85.
Miller, David Philip, 'The Royal Society of London 1800–1835: A Study in the Cultural Politics of Scientific Organization', PhD thesis, University of Pennsylvania 1981.
Miller, David Philip, 'Sir Joseph Banks: An Historiographical Perspective', in: *History of Science* 19 (1981), pp. 284–92.
Miller, David Philip, 'Between Hostile Camps: Sir Humphry Davy's Presidency of the Royal Society of London, 1820–1827', in: *British Journal for the History of Science* 16 (1983), pp. 1–47.
Miller, David Philip, 'The Revival of the Physical Sciences in Britain, 1815–1840', in: *Osiris* 2 (1986), pp. 107–34.
Miller, Peter N., *Peiresc's Europe: Learning and Virtue in the Seventeenth Century*, New Haven, CT: Yale University Press 2000.
Miller, Peter N., *Peiresc's Mediterranean World*, Cambridge, MA: Harvard University Press 2015.
Mills, Stella, 'Thomas Kirkman–The Mathematical Cleric of Croft', in: *Manchester Literary and Philosophical Society: Memoirs and Proceedings* 120 (1977–80), pp. 100–9.
Milne, Joshua, *A Treatise on the Valuation of Annuities and Assurances on Lives and Survivorships*, London: Longman, Hurst, Rees, Orme, and Brown 1815.
Molières, Joseph Privat de, *Mathematic Lessons, for the Use of Students in the Mathematics and Natural Philosophy*, trans. Thomas Haselden, London: printed for John Clarke 1730.
Monte, Guidobaldo Dal, *Le mechaniche* [...] *tradotte in volgare dal Sig. Filippo Pigafetta*, Venice: Francesco de' Franceschi 1581.
Moody, Jessica, *The Persistence of Memory: Remembering Slavery in Liverpool, 'Slaving Capital of the World'*, Liverpool: Liverpool University Press 2020.
Moore, John Hamilton, *The Practical Navigator, and Seaman's New Daily Assistant: Being Complete System of Practical Navigation*, London: Printed by W. and J. Richardson 1772; 9th edn 1791
Moote, A. Lloyd and Dorothy C. Moote, *The Great Plague. The Story of London's Most Deadly Year*, Baltimore: Johns Hopkins University Press 2004.
Morel, Thomas, '*De Re Geometrica*: Writing, Drawing, and Preaching Mathematics in Early Modern Mines', in: *Isis* 111/1 (2020), pp. 22–45.
Morel, Thomas, 'Le microcosme de la géométrie souterraine: échanges et transmissions en mathématiques pratiques', in: *Philosophia Scientiae* 19/2 (2015), pp. 17–36.

Morel, Thomas, *Underground Mathematics: Craft Culture and Knowledge Production in the Holy Roman Empire*, Cambridge: Cambridge University Press 2022.

Morel, Thomas, 'Mathematics and Technological Change: The Silent Rise of Practical Mathematics, in: *Bloomsbury Cultural History of Mathematics*, vol. 3, 2024, pp. 179–206.

Mores, Edward Rowe, *A List of Policies and Other Instruments of the Society as well General as Special*, London: [*s.n.*] 1764.

Morgan, Sylvanus, *Horologiographia optica. Dialling Universall and Particular*, London: R. & W. Leybourn for Andrew Kemb and Robert Boydell 1652.

Morrell, J. B., 'Brewster and the Early British Association for the Advancement of Science', in: *'Martyr of Science': Sir David Brewster 1781–1868*, ed. A. D. Morrison-Low and J. R. R. Christie, Edinburgh: Royal Scottish Museum 1984, pp. 25–9.

Morrell, Jack and Arnold Thackray, *Gentlemen of Science: Early Years of the British Association for the Advancement of Science*, Oxford: Clarendon Press 1981.

Morrell, Jack and Arnold Thackray (eds), *Gentlemen of Science: Early Correspondence of the British Association for the Advancement of Science*, Camden Fourth Series 30, London: Royal Historical Society 1984.

Morris, R. J., 'Clubs, Societies and Associations', Chapter 8 in: *The Cambridge Social History of Britain, 1750–1950*, ed. F. Thompson, Cambridge: Cambridge University Press 1990, pp. 395–444.

Mosley, Adam, 'Objects of Knowledge: Mathematics and Models in Sixteenth-Century Cosmology and Astronomy', in: *Transmitting Knowledge: Words, Images, and Instruments in Early Modern Europe*, ed. Sachiko Kusukawa and Ian MacLean, Oxford: Oxford University Press 2006, pp. 193–216.

Mosley, Adam, 'Objects, Texts and Images in the History of Science', in: *Studies in History and Philosophy of Science* 38 (2007), pp. 289–302.

Mota, Catarina Alexandra Pereira, *'A história do conceito de reta tangente em Portugal: um estudo desde o século XVIII até à matemática moderna'*, doctoral thesis, Braga: Escola de Ciências da Universidade do Minho 2018, http://hdl.handle.net/1822/58337.

Mountaine, William, *The Seaman's Vade-Mecum*, London: Printed for W. Mount and T. Page 1744.

Mountaine, William, *A Description of the Lines Drawn on Gunter's Scale, as Improved by Mr. John Robertson*, London: Nairne & Blunt 1778.

Mountaine, William and James Dodson, *An Account of the Methods Used to Describe Lines, on Dr. Halley's Chart of the Terraqueous Globe, Shewing the Variation of the Magnetic Needle*, London: William Mount and Thomas Page 1746.

Muccillo, Maria, 'La biblioteca greca di Francesco Patrizi', in: *Bibliothecae Selectae: Da Cusano a Leopardi*, ed. Eugenio Canone, Florence: Olschki 1993, pp. 73–118.

Mukerji, Chandra, *Impossible Engineering: Technology and Territoriality on the Canal Du Midi*, Princeton, NJ: Princeton University Press 2009.

Mulhern, James, 'Manuscript Schoolbooks', in: *Journal of Educational Research* 32 (1939), pp. 428–48.

Muller, E. and K. Zandvliet (ed.), *Admissies als landmeter in Nederland voor 1811*, Alphen aan den Rijn: Canaletto 1987.

Mydorge, Claude, *Claudii Mydorgii [...] Prodromi catoptricorum et dioptricorum: sive conicorum operis ad abdita radii reflexi et refracti mysteria praeuij & facem praeferentis*, Parisiis: Ex typographia I. Dedin 1639.

Nangle, B. C., *The Monthly Review, Second Series, 1790–1815: Indexes of Contributors and Articles*, Oxford: Clarendon Press 1955.

National Loan Fund Life Assurance and Deferred Annuity Society, *First Report of the Directors at the Annual Meeting of Proprietors, Held on the 13th day of May 1840, of the National Loan Fund Life Assurance and Deferred Annuity Society*, London 1840.

Naylor, Ron, 'Paolo Sarpi and the First Copernican Tidal Theory', in: *British Journal for the History of Science* 47 (2014), pp. 661–75.

Neerfeld, Christiane, *'Historia per forma di diaria': La cronachistica veneziana contemporanea a cavallo tra il Quattro e Cinquecento*, Venice: Istituto Veneto di Scienze, Lettere ed Arti 2006.

Nelson, G., *The Wonders of Nature Throughout the World Display'd, Both for Diversion and Instruction*, London: Watts 1740.

Neugebauer, Otto and Abraham Sachs, *Mathematical Cuneiform Texts*, American Oriental Series 29, New Haven, CT: American Oriental Society 1945.

Neumann, Peter M., *The Mathematical Writings of Évariste Galois*, Zürich: European Mathematical Society 2011.

Newhouse, Daniel, *The Whole Art of Navigation in Five Books*, London: Printed for the author 1685.

Newton, John, *Cosmographia, or a view of the Terrestrial and Coelestial Globes, in a Brief Explanation of the Principles of plain and solid Geometry, applied to surveying and gauging a cask*, London: for Thomas Passinger 1679.

Newton, John, *The English Academy, or, A Brief Introduction to the Seven Liberal Arts*, London: A. Milbourn for Tho. Passenger 1693.

Nicholson, Francis, 'The Literary and Philosophical Society 1781–1851', in: *Memoirs and Proceedings of the Manchester Literary and Philosophical Society (Manchester Memoirs)* 68 (1923–24), pp. 97–148.

Nieuwe uytgereckende Taafelen, Wegens de Son, Maan en Sterren, Als meede een Almanach, voor eenige toekoomende Jaeren, Amsterdam: Jacobus Robyn [1701].

Norman, Robert, *The Newe Attractive, Containyng a Short Discourse of the Magnes or Lodestone*, London: Iohn Kyngston for Richard Ballard 1581.

Norris, Adrian, 'Leeds City Museum—Its Natural History Collections', in: *Journal of Biological Curation* 1 (1993), pp. 29–39.

Norris, Richard, *The Manner of Finding of the True Sum of the Infinite Secants of an Arch, by an Infinite Series*, London: Thomas James for the author 1685.

Nothaft, C. P. E., 'A Sixteenth-Century Debate on the Jewish Calendar: Jacob Christmann and Joseph Justus Scaliger', in: *Jewish Quarterly Review* 103 (2013), pp. 47–73.

Nuovo, Angela, 'The Creation and Dispersal of the Library of Gian Vincenzo Pinelli', in: *Books on the Move: Tracking Copies through Collections and the Book Trade*, ed. Robin Myers, Michael Harris, and Giles Mandelbrote, London: British Library 2007, pp. 39–67.

Nuovo, Angela, 'Gian Vincenzo Pinelli's Collection of Catalogues of Private Libraries in Sixteenth-Century Europe', in: *Gutenberg-Jahrbuch* 82 (2007), pp. 129–44.

Nuovo, Angela, 'Manuscript Writings on Politics and Current Affairs in the Collection of Gian Vincenzo Pinelli (1535–1601)', in: *Italian Studies* 66 (2011), pp. 193–205.

Ogborn, Maurice Edward, *Equitable Assurances: The Story of Life Assurance in the Experience of The Equitable Life Assurance Society 1762–1962*, London: George Allen & Unwin 1962.

Ogilvie, Brian W., *The Science of Describing: Natural History in Renaissance Europe*, Chicago: University of Chicago Press 2006.

Oheim, Carl Eugen Pabst von, *Geschichte des 'Thurmhofer Hilfsstollens' bei Freiberg mit dem Bericht von C.E. Pabst von Ohain aus dem Jahr 1772*, ed. Jens Kugler, Schriftenreihe Akten und Berichte vom sächsischen Bergbau 4, Kleinvoigtsberg: Kugler Verlag 1998.

Oosterhoff, Richard J., 'Tutor, Antiquarian, and Almost a Practitioner: Brian Twyne's Readings of Mathematics', in: *Reading Mathematics in Early Modern Europe: Studies in the Production,*

Collection, and Use of Mathematical Books, ed. P. Beeley, Y. Nasifoglu, and B. Wardhaugh, London: Routledge 2021, pp. 151–66.

Oppel, Wilhelm von, Friedrich and Johann Gottlieb Kern, *Bericht von Bergbau*, Leipzig: Crusius 1772.

Orange, A. D., 'The British Association for the Advancement of Science: The Provincial Background', in: *Science Studies* 1 (1971), pp. 315–29.

Orange, A. D., 'The Origins of the British Association for the Advancement of Science', in: *British Journal for the History of Science* 6/2 (1972), pp. 152–76.

Orange, A. D., 'Hyaenas in Yorkshire: William Buckland and the Cave in Kirkdale', in: *History Today* 22/11 (1972), pp. 777–85.

Orange, A. D., *Philosophers and Provincials: The Yorkshire Philosophical Society from 1822 to 1844*, York: Yorkshire Philosophical Society 1973.

Orange, A. D., 'The Idols of the Theatre: The British Association and its Early Critics', in: *Annals of Science* 32/3 (1975), pp. 277–94.

Orange, Derek, 'Science in Early Nineteenth-Century York: The Yorkshire Philosophical Society and the British Association', in: *York 1831–1981: 150 Years of Scientific Endeavour and Social Change*, ed. C. H. Feinstein, York: William Sessions, Ebor Press in association with the British Association for the Advancement of Science (York Committee) 1981, pp. 1–29.

Orange, Derek, 'Rational Dissent and Provincial Science: William Turner and the Newcastle Literary and Philosophical Society', in: *Metropolis and Province: Science in British Culture, 1780–1850*, ed. Ian Inkster and Jack Morrell, London: Hutchinson 1983, pp. 205–30.

Ordonnance de Louis XIV, donnée à fontainebleau au mois d'Aoust 1681, touchant la Marine, Paris: Denys Thierry 1714.

Osburn, William, Jnr, *An Account of an Egyptian Mummy, Presented to the Museum of the Leeds Philosophical and Literary Society, by the Late John Bladys, Esq.*, Leeds: Leeds Philosophical and Literary Society 1828.

Otis, Jessica, '"Set Them to the Cyphering Schoole": Reading, Writing, and Arithmetical Education, circa 1540–1700', in: *Journal of British Studies* 56 (2017), pp. 453–82.

Ozanam, Jacques, *Cours de Mathematique*, vol. I, Paris: Jean Jombert 1693.

Ozanam, Jacques, *La Trigonometrie rectiligne et spherique*, Paris: Claude Jombert 1720.

Palmieri, Paolo, 'Mental Models in Galileo's Early Mathematization of Nature', in: *Studies in History and Philosophy of Science* 34 (2003), pp. 229–64.

Palumbo, Margherita, 'Books on the Run: The Case of Francesco Patrizi', in: *Fruits of Migration: Heterodox Italian Migrants and Central European Culture, 1550–1620*, ed. Cornel Zwierlein and Vincenzo Lavenia, Leiden: Brill 2018, pp. 45–71.

Panciera, W., 'Giulio Savorgnan e la costruzione della fortezza di Nicosia (1567–1570)', in: *La Serenissima a Cipro*, ed. E. Skoufari, Rome: Viella 2013, pp. 131–42.

Pangallo, Matteo A., 'Correction to Plomer's Biography of Thomas Harper', in: *Notes and Queries* 254 (2009), pp. 203–5.

Panteki, Maria, 'William Wallace and the Introduction of Continental Calculus to Britain: a Letter to George Peacock', in: *Historia Mathematica* 14 (1987), pp. 119–32.

Parker, Irene, *Dissenting Academies in England, Their Rise and Progress and Their Place among the Educational Systems of the Country*, Cambridge: at the University Press 1914.

Patoun, Archibald, *A Compleat Treatise of Practical Navigation Demonstrated from It's First Principles: Together with All the Necessary Tables. To Which Are Added, the Useful Theorems of Mensuration, Surveying, and Gauging; with Their Application to Practice. Written for the Use of the Academy in Tower-Street*, London: Printed for J. Brotherton [etc.] 1734.

Patton, William J., *Three Ballynahinch Boys: A New Year's Address to the Young for 1880*, Belfast: Archer & Sons, 1880.

Peacock, George, 'Report on the Recent Progress and Present State of Certain Branches of Analysis', in: *Report of the Third Meeting of the British Association for the Advancement of Science; held at Cambridge in 1833*, London: John Murray 1834, pp. 185–352.

Pearce, E. H., *Annals of Christ's Hospital*, London: Methuen & Co. 1901.

Pepys, Samuel, *Samuel Pepys's Naval Minutes*, ed. J. R. Tanner, London: Navy Records Society 1926.

Pepys, Samuel, *Private Correspondence and Miscellaneous Papers of Samuel Pepys 1679–1703*, ed. Joseph R. Tanner, 2 vols, London: G. Bell 1926.

Pepys, Samuel, *The Tangier Papers of Samuel Pepys*, ed. E. Chappell, London: Navy Records Society 1935.

Perl, Teri, '*The Ladies' Diary or Woman's Almanack* 1704–1841', in: *Historia Mathematica* 6 (1979), pp. 36–53.

Petri, Nicolaus, *Practicque, om te leeren rekenen, cijpheren ende boeckhouwen, met die regel coss ende geometrie seer profijtelijcken voor alle coopluyden*, Amsterdam: Cornelis Claesz 1583.

Petty, William, *Discourse made before the Royal Society the 26. November 1674. Concerning the Use of Duplicate Proportion in sundry important particulars*, London: for John Martyn 1674.

Petty, William, *Political Arithmetick or a discourse concerning, the extent and value of lands, people. Buildings, husbandry, manufactrure, commerce, fishery*, [. . .], London: for Robert Clavel 1690.

Philander, 'William Hilton, The Lancashire Mathematician', in: *Manchester City News Notes and Queries* 96 (1882), pp. 337–8.

Philo, J.-M., 'Henry Savile's Tacitus in Italy', in: *Renaissance Studies* 32 (2017), pp. 687–707.

Philo, J.-M., 'English and Scottish Scholars at the Library of Gian Vincenzo Pinelli (1565–1601)', in: *Renaissance and Reformation* 42 (2019), pp. 51–80.

Pickering, Paul and Alex Tyrell, *The People's Bread: A History of the Anti-Corn Law League*, London: Leicester University Press 2000.

Pigatto, Luisa, 'Tycho Brahe and the Republic of Venice: A Failed Project', in: *Tycho Brahe and Prague: Crossroads of European Science*, ed. J. R. Christianson et al., Frankfurt: H. Deutsch 2002, pp. 187–202.

Pimentel, Luís Serrão, *Methodo Lusitanico de desenhar as Fortificaçoens das Praças Regulares, & Irregulares, Fortes de Campanha, e outras obras pertencentes á Architectura Militar*, Lisbon: Antonio Craesbeeck de Mello 1680, https://purl.pt/24485.

Pizzorusso, Giovanni, 'Francesco Ingoli: Knowledge and Curial Service in 17th-Century Rome', in: *Copernicus Banned: The Entangled Matter of the anti-Copernican Decree of 1616*, ed. N. Fabbri and F. Favino, Florence: Olschki 2018, pp. 157–89.

Playfair, John, '*Traité de Méchanique Céleste* [review]', in: *The Edinburgh Review* 11 (1808), pp. 249–84.

Plumley, N., 'The Royal Mathematical School within Christ's Hospital: The Early Years—Its Aims and Achievements', in: *Vistas in Astronomy* 20 (1976), pp. 51–9.

Poelje, Otto van, 'Gunter Rules in Navigation', in: *Journal of the Oughtred Society* 13 (2004), pp. 11–22.

Pomata, Gianna and Nancy G. Siraisi (eds), *Historia: Empiricism and Erudition in Early Modern Europe*, Cambridge, MA: MIT Press 2005.

Poole, William, 'The Origin and Development of the Savilian Library', in: *Reading Mathematics in Early Modern Europe: Studies in the Production, Collection, and Use of Mathematical Books*, ed. P. Beeley, Y. Nasifoglu, and B. Wardhaugh, London: Routledge 2021, pp. 167–91.

Poole, William, 'Sir Henry Savile and the Early Professors', in: *Oxford's Savilian Professors of Geometry*, ed. Robin Wilson, Oxford: Oxford University Press 2022, pp. 2–27.

Popplow, Marcus, 'Why Draw Pictures of Machines? The Social Contexts of Early Modern Machine Drawings', in: *Picturing Machines, 1400–1700*, ed. W. Lefèvre, Cambridge, MA: MIT Press 2004, pp. 17–52.

Porter, William Smith, *Sheffield Literary and Philosophical Society: A Centenary Retrospect 1822–1922*, Sheffield: J. W. Northend 1922.

Prescot, Bar[tholomew], 'Letters on the Newtonian System', in: *Liverpool Apollonius*, no. 2 (1824), pp. 132–86.

Price, Richard, *Observations on Reversionary Payments: On Schemes for Providing Annuities for Widows, and for Persons in Old Age; on the Method of Calculating the Values of Assurances on Lives; and on the National Debt*, London: Cadell 1771.

Price, Richard, *Observations on Reversionary Payments: On Schemes for Providing Annuities for Widows, and for Persons in Old Age; on the Method of Calculating the Values of Assurances on Lives; and on the National Debt*, London: Cadell 1783.

Pugliano, Valentina, 'Specimen Lists: Artisanal Writing or Natural Historical Paperwork?', in: *Isis* 103 (2012), pp. 716–26.

Price, Richard, 'Ulisse Aldrovandi's Color Sensibility: Natural History, Language and the Lay Color Practices of Renaissance *Virtuosi*', in: *Early Science and Medicine* 20 (2015), pp. 358–96.

Pycior, Helena M., *Symbols, Impossible Numbers, and Geometric Entanglements; British Algebra through the Commentaries on Newton's Universal Arithmetick*, Cambridge: Cambridge University Press 1997.

Rahn, Johann Heinrich, *An Introduction to Algebra. Translated out of the High-Dutch into English by Thomas Brancker, M.A. Much altered and augmented by D[r] P[ell]*, London: William Godbid for Moses Pitt 1668.

Raines, Dorit, 'L'archivio familiare strumento di formazione politica del patriziato veneziano', in: *Accademie e biblioteche d'Italia* 64 (1996), pp. 5–36.

Rann, K. and R. S. Johnson, 'Chasing the Line: Hutton's Contribution to the Invention of Contours', in: *Journal of Maps* 15 (2019), pp. 48–56.

Raphael, Renée, *Reading Galileo: Scribal Technologies and the Two New Sciences*, Baltimore: Johns Hopkins University Press 2017.

Raugei, Anna Maria, *Gian Vincenzo Pinelli e la sua biblioteca*, Geneva: Droz 2018.

Ravid, Benjamin, 'The Venetian Government and the Jews', in: *The Jews of Early Modern Venice*, ed. Robert C. Davis and Benjamin C. Ravid, Baltimore: Johns Hopkins University Press 2001, pp. 3–30.

Raviola, Blythe Alice, *Giovanni Botero: Un profilo fra storia e storiografia*, Milan: Bruno Mondadori 2020.

Raynaud, Dominique, 'Introduction', in: *Géométrie pratique: Géomètres, ingénieurs et architectes, xvie–xviiie siècle*, ed. Dominique Raynaud, Besançon: Presses universitaires de Franche-Comté 2015, pp. 9–20.

Read, Nathan S., *An Astronomical Dictionary: Compiled from Hutton's Mathematical and Philosophical Dictionary*, New Haven: Hezekiah Howe 1817.

Recorde, Robert, *The Pathvvay to Knowledg*, London: Reynold Wolfe 1551.

Reden, Claus Friedrich von, *Rede bei dem feyerlichen Anfange des tiefen Georg-Stollen-Baues unweit der Bergstadt Grund*, Clausthal: Wendeborn 1777.

Reid, David A., 'Science and Pedagogy in the Dissenting Academies of Enlightenment Britain', PhD thesis, University of Wisconsin-Madison 1999.

Reimann, C., 'Ferdinando de' Medici and the *Typographia Medicea*', in: *Print and Power in Early Modern Europe, 1500–1800*, ed. Nina Lamal, Jamie Cumby, and Helmer J, Helmers, Leiden: Brill 2021, pp. 220–38.

[Remmelin, Johann], *Numerus Figuratus, sive Arithmetica Analytica Arte Mirabili Jnavdita Nova Constans. Hic Dn. Johannis Favlhaberi Logistae Vlmensis Ars, Qvam ex Biblicis hausit Numeris, detegitur, & simùl in Prooemio ipsius Antagonistae charta famosa refutatur*, [no place] 1614.

[Remmelin, Johann], *Mysterium Arithmeticum Sive, Cabalistica & Philosophica Inventio, nova admiranda & ardua, qua Numeri Ratione et Methodo computentur, Mortalibus à Mundi Primordio Abdita, et ad Finem non sine singulari omnipotentis Dei provisione revelata. Cum Illuminatissimis laudatissimisque Fraternitatis Roseae crucis Famae Viris humiliter & syncerè dicata*, [no place] 1615.

Reyersz, Heyndrick, *Jaep en Veer, of stuurmans praetjen [...]*, Amsterdam: Dirk Pietersz. 1622.

Ribeiro, Dulcyene Maria, 'A formação dos engenheiros militares: Azevedo Fortes, Matemática e ensino da Engenharia Militar no século XVIII em Portugal e no Brasil', doctoral thesis, São Paulo: Faculdade de Educação da Universidade de São Paulo 2009, https://teses.usp.br/teses/disponiveis/48/48134/tde-08122009-151638/en.php

Rice, Adrian C. and Robin J. Wilson, 'From National to International Society: The London Mathematical Society, 1867–1900', in: *Historia Mathematica* 25 (1995), pp. 185–217.

Rice, Adrian C., Robin J. Wilson, and J. Helen Gardner, 'From Student Club to National Society: The Founding of the London Mathematical Society in 1865', in: *Historia Mathematica* 22 (1995), pp. 402–21.

Rickey, V. Frederick and Amy Shell-Gellasch, 'Mathematics Education at West Point: The First Hundred Years', in: *Convergence* (2010), https://www.maa.org/press/periodicals/convergence/mathematics-education-at-west-point-the-first-hundred-years.

Rigaud, Stephen Jordan (ed.), *Correspondence of Scientific Men of the Seventeenth Century*, 2 vols, Oxford: at the University Press 1841.

Rivington, Charles A., 'Early Printers to the Royal Society 1663–1708', in: *Notes and Records of the Royal Society* 39 (1984), pp. 1–27.

Rivolta, Adolfo, *Catalogo dei Codici Pinelliani dell'Ambrosiana*, Milan: Tipografia Arcivescovile 1933.

Roberts, A., *The Adventures of (Mr T.S.) An English Merchant*, London: Moses Pitt 1670.

Robins, Benjamin, *New principles of Gunnery: Containing, the Determination of the Force of Gun-Powder, and an Investigation of the Difference in the Resisting Power of the Air to Swift and Slow Motions*, London: Printed for J. Nourse, without Temple-Bar 1742.

Robinson, Charles J., *A Register of the Scholars Admitted into Merchant Taylors' School*, vol. 2, Lewes: Farncombe and Co. 1883.

Robinson, Henry, *Englands safety in Trades Encreased*, London: E. P. for Nicholas Bourne 1641.

Robinson, Henry W. and Walter Adams (eds), *The Diary of Robert Hooke, 1672–1680*, London: Taylor & Francis 1935.

Robson, Eleanor, *Mathematics in Ancient Iraq: A Social History*, Princeton, NJ: Princeton University Press 2008.

Roche, J. J., 'Harriot, Thomas (*c.*1560–1621), Mathematician and Natural Philosopher', in: *Oxford Dictionary of National Biography*, 2006, https://doi.org/10.1093/ref:odnb/12379.

Rodella, Massimo, 'Fortuna e sfortuna della biblioteca di Gian Vincenzo Pinelli: la vendita a Federico Borromeo', in: *Bibliotheca* 2 (2003), pp. 87–125.

Roderick, Gordon W. and Michael D. Stephens, 'Approaches to Technical Education in 19th Century England, Part III: The Liverpool Literary and Philosophical Society', in: *The Vocational Aspect of Education* 23/54 (Spring 1971), pp. 49–54.

Rodet, Léon, 'L'algèbre d'Al-Khārizmi et les méthodes indienne et grecque', in: *Journal asiatique* 11 (1878), pp. 5–98.

Rodger, N. A. M., 'Commissioned Officers' Careers in the Royal Navy, 1690–1815', in: *Journal for Maritime Research* 3/1 (2001), pp. 85–129.

Rose, Jonathan, *The Intellectual Life of the British Working Classes*, New Haven, CT: Yale University Press 2001.

Rose, P. L., 'A Venetian Patron and Mathematician of the Sixteenth Century: Francesco Barozzi (1537–1604)', in: *Studi veneziani* 1 (1977), pp. 119–78.

Rose, Susan, 'Mathematics and the Art of Navigation: the Advance of Scientific Seamanship in Elizabethan England', in: *Transactions of the Royal Historical Society* 14 (2004), pp. 175–84.

Rösler, Balthasar, *Speculum metallurgiae politissimum, oder, Hell-polierter Berg-Bau-Spiegel*, Dresden: Winckler 1700.

Rossini, P., 'New Theories for New Instruments: Fabrizio Mordente's Proportional Compass and the Genesis of Giordano Bruno's Atomist Geometry', in: *Studies in History and Philosophy of Science* 76 (2019), pp. 60–8.

Roth, Peter, *Arithmetica Philosophica, Oder schöne newe wolgegründte Vberauß Kunstliche Rechnung der Coß oder Algebrae, In drey vnterschiedliche Theil getheilt. Im I. Theil werden deß hochgelehrten, fürtrefflichen vnd weitberühmbten Herrn D. Hieronymi Cardani, Mathematici, Philosophi vnd Medici dreyzehn Reguln (als der Schlüssel, nach welchen alle ratio. vnd irrational, wie auch binomi vnd residui cubicossische Exempla vnd aequationes zu solviren vnd auffzulösen) auffs trewlichst vnd fleissigst beschrieben vnd gesetzt. Deßgleichen noch drey andere newerfundene nützliche Reguln, zu den ersten drey cubiccossischen aequationib. (fürnemlich aber der andern vnd dritten Regul Cardani, wann der Cubus deß dritten theils der Zahlen Radicum grösser, als das Quadrat deß halben theils der ledigen Zahl) gehörig. Im II. Theil folget die allerkünstlichste Resolution deß gantzen Arithmetisch. Cubiccossischen Lustgartens, welcher von dem Wolerfahrnen Herrn Johann Faulhabern, Burgern vnd Rechenmeistern zu Vlm, mit 160. Bäumlein, das ist, außerlesenen kunstlichen Quaestionen gepflantzt worden, sampt deroselben nach notdurfft daran gehenckten erklärung, Vnd einer noch überaus schönen herrlichen incorporirten polygonalischen Regul, vnd der daraus componirten Taffeln, dardurch auff beede Weg leichtlich die Summa etlicher Polygonahlzahlen, vnd herwiderumben derselben Radices mögen gefunden werden. Vnd dann endlich im III. Theil, als zum Beschluß, eine anzahl wunderbarliche, newerfundene, künstliche, ja von vielen hochverstendigen diser Kunst gelehrten, für vnmüglich geachte Surdische, Zensizensi. Surdesoli. Zensicubi. Bsurdesoli. wie auch Longi. Plani. vnd Stereometrische Cossische Quaestiones vnnd Exempla, der gestalt vorhin in keiner Sprach gesehen worden. Calculirt, solvirt, auch auff das aller trewlichst den jenigen, so was mehrers in dieser edlen vnd sinnreichen Kunst zu erfahren begierig, beschrieben vnd an tag geben*, Nürnberg 1608.

Roux, S., 'Forms of Mathematization (14th–17th Centuries)', in: *Early Science and Medicine* 15 (2010), pp. 319–37.

The Record of the Royal Society of London, 3rd edn. London: for the Royal Society 1912.

Rubinstein, David, *The Nature of the World: The Yorkshire Philosophical Society, 1822–2000*, York: Quacks Books 2009.

Rudolff, Christoff, *Behend und Hubsch Rechnung durch die kunstreichen regeln Algebre, so gemeincklich die Coß geneñt werden*, Straßburg: Cephaleus 1525 [for a later edition see Stifel].

Russell, G. F., *Memoir of Richard Busby D.D. (1606–1695) with Some Account of Westminster School in the Seventeenth Century*, London: Lawrence and Bullen 1895.

Sachse, William L., 'The Journal of Nathan Prince, 1747', in: *The American Neptune*, 16/2 (1956), pp. 81–97.

Saldanha, Arun, 'The Itineraries of Geography: Jan Huygen van Linschoten's "Itinerario" and Dutch Expeditions to the Indian Ocean, 1594–1602', in: *Annals of the Association of American Geographers* 101/1 (2011), pp. 149–77.

Sander, C., 'Early-Modern Magnetism: Uncovering New Textual Links between Leonardo Garzoni SJ (1543–1592), Paolo Sarpi OSM (1552–1623), Giambattista Della Porta (1535–1615), and the Accademia dei Lincei', in: *Archivum Historicum Societatis Iesu* 85 (2016), pp. 303–63.

Sandman, Alison, 'Cosmographers vs. Pilots: Navigation, Cosmography, and the State in Early Modern Spain', PhD dissertation, University of Wisconsin–Madison 2001.

Sansovino, Francesco, *Venetia città nobilissima et singolare*, Venice: Sansovino 1581.

Sasaki, Chikara, *Descartes's Mathematical Thought*, Dordrecht: Springer 2003.

Savio, Andrea, *Tra spezie e spie: Filippo Pigafetta nel Mediterraneo del Cinquecento*, Rome: Viella 2020.

Schaefer, Christoph, '"Krygsvernuftelingen": Militäringenieure und Fortifikation in den Vereinigten Niederlanden', PhD dissertation, University of Giessen 2001.

Schellhas, Walter, *Der Rechenmeister Adam Ries (1492 bis 1559) und der Bergbau*, Freiberg: Bergakademie 1977.

Scheubel, Johann, *Algebrae compendiosa facilisque descriptio*, Paris: Cauellat 1551.

Schilder, Günter, *Cornelis Claesz (c.1551–1609): Stimulator and Driving Force of Dutch Cartography*, Monumenta Cartographica Neerlandica 7, Alphen aan den Rijn: Canaletto/Repro-Holland 2003.

Schmidt, Martin, *Die Wasserwirtschaft des Oberharzer Bergbaus*, Neuwied: Neuwieder Verlagsgesellschaft 1989.

Schneider, Ivo, *Johannes Faulhaber (1580–1635): Rechenmeister in einer Zeit des Umbruchs*, Basel 1993.

Schneider, Ivo, 'Ausbildung und fachliche Kontrolle der deutschen Rechenmeister vor dem Hintergrund ihrer Herkunft und ihres sozialen Status', in: *Verfasser und Herausgeber mathematischer Texte der frühen Neuzeit*, ed. Rainer Gebhardt, Schriften des Adam-Ries-Bundes Annaberg-Buchholz 14, Annaberg-Buchholz 2002, pp. 1–22.

Schneider, Ivo, 'Trends in German Mathematics at the Time of Descartes' Stay in Southern Germany', in: *Mathématiciens français du XVIIe siècle: Pascal, Descartes, Fermat*, ed. M. Serfati and D. Descotes, Clermont-Ferrand: Presses Universitaires Blaise Pascal 2008, pp. 45–67.

Schneider, Ivo, 'Between Rosicrucians and Cabbalah—Johannes Faulhaber's Mathematics of Biblical Numbers', in: *Mathematics and the Divine: A Historical Study*, ed. Teun Koetsier und Luc Bergmans, Amsterdam: Elsevier 2005, pp. 311–30.

Schneider, Ivo, 'The Concept of Algebra in the Publications of Johannes Faulhaber in the Context of the Activities of the Rechenmeister', in: *Pluralité de l'algèbre à la renaissance*, ed. Sabine Rommeveaux, Maryvonne Spiesser, and Veronica Gavagna, Paris: Honoré Champion 2012, pp. 311–29.

Schönberg, Abraham, *Ausführliche Berg-Information*, Leipzig und Zwickau: Fleischer und Büschel 1693.

Schotte, Margaret E., 'Expert Records: Nautical Logbooks from Columbus to Cook', in: *Information & Culture: A Journal of History* 48/3 (2013), pp. 281–322.

Schotte, Margaret E., *Sailing School: Navigating Science and Skill, 1550–1800*, Baltimore: Johns Hopkins University Press 2019.

Schotte, Margaret E., 'Sailors, States, and the Creation of Nautical Knowledge', in: *A World at Sea: Maritime Practices and Global History*, ed. Lauren Benton and Nathan Perl-Rosenthal, Philadelphia: University of Pennsylvania Press 2020, pp. 89–107.

Schotte, Margaret E., 'Nautical Manuals and Ships' Instruments, 1550–1800: Lessons in Two and Three Dimensions', in: *The Routledge Companion to Marine and Maritime Worlds 1400–1800*, ed. Claire Jowitt, Craig Lambert, and Steve Mentz, London: Macmillan 2020, pp. 273–97.

Scriba, Christoph J., *James Gregorys frühe Schriften zur Infinitesimalrechnung*, Giessen: Selbstverlag des Mathematischen Seminars 1957.

Scriba, Christoph J., 'Mercator's Kinckhuysen-Translation in the Bodleian Library at Oxford', in: *British Journal for the History of Science* 2 (1964), pp. 45–58.

Scriba, Christoph J., 'The Autobiography of John Wallis, F.R.S.', in: *Notes and Records of the Royal Society* 25 (1970), pp. 17–46.

Scriba, Christoph J., 'John Pell's English Edition of J. H. Rahn's Teutsche Algebra', in: *For Dirk Struik. Scientific, Historical and Political Essays in Honor of Dirk. J. Struik*, ed. Robert S. Cohen, John J. Stachel, and Marx W. Wartofsky, Dordrecht: D. Reidel 1974, pp. 261–74.

SDUK, *Report of the State of Literary, Scientific, and Mechanics' Institutions in England*, London: Society for the Diffusion of Useful Knowledge 1841.

Secord, Anne, 'Science in the Pub: Artisan Botanists in Early Nineteenth–Century Lancashire', in: *History of Science* 32/3 (1994), pp. 269–315.

Sedgwick, W. F., rev. Anita McConnell, 'Robertson John, (1707–1776), Mathematician', in: *Oxford Dictionary of National Biography*, 2008, https://doi.org/10.1093/ref:odnb/23802.

Sellers, David, 'An Early Astronomical Society: Highfield Astronomical and Meteorological Society, Halifax', in: *The Antiquarian Astronomer: Journal of the Society for the History of Astronomy* 13 (June 2019), pp. 42–9.

Sennewald, Rainer, 'Die Stipendiatenausbildung von 1702 bis zur Gründung der Bergakademie Freiberg 1765/66', in: *Technische Akademie Freiberg, Festgabe zur 300. Jahrestag der Gründung der Stipendienkasse für die akademische Ausbildung im Berg- und Hüttenfach zu Freiberg in Sachsen*, Freiberg: TU Bergakademie 2002, pp. 407–29.

Sepulveda, Christovam Ayres de Magalhães, *Historia Organica e Politica do Exercito Português: Provas*, 17 vols, Lisbon: Imprensa Nacional, Coimbra: Imprensa da Universidade 1902–1932, https://purl.pt/24869.

Shapin, Steven A., 'The Pottery Philosophical Society, 1819–1835: An Examination of the Cultural Uses of Provincial Science', in: *Science Studies* 2 (1972), pp. 311–36.

Shapin, Steven and Arnold Thackray, 'Prosopography as a Research Tool in History of Science: The British Scientific Community 1700–1900', in: *History of Science* 1 (1974), pp. 1–28.

Shapiro, Barbara J., *John Wilkins 1614–1672: An Intellectual Biography*, Berkeley and Los Angeles: University of California Press 1969.

Sharpe, Kevin, *Sir Robert Cotton, 1586–1631: History and Politics in Early Modern England*, Oxford: Oxford University Press 1979.

Sheehan, Donal, 'The Manchester Literary and Philosophical Society', in: *Isis* 33/4 (1941), pp. 519–23.

Sheils, Sarah, *From Cave to Cosmos : A history of the Yorkshire Philosophical Society*, York: Yorkshire Philosophical Society 2022.

Sherman, Arnold A., 'Pressure from Leadenhall: The East India Company Lobby, 1660–1678', in: *Business History Review* 50 (1976), pp. 329–55.

Sherwood, Marika, *After Abolition: Britain and the Slave Trade Since 1807*, London: I. B. Taurus & Co. 2007.

Simpson, Thomas, *The Doctrine of Annuities and Reversions: Deduced from General and Evident Principles*, London: J. Nourse 1742.

Simpson, Thomas, *The Doctrine and Application of Fluxions: Containing (besides What is Common on the Subject) a Number of New Improvements in the Theory: and the Solution of a Variety of New, and Very Interesting, Problems in Different Branches of the Mathematicks. Part I*, London: Printed for J. Nourse 1750.

Sitters, Matthijs H., 'Sybrandt Hansz. Cardinael (1578–1647) Meester in de meetkunde', in: *Niew Archief voor wiskunde*, ser. 5, 4/4 (2003), pp. 309–16.

Sitters, Matthijs H., 'Sybrandt Hansz. Cardinael 1578–1647: Rekenmeester en wiskundige: zijn leven en zijn werk', PhD dissertation, Groningen 2007.

Sitters, Matthijs H., *Sybrandt Hansz. Cardinael 1578–1647: Rekenmeester en wiskundige: zijn leven en zijn werk*, Hilversum: Verloren 2008.

Sitters, Matthijs H. and Albrecht Heeffer, *The Hundred Geometrical Questions by Sybrandt Hansz Cardinael: Early-Modern Dutch Geometry from the Surveyor's Tradition*, Boston: Docent Press 2016.

Slack, Paul, *The Invention of Improvement: Information and Material Progress in Seventeenth-Century England*, Oxford: Oxford University Press 2015.

Sloan, Kim, 'Thomas Weston and the Academy at Greenwich', in: *Transactions of the Greenwich and Lewisham Antiquarian Society* 9/6 (1984), pp. 313–33.

Sloan Kim, 'The Teaching of Non-Professional Artists in Eighteenth-Century England', PhD dissertation, University of London 1986.

Smith, Ashley, *The Birth of Modern Education: The Contribution of the Dissenting Academies, 1660–1800*, London 1954.

Smith, Crosbie and M. Norton Wise, *Energy & Empire: A Biographical Study of Lord Kelvin*, Cambridge: Cambridge University Press 1989.

Smith, R. Angus, *A Centenary of Science in Manchester: in A Series of Notes*, London: Taylor and Francis 1883.

Snell, Charles, *The Tradesman's Director; or, a short and easy Method of Keeping his Books of Accompts*, London: Richard Baldwin 1697.

Sørensen, Henrik Kragh, 'The Mathematics of Niels Henrik Abel: Continuation and New Approaches in Mathematics During the 1820s', PhD thesis, University of Aarhus 2002.

Speidell, Euclid, *Logarithmotechnia: or, the Making of Numbers Called Logarithms to Twenty five Places*, London: Henry Clark for the author 1688.

Speidell, John, *A Geometricall Extraction, or A Compendious Collection of the Chiefe and Choyse Problemes, Collected out of the Best, and Latest Writers: VVhereunto is Added, About 30. Problemes of the Authors Inuention, Being for the Most Part, Performed by a Better and Briefer Way, Then by Any Former writer*, London: Edward Allde 1617.

Speidell, John, *A Briefe Treatise for the Measuring of Glasse, Board, Timber, or Stone, Square or Round: Being Performed Only by Simple Addition and Substraction, and That in Whole Numbers, with[o]ut Any Multiplication, or Division At All*, London 164[0?].

Springer, Friedrich P., 'Über Kameralismus und Bergbau', in: *Der Anschnitt* 62 (5/6) (2010), pp. 230–41.

Stampioen, Jan, *Resolutie ende ontbindinghe der twee vraegh-stucken, in den jare 1632*, Rotterdam: op de weduwe van Matthiis Bastiaensz 1634.

Stampioen, Jan, *Algebra, ofte Nieuwe stel-regel, waer door alles ghevonden wordt inde wiskonst, wat vindtbaer is*, 's Graven-Hage: ghedruckt ten huyse vanden autheur 1639.

Steadman, Mark, 'A History of the Scientific Collections of the Leeds Philosophical and Literary Society's Museum in the Nineteenth Century: Acquiring, Interpreting & Presenting the Natural World in the English Industrial City', PhD thesis, University of Leeds 2019.

Stedall, Jacqueline A., 'Rob'd of Glories: The Misfortunes of Thomas Harriot and his Algebra', in: *Archive for History of Exact Sciences* 54 (2000), pp. 455–97.

Stedall, Jacqueline A., 'Ariadne's Thread: The Life and Times of Oughtred's Clavis', in: *Annals of Science* 57 (2000), pp. 27–60.

Stedall, Jacqueline A., *A Discourse Concerning Algebra: English Algebra to 1685*, Oxford: Oxford University Press 2007.

Stedall, Jacqueline A., 'Tracing Mathematical Networks in Seventeenth-Century England', in: *The Oxford Handbook of the History of Mathematics*, ed. Eleanor Robson and Jacqueline Stedall, Oxford: Oxford University Press 2009, pp. 133–52.

Steele, Brett D., 'Military "Progress" and Newtonian Science in the Age of Enlightenment', in: *The Heirs of Archimedes: Science and the Art of War through the Age of Enlightenment*, ed. Brett D. Steele and Tamera Dorland, Cambridge, MA: MIT Press, 2005, pp. 361–90.

Steele, E. D., 'The Leeds Patriciate and the Cultivation of Learning, 1819–1905: A Study of the Leeds Philosophical and Literary Society', in: *Proceedings of the Leeds Philosophical and Literary Society, Literary and Historical Section* 16/9 (1978), pp. 183–202.

Steen, Isaiah, *A Treatise on Mental Arithmetic in Theory and Practice*, Belfast: Henderson 1846; 2nd edn 1848; 3rd edn 1857.

Steenstra, Pybo, *Grond-beginzels der Stuurmans-kunst*, Amsterdam: G. Hulst Van Keulen 1779.

Steenstra, Pybo, *Uitgewerkt Examen*, Middelburg: P. Gillissen en Zoon 1781.

Stegeman, Saskia, *Patronage and Services in the Republic of Letters: The Networks of Theodorus Janssonius van Almeloveen* (1657–1712), Amsterdam: APA-Holland University Press 2005.

Stella, Aldo, 'Galileo, il circolo culturale di Gian Vincenzo Pinelli e la *Patavina libertas*', in: *Galileo e la cultura padovana*, ed. Giovanni Santinello, Padua: Cedam 1992, pp. 307–25.

Stenhouse, Brigitte, 'Mary Somerville's Early Contributions to the Circulation of Differential Calculus', in: *Historia Mathematica* 51 (2020), pp. 1–25.

Stevin, Simon, *Wisconstige gedachtenissen, Inhoudende t'ghene daer hem in gheoeffent heeft den [...] vorst ende heere, Mavrits prince van Oraengien*, Tot Leyden: Inde Druckerye van Ian Bouvvensz 1605–8.

Stevin, Simon, *L'arithmetique de Simon Stevin de Bruges: contenant les computations des nombres Arithmetiques ou vulgaires: aussi l'Algebre, auec les equations de cinc quantitez: ensemble les quatre premiers liures d'Algebre de Diophante d'Alexandrie, maintenant premierement traduicts en François*, Leyden: Christophle Plantin 1685.

Stewart, A. T. Q., *Belfast Royal Academy: The First Century 1785–1885*, Antrim: Greystone Press 1985.

Stewart, Larry, *The Rise of Public Science: Rhetoric, Technology, and Natural Philosophy in Newtonian Britain, 1660–1750*, Cambridge: Cambridge University Press 1992.

Stewart, Larry, 'Other Centres of Calculation, or, Where the Royal Society Didn't Count: Commerce, Coffee-Houses and Natural Philosophy in Early Modern London', in: *British Journal for the History of Science* 32 (1999), pp. 133–53.

[Stifel, Michael], *Ein Rechen Büchlin Vom End Christ. Apocalypsis in Apocalypsim*, Wittenberg 1532.

[Stifel, Michael], *Die Coß Christoffs Rudolffs. Die schönen Exempeln der Coß Durch Michael Stifel Gebessert vnd sehr vermehrt*, Königsberg 1553.

Stone, Lawrence, 'The Educational Revolution in England, 1560–1640', in: *Past & Present* 28 (1964), pp. 41–80.

Storey, Charles Ambrose, *Persian Literature: A Bio-Bibliographical Survey* 2/1 (1927), p. 19.

Strode, Thomas, *A Short Treatise of Combinations, Elections, Permutations & Composition of Quantities*, London: W. Godbid for Enoch Wyer 1678.

Strode, Thomas, *A new and easie method to the art of dyalling*, London: for J. Taylor and T. Newborough 1688.

Sturges, R. P., 'The Membership of the Derby Philosophical Society, 1783–1802', in: *Midland History* 4/3 (1978), pp. 212–29.

Sturmy, Samuel, *The Mariners Magazine*, London: Anne Goodbid for William Fisher, at the Postern-Gate near Tower-Hill and five others, 1679.

Sullivan, F. B., 'The Naval Schoolmaster During the Eighteenth Century and the Early Nineteenth Century', in: *The Mariner's Mirror*, 62/4 (1976), pp. 311–26.

Swale, J. H., 'Advertisement', in: *Liverpool Apollonius*, no. 1 (1823).

Sullivan, F. B., [Geometrical Construction 4], in: *Liverpool Apollonius*, no. 2 (1824), pp. 51–2.

Sullivan, F. B., 'To Correspondents &c.', in: *Liverpool Apollonius*, no. 2 (1824).

Swetz, Frank J., 'The Mystery of Robert Adrain', in: *Mathematics Magazine* 81 (2008), pp. 332–44.

Sullivan, F. B., *Mathematical Expeditions: Exploring Word Problems across the Ages*, Baltimore: Johns Hopkins University Press 2010.

[Swift, Jonathan], *Travels into Several Remote Nations of the World, By Lemuel Gulliver*, London: printed for Benj. Motte 1726.

Tacquet, Andreas, *Elementa geometriæ planæ ac solidæ*, 2nd edn, Antwerp: Jacob van Meurs 1665.

Tafuri, Manfredo, *Venice and the Renaissance*, Cambridge, MA: MIT Press 1989.

Tartaglia, Nicolò, *Nova scientia inventa da Nicolo Tartalea*, Venetia: Per Stephano da Sabio 1537.

Tartaglia Nicolò, *Quesiti et inventioni diverse de Nicolo Tartalea Brisciano*, Venetia: Ruffinelli, 1546.

Taylor, E. G. R., *The Mathematical Practitioners of Tudor and Stuart England*, Cambridge: Cambridge University Press, 1970; first edn 1954.

Taylor, E. G. R. (ed.), *A Regiment for the Sea and other Writings on Navigation by William Bourne of Gravesend, a Gunner (c.1535–1582)*, Cambridge: Haklyut Society 1963.

Taylor, E. G. R., *The Mathematical Practitioners of Hanoverian England, 1714–1840*, Cambridge: Cambridge University Press 1966.

Taylor, William, *The Diagrams of Euclid's Elements of Geometry, in an Embossed or Tangible Form, for the Use of Blind Persons Who Wish to Enter upon the Study of that Noble Science*, York: Wolstenhome 1828.

Tebay, Septimus, 'Show How Three Billiard Balls Must Be Struck Simultaneously, so That Each Ball Shall *Just* Touch the Other Two', in: *Lady's and Gentleman's Diary for 1853* (1852), p. 78.

Tebay, Septimus, 'To the Editor of the Mechanics' Magazine', in: *Mechanics' Magazine* 60 (1854), pp. 158–9.

Tebay, Septimus, 'To the Editor of the Mechanics' Magazine', in: *Mechanics' Magazine* 60 (1854), p. 200.

Tenfelde, Klaus, Stefan Berger, and Hans-Christoph Seidel (eds), *Geschichte des deutschen Bergbaus. Der alteuropäische Bergbau: von den Anfängen bis zur Mitte des 18.* Jahrhunderts, vol. 1, Münster: Aschendorff Verlag 2012.

[Tenneur, Jacques Alexandre le], *Traité des quantitez incommenssurables*, Paris: I. Dedin 1640.

Thbaut, Louis, 'Le mécanicien anobli Pierre-Joseph Laurent, 1713–1773: des mines d'Anzin au Canal de Saint-Quentin', thèse de 3e cycle, Lille III 1974.

Thackrah, Charles Turner, *An Introductory Discourse Delivered to the Leeds Philosophical and Literary Society, April 6, 1821*, Leeds: Philosophical and Literary Society 1821.

Thackray, Arnold, 'Natural Knowledge in Cultural Context: The Manchester Model', in: *American Historical Review* 79/3 (1974), pp. 672–709.

Thomas, Keith, 'Numeracy in Early Modern England: The Prothero Lecture', in: *Transactions of the Royal Historical Society* 37 (1987), pp. 103–32.

Thompson, Silvanus P., *The Life of William Thomson, Baron Kelvin of Largs*, London: Macmillan & Co. 1910.

Thomson, James, *A Treatise on Arithmetic in Theory and Practice*, Belfast: Simms & McIntyre 1819; 2nd edn 1825.

Thomson, James, *An Introduction to Modern Geography*, Belfast: Simms & McIntyre 1827.

Thomson, James, *Remarks on the Phenomena of the Heavens, as They Appear from Different Bodies in the Solar System*, Belfast: A. Mackay 1827.

Thomson, James, 'On the True and Extended Interpretations of Formulae in Spherical Trigonometry', in: *The London and Edinburgh Philosophical Magazine and Journal of Science* 10 (1837), pp. 18–24.

Thomson, James, 'A Geometrical Proposition', in: *The London and Edinburgh Philosophical Magazine and Journal of Science* 15 (1839), pp. 41–2.

Thomson, James, 'Investigation of a New Series for the Computation of Logarithms: with a New Investigation of a Series for the Rectification of the Circle', in: *Transactions of the Royal Society of Edinburgh* 14/1 (1840), pp. 217–23.

Toldervy, William, *Selected Epitaphs*, 2 vols, London: printed for W. Owen 1755.

Tomasini, Giacomo Filippo, *Bibliothecae Patavinae manuscriptae publicae et privatae*, Udine: Schiratti 1639.

Topham, Jonathan R., 'Science, Print, and Crossing Borders: Importing French Science Books into Britain, 1789–1815', in: *Geographies of Nineteenth-Century Science*, ed. David N. Livingstone, and Charles Withers, Chicago: University of Chicago Press 2011, pp. 119–52.

Toplis, John, 'On the Decline of Mathematical Studies, and the Sciences Dependent upon Them', in: *Philosophical Magazine* 20 (1805), pp. 25–31.

Trabucco, Oreste, 'Telesian Controversies on the Winds and Meteorology', in: *Bernardino Telesio and the Natural Sciences in the Renaissance*, ed. P. D. Omodeo, Leiden: Brill 2019, pp. 96–115.

Trollope, William, *A History of the Royal Foundation of Christ's Hospital*, London: William Pickering 1834.

Turnbull, G. H., 'Samuel Hartlib's Connection with Sir Francis Kynaston's "Musaeum Minervae"', in: *Notes and Queries* 97 (1952), pp. 33–7.

Turnbull, Herbert Westren (ed.), *James Gregory Tercentenary Volume*, London: G. Bell & Sons 1939.

Turnbull, Herbert Westren, Joseph Frederick Scott, A. Rupert Hall, and Laura Tilling (eds), *The Correspondence of Isaac Newton*, 7 vols, Cambridge: Cambridge University Press 1967–81.

Turner, A. J., 'Mathematical Instruments and the Education of Gentlemen', in: *Annals of Science* 30/1 (1973), pp. 51–88.

Turner, A. J., 'William Oughtred, Richard Delamain and the Horizontal Instrument in Seventeenth Century England', in: *Annali dell'Istituto e Museo di Storia della Scienza di Firenze* 6/2 (1981), pp. 99–125.

Turner, Gerard L'E., *Elizabethan Instrument Makers: The Origins of the London Trade in Precision Instrument Making*, Oxford: Oxford University Press 2000.

Tyacke, Sarah, 'All at Sea: Some Cartographical Problems in the North 1500–1700', in: *IMCoS Journal* 111 (2007), pp. 38–41.

Unguru, Sabetai, 'On the Need to Rewrite the History of Greek Mathematics', in: *Archive for History of Exact Sciences* 15 (1975), pp. 67–114.

Unguru, Sabetai, 'History of Ancient Mathematics: Some Reflections on the State of the Art', in: *Isis* 70 (1979), pp. 555–65.

Valleriani, Matteo, *Galileo Engineer*, Dordrecht: Springer 2010.

Ventrice, P., 'Ettore Ausonio matematico dell'Accademia veneziana della Fama', in: *Ethos e cultura: Studi in onore di Ezio Riondato*, ed. Ezio Rionato, Padua: Antenore 1991, pp. 1135–54.

Vergé-Franceschi, Michel, *Marine et éducation sous l'ancien régime*, Paris: Editions du Centre national de la recherche scientifique 1991.

Vesey, F., *Cases Argued and Determined in the High Court of Chancery: In the Time of Lord Chancellor Hardwicke, from the Year 1746–47, to 1755*, London: W. Strahan and M. Woodfall 1773.

Villefosse, Antoine-Marie Héron de, *Über den Mineral-Reichtum. Betractungen über die Berg-, Hütten- und Salzwerke verschiedener Staaten, sowohl hinsichtlich ihrer Production und Verwaltung, als auch des jetzigen Zustandes der Bergbau- und Hüttenkunde, Zweyter Band, Des technischen Theils erste und zweyte Abtheilung*, Sondershausen: Bernhard Friedrich Voigt 1822.

Vine, Angus, *In Defiance of Time: Antiquarian Writing in Early Modern England*, Oxford: Oxford University Press 2010.

Vine, Angus, *Miscellaneous Order: Manuscript Culture and the Early Modern Organization of Knowledge*, Oxford: Oxford University Press 2019.

Viterbo, Ariel, 'Gli ebrei a Padova nel Settecento', in: *Ramhal: Pensiero ebraico e kabbalah tra Padova ed Eretz Israel*, ed. Gadi Luzzatto Voghera and Mauro Perani, Padua: Esedra 2010, pp. 13–23.

Vivo, Filippo De, 'How to Read Venetian *Relazioni*', in: *Renaissance and Reformation* 32 (2011), pp. 25–59.

Vivo, Filippo De, 'Archives of Speech: Recording Diplomatic Negotiation in Late Medieval and Early Modern Italy', in: *European History Quarterly* 46 (2016), pp. 519–44.

Voogt, Claas Jansz., *De Zeemans Wegh-Wyser*, Amsterdam: J. van Keulen 1695.

Waard, Cornelis de, René Pintard, Bernard Rochot, et al. (eds), *Correspondance du P. Marin Mersenne, Religieux Minime*, 17 vols, Paris: PUF and CNRS 1933–88.

Wach, Howard M., 'Culture and the Middle Classes: Popular Knowledge in Industrial Manchester', in: *Journal of British Studies* 27/4 (1988), pp. 375–404.

Waerden, B. L. van der, 'Defense of a "Shocking" Point of View', in: *Archive for History of Exact Sciences* 15 (1976), pp. 199–210.

Waghenaer, Lucas Janszoon, *Spieghel der Zeevaerdt*, Leiden: Christophe Plantin 1584.

Walker, Martyn Austin, '"A Solid and Practical Education within Reach of the Humblest Means": The Growth and Development of the Yorkshire Union of Mechanics' Institutes 1838–1891', PhD thesis, University of Huddersfield 2010.

Walker, Martyn Austin, *The Development of the Mechanics' Institute Movement in Britain and Beyond: Supporting Further Education for the Adult Working Classes*, Routledge Research in Education, London: Routledge 2016.

Wallace, William A., 'Randall Redivivus: Galileo and the Paduan Aristotelians', in: *Journal of the History of Ideas* 49 (1988), pp. 133–49.

Wallis, Helen M., 'The First English Globe: a Recent Discovery', in: *The Geographical Journal*, 117/3 (1951), pp. 275–90.

Wallis, Helen M., 'England's Search for the Northern Passages in the Sixteenth and Early Seventeenth Centuries', in: *Arctic*, 37/4 (1984), pp. 453–72.

Wallis, John, *Mechanica: sive, de motu, tractatus geometricus*, 3 parts, London: William Godbid for Moses Pitt 1670–71.

Wallis, John, *A Treatise of Algebra, both Historical and Practical*, London: John Playford for Richard Davis 1685.

Wallis, John, *Opera mathematica*, 3 vols and suppl., Oxford: at the University Press 1693–99.

Wallis, Peter John, *Newcastle Mathematical Libraries: William Armstrong, Charles Hutton and Others*, Northern Notes 4 (1972: supplement), University of Durham 1972.

Wallis, Peter and Ruth Wallis, *Mathematical Tradition in the North of England*, Durham: NEBMA 1991.

Wallis, Peter and Ruth Wallis, *Index of British Mathematicians, Part III: 1701–1800*, Newcastle upon Tyne 1993.

Walsham, Alexandra, 'The Social History of the Archive: Record-Keeping in Early Modern Europe', in: *Past and Present*, Supplement 11 (2016), pp. 9–48.

Walters, Alice N., 'Conversation Pieces: Science and Politeness in Eighteenth-Century England', in: *History of Science* 35 (1997), pp. 121–54.

Waquet, Françoise and Hans Bots, *La république des lettres*, Paris: Belin; Brussels: De Boeck 1997.

Ward, Edward, *The London-Spy Compleat, in Eighteen-Parts*, London: Printed and sold by J. How 1703.

Ward, John, *The Lives of the Professors of Gresham College: to Which is Prefixed the Life of the Founder, Sir Thomas Gresham*, London: John Moore for the author 1741.

[Ward, Seth], *Vindiciae Academiarum Containing, Some Briefe Animadversions upon Mr. Websters Book, Stiled, The Examination of Academies*, Oxford: Leonard Lichfield for Thomas Robinson 1654.

Wardhaugh, Benjamin, 'Charles Hutton and the "Dissensions" of 1783–84: Scientific Networking and its Failures', in: *Notes and Records of the Royal Society* 71/1 (2017), pp. 41–59.

Wardhaugh, Benjamin (ed.), *The Correspondence of Charles Hutton (1737–1823): Mathematical Networks in Georgian Britain*, Oxford: Oxford University Press 2017.

Wardhaugh, Benjamin, *Gunpowder and Geometry: The Life of Charles Hutton, Pit Boy, Mathematician and Scientific Rebel*, London: William Collins 2019.

Wardhaugh, Benjamin, *The Book of Wonders: The Many Lives of Euclid's Elements*, Glasgow: HarperCollins 2020.

Wardhaugh, Benjamin, 'Rehearsing in the Margins: Mathematical Print and Mathematical Learning in the Early Modern Period', in: *The Palgrave Handbook of Literature and Mathematics*, ed. Robert Tubbs, Alice Jenkins, and Nina Engelhardt, Cham: Palgrave Macmillan 2021, pp. 553–67.

Warwick, Andrew, *Masters of Theory: Cambridge and the Rise of Mathematical Physics*, Chicago: University of Chicago Press 2003.

Waters, David W., *The Art of Navigation in England in Elizabethan and Early Stuart Times*, New Haven, CT: Yale University Press 1958; 2nd edn, Greenwich: National Maritime Museum 1978.

Watson, Robert Spence, *The History of the Literary and Philosophical Society of Newcastle-upon-Tyne (1793–1896)*, London: Walter Scott Ltd 1897.

Weale, John, *A Catalogue of Books, on the Sciences: Astronomy, Mathematics, Natural Philosophy, &c; With Some Added That are Curious and Miscellaneous; Chiefly from the Libraries of Rev. Nevile Maskelyne, D.D., Astron. Royal and F.R.S.; Bishop Horsley, F.R.S., &c.; Dr. Charles Hutton, LL.D. F.R.S., &c.; William Phillips, F.R.L. and G.SS.; and Richard Heber, esq. On Sale, By John Weale, (Scientific and Architectural Bookseller, 59, High Holborn, London.)*, [London] 1835.

Webster, Charles, *The Great Instauration: Science, Medicine, and Reform, 1626–1660*, London: Duckworths 1975.

[Wehe, Zimbertus] alias Hisaias sub cruce, *Expolitio famae sidereae novae Faulhaberianae*, Ulm: Parnasische Truckerey 1619.

Weil, André, 'Who Betrayed Euclid?', in: *Archive for History of Exact Sciences* 19 (1978), pp. 91–3.

Weisbach, Julius, *Die neue Markscheidekunst und ihre Anwendung auf bergmännische Anlage. Erste Abtheilung: die trigonometrischen und Nivellir-Arbeiten über Tage, sowie die Anwendung auf die Anlage des Rothschönberger Stollns*, Braunschweig: Vieweg 1851.

Westman, Robert, *The Copernican Question: Prognostication, Skepticism and Celestial Order*, Berkeley: University of California Press 2011.

Weston, Thomas, *A Copy-Book Written for the Use of the Young-Gentlemen at the Academy in Greenwich*, [London] 1726.

Weston, Thomas, *Drawing-Book Compos'd for the Use of the Young Gentlemen at the Academy in Greenwich*, [London 1726].

Weston, Thomas, *Veteris arithmeticae elementa: sive De symbolicis et practicis partibus arithmeticae, ab antiquis hebraesis, graecis et romanis usurpatae [...] tractatus: in usum studiosae juventutis in Academiâ Grenovici [...]* [London 1726].

Weston, Thomas, *A Treatise of Arithmetic, in Whole Numbers and Fractions*, London: published by John Weston, printed for J. Hooke 1729.

Weston, Thomas, *Mathematical Discourses Concerning Two New Sciences*, trans. Thomas Weston, London: printed for J. Hooke 1730.

Whiteside, Derek Thomas (ed.), *The Mathematical Papers of Isaac Newton*, 8 vols, Cambridge: Cambridge University Press 1967–81.

Whyman, Susan E., *Sociability and Power in Late Stuart England: The cultural worlds of the Verneys, 1660–1720*, Oxford: Oxford University Press 1999.

Wilding, Nick, *Galileo's Idol: Gianfrancesco Sagredo and the Politics of Knowledge*, Chicago: University of Chicago Press 2014.

Wilkinson, Thomas Turner, 'Mathematical Periodicals (continued)', in: *Mechanics' Magazine* 50 (1849), pp. 5–9.

Wilkinson, Thomas Turner, 'Memoir of James Wolfenden, of Hollinwood', in: *Mechanics' Magazine* 50 (1849), pp. 387–93.

Wilkinson, Thomas Turner, 'On the origin and progress of the study of geometry in Lancashire', in: *Notes and Queries* 34 (1850), pp. 57–60.

Wilkinson, Thomas Turner, 'Memoir of the Late J. H. Swale', in: *Mechanics' Magazine* 56 (1852), pp. 194–6, 206–9, 224–6.

Wilkinson, Thomas Turner, 'Mathematical Periodicals', in: *Mechanics' Magazine* 58 (1853), pp. 306–7, 327–8.

Wilkinson, Thomas Turner, 'Mathematical Periodicals', in: *Mechanics' Magazine* 59 (1853), pp. 528–9.

Wilkinson, Thomas Turner, 'Mathematics and Mathematicians, the Journals of the late Reuben Burrow', in: *London, Edinburgh, and Dublin Philosophical Magazine*, 4th series, 5 (1853), pp. 185–93; 6 (1853), pp. 196–204.

Wilkinson, Thomas Turner, 'Mathematical Periodicals', in: *Mechanics' Magazine* 60 (1854), pp. 182–3.

Wilkinson, Thomas Turner, 'The Lancashire Geometers and their Writings', in: *Memoirs of the Literary and Philosophical Society of Manchester* 11 (1854), pp. 123–157.

Wilkinson, Thomas Turner, 'Biographical Notices of some Liverpool Mathematicians', in: *Transactions of the Historic Society of Lancashire and Cheshire*, new ser. 2 (1862), pp. 29–40.

Wilkinson, Thomas Turner, 'An account of the Life and Writings of the Late Henry Buckley', in: *Transactions of the Historic Society of Lancashire and Cheshire*, new ser. 3 (1863), pp. 115–28.

Willmoth, Frances, *Sir Jonas Moore: Practical Mathematics and Restoration Science*, Woodbridge: The Boydell Press 1993.

Wilson, Arline, 'The Cultural Identity of Liverpool, 1790–1850: The Early Learned Societies', in: *Transactions of the Historic Society of Lancashire and Cheshire* 147 (1997), pp. 55–80.

Wilson, Charles, *England's Apprenticeship, 1603–1763*, London: Longmans, Green & Co 1965.

Wilson, James, *Biography of the Blind: or the Lives of Such as Have Distinguished Themselves as Poets, Philosophers, Artists, &c.*, Birmingham: J. W. Showell 1838.

Wilson, J. I., *A Brief History of Christ's Hospital*, London: John van Voorst 1842.

Wilson, Robin J., 'Kirkman, Thomas Penyngton (1806–1895), Mathematician and Philosopher', in: *Oxford Dictionary of National Biography*, 2004, https://doi.org/10.1093/ref:odnb/51577.

Wingate, Edmund, *Arithmetique made easie, or, A perfect Methode for the true knowledge and practice of Natural Arithmetique., according to the ancient vulgar way, without dependence upon any other Author for the grounds thereof.* Second Edition by John Kersey. London: J. Flesher for Philemon Stephens 1650.

Wood, Anthony, *Athenae Oxonienses. An Exact History of All the Writers and Bishops Who Have Had Their Education in the Most Ancient and Famous University of Oxford*, 2 vols, London: for Thomas Bennet 1691–92.

Woodhouse, Robert, [Review of Lagrange, *Theorie des fonctions*], in: *The Monthly Review* 28 (1799), pp. 481–99.

Woolhouse, Wesley S. B., *Investigation of Mortality in the Indian Army*, London 1839.

Wright, Edward, *Certaine Errors in Nauigation*, London: Valentine Sims [and W. White] 1599.

Wussing, Hans, *Die Genesis des abstrakten Gruppenbegriffes*, Berlin: Deutscher Verlag der Wissenschaften 1969; English translation by Abe Schenitzer: *The Genesis of the Abstract Group Concept*, Cambridge, MA: MIT Press 1984.

Yale, Elizabeth, *Sociable Knowledge: Natural History and the Nation in Early Modern Britain*, Philadelphia: University of Pennsylvania Press 2016.

Yale, Elizabeth, 'The Book and the Archive in the History of Science', in: *Isis* 107 (2016), pp. 106–15.

Yates, Frances A., 'Giordano Bruno: Some New Documents', in: *Revue Internationale de Philosophie* 16 (1951), pp: 174–99.

Yates, Frances A., *The Rosicrucian Enlightenment*, London: Routledge and Kegan Paul 1972.

Yeo, R., *Notebooks, English Virtuosi, and Early Modern Science*, Chicago: University of Chicago Press 2014.

Zitarelli, David E., *A History of Mathematics in the United States and Canada*, vol. 1: *1492–1900*, Providence, RI: MAA Press 2019.

Zweckbronner, G., 'Rechenmeister, Ingenieur und Bürger zu Ulm—Johann Faulhaber (1580–1635) in seiner Zeit', *Technikgeschichte* 47/2 (1980), pp. 114–32.

Figure Sources and Credits

3.1 Courtesy of the John Carter Brown Library, https://jcblibrary.org/collection/digital-images.
3.2 Collection Maritiem Museum Rotterdam. MMR Image reproduced by permission of curator Marcel Kroon
3.3 Public domain
https://www.nationaalarchief.nl/onderzoeken/open data/open-data-archiefinventarissen-en-scans-van-archieven; see section 'Open data scans en andere digitale objecten', notes 'Publiek domein' and 'CCO'.
3.4 Nationaal Archief, Den Haag, Adm. coll., 1.01.47.11, inv. nr. 13. Public domain. See Figure 3.3 above.
4.1 British Museum: 1848,0911.395.
4.2 Wellcome Library no. 9661i. Wellcome Collection.
4.3 The Trustees, Admiral Long's Foundation.
4.4 Wellcome Library no. 6080i. Wellcome Collection.
4.5 National Maritime Museum, MKY/9/11, via Cambridge Digital Library, https://cudl.lib.cam.ac.uk/view/MS-MKY-00009/11).
5.1 Biblioteca Nacional de Portugal.
5.2 Biblioteca Nacional de Portugal.
5.3 Biblioteca Nacional de Portugal.
5.4 Biblioteca Pública de Évora, cód. Manizola 258.
5.5 Biblioteca Pública de Évora, cód. Manizola 258.
5.6 Biblioteca Nacional de Portugal, cód. 5194, fl. 333.
10.1 © Visual elaboration of the author.
10.2 Oxford, Bodleian Library, MS Canon. Ital. 145, fols 43^{r} and 45^{r}. © Bodleian Library, Oxford.
10.3 Milan, Ambrosian Library, MS S 105 sup., fol. 241^{r-v}. © Ambrosian Library, Milan.
10.4 Milan, Ambrosian Library, MS R 99 sup., fol. 83^{r}. © Ambrosian Library, Milan.
10.5 Florence, Biblioteca Nazionale Centrale, MS Gal. 83, fol. 4^{r}. © Museo Galileo, Florence; public domain.
10.6 Florence, Biblioteca Nazionale Centrale, MS Gal. 72, fol. 134^{r}. © Museo Galileo, Florence; public domain.
11.1 Oxford, Bodleian Library, MS Don. d. 45, f. 279^{v}. © Bodleian Library., Oxford.
11.2 Oxford, Bodleian Library Savile H 18 (6). © Bodleian Library., Oxford.
11.3 Oxford, History of Science Museum, Evans Collection, shelf-mark: LE/COL.
11.4 Oxford, Christ Church Library, shelf-mark ON.6.19 (5).
11.5 Oxford, Bodleian Library, MS Savile G 20 (4). © Bodleian Library., Oxford.
14.1 From copy owned by A. Heeffer.
14.2 From copy owned by A. Heeffer.
14.3 University of Oklahoma Libraries History of Science Collections.
14.4 Sächische Landesbibliothek, Dresden.
14.5 Early English Books Online.
14.6 Cuneiform Digital Library Initiative, https://cdli.mpiwg-berlin.mpg.de/
15.1 Lincoln, Lincolnshire Archives, Monson Papers, classmark: MON 7/21.

15.2 Copyright the Board of Trustees of the Science Museum.
15.3 Penn Libraries, University of Pennsylvania: Kislak Center for Special Collections, Rare Books and Manuscripts University of Pennsylvania Oversize LJS 321.
15.4 Whipple Museum of the History of Science, inv. no. Wh.2255.

Index

Please note: Indexed terms that span two pages (e.g. 52–53) may occasionally only appear on one of those pages.